ANATOLY ROZENBLAT

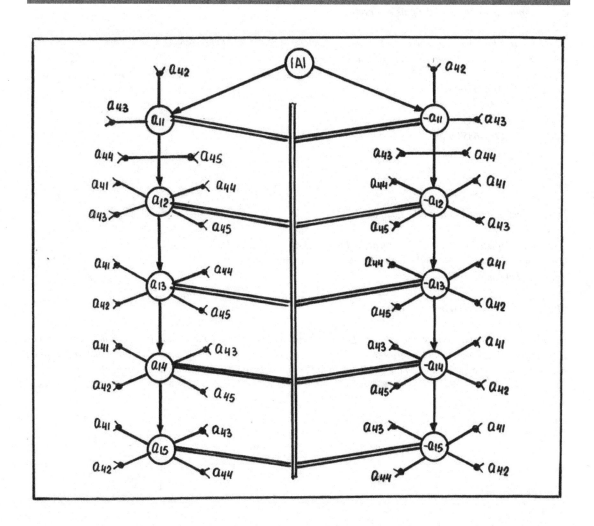

SOLVING OF DETERMINANTS
with functional graphs

authorHOUSE®

AuthorHouse™
1663 Liberty Drive
Bloomington, IN 47403
www.authorhouse.com
Phone: 1 (800) 839-8640

Published by AuthorHouse 02/17/2015

ISBN: 978-1-4969-6858-6 (sc)
ISBN: 978-1-4969-6857-9 (e)

Dedicated to my lovely grandchild Vlada and her family

Preface to the first edition

In last ten years' period rather widely appeared application of some cardinal methods in operative activity and also in planning of economy, industry and other military fields advantageously in distribution of equipments and weapons. And considerable role in this question plays the Linear Algebra conformably to theory of determinants and progressive Linear Programming methods(mathematical base, theoretical and calculation processes).

Therefore *in the first chapter* are considered the matrix decisions for determinants, having matrices of size with $2 \leq n \leq 3$,as the most lightly matrices solved by Sarrus rule and others known methods, as Expansion and Laplace's methods. And also considered some applied methods for solving the system of linear equations with application of Cramer's rule and functional graphs in comparative form.

In the second chapter "General algorithms for evaluation of determinants with 4x4 matrices" considered of some problems for solving of such matrices determinants. And also in comparative analysis considered the Expansion and Laplace's methods with using of the functional graphs which considerably reduce the calculation processes of the different matrices determinants particularly in numerical data. And besides admitted that in general view could be fitted of some formulas for evaluation of matrices determinants for calculation procedures.

In the third and fourth chapters "Algorithms and graphs for solving determinants with sizes of $5 \leq n \leq 7$ " are considered in general view the different algorithms and designed formulas for evaluation the matrices determinants with application of Expansion ,Laplace's methods .And also shown the different applied examples in question of evaluation the matrices determinants in numerical data with application of the functional graphs.

In the fifth chapter "Functional graphs and algorithms for symmetric and oblique(skew) determinants " are considered the peculiarities of above-named matrices determinants advantageously for sizes $3 \leq n \leq 5$ applicably to Expansion and Laplace's methods with application of the functional graphs for the given conditions in question of evaluation the value of the different matrices determinants in general view and numerical data.

And also in this book is given of some practical recommendations for the students and some people who is interested by the questions of evaluation the matrices determinants used for solving the system of linear equations with goal that further to discover the possibility of the Cramer's rule in practical conditions for given numerical data of matrix determinants ,having sizes $2 \leq n \leq 6$ by the mathematical methods with application of the functional graphs, reducing the calculation procedures.

November 20 ,2014

Anatoly Rozenblat,
Independent Scientist &Inventor
Chicago, Illinois , USA

CONTENTS

CHAPTER FIVE FUNCTIONAL GRAPHS AND ALGORITHMS FOR SYMMETRIC AND OBLIQUE (SKEW) DETERMINANTS

GUIDE TO THE USE OF GRAPHS FOR MATRICES DETERMINANTS

2x 2 matrices

1) The determinant $|A|$ of a 2x2 matrix is defined by analytic method(a) and functional graph (b) as:

$$|A| = \begin{vmatrix} a_{11} & a_{12} \\ a_{21} & a_{22} \end{vmatrix} = a_{11}a_{22} - a_{12}a_{21}$$

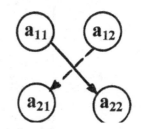

a) Analytic method

b) Functional graph
[----- negative area]

2) Sample problems:

$$|A| = \begin{vmatrix} 2 & -5 \\ 0 & 4 \end{vmatrix} = 8$$

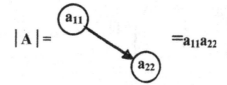

$$|A| = a_{11}a_{22}$$

$$|A| = \begin{vmatrix} 0 & -3 \\ 1 & 4 \end{vmatrix} = 3$$

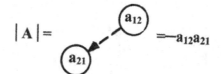

$$|A| = -a_{12}a_{21}$$

3x 3 matrices

1) The determinant $|A|$ of a 3x3 matrix is defined by analytic method as:

$$|A| = \begin{vmatrix} a_{11} & a_{12} & a_{13} \\ a_{21} & a_{22} & a_{23} \\ a_{31} & a_{32} & a_{33} \end{vmatrix} = a_{21}(a_{13}a_{32} - a_{12}a_{33}) + a_{22}(a_{11}a_{33} - a_{13}a_{31}) + a_{23}(a_{12}a_{31} - a_{11}a_{32})$$

2) This 3x3 matrix determinant is introduced by the functional graph in view of:

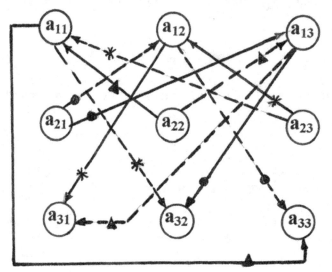

Functional graph for 3x3 matrix determinant (---- negative area)

3) Sample problems:

$$A \mid = \begin{vmatrix} 0 & 2 & -3 \\ 0 & 0 & 4 \\ 5 & -1 & 0 \end{vmatrix} = 40$$

$$|A| = \quad = a_{23}a_{12}a_{31}$$

$$|A| = \begin{vmatrix} 4 & 5 & 0 \\ 0 & 0 & -1 \\ 0 & -2 & 3 \end{vmatrix} = -8$$

$$|A| = \quad = -a_{23}a_{11}a_{32}$$

4x4 matrices

1) The determinant $|A|$ of a 4x4 matrix is defined by analytic method as ,for a instance , at expansion of its elements to the first row in general view:

$$|A| = a_{11}[a_{22}(a_{33}a_{44}-a_{34}a_{43})+a_{23}(a_{34}a_{42}-a_{32}a_{44})+a_{24}(a_{32}a_{43}-a_{33}a_{42})]+$$
$$+a_{12}[a_{21}(a_{34}a_{43}-a_{33}a_{44})+a_{23}(a_{31}a_{44}-a_{34}a_{41})+a_{24}(a_{33}a_{41}-a_{31}a_{43})]+$$
$$+a_{13}[a_{21}(a_{32}a_{44}-a_{34}a_{42})+a_{22}(a_{34}a_{41}-a_{31}a_{44})+a_{24}(a_{31}a_{42}-a_{32}a_{41})]+$$
$$+a_{14}[a_{21}(a_{33}a_{42}-a_{32}a_{43})+a_{22}(a_{31}a_{43}-a_{33}a_{41})+a_{23}(a_{32}a_{41}-a_{31}a_{42})]$$

X

2) This 4x4 matrix determinant is introduced by the functional graph for positive area (a) and negative area (b) in view of:

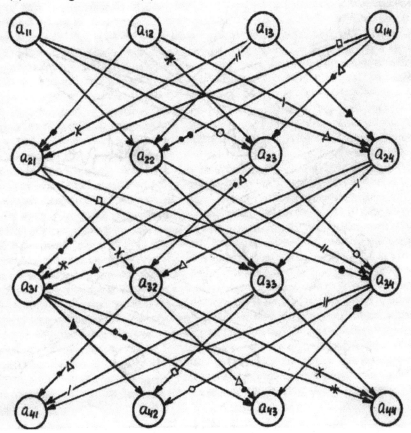

a) Functional graph for the determinant of the 4x4 matrix elements of which have expansion to the first row for the positive area

3) Sample problems:

$$|A| = \begin{vmatrix} 0 & 0 & 0 & 2 \\ -3 & 0 & 1 & 4 \\ 5 & -6 & -1 & 3 \\ 0 & 0 & -4 & 2 \end{vmatrix} = 144$$

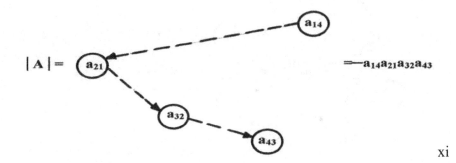

$|A| = $ $= -a_{14}a_{21}a_{32}a_{43}$

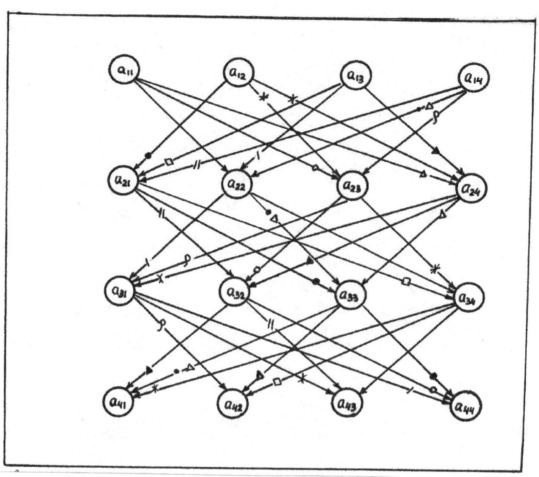

b) Functional graph for the determinant of the 4x4 matrix elements of which
    have expansion to the first row for the negative area

$$|A| = \begin{vmatrix} 0 & 2 & -3 & 0 \\ 4 & 0 & 0 & 1 \\ 3 & 0 & 0 & -2 \\ 0 & -1 & 5 & 0 \end{vmatrix} = -86$$

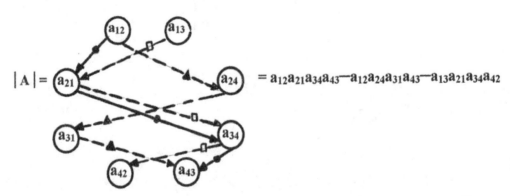

$|A| =$ $= a_{12}a_{21}a_{34}a_{43} - a_{12}a_{24}a_{31}a_{43} - a_{13}a_{21}a_{34}a_{42}$

CHAPTER ONE   SOLVING OF DETERMINANTS FOR MATRICES OF SIZE 2≤n≤3

1. Introduction

At the last ten years a matrix determinant discovered itself in wide use advantageously in linear programming at solving of extreme tasks joined with planned of manufacture, effective management in industry and optimization of freightage transportation. And besides the above-named parameter presents in some other fields of mathematical research works in designing of some algorithms and formulas for evaluation of the matrices determinants which could be fit for solving of the system linear algebraic equations and also the multiple regression models advantageously in such fields ,as engineering ,education ,economics, food research ,psychology, sociology and statistics.

And for this reason today appeared necessity of analysis the well-known methods which are fit and useful for evaluation of the matrices determinants and also of their properties with goal to minimize the calculation procedures . This could be achieved by using the special algorithms, formulas and functional graphs for the given matrix in solving of any determinants that then to definite the unknown coefficients ,incoming to the different mathematical models.

So, the applied methods of determinants and its application in linear algebra is main object of this work , but in question of historical information about theory of that subject should addresses to the reference book written by Thomas Muir **[1]**. In this book author has expressed in general view some works the well-known mathematics of XVI to XVIII centuries, such as Gauss, Cramer, Jacobi and many others.

However, today is XXI century and for this lengthy period has appeared many new scientific works in question of studying the applied methods for solving of the matrices determinants applicably to n≤5 ,such as **[2],[3],[4],[5],[6],[7],[8]** and many others. And now we are going in detail to consider some well-known applied methods for definition of the matrices determinants for some people who interested in this problem for solving of the considered problems.

2. Definition of the determinant of the matrix with sizes of n≤3

2.1 For the 2x2 matrix

Consider the square (n x n) matrix 2x2 with n=2 rows and n=2 columns in general view:

$$A=\begin{vmatrix} a_{11} & a_{12} \\ a_{21} & a_{22} \end{vmatrix}$$

where,

a_{11}- the first element of the first row or the first element of the first column;

a_{12}-the second element of the first row or the first element of the second column;

a_{21}-the first element of the second row or the second element of the first column;

a_{22}-the second element of the second row or the second element of the second column .

So, the determinant of the 2x2 matrix is defined in general form by formula in view of:

$$|A|=a_{11}a_{22}-a_{12}a_{21} \qquad (1)$$

where,

$a_{11}a_{22}$- positive product of two elements of the 2x2 matrix;

$a_{12}a_{21}$-negative product of two elements of the 2x2 matrix.

The functional ties between elements of the determinant of the 2x2 matrix is shown in Figure 1 .

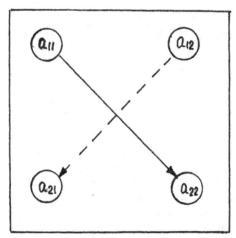

Figure 1 Functional ties between elements of the 2x2 matrix
(-------- negative ties between elements)

In   Table 1is shown the parameters of considered 2x2 matrices and Figures 2÷21 are shown by vector analysis the value of the determinants in view *of the area of the parallelogram* at the different combinations of given elements of the 2x2 matrices which present the sides of the above-named parallelogram in view of vectors in general view:

Table 1 Parameters of considered 2x2 matrices

$$A=\begin{vmatrix} a_{11} & a_{12} \\ a_{21} & a_{22} \end{vmatrix} \text{ for Figures 2 and 3 ;} \quad A=\begin{vmatrix} -a_{11} & -a_{12} \\ -a_{21} & -a_{22} \end{vmatrix} \text{ for Figures 4 and 5;}$$

$$A=\begin{vmatrix} -a_{11} & a_{12} \\ -a_{21} & a_{22} \end{vmatrix} \text{ for Figures 6 and 7 ;} \quad A=\begin{vmatrix} a_{11} & -a_{12} \\ a_{21} & -a_{22} \end{vmatrix} \text{ for Figures 8 and 9;}$$

$$A=\begin{vmatrix} -a_{11} & -a_{12} \\ a_{21} & a_{22} \end{vmatrix} \text{ for Figures 10 and 11;} \quad A=\begin{vmatrix} a_{11} & a_{12} \\ -a_{21} & -a_{22} \end{vmatrix} \text{ for Figures 12 and 13;}$$

$$A=\begin{vmatrix} -a_{11} & a_{12} \\ a_{21} & a_{22} \end{vmatrix} \text{ for Figures 14 and 15 ;} \quad A=\begin{vmatrix} a_{11} & -a_{12} \\ a_{21} & a_{22} \end{vmatrix} \text{ for Figures 16 and 17;}$$

$$A=\begin{vmatrix} a_{11} & a_{12} \\ -a_{21} & a_{22} \end{vmatrix} \text{ for Figures 18 and 19 ;} \quad A=\begin{vmatrix} a_{11} & a_{12} \\ a_{21} & -a_{22} \end{vmatrix} \text{ for Figures 20 and 21}$$

In Figure 2 is shown the area of the parallelogram on base of positive vectors (v_1) and (v_2) from elements of the 2x2 matrix and in Figure 3 is shown the area of parallelogram which is value for the determinant formed on base vectors(v_1) and (v_2) ,as product between elements of the 2x2 matrix which is equal $S_2=a_{11}a_{22}-a_{12}a_{21}$.

In Figure 4 is shown the area of the parallelogram formed on base of negative vectors($-v_1$) and ($-v_2$) from elements of the 2x2 matrix and in Figure 5 is shown the area of the parallelogram which is value for the determinant formed on base of vectors (v_1) and ($-v_2$) ,as product between elements of the 2x2 matrix which is equal $S_4=a_{11}a_{22}-a_{12}a_{21}$. In Figure 6 is shown the area of the parallelogram formed on base of negative vectors ($-v_1$) and ($-v_2$) from elements of the 2x2 matrix and in Figure 7 is shown the area of the parallelogram which is value for the determinant formed on base of vectors ($-v_1$) and (v_2) ,as product between elements of the 2x2 matrix which is equal $S_6=a_{12}a_{21}-a_{11}a_{22}$.

2

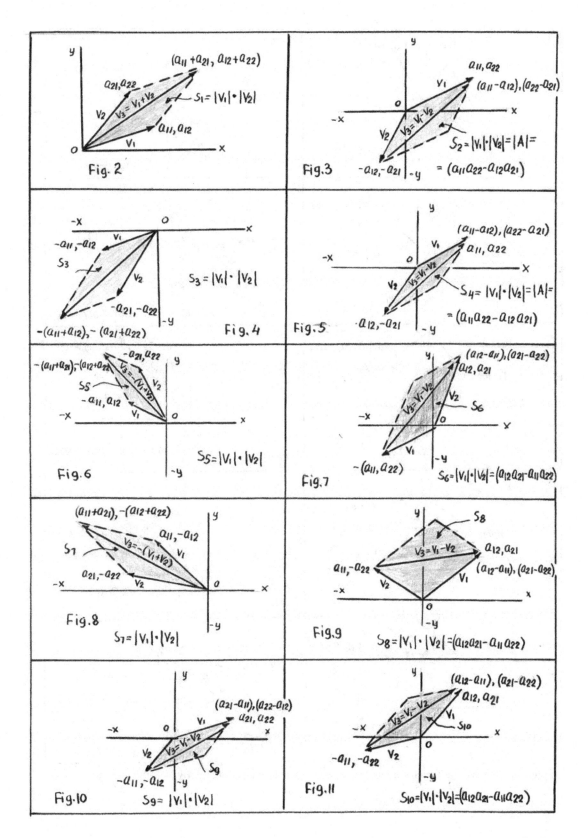

Fig. 2

$(a_{11}+a_{21}, a_{12}+a_{22})$

a_{21}, a_{22}

V_2 $V_3=V_1+V_2$

$S_1=|V_1|\cdot|V_2|$

a_{11}, a_{12}

V_1

Fig. 3

a_{11}, a_{22}

V_1 $(a_{11}-a_{12}), (a_{22}-a_{21})$

$V_3=V_1-V_2$

V_2

$S_2=|V_1|\cdot|V_2|=|A|=$

$-a_{12}, -a_{21}$ $= (a_{11}a_{22}-a_{12}a_{21})$

Fig. 4

$-a_{11}, -a_{12}$ V_1

S_3

V_2

$-a_{21}, -a_{22}$

$S_3=|V_1|\cdot|V_2|$

$-(a_{11}+a_{12}), -(a_{21}+a_{22})$

Fig. 5

$(a_{11}-a_{12}), (a_{22}-a_{21})$

V_1 a_{11}, a_{22}

$V_3=V_1-V_2$

V_2

$S_4=|V_1|\cdot|V_2|=|A|=$

$a_{12}, -a_{21}$ $= (a_{11}a_{22}-a_{12}a_{21})$

Fig. 6

$-a_{21}, a_{22}$ $V_3=-(V_1+V_2)$

$-(a_{11}+a_{21}), -(a_{12}+a_{22})$

S_5 V_2

$-a_{11}, a_{12}$ V_1

$S_5=|V_1|\cdot|V_2|$

Fig. 7

$(a_{12}-a_{11}), (a_{21}-a_{22})$

a_{12}, a_{21}

$V_3=V_1-V_2$ V_2 S_6

V_1

$-(a_{11}, a_{22})$ $S_6=|V_1|\cdot|V_2|=(a_{12}a_{21}-a_{11}a_{22})$

Fig. 8

$(a_{11}+a_{21}), -(a_{12}+a_{22})$

$a_{11}, -a_{12}$

S_7 $V_3=-(V_1+V_2)$ V_1

$a_{21}, -a_{22}$ V_2

$S_7=|V_1|\cdot|V_2|$

Fig. 9

S_8

$V_3=V_1-V_2$ a_{12}, a_{21}

$a_{11}, -a_{22}$ $(a_{12}-a_{11}), (a_{21}-a_{22})$

V_2 V_1

$S_8=|V_1|\cdot|V_2|=(a_{12}a_{21}-a_{11}a_{22})$

Fig. 10

$(a_{21}-a_{11}), (a_{22}-a_{12})$

V_1 a_{21}, a_{22}

V_2 $V_3=V_1-V_2$

$-a_{11}, -a_{12}$ S_9

$S_9=|V_1|\cdot|V_2|$

Fig. 11

$(a_{12}-a_{11}), (a_{21}-a_{22})$

a_{12}, a_{21}

$V_3=V_1-V_2$ V_1

S_{10}

V_2

$-a_{11}, -a_{22}$

$S_{10}=|V_1|\cdot|V_2|=(a_{12}a_{21}-a_{11}a_{22})$

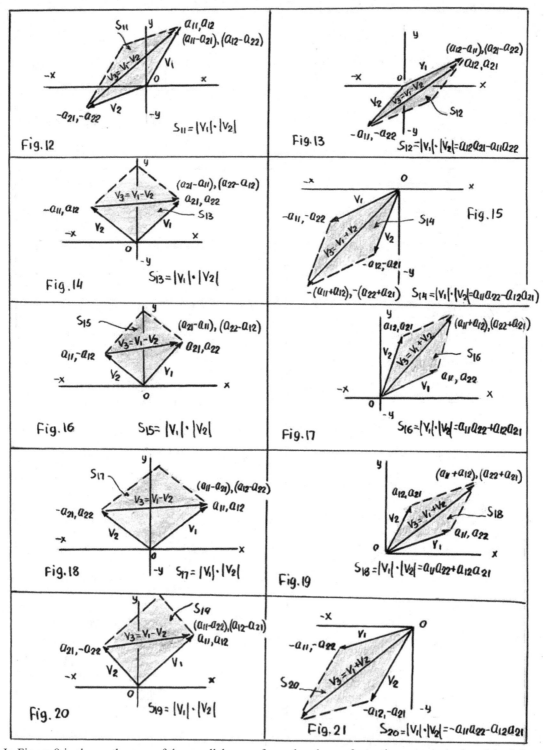

Fig. 12 $S_{11} = |V_1| \cdot |V_2|$

Fig. 13 $S_{12} = |V_1| \cdot |V_2| = a_{12}a_{21} - a_{11}a_{22}$

Fig. 14 $S_{13} = |V_1| \cdot |V_2|$

Fig. 15 $S_{14} = |V_1| \cdot |V_2| = a_{11}a_{22} - a_{12}a_{21}$

Fig. 16 $S_{15} = |V_1| \cdot |V_2|$

Fig. 17 $S_{16} = |V_1| \cdot |V_2| = a_{11}a_{22} + a_{12}a_{21}$

Fig. 18 $S_{17} = |V_1| \cdot |V_2|$

Fig. 19 $S_{18} = |V_1| \cdot |V_2| = a_{11}a_{22} + a_{12}a_{21}$

Fig. 20 $S_{19} = |V_1| \cdot |V_2|$

Fig. 21 $S_{20} = |V_1| \cdot |V_2| = -a_{11}a_{22} - a_{12}a_{21}$

In Figure 8 is shown the area of the parallelogram formed on base of negative vectors ($-v_1$) and ($-v_2$) from elements of the 2x2 matrix and in Figure 9 is shown the area of the parallelogram which is value for the determinant formed on base of vectors (v_1) and ($-v_2$), as product between elements of the 2x2 matrix which is equal $S_8 = a_{12}a_{21} - a_{11}a_{22}$.

4

In Figure 10 is shown the area of the parallelogram formed on base of vectors (v_1) and $(-v_2)$ from elements of the 2x2 matrix and in Figure 11 is shown the area of the parallelogram which is value for the determinant formed on base of vectors (v_1) and $(-v_2)$,as product between elements of the 2x2 matrix which is equal $S_{10}=a_{12}a_{21}-a_{11}a_{22}$.

In Figure 12 is shown the area of the parallelogram formed on base vectors (v_1) and $(-v_2)$ from elements of the 2x2 matrix and in Figure 13 is shown the area of the parallelogram which is value for the determinant formed on base of vectors (v_1) and $(-v_2)$,as product between elements of the 2x2 matrix which is equal $S_{12}=a_{12}a_{21}-a_{11}a_{22}$.

In Figure 14 is shown the area of the parallelogram formed on base of vectors (v_1) and $(-v_2)$ from elements of the 2x2 matrix and in Figure 15 is shown the area of the parallelogram which is value for the determinant formed on base of negative vectors $(-v_1)$ and $(-v_2)$,as product between elements of the 2x2 matrix which is equal $S_{14}=-(a_{11}a_{22}+a_{12}a_{21})$. In Figure 16 is shown the area of the parallelogram formed on base of vectors (v_1) and $(-v_2)$ from elements of the 2x2 matrix and in Figure 17 is shown the area of the parallelogram which is value for the determinant formed on base of vectors (v_1) and (v_2) ,as product between elements of the 2x2 matrix which is equal $S_{16}=a_{11}a_{22}+a_{12}a_{21}$.

In Figure 18 is shown the area of the parallelogram formed on base of vectors (v_1) and $(-v_2)$ from elements of the 2x2 matrix and in Figure19 is shown the area of the parallelogram which is value for the determinant formed on base of vectors (v_1) and (v_2) ,as product between elements of the 2x2 matrix which is equal $S_{18}=a_{11}a_{22}+a_{12}a_{21}$. In Figure 20 is shown the area of the parallelogram formed on base of vectors (v_2) and $(-v_2)$ from elements of the 2x2 matrix and in Figure 21 is shown the area of the parallelogram which is value for the determinant formed on base of negative vectors $(-v_1)$ and $(-v_2)$,as product between elements of the 2x2 matrix which is equal $S_{20}=-(a_{11}a_{22}+a_{12}a_{21})$.

In Table 2 is shown evaluation of the determinants of the 2x2 matrices at the different combinations of its elements with zero value.

Table 2 Evaluation of the determinants of the 2x2 matrices at the different combinations of its elements with zero value

n/n	Zero value for elements of the 2x2 matrix				Result of determinant in general view
	a_{11}	a_{12}	a_{21}	a_{22}	
1	0	0	0	0	0
2	0	-	0	-	0
3	-	0	-	0	0
4	0	0	-	-	0
5	-	-	0	0	0
6	0	-	-	-	$\lvert A \rvert = -a_{12}a_{21}$
7	-	0	-	-	$\lvert A \rvert = a_{11}a_{22}$
8	-	-	0	0	$\lvert A \rvert = a_{11}a_{22}$
9	-	-	-	0	$\lvert A \rvert = -a_{12}a_{21}$
10	0	-	-	0	$\lvert A \rvert = -a_{12}a_{21}$
11	-	0	0	-	$\lvert A \rvert = a_{11}a_{22}$
12	0	0	0	-	0
13	0	0	-	0	0
14	-	0	0	0	0
15	0	-	0	0	0

2.2 For the 3x3 matrix

The determinant $|A|$ of the 3x3 matrix is expressed by the following view:

$$A = \begin{vmatrix} a_{11} & a_{12} & a_{13} \\ a_{21} & a_{22} & a_{23} \\ a_{31} & a_{32} & a_{33} \end{vmatrix}$$

where,

a_{11}-the first element of the first row or the first element of the first column;

a_{12}-the second element of the first row or the first element of the second column;

a_{13}-the third element of the first row or the first element of the third column;

a_{21}-the first element of the second row or the second element of the first column;

a_{22}-the second element of the second row or the second element of the second column;

a_{23}-the third element of the second row or the second element of the third column;

a_{31}-the first element of the third row or the third element of the first column;

a_{32}-the second element of the third row or the third element of the second column;

a_{33}-the third element of the third row or the third element of the third column .

The determinant $|A|$ of the 3x3 matrix can be evaluated by the following method:

A. Sarrus rule(diagonal method) [9]

The Sarrus rule is a mnemonic for this formula:

❖ To the matrix determinant at right side should to write two the first columns ,as this shown in the illustration:

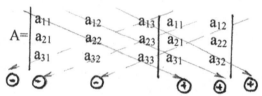

❖ To form the sum of the products of three main positive(+) diagonal vectors ,presenting the elements of 3x3 matrix and then

❖ To form   the sum of three negative (-) diagonal vectors, presenting the elements of 3x3 matrix.

So, for the above-considered conditions the determinant of the 3x3 matrix in general view is equal:

$$|A| = (a_{11}a_{22}a_{33} + a_{12}a_{23}a_{31}) - (a_{13}a_{22}a_{31} + a_{11}a_{23}a_{32} + a_{12}a_{21}a_{33}) =$$
$$= a_{21}(a_{13}a_{32} - a_{12}a_{33}) + a_{22}(a_{11}a_{33} - a_{13}a_{31}) + a_{23}(a_{12}a_{31} - a_{11}a_{32}) \quad (2)$$

But necessary to admit that Sarrus rule is fit only for the 2x2 and 3x3 matrices and could not be used on over higher size of the matrices.

B. Expansion (cofactor) method [2]

The essential of this expansion method(theorem) consists in such form: "….If we take any row or column of the determinant than we see that the value of the determinant $|A|$ always will be equal of sum product of taken elements row or column on their algebraic supplemental (cofactor)…."i.e we can this method to mark in the following form in general form for any row in view of:

$$|A| = a_{11}A_{11} + a_{12}A_{12} + \dots\ a_{1n}A_{1n} = a_{21}A_{21} + a_{22}A_{22} + \dots a_{2n}A_{2n} =$$

$$\dots\dots\dots\dots\dots\dots\dots\dots\dots\dots\dots\dots\dots\dots\dots\dots\dots$$

$$a_{n1}A_{n1} + a_{n2}A_{n2} + \dots\dots a_{nn}A_{nn} = \sum_{i=1}^{n} a_{in}A_{in} = \sum_{i=1}^{n} a_{in}(-1)^{i+j} M_{ij} \quad (3)$$

or to any column:

$$\left|A\right| = a_{11}A_{11}+a_{21}A_{21}+\ldots\ldots a_{n1}A_{n1}=a_{12}A_{12}+a_{22}A_{22}+\ldots.a_{n2}A_{n2}=$$

$$\ldots\ldots\ldots\ldots\ldots\ldots\ldots\ldots\ldots\ldots\ldots\ldots\ldots\ldots\ldots\ldots\ldots$$

$$a_{1i}A_{1i}+a_{2i}A_{2i}+\ldots\ldots.a_{ni}A_{ni}=\sum_{i=1}^{n}a_{ni}A_{ni}=\sum_{i=1}^{n}a_{ni}(-1)^{j+i}M_{ji} \quad (4)$$

$$(i=1,2,3\ldots..n;\ j=1,2,3..n)$$

where,

M_{ij}-minor of element (a_{ij}) for any row;

M_{ji}-minor of element (a_{ji}) for any column;

A_{in} -cofactor element for row ;

A_{ni}-cofactor element for column

Applicably to the determinant of the 3x3 matrix we have in general view:

❖ For the rows of the 3x3 matrix

$$A=\begin{vmatrix} a_{11} & a_{12} & a_{13} \\ a_{21} & a_{22} & a_{23} \\ a_{31} & a_{32} & a_{33} \end{vmatrix}$$

- If the first row of A is selected, then we have the determinant equal $\left|A\right|=a_{11}A_{11}+a_{12}A_{12}+a_{13}A_{13}$

where,

$A_{11}=(-1)^{1+1}M_{11}=(a_{22}a_{33}-a_{23}a_{32})$

$A_{12}=(-1)^{1+2}M_{12}=-(a_{21}a_{33}-a_{23}a_{31})$

$A_{13}=(-1)^{1+3}M_{13}=(a_{21}a_{32}-a_{22}a_{31})$ and the determinant $\left|A\right|$ of the 3x3 matrix in general view is equal:

$$\left|A\right| = a_{11}(a_{22}a_{33}-a_{23}a_{32})-a_{12}(a_{21}a_{33}-a_{23}a_{31})+a_{13}(a_{21}a_{32}-a_{22}a_{31})=$$
$$=(a_{11}a_{22}a_{33}+a_{12}a_{23}a_{31}+a_{13}a_{21}a_{32})-(a_{11}a_{23}a_{32}+a_{12}a_{21}a_{33}+a_{13}a_{22}a_{31}) \quad (5)$$

In Figure 22 is shown the functional ties between elements of the 3x3 matrix at expansion of its elements to the first row.

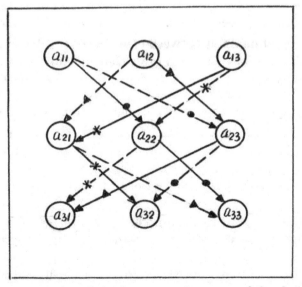

Figure 22 Functional ties between elements of the determinant of the 3x3 matrix for expansion of its elements to the first row

(------- negative ties)

7

- If the second row of A is selected, then we have the following data in general view:
$$|A|=a_{21}A_{21}+a_{22}A_{22}+a_{23}A_{23}$$
where,
$A_{21}=(-1)^{2+1}M_{21}=-(a_{12}a_{33}-a_{13}a_{32})$
$A_{22}=(-1)^{2+2}M_{22}=(a_{11}a_{33}-a_{13}a_{31})$
$A_{23}=(-1)^{2+3}M_{23}=-(a_{11}a_{32}-a_{12}a_{31})$ and the determinant $|A|$ of the 3x3 matrix is equal in general view:

$$|A|=-a_{21}(a_{12}a_{33}-a_{13}a_{32})+a_{22}(a_{11}a_{33}-a_{13}a_{31})-a_{23}(a_{11}a_{32}-a_{12}a_{31})=$$
$$=(a_{21}a_{13}a_{32}+a_{22}a_{11}a_{33}+a_{23}a_{12}a_{31})-(a_{21}a_{12}a_{33}+a_{22}a_{13}a_{31}+a_{23}a_{11}a_{32}) \quad (6)$$

In Figure 23 is shown the functional ties between elements of the determinant of the 3x3 matrix for expansion of its elements to the second row.

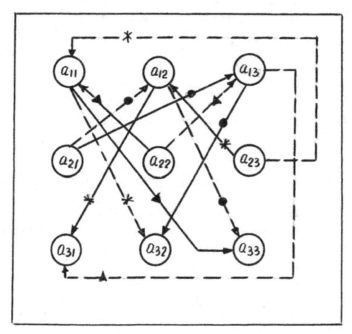

**Figure 23 Functional ties between elements of the determinant
of the 3x3 matrix for expansion of its elements to the second row**
(------negative ties)

- If the third row of A is selected ,then we have the following data in general view:
$$|A|=a_{31}A_{31}+a_{32}A_{32}+a_{33}A_{33}$$
where,
$A_{31}=(-1)^{3+1}M_{31}=(a_{12}a_{23}-a_{13}a_{22})$
$A_{32}=(-1)^{3+2}M_{32}=-(a_{11}a_{23}-a_{13}a_{21})$
$A_{33}=(-1)^{3+3}M_{33}=(a_{11}a_{22}-a_{12}a_{21})$ and the determinant $|A|$ of the 3x3 matrix is equal in general view:

$$|A|=a_{31}(a_{12}a_{23}-a_{13}a_{22})-a_{32}(a_{11}a_{23}-a_{13}a_{21})+a_{33}(a_{11}a_{22}-a_{12}a_{21})=$$
$$=(a_{31}a_{12}a_{23}+a_{32}a_{13}a_{21}+a_{33}a_{11}a_{22})-(a_{31}a_{13}a_{22}+a_{32}a_{11}a_{23}+a_{33}a_{12}a_{21}) \quad (7)$$

In Figure 24 is shown the functional ties between elements of the determinant of the 3x3 matrix for expansion of its elements to the third row.

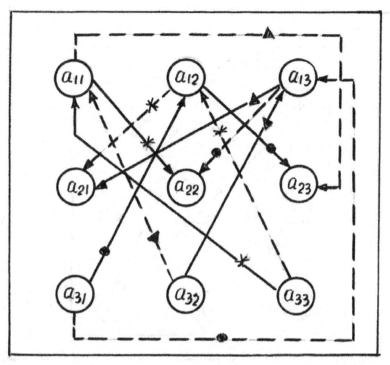

**Figure 24 Functional ties between elements of the determinant
of the 3x3 matrix for expansion of its elements to the third row**
(-------negative ties)

❖ For the columns of the 3x3 matrix

- If the first column of A is selected ,then we have the following data in general view:
$$|A|=a_{11}A_{11}+a_{21}A_{21}+a_{31}A_{31}$$
where,
$A_{11}=(-1)^{1+1}M_{11}=(a_{22}a_{33}-a_{23}a_{32})$
$A_{21}=(-1)^{2+1}M_{21}=-(a_{12}a_{33}-a_{13}a_{32})$
$A_{31}=(-1)^{3+1}M_{31}=(a_{12}a_{23}-a_{13}a_{22})$ and the determinant $|A|$ of the 3x3 matrix is equal in general view:

$$|A|=a_{11}(a_{22}a_{33}-a_{23}a_{32})-a_{21}(a_{12}a_{33}-a_{13}a_{32})+a_{31}(a_{12}a_{23}-a_{13}a_{22})=$$
$$=(a_{11}a_{22}a_{33}+a_{21}a_{13}a_{32}+a_{31}a_{12}a_{23})-(a_{11}a_{23}a_{32}+a_{21}a_{12}a_{33}+a_{31}a_{13}a_{22}) \quad (8)$$

In Figure 25 is shown the functional ties between elements of the determinant of the 3x3 matrix for expansion of its elements to the first column.
- If the second column of A is selected ,then we have the following data in general view:
$$|A|=a_{12}A_{12}+a_{22}A_{22}+a_{32}A_{32}$$
where,
$A_{12}=(-1)^{2+1}M_{12}=-(a_{21}a_{33}-a_{23}a_{31})$
$A_{22}=(-1)^{2+2}M_{22}=(a_{11}a_{33}-a_{13}a_{31})$
$A_{32}=(-1)^{2+3}M_{32}=-(a_{11}a_{23}-a_{13}a_{21})$

and the determinant $|A|$ of the 3x3 matrix is equal in general view:

9

$$|A| = -a_{12}(a_{21}a_{33}-a_{23}a_{31})+a_{22}(a_{11}a_{33}-a_{13}a_{31})-a_{32}(a_{11}a_{23}-a_{13}a_{21})=$$
$$=(a_{12}a_{23}a_{31}+a_{22}a_{11}a_{33}+a_{32}a_{13}a_{21})-(a_{12}a_{21}a_{33}+a_{22}a_{13}a_{31}+a_{32}a_{11}a_{23}) \quad (9)$$

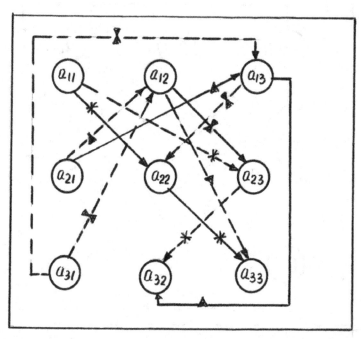

**Figure 25 Functional ties between elements of the determinant
of the 3x3 matrix for expansion of its elements to the first column**
(-------- negative ties)

In Figure 26 is shown the functional ties between elements of the determinant of the 3x3 matrix for expansion of its elements to the second column.

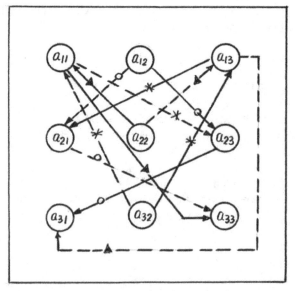

**Figure 26   Functional ties between elements of the determinant
of the 3x3 matrix for expansion of its elements to the second column**
(-------- negative ties)

- If the third column of A is selected ,then we have the following data in general view:
$$|A|=a_{13}A_{13}+a_{23}A_{23}+a_{33}A_{33}$$

where,

$A_{13}=(-1)^{3+1}M_{13}=(a_{21}a_{32}-a_{22}a_{31})$

$A_{23}=(-1)^{3+2}M_{23}=-(a_{11}a_{32}-a_{12}a_{31})$

$A_{33}=(-1)^{3+3}M_{33}=(a_{11}a_{22}-a_{12}a_{21})$ and the determinant $|A|$ of the 3x3 matrix is equal in general view:

$$|A|=a_{13}(a_{21}a_{32}-a_{22}a_{31})-a_{23}(a_{11}a_{32}-a_{12}a_{31})+a_{33}(a_{11}a_{22}-a_{12}a_{21})=$$
$$=(a_{13}a_{21}a_{32}+a_{23}a_{12}a_{31}+a_{33}a_{11}a_{22})-(a_{13}a_{22}a_{31}+a_{23}a_{11}a_{32}+a_{33}a_{12}a_{21}) \quad (10)$$

In Figure 27 is shown the functional ties between elements of the determinant of the 3x3 matrix for expansion of its elements to the third column.

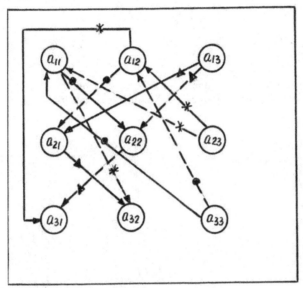

Figure 27 Functional ties between elements of the determinant of the 3x3 matrix for expansion of its elements to the third column
(-------- negative ties)

C. **Laplace's theorem (method) [2]**

The essential aspects of Laplace's theorem are expressed in the following form:
".....If we in the determinant of the nxn matrix mark out randomly k-rows (or columns) [$1\leq k\leq n-1$] ,then value of the determinant is equal of sum product all minors of k-order elements formed from the above-marked out rows(or columns)on their algebraic supplemental (cofactors)........'.
Applicably to the determinant of the 3x3 matrix in general view we have the following data:

$$A=\begin{vmatrix} a_{11} & a_{12} & a_{13} \\ a_{21} & a_{22} & a_{23} \\ a_{31} & a_{32} & a_{33} \end{vmatrix}$$

a) **For the first and second rows**
- In the 3x3 matrix we mark out randomly two paired rows ,for a instance the first(i_1) and second (i_2) rows and write all kinds of minors the second order (M_{ij}) and algebraic supplemental (A_{ij}) ,as this shown in sample(a);

11

- And then in the 3x3 matrix we mark out the first (j_1) and second (j_2) two paired columns and write the minor (M_1) and algebraic supplemental(cofactor) (A_1) ,as this shown in sample (a):

(a)

$M_1=(a_{11}a_{22}-a_{12}a_{21})$; $A_1=(-1)^{(1+2)+(1+2)}(a_{33})=a_{33}$

- And then in the 3x3 matrix we marked out the first(j_1) and third(j_3) two paired columns and write the minor (M_2) and algebraic supplemental (cofactor) A_2,as this shown in sample (b):

(b)

$M_2=(a_{11}a_{23}-a_{13}a_{21})$; $A_2=(-1)^{(1+2)+(1+3)}(a_{32})=-a_{32}$

- And then in the 3x3 matrix we mark out the second (j_2) and third (j_3) two paired columns and write the minor (M_3) and algebraic supplemental (cofactor) A_3,as this shown in sample (c):

(c)

$M_3=(a_{12}a_{23}-a_{13}a_{22})$; $A_3=(-1)^{(1+2)+(2+3)}(a_{31})=a_{31}$

So, the determinant of the 3x3 matrix is defined by formula:

$$|A|=a_{33}(a_{11}a_{22}-a_{12}a_{21})-a_{32}(a_{11}a_{23}-a_{13}a_{21})+a_{31}(a_{12}a_{23}-a_{13}a_{22}) \quad (11)$$

In Table 3 is shown evaluation of the determinants of the 3x3 matrices at combination of its elements to the zero value

Table 3 Evaluation of the determinants of the 3x3 matrices at combination of its elements to the zero value

n/n	Zero value for elements									Value of determinant $|A|$ in general view
	a_{11}	a_{12}	a_{13}	a_{21}	a_{22}	a_{23}	a_{31}	a_{32}	a_{33}	
1	0	0	0	0	0	0	0	0	0	0
2	0	0	0	-	-	-	-	-	-	0
3	-	-	-	0	0	0	-	-	-	0
4	-	-	-	-	-	-	0	0	0	0
5	0	-	-	0	-	-	0	-	-	0
6	-	0	-	-	0	-	-	0	-	0
7	-	-	0	-	-	0	-	-	0	0
8	0	-	-	-	0	-	-	-	0	$|A|=a_{12}a_{23}a_{31}+a_{13}a_{21}a_{32}$
9	-	-	0	-	0	-	0	-	-	$|A|=-(a_{11}a_{23}a_{32}+a_{12}a_{21}a_{33})$
10	0	0	-	0	0	-	0	0	-	0
11	0	-	0	0	-	0	0	-	0	0
12	-	0	0	-	0	0	-	0	0	0
13	-	-	-	0	0	0	0	0	0	0
14	0	0	0	0	0	0	-	-	-	0
15	0	-	-	-	-	-	-	-	-	$|A|=(a_{12}a_{23}a_{31}+a_{13}a_{21}a_{32})-(a_{13}a_{22}a_{31}+a_{12}a_{21}a_{33})$
16	-	0	-	-	-	-	-	-	-	$|A|=(a_{11}a_{22}a_{33}+a_{13}a_{21}a_{32})-(a_{13}a_{22}a_{31}+a_{11}a_{23}a_{32})$
17	-	-	0	-	-	-	-	-	-	$|A|=(a_{11}a_{22}a_{33}+a_{12}a_{23}a_{31})-(a_{11}a_{23}a_{32}+a_{12}a_{21}a_{33})$
18	-	-	-	0	-	-	-	-	-	$|A|=(a_{11}a_{22}a_{33}+a_{12}a_{23}a_{31})-(a_{13}a_{22}a_{31}+a_{11}a_{23}a_{32})$
19	-	-	-	-	0	-	-	-	-	$|A|=(a_{12}a_{23}a_{31}+a_{13}a_{21}a_{32})-(a_{11}a_{23}a_{32}+a_{12}a_{21}a_{33})$
20	-	-	-	-	-	0	-	-	-	$|A|=(a_{11}a_{22}a_{33}+a_{13}a_{21}a_{32})-(a_{13}a_{22}a_{31}+a_{12}a_{21}a_{33})$
21	-	-	-	-	-	-	0	-	-	$|A|=(a_{11}a_{22}a_{33}+a_{13}a_{21}a_{32})-(a_{11}a_{23}a_{32}+a_{12}a_{21}a_{33})$
22	-	-	-	-	-	-	-	0	-	$|A|=(a_{11}a_{22}a_{33}+a_{12}a_{23}a_{31})-(a_{13}a_{22}a_{31}+a_{12}a_{21}a_{33})$
23	-	-	-	-	-	-	-	-	0	$|A|=(a_{12}a_{23}a_{31}+a_{13}a_{21}a_{32})-(a_{13}a_{22}a_{31}+a_{11}a_{23}a_{32})$
24	0	0	0	0	-	-	-	-	-	0
25	0	0	0	-	0	-	-	-	-	0
26	0	0	0	-	-	0	-	-	-	0
27	0	0	0	-	-	-	0	-	-	0
28	0	0	0	-	-	-	-	0	-	0
29	0	0	0	-	-	-	-	-	0	0
30	0	-	-	0	0	0	-	-	-	0
31	-	0	-	0	0	0	-	-	-	0
32	-	-	0	0	0	0	-	-	-	0
33	0	-	-	0	0	0	-	-	-	0
34	-	-	-	0	0	0	-	0	-	0
35	-	-	-	0	0	0	-	-	0	0
36	0	-	-	-	-	-	0	0	0	0
37	-	0	-	-	-	-	0	0	0	0

38	-	-	0	-	-	-	0	0	0	0
39	-	-	-	0	-	-	0	0	0	0
40	-	-	-	-	0	-	0	0	0	0
41	-	-	-	-	-	0	0	0	0	0
42	0	0	0	0	0	-	-	-	-	0
43	0	0	0	-	0	0	-	-	-	0
44	0	0	0	0	-	0	-	-	-	0
45	0	0	0	-	-	-	0	0	-	0
46	0	0	0	-	-	-	0	-	0	0
47	0	0	0	-	-	-	-	0	0	0
48	0	0	0	-	-	-	-	0	0	0
49	0	0	-	0	0	0	-	-	-	0
50	-	0	0	0	0	0	-	-	-	0
51	0	-	0	0	0	0	-	-	-	0
52	-	-	-	0	0	-	0	0	0	0
53	-	-	-	-	0	0	0	0	0	0
54	-	-	-	0	-	0	0	0	0	0
55	-	-	-	0	0	-	0	0	0	0

Example 1 :

Calculate the determinant of the 3x3 matrix by *Sarrus rule, Expansion (cofactor) and Laplace's methods:*

$$A = \begin{vmatrix} 0 & 2 & -6 \\ 4 & -5 & 1 \\ 3 & -2 & -1 \end{vmatrix}$$

- **By Sarrus rule** we have the value of determinant of the 3x3 matrix:
  $|A| = a_{21}(a_{13}a_{32} - a_{12}a_{33}) + a_{22}(a_{11}a_{33} - a_{13}a_{31}) + a_{23}(a_{12}a_{31} - a_{11}a_{32}); \; |A| = -28$
- **By Expansion (cofactor) method**, for a instance, the third column we have the value of determinant of the 3x3 matrix:
  $|A| = a_{13}(a_{21}a_{32} - a_{22}a_{31}) - a_{23}(a_{11}a_{32} - a_{12}a_{31}) + a_{33}(a_{11}a_{22} - a_{12}a_{21}); \; |A| = -28$
- **By Laplace's theorem(method)**, we have the value of determinant of the 3x3 matrix:
  $|A| = a_{33}(a_{11}a_{22} - a_{12}a_{21}) - a_{32}(a_{11}a_{23} - a_{13}a_{21}) + a_{31}(a_{12}a_{23} - a_{13}a_{22}); \; |A| = -28$

The same value for the determinant of the 3x3 matrix is obtained in above-indicated calculations for three considered methods.

D. Vector analysis for the determinant of the 3x3 matrix

Now we consider of some combinations the 3x3 matrices for evaluation of the determinants by vector analysis.

For Figure 28

$$A = \begin{vmatrix} a_{11} & a_{12} & a_{13} \\ a_{21} & a_{22} & a_{23} \\ a_{31} & a_{32} & a_{33} \end{vmatrix}$$

For Figure 29

$$A = \begin{vmatrix} -a_{11} & -a_{12} & -a_{13} \\ -a_{21} & -a_{22} & -a_{23} \\ -a_{31} & -a_{32} & -a_{33} \end{vmatrix}$$

For Figure 30

$$A = \begin{vmatrix} a_{11} & a_{12} & a_{13} \\ a_{21} & a_{22} & a_{23} \\ a_{31} & a_{32} & -a_{33} \end{vmatrix}$$

For Figure 31

$$A = \begin{vmatrix} a_{11} & a_{12} & a_{13} \\ -a_{21} & -a_{22} & -a_{23} \\ -a_{31} & -a_{32} & -a_{33} \end{vmatrix}$$

14

In Figure 28 ÷31 are shown by vector analysis the value of determinant in view of volume of the parallelepiped for some combinations of the given elements of the 3x3 matrices , presenting the sides of the above-named parallelepiped by the vectors in general view.

Analysis of Figure 28 shows the following results :
- All elements of the 3x3 matrix have the positive values
- The value of vectors V_1, V_2 and V_3 also have the positive values
- The determinant of the 3x3 matrix is expressed in view of volume (V) of parallelepiped which is formed on base of positive vectors V_1, V_2 and V_3 and is equal ,as $V = |S| * |H|$, where :

S- area of parallelepiped is in absolute value
H- height of parallelepiped is also in absolute value

Analysis of Figure 29 shows the following results :
- All elements of the 3x3 matrix have the negative values
- The value of vectors V_1, V_2 and V_3 also have the negative values
- The determinant of the 3x3 matrix is expressed in view of volume(V) of parallelepiped which is formed on base of negative vectors V_1, V_2 and V_3 and it volume is equal ,as $V = |S| * |H|$.

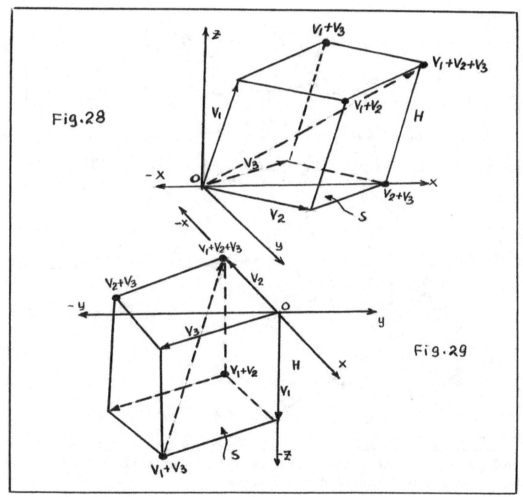

Figures 28÷29 Vector analysis the value of determinant of the 3x3 matrices at positive (Fig.28) and negative (Fig.29) values of all its elements

Analysis of Figure 30 shows the following results:
- All elements of the 3x3 matrix have the positive values ,besides of element(a_{33}) which has the negative value
- The vectors V_1 and V_2 have the positive values
- The vector V_3 has the negative value
- The determinant of the 3x3 matrix also is expressed in view of volume(V) of parallelepiped which is formed on base positive vectors V_1 and V_2 and negative vector V_3 and its volume is equal as $V=|S|*|H|$.

Analysis of Figure 31 shows the following results:
- All elements of the second and third rows of the 3x3 matrix have the negative values
- All elements of the first row of the 3x3 matrix has the positive values
- The value of the first vector V_1 has the positive value
- The value of the second V_2 and third V_3 vectors have the negative values
- The determinant of the 3x3 matrix also is expressed in view of volume (V) of parallelepiped which is formed on base of positive vector V_1 and negative vectors V_2 and V_3 and its volume is equal as $V=|S|*|H|$.

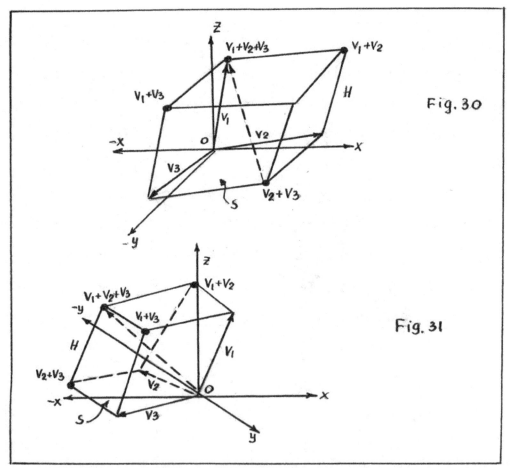

Figure 30÷31 Vector analysis the value of the determinant of the 3x3 matrices at combined values of its elements and formed vectors

16

E. Comparative analysis in evaluation of the determinant of the 3x3 matrix by the different methods

Referring to above-indicated three methods in question of evaluation the determinant of the 3x3 matrix, we can now in comparative form to analyze the advantages of each method. In Figure 32 is shown the diagrams for evaluation of the determinant of the 3x3 matrix obtained by each method in general view.

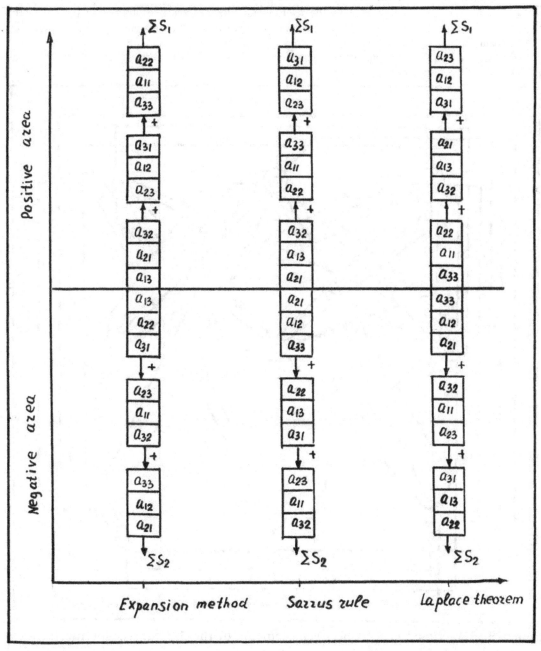

Figure 32 Comparative diagrams in evaluation of the determinant of the 3x3 matrix by the different methods ($\sum S_1$, $\sum S_2$ -sums of elements for positive and negative areas)

Analysis of data shown in Figure 32 shows the following results:

- Total quantity of elements ,participation in evaluation of the determinant of the 3x3 matrix are equal 18(9-for positive area and 9-for negative area);
- For each method presents the following similar elements, participating in functional ties ,as for positive and also for negative areas and these elements are such as:

    $a_{11}, a_{12}, a_{13}, a_{21}, a_{22}, a_{23}, a_{31}, a_{32}, a_{33}$

- However, the order of functional ties between elements of the 3x3 matrix for each method is the different, as this is shown in Example and diagrams in Figure 32;
- And besides the result of value for the determinant of the 3x3 matrix is obtained by each method is the same, i.e the value of the determinant is constant parameter for given 3x3 matrix in numerical data and equal $\lvert A \rvert = -28$.

In Figure 33 is shown the functional graphs between elements of the 3x3 matrix for the different methods in evaluation of the determinant.

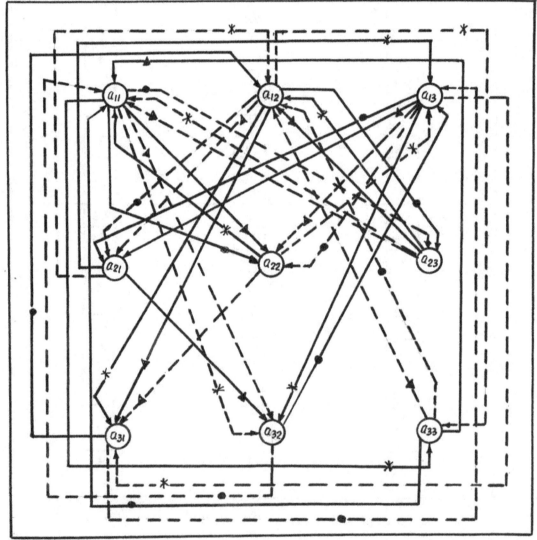

Figure 33 Functional graphs between elements of the 3x3 matrix for the different methods in evaluation of the determinant

(▲ Expansion method; * Sarrus rule; ●Laplace's theorem ; --------------negative area)

18

Analysis of data shown in Figure 33 , considering the functional graphs between elements of the 3x3 matrix for the different methods in evaluation of the determinant indicates on the following results:

a) For Expansion method of the 3x3 matrices

- The third element (a_{13})of the first row has ties with the first element(a_{21}) of the second row and the second element(a_{32}) of the third row for positive area; and also with the second element(a_{22}) of the second row and the first element (a_{31}) of the third row for negative area;
- The third element (a_{23}) of the second row has ties with the first element(a_{11}) of the first row and the second element(a_{32})of the third row for negative area; and also with the second element (a_{12}) of the first row and the first element(a_{31}) of the third row for positive area;
- The third element(a_{33}) of the third row has ties with the first element(a_{11}) of the first row and the second element (a_{22}) of the second row for positive area; and also with the second element(a_{12}) of the first row and the first element(a_{21})of the second row for negative area.

b) For Sarrus method of the 3x3 matrices

- The first element(a_{21}) of the second row has ties with the third element (a_{13}) of the first row and the second element(a_{32}) of the third row for positive area; and also with the second element(a_{12}) of the first row and the third element(a_{33}) of the third row for negative area;
- The second element (a_{22}) of the second row has ties with the first element (a_{11}) of the first row and the third element(a_{33}) of the third row for positive area; and also with the third element (a_{13}) of the first row and the first element(a_{31}) of the third row for negative area;
- The third element (a_{23}) of the second row has ties with the second element(a_{12}) of the first row and the first element(a_{31}) of the third row for positive area; and also with the first element(a_{11}) of the first row and the second element (a_{32}) of the third row for negative area.

c) For Laplace's theorem of the 3x3 matrices

- The third element(a_{33}) of the third row has ties with the first element(a_{11}) of the first row and the second element(a_{22}) of the second row for positive area; and also with the second element (a_{12})of the first row and the first element(a_{21}) of the second row for negative area;
- The second element(a_{32}) of the third row has ties with the first element (a_{11}) of the first row and the third element(a_{23}) of the second row for positive area; and also with the third element(a_{13}) of the first row and the first element (a_{21}) of the second row for negative area;
- The first element (a_{31})of the third row has ties with the second row(a_{12})of the first row and the third element (a_{23}) of the second row for positive area; and also with the third element (a_{13}) of the first row and the second element(a_{22}) of the second row for negative area.

So, from Figure 33 we see that despite on the difference of functional order ties between elements of the 3x3 matrix, the value of determinant is the same. And this fact means that value of the determinant of the 3x3 matrix can be evaluated by any convenience above-indicated method for given matrix in numerical data.

CHAPTER TWO GENERAL ALGORITHMS FOR EVALUATION OF DETERMINANTS WITH 4X4 MATRICES

1. Problems for solving of the determinants of the matrices, having size of n ≥4

As we see that evaluation of the determinants of the matrices, having sizes of n≤3 does not have considerable difficulties. However, with increasing of sizes of matrices with n≥3 considerable increases the operational calculations for solving of any determinant and this fact confirms by data of author **[5]**.

And this reason can explain by the following fact that in process of calculation of the determinant appears a lot of the *incoming (primary) elements (N_1)* of the given matrix, and also forms the *block (sectional) elements (N_2) and single elements (N_3)*, incoming to positive and negative areas in process of evaluation the determinant on any matrix. In Table 4 is shown the distribution of used elements for any considered matrix at evaluation of its determinant.

Table 4 Distribution of used elements for any considered matrix at evaluation of its determinant

Size of matrix (n x n)	Used elements at evaluation of the determinant		
	Incoming(primary) elements ,N_1	Calculations elements	
		Block (sectional),N_2 [1]	Single elements,N_3
2x2	4	2	4
3x3	9	6	18
4x4	16	24	96 [8]
5x5	25	120	600
6x6	36	720	2230*
7x7	49	840	6750*
8x8	64	40320	17400*
9x9	81	362880	40100*
10x10	100	3628800	86320*

where,

$N_2 = n! = 1 \times 2 \times 3 \ldots n$; * forecasting data

In Table 5 is shown the data for incoming (primary) elements (N_1) in process of evaluation the determinants of the different matrices. In Table 6 is shown the data for definition of block (sectional) elements (N_2) in process of evaluation the determinants of the different matrices. And in Table 7 is shown the data for definition of single elements (N_3), incoming to the positive and negative areas at evaluation of the determinant of the different matrices.

Table 5 Data for definition of the incoming (primary) elements (N_1)

y(N_1)	x(n)	logx	logy	(logx)(logy)	$(logx)^2$	$(logy)^2$	Y_c	$(logy_c)^2$	$(y-y_c)^2$	$(y-y')^2$
4	2	0.301	0.602	0.181	0.091	0.362	4.05	0.369	0.003	1495.3
9	3	0.477	0.954	0.455	0.228	0.911	9.06	0.916	0.004	1133.7
16	4	0.602	1.204	0.724	0.362	1.449	16.1	1.453	0.003	711.3
25	5	0.699	1.398	0.977	0.489	1.954	25.01	1.954	0	312.2
36	6	0.778	1.556	1.211	0.606	2.422	36.92	2.419	0.006	44.4
49	7	0.845	1.690	1.428	0.714	2.856	48.8	2.851	0.040	40.1
64	8	0.903	1.806	1.631	0.816	3.262	63.59	3.252	0.168	454.9
81	9	0.954	1.908	1.820	0.911	3.642	80.35	3.629	0.423	1469
100	10	1.000	2.000	2.000	1.000	4.000	99.05	3.983	0.903	3287
Total: 384	54	6.559	13.12	10.427	5.217	20.86		20.83	1.55	8948

Mean: y'=42.67;x'=6	
Standard error σ=2.31	1.$\sum logy = nloga + b\sum logx$
Coeff. determination R^2=0.99	2. $\sum(logx)(logy) = loga\sum logx + b\sum(logx)^2$ (12)
Coeff. of correlation r=0.99	

The data taken from Table 5 permits to solve two normal equations (12) and find the non-linear regression equation. After of solving the above-named two normal equations, we have the following data: loga=0.011 ;b=1.986 and regression equation has view:

$$logY_c = 0.011 + 1.986 logX \quad \text{or}$$
$$Y_c = 1.023(X)^{1.986} \quad \text{or}$$
$$N_1 = 1.023(n)^{1.986} \quad (13)$$

Analyzing the data from Table 5 and regression equation in view of $N_1=1.023(n)^{1.986}$, we can determine the statistical parameters of this regression model:

1) Standard error is calculated by formula
$\sigma= \sum(\log y_c)^2/n=[\sum(\log y)^2-\sum(\log y_c)^2]/n$; $\sigma= 2.314$
2) Coefficients of determination $(R^2)=[\sum(y-y')^2-\sum(y-y_c)^2]/[\sum(y-y')^2]$; $R^2=0.999$
3) Coefficient of correlation is equal $r=(R^2)^{0.5}$; $r=0.999$

Table 6 Data for definition of the block (sectional) elements (N_2)

$y(N_2)$	$x(n)$	$\log x$	$\log y$	$(\log x)^2$	$(\log y)^2$	$(\log x)(\log y)$	y_c	$(\log y_c)^2$	$(y-y')^2$	$(y-y_c)^2$
2	2	0.301	0.301	0.091	0.091	0.091	2	0.091	$2\cdot10^{11}$	0
6	3	0.477	0.778	0.227	0.606	0.371	74	3.494	$2\cdot10^{11}$	0
24	4	0.602	1.380	0.362	1.905	0.831	933	8.820	$2\cdot10^{11}$	0
120	5	0.698	2.079	0.489	4.323	1.451	$6.6\cdot10^3$	14.615	$2\cdot10^{11}$	0
720	6	0.778	2.857	0.606	8.164	2.222	$3.3\cdot10^4$	20.428	$2\cdot10^{11}$	10^{11}
5040	7	0.845	3.702	0.714	13.708	3.128	$1.3\cdot10^5$	26.1	$2\cdot10^{11}$	10^{11}
$40\cdot10^3$	8	0.903	4.606	0.816	21.211	4.159	$4.2\cdot10^5$	31.6	$2\cdot10^{11}$	10^{11}
$36\cdot10^4$	9	0.954	5.559	0.911	30.911	5.303	$11\cdot10^5$	36.838	10^{11}	$7\cdot10^{11}$
$36\cdot10^5$	10	1.000	6.559	1.000	43.031	6.559	$3\cdot10^7$	41.888	10^{13}	$5\cdot10^{11}$
Total: $4\cdot10^6$	54	6.558	27.82	5.216	123.95	24.115			183.51	$116\cdot10^{11}$	$13\cdot10^{11}$

Mean: $y'=4.5\cdot10^5$;$x'=6$	
Standard error: $\sigma=20.4$	1) $\sum\log y=n\log a+ b\sum\log x$
Coeff.of determination $R^2=0.887$	2) $\sum(\log x\cdot\log y)=\log a\sum\log x+b\sum(\log x)^2$
Coeff.of correlation r=0.942	

The data taken from Table 6 permits to solve the above-indicated two normal equations and find the non-linear regression equation. After of solving the above-named two normal equations , we have the following data : $\log a= -3.326$;$b=8.8$ and regression equation has view

$$\log Y_c= -3.326+8.8\log X \text{ and}$$
$$Y_c=4.7\cdot10^{-3}(X)^{8.8} \quad (14) \text{ or}$$
$$N_2=4.7\cdot10^{-3}(n)^{8.8}$$

Analyzing of data from Table 6 and regression equation in view of $N_2=4.7\cdot10^{-3}(n)^{8.8}$,we can determine the statistical parameters of this regression model:

1) Standard error is calculated by formula:
$\sigma=\sum(\log Y_c)^2/n=[\sum(\log y)^2-\sum(\log y_c)^2]/n$; $\sigma=20.4$
2) Coefficient of determination (R^2) is equal:
$R^2=[\sum(y-y')^2-\sum(y-y_c)^2/\sum(y-y')^2$; $R^2=0.887$
3) Coefficient of correlation (r) is equal: $r=(R^2)^{0.5}$; $r=0.942$

In Figure 34 is shown the comparative analysis of quantity used elements for matrix determinant , having sizes of $2\leq n\leq10$,but the considered non-linear models can be fit for receiving the approximate forecasting data in question of evaluation of any determinant ,having square nxn matrix.

Table 7 Data for definition of the single elements N_3

$y(N_3)$	$x(n)$	$\log x$	$\log y$	$(\log x\cdot\log y)$	$(\log x)^2$	$(\log y)^2$	y_c	$(\log y_c)^2$	$(y-y')^2$	$(y-y_c)^2$
4	2	0.301	0.602	0.181	0.091	0.362	2	0.091	$3\cdot10^8$	10^3
18	3	0.477	1.255	0.598	0.228	1.576	27	2.049	$3\cdot10^8$	10^3
96	4	0.602	1.982	1.193	0.362	3.929	174	5.021	$3\cdot10^8$	$6\cdot10^3$
600	5	0.699	2.778	1.942	0.489	7.718	740	8.232	$3\cdot10^8$	$20\cdot10^3$
$2\cdot10^3$	6	0.778	3.348	2.605	0.606	11.211	2409	11.437	$2\cdot10^8$	$32\cdot10^3$
$7\cdot10^3$	7	0.845	3.829	3.235	0.714	14.663	6529	14.553	10^8	$48\cdot10^3$

17·10³	8	0.903	4.241	3.829	0.816	17.982	15·10³	17.556	10⁶	0.1·10⁸
40·10³	9	0.954	4.603	4.391	0.911	21.188	33·10³	20.439	5·10⁸	0.5·10⁸
86·10³	10	1.000	4.936	4.936	1.000	24.365	65·10³	23.203	48·10⁸	3·10⁸
Total: 152·10³	54	6.559	27.574	22.91	5.217	102.99		102.6	69·10⁸	3.6·10⁸

Mean: y'=17058; x'=6	
Standard error: σ= 11.39	1) $\sum logy = nloga + b\sum logx$
Coeff. of determination R^2=0.949	2) $\sum(logx \cdot logy) = loga\sum logx + b\sum(logx)^2$
Coeff. of correlation r=0.974	

The data taken from Table 7 permits to solve the above-indicated two normal equations and find the regression non-linear equation. So , we have the following data after of solving two equations:
loga = —1.652; b=6.469 and regression model has the following view

$$logY_c = -1.652 + 6.469 logX \quad \text{and}$$
$$Y_c = 0.022(X)^{6.469} \quad (15) \text{ or}$$
$$N_3 = 0.022(n)^{6.469}$$

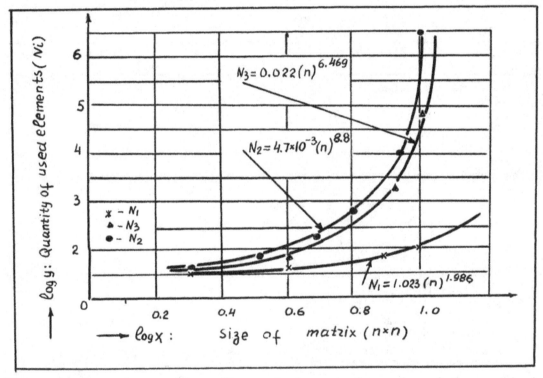

Figure 34 Dependence of quantity used elements from size of the matrix in evaluation of the determinant

Analysis of data indicated in Figure 34 shows fact that with increasing of size the matrix determinant considerably increases the quantity of used elements .And this value is more increasing for the quantity block elements (N_2) which participates in process of evaluation the matrix determinant in compare with the primary (incoming elements (N_1) of considered matrix ,where the quantity is considerably less.

The quantity of single elements (N_3) ,participating in process of evaluation the matrix determinant forms the average position between the values of quantity (N_1) and (N_2).From Figure 32 we also see that with increasing larger size of matrix determinant ,for a instance, of the 5x5 matrix and more ,the quantity of any above-said elements considerably increases and this factor makes the process of estimation of the matrix determinant rather complex and unpredictable and further demands the additional special algorithms and computer programs.

And these conclusions are confirmed by the following calculation procedures, on base of the square matrices, having sizes of $4 \leq n \leq 6$, where author tries with mathematical formulas and functional graphs to show the complexity of whole computational processes of evaluation of the matrix determinant.

2. The general algorithms and functional graphs for definition of the determinant of the 4x4 matrix

2.1 By expansion (cofactor)method for the rows and columns

Theory of determinants advantageously is joined with studying of a system linear equation which can be introduced in general view as:

$$a_{11}X_1+a_{12}X_2+a_{13}X_3+....a_{1n}X_1=b_1$$
$$a_{21}X_1+a_{22}X_2+a_{23}X_3+... a_{2n}X_n=b_2$$
$$a_{31}X_1+a_{32}X_2+a_{33}X_3+....a_{3n}X_n=b_3$$
$$.......................................$$
$$a_{n1}X_1+a_{n2}X_2+a_{n3}X_3+....a_{nn}X_n=b_n$$

where,

$a_{11},a_{12},a_{13}......a_{1n}$—elements of the first row
$a_{21},a_{22},a_{23}......a_{2n}$-elements of the second row
$a_{31},a_{32},a_{33}......a_{3n}$-elements of the third row
....
$a_{n1},a_{n2},a_{n3}......a_{nn}$-elements of n-rows
$X_1,X_2,X_3......X_n$-unknown parameters of system
$b_1,b_2,b_3........b_n$-free elements ($b_n \neq 0$ for not uniform system)
So, the above-indicated linear system can be expressed in general view as square n-by -n matrix:

$$A = \begin{vmatrix} a_{11} & a_{12} & a_{13}.......a_{1n} \\ a_{21} & a_{22} & a_{23}.......a_{2n} \\ a_{31} & a_{32} & a_{33}.......a_{3n} \\ .... \\ a_{n1} & a_{n2} & a_{n3}.......a_{nn} \end{vmatrix}$$

a) For the rows

- ### To the first row

Applicably to the determinant of the 4x4 matrix, we have the following conditions when elements of the matrix determinant have expansion to the first row:

$$|A| = \sum_{j=1}^{4} a_{1j}A_{1j} = \sum_{j=1}^{4} a_{1j}(-1)^{1+j}M_{1j} \qquad (16)$$

where,

a_{1j}— elements of the first row
M_{1j}—minor of algebraic supplemental (cofactor) A_{1j} of element a_{1j} taken with $(-1)^{1+j}$.
Applicably to the first row, we have the value of determinant in general view as:

$$|A| = a_{11}A_{11}+a_{12}A_{12}+a_{13}A_{13}+a_{14}A_{14}$$

where,

$$A_{11}=(-1)^{1+1}\begin{vmatrix} a_{22} & a_{23} & a_{24} \\ a_{32} & a_{33} & a_{34} \\ a_{42} & a_{43} & a_{44} \end{vmatrix} \qquad A_{12}=(-1)^{1+2}\begin{vmatrix} a_{21} & a_{23} & a_{24} \\ a_{31} & a_{33} & a_{34} \\ a_{41} & a_{43} & a_{44} \end{vmatrix}$$

$$A_{13}=(-1)^{1+3}\begin{vmatrix} a_{21} & a_{22} & a_{24} \\ a_{31} & a_{32} & a_{34} \\ a_{41} & a_{42} & a_{44} \end{vmatrix} \qquad A_{14}=(-1)^{1+4}\begin{vmatrix} a_{21} & a_{22} & a_{23} \\ a_{31} & a_{32} & a_{33} \\ a_{41} & a_{42} & a_{43} \end{vmatrix}$$

After of some transformations, we finally have the value for determinant of the 4x4 matrix elements of which have *expansion to the first row* in general view:

$$\begin{aligned}
|A| = & a_{11}[a_{22}(a_{33}a_{44}-a_{34}a_{43})+a_{23}(a_{34}a_{42}-a_{32}a_{44})+a_{24}(a_{32}a_{43}-a_{33}a_{42})]+ \\
& +a_{12}[a_{21}(a_{34}a_{43}-a_{33}a_{44})+a_{23}(a_{31}a_{44}-a_{34}a_{41})+a_{24}(a_{33}a_{41}-a_{31}a_{43})]+ \\
& +a_{13}[a_{21}(a_{32}a_{44}-a_{34}a_{42})+a_{22}(a_{34}a_{41}-a_{31}a_{44})+a_{24}(a_{31}a_{42}-a_{32}a_{41})]+ \\
& +a_{14}[a_{21}(a_{33}a_{42}-a_{32}a_{43})+a_{22}(a_{31}a_{43}-a_{33}a_{41})+a_{23}(a_{32}a_{41}-a_{31}a_{42})] \quad (17)
\end{aligned}$$

In Figure 35 is shown the functional graphs for the determinant of the 4x4 matrix in general view *for the positive area* and in Figure 36-*for the negative area.*

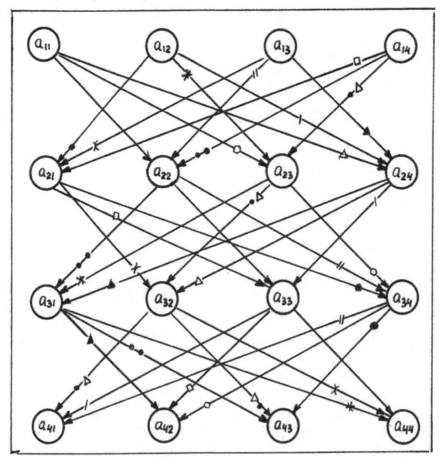

Figure 35 Functional graphs for the determinant of the 4x4 matrix elements of which have expansion to the first row for the positive area

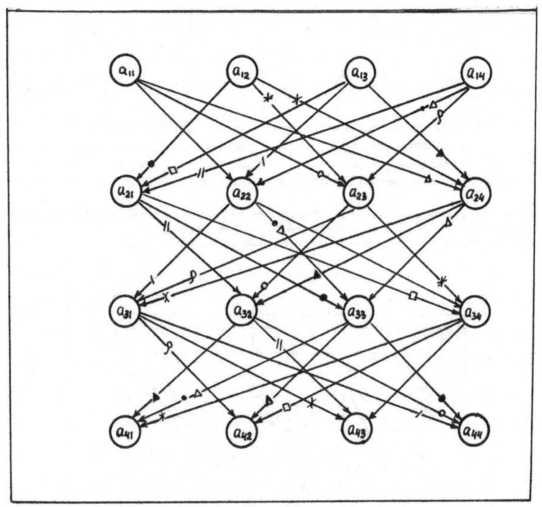

Figure 36 Functional graphs for the determinant of the 4x4 matrix elements of which have expansion to the first row for the negative area

- **To the second row**

The value of determinant of the 4x4 matrix elements of which have expansion to the second row can be expressed by the following formula:

$$|A| = \sum_{j=1}^{4} a_{2j}A_{2j} = \sum_{j=1}^{4} a_{2j}(-1)^{2+j}M_{2j} \qquad (18)$$

For this case , the value of determinant is equal:

$$|A| = a_{21}A_{21} + a_{22}A_{22} + a_{23}A_{23} + a_{24}A_{24}$$

where,

$$A_{21} = (-1)^{2+1} \begin{vmatrix} a_{12} & a_{13} & a_{14} \\ a_{32} & a_{33} & a_{34} \\ a_{42} & a_{43} & a_{44} \end{vmatrix} \qquad A_{22} = (-1)^{2+2} \begin{vmatrix} a_{11} & a_{13} & a_{14} \\ a_{31} & a_{33} & a_{34} \\ a_{41} & a_{43} & a_{44} \end{vmatrix}$$

25

$$A_{23}=(-1)^{2+3}\begin{vmatrix} a_{11} & a_{12} & a_{14} \\ a_{31} & a_{32} & a_{34} \\ a_{41} & a_{42} & a_{44} \end{vmatrix} \qquad A_{24}=(-1)^{2+4}\begin{vmatrix} a_{11} & a_{12} & a_{13} \\ a_{31} & a_{32} & a_{33} \\ a_{41} & a_{42} & a_{43} \end{vmatrix}$$

After of some transformations, we finally have the value for determinant of the 4x4 matrix elements of which have expansion to the second row in general view:

$$\begin{aligned} |A| =\ & a_{21}[a_{12}(a_{34}a_{43}-a_{33}a_{44})+a_{13}(a_{32}a_{44}-a_{34}a_{42})+a_{14}(a_{33}a_{42}-a_{32}a_{43})]+ \\ &+a_{22}[a_{11}(a_{33}a_{44}-a_{34}a_{43})+a_{13}(a_{34}a_{41}-a_{31}a_{44})+a_{14}(a_{31}a_{43}-a_{33}a_{41})]+ \\ &+a_{23}[a_{11}(a_{34}a_{42}-a_{32}a_{44})+a_{12}(a_{31}a_{44}-a_{34}a_{41})+a_{14}(a_{32}a_{41}-a_{31}a_{42})]+ \\ &+a_{24}[a_{11}(a_{32}a_{43}-a_{33}a_{42})+a_{12}(a_{33}a_{41}-a_{31}a_{43})+a_{13}(a_{31}a_{42}-a_{32}a_{41})] \quad (19) \end{aligned}$$

In Figure 37 is shown the functional graphs between elements of the matrix determinant 4x4 at expansion of its elements *to the second row for the positive area* and in Figure 38 *–for the negative area.*

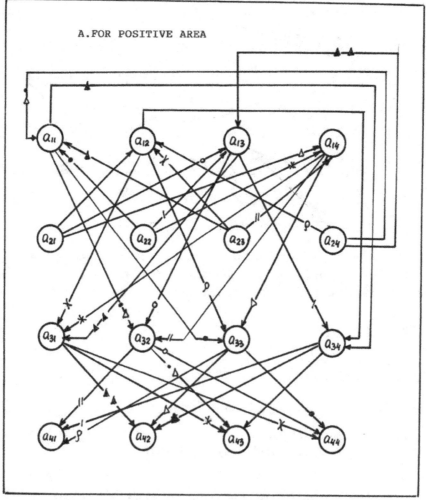

Figure 37 Functional graphs between elements of the determinant of the 4x4 matrix at expansion of its elements to the second row for the positive area

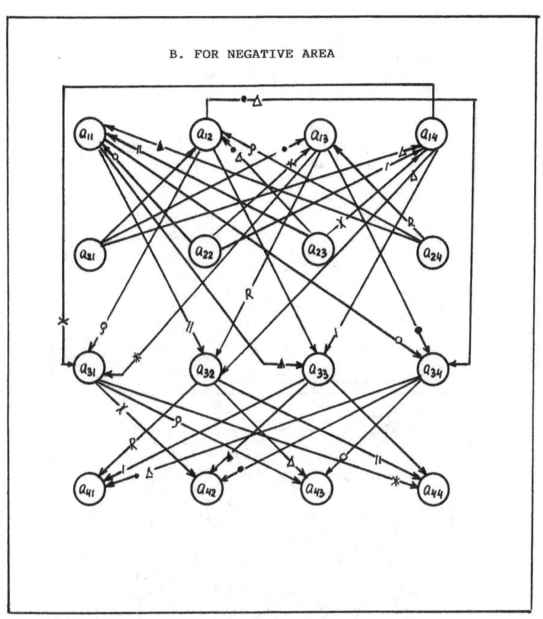

Figure 38 Functional graphs between elements of the determinant of the 4x4 matrix at expansion of its elements to the second row for the negative area

- **To the third row**

The value of determinant of the 4x4 matrix elements of which have expansion to the third row can be expressed by the following formula:

$$|A| = \sum_{j=1}^{4} a_{3j}A_{3j} = \sum_{j=1}^{4} a_{3j}(-1)^{3+j}M_{3j} \qquad (20)$$

In this case , the value of determinant is equal:

$$|A| = a_{31}A_{31} + a_{32}A_{32} + a_{33}A_{33} + a_{34}A_{34}$$

where,

$a_{31}, a_{32}, a_{33}, a_{34}$-elements of the 4x4 matrix

A_{31}-algebraic supplemental (cofactor) taken to the third row and the first column

A_{32}-algebraic supplemental taken to the third row and the second column

A_{33}-algebraic supplemental taken to the third row and the third column

A_{34}-algebraic supplemental taken to the third row and the fourth column

$$A_{31}=(-1)^{3+1}\begin{vmatrix} a_{12} & a_{13} & a_{14} \\ a_{22} & a_{23} & a_{24} \\ a_{42} & a_{43} & a_{44} \end{vmatrix} \qquad A_{32}=(-1)^{3+2}\begin{vmatrix} a_{11} & a_{13} & a_{14} \\ a_{21} & a_{23} & a_{24} \\ a_{41} & a_{43} & a_{44} \end{vmatrix}$$

$$A_{33}=(-1)^{3+3}\begin{vmatrix} a_{11} & a_{12} & a_{14} \\ a_{21} & a_{22} & a_{24} \\ a_{41} & a_{42} & a_{44} \end{vmatrix} \qquad A_{34}=(-1)^{3+4}\begin{vmatrix} a_{11} & a_{12} & a_{13} \\ a_{21} & a_{22} & a_{23} \\ a_{41} & a_{42} & a_{43} \end{vmatrix}$$

In Figure 39 is shown the functional graphs between elements of the determinant of the 4x4 matrix at expansion of its elements to the third row *for positive area* and in Figure 40-*for negative area.*

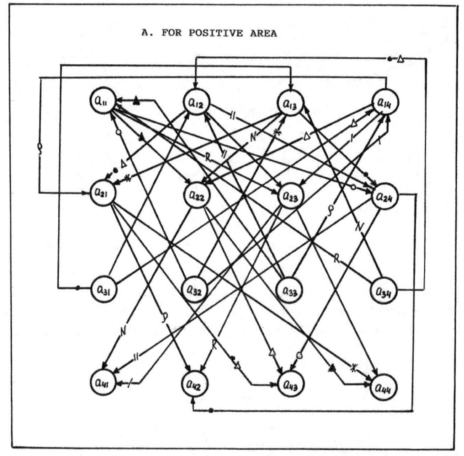

Figure 39 Functional graphs between elements of the determinant of the 4x4 matrix at expansion of its elements to the third row for positive area

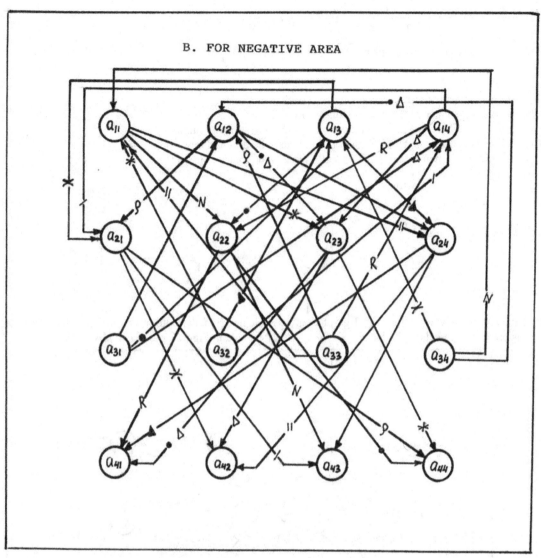

Figure 40 Functional graphs between elements of the determinant of the 4x4 matrix at expansion of its elements to the third row for negative area

After of some transformations, we finally have the value for determinant of the 4x4 matrix elements of which have the expansion to the third row in general view:

$$|A| = a_{31}[a_{12}(a_{23}a_{44}-a_{24}a_{43})+a_{13}(a_{24}a_{42}-a_{22}a_{44})+a_{14}(a_{22}a_{43}-a_{23}a_{42})]+$$
$$+a_{32}[a_{11}(a_{24}a_{43}-a_{23}a_{44})+a_{13}(a_{21}a_{44}-a_{24}a_{41})+a_{14}(a_{23}a_{41}-a_{21}a_{43})]+$$
$$+a_{33}[a_{11}(a_{22}a_{44}-a_{24}a_{42})+a_{12}(a_{24}a_{41}-a_{21}a_{44})+a_{14}(a_{21}a_{42}-a_{22}a_{41})]+$$
$$+a_{34}[a_{11}(a_{23}a_{42}-a_{22}a_{43})+a_{12}(a_{21}a_{43}-a_{23}a_{41})+a_{13}(a_{22}a_{41}-a_{21}a_{42})] \quad (21)$$

- **To the fourth row**

The value of determinant of the 4x4 matrix elements of which have expansion to the fourth row can be expressed by the following formula in general view:

29

$$\left|\mathbf{A}\right|=\sum_{j=1}^{4}\mathbf{a_{4j}A_{4j}}=\sum_{j=1}^{4}\mathbf{a_{4j}(-1)^{4+j}M_{4j}} \qquad (22)$$

In this case , the value of determinant of the 4x4 matrix is equal:

$$\left|\mathbf{A}\right|=\mathbf{a_{41}A_{41}+a_{42}A_{42}+a_{43}A_{43}+a_{44}A_{44}}$$

where,

$a_{41}, a_{42}, a_{43}, a_{44}$-elements of the 4x4 matrix

A_{41}-algebraic supplemental taken to the fourth row and the first column

A_{42}-algebraic supplemental taken to the fourth row and the second column

A_{43}-algebraic supplemental taken to the fourth row and the third column

A_{44}-algebraic supplemental taken to the fourth row and the fourth column

$$A_{41}=(-1)^{4+1}\begin{vmatrix} a_{12} & a_{13} & a_{14} \\ a_{22} & a_{23} & a_{24} \\ a_{32} & a_{33} & a_{34} \end{vmatrix} \qquad A_{42}=(-1)^{4+2}\begin{vmatrix} a_{11} & a_{13} & a_{14} \\ a_{21} & a_{23} & a_{24} \\ a_{31} & a_{33} & a_{34} \end{vmatrix}$$

$$A_{43}=(-1)^{4+3}\begin{vmatrix} a_{11} & a_{12} & a_{14} \\ a_{21} & a_{22} & a_{24} \\ a_{31} & a_{32} & a_{34} \end{vmatrix} \qquad A_{44}=(-1)^{4+4}\begin{vmatrix} a_{11} & a_{12} & a_{13} \\ a_{21} & a_{22} & a_{23} \\ a_{31} & a_{32} & a_{33} \end{vmatrix}$$

In Figure 41 is shown the functional graphs between elements of the determinant of the 4x4 matrix at expansion of its elements to the fourth row *for positive area* and in Figure 42 *–for negative area.*

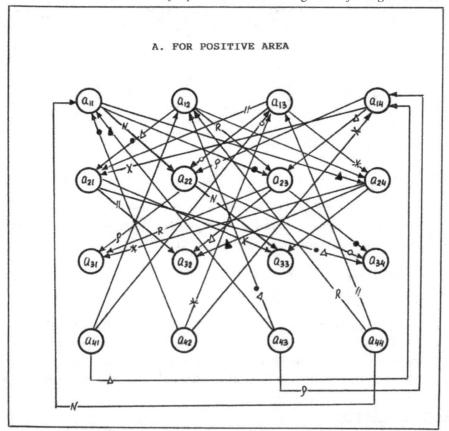

Figure 41 Functional graphs between elements of the determinant of the 4x4 matrix at expansion of its elements to the fourth row for positive area

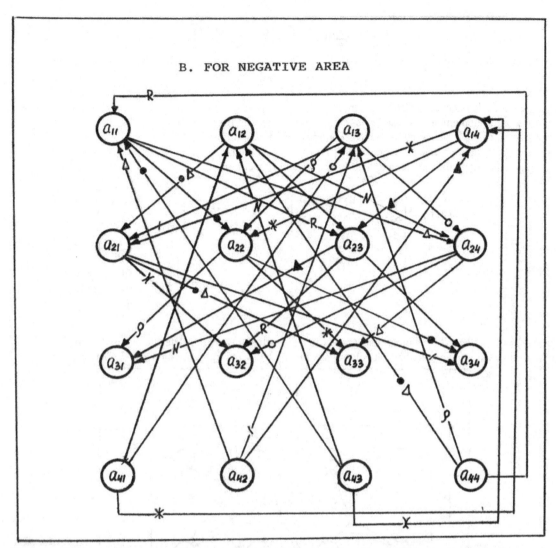

B. FOR NEGATIVE AREA

Figure 42 Functional graphs between elements of the determinant of the 4x4 matrix at expansion of its elements to the fourth row for negative area

After of some transformations, we finally have the value for determinant of the 4x4 matrix elements of which have expansion to the fourth row in general view:

$$|A| = a_{41}[a_{12}(a_{24}a_{33}-a_{23}a_{34})+a_{13}(a_{22}a_{34}-a_{24}a_{32})+a_{14}(a_{23}a_{32}-a_{22}a_{33})]+$$
$$+a_{42}[a_{11}(a_{23}a_{34}-a_{24}a_{33})+a_{13}(a_{24}a_{31}-a_{21}a_{34})+a_{14}(a_{21}a_{33}-a_{23}a_{31})]+$$
$$+a_{43}[a_{11}(a_{24}a_{32}-a_{22}a_{34})+a_{12}(a_{21}a_{34}-a_{24}a_{31})+a_{14}(a_{22}a_{31}-a_{21}a_{32})]+$$
$$+a_{44}[a_{11}(a_{22}a_{33}-a_{23}a_{32})+a_{12}(a_{23}a_{31}-a_{21}a_{33})+a_{13}(a_{21}a_{32}-a_{22}a_{31})] \quad (23)$$

b) *For the columns*
- ### To the first column

The value of determinant of the 4x4 matrix at expansion of its elements to the first column can be expressed by the following formula:

31

$$\left| A \right| = \sum_{i=1}^{4} a_{ai}A_{i1} = \sum_{i=1}^{4} a_{i1}(-1)^{i+1}M_{i1} \qquad (24)$$

In this case, the value of determinant is equal:

$$\left| A \right| = a_{11}A_{11} + a_{21}A_{21} + a_{31}A_{31} + a_{41}A_{41}$$

where,

$a_{11}, a_{21}, a_{31}, a_{41}$-elements of the first column

A_{11}-algebraic supplemental (cofactor) taken to the first row and first column

A_{21}-algebraic supplemental taken to the second row and first column

A_{31}-algebraic supplemental taken to the third row and first column

A_{41}-algebraic supplemental taken to the fourth row and first column

$$A_{11}=(-1)^{1+1}\begin{vmatrix} a_{22} & a_{23} & a_{24} \\ a_{32} & a_{33} & a_{34} \\ a_{42} & a_{43} & a_{44} \end{vmatrix} \qquad A_{21}=(-1)^{2+1}\begin{vmatrix} a_{12} & a_{13} & a_{14} \\ a_{32} & a_{33} & a_{34} \\ a_{42} & a_{43} & a_{44} \end{vmatrix}$$

$$A_{31}=(-1)^{3+1}\begin{vmatrix} a_{12} & a_{13} & a_{14} \\ a_{22} & a_{23} & a_{24} \\ a_{42} & a_{43} & a_{44} \end{vmatrix} \qquad A_{41}=(-1)^{4+1}\begin{vmatrix} a_{12} & a_{13} & a_{14} \\ a_{22} & a_{23} & a_{24} \\ a_{32} & a_{33} & a_{34} \end{vmatrix}$$

In Figure 43 is shown the functional graphs between elements of the determinant of the 4x4 matrix at expansion of its elements to the first column *for positive area* and in Figure 44-*for negative area.*

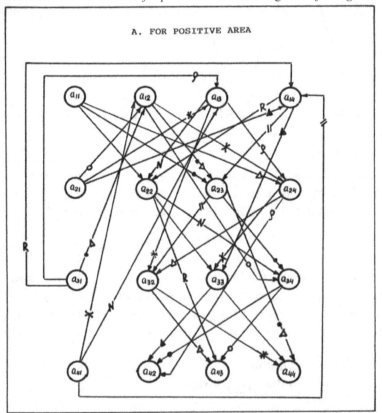

Figure 43 Functional graphs between elements of the determinant of the 4x4 matrix at expansion of its elements to the first column for positive area

32

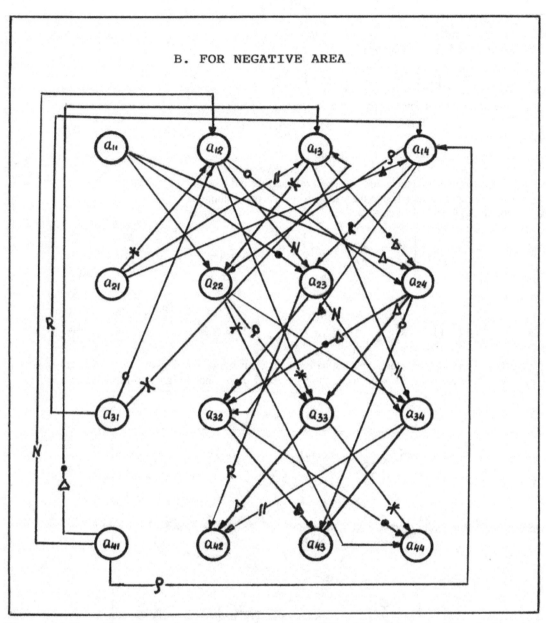

Figure 44 Functional graphs between elements of the determinant of the 4x4 matrix at expansion of its elements to the first column for negative area

After of some transformations, we finally have the value for determinant of the 4x4 matrix at expansion of its elements to the first column in general view:

$$|A| = a_{11}[a_{22}(a_{33}a_{44}-a_{34}a_{43})+a_{23}(a_{34}a_{42}-a_{32}a_{44})+a_{24}(a_{32}a_{43}-a_{33}a_{42})]+$$
$$+a_{21}[a_{12}(a_{34}a_{43}-a_{33}a_{44})+a_{13}(a_{32}a_{44}-a_{34}a_{42})+a_{14}(a_{33}a_{42}-a_{32}a_{43})]+$$
$$+a_{31}[a_{12}(a_{23}a_{44}-a_{24}a_{43})+a_{13}(a_{24}a_{42}-a_{22}a_{44})+a_{14}(a_{22}a_{43}-a_{23}a_{42})]+$$
$$+a_{41}[a_{12}(a_{24}a_{33}-a_{23}a_{34})+a_{13}(a_{22}a_{34}-a_{24}a_{32})+a_{14}(a_{23}a_{32}-a_{22}a_{33})] \quad (25)$$

33

- **To the second column**

The value of determinant of the 4x4 matrix elements of which have expansion to the second column can be expressed by the following formula in general view:

$$|A|=\sum_{i=1}^{4}a_{i2}A_{i2}=\sum_{i=1}^{4}a_{i2}(-1)^{i+2}M_{i2} \qquad (26)$$

In this case , the value of determinant of the 4x4 matrix is equal:

$$|A|=a_{12}A_{12}+a_{22}A_{22}+a_{32}A_{32}+a_{42}A_{42}$$

where,

$a_{12},a_{22},a_{32},a_{42}$-elements of the second column

A_{12}-algebraic supplemental taken to the first row and second column

A_{22}-algebraic supplemental taken to the second row and second column

A_{32}-algebraic supplemental taken to the third row and second column

A_{42}-algebraic supplemental taken to the fourth row and second column.

$$A_{12}=(-1)^{1+2}\begin{vmatrix} a_{21} & a_{23} & a_{24} \\ a_{31} & a_{33} & a_{34} \\ a_{41} & a_{43} & a_{44} \end{vmatrix} \qquad A_{22}=(-1)^{2+2}\begin{vmatrix} a_{11} & a_{13} & a_{14} \\ a_{31} & a_{33} & a_{34} \\ a_{41} & a_{43} & a_{44} \end{vmatrix}$$

$$A_{32}=(-1)^{3+2}\begin{vmatrix} a_{11} & a_{13} & a_{14} \\ a_{21} & a_{23} & a_{24} \\ a_{41} & a_{43} & a_{44} \end{vmatrix} \qquad A_{42}=(-1)^{4+2}\begin{vmatrix} a_{11} & a_{13} & a_{14} \\ a_{21} & a_{23} & a_{24} \\ a_{31} & a_{33} & a_{34} \end{vmatrix}$$

In Figure 45 is shown the functional graphs between elements of the determinant of the 4x4 matrix at expansion of its elements to the second column *for positive area* and in Figure 46-*for negative area.*

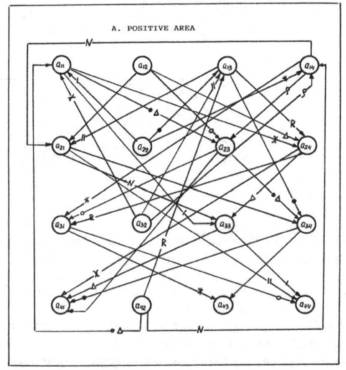

Figure 45 Functional graphs between elements of the determinant of the 4x4 matrix at expansion of its elements to the second column for positive area

34

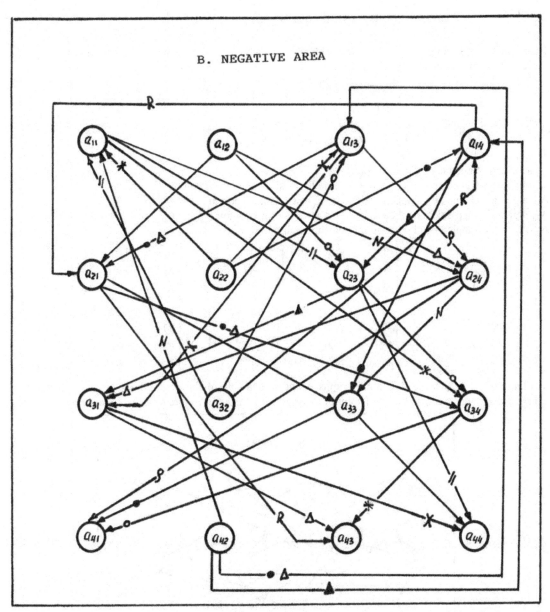

Figure 46 Functional graphs between elements of the determinant of the 4x4 matrix at expansion of its elements to the second column for negative area

After of some transformations, we finally have the value for determinant of the 4x4 matrix at expansion of its elements to the second column in general view:

$$|A| = a_{12}[a_{21}(a_{34}a_{43}-a_{33}a_{44})+a_{23}(a_{31}a_{44}-a_{34}a_{41})+a_{24}(a_{33}a_{41}-a_{31}a_{43})]+$$
$$+a_{22}[a_{11}(a_{33}a_{44}-a_{34}a_{43})+a_{13}(a_{34}a_{41}-a_{31}a_{44})+a_{14}(a_{31}a_{43}-a_{33}a_{41})]+$$
$$+a_{32}[a_{11}(a_{24}a_{41}-a_{23}a_{44})+a_{13}(a_{21}a_{44}-a_{24}a_{41})+a_{14}(a_{23}a_{41}-a_{21}a_{43})]+$$
$$+a_{42}[a_{11}(a_{23}a_{34}-a_{24}a_{33})+a_{13}(a_{24}a_{31}-a_{21}a_{34})+a_{14}(a_{21}a_{33}-a_{23}a_{31}) \quad (27)$$

- **To the third column**

The value of determinant of the 4x4 matrix elements of which have expansion to the third column can be expressed by the following formula in general view:

$$\left|A\right| = \sum_{i=1}^{4} a_{i3}A_{i3} = \sum_{i=1}^{4} a_{i3}(-1)^{i+3}M_{i3} \qquad (28)$$

In this case, the value of determinant is equal in general view:

$$\left|A\right| = a_{13}A_{13} + a_{23}A_{23} + a_{33}A_{33} + a_{43}A_{43}$$

where,

$a_{13}, a_{23}, a_{33}, a_{43}$-elements of the third column

A_{13}-algebraic supplemental (cofactor) taken to the first row and third column

A_{23}-algebraic supplemental taken to the second row and third column

A_{33}-algebraic supplemental taken to the third row and third column

A_{43}-algebraic supplemental taken to the fourth row and third column.

$$A_{13} = (-1)^{1+3} \begin{vmatrix} a_{21} & a_{22} & a_{24} \\ a_{31} & a_{32} & a_{34} \\ a_{41} & a_{42} & a_{44} \end{vmatrix} \qquad A_{23} = (-1)^{2+3} \begin{vmatrix} a_{11} & a_{12} & a_{14} \\ a_{21} & a_{22} & a_{24} \\ a_{41} & a_{42} & a_{44} \end{vmatrix}$$

$$A_{33} = (-1)^{3+3} \begin{vmatrix} a_{11} & a_{12} & a_{14} \\ a_{21} & a_{22} & a_{24} \\ a_{41} & a_{42} & a_{44} \end{vmatrix} \qquad A_{43} = (-1)^{4+3} \begin{vmatrix} a_{11} & a_{12} & a_{14} \\ a_{21} & a_{22} & a_{24} \\ a_{31} & a_{32} & a_{34} \end{vmatrix}$$

In Figure 47 is shown the functional graphs between elements of the determinant of the 4x4 matrix at expansion of its elements to the third column *for positive area* and in Figure 48-*for negative area.*

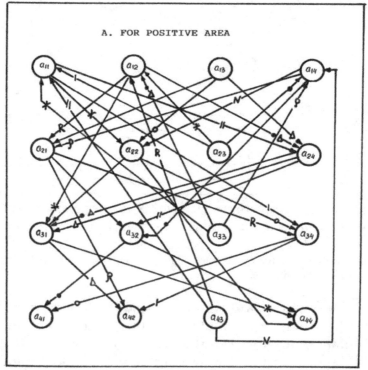

Figure 47 Functional graphs between elements of the determinant of the 4x4 matrix at expansion of its elements to the third row for positive area

36

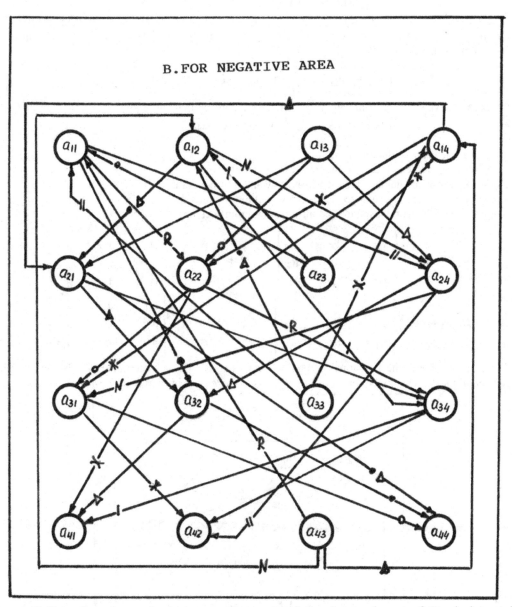

B. FOR NEGATIVE AREA

Figure 48 Functional graphs between elements of the determinant of the 4x4 matrix at expansion of its elements to the third row for negative area

After of some transformation, we finally have the value for determinant of the 4x4 matrix at expansion of its elements to the third column in general view:

$$|A| = a_{13}[a_{21}(a_{32}a_{44}-a_{34}a_{42})+a_{22}(a_{34}a_{41}-a_{31}a_{44})+a_{24}(a_{31}a_{42}-a_{32}a_{41})]+$$
$$+a_{23}[a_{11}(a_{34}a_{42}-a_{32}a_{44})+a_{12}(a_{31}a_{44}-a_{34}a_{41})+a_{14}(a_{32}a_{41}-a_{31}a_{42})]+$$
$$+a_{33}[a_{11}(a_{22}a_{44}-a_{24}a_{42})+a_{12}(a_{24}a_{41}-a_{21}a_{44})+a_{14}(a_{21}a_{42}-a_{22}a_{41})]+$$
$$+a_{43}[a_{11}(a_{24}a_{32}-a_{22}a_{34})+a_{12}(a_{21}a_{34}-a_{24}a_{31})+a_{14}(a_{22}a_{31}-a_{21}a_{32})] \quad (29)$$

- **To the fourth column**

The value of determinant of the 4x4 matrix at expansion of its elements to the fourth column can be expressed by the following formula in general view:

$$|A|=\sum_{i=1}^{4}a_{i4}A_{i4}=\sum_{i=1}^{4}a_{i4}(-1)^{i+4}M_{i4} \qquad (30)$$

In this case , the value of determinant of the 4x4 matrix in general view is equal:

$$|A|=a_{14}A_{14}+a_{24}A_{24}+a_{34}A_{34}+a_{44}A_{44}$$

where,

$a_{14},a_{24},a_{34},a_{44}$-elements of the fourth column

A_{14}-algebraic supplemental taken to the first row and fourth column

A_{24}-algebraic supplemental taken to the second row and fourth column

A_{34}-algebraic supplemental taken to the third row and fourth column

A_{44}-algebraic supplemental taken to the fourth row and fourth column.

$$A_{14}=(-1)^{1+4}\begin{vmatrix} a_{21} & a_{22} & a_{23} \\ a_{31} & a_{32} & a_{33} \\ a_{41} & a_{42} & a_{43} \end{vmatrix} \qquad A_{24}=(-1)^{2+4}\begin{vmatrix} a_{11} & a_{12} & a_{13} \\ a_{31} & a_{32} & a_{33} \\ a_{41} & a_{42} & a_{43} \end{vmatrix}$$

$$A_{34}=(-1)^{3+4}\begin{vmatrix} a_{11} & a_{12} & a_{13} \\ a_{21} & a_{22} & a_{23} \\ a_{41} & a_{42} & a_{43} \end{vmatrix} \qquad A_{44}=(-1)^{4+4}\begin{vmatrix} a_{11} & a_{12} & a_{13} \\ a_{21} & a_{22} & a_{23} \\ a_{31} & a_{32} & a_{33} \end{vmatrix}$$

In Figure 49 is shown the functional graphs between elements of the determinant of the 4x4 matrix at expansion of its elements to the fourth column *for positive area* and in Figure 50-*for negative area.*

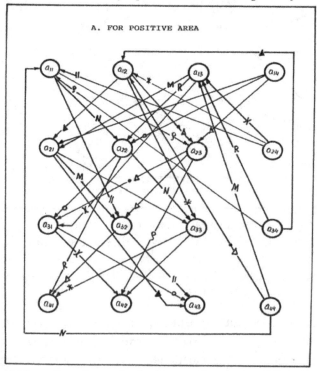

Figure 49 Functional graphs between elements of the determinant of the 4x4 matrix at expansion of its elements to the fourth column for positive area

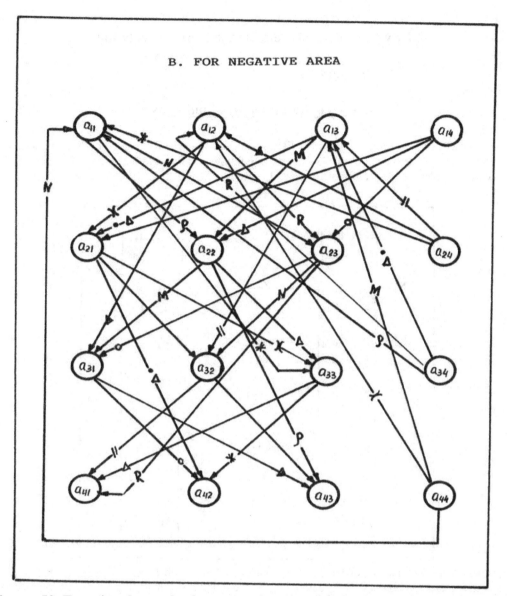

Figure 50 Functional graphs between elements of the determinant of the 4x4 matrix at expansion of its elements to the fourth column for negative area

After of some transformations , we finally have the value for determinant of the 4x4 matrix at expansion of its elements to the fourth column in general view:

$$|A| = a_{14}[a_{21}(a_{33}a_{42}-a_{32}a_{43})+a_{22}(a_{31}a_{43}-a_{33}a_{41})+a_{23}(a_{32}a_{41}-a_{31}a_{42})]+$$
$$+a_{24}[a_{11}(a_{32}a_{43}-a_{33}a_{42})+a_{12}(a_{33}a_{41}-a_{31}a_{43})+a_{13}(a_{31}a_{42}-a_{32}a_{41})]+$$
$$+a_{34}[a_{11}(a_{23}a_{42}-a_{22}a_{43})+a_{12}(a_{21}a_{43}-a_{23}a_{41})+a_{13}(a_{22}a_{41}-a_{21}a_{42})]+$$
$$+a_{44}[a_{11}(a_{22}a_{33}-a_{23}a_{32})+a_{12}(a_{23}a_{31}-a_{21}a_{33})+a_{13}(a_{21}a_{32}-a_{22}a_{31})] \qquad (31)$$

In Appendix 1 is shown the evaluation of determinants of the 4x4 matrices at expansion of its elements to the rows and columns in numerical data applicably to expansion of its elements to the first row and formula (17)

.

2.2 By Laplace's theorem for the rows and columns

a) *For the rows*

- ## To the first and second rows

Referring to well-known Laplace's theorem [2], we can now to write for marked out jointly the first and second rows of the determinant of the 4x4 matrix the following minors(M_{ij}) of the second order and also the algebraic supplemental A_{ij} (cofactor) in general view:

$$M_1 = \begin{vmatrix} a_{11} & a_{12} \\ a_{21} & a_{22} \end{vmatrix} \quad A_1 = (-1)^{(1+2)+(1+2)} \begin{vmatrix} a_{33} & a_{34} \\ a_{43} & a_{44} \end{vmatrix}$$

$$M_2 = \begin{vmatrix} a_{11} & a_{13} \\ a_{21} & a_{23} \end{vmatrix} \quad A_2 = (-1)^{(1+2)+(1+3)} \begin{vmatrix} a_{32} & a_{34} \\ a_{42} & a_{44} \end{vmatrix}$$

$$M_3 = \begin{vmatrix} a_{11} & a_{14} \\ a_{21} & a_{24} \end{vmatrix} \quad A_3 = (-1)^{(1+2)+(1+4)} \begin{vmatrix} a_{32} & a_{33} \\ a_{42} & a_{43} \end{vmatrix}$$

$$M_4 = \begin{vmatrix} a_{12} & a_{13} \\ a_{22} & a_{23} \end{vmatrix} \quad A_4 = (-1)^{(1+2)+(2+3)} \begin{vmatrix} a_{31} & a_{34} \\ a_{41} & a_{44} \end{vmatrix}$$

$$M_5 = \begin{vmatrix} a_{12} & a_{14} \\ a_{22} & a_{24} \end{vmatrix} \quad A_5 = (-1)^{(1+2)+(2+4)} \begin{vmatrix} a_{31} & a_{33} \\ a_{41} & a_{43} \end{vmatrix}$$

$$M_6 = \begin{vmatrix} a_{13} & a_{14} \\ a_{23} & a_{24} \end{vmatrix} \quad A_6 = (-1)^{(1+2)+(3+4)} \begin{vmatrix} a_{31} & a_{32} \\ a_{41} & a_{42} \end{vmatrix}$$

So, the value of determinant of the 4x4 matrix in accordance with Laplace's theorem for marked out jointly the first and second rows can be determined by the following formula in general view:

$$|A| = M_1 A_1 - M_2 A_2 + M_3 A_3 + M_4 A_4 - M_5 A_5 + M_6 A_6 \quad (32)$$

After of some transformation formula (32) has the following view in general form:

$$|A| = a_{11}[a_{22}(a_{33}a_{44} - a_{34}a_{43}) + a_{23}(a_{34}a_{42} - a_{32}a_{44}) + a_{24}(a_{32}a_{43} - a_{33}a_{42})] +$$
$$+ a_{12}[a_{21}(a_{34}a_{43} - a_{33}a_{44}) + a_{23}(a_{31}a_{44} - a_{34}a_{41}) + a_{24}(a_{33}a_{41} - a_{31}a_{43})] +$$
$$+ a_{13}[a_{21}(a_{32}a_{44} - a_{34}a_{42}) + a_{22}(a_{34}a_{41} - a_{31}a_{44}) + a_{24}(a_{31}a_{42} - a_{32}a_{41})] +$$
$$+ a_{14}[a_{21}(a_{33}a_{42} - a_{32}a_{43}) + a_{22}(a_{31}a_{43} - a_{33}a_{41}) + a_{23}(a_{32}a_{41} - a_{31}a_{42})] \quad (33)$$

In Figure 51 is shown the functional graphs for the determinant of the 4x4 matrix for the first and second rows applicably to Laplace's theorem *for positive area* and in Figure 52- *for negative area*.

40

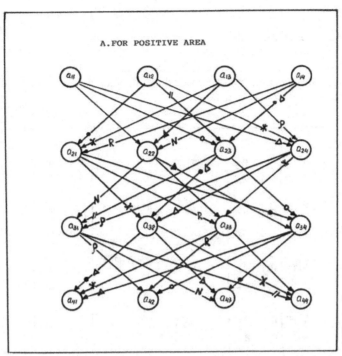

Figure 51 Functional graphs for the determinant of the 4x4 matrix for the first and second rows for positive area

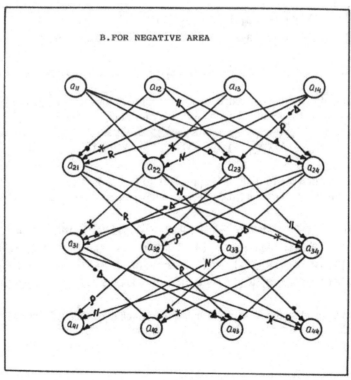

Figure 52 Functional graphs for the determinant of the 4x4 matrix for the first and second rows for negative area

- ### To the first and third rows

Considering the determinant of the 4x4 matrix with marked out jointly the first and third rows, we can now to write in accordance with Laplace;s theorem the following minors of the second order (M_{ij}) and also the algebraic supplemental A_{ij}(cofactor) in general view:

$$M_1 = \begin{vmatrix} a_{11} & a_{12} \\ a_{31} & a_{32} \end{vmatrix} \qquad A_1 = (-1)^{(1+3)+(1+2)} \begin{vmatrix} a_{23} & a_{24} \\ a_{43} & a_{44} \end{vmatrix}$$

$$M_2 = \begin{vmatrix} a_{11} & a_{13} \\ a_{31} & a_{33} \end{vmatrix} \qquad A_2 = (-1)^{(1+3)+(1+3)} \begin{vmatrix} a_{22} & a_{24} \\ a_{42} & a_{44} \end{vmatrix}$$

$$M_3 = \begin{vmatrix} a_{11} & a_{14} \\ a_{31} & a_{34} \end{vmatrix} \qquad A_3 = (-1)^{(1+3)+(1+4)} \begin{vmatrix} a_{22} & a_{23} \\ a_{42} & a_{43} \end{vmatrix}$$

$$M_4 = \begin{vmatrix} a_{12} & a_{13} \\ a_{32} & a_{33} \end{vmatrix} \qquad A_4 = (-1)^{(1+3)+(2+3)} \begin{vmatrix} a_{21} & a_{24} \\ a_{41} & a_{44} \end{vmatrix}$$

$$M_5 = \begin{vmatrix} a_{12} & a_{14} \\ a_{32} & a_{34} \end{vmatrix} \qquad A_5 = (-1)^{(1+3)+(2+4)} \begin{vmatrix} a_{21} & a_{23} \\ a_{41} & a_{43} \end{vmatrix}$$

$$M_6 = \begin{vmatrix} a_{13} & a_{14} \\ a_{33} & a_{34} \end{vmatrix} \qquad A_6 = (-1)^{(1+3)+(3+4)} \begin{vmatrix} a_{21} & a_{22} \\ a_{41} & a_{42} \end{vmatrix}$$

So, the value of determinant of the 4x4 matrix in accordance with Laplace's theorem for marked out jointly the first and third rows can be determined by the following formula in general view:

$$|A| = -M_1A_1 + M_2A_2 - M_3A_3 - M_4A_4 + M_5A_5 - M_6A_6 \qquad (34)$$

After of some transformation the formula (34) applicably to the marked out jointly the first and third rows for the determinant of the 4x4 matrix has the following view in general view:

$$
\begin{aligned}
|A| = &a_{11}[a_{32}(a_{24}a_{43} - a_{23}a_{44}) + a_{33}(a_{22}a_{44} - a_{24}a_{42}) + a_{34}(a_{23}a_{42} - a_{22}a_{43})] + \\
&+ a_{12}[a_{31}(a_{23}a_{44} - a_{24}a_{43}) + a_{33}(a_{24}a_{41} - a_{21}a_{44}) + a_{34}(a_{21}a_{43} - a_{23}a_{41})] + \\
&+ a_{13}[a_{31}(a_{24}a_{42} - a_{22}a_{44}) + a_{32}(a_{21}a_{44} - a_{24}a_{41}) + a_{34}(a_{22}a_{41} - a_{21}a_{42})] + \\
&+ a_{14}[a_{31}(a_{22}a_{43} - a_{23}a_{42}) + a_{32}(a_{23}a_{41} - a_{21}a_{43}) + a_{33}(a_{21}a_{42} - a_{22}a_{41})] \qquad (35)
\end{aligned}
$$

In Figure 53 is shown the functional graphs for the determinant of the 4x4 matrix in general view for marked out jointly the first and third rows applicably to Laplace's theorem *for positive area* and in Figure 54 –*for negative area.*

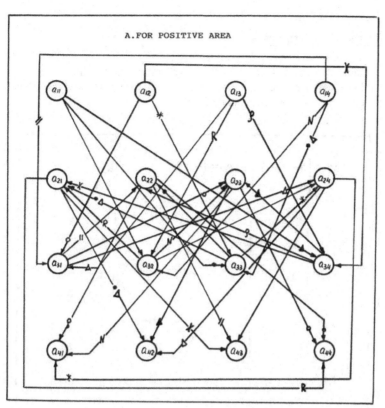

Figure 53 Functional graphs of the first and third rows for positive area

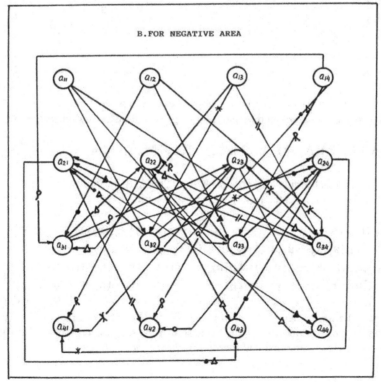

Figure 54 Functional graphs of the first and third rows for negative area

- ## To the first and fourth rows

Referring to the determinant of the 4x4 matrix with marked out jointly the first and fourth rows ,we can now to write in accordance with Laplace's theorem the following minors (M_{ij}) of the second order and also the algebraic supplemental A_{ij}(cofactor) in general view:

$$M_1 = \begin{vmatrix} a_{11} & a_{12} \\ a_{41} & a_{42} \end{vmatrix} \qquad A_1 = (-1)^{(1+4)+(1+2)} \begin{vmatrix} a_{23} & a_{24} \\ a_{33} & a_{34} \end{vmatrix}$$

$$M_2 = \begin{vmatrix} a_{11} & a_{13} \\ a_{41} & a_{43} \end{vmatrix} \qquad A_2 = (-1)^{(1+4)+(1+3)} \begin{vmatrix} a_{22} & a_{24} \\ a_{32} & a_{34} \end{vmatrix}$$

$$M_3 = \begin{vmatrix} a_{11} & a_{14} \\ a_{41} & a_{44} \end{vmatrix} \qquad A_3 = (-1)^{(1+4)+(1+4)} \begin{vmatrix} a_{22} & a_{23} \\ a_{32} & a_{33} \end{vmatrix}$$

$$M_4 = \begin{vmatrix} a_{12} & a_{13} \\ a_{42} & a_{43} \end{vmatrix} \qquad A_4 = (-1)^{(1+4)+(2+3)} \begin{vmatrix} a_{21} & a_{24} \\ a_{31} & a_{34} \end{vmatrix}$$

$$M_5 = \begin{vmatrix} a_{12} & a_{14} \\ a_{42} & a_{44} \end{vmatrix} \qquad A_5 = (-1)^{(1+4)+(2+4)} \begin{vmatrix} a_{21} & a_{23} \\ a_{31} & a_{33} \end{vmatrix}$$

$$M_6 = \begin{vmatrix} a_{13} & a_{14} \\ a_{43} & a_{44} \end{vmatrix} \qquad A_6 = (-1)^{(1+4)+(3+4)} \begin{vmatrix} a_{21} & a_{22} \\ a_{31} & a_{32} \end{vmatrix}$$

So, the value of determinant of the 4x4 matrix in accordance with Lapalce's theorem for marked out jointly the first and fourth rows can be determined by the following formula in general view:

$$|A| = M_1A_1 - M_2A_2 + M_3A_3 + M_4A_4 - M_5A_5 + M_6A_6 \quad (36)$$

After of some transformations formula (36) applicably to the marked out jointly the first and fourth rows of the determinant of the 4x4 matrix has the following view in general view:

$$\begin{aligned}
|A| = {} & a_{11}[a_{42}(a_{23}a_{34} - a_{24}a_{33}) + a_{43}(a_{24}a_{32} - a_{22}a_{34}) + a_{44}(a_{22}a_{33} - a_{23}a_{32})] + \\
& + a_{12}[a_{41}(a_{24}a_{33} - a_{23}a_{34}) + a_{43}(a_{21}a_{34} - a_{24}a_{31}) + a_{44}(a_{23}a_{31} - a_{21}a_{33})] + \\
& + a_{13}[a_{41}(a_{22}a_{34} - a_{24}a_{32}) + a_{42}(a_{24}a_{31} - a_{21}a_{34}) + a_{44}(a_{21}a_{32} - a_{22}a_{31})] + \\
& + a_{14}[a_{41}(a_{23}a_{32} - a_{22}a_{33}) + a_{42}(a_{21}a_{33} - a_{23}a_{31}) + a_{43}(a_{22}a_{31} - a_{21}a_{32})]
\end{aligned}$$

$$(37)$$

In Figure 55 is shown the functional graphs for the determinant of the 4x4 matrix in general view of the first and fourth rows *for positive area*. In Figure 56 is shown the functional graphs for the determinant of the 4x4 matrix in general view of the first and fourth rows *for negative area*.

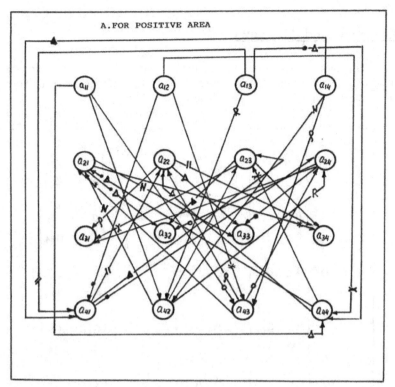

Figure 55 Functional graphs of the first and fourth rows for positive area

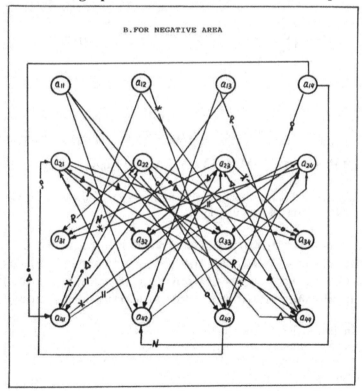

Figure 56 Functional graphs of the first and fourth rows for negative area

• To the second and third rows

Referring to the determinant of the 4x4 matrix with marked out jointly the second and third rows, we can now to write in accordance with Laplace's theorem the following minors of the second order (M_{ij}) and also the algebraic supplemental A_{ij}(cofactor) in general view:

$$M_1 = \begin{vmatrix} a_{21} & a_{22} \\ a_{31} & a_{32} \end{vmatrix} \qquad A_1 = (-1)^{(2+3)+(1+2)} \begin{vmatrix} a_{13} & a_{14} \\ a_{43} & a_{44} \end{vmatrix}$$

$$M_2 = \begin{vmatrix} a_{21} & a_{23} \\ a_{31} & a_{33} \end{vmatrix} \qquad A_2 = (-1)^{(2+3)+(1+3)} \begin{vmatrix} a_{12} & a_{14} \\ a_{42} & a_{44} \end{vmatrix}$$

$$M_3 = \begin{vmatrix} a_{21} & a_{24} \\ a_{31} & a_{34} \end{vmatrix} \qquad A_3 = (-1)^{(2+3)+(1+4)} \begin{vmatrix} a_{12} & a_{13} \\ a_{42} & a_{43} \end{vmatrix}$$

$$M_4 = \begin{vmatrix} a_{22} & a_{23} \\ a_{32} & a_{33} \end{vmatrix} \qquad A_4 = (-1)^{(2+3)+(2+3)} \begin{vmatrix} a_{11} & a_{14} \\ a_{41} & a_{44} \end{vmatrix}$$

$$M_5 = \begin{vmatrix} a_{22} & a_{24} \\ a_{32} & a_{34} \end{vmatrix} \qquad A_5 = (-1)^{(2+3)+(2+4)} \begin{vmatrix} a_{11} & a_{13} \\ a_{41} & a_{43} \end{vmatrix}$$

$$M_6 = \begin{vmatrix} a_{23} & a_{24} \\ a_{33} & a_{34} \end{vmatrix} \qquad A_6 = (-1)^{(2+3)+(3+4)} \begin{vmatrix} a_{11} & a_{12} \\ a_{41} & a_{42} \end{vmatrix}$$

So, the value of determinant of the 4x4 matrix in accordance with Laplace's theorem for marked out jointly the second and third rows can be determined by the following formula in general view:

$$|A| = M_1A_1 - M_2A_2 + M_3A_3 + M_4A_4 - M_5A_5 + M_6A_6 \quad (38)$$

After of some transformations formula (38) applicably for marked out jointly the second and third rows of the determinant of the 4x4 matrix has the following view:

$$|A| = a_{21}[a_{32}(a_{13}a_{44} - a_{14}a_{43}) + a_{33}(a_{14}a_{42} - a_{12}a_{44}) + a_{34}(a_{12}a_{43} - a_{13}a_{42})] +$$
$$+ a_{22}[a_{31}(a_{14}a_{43} - a_{13}a_{44}) + a_{33}(a_{11}a_{44} - a_{14}a_{41}) + a_{34}(a_{13}a_{41} - a_{11}a_{43})] +$$
$$+ a_{23}[a_{31}(a_{12}a_{44} - a_{14}a_{42}) + a_{32}(a_{14}a_{41} - a_{11}a_{44}) + a_{34}(a_{11}a_{42} - a_{12}a_{41})] +$$
$$+ a_{24}[a_{31}(a_{13}a_{42} - a_{12}a_{43}) + a_{32}(a_{11}a_{43} - a_{13}a_{41}) + a_{33}(a_{12}a_{41} - a_{11}a_{42})]$$

$$(39)$$

In Figure 57 is shown the functional graphs for the determinant of the 4x4 matrix at marked out jointly the second and third rows *for positive area* in general view.
And in Figure 58 is shown the functional graphs for the determinant of the 4x4 matrix at marked out jointly the second and third rows *for negative area* in general view.

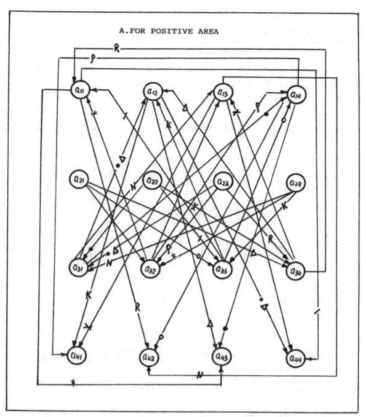

Figure 57 Functional graphs of the second and third rows for positive area

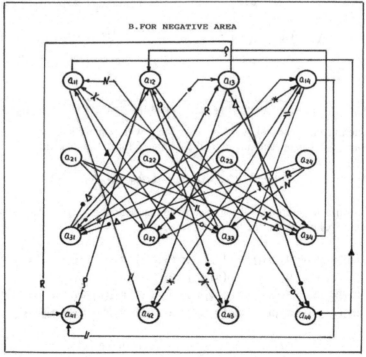

Figure 58 Functional graphs of the second and third rows for negative area

47

- **To the second and fourth rows**

Referring to Laplace's theorem for the determinant of the 4x4 matrix , we can now to write for marked out jointly the second and fourth rows the following minors of the second order (M_{ij}) and also the algebraic supplemental A_{ij}(cofactor) in general view:

$$M_1 = \begin{vmatrix} a_{21} & a_{22} \\ a_{41} & a_{42} \end{vmatrix} \qquad A_1 = (-1)^{(2+4)+(1+2)} \begin{vmatrix} a_{13} & a_{14} \\ a_{33} & a_{34} \end{vmatrix}$$

$$M_2 = \begin{vmatrix} a_{21} & a_{23} \\ a_{41} & a_{43} \end{vmatrix} \qquad A_2 = (-1)^{(2+4)+(1+3)} \begin{vmatrix} a_{12} & a_{14} \\ a_{32} & a_{34} \end{vmatrix}$$

$$M_3 = \begin{vmatrix} a_{21} & a_{24} \\ a_{41} & a_{44} \end{vmatrix} \qquad A_3 = (-1)^{(2+4)+(1+4)} \begin{vmatrix} a_{12} & a_{13} \\ a_{32} & a_{33} \end{vmatrix}$$

$$M_4 = \begin{vmatrix} a_{22} & a_{23} \\ a_{42} & a_{43} \end{vmatrix} \qquad A_4 = (-1)^{(2+4)+(2+3)} \begin{vmatrix} a_{11} & a_{14} \\ a_{31} & a_{34} \end{vmatrix}$$

$$M_5 = \begin{vmatrix} a_{22} & a_{24} \\ a_{42} & a_{44} \end{vmatrix} \qquad A_5 = (-1)^{(2+4)+(2+4)} \begin{vmatrix} a_{11} & a_{13} \\ a_{31} & a_{33} \end{vmatrix}$$

$$M_6 = \begin{vmatrix} a_{23} & a_{24} \\ a_{43} & a_{44} \end{vmatrix} \qquad A_6 = (-1)^{(2+4)+(3+4)} \begin{vmatrix} a_{11} & a_{12} \\ a_{31} & a_{32} \end{vmatrix}$$

So, the value of determinant of the 4x4 matrix in accordance with Laplace's theorem for marked out jointly the second and fourth rows can be determined by the following formula in general view:

$$|A| = -M_1A_1 + M_2A_2 - M_3A_3 - M_4A_4 + M_5A_5 - M_6A_6 \quad (40)$$

After of some transformations formula (40) applicably to marked out jointly the second and fourth rows of the determinant of the 4x4 matrix has the following view:

$$|A| = a_{21}[a_{42}(a_{14}a_{33} - a_{13}a_{34}) + a_{43}(a_{12}a_{34} - a_{14}a_{32}) + a_{44}(a_{13}a_{32} - a_{12}a_{33})] +$$
$$+ a_{22}[a_{41}(a_{13}a_{34} - a_{14}a_{33}) + a_{43}(a_{14}a_{31} - a_{11}a_{34}) + a_{44}(a_{11}a_{33} - a_{13}a_{31})] +$$
$$+ a_{23}[a_{41}(a_{14}a_{32} - a_{12}a_{34}) + a_{42}(a_{11}a_{34} - a_{14}a_{31}) + a_{44}(a_{12}a_{31} - a_{11}a_{32})] +$$
$$+ a_{24}[a_{41}(a_{12}a_{33} - a_{13}a_{32}) + a_{42}(a_{13}a_{31} - a_{11}a_{33}) + a_{43}(a_{11}a_{32} - a_{12}a_{31})]$$

$$(41)$$

In Figure 59 is shown the functional graphs for the determinant of the 4x4 matrix for the second and fourth rows in general view *for positive area* .And in Figure 60 is shown the functional graphs for the determinant of the 4x4 matrix for the second and fourth rows in general view *for negative area.*

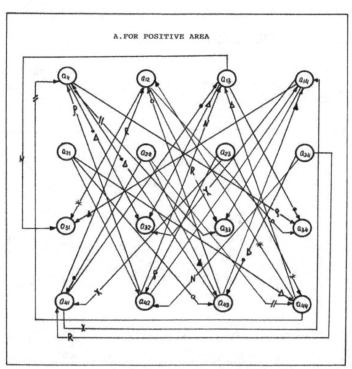

Figure 59 Functional graphs of the second and fourth rows for positive area

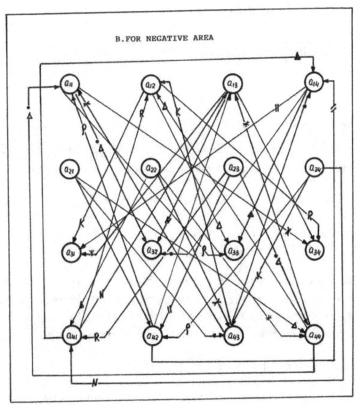

Figure 60 Functional graphs of the second and fourth rows for negative area

• To the third and fourth rows

Considering the essential of Laplace's theorem we can now to write for marked out jointly the third and fourth rows for the determinant of the 4x4 matrix the following minors of the second order (M_{ij}) and also the algebraic supplemental A_{ij}(cofactor) in general view:

$$M_1 = \begin{vmatrix} a_{31} & a_{32} \\ a_{41} & a_{42} \end{vmatrix} \qquad A_1 = (-1)^{(3+4)+(1+2)} \begin{vmatrix} a_{13} & a_{14} \\ a_{23} & a_{24} \end{vmatrix}$$

$$M_2 = \begin{vmatrix} a_{31} & a_{33} \\ a_{41} & a_{43} \end{vmatrix} \qquad A_2 = (-1)^{(3+4)+(1+3)} \begin{vmatrix} a_{12} & a_{14} \\ a_{22} & a_{24} \end{vmatrix}$$

$$M_3 = \begin{vmatrix} a_{31} & a_{34} \\ a_{41} & a_{44} \end{vmatrix} \qquad A_3 = (-1)^{(3+4)+(1+4)} \begin{vmatrix} a_{12} & a_{13} \\ a_{22} & a_{23} \end{vmatrix}$$

$$M_4 = \begin{vmatrix} a_{32} & a_{33} \\ a_{42} & a_{43} \end{vmatrix} \qquad A_4 = (-1)^{(3+4)+(2+3)} \begin{vmatrix} a_{11} & a_{14} \\ a_{21} & a_{24} \end{vmatrix}$$

$$M_5 = \begin{vmatrix} a_{32} & a_{34} \\ a_{42} & a_{44} \end{vmatrix} \qquad A_5 = (-1)^{(3+4)+(2+4)} \begin{vmatrix} a_{11} & a_{13} \\ a_{21} & a_{23} \end{vmatrix}$$

$$M_6 = \begin{vmatrix} a_{33} & a_{34} \\ a_{43} & a_{44} \end{vmatrix} \qquad A_6 = (-1)^{(3+4)+(3+4)} \begin{vmatrix} a_{11} & a_{12} \\ a_{21} & a_{22} \end{vmatrix}$$

So, the value of determinant of the 4x4 matrix in accordance with Laplace's theorem for marked out jointly the third and fourth rows can be determined by the following formula in general view:

$$|A| = M_1A_1 - M_2A_2 + M_3A_3 + M_4A_4 - M_5A_5 + M_6A_6 \qquad (42)$$

After of some transformations formula (42) applicably to the marked out jointly the third and fourth rows of the determinant of the 4x4 matrix has the following view in general view:

$$\begin{aligned}
|A| = &a_{31}[a_{42}(a_{13}a_{24}-a_{14}a_{23})+a_{43}(a_{14}a_{22}-a_{12}a_{24})+a_{44}(a_{12}a_{23}-a_{13}a_{22})]+ \\
&+a_{32}[a_{41}(a_{14}a_{23}-a_{13}a_{24})+a_{43}(a_{11}a_{24}-a_{14}a_{21})+a_{44}(a_{13}a_{21}-a_{11}a_{23})]+ \\
&+a_{33}[a_{41}(a_{12}a_{24}-a_{14}a_{22})+a_{42}(a_{14}a_{21}-a_{11}a_{24})+a_{44}(a_{11}a_{22}-a_{12}a_{21})]+ \\
&+a_{34}[a_{41}(a_{13}a_{22}-a_{12}a_{23})+a_{42}(a_{11}a_{23}-a_{13}a_{21})+a_{43}(a_{12}a_{21}-a_{11}a_{22})]
\end{aligned}$$

$$(43)$$

In Figure 61 is shown the functional graphs for the determinant of the 4x4 matrix at marked out jointly the third and fourth rows *for positive area* in general view. In Figure 62 is shown the functional graphs for the determinant of the 4x4 matrix at marked out jointly the third and fourth rows *for negative area* in general view.

50

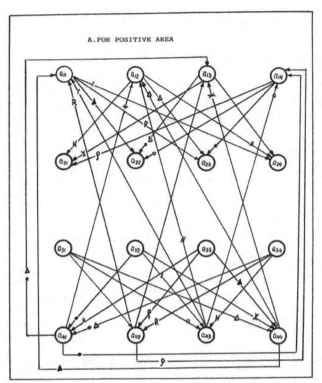

Figure 61 Functional graphs of the third and fourth rows for positive area

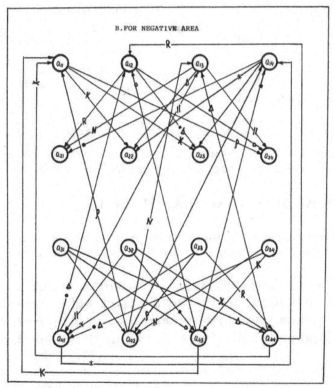

Figure 62 Functional graphs of the third and fourth rows for negative area

51

b) *For the columns*

- ## To the first and second columns

Referring to Laplace's theorem, we can now to write for marked out two jointly columns in view of the first and second columns for the determinant of the 4x4 matrix the following minors of the second order (M_{ij}) and also the algebraic supplemental A_{ij}(cofactor) in general view:

$$M_1 = \begin{vmatrix} a_{11} & a_{12} \\ a_{21} & a_{22} \end{vmatrix} \qquad A_1 = (-1)^{(1+2)+(1+2)} \begin{vmatrix} a_{33} & a_{34} \\ a_{43} & a_{44} \end{vmatrix}$$

$$M_2 = \begin{vmatrix} a_{11} & a_{12} \\ a_{31} & a_{32} \end{vmatrix} \qquad A_2 = (-1)^{(1+2)+(1+3)} \begin{vmatrix} a_{23} & a_{24} \\ a_{43} & a_{44} \end{vmatrix}$$

$$M_3 = \begin{vmatrix} a_{11} & a_{12} \\ a_{41} & a_{42} \end{vmatrix} \qquad A_3 = (-1)^{(1+2)+(1+4)} \begin{vmatrix} a_{23} & a_{24} \\ a_{33} & a_{34} \end{vmatrix}$$

$$M_4 = \begin{vmatrix} a_{21} & a_{22} \\ a_{31} & a_{32} \end{vmatrix} \qquad A_4 = (-1)^{(1+2)+(2+3)} \begin{vmatrix} a_{13} & a_{14} \\ a_{43} & a_{44} \end{vmatrix}$$

$$M_5 = \begin{vmatrix} a_{21} & a_{22} \\ a_{41} & a_{42} \end{vmatrix} \qquad A_5 = (-1)^{(1+2)+(2+4)} \begin{vmatrix} a_{13} & a_{14} \\ a_{33} & a_{34} \end{vmatrix}$$

$$M_6 = \begin{vmatrix} a_{31} & a_{32} \\ a_{41} & a_{42} \end{vmatrix} \qquad A_6 = (-1)^{(1+2)+(3+4)} \begin{vmatrix} a_{13} & a_{14} \\ a_{23} & a_{24} \end{vmatrix}$$

So, the value of determinant of the 4x4 matrix in accordance with Laplace's theorem for marked out jointly the first and second columns can be determined by the following formula in general view:

$$\left| A \right| = M_1A_1 - M_2A_2 + M_3A_3 + M_4A_4 - M_5A_5 + M_6A_6 \quad (44)$$

After of some transformations formula (44) applicably to the marked out jointly the first and second columns of the determinant of the 4x4 matrix has the following view in general form:

$$\left| A \right| = a_{11}[a_{22}(a_{33}a_{44} - a_{34}a_{43}) + a_{32}(a_{24}a_{43} - a_{23}a_{44}) + a_{42}(a_{23}a_{34} - a_{24}a_{33})] +$$
$$+ a_{12}[a_{21}(a_{34}a_{43} - a_{33}a_{44}) + a_{31}(a_{23}a_{44} - a_{24}a_{43}) + a_{41}(a_{24}a_{33} - a_{23}a_{34})] +$$
$$+ a_{13}[a_{24}(a_{31}a_{42} - a_{32}a_{41}) + a_{34}(a_{22}a_{41} - a_{21}a_{42}) + a_{44}(a_{21}a_{32} - a_{22}a_{31})] +$$
$$+ a_{14}[a_{23}(a_{32}a_{41} - a_{31}a_{42}) + a_{33}(a_{21}a_{42} - a_{22}a_{41}) + a_{43}(a_{22}a_{31} - a_{21}a_{32})]$$

$$(45)$$

In Figure 63 is shown the functional graphs for the determinant of the 4x4 matrix at the first and second columns *for positive area* in general view. In Figure 64 is shown the functional graphs for the determinant of the 4x4 matrix at the first and second columns *for negative area* in general view.

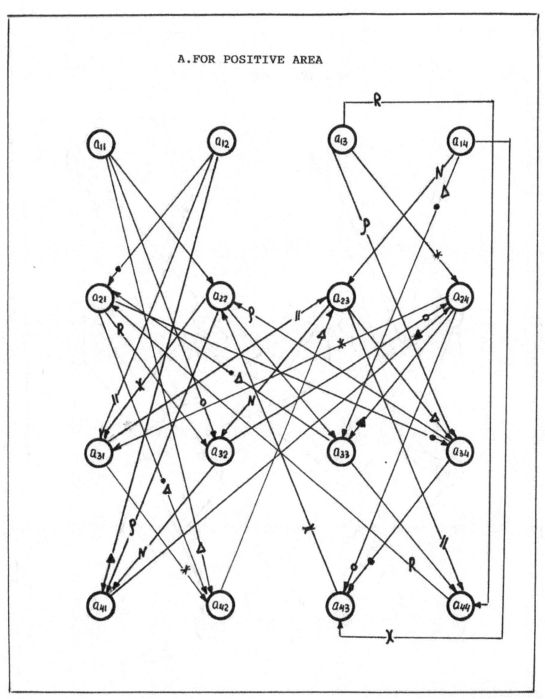

A.FOR POSITIVE AREA

Figure 63 Functional graphs between elements of the determinant of the 4x4 matrix at marked out jointly the first and second columns for positive area applicably to Laplace's theorem

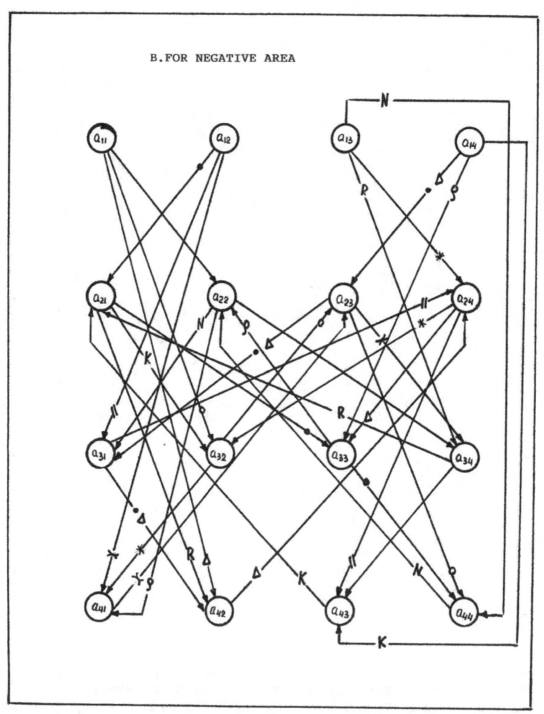

Figure 64 Functional graphs between elements of the determinant of the 4x4 matrix at marked out jointly the first and second columns for negative area applicably to Laplace's theorem

54

- ## To the first and third columns

Referring to Laplace's theorem, we can now to write for marked out jointly two columns in view of the first and third columns for the determinant of the 4x4 matrix the following minors of the second order (M_{ij}) and also the algebraic supplemental in general view:

$$M_1 = \begin{vmatrix} a_{11} & a_{13} \\ a_{21} & a_{23} \end{vmatrix} \qquad A_1 = (-1)^{(1+3)+(1+2)} \begin{vmatrix} a_{32} & a_{34} \\ a_{42} & a_{44} \end{vmatrix}$$

$$M_2 = \begin{vmatrix} a_{11} & a_{13} \\ a_{31} & a_{33} \end{vmatrix} \qquad A_2 = (-1)^{(1+3)+(1+3)} \begin{vmatrix} a_{22} & a_{24} \\ a_{42} & a_{44} \end{vmatrix}$$

$$M_3 = \begin{vmatrix} a_{11} & a_{13} \\ a_{41} & a_{43} \end{vmatrix} \qquad A_3 = (-1)^{(1+3)+(1+4)} \begin{vmatrix} a_{22} & a_{24} \\ a_{32} & a_{34} \end{vmatrix}$$

$$M_4 = \begin{vmatrix} a_{21} & a_{23} \\ a_{31} & a_{33} \end{vmatrix} \qquad A_4 = (-1)^{(1+3)+(2+3)} \begin{vmatrix} a_{12} & a_{14} \\ a_{42} & a_{44} \end{vmatrix}$$

$$M_5 = \begin{vmatrix} a_{21} & a_{23} \\ a_{41} & a_{43} \end{vmatrix} \qquad A_5 = (-1)^{(1+3)+(2+4)} \begin{vmatrix} a_{12} & a_{14} \\ a_{32} & a_{34} \end{vmatrix}$$

$$M_6 = \begin{vmatrix} a_{31} & a_{33} \\ a_{41} & a_{43} \end{vmatrix} \qquad A_6 = (-1)^{(1+3)+(3+4)} \begin{vmatrix} a_{12} & a_{14} \\ a_{22} & a_{24} \end{vmatrix}$$

So, the value of determinant of the 4x4 matrix in accordance with Laplace's theorem for marked out jointly the first and third columns can be determined by the following formula in general view:

$$|A| = -M_1A_1 + M_2A_2 - M_3A_3 - M_4A_4 + M_5A_5 - M_6A_6 \qquad (46)$$

After of some transformations formula (46) applicably to the marked out jointly the first and third columns of the determinant of the 4x4 matrix has the following view in general form:

$$|A| = a_{11}[a_{23}(a_{34}a_{42} - a_{32}a_{44}) + a_{33}(a_{22}a_{44} - a_{24}a_{42}) + a_{43}(a_{24}a_{32} - a_{22}a_{34})] +$$
$$+ a_{12}[a_{24}(a_{33}a_{41} - a_{31}a_{43}) + a_{34}(a_{21}a_{43} - a_{23}a_{41}) + a_{44}(a_{23}a_{31} - a_{21}a_{33})] +$$
$$+ a_{13}[a_{21}(a_{32}a_{44} - a_{34}a_{42}) + a_{31}(a_{24}a_{42} - a_{22}a_{44}) + a_{41}(a_{22}a_{34} - a_{24}a_{32})] +$$
$$+ a_{14}[a_{22}(a_{31}a_{43} - a_{33}a_{41}) + a_{32}(a_{23}a_{41} - a_{21}a_{43}) + a_{42}(a_{21}a_{33} - a_{23}a_{31})]$$

$$(47)$$

In Figure 65 is shown the functional graphs for the determinant of the 4x4 matrix for the first and third columns *for positive area* in general view. In Figure 66 is shown the functional graphs for the determinant of the 4x4 matrix for the first and third columns *for negative area.*

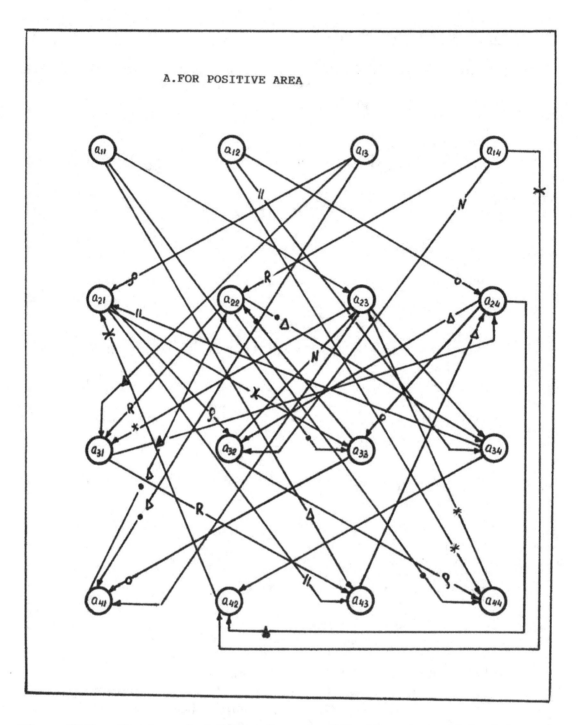

A.FOR POSITIVE AREA

Figure 65 Functional graphs between elements of the determinant of the 4x4 matrix for the first and third columns of positive area applicably to Laplace's theorem

56

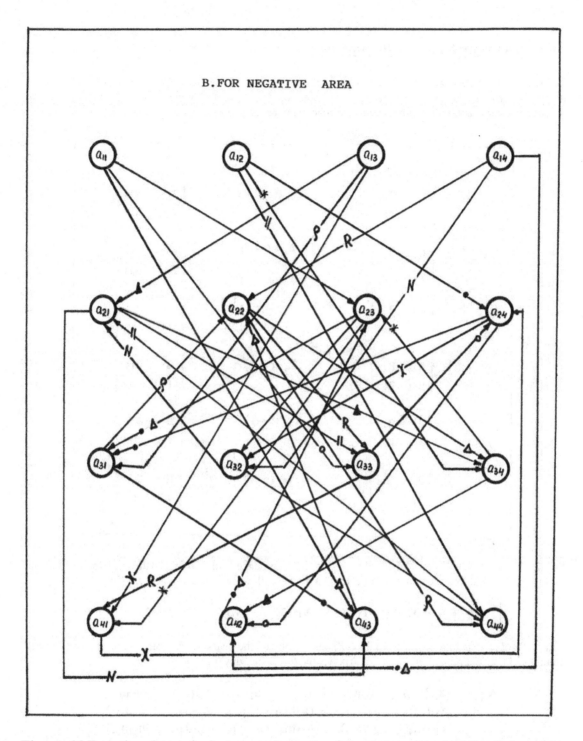

Figure 66 Functional graphs between elements of the determinant of the 4x4 matrix at marked out jointly the first and third columns for negative area applicably to Laplace's theorem

57

- ## To the first and fourth columns

Referring to Laplace's theorem, we can now to write for marked out two jointly columns in view of the first and fourth columns for the determinant of the 4x4 matrix the following minors (M_{ij}) of the second order and also the algebraic supplemental A_{ij}(cofactor) in general view:

$$M_1 = \begin{vmatrix} a_{11} & a_{14} \\ a_{21} & a_{24} \end{vmatrix} \qquad A_1 = (-1)^{(1+4)+(1+2)} \begin{vmatrix} a_{32} & a_{33} \\ a_{42} & a_{43} \end{vmatrix}$$

$$M_2 = \begin{vmatrix} a_{11} & a_{14} \\ a_{31} & a_{34} \end{vmatrix} \qquad A_2 = (-1)^{(1+4)+(1+3)} \begin{vmatrix} a_{22} & a_{23} \\ a_{42} & a_{43} \end{vmatrix}$$

$$M_3 = \begin{vmatrix} a_{11} & a_{14} \\ a_{41} & a_{44} \end{vmatrix} \qquad A_3 = (-1)^{(1+4)+(1+4)} \begin{vmatrix} a_{22} & a_{23} \\ a_{32} & a_{33} \end{vmatrix}$$

$$M_4 = \begin{vmatrix} a_{21} & a_{24} \\ a_{31} & a_{34} \end{vmatrix} \qquad A_4 = (-1)^{(1+4)+(2+3)} \begin{vmatrix} a_{12} & a_{13} \\ a_{42} & a_{43} \end{vmatrix}$$

$$M_5 = \begin{vmatrix} a_{21} & a_{24} \\ a_{41} & a_{42} \end{vmatrix} \qquad A_5 = (-1)^{(1+4)+(2+4)} \begin{vmatrix} a_{12} & a_{13} \\ a_{32} & a_{33} \end{vmatrix}$$

$$M_6 = \begin{vmatrix} a_{31} & a_{34} \\ a_{41} & a_{44} \end{vmatrix} \qquad A_6 = (-1)^{(1+4)+(3+4)} \begin{vmatrix} a_{12} & a_{13} \\ a_{22} & a_{23} \end{vmatrix}$$

So, the value of determinant of the 4x4 matrix in accordance with Laplace's theorem for marked out jointly the first and fourth columns can be defined by the following formula in general view:

$$|A| = M_1A_1 - M_2A_2 + M_3A_3 + M_4A_4 - M_5A_5 + M_6A_6 \qquad (48)$$

After of some transformations formula (48) applicably to the marked out jointly the first and fourth columns of the determinant of the 4x4 matrix has the following view in general form:

$$|A| = a_{11}[a_{24}(a_{32}a_{43} - a_{33}a_{42}) + a_{34}(a_{23}a_{42} - a_{22}a_{43}) + a_{44}(a_{22}a_{33} - a_{23}a_{32})] + \\ + a_{12}[a_{23}(a_{31}a_{44} - a_{34}a_{41}) + a_{33}(a_{24}a_{41} - a_{21}a_{44}) + a_{43}(a_{21}a_{34} - a_{24}a_{31})] + \\ + a_{13}[a_{22}(a_{34}a_{41} - a_{31}a_{44}) + a_{32}(a_{21}a_{44} - a_{24}a_{41}) + a_{42}(a_{24}a_{31} - a_{21}a_{34})] + \\ + a_{14}[a_{21}(a_{33}a_{42} - a_{32}a_{43}) + a_{31}(a_{22}a_{43} - a_{23}a_{42}) + a_{41}(a_{23}a_{32} - a_{22}a_{33})] \qquad (49)$$

In Figure 67 is shown the functional graphs between elements of the 4x4 matrix for the first and fourth columns *for positive area* in general view applicably to Laplace's theorem. In Figure 68 is shown the functional graphs between elements of the 4x4 matrix for the first and fourth columns *for negative area* in general view applicably to Laplace's theorem.

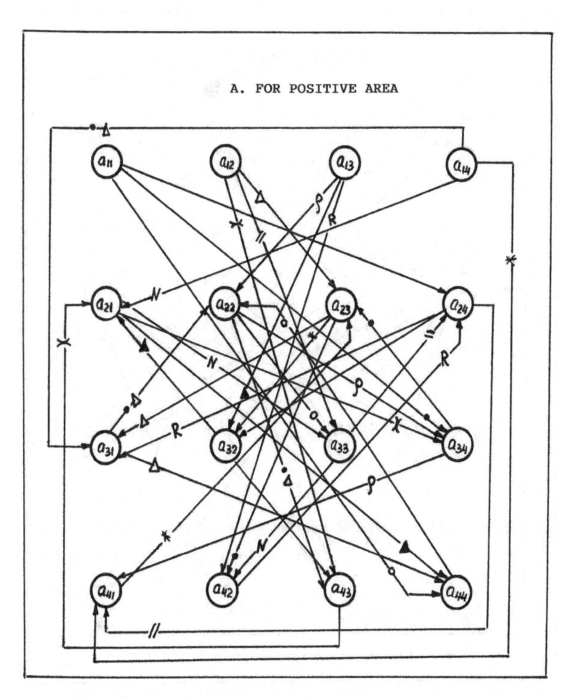

Figure 67 Functional graphs between elements of the determinant of the 4x4 matrix for the first and fourth columns for positive area in general view applicably to Laplace's theorem

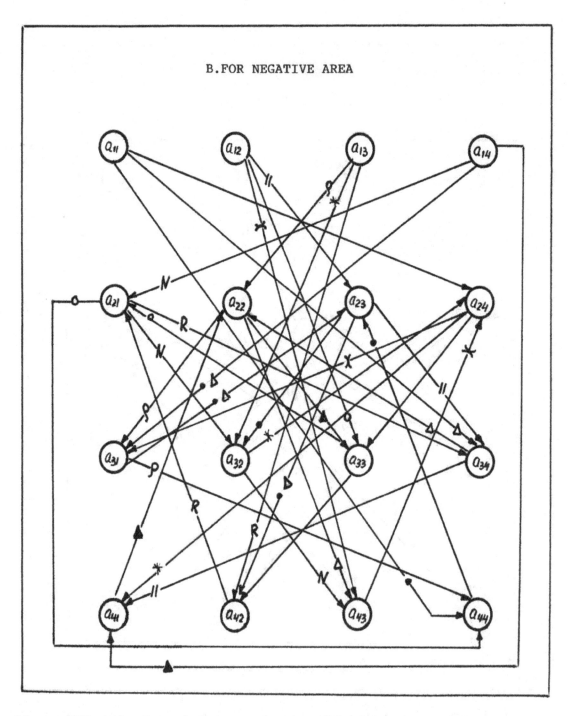

Figure 68 Functional graphs between elements of the determinant of the 4x4 matrix for the first and fourth columns for negative area in general view applicably to Laplace's theorem

- ## To the second and third columns

Referring to Laplace's theorem ,we can now to write for marked out two jointly columns in view of the second and third columns for the determinant of the 4x4 matrix the following minors of the second order (M_{ij}) and also the algebraic supplemental A_{ii}(cofactor) in general view:

$$M_1=\begin{vmatrix} a_{12} & a_{13} \\ a_{22} & a_{23} \end{vmatrix} \qquad A_1=(-1)^{(2+3)+(1+2)}\begin{vmatrix} a_{31} & a_{34} \\ a_{41} & a_{44} \end{vmatrix}$$

$$M_2=\begin{vmatrix} a_{12} & a_{13} \\ a_{32} & a_{33} \end{vmatrix} \qquad A_2=(-1)^{(2+3)+(1+3)}\begin{vmatrix} a_{21} & a_{24} \\ a_{41} & a_{44} \end{vmatrix}$$

$$M_3=\begin{vmatrix} a_{12} & a_{13} \\ a_{42} & a_{43} \end{vmatrix} \qquad A_3=(-1)^{(2+3)+(1+4)}\begin{vmatrix} a_{21} & a_{24} \\ a_{31} & a_{34} \end{vmatrix}$$

$$M_4=\begin{vmatrix} a_{22} & a_{23} \\ a_{32} & a_{33} \end{vmatrix} \qquad A_4=(-1)^{(2+3)+(2+3)}\begin{vmatrix} a_{11} & a_{14} \\ a_{41} & a_{44} \end{vmatrix}$$

$$M_5=\begin{vmatrix} a_{22} & a_{23} \\ a_{42} & a_{43} \end{vmatrix} \qquad A_5=(-1)^{(2+3)+(2+4)}\begin{vmatrix} a_{11} & a_{14} \\ a_{31} & a_{34} \end{vmatrix}$$

$$M_6=\begin{vmatrix} a_{32} & a_{33} \\ a_{42} & a_{43} \end{vmatrix} \qquad A_6=(-1)^{(2+3)+(3+4)}\begin{vmatrix} a_{11} & a_{14} \\ a_{21} & a_{24} \end{vmatrix}$$

So ,the value of determinant of the 4x4 matrix in accordance with Laplace's theorem for marked out jointly the second and third columns can be determined by the following formula in general view:

$$|A|=M_1A_1-M_2A_2+M_3A_3+M_4A_4-M_5A_5+M_6A_6 \qquad (50)$$

After of some transformations formula (50) applicably to the marked out jointly the second and third columns of the determinant of the 4x4 matrix has the following view in general form:

$$\begin{aligned}|A|=&a_{21}[a_{32}(a_{13}a_{44}-a_{14}a_{43})+a_{33}(a_{14}a_{42}-a_{12}a_{44})+a_{34}(a_{12}a_{43}-a_{13}a_{42})]+\\ &+a_{22}[a_{31}(a_{14}a_{43}-a_{13}a_{44})+a_{33}(a_{11}a_{44}-a_{14}a_{41})+a_{34}(a_{13}a_{41}-a_{11}a_{43})]+\\ &+a_{23}[a_{31}(a_{12}a_{44}-a_{14}a_{42})+a_{32}(a_{14}a_{41}-a_{11}a_{44})+a_{34}(a_{11}a_{42}-a_{12}a_{41})]+\\ &+a_{24}[a_{31}(a_{13}a_{42}-a_{12}a_{43})+a_{32}(a_{11}a_{43}-a_{13}a_{41})+a_{33}(a_{12}a_{41}-a_{11}a_{42})] \quad (51)\end{aligned}$$

In Figure 69 is shown the functional graphs for the determinant of the 4x4 matrix at the second and third columns *for positive area* in general view applicably to Laplace's theorem. In Figure 70 is shown the functional graphs for the determinant of the 4x4 matrix at the second and third columns *for negative area* in general view applicably to Laplace's theorem.

61

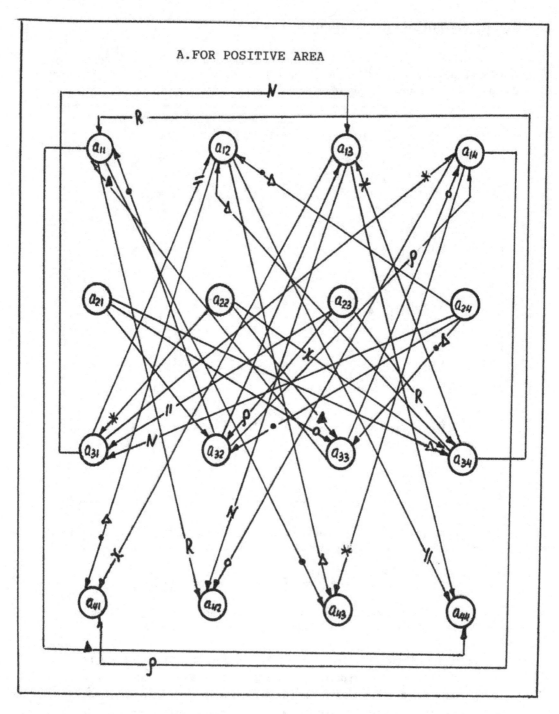

Figure 69 Functional graphs between elements of the determinant of the 4x4 matrix for the second and third columns for positive area in general view applicably to Laplace's theorem

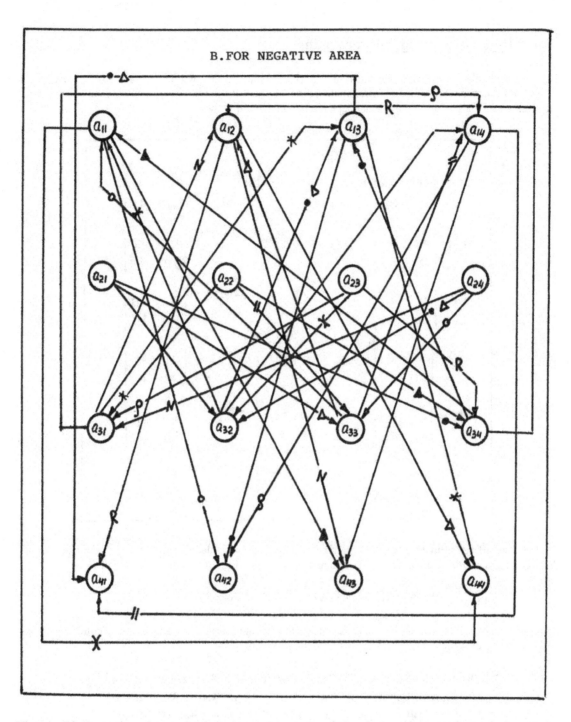

Figure 70 Functional graphs between elements of the determinant of the 4x4 matrix for the second and third columns for negative area in general view applicably to Laplace's theorem

63

- ## To the second and fourth columns

Referring to Laplace's theorem, we can now to write for marked out two jointly columns in view of the second and fourth columns for the determinant of the 4x4 matrix the following minors of the second order (M_{ij}) and also the algebraic supplemental A_{ij}(cofactor) in general view:

$$M_1 = \begin{vmatrix} a_{12} & a_{14} \\ a_{22} & a_{24} \end{vmatrix} \qquad A_1 = (-1)^{(2+4)+(1+2)} \begin{vmatrix} a_{31} & a_{33} \\ a_{41} & a_{43} \end{vmatrix}$$

$$M_2 = \begin{vmatrix} a_{12} & a_{14} \\ a_{32} & a_{34} \end{vmatrix} \qquad A_2 = (-1)^{(2+4)+(1+3)} \begin{vmatrix} a_{21} & a_{23} \\ a_{41} & a_{43} \end{vmatrix}$$

$$M_3 = \begin{vmatrix} a_{12} & a_{14} \\ a_{42} & a_{44} \end{vmatrix} \qquad A_3 = (-1)^{(2+4)+(1+4)} \begin{vmatrix} a_{21} & a_{23} \\ a_{31} & a_{33} \end{vmatrix}$$

$$M_4 = \begin{vmatrix} a_{22} & a_{24} \\ a_{32} & a_{34} \end{vmatrix} \qquad A_4 = (-1)^{(2+4)+(2+3)} \begin{vmatrix} a_{11} & a_{13} \\ a_{41} & a_{43} \end{vmatrix}$$

$$M_5 = \begin{vmatrix} a_{22} & a_{24} \\ a_{42} & a_{44} \end{vmatrix} \qquad A_5 = (-1)^{(2+4)+(2+4)} \begin{vmatrix} a_{11} & a_{13} \\ a_{31} & a_{33} \end{vmatrix}$$

$$M_6 = \begin{vmatrix} a_{32} & a_{34} \\ a_{42} & a_{44} \end{vmatrix} \qquad A_6 = (-1)^{(2+4)+(3+4)} \begin{vmatrix} a_{11} & a_{13} \\ a_{21} & a_{23} \end{vmatrix}$$

So, the value of determinant of the 4x4 matrix in accordance of Laplace's theorem for marked out jointly the second and fourth columns can be determined by the following formula in general view:

$$|A| = -M_1 A_1 + M_2 A_2 - M_3 A_3 - M_4 A_4 + M_5 A_5 - M_6 A_6 \qquad (52)$$

After of some transformations formula (52) applicably to the marked out jointly the second and fourth columns of the determinant of the 4x4 matrix has the following view:

$$\begin{aligned} |A| = &a_{11}[a_{23}(a_{34}a_{42}-a_{32}a_{44})+a_{33}(a_{22}a_{44}-a_{24}a_{42})+a_{43}(a_{24}a_{32}-a_{22}a_{34})]+ \\ &+a_{12}[a_{24}(a_{33}a_{41}-a_{31}a_{43})+a_{34}(a_{21}a_{43}-a_{23}a_{41})+a_{44}(a_{23}a_{31}-a_{21}a_{33})]+ \\ &+a_{13}[a_{22}(a_{34}a_{41}-a_{31}a_{44})+a_{32}(a_{21}a_{44}-a_{24}a_{41})+a_{42}(a_{24}a_{31}-a_{21}a_{34})]+ \\ &+a_{14}[a_{22}(a_{31}a_{43}-a_{33}a_{41})+a_{32}(a_{23}a_{41}-a_{21}a_{43})+a_{42}(a_{21}a_{33}-a_{23}a_{31})] \end{aligned}$$

$$(53)$$

In Figure 71 is shown the functional graphs between elements of the determinant of the 4x4 matrix applicably to Laplace's theorem for *positive area* in general view. In Figure 72 is shown the functional graphs between elements of the determinant of the 4x4 matrix applicably to Laplace's theorem *for negative area* in general view.

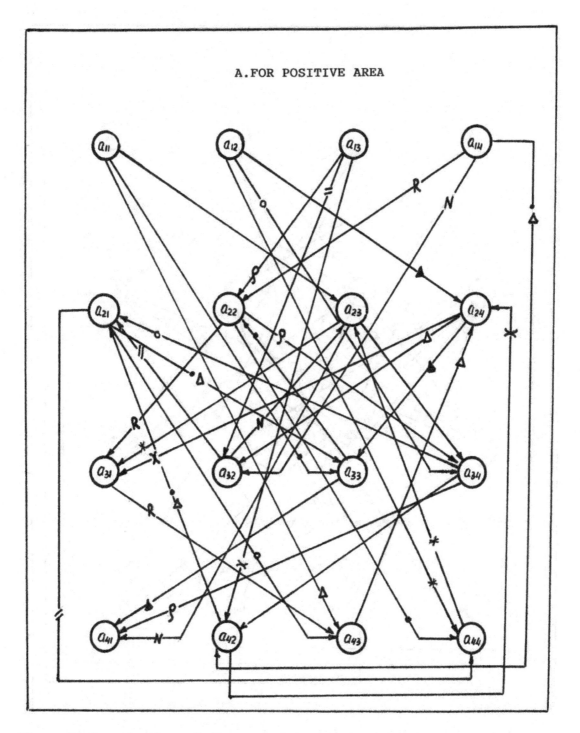

A. FOR POSITIVE AREA

Figure 71 Functional graphs between elements of the 4x4 matrix for second and fourth columns applicably to Laplace's theorem for positive area in general view

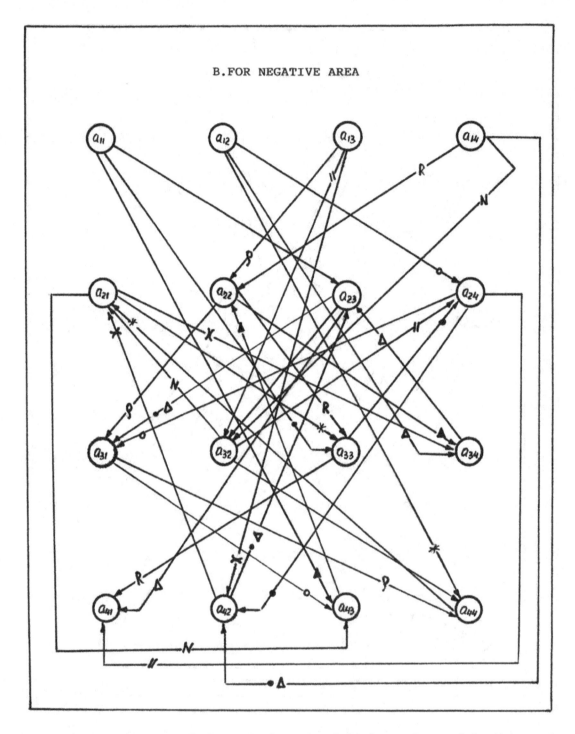

B. FOR NEGATIVE AREA

Figure 72 Functional graphs between elements of the determinant of the 4x4 matrix applicably to Laplace's theorem for the second and fourth columns for negative area in general view

• To the third and fourth columns

Referring to Laplace's theorem ,we can now to write for marked out two jointly columns in view of the third and fourth columns for the determinant of the 4x4 matrix the following minors of the second order (M_{ij}) and also the algebraic supplemental A_{ij}(cofactor) in general view:

$$M_1 = \begin{vmatrix} a_{13} & a_{14} \\ a_{23} & a_{24} \end{vmatrix} \qquad A_1 = (-1)^{(3+4)+(1+2)} \begin{vmatrix} a_{31} & a_{32} \\ a_{41} & a_{42} \end{vmatrix}$$

$$M_2 = \begin{vmatrix} a_{13} & a_{14} \\ a_{33} & a_{34} \end{vmatrix} \qquad A_2 = (-1)^{(3+4)+(1+3)} \begin{vmatrix} a_{21} & a_{22} \\ a_{41} & a_{42} \end{vmatrix}$$

$$M_3 = \begin{vmatrix} a_{13} & a_{14} \\ a_{43} & a_{44} \end{vmatrix} \qquad A_3 = (-1)^{(3+4)+(1+4)} \begin{vmatrix} a_{21} & a_{22} \\ a_{31} & a_{32} \end{vmatrix}$$

$$M_4 = \begin{vmatrix} a_{23} & a_{24} \\ a_{33} & a_{34} \end{vmatrix} \qquad A_4 = (-1)^{(3+4)+(2+3)} \begin{vmatrix} a_{11} & a_{12} \\ a_{41} & a_{42} \end{vmatrix}$$

$$M_5 = \begin{vmatrix} a_{23} & a_{24} \\ a_{43} & a_{44} \end{vmatrix} \qquad A_5 = (-1)^{(3+4)+(2+4)} \begin{vmatrix} a_{11} & a_{12} \\ a_{31} & a_{32} \end{vmatrix}$$

$$M_6 = \begin{vmatrix} a_{33} & a_{34} \\ a_{43} & a_{44} \end{vmatrix} \qquad A_6 = (-1)^{(3+4)+(3+4)} \begin{vmatrix} a_{11} & a_{12} \\ a_{21} & a_{22} \end{vmatrix}$$

So, the value of determinant of the 4x4 matrix in accordance with Laplace's theorem for marked out jointly the third and fourth columns can be determined by the following formula in general view:

$$|A| = M_1A_1 - M_2A_2 + M_3A_3 + M_4A_4 - M_5A_5 + M_6A_6 \qquad (54)$$

After of some transformations formula (54) applicably to the marked out jointly the third and fourth columns of the determinant of the 4x4 matrix has the following view:

$$\begin{aligned} |A| = & a_{11}[a_{22}(a_{33}a_{44}-a_{34}a_{43})+a_{23}(a_{34}a_{42}-a_{32}a_{44})+a_{24}(a_{32}a_{43}-a_{33}a_{42})]+ \\ & +a_{12}[a_{31}(a_{23}a_{44}-a_{24}a_{43})+a_{33}(a_{24}a_{41}-a_{21}a_{44})+a_{34}(a_{21}a_{43}-a_{23}a_{41})]+ \\ & +a_{13}[a_{21}(a_{32}a_{44}-a_{34}a_{42})+a_{22}(a_{34}a_{41}-a_{31}a_{44})+a_{24}(a_{31}a_{42}-a_{32}a_{41})]+ \\ & +a_{14}[a_{23}(a_{32}a_{41}-a_{31}a_{42})+a_{33}(a_{21}a_{42}-a_{22}a_{41})+a_{43}(a_{22}a_{31}-a_{21}a_{32})] \qquad (55) \end{aligned}$$

In Figure 73 is shown the functional graphs between elements of the determinant of the 4x4 matrix for the third and fourth columns applicably to Laplace's theorem *for positive area* in general view. In Figure 74 is shown the functional graphs between elements of the determinant of the 4x4 matrix for the third and fourth columns applicably to Laplace's theorem *for negative area* in general view.

67

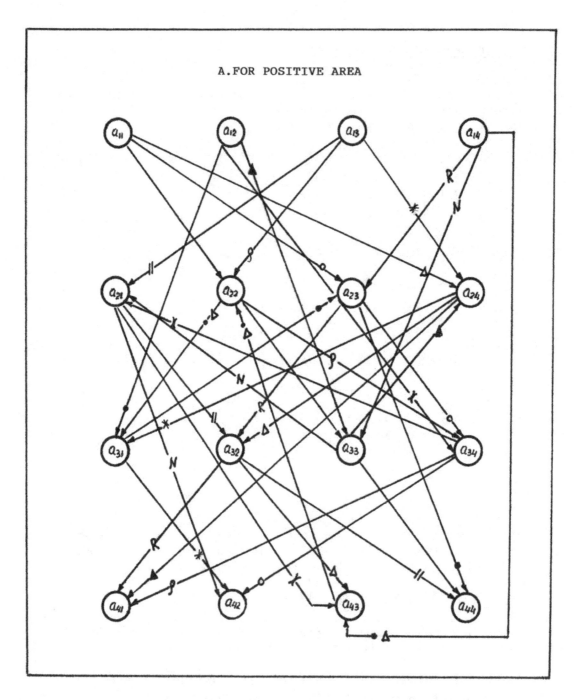

Figure 73 Functional graphs between elements of the determinant of the 4x4 matrix for the third and fourth columns applicably to Laplace's theorem for positive area in general view

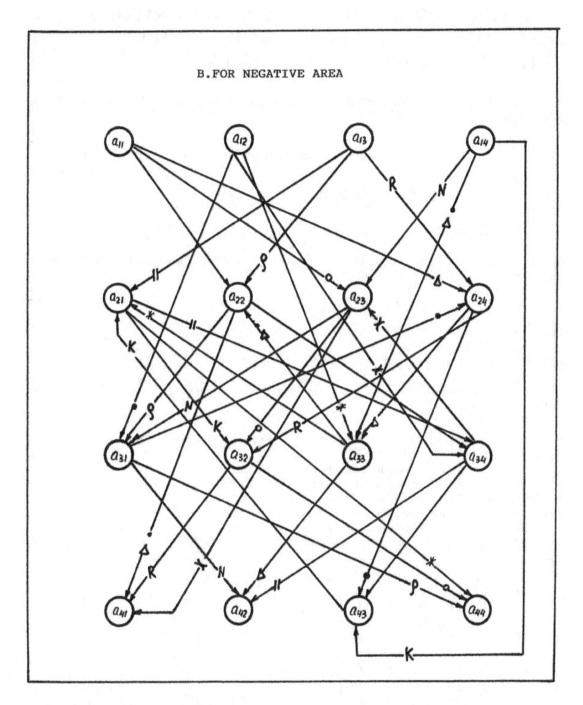

Figure 74 Functional graphs between elements of the determinant of the 4x4 matrix for the third and fourth columns applicably to Laplace's theorem for negative area in general view

In Appendix 2 is shown evaluation of the determinant of the 4x4 matrix with using Laplace's theorem for jointly rows and columns.

69

CHAPTER THREE SOME MAIN ALGORITHMS AND FUNCTIONAL GRAPHS FOR SOLVING OF DETERMINANT WITH 5x5 MATRICES

1. Principal equations and functional graphs for solving of the determinant of the 5x5 matrix

1.1 By expansion(cofactor) method for the rows and columns

❖ *For the rows:*

a) For the first row

The value of determinant of the 5x5 matrix at conditions when its elements have expansion to the first row can be expressed by the following formula:

$$|A| = \sum_{j=1}^{5} a_{1j}A_{1j} = \sum_{j=1}^{5} a_{1j}(-1)^{1+j} M_{1j} \qquad (56)$$

where,

a_{1j}- element of the first row

M_{1j}-minor of algebraic supplemental A_{1j} of element a_{1j} taken with $(-1)^{1+j}$.

And now we can evaluate the determinant of the 5x5 matrix *at expansion of its elements to the first row* by the following formula:

$$|A| = a_{11}A_{11} - a_{12}A_{12} + a_{13}A_{13} - a_{14}A_{14} + a_{15}A_{15} \qquad (57)$$

where,

A_{11}-algebraic supplemental, belonging to the first row and the first column

A_{12}-algebraic supplemental, belonging to the first row and the second column

A_{13}-algebraic supplemental, belonging to the first row and the third column

A_{14}-algebraic supplemental, belonging to the first row and fourth column

A_{15}-algebraic supplemental, belonging to the first row and fifth column .

$$A_{11}=(-1)^{1+1}\begin{vmatrix} a_{22} & a_{23} & a_{24} & a_{25} \\ a_{32} & a_{33} & a_{34} & a_{35} \\ a_{42} & a_{43} & a_{44} & a_{45} \\ a_{52} & a_{53} & a_{54} & a_{55} \end{vmatrix} \qquad A_{12}=(-1)^{1+2}\begin{vmatrix} a_{21} & a_{23} & a_{24} & a_{25} \\ a_{31} & a_{33} & a_{34} & a_{35} \\ a_{41} & a_{43} & a_{44} & a_{45} \\ a_{51} & a_{53} & a_{54} & a_{55} \end{vmatrix}$$

$$A_{13}=(-1)^{1+3}\begin{vmatrix} a_{21} & a_{22} & a_{24} & a_{25} \\ a_{31} & a_{32} & a_{34} & a_{35} \\ a_{41} & a_{42} & a_{44} & a_{45} \\ a_{51} & a_{52} & a_{54} & a_{55} \end{vmatrix} \qquad A_{14}=(-1)^{1+4}\begin{vmatrix} a_{21} & a_{22} & a_{23} & a_{25} \\ a_{31} & a_{32} & a_{33} & a_{35} \\ a_{41} & a_{42} & a_{43} & a_{45} \\ a_{51} & a_{52} & a_{53} & a_{55} \end{vmatrix}$$

$$A_{15}=(-1)^{1+5}\begin{vmatrix} a_{21} & a_{22} & a_{23} & a_{24} \\ a_{31} & a_{32} & a_{33} & a_{34} \\ a_{41} & a_{42} & a_{43} & a_{44} \\ a_{51} & a_{52} & a_{53} & a_{54} \end{vmatrix}$$

70

In Figure 75 is shown the tied graph (tree) for finding in general view the value of determinant of the 5x5 matrix *at expansion of its elements to the first row*.

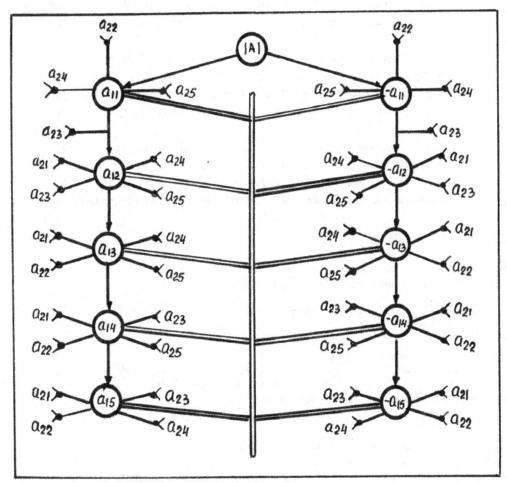

Figure 75 Tied graph (tree) for elements of the determinant of the 5x5 matrix at expansion of its elements to the first row in general view

So, the value of determinant of the 5x5 matrix *at expansion of its elements to the first row*, after of some transformations, can be defined by the following formula in general view:

$$|A| = a_{11}a_{22}[a_{33}(a_{44}a_{55}-a_{45}a_{54})+a_{34}(a_{45}a_{53}-a_{43}a_{55})+a_{35}(a_{43}a_{54}-a_{44}a_{53})]+$$
$$+a_{11}a_{23}[a_{32}(a_{45}a_{54}-a_{44}a_{55})+a_{34}(a_{42}a_{55}-a_{45}a_{52})+a_{35}(a_{44}a_{52}-a_{42}a_{54})]+$$
$$+a_{11}a_{24}[a_{32}(a_{43}a_{55}-a_{45}a_{53})+a_{33}(a_{45}a_{52}-a_{42}a_{55})+a_{35}(a_{42}a_{53}-a_{43}a_{52})]+$$
$$+a_{11}a_{25}[a_{32}(a_{44}a_{53}-a_{43}a_{54})+a_{33}(a_{42}a_{54}-a_{44}a_{52})+a_{34}(a_{43}a_{52}-a_{42}a_{53})]+$$

$$+a_{12}a_{21}[a_{33}(a_{45}a_{54}-a_{44}a_{55})+a_{34}(a_{43}a_{55}-a_{45}a_{53})+a_{35}(a_{44}a_{53}-a_{43}a_{54})]+$$
$$+a_{12}a_{23}[a_{31}(a_{44}a_{55}-a_{45}a_{54})+a_{34}(a_{45}a_{51}-a_{41}a_{55})+a_{35}(a_{41}a_{54}-a_{44}a_{51})]+$$
$$+a_{12}a_{24}[a_{31}(a_{45}a_{53}-a_{43}a_{55})+a_{33}(a_{41}a_{55}-a_{45}a_{51})+a_{35}(a_{43}a_{51}-a_{41}a_{53})]+$$
$$+a_{12}a_{25}[a_{31}(a_{43}a_{54}-a_{44}a_{53})+a_{33}(a_{44}a_{51}-a_{41}a_{54})+a_{34}(a_{41}a_{53}-a_{43}a_{51})]+$$

71

$+a_{13}a_{21}[a_{32}(a_{44}a_{55}-a_{45}a_{54})+a_{34}(a_{45}a_{52}-a_{42}a_{55})+a_{35}(a_{42}a_{54}-a_{44}a_{52})]+$
$+a_{13}a_{22}[a_{31}(a_{45}a_{54}-a_{44}a_{55})+a_{34}(a_{41}a_{55}-a_{45}a_{51})+a_{35}(a_{44}a_{51}-a_{41}a_{54})]+$
$+a_{13}a_{24}[a_{31}(a_{42}a_{55}-a_{45}a_{52})+a_{32}(a_{45}a_{51}-a_{41}a_{55})+a_{35}(a_{41}a_{52}-a_{42}a_{51})]+$
$+a_{13}a_{25}[a_{31}(a_{44}a_{52}-a_{42}a_{54})+a_{32}(a_{41}a_{54}-a_{44}a_{51})+a_{34}(a_{42}a_{51}-a_{41}a_{52})]+$

$+a_{14}a_{21}[a_{32}(a_{45}a_{53}-a_{43}a_{55})+a_{33}(a_{42}a_{55}-a_{45}a_{52})+a_{35}(a_{43}a_{52}-a_{42}a_{53})]+$
$+a_{14}a_{22}[a_{31}(a_{43}a_{55}-a_{45}a_{53})+a_{33}(a_{45}a_{51}-a_{41}a_{55})+a_{35}(a_{41}a_{53}-a_{43}a_{51})]+$
$+a_{14}a_{23}[a_{31}(a_{45}a_{52}-a_{42}a_{55})+a_{32}(a_{41}a_{55}-a_{45}a_{51})+a_{35}(a_{42}a_{51}-a_{41}a_{52})]+$
$+a_{14}a_{25}[a_{31}(a_{42}a_{53}-a_{43}a_{52})+a_{32}(a_{43}a_{51}-a_{41}a_{53})+a_{33}(a_{41}a_{52}-a_{42}a_{51})]+$

$+a_{15}a_{21}[a_{32}(a_{43}a_{54}-a_{44}a_{53})+a_{33}(a_{44}a_{52}-a_{42}a_{54})+a_{34}(a_{42}a_{53}-a_{43}a_{52})]+$
$+a_{15}a_{22}[a_{31}(a_{44}a_{53}-a_{43}a_{54})+a_{33}(a_{41}a_{54}-a_{44}a_{51})+a_{34}(a_{43}a_{51}-a_{41}a_{53})]+$
$+a_{15}a_{23}[a_{31}(a_{42}a_{54}-a_{44}a_{52})+a_{32}(a_{44}a_{51}-a_{41}a_{54})+a_{34}(a_{41}a_{52}-a_{42}a_{51})]+$
$+a_{15}a_{24}[a_{31}(a_{43}a_{52}-a_{42}a_{53})+a_{32}(a_{41}a_{53}-a_{43}a_{51})+a_{33}(a_{42}a_{51}-a_{41}a_{52})]$ (58)

b) For the second row

Applicably to the 5x5 matrix at conditions when its elements have expansion to the second row , we now can to evaluate the determinant by the following formula:

$$\left|A\right|=\sum_{j=1}^{5}a_{2j}A_{2j}=\sum_{j=1}^{5}a_{2j}(-1)^{2+j}M_{2j}$$ (59)

where,

a_{2j}- element of the second row

M_{2j}-minor of algebraic supplemental (cofactor) A_{2j} of element a_{2j} taken with $(-1)^{2+j}$.

And now for practical conditions, we can evaluate the determinant of the 5x5 matrix at expansion of its elements to the second row applicably to suggested formula:

$$\left|A\right|=-a_{21}A_{21}+a_{22}A_{22}-a_{23}A_{23}+a_{24}A_{24}-a_{25}A_{25}$$ (60)

where,

$a_{21},a_{22},a_{23},a_{24},a_{25}$-elements of the second row

A_{21}-algebraic supplemental, belonging to the second row and the first column

A_{22}-algebraic supplemental, belonging to the second row and the second column

A_{23}-algebraic supplemental, belonging to the second row and the third column

A_{24}-algebraic supplemental, belonging to the second row and the fourth column

A_{25}-algebraic supplemental, belonging to the second row and the fifth column .

$$A_{21}=(-1)^{2+1}\begin{vmatrix} a_{12} & a_{13} & a_{14} & a_{15} \\ a_{32} & a_{33} & a_{34} & a_{35} \\ a_{42} & a_{43} & a_{44} & a_{45} \\ a_{52} & a_{53} & a_{54} & a_{55} \end{vmatrix} \qquad A_{22}=(-1)^{2+2}\begin{vmatrix} a_{11} & a_{13} & a_{14} & a_{15} \\ a_{31} & a_{33} & a_{34} & a_{35} \\ a_{41} & a_{43} & a_{44} & a_{45} \\ a_{51} & a_{53} & a_{54} & a_{55} \end{vmatrix}$$

$$A_{23}=(-1)^{2+3}\begin{vmatrix} a_{11} & a_{12} & a_{14} & a_{15} \\ a_{31} & a_{32} & a_{34} & a_{35} \\ a_{41} & a_{42} & a_{44} & a_{45} \\ a_{51} & a_{52} & a_{54} & a_{55} \end{vmatrix} \qquad A_{24}=(-1)^{2+4}\begin{vmatrix} a_{11} & a_{12} & a_{13} & a_{15} \\ a_{31} & a_{32} & a_{33} & a_{35} \\ a_{41} & a_{42} & a_{43} & a_{45} \\ a_{51} & a_{52} & a_{53} & a_{55} \end{vmatrix}$$

$$A_{25}=(-1)^{2+5}\begin{vmatrix} a_{11} & a_{12} & a_{13} & a_{14} \\ a_{31} & a_{32} & a_{33} & a_{34} \\ a_{41} & a_{42} & a_{43} & a_{44} \\ a_{51} & a_{52} & a_{53} & a_{54} \end{vmatrix}$$

In Figure 76 is shown the tied graph (tree) for finding in general view the value of the determinant of the 5x5 matrix *at expansion of its elements to the second row.*

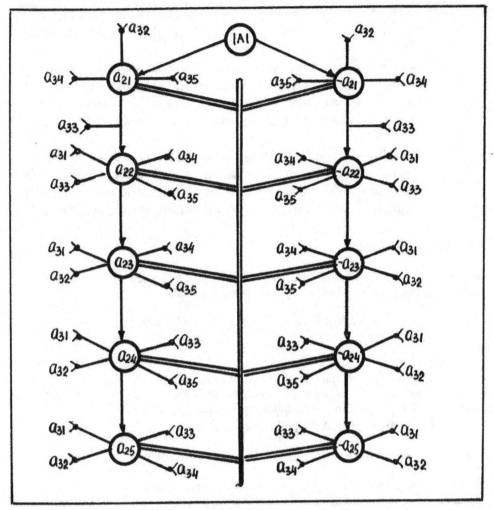

Figure 76 Tied graph (tree) for determinant of the 5x5 matrix at expansion of its elements to the second row

So, the value of determinant of the 5x5 matrix at expansion of its elements to the second row , after of some transformations, is equal in general view:

$$|A| = a_{21}a_{32}[a_{13}(a_{44}a_{55}-a_{45}a_{54})+a_{14}(a_{45}a_{53}-a_{43}a_{55})+a_{15}(a_{43}a_{54}-a_{44}a_{53})]+$$
$$+a_{21}a_{33}[a_{12}(a_{45}a_{54}-a_{44}a_{55})+a_{14}(a_{42}a_{55}-a_{45}a_{52})+a_{15}(a_{44}a_{52}-a_{42}a_{54})]+$$
$$+a_{21}a_{34}[a_{12}(a_{43}a_{55}-a_{45}a_{53})+a_{13}(a_{45}a_{52}-a_{42}a_{55})+a_{15}(a_{42}a_{53}-a_{43}a_{52})]+$$
$$+a_{21}a_{35}[a_{12}(a_{44}a_{53}-a_{43}a_{54})+a_{13}(a_{42}a_{54}-a_{44}a_{52})+a_{14}(a_{43}a_{52}-a_{42}a_{53})]+$$

$$+a_{22}a_{31}[a_{13}(a_{45}a_{54}-a_{44}a_{55})+a_{14}(a_{43}a_{55}-a_{45}a_{53})+a_{15}(a_{44}a_{53}-a_{43}a_{54})]+$$
$$+a_{22}a_{33}[a_{11}(a_{44}a_{55}-a_{45}a_{54})+a_{14}(a_{45}a_{51}-a_{41}a_{55})+a_{15}(a_{41}a_{54}-a_{44}a_{51})]+$$
$$+a_{22}a_{34}[a_{11}(a_{45}a_{53}-a_{43}a_{55})+a_{13}(a_{41}a_{55}-a_{45}a_{51})+a_{15}(a_{43}a_{51}-a_{41}a_{53})]+$$
$$+a_{22}a_{35}[a_{11}(a_{43}a_{54}-a_{44}a_{53})+a_{13}(a_{44}a_{51}-a_{41}a_{54})+a_{14}(a_{41}a_{53}-a_{43}a_{51})]+$$

73

$+a_{23}a_{31}[a_{12}(a_{44}a_{55}-a_{45}a_{54})+a_{14}(a_{45}a_{52}-a_{42}a_{55})+a_{15}(a_{42}a_{54}-a_{44}a_{52})]+$
$+a_{23}a_{32}[a_{11}(a_{45}a_{54}-a_{44}a_{55})+a_{14}(a_{41}a_{55}-a_{45}a_{51})+a_{15}(a_{44}a_{51}-a_{41}a_{54})]+$
$+a_{23}a_{34}[a_{11}(a_{42}a_{55}-a_{45}a_{52})+a_{12}(a_{45}a_{51}-a_{41}a_{55})+a_{15}(a_{41}a_{52}-a_{42}a_{51})]+$
$+a_{23}a_{35}[a_{11}(a_{44}a_{52}-a_{42}a_{54})+a_{12}(a_{41}a_{54}-a_{44}a_{51})+a_{14}(a_{42}a_{51}-a_{41}a_{52})]+$

$+a_{24}a_{31}[a_{12}(a_{45}a_{53}-a_{43}a_{55})+a_{13}(a_{42}a_{55}-a_{45}a_{52})+a_{15}(a_{43}a_{52}-a_{42}a_{53})]+$
$+a_{24}a_{32}[a_{11}(a_{43}a_{55}-a_{45}a_{53})+a_{13}(a_{45}a_{51}-a_{41}a_{55})+a_{15}(a_{41}a_{53}-a_{43}a_{51})]+$
$+a_{24}a_{33}[a_{11}(a_{45}a_{52}-a_{42}a_{55})+a_{12}(a_{41}a_{55}-a_{45}a_{51})+a_{15}(a_{42}a_{51}-a_{41}a_{52})]+$
$+a_{24}a_{35}[a_{11}(a_{42}a_{53}-a_{43}a_{52})+a_{12}(a_{43}a_{51}-a_{41}a_{53})+a_{13}(a_{41}a_{52}-a_{42}a_{51})]+$

$+a_{25}a_{31}[a_{12}(a_{43}a_{54}-a_{44}a_{53})+a_{13}(a_{44}a_{52}-a_{42}a_{54})+a_{14}(a_{42}a_{53}-a_{43}a_{52})]+$
$+a_{25}a_{32}[a_{11}(a_{44}a_{53}-a_{43}a_{54})+a_{13}(a_{41}a_{54}-a_{44}a_{51})+a_{14}(a_{43}a_{51}-a_{41}a_{53})]+$
$+a_{25}a_{33}[a_{11}(a_{42}a_{54}-a_{44}a_{52})+a_{12}(a_{44}a_{51}-a_{41}a_{54})+a_{14}(a_{41}a_{52}-a_{42}a_{51})]+$
$+a_{25}a_{34}[a_{11}(a_{43}a_{52}-a_{42}a_{53})+a_{12}(a_{41}a_{53}-a_{43}a_{51})+a_{13}(a_{42}a_{51}-a_{41}a_{52})]$ (61)

c) For the third row

Considering the square 5x5 matrix , we can now to evaluate the determinant at conditions when its elements have expansion to the third row by the following formula:

$$|A| = \sum_{j=1}^{5} a_{3j}A_{3j} = \sum_{j=1}^{5} a_{3j}(-1)^{3+j}M_{3j}$$ (62)

where,

a_{3j}-element of the third row

M_{3j}-minor of algebraic supplemental (cofactor) A_{3j} of element a_{3j} taken with $(-1)^{3+j}$.

And now for practical conditions we can evaluate the determinant of the 5x5 matrix *at expansion of its elements to the third row* applicably to the following formula:

$$|A| = a_{31}A_{31} - a_{32}A_{32} + a_{33}A_{33} - a_{34}A_{34} + a_{35}A_{35}$$ (63)

where,

$a_{31},a_{32},a_{33},a_{34},a_{35}$-elements of the 5x5 matrix at expansion of its elements to the third row and the next all of its columns;

A_{31}-algebraic supplemental, belonging to the third row and first column

A_{32}-algebraic supplemental, belonging to the third row and second column

A_{33}-algebraic supplemental, belonging to the third row and third column

A_{34}-algebraic supplemental, belonging to the third row and fourth column

A_{35}-algebraic supplemental, belonging to the third row and fifth column .

$$A_{31}=(-1)^{3+1}\begin{vmatrix} a_{12} & a_{13} & a_{14} & a_{15} \\ a_{22} & a_{23} & a_{24} & a_{25} \\ a_{42} & a_{43} & a_{44} & a_{45} \\ a_{52} & a_{53} & a_{54} & a_{55} \end{vmatrix} \qquad A_{32}=(-1)^{3+2}\begin{vmatrix} a_{11} & a_{13} & a_{14} & a_{15} \\ a_{21} & a_{23} & a_{24} & a_{25} \\ a_{41} & a_{43} & a_{44} & a_{45} \\ a_{51} & a_{53} & a_{54} & a_{55} \end{vmatrix}$$

$$A_{33}=(-1)^{3+3}\begin{vmatrix} a_{11} & a_{12} & a_{14} & a_{15} \\ a_{21} & a_{22} & a_{24} & a_{25} \\ a_{41} & a_{42} & a_{44} & a_{45} \\ a_{51} & a_{52} & a_{54} & a_{55} \end{vmatrix} \qquad A_{34}=(-1)^{3+4}\begin{vmatrix} a_{11} & a_{12} & a_{13} & a_{15} \\ a_{21} & a_{22} & a_{23} & a_{25} \\ a_{41} & a_{42} & a_{43} & a_{45} \\ a_{51} & a_{52} & a_{53} & a_{55} \end{vmatrix}$$

$$A_{35}=(-1)^{3+5}\begin{vmatrix} a_{11} & a_{12} & a_{13} & a_{14} \\ a_{21} & a_{22} & a_{23} & a_{24} \\ a_{41} & a_{42} & a_{43} & a_{44} \\ a_{51} & a_{52} & a_{53} & a_{54} \end{vmatrix}$$

In Figure 77 is shown the tied graph(tree) for finding in general vie the value of determinant of the 5x5 matrix at expansion of its elements to the third row.

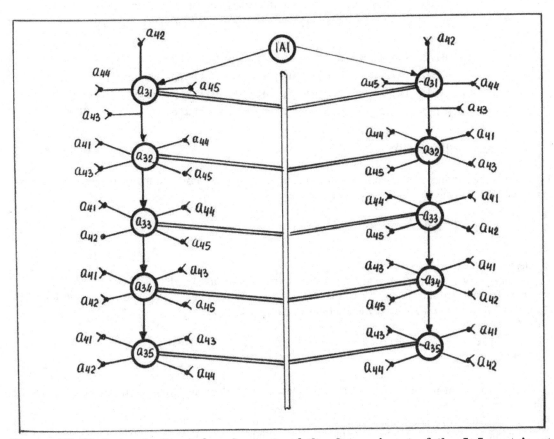

Figure 77 Tied graph (tree) for elements of the determinant of the 5x5 matrix at expansion of its elements to the third row in general view

So, the value of determinant of the 5x5 matrix *at expansion of its elements to the third row,* after of some transformations, is equal in general view:

$$|A| = a_{31}a_{42}[a_{13}(a_{24}a_{55} - a_{25}a_{54}) + a_{14}(a_{25}a_{53} - a_{23}a_{55}) + a_{15}(a_{23}a_{54} - a_{24}a_{53})] +$$
$$+ a_{31}a_{43}[a_{12}(a_{25}a_{54} - a_{24}a_{55}) + a_{14}(a_{22}a_{55} - a_{25}a_{52}) + a_{15}(a_{24}a_{52} - a_{22}a_{54})] +$$
$$+ a_{31}a_{44}[a_{12}(a_{23}a_{55} - a_{25}a_{53}) + a_{13}(a_{25}a_{52} - a_{22}a_{55}) + a_{15}(a_{22}a_{53} - a_{23}a_{52})] +$$
$$+ a_{31}a_{45}[a_{12}(a_{24}a_{53} - a_{23}a_{54}) + a_{13}(a_{22}a_{54} - a_{24}a_{52}) + a_{14}(a_{23}a_{52} - a_{22}a_{53})] +$$

$$+ a_{32}a_{41}[a_{13}(a_{25}a_{54} - a_{24}a_{55}) + a_{14}(a_{23}a_{55} - a_{25}a_{53}) + a_{15}(a_{24}a_{53} - a_{23}a_{54})] +$$
$$+ a_{32}a_{43}[a_{11}(a_{24}a_{55} - a_{25}a_{54}) + a_{14}(a_{25}a_{51} - a_{21}a_{55}) + a_{15}(a_{21}a_{54} - a_{24}a_{51})] +$$
$$+ a_{32}a_{44}[a_{11}(a_{25}a_{53} - a_{23}a_{55}) + a_{13}(a_{21}a_{55} - a_{25}a_{51}) + a_{15}(a_{23}a_{51} - a_{21}a_{53})] +$$
$$+ a_{32}a_{45}[a_{11}(a_{23}a_{54} - a_{24}a_{53}) + a_{13}(a_{24}a_{51} - a_{21}a_{54}) + a_{14}(a_{21}a_{53} - a_{23}a_{51})] +$$

$$+ a_{33}a_{41}[a_{12}(a_{24}a_{55} - a_{25}a_{54}) + a_{14}(a_{25}a_{52} - a_{22}a_{55}) + a_{15}(a_{22}a_{54} - a_{24}a_{52})] +$$
$$+ a_{33}a_{42}[a_{11}(a_{25}a_{54} - a_{24}a_{55}) + a_{14}(a_{21}a_{55} - a_{25}a_{51}) + a_{15}(a_{24}a_{51} - a_{21}a_{54})] +$$
$$+ a_{33}a_{44}[a_{11}(a_{22}a_{55} - a_{25}a_{52}) + a_{12}(a_{25}a_{51} - a_{21}a_{55}) + a_{15}(a_{21}a_{52} - a_{22}a_{51})] +$$
$$+ a_{33}a_{45}[a_{11}(a_{24}a_{52} - a_{22}a_{54}) + a_{12}(a_{21}a_{54} - a_{24}a_{51}) + a_{14}(a_{22}a_{51} - a_{21}a_{52})] +$$

$$+a_{34}a_{41}[a_{12}(a_{25}a_{53}-a_{23}a_{55})+a_{13}(a_{22}a_{55}-a_{25}a_{52})+a_{15}(a_{23}a_{52}-a_{22}a_{53})]+$$
$$+a_{34}a_{42}[a_{11}(a_{23}a_{55}-a_{25}a_{53})+a_{13}(a_{25}a_{51}-a_{21}a_{55})+a_{15}(a_{21}a_{53}-a_{23}a_{51})]+$$
$$+a_{34}a_{43}[a_{11}(a_{25}a_{52}-a_{22}a_{55})+a_{12}(a_{21}a_{55}-a_{25}a_{51})+a_{15}(a_{22}a_{51}-a_{21}a_{52})]+$$
$$+a_{34}a_{45}[a_{11}(a_{22}a_{53}-a_{23}a_{52})+a_{12}(a_{23}a_{51}-a_{21}a_{53})+a_{13}(a_{21}a_{52}-a_{22}a_{51})]+$$

$$+a_{35}a_{41}[a_{12}(a_{23}a_{54}-a_{24}a_{53})+a_{13}(a_{24}a_{52}-a_{22}a_{54})+a_{14}(a_{22}a_{53}-a_{23}a_{52})]+$$
$$+a_{35}a_{42}[a_{11}(a_{24}a_{53}-a_{23}a_{54})+a_{13}(a_{21}a_{54}-a_{24}a_{51})+a_{14}(a_{23}a_{51}-a_{21}a_{53})]+$$
$$+a_{35}a_{43}[a_{11}(a_{22}a_{54}-a_{24}a_{52})+a_{12}(a_{24}a_{51}-a_{21}a_{54})+a_{14}(a_{21}a_{52}-a_{22}a_{51})]+$$
$$+a_{35}a_{44}[a_{11}(a_{23}a_{52}-a_{22}a_{53})+a_{12}(a_{21}a_{53}-a_{23}a_{51})+a_{13}(a_{22}a_{51}-a_{21}a_{52})] \quad (64)$$

d) **For the fourth row**

Considering the 5x5 matrix at conditions when its elements have expansion to the fourth row, we can now to evaluate the determinant by the following formula:

$$|A| = \sum_{j=1}^{5} a_{4j}A_{4j} = \sum_{j=1}^{5} a_{4j}(-1)^{4+j} M_{4j} \quad (65)$$

where,

a_{4j}- element of the fourth row

M_{4j} —minor of algebraic supplemental (cofactor) $A_{4j}$of element a_{4j} taken with $(-1)^{4+j}$.

And now for practical conditions, we can evaluate the determinant of the 5x5 matrix *at expansion of its elements to the fourth row* applicably to the following formula:

$$|A| = -a_{41}A_{41}+a_{42}A_{42}-a_{43}A_{43}+a_{44}A_{44}-a_{45}A_{45} \quad (66)$$

where,

$a_{41}, a_{42}, a_{43}, a_{44}, a_{45}$-elements of the 5x5 matrix at expansion of its elements to the fourth row and the next all of its columns

A_{41}-algebraic supplemental, belonging to the fourth row and the first column

A_{42}-algebraic supplemental, belonging to the fourth row and second column

A_{43}-algebraic supplemental, belonging to the fourth row and third column

A_{44}-algebraic supplemental, belonging to the fourth row and fourth column

A_{45}-algebraic supplemental, belonging to the fourth row and fifth column .

$$A_{41}=(-1)^{4+1}\begin{vmatrix} a_{12} & a_{13} & a_{14} & a_{15} \\ a_{22} & a_{23} & a_{24} & a_{25} \\ a_{32} & a_{33} & a_{34} & a_{35} \\ a_{52} & a_{53} & a_{54} & a_{55} \end{vmatrix} \qquad A_{42}=(-1)^{4+2}\begin{vmatrix} a_{11} & a_{13} & a_{14} & a_{15} \\ a_{21} & a_{23} & a_{24} & a_{25} \\ a_{31} & a_{33} & a_{34} & a_{35} \\ a_{51} & a_{53} & a_{54} & a_{55} \end{vmatrix}$$

$$A_{43}=(-1)^{4+3}\begin{vmatrix} a_{11} & a_{12} & a_{14} & a_{15} \\ a_{21} & a_{22} & a_{24} & a_{25} \\ a_{31} & a_{32} & a_{34} & a_{35} \\ a_{51} & a_{52} & a_{54} & a_{55} \end{vmatrix} \qquad A_{44}=(-1)^{4+4}\begin{vmatrix} a_{11} & a_{12} & a_{13} & a_{15} \\ a_{21} & a_{22} & a_{23} & a_{25} \\ a_{31} & a_{32} & a_{33} & a_{35} \\ a_{51} & a_{52} & a_{53} & a_{55} \end{vmatrix}$$

$$A_{45}=(-1)^{4+5}\begin{vmatrix} a_{11} & a_{12} & a_{13} & a_{14} \\ a_{21} & a_{22} & a_{23} & a_{24} \\ a_{31} & a_{32} & a_{33} & a_{34} \\ a_{51} & a_{52} & a_{53} & a_{54} \end{vmatrix}$$

In Figure 78 is shown the tied graph (tree) for finding in general view the value of determinant of the 5x5 matrix at expansion of its elements to the fourth row.

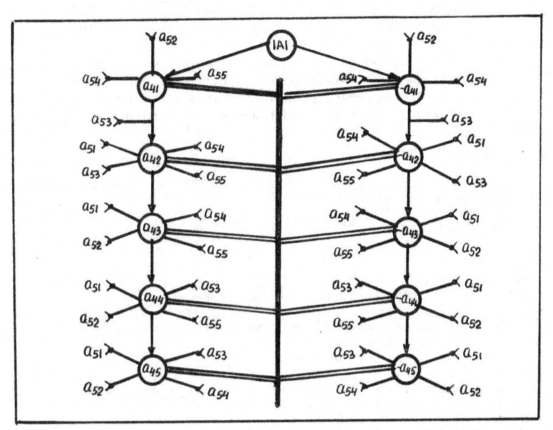

Figure 78 Tied graph (tree) for elements of the determinant of the 5x5 matrix at expansion of its elements to the fourth row in general view

So, the value of determinant of the 5x5 matrix *at expansion of its elements to the fourth row,* after of some transformations, is equal in general view:

$$\begin{aligned}
|A| =\ & a_{41}a_{52}[a_{13}(a_{24}a_{35}-a_{25}a_{34})+a_{14}(a_{25}a_{33}-a_{23}a_{35})+a_{15}(a_{23}a_{34}-a_{24}a_{33})]+ \\
& +a_{41}a_{53}[a_{12}(a_{25}a_{34}-a_{24}a_{35})+a_{14}(a_{22}a_{35}-a_{25}a_{32})+a_{15}(a_{24}a_{32}-a_{22}a_{34})]+ \\
& +a_{41}a_{54}[a_{12}(a_{23}a_{35}-a_{25}a_{33})+a_{13}(a_{25}a_{32}-a_{22}a_{35})+a_{15}(a_{22}a_{33}-a_{23}a_{32})]+ \\
& +a_{41}a_{55}[a_{12}(a_{24}a_{33}-a_{23}a_{34})+a_{13}(a_{22}a_{34}-a_{24}a_{32})+a_{14}(a_{23}a_{32}-a_{22}a_{33})]+ \\
\\
& +a_{42}a_{51}[a_{13}(a_{25}a_{34}-a_{24}a_{35})+a_{14}(a_{23}a_{35}-a_{25}a_{33})+a_{15}(a_{24}a_{33}-a_{23}a_{34})]+ \\
& +a_{42}a_{53}[a_{11}(a_{24}a_{35}-a_{25}a_{34})+a_{14}(a_{25}a_{31}-a_{21}a_{35})+a_{15}(a_{21}a_{34}-a_{24}a_{31})]+ \\
& +a_{42}a_{54}[a_{11}(a_{25}a_{33}-a_{23}a_{35})+a_{13}(a_{21}a_{35}-a_{25}a_{31})+a_{15}(a_{23}a_{31}-a_{21}a_{33})]+ \\
& +a_{42}a_{55}[a_{11}(a_{23}a_{34}-a_{24}a_{33})+a_{13}(a_{24}a_{31}-a_{21}a_{34})+a_{14}(a_{21}a_{33}-a_{23}a_{31})]+ \\
\\
& +a_{43}a_{51}[a_{12}(a_{24}a_{35}-a_{25}a_{34})+a_{14}(a_{25}a_{32}-a_{22}a_{35})+a_{15}(a_{22}a_{34}-a_{24}a_{32})]+ \\
& +a_{43}a_{52}[a_{11}(a_{25}a_{34}-a_{24}a_{35})+a_{14}(a_{21}a_{35}-a_{25}a_{31})+a_{15}(a_{24}a_{31}-a_{21}a_{34})]+ \\
& +a_{43}a_{54}[a_{11}(a_{22}a_{35}-a_{25}a_{32})+a_{12}(a_{25}a_{31}-a_{21}a_{35})+a_{15}(a_{21}a_{32}-a_{22}a_{31})]+ \\
& +a_{43}a_{55}[a_{11}(a_{24}a_{32}-a_{22}a_{34})+a_{12}(a_{21}a_{34}-a_{24}a_{31})+a_{14}(a_{22}a_{31}-a_{21}a_{32})]
\end{aligned}$$

$$+a_{44}a_{51}[a_{12}(a_{25}a_{33}-a_{23}a_{35})+a_{13}(a_{22}a_{35}-a_{25}a_{32})+a_{15}(a_{23}a_{32}-a_{22}a_{33})]+$$
$$+a_{44}a_{52}[a_{11}(a_{23}a_{35}-a_{25}a_{33})+a_{13}(a_{25}a_{31}-a_{21}a_{35})+a_{15}(a_{21}a_{33}-a_{23}a_{31})]+$$
$$+a_{44}a_{53}[a_{11}(a_{25}a_{32}-a_{22}a_{35})+a_{12}(a_{21}a_{35}-a_{25}a_{31})+a_{15}(a_{22}a_{31}-a_{21}a_{32})]+$$
$$+a_{44}a_{55}[a_{11}(a_{22}a_{33}-a_{23}a_{32})+a_{12}(a_{23}a_{31}-a_{21}a_{33})+a_{13}(a_{21}a_{32}-a_{22}a_{31})]+$$

$$+a_{45}a_{51}[a_{12}(a_{23}a_{34}-a_{24}a_{33})+a_{13}(a_{24}a_{32}-a_{22}a_{34})+a_{14}(a_{22}a_{33}-a_{23}a_{32})]+$$
$$+a_{45}a_{52}[a_{11}(a_{24}a_{33}-a_{23}a_{34})+a_{13}(a_{21}a_{34}-a_{24}a_{31})+a_{14}(a_{23}a_{31}-a_{21}a_{33})]+$$
$$+a_{45}a_{53}[a_{11}(a_{22}a_{34}-a_{24}a_{32})+a_{12}(a_{24}a_{31}-a_{21}a_{34})+a_{14}(a_{21}a_{32}-a_{22}a_{31})]+$$
$$+a_{45}a_{54}[a_{11}(a_{23}a_{32}-a_{22}a_{33})+a_{12}(a_{21}a_{33}-a_{23}a_{31})+a_{13}(a_{22}a_{31}-a_{21}a_{32})] \quad (67)$$

e) For the fifth row

Considering the 5x5 matrix at conditions when its elements have expansion to the fifth row ,we can now to evaluate the determinant by the following formula:

$$|A|=\sum_{j=1}^{5}a_{5j}A_{5j}=\sum_{j=1}^{5}a_{5j}(-1)^{5+j}M_{5j} \quad (68)$$

where,

a_{5j}— element of the fifth row

M_{5j}- minor of algebraic supplemental (cofactor) A_{5j} of element a_{5j} taken with $(-1)^{5+j}$.

And now for practical conditions, we can evaluate the determinant of the 5x5 matrix *at expansion of its elements to the fifth row* applicably to suggested formula:

$$|A|=a_{51}A_{51}-a_{52}A_{52}+a_{53}A_{53}-a_{54}A_{54}+a_{55}A_{55} \quad (69)$$

where,

$a_{51},a_{52},a_{53},a_{54},a_{55}$-elemeents of the 5x5 matrix at expansion of its elements to the fifth row and the next all of its columns

A_{51}-algebraic supplemental, belonging to the fifth row and first column

A_{52}-algebraic supplemental, belonging to the fifth row and second column

A_{53}-algebraic supplemental, belonging to the fifth row and third column

A_{54}-algebraic supplemental, belonging to the fifth row and fourth column

A_{55}-algebraic supplemental, belonging to the fifth row and fifth column .

$$A_{51}=(-1)^{5+1}\begin{vmatrix} a_{12} & a_{13} & a_{14} & a_{15} \\ a_{22} & a_{23} & a_{24} & a_{25} \\ a_{32} & a_{33} & a_{34} & a_{35} \\ a_{42} & a_{43} & a_{44} & a_{45} \end{vmatrix} \qquad A_{52}=(-1)^{5+2}\begin{vmatrix} a_{11} & a_{13} & a_{14} & a_{15} \\ a_{21} & a_{23} & a_{24} & a_{25} \\ a_{31} & a_{33} & a_{34} & a_{35} \\ a_{41} & a_{43} & a_{44} & a_{45} \end{vmatrix}$$

$$A_{53}=(-1)^{5+3}\begin{vmatrix} a_{11} & a_{12} & a_{14} & a_{15} \\ a_{21} & a_{22} & a_{24} & a_{25} \\ a_{31} & a_{32} & a_{34} & a_{35} \\ a_{41} & a_{42} & a_{44} & a_{45} \end{vmatrix} \qquad A_{54}=(-1)^{5+4}\begin{vmatrix} a_{11} & a_{12} & a_{13} & a_{15} \\ a_{21} & a_{22} & a_{23} & a_{25} \\ a_{31} & a_{32} & a_{33} & a_{35} \\ a_{41} & a_{42} & a_{43} & a_{45} \end{vmatrix}$$

$$A_{55}=(-1)^{5+5}\begin{vmatrix} a_{11} & a_{12} & a_{13} & a_{14} \\ a_{21} & a_{22} & a_{23} & a_{24} \\ a_{31} & a_{32} & a_{33} & a_{34} \\ a_{41} & a_{42} & a_{43} & a_{44} \end{vmatrix}$$

In Figure 79 is shown the tied graph (tree) for finding in general view the value of determinant of the 5x5 matrix at expansion of its elements to the fifth row.

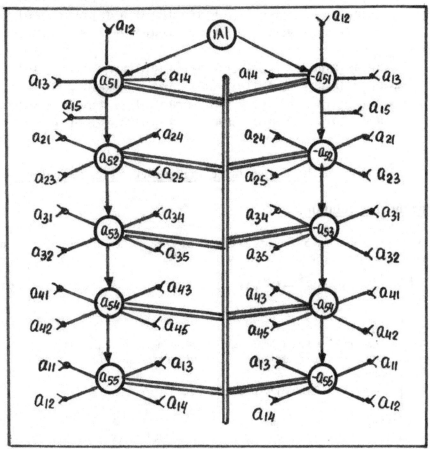

Figure 79 Tied graph (tree) for elements of the determinant of the 5x5 matrix at expansion of its elements to the fifth row in general view

So, the value of determinant of the 5x5 matrix *at expansion of its elements to the fifth row,* after of some transformations, is equal in general view:

$$|A| = a_{51}a_{12}[a_{23}(a_{34}a_{45}-a_{35}a_{44})+a_{24}(a_{35}a_{43}-a_{33}a_{45})+a_{25}(a_{33}a_{44}-a_{34}a_{43})]+$$
$$+a_{51}a_{13}[a_{22}(a_{35}a_{44}-a_{34}a_{45})+a_{24}(a_{32}a_{45}-a_{35}a_{42})+a_{25}(a_{34}a_{42}-a_{32}a_{44})]+$$
$$+a_{51}a_{14}[a_{22}(a_{33}a_{45}-a_{35}a_{43})+a_{23}(a_{35}a_{42}-a_{32}a_{45})+a_{25}(a_{32}a_{43}-a_{33}a_{42})]+$$
$$+a_{51}a_{15}[a_{22}(a_{34}a_{43}-a_{33}a_{44})+a_{23}(a_{32}a_{44}-a_{34}a_{42})+a_{24}(a_{33}a_{42}-a_{32}a_{43})]+$$

$$+a_{52}a_{21}[a_{13}(a_{34}a_{45}-a_{35}a_{44})+a_{14}(a_{35}a_{43}-a_{33}a_{45})+a_{15}(a_{33}a_{44}-a_{34}a_{43})]+$$
$$+a_{52}a_{23}[a_{11}(a_{35}a_{44}-a_{34}a_{45})+a_{14}(a_{31}a_{45}-a_{35}a_{41})+a_{15}(a_{34}a_{41}-a_{31}a_{44})]+$$
$$+a_{52}a_{24}[a_{11}(a_{33}a_{45}-a_{35}a_{43})+a_{13}(a_{35}a_{41}-a_{31}a_{45})+a_{15}(a_{31}a_{43}-a_{33}a_{41})]+$$
$$+a_{52}a_{25}[a_{11}(a_{34}a_{43}-a_{33}a_{44})+a_{13}(a_{31}a_{44}-a_{34}a_{41})+a_{14}(a_{33}a_{41}-a_{31}a_{43})]+$$

$$+a_{53}a_{31}[a_{12}(a_{24}a_{45}-a_{25}a_{44})+a_{14}(a_{25}a_{42}-a_{22}a_{45})+a_{15}(a_{22}a_{44}-a_{24}a_{42})]+$$
$$+a_{53}a_{32}[a_{11}(a_{25}a_{44}-a_{24}a_{45})+a_{14}(a_{21}a_{45}-a_{25}a_{41})+a_{15}(a_{24}a_{41}-a_{21}a_{44})]+$$
$$+a_{53}a_{34}[a_{11}(a_{22}a_{45}-a_{25}a_{42})+a_{12}(a_{25}a_{41}-a_{21}a_{45})+a_{15}(a_{21}a_{42}-a_{22}a_{41})]+$$
$$+a_{53}a_{35}[a_{11}(a_{24}a_{42}-a_{22}a_{44})+a_{12}(a_{21}a_{44}-a_{24}a_{41})+a_{14}(a_{22}a_{41}-a_{21}a_{42})]+$$

79

$$+a_{54}a_{41}[a_{12}(a_{23}a_{35}-a_{25}a_{33})+a_{13}(a_{25}a_{32}-a_{22}a_{35})+a_{15}(a_{22}a_{33}-a_{23}a_{32})]+$$
$$+a_{54}a_{42}[a_{11}(a_{25}a_{33}-a_{23}a_{35})+a_{13}(a_{21}a_{35}-a_{25}a_{31})+a_{13}(a_{23}a_{31}-a_{21}a_{33})]+$$
$$+a_{54}a_{43}[a_{11}(a_{22}a_{35}-a_{25}a_{32})+a_{12}(a_{25}a_{31}-a_{21}a_{35})+a_{15}(a_{21}a_{32}-a_{22}a_{31})]+$$
$$+a_{54}a_{45}[a_{11}(a_{23}a_{32}-a_{22}a_{33})+a_{12}(a_{21}a_{33}-a_{23}a_{31})+a_{13}(a_{22}a_{31}-a_{21}a_{32})]+$$

$$+a_{55}a_{11}[a_{22}(a_{33}a_{44}-a_{34}a_{43})+a_{23}(a_{34}a_{42}-a_{32}a_{44})+a_{24}(a_{32}a_{43}-a_{33}a_{42})]+$$
$$+a_{55}a_{12}[a_{21}(a_{34}a_{43}-a_{33}a_{44})+a_{23}(a_{31}a_{44}-a_{34}a_{41})+a_{24}(a_{33}a_{41}-a_{31}a_{43})]+$$
$$+a_{55}a_{13}[a_{21}(a_{32}a_{44}-a_{34}a_{42})+a_{22}(a_{34}a_{41}-a_{31}a_{44})+a_{24}(a_{31}a_{42}-a_{32}a_{41})]+$$
$$+a_{55}a_{14}[a_{21}(a_{33}a_{42}-a_{32}a_{43})+a_{22}(a_{31}a_{43}-a_{33}a_{41})+a_{23}(a_{32}a_{41}-a_{31}a_{42})] \qquad (70)$$

❖ By Expansion(cofactor)method for the columns

a) For the first column

The value of determinant of the 5x5 matrix at conditions when its elements have expansion to the first column can be expressed by the following formula:

$$|A| = \sum_{i=1}^{5} a_{i1}A_{i1} = \sum_{i=1}^{5} a_{i1}(-1)^{i+1} M_{i1} \qquad (71)$$

where,

a_{i1}- elements of the first column

M_{i1}-minor of algebraic supplemental (cofactor) A_{i1} of element (a_{i1}) taken with $(-1)^{i+1}$.

And now we can evaluate the determinant of the 5x5 matrix at expansion of its elements to the first column by the following formula:

$$|A| = a_{11}A_{11}-a_{21}A_{21}+a_{31}A_{31}-a_{41}A_{41}+a_{51}A_{51} \qquad (72)$$

where,

$a_{11},a_{21},a_{31},a_{41},a_{51}$-elements of the first column

A_{11}-algebraic supplemental, belonging to the first column and the first row

A_{21}-algebraic supplemental, belonging to the first column and the second row

A_{31}-algebraic supplemental, belonging to the first column and the third row

A_{41}-algebraic supplemental, belonging to the first column and the fourth row

A_{51}-algebraic supplemental, belonging to the first column and the fifth row .

$$A_{11}=(-1)^{1+1} \cdot \begin{vmatrix} a_{22} & a_{23} & a_{24} & a_{25} \\ a_{32} & a_{33} & a_{34} & a_{35} \\ a_{42} & a_{43} & a_{44} & a_{45} \\ a_{52} & a_{53} & a_{54} & a_{55} \end{vmatrix} \qquad A_{21}=(-1)^{2+1} \cdot \begin{vmatrix} a_{12} & a_{13} & a_{14} & a_{15} \\ a_{32} & a_{33} & a_{34} & a_{35} \\ a_{42} & a_{43} & a_{44} & a_{45} \\ a_{52} & a_{53} & a_{54} & a_{55} \end{vmatrix}$$

$$A_{31}=(-1)^{3+1} \begin{vmatrix} a_{12} & a_{13} & a_{14} & a_{15} \\ a_{22} & a_{23} & a_{24} & a_{25} \\ a_{42} & a_{43} & a_{44} & a_{45} \\ a_{52} & a_{53} & a_{54} & a_{55} \end{vmatrix} \qquad A_{41}=(-1)^{4+1} \begin{vmatrix} a_{12} & a_{13} & a_{14} & a_{15} \\ a_{22} & a_{23} & a_{24} & a_{25} \\ a_{32} & a_{33} & a_{34} & a_{35} \\ a_{52} & a_{53} & a_{54} & a_{55} \end{vmatrix}$$

$$A_{51}=(-1)^{5+1} \begin{vmatrix} a_{12} & a_{13} & a_{14} & a_{15} \\ a_{22} & a_{23} & a_{24} & a_{25} \\ a_{32} & a_{33} & a_{34} & a_{35} \\ a_{42} & a_{43} & a_{44} & a_{45} \end{vmatrix}$$

In Figure 80 is shown the tied graph (tree) for finding in general view the value of determinant of the 5x5 matrix at expansion of its elements to the first column.

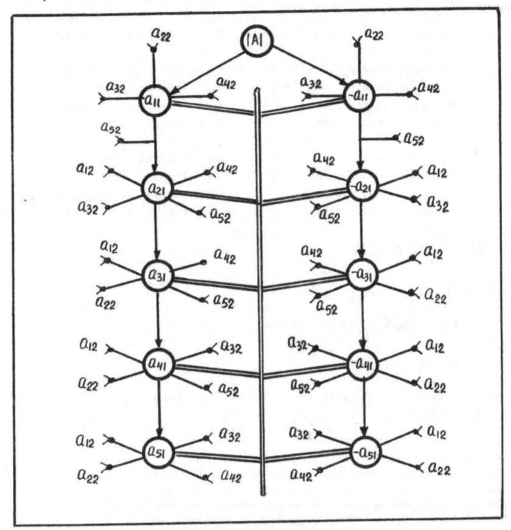

Figure 80 Tied graph (tree) for elements of the determinant of the 5x5 matrix at expansion of its elements to the first column in general view

So, the value of determinant of the 5x5 matrix *at expansion of its elements to the first column,* after of some transformations, is equal in general view:

$$|A| = a_{11}a_{22}[a_{33}(a_{44}a_{55}-a_{45}a_{54})+a_{34}(a_{45}a_{53}-a_{43}a_{55})+a_{35}(a_{43}a_{54}-a_{44}a_{53})]+$$
$$+a_{11}a_{32}[a_{23}(a_{45}a_{54}-a_{44}a_{55})+a_{24}(a_{43}a_{55}-a_{45}a_{53})+a_{25}(a_{44}a_{53}-a_{43}a_{54})]+$$
$$+a_{11}a_{42}[a_{23}(a_{34}a_{55}-a_{35}a_{54})+a_{24}(a_{35}a_{53}-a_{33}a_{55})+a_{25}(a_{33}a_{54}-a_{34}a_{53})]+$$
$$+a_{11}a_{52}[a_{23}(a_{35}a_{44}-a_{34}a_{45})+a_{24}(a_{33}a_{45}-a_{35}a_{43})+a_{25}(a_{34}a_{43}-a_{33}a_{44})]+$$

$$+a_{21}a_{12}[a_{33}(a_{45}a_{54}-a_{44}a_{55})+a_{34}(a_{43}a_{55}-a_{45}a_{53})+a_{35}(a_{44}a_{53}-a_{43}a_{54})]+$$
$$+a_{21}a_{32}[a_{13}(a_{44}a_{55}-a_{45}a_{54})+a_{14}(a_{45}a_{53}-a_{43}a_{55})+a_{15}(a_{43}a_{54}-a_{44}a_{53})]+$$
$$+a_{21}a_{42}[a_{13}(a_{35}a_{54}-a_{34}a_{55})+a_{14}(a_{33}a_{55}-a_{35}a_{53})+a_{15}(a_{34}a_{53}-a_{33}a_{54})]+$$
$$+a_{21}a_{52}[a_{13}(a_{34}a_{45}-a_{35}a_{44})+a_{14}(a_{35}a_{43}-a_{33}a_{45})+a_{15}(a_{33}a_{44}-a_{34}a_{43})]+$$

81

$$+a_{31}a_{12}[a_{23}(a_{44}a_{55}-a_{45}a_{54})+a_{24}(a_{45}a_{53}-a_{43}a_{55})+a_{25}(a_{43}a_{54}-a_{44}a_{53})]+$$
$$+a_{31}a_{22}[a_{13}(a_{45}a_{54}-a_{44}a_{55})+a_{14}(a_{43}a_{55}-a_{45}a_{53})+a_{15}(a_{44}a_{53}-a_{43}a_{54})]+$$
$$+a_{31}a_{42}[a_{13}(a_{24}a_{55}-a_{25}a_{54})+a_{14}(a_{25}a_{53}-a_{23}a_{55})+a_{15}(a_{23}a_{54}-a_{24}a_{53})]+$$
$$+a_{31}a_{52}[a_{13}(a_{25}a_{44}-a_{24}a_{45})+a_{14}(a_{23}a_{45}-a_{25}a_{43})+a_{15}(a_{24}a_{43}-a_{23}a_{44})]+$$

$$+a_{41}a_{12}[a_{23}(a_{35}a_{54}-a_{34}a_{55})+a_{24}(a_{33}a_{55}-a_{35}a_{53})+a_{25}(a_{34}a_{53}-a_{33}a_{54})]+$$
$$+a_{41}a_{22}[a_{13}(a_{34}a_{55}-a_{35}a_{54})+a_{14}(a_{35}a_{53}-a_{33}a_{55})+a_{15}(a_{33}a_{54}-a_{34}a_{53})]+$$
$$+a_{41}a_{32}[a_{13}(a_{25}a_{54}-a_{24}a_{55})+a_{14}(a_{23}a_{55}-a_{25}a_{53})+a_{15}(a_{24}a_{53}-a_{23}a_{54})]+$$
$$+a_{41}a_{52}[a_{13}(a_{24}a_{35}-a_{25}a_{34})+a_{14}(a_{25}a_{33}-a_{23}a_{35})+a_{15}(a_{23}a_{34}-a_{24}a_{33})]+$$

$$+a_{51}a_{12}[a_{23}(a_{34}a_{45}-a_{35}a_{44})+a_{24}(a_{35}a_{43}-a_{33}a_{45})+a_{25}(a_{33}a_{44}-a_{34}a_{43})]+$$
$$+a_{51}a_{22}[a_{13}(a_{35}a_{44}-a_{34}a_{45})+a_{14}(a_{33}a_{45}-a_{35}a_{43})+a_{15}(a_{34}a_{43}-a_{33}a_{44})]+$$
$$+a_{51}a_{32}[a_{13}(a_{24}a_{45}-a_{25}a_{44})+a_{14}(a_{25}a_{43}-a_{23}a_{45})+a_{15}(a_{23}a_{44}-a_{24}a_{43})]+$$
$$+a_{51}a_{42}[a_{13}(a_{25}a_{34}-a_{24}a_{35})+a_{14}(a_{23}a_{35}-a_{25}a_{33})+a_{15}(a_{24}a_{33}-a_{23}a_{34})] \qquad (73)$$

b) For the second column

The value of determinant of the 5x5 matrix at conditions when its elements have expansion to the second column can be expressed by the following formula:

$$\left| A \right| = \sum_{i=1}^{5} a_{i2}A_{i2} = \sum_{i=1}^{5} a_{i2}(-1)^{i+2} M_{i2} \qquad (74)$$

where,

a_{i2}- elements of the second column

M_{i2}-minor of algebraic supplemental (cofactor) A_{i2} of element a_{i2} taken with $(-1)^{i+2}$.

And now we can evaluate the determinant of the 5x5 matrix at expansion of its elements to the second column by the following formula:

$$\left| A \right| = -a_{12}A_{12}+a_{22}A_{22}-a_{32}A_{32}+a_{42}A_{42}-a_{52}A_{52} \qquad (75)$$

where,

$a_{12},a_{22},a_{32},a_{42},a_{52}$-elemeents of the second column of the 5x5 matrix

A_{12}-algebraic supplemental, belonging to the second column and the first row

A_{22}-algebraic supplemental, belonging to the second column and the second row

A_{32}-algebraic supplemental, belonging to the second column and third row

A_{42}-algebraic supplemental, belonging to the second column and fourth row

A_{52}-algebraic supplemental, belonging to the second column and fifth row .

$$A_{12}=(-1)^{1+2}\begin{vmatrix} a_{21} & a_{23} & a_{24} & a_{25} \\ a_{31} & a_{33} & a_{34} & a_{35} \\ a_{41} & a_{43} & a_{44} & a_{45} \\ a_{51} & a_{53} & a_{54} & a_{55} \end{vmatrix} \qquad A_{22}=(-1)^{2+2}\begin{vmatrix} a_{11} & a_{13} & a_{14} & a_{15} \\ a_{31} & a_{33} & a_{34} & a_{35} \\ a_{41} & a_{43} & a_{44} & a_{45} \\ a_{51} & a_{53} & a_{54} & a_{55} \end{vmatrix}$$

$$A_{32}=(-1)^{3+2}\begin{vmatrix} a_{11} & a_{13} & a_{14} & a_{15} \\ a_{21} & a_{23} & a_{24} & a_{25} \\ a_{41} & a_{43} & a_{44} & a_{45} \\ a_{51} & a_{53} & a_{54} & a_{55} \end{vmatrix} \qquad A_{42}=(-1)^{4+2}\begin{vmatrix} a_{11} & a_{13} & a_{14} & a_{15} \\ a_{21} & a_{23} & a_{24} & a_{25} \\ a_{31} & a_{33} & a_{34} & a_{35} \\ a_{51} & a_{53} & a_{54} & a_{55} \end{vmatrix}$$

$$A_{52}=(-1)^{5+2}\begin{vmatrix} a_{11} & a_{13} & a_{14} & a_{15} \\ a_{21} & a_{23} & a_{24} & a_{25} \\ a_{31} & a_{33} & a_{34} & a_{35} \\ a_{41} & a_{43} & a_{44} & a_{45} \end{vmatrix}$$

In Figure 81 is shown the tied graph (tree) for finding in general view the value of determinant of the 5x5 matrix at expansion of its elements to the second column.

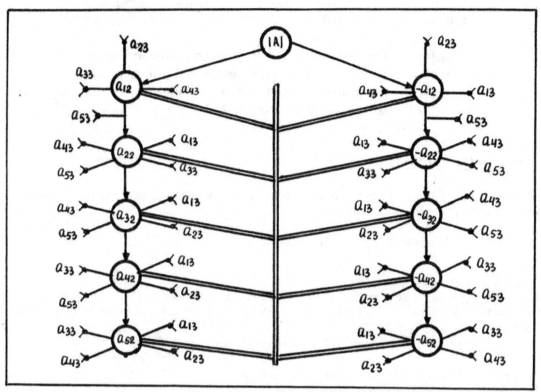

Figure 81 Tied graph (tree) for elements of the determinant of the 5x5 matrix at expansion of its elements to the second column in general view

So, the value of determinant of the 5x5 matrix *at expansion of its elements to the second column,* after of some transformations, is equal in general view:

$$|A| = a_{12}a_{23}[a_{31}(a_{44}a_{55}-a_{45}a_{54})+a_{34}(a_{45}a_{51}-a_{41}a_{55})+a_{35}(a_{41}a_{54}-a_{44}a_{51})]+$$
$$+a_{12}a_{33}[a_{21}(a_{45}a_{54}-a_{44}a_{55})+a_{24}(a_{41}a_{55}-a_{45}a_{51})+a_{25}(a_{44}a_{51}-a_{41}a_{54})]+$$
$$+a_{12}a_{43}[a_{21}(a_{34}a_{55}-a_{35}a_{54})+a_{24}(a_{35}a_{51}-a_{31}a_{55})+a_{25}(a_{31}a_{54}-a_{34}a_{51})]+$$
$$+a_{12}a_{53}[a_{21}(a_{35}a_{44}-a_{34}a_{45})+a_{24}(a_{31}a_{45}-a_{35}a_{41})+a_{25}(a_{34}a_{41}-a_{31}a_{44})]+$$

$$+a_{22}a_{13}[a_{31}(a_{45}a_{54}-a_{44}a_{55})+a_{34}(a_{41}a_{55}-a_{45}a_{51})+a_{35}(a_{44}a_{51}-a_{41}a_{54})]+$$
$$+a_{22}a_{33}[a_{11}(a_{44}a_{55}-a_{45}a_{54})+a_{14}(a_{45}a_{51}-a_{41}a_{55})+a_{15}(a_{41}a_{54}-a_{44}a_{51})]+$$
$$+a_{22}a_{43}[a_{11}(a_{35}a_{54}-a_{34}a_{55})+a_{14}(a_{31}a_{55}-a_{35}a_{51})+a_{15}(a_{34}a_{51}-a_{31}a_{54})]+$$
$$+a_{22}a_{53}[a_{11}(a_{34}a_{45}-a_{35}a_{44})+a_{14}(a_{35}a_{41}-a_{31}a_{45})+a_{15}(a_{31}a_{44}-a_{34}a_{41})]+$$

$$+a_{32}a_{13}[a_{21}(a_{44}a_{55}-a_{45}a_{54})+a_{24}(a_{45}a_{51}-a_{41}a_{55})+a_{25}(a_{41}a_{54}-a_{44}a_{51})]+$$
$$+a_{32}a_{23}[a_{11}(a_{45}a_{54}-a_{44}a_{55})+a_{14}(a_{41}a_{55}-a_{45}a_{51})+a_{15}(a_{44}a_{51}-a_{41}a_{54})]+$$
$$+a_{32}a_{43}[a_{11}(a_{24}a_{55}-a_{25}a_{54})+a_{14}(a_{25}a_{51}-a_{21}a_{55})+a_{15}(a_{21}a_{54}-a_{24}a_{51})]+$$
$$+a_{32}a_{53}[a_{11}(a_{25}a_{44}-a_{24}a_{45})+a_{14}(a_{21}a_{45}-a_{25}a_{41})+a_{15}(a_{24}a_{41}-a_{21}a_{44})]+$$

83

$+a_{42}a_{13}[a_{21}(a_{35}a_{54}-a_{34}a_{55})+a_{24}(a_{31}a_{55}-a_{35}a_{51})+a_{25}(a_{34}a_{51}-a_{31}a_{54})]+$

$+a_{42}a_{23}[a_{11}(a_{34}a_{55}-a_{35}a_{54})+a_{14}(a_{35}a_{51}-a_{31}a_{55})+a_{15}(a_{31}a_{54}-a_{34}a_{51})]+$

$+a_{42}a_{33}[a_{11}(a_{25}a_{54}-a_{24}a_{55})+a_{14}(a_{21}a_{55}-a_{25}a_{51})+a_{15}(a_{24}a_{51}-a_{21}a_{54})]+$

$+a_{42}a_{53}[a_{11}(a_{24}a_{35}-a_{25}a_{34})+a_{14}(a_{25}a_{31}-a_{21}a_{35})+a_{15}(a_{21}a_{34}-a_{24}a_{31})]+$

$+a_{52}a_{13}[a_{21}(a_{34}a_{45}-a_{35}a_{44})+a_{24}(a_{35}a_{41}-a_{31}a_{45})+a_{25}(a_{31}a_{44}-a_{34}a_{41})]+$

$+a_{52}a_{23}[a_{11}(a_{35}a_{44}-a_{34}a_{45})+a_{14}(a_{31}a_{45}-a_{35}a_{41})+a_{15}(a_{34}a_{41}-a_{31}a_{44})]+$

$+a_{52}a_{33}[a_{11}(a_{24}a_{45}-a_{25}a_{44})+a_{14}(a_{25}a_{41}-a_{21}a_{45})+a_{15}(a_{21}a_{44}-a_{24}a_{41})]+$

$+a_{52}a_{43}[a_{11}(a_{25}a_{34}-a_{24}a_{35})+a_{14}(a_{21}a_{35}-a_{25}a_{31})+a_{15}(a_{24}a_{31}-a_{21}a_{34})]$ (76)

c) **For the third column**

The value of determinant of the 5x5 matrix at conditions when its elements have expansion to the third column can be expressed by the following formula:

$$|A| = \sum_{i=1}^{5} a_{i3}A_{i3} = \sum_{i=1}^{5} a_{i3}(-1)^{i+3} M_{i3} \quad (77)$$

where,

a_{i3}- elements of the third column

M_{i3}-minor of algebraic supplemental (cofactor) A_{i3} of element a_{i3} taken with $(-1)^{i+3}$.

And now we can evaluate the determinant of the 5x5 matrix at expansion of its elements to the third column by the following formula:

$$|A| = a_{13}A_{13} - a_{23}A_{23} + a_{33}A_{33} - a_{43}A_{43} + a_{53}A_{53} \quad (78)$$

where,

$a_{13}, a_{23}, a_{33}, a_{43}, a_{53}$-elements of the third column of the 5x5 matrix

A_{13}-algebraic supplemental, belonging to the first row and third column

A_{23}-algebraic supplemental, belonging to the second row and third column

A_{33}-algebraic supplemental, belonging to the third row and third column

A_{43}-algebraic supplemental, belonging to the fourth row and third column

A_{53}-algebraic supplemental , belonging to the fifth row and third column.

$$A_{13}=(-1)^{1+3} \begin{vmatrix} a_{21} & a_{22} & a_{24} & a_{25} \\ a_{31} & a_{32} & a_{34} & a_{35} \\ a_{41} & a_{42} & a_{44} & a_{45} \\ a_{51} & a_{52} & a_{54} & a_{55} \end{vmatrix} \qquad A_{23}=(-1)^{2+3} \begin{vmatrix} a_{11} & a_{12} & a_{14} & a_{15} \\ a_{31} & a_{32} & a_{34} & a_{35} \\ a_{41} & a_{42} & a_{44} & a_{45} \\ a_{51} & a_{52} & a_{54} & a_{55} \end{vmatrix}$$

$$A_{33}=(-1)^{3+3} \begin{vmatrix} a_{11} & a_{12} & a_{14} & a_{15} \\ a_{21} & a_{22} & a_{24} & a_{25} \\ a_{41} & a_{42} & a_{44} & a_{45} \\ a_{51} & a_{52} & a_{54} & a_{55} \end{vmatrix} \qquad A_{43}=(-1)^{4+3} \begin{vmatrix} a_{11} & a_{12} & a_{14} & a_{15} \\ a_{21} & a_{22} & a_{24} & a_{25} \\ a_{31} & a_{32} & a_{34} & a_{35} \\ a_{51} & a_{52} & a_{54} & a_{55} \end{vmatrix}$$

$$A_{53}=(-1)^{5+3} \begin{vmatrix} a_{11} & a_{12} & a_{14} & a_{15} \\ a_{21} & a_{22} & a_{24} & a_{25} \\ a_{31} & a_{32} & a_{34} & a_{35} \\ a_{41} & a_{42} & a_{44} & a_{45} \end{vmatrix}$$

In Figure 82 is shown the tied graph (tree) for finding in general view the value of determinant of the 5x5 matrix at expansion of its elements to the third column.

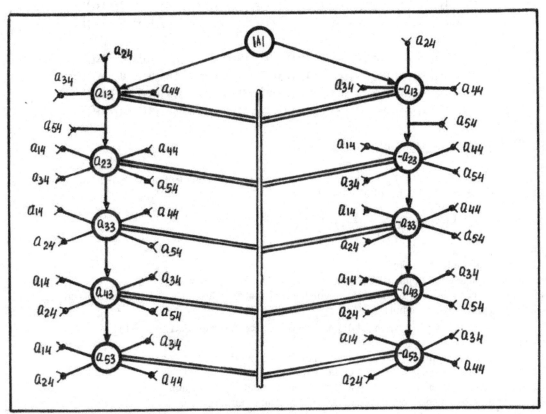

Figure 82 Tied graph (tree) for elements of the determinant of the 5x5 matrix at expansion of its elements to the third column in general view

So, the value of determinant of the 5x5 matrix *at expansion of its elements to the third column, after of some transformations,* is equal in general view:

$$|A| = a_{13}a_{24}[a_{31}(a_{42}a_{55}-a_{45}a_{52})+a_{32}(a_{45}a_{51}-a_{41}a_{55})+a_{35}(a_{41}a_{52}-a_{42}a_{51})]+$$
$$+a_{13}a_{34}[a_{21}(a_{45}a_{52}-a_{42}a_{55})+a_{22}(a_{41}a_{55}-a_{45}a_{51})+a_{25}(a_{42}a_{51}-a_{41}a_{52})]+$$
$$+a_{13}a_{44}[a_{21}(a_{32}a_{55}-a_{35}a_{52})+a_{22}(a_{35}a_{51}-a_{31}a_{55})+a_{25}(a_{31}a_{52}-a_{32}a_{51})]+$$
$$+a_{13}a_{54}[a_{21}(a_{35}a_{42}-a_{32}a_{45})+a_{22}(a_{31}a_{45}-a_{35}a_{41})+a_{25}(a_{32}a_{41}-a_{31}a_{42})]+$$

$$+a_{23}a_{14}[a_{31}(a_{45}a_{52}-a_{42}a_{55})+a_{32}(a_{41}a_{55}-a_{45}a_{51})+a_{35}(a_{42}a_{51}-a_{41}a_{52})]+$$
$$+a_{23}a_{34}[a_{11}(a_{42}a_{55}-a_{45}a_{52})+a_{12}(a_{45}a_{51}-a_{41}a_{55})+a_{15}(a_{41}a_{52}-a_{42}a_{51})]+$$
$$+a_{23}a_{44}[a_{11}(a_{35}a_{52}-a_{32}a_{55})+a_{12}(a_{31}a_{55}-a_{35}a_{51})+a_{15}(a_{32}a_{51}-a_{31}a_{52})]+$$
$$+a_{23}a_{54}[a_{11}(a_{32}a_{45}-a_{35}a_{42})+a_{12}(a_{35}a_{41}-a_{31}a_{45})+a_{15}(a_{31}a_{42}-a_{32}a_{41})]+$$

$$+a_{33}a_{14}[a_{21}(a_{42}a_{55}-a_{45}a_{52})+a_{22}(a_{45}a_{51}-a_{41}a_{55})+a_{25}(a_{41}a_{52}-a_{42}a_{51})]+$$
$$+a_{33}a_{24}[a_{11}(a_{45}a_{52}-a_{42}a_{55})+a_{12}(a_{41}a_{55}-a_{45}a_{51})+a_{15}(a_{42}a_{51}-a_{41}a_{52})]+$$
$$+a_{33}a_{44}[a_{11}(a_{22}a_{55}-a_{25}a_{52})+a_{12}(a_{25}a_{51}-a_{21}a_{55})+a_{15}(a_{21}a_{52}-a_{22}a_{51})]+$$
$$+a_{33}a_{54}[a_{11}(a_{25}a_{42}-a_{22}a_{45})+a_{12}(a_{21}a_{45}-a_{25}a_{41})+a_{15}(a_{22}a_{41}-a_{21}a_{42})]+$$

85

$+a_{43}a_{14}[a_{21}(a_{35}a_{52}-a_{32}a_{55})+a_{22}(a_{31}a_{55}-a_{35}a_{51})+a_{25}(a_{32}a_{51}-a_{31}a_{52})]+$
$+a_{43}a_{24}[a_{11}(a_{32}a_{55}-a_{35}a_{52})+a_{12}(a_{35}a_{51}-a_{31}a_{55})+a_{15}(a_{31}a_{52}-a_{32}a_{51})]+$
$+a_{43}a_{34}[a_{11}(a_{25}a_{52}-a_{22}a_{55})+a_{12}(a_{21}a_{55}-a_{25}a_{51})+a_{15}(a_{22}a_{51}-a_{21}a_{52})]+$
$+a_{43}a_{54}[a_{11}(a_{22}a_{35}-a_{25}a_{32})+a_{12}(a_{25}a_{31}-a_{21}a_{35})+a_{15}(a_{21}a_{32}-a_{22}a_{31})]+$

$+a_{53}a_{14}[a_{21}(a_{32}a_{45}-a_{35}a_{42})+a_{22}(a_{35}a_{41}-a_{31}a_{45})+a_{25}(a_{31}a_{42}-a_{32}a_{41})]+$
$+a_{53}a_{24}[a_{11}(a_{35}a_{42}-a_{32}a_{45})+a_{12}(a_{31}a_{45}-a_{35}a_{41})+a_{15}(a_{32}a_{41}-a_{31}a_{42})]+$
$+a_{53}a_{34}[a_{11}(a_{22}a_{45}-a_{25}a_{42})+a_{12}(a_{25}a_{41}-a_{21}a_{45})+a_{15}(a_{21}a_{42}-a_{22}a_{41})]+$
$+a_{53}a_{44}[a_{11}(a_{25}a_{32}-a_{22}a_{35})+a_{12}(a_{21}a_{35}-a_{25}a_{31})+a_{15}(a_{22}a_{31}-a_{21}a_{32})]$ (79)

d) **For the fourth column**

The value of determinant of the 5x5 matrix at conditions when its elements have expansion to the fourth column can be expressed by the following formula:

$$|A|=\sum_{i=1}^{5}a_{i4}A_{i4}=\sum_{i=1}^{5}a_{i4}(-1)^{i+4}M_{i4} \qquad (80)$$

where,

a_{i4}- elements of the fourth column

M_{i4}- minor of algebraic supplemental (cofactor) A_{i4} of element a_{i4} taken with $(-1)^{i+4}$.

And now we can evaluate the determinant of the 5x5 matrix at expansion of its elements to the fourth column by the following formula:

$$|A|=-a_{14}A_{14}+a_{24}A_{24}-a_{34}A_{34}+a_{44}A_{44}-a_{54}A_{54} \qquad (81)$$

where,

$a_{14},a_{24},a_{34},a_{44},a_{54}$-elements of the fourth column of the 5x5 matrix

A_{14}-algebraic supplemental, belonging to the first row and fourth column

A_{24}-algebraic supplemental, belonging to the second row and fourth column

A_{34}-algebraic supplemental, belonging to the third row and fourth column

A_{44}-algebraic supplemental, belonging to the fourth row and fourth column

A_{54}-algebraic supplemental, belonging to the fifth row and fourth column .

$$A_{14}=(-1)^{1+4}\begin{vmatrix} a_{21} & a_{22} & a_{23} & a_{25} \\ a_{31} & a_{32} & a_{33} & a_{35} \\ a_{41} & a_{42} & a_{43} & a_{45} \\ a_{51} & a_{52} & a_{53} & a_{55} \end{vmatrix} \qquad A_{24}=(-1)^{2+4}\begin{vmatrix} a_{11} & a_{12} & a_{13} & a_{15} \\ a_{31} & a_{32} & a_{33} & a_{35} \\ a_{41} & a_{42} & a_{43} & a_{45} \\ a_{51} & a_{52} & a_{53} & a_{55} \end{vmatrix}$$

$$A_{34}=(-1)^{3+4}\begin{vmatrix} a_{11} & a_{12} & a_{13} & a_{15} \\ a_{21} & a_{22} & a_{23} & a_{25} \\ a_{41} & a_{42} & a_{43} & a_{45} \\ a_{51} & a_{52} & a_{53} & a_{55} \end{vmatrix} \qquad A_{44}=(-1)^{4+4}\begin{vmatrix} a_{11} & a_{12} & a_{13} & a_{15} \\ a_{21} & a_{22} & a_{23} & a_{25} \\ a_{31} & a_{32} & a_{33} & a_{35} \\ a_{51} & a_{52} & a_{53} & a_{55} \end{vmatrix}$$

$$A_{54}=(-1)^{5+4}\begin{vmatrix} a_{11} & a_{12} & a_{13} & a_{15} \\ a_{21} & a_{22} & a_{23} & a_{25} \\ a_{31} & a_{32} & a_{33} & a_{35} \\ a_{41} & a_{42} & a_{43} & a_{45} \end{vmatrix}$$

In Figure 83 is shown the tied graph (tree) for finding in general view the value of determinant of the 5x5 matrix at expansion of its elements to the fourth column.

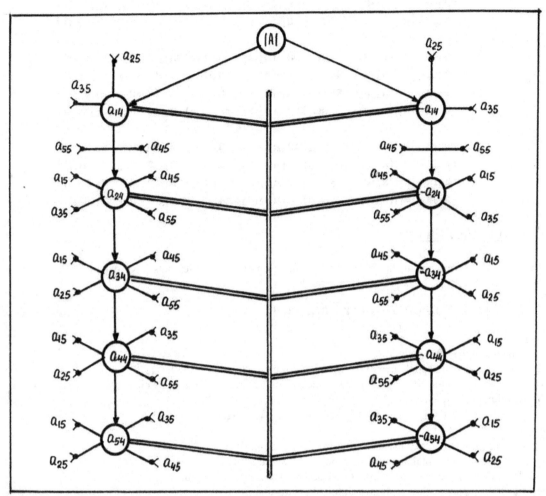

Figure 83 Tied graph (tree) for elements of the determinant of the 5x5 matrix at expansion of its elements to the fourth column in general view

So, the determinant of the 5x5 matrix *at expansion of its elements to the fourth column,* after of some transformations, is equal in general view:

$$|A| = a_{14}a_{25}[a_{31}(a_{42}a_{53}-a_{43}a_{52})+a_{32}(a_{43}a_{51}-a_{41}a_{53})+a_{33}(a_{41}a_{52}-a_{42}a_{51})]+$$
$$+a_{14}a_{35}[a_{21}(a_{43}a_{52}-a_{42}a_{53})+a_{22}(a_{41}a_{53}-a_{43}a_{51})+a_{23}(a_{42}a_{51}-a_{41}a_{52})]+$$
$$+a_{14}a_{45}[a_{21}(a_{32}a_{53}-a_{33}a_{52})+a_{22}(a_{33}a_{51}-a_{31}a_{53})+a_{23}(a_{31}a_{52}-a_{32}a_{51})]+$$
$$+a_{14}a_{55}[a_{21}(a_{33}a_{42}-a_{32}a_{43})+a_{22}(a_{31}a_{43}-a_{33}a_{41})+a_{23}(a_{32}a_{41}-a_{31}a_{42})]+$$

$$+a_{24}a_{15}[a_{31}(a_{43}a_{52}-a_{42}a_{53})+a_{32}(a_{41}a_{53}-a_{43}a_{51})+a_{33}(a_{42}a_{51}-a_{41}a_{52})]+$$
$$+a_{24}a_{35}[a_{11}(a_{42}a_{53}-a_{43}a_{52})+a_{12}(a_{43}a_{51}-a_{41}a_{53})+a_{13}(a_{41}a_{52}-a_{42}a_{51})]+$$
$$+a_{24}a_{45}[a_{11}(a_{33}a_{52}-a_{32}a_{53})+a_{12}(a_{31}a_{53}-a_{33}a_{51})+a_{13}(a_{32}a_{51}-a_{31}a_{52})]+$$
$$+a_{24}a_{55}[a_{11}(a_{32}a_{43}-a_{33}a_{42})+a_{12}(a_{33}a_{41}-a_{31}a_{43})+a_{13}(a_{31}a_{42}-a_{32}a_{41})+$$

87

$+a_{34}a_{15}[a_{21}(a_{42}a_{53}-a_{43}a_{52})+a_{22}(a_{43}a_{51}-a_{41}a_{53})+a_{23}(a_{41}a_{52}-a_{42}a_{51})]+$

$+a_{34}a_{25}[a_{11}(a_{43}a_{52}-a_{42}a_{53})+a_{12}(a_{41}a_{53}-a_{43}a_{51})+a_{13}(a_{42}a_{51}-a_{41}a_{52})]+$

$+a_{34}a_{45}[a_{11}(a_{22}a_{53}-a_{23}a_{52})+a_{12}(a_{23}a_{51}-a_{21}a_{53})+a_{13}(a_{21}a_{52}-a_{22}a_{51})]+$

$+a_{34}a_{55}[a_{11}(a_{23}a_{42}-a_{22}a_{43})+a_{12}(a_{21}a_{43}-a_{23}a_{41})+a_{13}(a_{22}a_{41}-a_{21}a_{42})]+$

$+a_{44}a_{15}[a_{21}(a_{33}a_{52}-a_{32}a_{53})+a_{22}(a_{31}a_{53}-a_{33}a_{51})+a_{23}(a_{32}a_{51}-a_{31}a_{52})]+$

$+a_{44}a_{25}[a_{11}(a_{32}a_{53}-a_{33}a_{52})+a_{12}(a_{33}a_{51}-a_{31}a_{53})+a_{13}(a_{31}a_{52}-a_{32}a_{51})]+$

$+a_{44}a_{35}[a_{11}(a_{23}a_{52}-a_{22}a_{53})+a_{12}(a_{21}a_{53}-a_{23}a_{51})+a_{13}(a_{22}a_{51}-a_{21}a_{52})]+$

$+a_{44}a_{55}[a_{11}(a_{22}a_{33}-a_{23}a_{32})+a_{12}(a_{23}a_{31}-a_{21}a_{33})+a_{13}(a_{21}a_{32}-a_{22}a_{31})]+$

$+a_{54}a_{15}[a_{21}(a_{32}a_{43}-a_{33}a_{42})+a_{22}(a_{33}a_{41}-a_{31}a_{43})+a_{23}(a_{31}a_{42}-a_{32}a_{41})]+$

$+a_{54}a_{25}[a_{11}(a_{33}a_{42}-a_{32}a_{43})+a_{12}(a_{31}a_{43}-a_{33}a_{41})+a_{13}(a_{32}a_{41}-a_{31}a_{41})]+$

$+a_{54}a_{35}[a_{11}(a_{22}a_{43}-a_{23}a_{42})+a_{12}(a_{23}a_{41}-a_{21}a_{43})+a_{13}(a_{21}a_{42}-a_{22}a_{41})]+$

$+a_{54}a_{45}[a_{11}(a_{23}a_{32}-a_{22}a_{33})+a_{12}(a_{21}a_{33}-a_{23}a_{31})+a_{13}(a_{22}a_{31}-a_{21}a_{32})]$ (82)

e) For the fifth row

The value of determinant of the 5x5 matrix at conditions when its elements have expansion to the fifth column can be expressed by the following formula:

$$|A|=\sum_{i=1}^{5}a_{i5}A_{i5}=\sum_{i=1}^{5}a_{i5}(-1)^{i+5}M_{i5} \quad (83)$$

where,

a_{i5}- elements of the fifth column

M_{i5}-minor of algebraic supplemental (cofactor) A_{i5} of element a_{i5} taken with $(-1)^{i+5}$.

And now we can evaluate the determinant of the 5x5 matrix at expansion of its elements to the fifth column by the following formula:

$$|A|=a_{15}A_{15}-a_{25}A_{25}+a_{35}A_{35}-a_{45}A_{45}+a_{55}A_{55} \quad (84)$$

where,

$a_{15},a_{25},a_{35},a_{45},a_{55}$-elements of the fifth column of the 5x5 matrix

A_{15}-algebraic supplemental, belonging to the first row and fifth column

A_{25}-algebraic supplemental, belonging to the second row and fifth column

A_{35}-algebraic supplemental, belonging to the third row and fifth column

A_{45}-algebraic supplemental, belonging to the fourth row and fifth column

A_{55}-algebraic supplemental, belonging to the fifth row and fifth column .

$$A_{15}=(-1)^{1+5}\begin{vmatrix} a_{21} & a_{22} & a_{23} & a_{24} \\ a_{31} & a_{32} & a_{33} & a_{34} \\ a_{41} & a_{42} & a_{43} & a_{44} \\ a_{51} & a_{52} & a_{53} & a_{54} \end{vmatrix} \qquad A_{25}=(-1)^{2+5}\begin{vmatrix} a_{11} & a_{12} & a_{13} & a_{14} \\ a_{31} & a_{32} & a_{33} & a_{34} \\ a_{41} & a_{42} & a_{43} & a_{44} \\ a_{51} & a_{52} & a_{53} & a_{54} \end{vmatrix}$$

$$A_{35}=(-1)^{3+5}\begin{vmatrix} a_{11} & a_{12} & a_{13} & a_{14} \\ a_{21} & a_{22} & a_{23} & a_{24} \\ a_{41} & a_{42} & a_{43} & a_{44} \\ a_{51} & a_{52} & a_{53} & a_{54} \end{vmatrix} \qquad A_{45}=(-1)^{4+5}\begin{vmatrix} a_{11} & a_{12} & a_{13} & a_{14} \\ a_{21} & a_{22} & a_{23} & a_{24} \\ a_{31} & a_{32} & a_{33} & a_{34} \\ a_{51} & a_{52} & a_{53} & a_{54} \end{vmatrix}$$

$$A_{55}=(-1)^{5+5}\begin{vmatrix} a_{11} & a_{12} & a_{13} & a_{14} \\ a_{21} & a_{22} & a_{23} & a_{24} \\ a_{31} & a_{32} & a_{33} & a_{34} \\ a_{41} & a_{42} & a_{43} & a_{44} \end{vmatrix}$$

In Figure 84 is shown the tied graph(tree) for finding in general view the value of determinant of the 5x5 matrix at expansion of its elements to the fifth column.

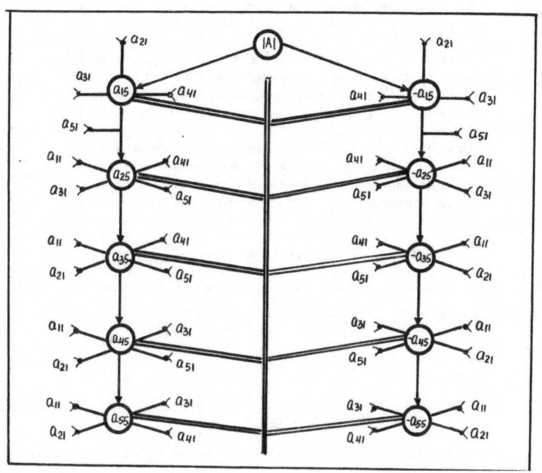

Figure 84 Tied graph (tree) for elements of the determinant of the 5x5 matrix at expansion of its elements to the fifth column in general view

So, the value of determinant of the 5x5 matrix *at expansion of its elements to the fifth column,* after of some transformations , is equal in general view:

$$|A| = a_{15}a_{21}[a_{32}(a_{43}a_{54}-a_{44}a_{53})+a_{33}(a_{44}a_{52}-a_{42}a_{54})+a_{34}(a_{42}a_{53}-a_{43}a_{52})]+$$
$$+a_{15}a_{31}[a_{22}(a_{44}a_{53}-a_{43}a_{54})+a_{23}(a_{42}a_{54}-a_{44}a_{52})+a_{24}(a_{43}a_{52}-a_{42}a_{53})]+$$
$$+a_{15}a_{41}[a_{22}(a_{33}a_{54}-a_{34}a_{53})+a_{23}(a_{34}a_{52}-a_{32}a_{54})+a_{24}(a_{32}a_{53}-a_{33}a_{52})]+$$
$$+a_{15}a_{51}[a_{22}(a_{34}a_{43}-a_{33}a_{44})+a_{23}(a_{32}a_{44}-a_{34}a_{42})+a_{24}(a_{33}a_{42}-a_{32}a_{43})]+$$

$$+a_{25}a_{11}[a_{32}(a_{44}a_{53}-a_{43}a_{54})+a_{33}(a_{42}a_{54}-a_{44}a_{52})+a_{34}(a_{43}a_{52}-a_{42}a_{53})]+$$
$$+a_{25}a_{31}[a_{12}(a_{43}a_{54}-a_{44}a_{53})+a_{13}(a_{44}a_{52}-a_{42}a_{54})+a_{14}(a_{42}a_{53}-a_{43}a_{52})]+$$
$$+a_{25}a_{41}[a_{12}(a_{34}a_{53}-a_{33}a_{54})+a_{13}(a_{32}a_{54}-a_{34}a_{52})+a_{14}(a_{33}a_{52}-a_{32}a_{53})]+$$
$$+a_{25}a_{51}[a_{12}(a_{33}a_{44}-a_{34}a_{43})+a_{13}(a_{34}a_{42}-a_{32}a_{44})+a_{14}(a_{32}a_{43}-a_{33}a_{42})]+$$

$+a_{35}a_{11}[a_{22}(a_{43}a_{54}{-}a_{44}a_{53}){+}a_{23}(a_{44}a_{52}{-}a_{42}a_{54}){+}a_{24}(a_{42}a_{53}{-}a_{43}a_{52})]+$
$+a_{35}a_{21}[a_{12}(a_{44}a_{53}{-}a_{43}a_{54}){+}a_{13}(a_{42}a_{54}{-}a_{44}a_{52}){+}a_{14}(a_{43}a_{52}{-}a_{42}a_{53})]+$
$+a_{35}a_{41}[a_{12}(a_{23}a_{54}{-}a_{24}a_{53}){+}a_{13}(a_{24}a_{52}{-}a_{22}a_{54}){+}a_{14}(a_{22}a_{53}{-}a_{23}a_{52})]+$
$+a_{35}a_{51}[a_{12}(a_{24}a_{43}{-}a_{23}a_{44}){+}a_{13}(a_{22}a_{44}{-}a_{24}a_{42}){+}a_{14}(a_{23}a_{42}{-}a_{22}a_{43})]+$

$+a_{45}a_{11}[a_{22}(a_{34}a_{53}{-}a_{33}a_{54}){+}a_{23}(a_{32}a_{54}{-}a_{34}a_{52}){+}a_{24}(a_{33}a_{52}{-}a_{32}a_{53})]+$
$+a_{45}a_{21}[a_{12}(a_{33}a_{54}{-}a_{34}a_{53}){+}a_{13}(a_{34}a_{52}{-}a_{32}a_{54}){+}a_{14}(a_{32}a_{53}{-}a_{33}a_{52})]+$
$+a_{45}a_{31}[a_{12}(a_{24}a_{53}{-}a_{23}a_{54}){+}a_{13}(a_{22}a_{54}{-}a_{24}a_{52}){+}a_{14}(a_{23}a_{52}{-}a_{22}a_{53})]+$
$+a_{45}a_{51}[a_{12}(a_{23}a_{34}{-}a_{24}a_{33}){+}a_{13}(a_{24}a_{32}{-}a_{22}a_{34}){+}a_{14}(a_{22}a_{33}{-}a_{23}a_{32})]+$

$+a_{55}a_{11}[a_{22}(a_{33}a_{44}{-}a_{34}a_{43}){+}a_{23}(a_{34}a_{42}{-}a_{32}a_{44}){+}a_{24}(a_{32}a_{43}{-}a_{33}a_{42})]+$
$+a_{55}a_{21}[a_{12}(a_{34}a_{43}{-}a_{33}a_{44}){+}a_{13}(a_{32}a_{44}{-}a_{34}a_{42}){+}a_{14}(a_{33}a_{42}{-}a_{32}a_{43})]+$
$+a_{55}a_{31}[a_{12}(a_{23}a_{44}{-}a_{24}a_{43}){+}a_{13}(a_{24}a_{42}{-}a_{22}a_{44}){+}a_{14}(a_{22}a_{43}{-}a_{23}a_{42})]+$
$+a_{55}a_{41}[a_{12}(a_{24}a_{33}{-}a_{23}a_{34}){+}a_{13}(a_{22}a_{34}{-}a_{24}a_{32}){+}a_{14}(a_{23}a_{32}{-}a_{22}a_{33})]$ (85)

1.2 **By Laplace's theorem for the rows and columns**

The main features of Laplace's theorem we consider in general view for the determinant of the 5x5 matrix:

$$A=\begin{vmatrix} a_{11} & a_{12} & a_{13} & a_{14} & a_{15} \\ a_{21} & a_{22} & a_{23} & a_{24} & a_{25} \\ a_{31} & a_{32} & a_{33} & a_{34} & a_{35} \\ a_{41} & a_{42} & a_{43} & a_{44} & a_{45} \\ a_{51} & a_{52} & a_{53} & a_{54} & a_{55} \end{vmatrix}$$

a) **For the first and second rows**

The value of determinant of the 5x5 matrix at conditions when marked out jointly the first and second rows is expressed in the following steps:
1) Primary should primary to mark out the first and second rows and then to mark jointly the first and second columns, writing the minor(M_{12}) and algebraic supplemental(A_{12}) and then;
2) In the 5x5 matrix should to mark out jointly the first and third columns, writing the minor (M_{13}) and algebraic supplemental(cofactor) A_{13} ;
3) And then in the 5x5 matrix should to mark out jointly the first and fourth columns, writing the minor (M_{14}) and algebraic supplemental (A_{14}) and further;
4) In the 5x5 matrix should to mark out jointly the first and fifth jointly two columns, writing the minor(M_{15}) and algebraic supplemental(A_{15}) and then;
5) Should in the 5x5 matrix to mark out jointly the second and third two columns, writing the minor(M_{23}) and algebraic supplemental (A_{23}) and then;
6) In the 5x5 matrix should to mark out jointly the second and fourth two columns, writing the minor(M_{24}) and algebraic supplemental (A_{24});
7) And then should to mark out jointly in the 5x5 matrix the second and fifth two columns, writing the minor(M_{25}) and algebraic supplemental (A_{25}) and then;
8) In the 5x5 matrix should to mark out jointly the third and fourth two columns, writing the minor (M_{34}) and algebraic supplemental (A_{34}) and then;
9) Should in the 5x5 matrix to mark out jointly the third and fifth two columns, writing the minor(M_{35}) and algebraic supplemental (A_{35}) and finally;
10) In the 5x5 matrix should to mark out jointly the fourth and fifth two columns, writing the minor (M_{45}) and algebraic supplemental (A_{45}).

After of all above-indicated procedures we have the following minors and algebraic supplemental (cofactors) applicably to Laplace's theorem for marked out jointly the first and second rows:

$$M_{12}=(a_{11}a_{22}-a_{12}a_{21}); \quad A_{12}=(-1)^{(1+2)+(1+2)}\begin{vmatrix} a_{33} & a_{34} & a_{35} \\ a_{43} & a_{44} & a_{45} \\ a_{53} & a_{54} & a_{55} \end{vmatrix}$$

$$M_{13}=(a_{11}a_{23}-a_{13}a_{21}); \quad A_{13}=(-1)^{(1+2)+(1+3)}\begin{vmatrix} a_{32} & a_{34} & a_{35} \\ a_{42} & a_{44} & a_{45} \\ a_{52} & a_{54} & a_{55} \end{vmatrix}$$

$$M_{14}=(a_{11}a_{24}-a_{14}a_{21}); \quad A_{14}=(-1)^{(1+2)+(1+4)}\begin{vmatrix} a_{32} & a_{33} & a_{35} \\ a_{42} & a_{43} & a_{45} \\ a_{52} & a_{53} & a_{55} \end{vmatrix}$$

$$M_{15}=(a_{11}a_{25}-a_{15}a_{21}); \quad A_{15}=(-1)^{(1+2)+(1+5)}\begin{vmatrix} a_{32} & a_{33} & a_{34} \\ a_{42} & a_{43} & a_{44} \\ a_{52} & a_{53} & a_{54} \end{vmatrix}$$

$$M_{23}=(a_{12}a_{23}-a_{13}a_{22}); \quad A_{23}=(-1)^{(1+2)+(2+3)}\begin{vmatrix} a_{31} & a_{34} & a_{35} \\ a_{41} & a_{44} & a_{45} \\ a_{51} & a_{54} & a_{55} \end{vmatrix}$$

$$M_{24}=(a_{12}a_{24}-a_{14}a_{22}); \quad A_{24}=(-1)^{(1+2)+(2+4)}\begin{vmatrix} a_{31} & a_{33} & a_{35} \\ a_{41} & a_{43} & a_{45} \\ a_{51} & a_{53} & a_{55} \end{vmatrix}$$

$$M_{25}=(a_{12}a_{25}-a_{15}a_{22}); \quad A_{25}=(-1)^{(1+2)+(2+5)}\begin{vmatrix} a_{31} & a_{33} & a_{34} \\ a_{41} & a_{43} & a_{44} \\ a_{51} & a_{53} & a_{54} \end{vmatrix}$$

$$M_{34}=(a_{13}a_{24}-a_{14}a_{23}); \quad A_{34}=(-1)^{(1+2)+(3+4)}\begin{vmatrix} a_{31} & a_{32} & a_{35} \\ a_{41} & a_{42} & a_{45} \\ a_{51} & a_{52} & a_{55} \end{vmatrix}$$

$$M_{35}=(a_{13}a_{25}-a_{15}a_{23}); \quad A_{35}=(-1)^{(1+2)+(3+5)}\begin{vmatrix} a_{31} & a_{32} & a_{34} \\ a_{41} & a_{42} & a_{44} \\ a_{51} & a_{52} & a_{54} \end{vmatrix}$$

$$M_{45}=(a_{14}a_{25}-a_{15}a_{24}); \quad A_{45}=(-1)^{(1+2)+(4+5)}\begin{vmatrix} a_{31} & a_{32} & a_{33} \\ a_{41} & a_{42} & a_{43} \\ a_{51} & a_{52} & a_{53} \end{vmatrix}$$

So, the value of determinant of the 5x5 matrix applicably to the Laplace's theorem for marked out jointly the first and second rows can be determined by the following formula in general view:

$$|A| = M_{12}A_{12} - M_{13}A_{13} + M_{14}A_{14} - M_{15}A_{15} + M_{23}A_{23} - M_{24}A_{24} +$$
$$+ M_{25}A_{25} + M_{34}A_{34} - M_{35}A_{35} + M_{45}A_{45}$$

In Figure 85 is shown the tied graphs (tree) for elements of the determinant of the 5x5 matrix at marked out jointly the first and second rows applicably to Laplace's theorem.

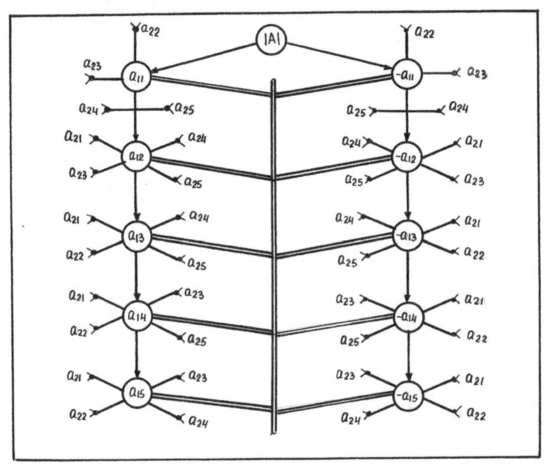

Figure 85 Tied graphs (tree) for elements of the determinant of the 5x5 matrix at marked out jointly the first and second rows applicably to Laplace's theorem

So, the value of the determinant of the 5x5 matrix at conditions when marked out jointly the first and second rows can be determined by the following formula in general view applicably to Laplace's theorem:

$$|A| = a_{11}a_{22}[a_{33}(a_{44}a_{55} - a_{45}a_{54}) + a_{34}(a_{45}a_{53} - a_{43}a_{55}) + a_{35}(a_{43}a_{54} - a_{44}a_{53})] +$$
$$+ a_{11}a_{23}[a_{32}(a_{45}a_{54} - a_{44}a_{55}) + a_{34}(a_{42}a_{55} - a_{45}a_{52}) + a_{35}(a_{44}a_{52} - a_{42}a_{54})] +$$
$$+ a_{11}a_{24}[a_{32}(a_{43}a_{55} - a_{45}a_{53}) + a_{33}(a_{45}a_{52} - a_{42}a_{55}) + a_{35}(a_{42}a_{53} - a_{43}a_{52})] +$$
$$+ a_{11}a_{25}[a_{32}(a_{44}a_{53} - a_{43}a_{54}) + a_{33}(a_{42}a_{54} - a_{44}a_{52}) + a_{34}(a_{43}a_{52} - a_{42}a_{53})] +$$

$$+ a_{12}a_{21}[a_{33}(a_{45}a_{54} - a_{44}a_{55}) + a_{34}(a_{43}a_{55} - a_{45}a_{53}) + a_{35}(a_{44}a_{53} - a_{43}a_{54})] +$$
$$+ a_{12}a_{23}[a_{31}(a_{44}a_{55} - a_{45}a_{54}) + a_{34}(a_{45}a_{51} - a_{41}a_{55}) + a_{35}(a_{41}a_{54} - a_{44}a_{51})] +$$
$$+ a_{12}a_{24}[a_{31}(a_{45}a_{53} - a_{43}a_{55}) + a_{33}(a_{41}a_{55} - a_{45}a_{51}) + a_{35}(a_{43}a_{51} - a_{41}a_{53})] +$$
$$+ a_{12}a_{25}[a_{31}(a_{43}a_{54} - a_{44}a_{53}) + a_{33}(a_{44}a_{51} - a_{41}a_{54}) + a_{34}(a_{41}a_{53} - a_{43}a_{51})] +$$

$+a_{13}a_{21}[a_{32}(a_{44}a_{55}-a_{45}a_{54})+a_{34}(a_{45}a_{52}-a_{42}a_{55})+a_{35}(a_{42}a_{54}-a_{44}a_{52})]+$
$+a_{13}a_{22}[a_{31}(a_{45}a_{54}-a_{44}a_{55})+a_{34}(a_{41}a_{55}-a_{45}a_{51})+a_{35}(a_{44}a_{51}-a_{41}a_{54})]+$
$+a_{13}a_{24}[a_{31}(a_{42}a_{55}-a_{45}a_{52})+a_{32}(a_{45}a_{51}-a_{41}a_{55})+a_{35}(a_{41}a_{52}-a_{42}a_{51})]+$
$+a_{13}a_{25}[a_{31}(a_{44}a_{52}-a_{42}a_{54})+a_{32}(a_{41}a_{54}-a_{44}a_{51})+a_{34}(a_{42}a_{51}-a_{41}a_{52})]+$

$+a_{14}a_{21}[a_{32}(a_{45}a_{53}-a_{43}a_{55})+a_{33}(a_{42}a_{55}-a_{45}a_{52})+a_{35}(a_{43}a_{52}-a_{42}a_{53})]+$
$+a_{14}a_{22}[a_{31}(a_{43}a_{55}-a_{45}a_{53})+a_{33}(a_{45}a_{51}-a_{41}a_{55})+a_{35}(a_{41}a_{53}-a_{43}a_{51})]+$
$+a_{14}a_{23}[a_{31}(a_{45}a_{52}-a_{42}a_{55})+a_{32}(a_{41}a_{55}-a_{45}a_{51})+a_{35}(a_{42}a_{51}-a_{41}a_{52})]+$
$+a_{14}a_{25}[a_{31}(a_{42}a_{53}-a_{43}a_{52})+a_{32}(a_{43}a_{51}-a_{41}a_{53})+a_{33}(a_{41}a_{52}-a_{42}a_{51})]+$

$+a_{15}a_{21}[a_{32}(a_{43}a_{54}-a_{44}a_{53})+a_{33}(a_{44}a_{52}-a_{42}a_{54})+a_{34}(a_{42}a_{53}-a_{43}a_{52})]+$
$+a_{15}a_{22}[a_{31}(a_{44}a_{53}-a_{43}a_{54})+a_{33}(a_{41}a_{54}-a_{44}a_{51})+a_{34}(a_{43}a_{51}-a_{41}a_{53})]+$
$+a_{15}a_{23}[a_{31}(a_{42}a_{54}-a_{44}a_{52})+a_{32}(a_{44}a_{51}-a_{41}a_{54})+a_{34}(a_{41}a_{52}-a_{42}a_{51})]+$
$+a_{15}a_{24}[a_{31}(a_{43}a_{52}-a_{42}a_{53})+a_{32}(a_{41}a_{53}-a_{43}a_{51})+a_{33}(a_{42}a_{51}-a_{41}a_{52})]$ (86)

In Appendix 3 is shown evaluation of the determinant of the 5x5 matrix for marked out jointly the first and second rows applicably to Laplace's theorem.

b) For the first and third rows

Referring to well-known Laplace's theorem **[2]** ,we can now to write for marked out jointly rows, in view of the first and third rows, of the determinant of the 5x5 matrix the following minors of the second order and also the algebraic supplemental (cofactors) in general view.

And value of the determinant $|A|$ of the 5x5 matrix in accordance with Laplace's theorem *for marked out jointly the first and third rows* can be determined by the following formula:

$$|A| = -M_1A_1+M_2A_2-M_3A_3+M_4A_4-M_5A_5+M_6A_6-M_7A_7-M_8A_8+M_9A_9-M_{10}A_{10}$$

or in detail the above-indicated formula ,after of some transformations ,has such view in general form:

$|A| = a_{11}a_{32}[a_{23}(a_{45}a_{54}-a_{44}a_{55})+a_{24}(a_{43}a_{55}-a_{45}a_{53})+a_{25}(a_{44}a_{53}-a_{43}a_{54})]+$
$+a_{11}a_{33}[a_{22}(a_{44}a_{55}-a_{45}a_{54})+a_{24}(a_{45}a_{52}-a_{42}a_{55})+a_{25}(a_{42}a_{54}-a_{44}a_{52})]+$
$+a_{11}a_{34}[a_{22}(a_{45}a_{53}-a_{43}a_{55})+a_{23}(a_{42}a_{55}-a_{45}a_{52})+a_{25}(a_{43}a_{52}-a_{42}a_{53})]+$
$+a_{11}a_{35}[a_{22}(a_{43}a_{54}-a_{44}a_{53})+a_{23}(a_{44}a_{52}-a_{42}a_{54})+a_{24}(a_{42}a_{53}-a_{43}a_{52})]+$

$+a_{12}a_{31}[a_{23}(a_{44}a_{55}-a_{45}a_{54})+a_{24}(a_{45}a_{53}-a_{43}a_{55})+a_{25}(a_{43}a_{54}-a_{44}a_{53})]+$
$+a_{12}a_{33}[a_{21}(a_{45}a_{54}-a_{44}a_{55})+a_{24}(a_{41}a_{55}-a_{45}a_{51})+a_{25}(a_{44}a_{51}-a_{41}a_{54})]+$
$+a_{12}a_{34}[a_{21}(a_{43}a_{55}-a_{45}a_{53})+a_{23}(a_{45}a_{51}-a_{41}a_{55})+a_{25}(a_{41}a_{53}-a_{43}a_{51})]+$
$+a_{12}a_{35}[a_{21}(a_{44}a_{53}-a_{43}a_{54})+a_{23}(a_{41}a_{54}-a_{44}a_{51})+a_{24}(a_{43}a_{51}-a_{41}a_{53})]+$

$+a_{13}a_{31}[a_{22}(a_{45}a_{54}-a_{44}a_{55})+a_{24}(a_{42}a_{55}-a_{45}a_{52})+a_{25}(a_{44}a_{52}-a_{42}a_{54})]+$
$+a_{13}a_{32}[a_{21}(a_{44}a_{55}-a_{45}a_{54})+a_{24}(a_{45}a_{51}-a_{41}a_{55})+a_{25}(a_{41}a_{54}-a_{44}a_{51})]+$
$+a_{13}a_{34}[a_{21}(a_{45}a_{52}-a_{42}a_{55})+a_{22}(a_{41}a_{55}-a_{45}a_{51})+a_{25}(a_{42}a_{51}-a_{41}a_{52})]+$
$+a_{13}a_{35}[a_{21}(a_{42}a_{54}-a_{44}a_{52})+a_{22}(a_{44}a_{51}-a_{41}a_{54})+a_{24}(a_{41}a_{52}-a_{42}a_{51})]+$

$+a_{14}a_{31}[a_{22}(a_{43}a_{55}-a_{45}a_{53})+a_{23}(a_{45}a_{52}-a_{42}a_{55})+a_{25}(a_{42}a_{53}-a_{43}a_{52})]+$
$+a_{14}a_{32}[a_{21}(a_{45}a_{53}-a_{43}a_{55})+a_{23}(a_{41}a_{55}-a_{45}a_{51})+a_{25}(a_{43}a_{51}-a_{41}a_{53})]+$
$+a_{14}a_{33}[a_{21}(a_{42}a_{55}-a_{45}a_{52})+a_{22}(a_{45}a_{51}-a_{41}a_{55})+a_{25}(a_{41}a_{52}-a_{42}a_{51})]+$
$+a_{14}a_{35}[a_{21}(a_{43}a_{52}-a_{42}a_{53})+a_{22}(a_{41}a_{53}-a_{43}a_{51})+a_{23}(a_{42}a_{51}-a_{41}a_{52})]+$

$$+a_{15}a_{31}[a_{22}(a_{44}a_{53}-a_{43}a_{54})+a_{23}(a_{42}a_{54}-a_{44}a_{52})+a_{24}(a_{43}a_{52}-a_{42}a_{53})]+$$
$$+a_{15}a_{32}[a_{21}(a_{43}a_{54}-a_{44}a_{53})+a_{23}(a_{44}a_{51}-a_{41}a_{54})+a_{24}(a_{41}a_{53}-a_{43}a_{51})]+$$
$$+a_{15}a_{33}[a_{21}(a_{44}a_{52}-a_{42}a_{54})+a_{22}(a_{41}a_{54}-a_{44}a_{51})+a_{24}(a_{42}a_{51}-a_{41}a_{52})]+$$
$$+a_{15}a_{34}[a_{21}(a_{42}a_{53}-a_{43}a_{52})+a_{22}(a_{43}a_{51}-a_{41}a_{53})+a_{23}(a_{41}a_{52}-a_{42}a_{51})] \quad (87)$$

$$M_1=(a_{11}a_{32}-a_{12}a_{31}); \quad A_1=(-1)^{(1+3)+(1+2)}\begin{vmatrix} a_{23} & a_{24} & a_{25} \\ a_{43} & a_{44} & a_{45} \\ a_{53} & a_{54} & a_{55} \end{vmatrix}$$

$$M_2=(a_{11}a_{33}-a_{13}a_{31}); \quad A_2=(-1)^{(1+3)+(1+3)}\begin{vmatrix} a_{22} & a_{24} & a_{25} \\ a_{42} & a_{44} & a_{45} \\ a_{52} & a_{54} & a_{55} \end{vmatrix}$$

$$M_3=(a_{11}a_{34}-a_{14}a_{31}); \quad A_3=(-1)^{(1+3)+(1+4)}\begin{vmatrix} a_{22} & a_{23} & a_{25} \\ a_{42} & a_{43} & a_{45} \\ a_{52} & a_{53} & a_{55} \end{vmatrix}$$

$$M_4=(a_{11}a_{35}-a_{15}a_{31}); \quad A_4=(-1)^{(1+3)+(1+5)}\begin{vmatrix} a_{22} & a_{23} & a_{24} \\ a_{42} & a_{43} & a_{44} \\ a_{52} & a_{53} & a_{54} \end{vmatrix}$$

$$M_5=(a_{12}a_{33}-a_{13}a_{32}); \quad A_5=(-1)^{(1+3)+(2+3)}\begin{vmatrix} a_{21} & a_{24} & a_{25} \\ a_{41} & a_{44} & a_{45} \\ a_{51} & a_{54} & a_{55} \end{vmatrix}$$

$$M_6=(a_{12}a_{34}-a_{14}a_{32}); \quad A_6=(-1)^{(1+3)+(2+4)}\begin{vmatrix} a_{21} & a_{23} & a_{25} \\ a_{41} & a_{43} & a_{45} \\ a_{51} & a_{53} & a_{55} \end{vmatrix}$$

$$M_7=(a_{12}a_{35}-a_{15}a_{32}); \quad A_7=(-1)^{(1+3)+(2+5)}\begin{vmatrix} a_{21} & a_{23} & a_{24} \\ a_{41} & a_{43} & a_{44} \\ a_{51} & a_{53} & a_{54} \end{vmatrix}$$

$$M_8=(a_{13}a_{34}-a_{14}a_{33}); \quad A_8=(-1)^{(1+3)+(3+4)}\begin{vmatrix} a_{21} & a_{22} & a_{25} \\ a_{41} & a_{42} & a_{45} \\ a_{51} & a_{52} & a_{55} \end{vmatrix}$$

$$M_9=(a_{13}a_{35}-a_{15}a_{33}); \quad A_9=(-1)^{(1+3)+(3+5)}\begin{vmatrix} a_{21} & a_{22} & a_{24} \\ a_{41} & a_{42} & a_{44} \\ a_{51} & a_{52} & a_{54} \end{vmatrix}$$

$$M_{10}=(a_{14}a_{35}-a_{15}a_{34}); \quad A_{10}=(-1)^{(1+3)+(4+5)}\begin{vmatrix} a_{21} & a_{22} & a_{23} \\ a_{41} & a_{42} & a_{43} \\ a_{51} & a_{52} & a_{53} \end{vmatrix}$$

Analysis of formula (87) indicates on fact that elements $(a_{31}, a_{32}, a_{33}, a_{34}, a_{35})$ of the third row present as the linking elements with other elements, participating in process of evaluation the determinant of the 5x5 matrix with marked out jointly the first and third rows. And these conclusions can be confirmed by the data shown in Figure 86 in general view.

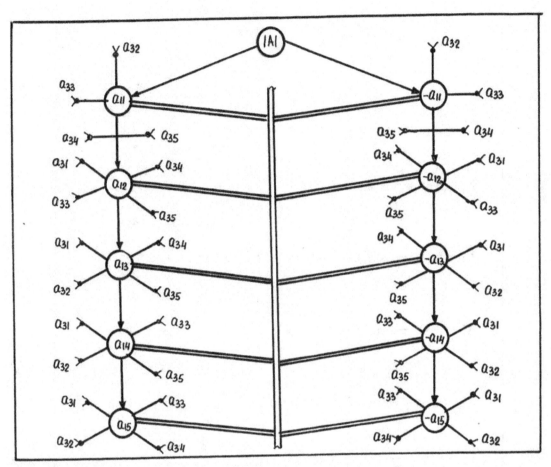

Figure 86 Tied graph (tree) for elements of the determinant of the 5x5 matrix at marked out jointly the first and third rows applicably to Laplace's theorem

In Appendix 4 is shown evaluation of the determinant of the 5x5 matrix for marked out the first and third rows applicably to Laplace's theorem.

c) **For the first and fourth rows**

Considering the general features of Laplace's theorem applicably to evaluation of the determinant of the 5x5 matrix *for marked out jointly the first and fourth rows,* we can now to write the following minors of the second order (M_{ij}) and also the algebraic(cofactors) supplemental (A_{ij}) and also to express the determinant by the following formula:

$$|A| = M_1A_1 - M_2A_2 + M_3A_3 - M_4A_4 + M_5A_5 - M_6A_6 + M_7A_7 + M_8A_8 - M_9A_9 + M_{10}A_{10}$$

where,

$M_1, M_2, M_3 \ldots\ldots\ldots M_{10}$ —minors of the second order;

$A_1, A_2, A_3 \ldots\ldots\ldots A_{10}$ —algebraic(cofactors) supplemental.

95

$$M_1 = (a_{11}a_{42} - a_{12}a_{41}); \quad A_1 = (-1)^{(1+4)+(1+2)} \begin{vmatrix} a_{23} & a_{24} & a_{25} \\ a_{33} & a_{34} & a_{35} \\ a_{53} & a_{54} & a_{55} \end{vmatrix}$$

$$M_2 = (a_{11}a_{42} - a_{12}a_{41}); \quad A_2 = (-1)^{(1+4)+(1+3)} \begin{vmatrix} a_{22} & a_{24} & a_{25} \\ a_{32} & a_{34} & a_{35} \\ a_{52} & a_{54} & a_{55} \end{vmatrix}$$

$$M_3 = (a_{11}a_{44} - a_{14}a_{41}); \quad A_3 = (-1)^{(1+4)+(1+4)} \begin{vmatrix} a_{22} & a_{23} & a_{25} \\ a_{32} & a_{33} & a_{35} \\ a_{52} & a_{53} & a_{55} \end{vmatrix}$$

$$M_4 = (a_{11}a_{45} - a_{15}a_{41}); \quad A_4 = (-1)^{(1+4)+(1+5)} \begin{vmatrix} a_{22} & a_{23} & a_{24} \\ a_{32} & a_{33} & a_{34} \\ a_{52} & a_{53} & a_{54} \end{vmatrix}$$

$$M_5 = (a_{12}a_{43} - a_{13}a_{42}); \quad A_5 = (-1)^{(1+4)+(2+3)} \begin{vmatrix} a_{21} & a_{24} & a_{25} \\ a_{31} & a_{34} & a_{35} \\ a_{51} & a_{54} & a_{55} \end{vmatrix}$$

$$M_6 = (a_{12}a_{44} - a_{14}a_{42}); \quad A_6 = (-1)^{(1+4)+(2+4)} \begin{vmatrix} a_{21} & a_{23} & a_{25} \\ a_{31} & a_{33} & a_{35} \\ a_{51} & a_{53} & a_{55} \end{vmatrix}$$

$$M_7 = (a_{12}a_{45} - a_{15}a_{42}); \quad A_7 = (-1)^{(1+4)+(2+5)} \begin{vmatrix} a_{21} & a_{23} & a_{24} \\ a_{31} & a_{33} & a_{34} \\ a_{51} & a_{53} & a_{54} \end{vmatrix}$$

$$M_8 = (a_{13}a_{44} - a_{14}a_{43}); \quad A_8 = (-1)^{(1+4)+(3+4)} \begin{vmatrix} a_{21} & a_{22} & a_{25} \\ a_{31} & a_{32} & a_{35} \\ a_{51} & a_{52} & a_{55} \end{vmatrix}$$

$$M_9 = (a_{13}a_{45} - a_{15}a_{43}); \quad A_9 = (-1)^{(1+4)+(3+5)} \begin{vmatrix} a_{21} & a_{22} & a_{24} \\ a_{31} & a_{32} & a_{34} \\ a_{51} & a_{52} & a_{54} \end{vmatrix}$$

$$M_{10} = (a_{14}a_{45} - a_{15}a_{44}); \quad A_{10} = (-1)^{(1+4)+(4+5)} \begin{vmatrix} a_{21} & a_{22} & a_{23} \\ a_{31} & a_{32} & a_{33} \\ a_{51} & a_{52} & a_{53} \end{vmatrix}$$

So, the value of determinant of the 5x5 matrix in accordance with Laplace's theorem *for marked out jointly the first and fourth rows* in general view can be determined by the following formula:

And besides analysis of above-indicated formula indicates on fact that elements (a$_{41}$,a$_{42}$,a$_{43}$,a$_{44}$,a$_{45}$)of the fourth row present as the linking elements with other elements ,participating in process of evaluation the determinant of the 5x5 matrix for marked out jointly the first and fourth rows. And these conclusions can be confirmed by data shown in Figure 87.

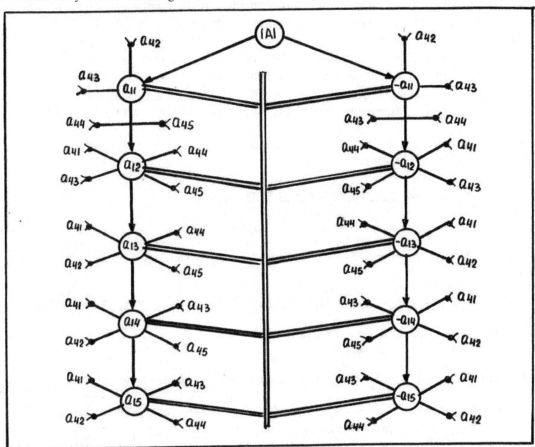

Figure 87 Tied graph (tree) for elements of the determinant of the 5x5 matrix at marked out jointly the first and fourth rows applicably to Laplace's theorem

$$
\begin{aligned}
|A| = &a_{11}a_{42}[a_{23}(a_{34}a_{55}-a_{35}a_{54})+a_{24}(a_{35}a_{53}-a_{33}a_{55})+a_{25}(a_{33}a_{54}-a_{34}a_{53})]+ \\
&+a_{11}a_{43}[a_{22}(a_{35}a_{54}-a_{34}a_{55})+a_{24}(a_{32}a_{55}-a_{35}a_{52})+a_{25}(a_{34}a_{52}-a_{32}a_{54})]+ \\
&+a_{11}a_{44}[a_{22}(a_{33}a_{55}-a_{35}a_{53})+a_{23}(a_{35}a_{52}-a_{32}a_{55})+a_{25}(a_{32}a_{53}-a_{33}a_{52})]+ \\
&+a_{11}a_{45}[a_{22}(a_{34}a_{53}-a_{33}a_{54})+a_{23}(a_{32}a_{54}-a_{34}a_{52})+a_{24}(a_{33}a_{52}-a_{32}a_{53})]+ \\
\\
&+a_{12}a_{41}[a_{23}(a_{35}a_{54}-a_{34}a_{55})+a_{24}(a_{33}a_{55}-a_{35}a_{53})+a_{25}(a_{34}a_{53}-a_{33}a_{54})]+ \\
&+a_{12}a_{43}[a_{21}(a_{34}a_{55}-a_{35}a_{54})+a_{24}(a_{35}a_{51}-a_{31}a_{55})+a_{25}(a_{31}a_{54}-a_{34}a_{51})]+ \\
&+a_{12}a_{44}[a_{21}(a_{35}a_{53}-a_{33}a_{55})+a_{23}(a_{31}a_{55}-a_{35}a_{51})+a_{25}(a_{33}a_{51}-a_{31}a_{53})]+ \\
&+a_{12}a_{45}[a_{21}(a_{33}a_{54}-a_{34}a_{53})+a_{23}(a_{34}a_{51}-a_{31}a_{54})+a_{24}(a_{31}a_{53}-a_{33}a_{51})]+ \\
\\
&+a_{13}a_{41}[a_{22}(a_{34}a_{55}-a_{35}a_{54})+a_{24}(a_{35}a_{52}-a_{32}a_{55})+a_{25}(a_{32}a_{54}-a_{34}a_{52})]+ \\
&+a_{13}a_{42}[a_{21}(a_{35}a_{54}-a_{34}a_{55})+a_{24}(a_{31}a_{55}-a_{35}a_{51})+a_{25}(a_{34}a_{51}-a_{31}a_{54})]+ \\
&+a_{13}a_{44}[a_{21}(a_{32}a_{55}-a_{35}a_{52})+a_{22}(a_{35}a_{51}-a_{31}a_{55})+a_{25}(a_{31}a_{52}-a_{32}a_{51})]+ \\
&+a_{13}a_{45}[a_{21}(a_{34}a_{52}-a_{32}a_{54})+a_{22}(a_{31}a_{54}-a_{34}a_{51})+a_{24}(a_{32}a_{51}-a_{31}a_{52})]+
\end{aligned}
$$

$$+a_{14}a_{41}[a_{22}(a_{35}a_{53}-a_{33}a_{55})+a_{23}(a_{32}a_{55}-a_{35}a_{52})+a_{25}(a_{33}a_{52}-a_{32}a_{53})]+$$
$$+a_{14}a_{42}[a_{21}(a_{33}a_{55}-a_{35}a_{53})+a_{23}(a_{35}a_{51}-a_{31}a_{55})+a_{25}(a_{31}a_{53}-a_{33}a_{51})]+$$
$$+a_{14}a_{43}[a_{21}(a_{35}a_{52}-a_{32}a_{55})+a_{22}(a_{31}a_{55}-a_{35}a_{51})+a_{25}(a_{32}a_{51}-a_{31}a_{52})]+$$
$$+a_{14}a_{45}[a_{21}(a_{32}a_{53}-a_{33}a_{52})+a_{22}(a_{33}a_{51}-a_{31}a_{53})+a_{23}(a_{31}a_{52}-a_{32}a_{51})]+$$

$$+a_{15}a_{41}[a_{22}(a_{33}a_{54}-a_{34}a_{53})+a_{23}(a_{34}a_{52}-a_{32}a_{54})+a_{24}(a_{32}a_{53}-a_{33}a_{52})]+$$
$$+a_{15}a_{42}[a_{21}(a_{34}a_{53}-a_{33}a_{54})+a_{23}(a_{31}a_{54}-a_{34}a_{51})+a_{24}(a_{33}a_{51}-a_{31}a_{53})]+$$
$$+a_{15}a_{43}[a_{21}(a_{32}a_{54}-a_{34}a_{52})+a_{22}(a_{34}a_{51}-a_{31}a_{54})+a_{24}(a_{31}a_{52}-a_{32}a_{51})]+$$
$$+a_{15}a_{44}[a_{21}(a_{33}a_{52}-a_{32}a_{53})+a_{22}(a_{31}a_{53}-a_{33}a_{51})+a_{23}(a_{32}a_{51}-a_{31}a_{52})] \qquad (88)$$

In Appendix 5 is shown evaluation of the determinant of the 5x5 matrix for marked out jointly the first and fourth rows applicably to Laplace's theorem.

d) For the first and fifth rows

Referring to the general features of Laplace's theorem applicably to evaluation of the determinant of the 5x5 matrix *for marked out jointly the first and fifth rows,* we can now to write the following minors of the second order (M_{ij}) and also the algebraic(cofactors) supplemental (A_{ij}) and also to express the determinant $|A|$ by the following formula:

$$|A| = - M_1A_1+M_2A_2-M_3A_3+M_4A_4-M_5A_5+M_6A_6-M_7A_7-M_8A_8+M_9A_9-M_{10}A_{10}$$
where,
M_{ij}-minors of the second order of i-row and j-column
A_{ij}-algebraic (cofactors) supplemental of i-row and j-column .

And in detail the above-indicated formula, after of some transformation, has the following view in general form:

$$|A| =a_{11}a_{52}[a_{23}(a_{35}a_{44}-a_{34}a_{45})+a_{24}(a_{33}a_{45}-a_{35}a_{43})+a_{25}(a_{34}a_{43}-a_{33}a_{44})]+$$
$$+a_{11}a_{53}[a_{22}(a_{34}a_{45}-a_{35}a_{44})+a_{24}(a_{35}a_{42}-a_{32}a_{45})+a_{25}(a_{32}a_{44}-a_{34}a_{42})]+$$
$$+a_{11}a_{54}[a_{22}(a_{35}a_{43}-a_{33}a_{45})+a_{23}(a_{32}a_{45}-a_{35}a_{42})+a_{25}(a_{33}a_{42}-a_{32}a_{43})]+$$
$$+a_{11}a_{55}[a_{22}(a_{33}a_{44}-a_{34}a_{43})+a_{23}(a_{34}a_{42}-a_{32}a_{44})+a_{24}(a_{32}a_{43}-a_{33}a_{42})]+$$

$$+a_{12}a_{51}[a_{23}(a_{34}a_{45}-a_{35}a_{44})+a_{24}(a_{35}a_{43}-a_{33}a_{45})+a_{25}(a_{33}a_{44}-a_{34}a_{43})]+$$
$$+a_{12}a_{53}[a_{21}(a_{35}a_{44}-a_{34}a_{45})+a_{24}(a_{31}a_{45}-a_{35}a_{41})+a_{25}(a_{34}a_{41}-a_{31}a_{44})]+$$
$$+a_{12}a_{54}[a_{21}(a_{33}a_{45}-a_{35}a_{43})+a_{23}(a_{35}a_{41}-a_{31}a_{45})+a_{25}(a_{31}a_{43}-a_{33}a_{41})]+$$
$$+a_{12}a_{55}[a_{21}(a_{34}a_{43}-a_{33}a_{44})+a_{23}(a_{31}a_{44}-a_{34}a_{41})+a_{24}(a_{33}a_{41}-a_{31}a_{43})]+$$

$$+a_{13}a_{51}[a_{22}(a_{35}a_{44}-a_{34}a_{45})+a_{24}(a_{32}a_{45}-a_{35}a_{42})+a_{25}(a_{34}a_{42}-a_{32}a_{44})]+$$
$$+a_{13}a_{52}[a_{21}(a_{34}a_{45}-a_{35}a_{44})+a_{24}(a_{35}a_{41}-a_{31}a_{45})+a_{25}(a_{31}a_{44}-a_{34}a_{41})+$$
$$+a_{13}a_{54}[a_{21}(a_{35}a_{42}-a_{32}a_{45})+a_{22}(a_{31}a_{45}-a_{35}a_{41})+a_{25}(a_{32}a_{41}-a_{31}a_{42})]+$$
$$+a_{13}a_{55}[a_{21}(a_{32}a_{44}-a_{34}a_{42})+a_{22}(a_{34}a_{41}-a_{31}a_{44})+a_{24}(a_{31}a_{42}-a_{32}a_{41})]+$$

$$+a_{14}a_{51}[a_{22}(a_{33}a_{45}-a_{35}a_{43})+a_{23}(a_{35}a_{42}-a_{32}a_{45})+a_{25}(a_{32}a_{43}-a_{33}a_{42})+$$
$$+a_{14}a_{52}[a_{21}(a_{35}a_{43}-a_{33}a_{45})+a_{23}(a_{31}a_{45}-a_{35}a_{41})+a_{25}(a_{33}a_{41}-a_{31}a_{43})+$$
$$+a_{14}a_{53}[a_{21}(a_{32}a_{45}-a_{35}a_{42})+a_{22}(a_{35}a_{41}-a_{31}a_{45})+a_{25}(a_{31}a_{42}-a_{32}a_{41})+$$
$$+a_{14}a_{55}[a_{21}(a_{33}a_{42}-a_{32}a_{43})+a_{22}(a_{31}a_{43}-a_{33}a_{41})+a_{23}(a_{32}a_{41}-a_{31}a_{42})]+$$

$$+a_{15}a_{51}[a_{22}(a_{34}a_{43}-a_{33}a_{44})+a_{23}(a_{32}a_{44}-a_{34}a_{42})+a_{24}(a_{33}a_{42}-a_{32}a_{43})]+$$
$$+a_{15}a_{52}[a_{21}(a_{33}a_{44}-a_{34}a_{43})+a_{23}(a_{34}a_{41}-a_{31}a_{44})+a_{24}(a_{31}a_{43}-a_{33}a_{41})]+$$
$$+a_{15}a_{53}[a_{21}(a_{34}a_{42}-a_{32}a_{44})+a_{22}(a_{31}a_{44}-a_{34}a_{41})+a_{24}(a_{32}a_{41}-a_{31}a_{42})]+$$
$$+a_{15}a_{54}[a_{21}(a_{32}a_{43}-a_{33}a_{42})+a_{22}(a_{33}a_{41}-a_{31}a_{43})+a_{23}(a_{31}a_{42}-a_{32}a_{41})] \quad (89)$$

$$M_1=(a_{11}a_{52}-a_{12}a_{51}); \quad A_1=(-1)^{(1+5)+(1+2)}\begin{vmatrix} a_{23} & a_{24} & a_{25} \\ a_{33} & a_{34} & a_{35} \\ a_{43} & a_{44} & a_{45} \end{vmatrix}$$

$$M_2=(a_{11}a_{53}-a_{13}a_{51}); \quad A_2=(-1)^{(1+5)+(1+3)}\begin{vmatrix} a_{22} & a_{24} & a_{25} \\ a_{32} & a_{34} & a_{35} \\ a_{42} & a_{44} & a_{45} \end{vmatrix}$$

$$M_3=(a_{11}a_{54}-a_{14}a_{51}); \quad A_3=(-1)^{(1+5)+(1+4)}\begin{vmatrix} a_{22} & a_{23} & a_{25} \\ a_{32} & a_{33} & a_{35} \\ a_{42} & a_{43} & a_{45} \end{vmatrix}$$

$$M_4=(a_{11}a_{55}-a_{15}a_{51}); \quad A_4=(-1)^{(1+5)+(1+5)}\begin{vmatrix} a_{22} & a_{23} & a_{24} \\ a_{32} & a_{33} & a_{34} \\ a_{42} & a_{43} & a_{44} \end{vmatrix}$$

$$M_5=(a_{12}a_{53}-a_{13}a_{52}); \quad A_5=(-1)^{(1+5)+(2+3)}\begin{vmatrix} a_{21} & a_{24} & a_{25} \\ a_{31} & a_{34} & a_{35} \\ a_{41} & a_{44} & a_{45} \end{vmatrix}$$

$$M_6=(a_{12}a_{54}-a_{14}a_{52}); \quad A_6=(-1)^{(1+5)+(2+4)}\begin{vmatrix} a_{21} & a_{23} & a_{25} \\ a_{31} & a_{33} & a_{35} \\ a_{41} & a_{43} & a_{45} \end{vmatrix}$$

$$M_7=(a_{12}a_{55}-a_{15}a_{52}); \quad A_7=(-1)^{(1+5)+(2+5)}\begin{vmatrix} a_{21} & a_{23} & a_{24} \\ a_{31} & a_{33} & a_{34} \\ a_{41} & a_{43} & a_{44} \end{vmatrix}$$

$$M_8=(a_{13}a_{54}-a_{14}a_{53}); \quad A_8=(-1)^{(1+5)+(3+4)}\begin{vmatrix} a_{21} & a_{22} & a_{25} \\ a_{31} & a_{32} & a_{35} \\ a_{41} & a_{42} & a_{45} \end{vmatrix}$$

$$M_9=(a_{13}a_{55}-a_{15}a_{53}); \quad A_9=(-1)^{(1+5)+(3+5)}\begin{vmatrix} a_{21} & a_{22} & a_{24} \\ a_{31} & a_{32} & a_{34} \\ a_{41} & a_{42} & a_{44} \end{vmatrix}$$

$$M_{10}=(a_{14}a_{55}-a_{15}a_{54}); \quad A_{10}=(-1)^{(1+5)+(4+5)}\begin{vmatrix} a_{21} & a_{22} & a_{23} \\ a_{31} & a_{32} & a_{33} \\ a_{41} & a_{42} & a_{43} \end{vmatrix}$$

And besides analysis of formula (89) indicates on fact that elements $(a_{51}, a_{52}, a_{53}, a_{54}, a_{55})$ of the fifth row present as the linking elements with other elements, participating in process of evaluation the determinant of the 5x5 matrix for marked out jointly the first and fifth rows. And these conclusions can be confirmed by data shown in Figure 88.

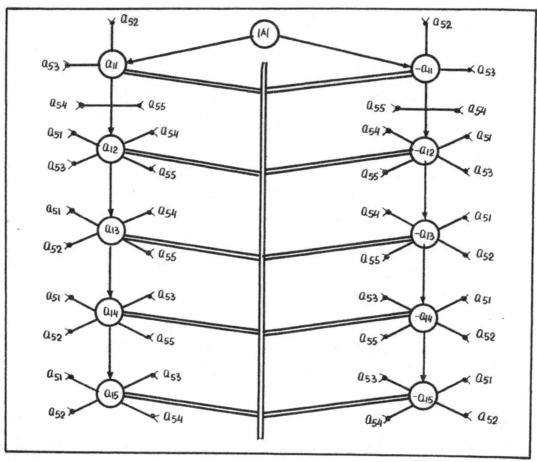

Figure 88 Tied graph (tree) for elements of the determinant of the 5x5 matrix for marked out jointly the first and fifth rows applicably to Laplace's theorem in general view

In Appendix 6 is shown evaluation of the determinant of the 5x5 matrix for marked out jointly the first and fifth rows applicably to Laplace's theorem.

e) For the second and third rows

Referring to Laplace's theorem, we can now to write for marked out two jointly rows, in view of the second and third rows, of the determinant of the 5x5 matrix the following minors of the second order (M_{ij}) and also the algebraic (cofactors) supplemental (A_{ij}) in general form:

$$|A| = M_1A_1 - M_2A_2 + M_3A_3 - M_4A_4 + M_5A_5 - M_6A_6 + M_7A_7 + M_8A_8 - M_9A_9 + M_{10}A_{10} \quad (*)$$

where,

M_{ij}- minors of the second order;

A_{ij}-algebraic(cofactors) supplemental .

100

$M_1 = (a_{21}a_{32} - a_{22}a_{31});$ $A_1 = (-1)^{(2+3)+(1+2)} \begin{vmatrix} a_{13} & a_{14} & a_{15} \\ a_{43} & a_{44} & a_{45} \\ a_{53} & a_{54} & a_{55} \end{vmatrix}$

$M_2 = (a_{21}a_{33} - a_{23}a_{31});$ $A_2 = (-1)^{(2+3)+(1+3)} \begin{vmatrix} a_{12} & a_{14} & a_{15} \\ a_{42} & a_{44} & a_{45} \\ a_{52} & a_{54} & a_{55} \end{vmatrix}$

$M_3 = (a_{21}a_{34} - a_{24}a_{31});$ $A_3 = (-1)^{(2+3)+(1+4)} \begin{vmatrix} a_{12} & a_{13} & a_{15} \\ a_{42} & a_{43} & a_{45} \\ a_{52} & a_{53} & a_{55} \end{vmatrix}$

$M_4 = (a_{21}a_{35} - a_{25}a_{31});$ $A_4 = (-1)^{(2+3)+(1+5)} \begin{vmatrix} a_{12} & a_{13} & a_{14} \\ a_{42} & a_{43} & a_{44} \\ a_{52} & a_{53} & a_{54} \end{vmatrix}$

$M_5 = (a_{22}a_{33} - a_{23}a_{32});$ $A_5 = (-1)^{(2+3)+(2+3)} \begin{vmatrix} a_{11} & a_{14} & a_{15} \\ a_{41} & a_{44} & a_{45} \\ a_{51} & a_{54} & a_{55} \end{vmatrix}$

$M_6 = (a_{22}a_{34} - a_{24}a_{32});$ $A_6 = (-1)^{(2+3)+(2+4)} \begin{vmatrix} a_{11} & a_{13} & a_{15} \\ a_{41} & a_{43} & a_{45} \\ a_{51} & a_{53} & a_{55} \end{vmatrix}$

$M_7 = (a_{22}a_{35} - a_{25}a_{32});$ $A_7 = (-1)^{(2+3)+(2+5)} \begin{vmatrix} a_{11} & a_{13} & a_{14} \\ a_{41} & a_{43} & a_{44} \\ a_{51} & a_{53} & a_{54} \end{vmatrix}$

$M_8 = (a_{23}a_{34} - a_{24}a_{33});$ $A_8 = (-1)^{(2+3)+(3+4)} \begin{vmatrix} a_{11} & a_{12} & a_{15} \\ a_{41} & a_{42} & a_{45} \\ a_{51} & a_{52} & a_{55} \end{vmatrix}$

$M_9 = (a_{23}a_{35} - a_{25}a_{33});$ $A_9 = (-1)^{(2+3)+(3+5)} \begin{vmatrix} a_{11} & a_{12} & a_{14} \\ a_{41} & a_{42} & a_{44} \\ a_{51} & a_{52} & a_{54} \end{vmatrix}$

$M_{10} = (a_{24}a_{35} - a_{25}a_{34});$ $A_{10} = (-1)^{(2+3)+(4+5)} \begin{vmatrix} a_{11} & a_{12} & a_{13} \\ a_{41} & a_{42} & a_{43} \\ a_{51} & a_{52} & a_{53} \end{vmatrix}$

In Figure 89 is shown the tied graph (tree) for elements of the determinant of the 5x5 matrix at marked out jointly the second and third rows applicably to Laplace's theorem.

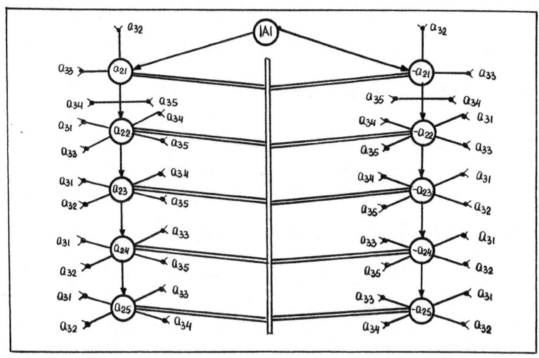

Figure 89 Tied graph (tree) for elements of the determinant of the 5x5 matrix at marked out jointly the second and third rows applicably to Laplace's theorem

In detail the above-indicated formula(*) for evaluation of the determinant of the 5x5 matrix at conditions when its elements *marked out jointly to the second and third rows* has the following view in general form:

$$\left| A \right| = a_{21}a_{32}[a_{13}(a_{44}a_{55}-a_{45}a_{54})+a_{14}(a_{45}a_{53}-a_{43}a_{55})+a_{15}(a_{43}a_{54}-a_{44}a_{53})]+$$
$$+a_{21}a_{33}[a_{12}(a_{45}a_{54}-a_{44}a_{55})+a_{14}(a_{42}a_{55}-a_{45}a_{52})+a_{15}(a_{44}a_{52}-a_{42}a_{54})]+$$
$$+a_{21}a_{34}[a_{12}(a_{43}a_{55}-a_{45}a_{53})+a_{13}(a_{45}a_{52}-a_{42}a_{55})+a_{15}(a_{42}a_{53}-a_{43}a_{52})]+$$
$$+a_{21}a_{35}[a_{12}(a_{44}a_{53}-a_{43}a_{54})+a_{13}(a_{42}a_{54}-a_{44}a_{52})+a_{14}(a_{43}a_{52}-a_{42}a_{53})]+$$

$$+a_{22}a_{31}[a_{13}(a_{45}a_{54}-a_{44}a_{55})+a_{14}(a_{43}a_{55}-a_{45}a_{53})+a_{15}(a_{44}a_{53}-a_{43}a_{54})]+$$
$$+a_{22}a_{33}[a_{11}(a_{44}a_{55}-a_{45}a_{54})+a_{14}(a_{45}a_{51}-a_{41}a_{55})+a_{15}(a_{41}a_{54}-a_{44}a_{51})]+$$
$$+a_{22}a_{34}[a_{11}(a_{45}a_{53}-a_{43}a_{55})+a_{13}(a_{41}a_{55}-a_{45}a_{51})+a_{15}(a_{43}a_{51}-a_{41}a_{53})]+$$
$$+a_{22}a_{35}[a_{11}(a_{43}a_{54}-a_{44}a_{53})+a_{13}(a_{44}a_{51}-a_{41}a_{54})+a_{14}(a_{41}a_{53}-a_{43}a_{51})]+$$

$$+a_{23}a_{31}[a_{12}(a_{44}a_{55}-a_{45}a_{54})+a_{14}(a_{45}a_{52}-a_{42}a_{55})+a_{15}(a_{42}a_{54}-a_{44}a_{52})]+$$
$$+a_{23}a_{32}[a_{11}(a_{45}a_{54}-a_{44}a_{55})+a_{14}(a_{41}a_{55}-a_{45}a_{51})+a_{15}(a_{44}a_{51}-a_{41}a_{55})]+$$
$$+a_{23}a_{34}[a_{11}(a_{42}a_{55}-a_{45}a_{52})+a_{12}(a_{15}a_{51}-a_{41}a_{54})+a_{15}(a_{41}a_{52}-a_{42}a_{51})]+$$
$$+a_{23}a_{35}[a_{11}(a_{44}a_{52}-a_{42}a_{54})+a_{12}(a_{41}a_{54}-a_{44}a_{51})+a_{14}(a_{42}a_{51}-a_{41}a_{52})]+$$

$$+a_{24}a_{31}[a_{12}(a_{45}a_{53}-a_{43}a_{55})+a_{13}(a_{42}a_{55}-a_{45}a_{52})+a_{15}(a_{43}a_{52}-a_{42}a_{53})]+$$
$$+a_{24}a_{32}[a_{11}(a_{43}a_{55}-a_{45}a_{53})+a_{13}(a_{45}a_{51}-a_{41}a_{55})+a_{15}(a_{41}a_{53}-a_{43}a_{51})]+$$
$$+a_{24}a_{33}[a_{11}(a_{45}a_{52}-a_{42}a_{55})+a_{12}(a_{41}a_{55}-a_{45}a_{51})+a_{15}(a_{42}a_{51}-a_{41}a_{52})]+$$
$$+a_{24}a_{35}[a_{11}(a_{42}a_{53}-a_{43}a_{52})+a_{12}(a_{43}a_{51}-a_{41}a_{53})+a_{13}(a_{41}a_{52}-a_{42}a_{51})]+$$

$+a_{25}a_{31}[a_{12}(a_{43}a_{54}-a_{44}a_{53})+a_{13}(a_{44}a_{52}-a_{42}a_{54})+a_{14}(a_{42}a_{53}-a_{43}a_{52})+$

$+a_{25}a_{32}[a_{11}(a_{44}a_{53}-a_{43}a_{54})+a_{13}(a_{41}a_{54}-a_{44}a_{51})+a_{14}(a_{43}a_{51}-a_{41}a_{53})+$

$+a_{25}a_{33}[a_{11}(a_{42}a_{54}-a_{44}a_{52})+a_{12}(a_{44}a_{51}-a_{41}a_{54})+a_{14}(a_{41}a_{52}-a_{42}a_{51})+$

$+a_{25}a_{34}[a_{11}(a_{43}a_{52}-a_{42}a_{53})+a_{12}(a_{41}a_{53}-a_{43}a_{51})+a_{13}(a_{42}a_{51}-a_{41}a_{52})$ (90)

In Appendix 7 is shown evaluation of the determinant of the 5x5 matrix for marked out jointly the second and third rows applicably to Laplace's theorem.

f) For the second and fourth rows

Considering the general features of Laplace theorem applicably to evaluation of the determinant of the 5x5 matrix *for marked out jointly the second and fourth rows* ,we can now to write the following minors of the second order (M_{ij}) and also the algebraic (cofactors) supplemental (A_{ij}) in general view:

$$M_1=(a_{21}a_{42}-a_{22}a_{41}); \quad A_1=(-1)^{(2+4)+(1+2)} \begin{vmatrix} a_{13} & a_{14} & a_{15} \\ a_{33} & a_{34} & a_{35} \\ a_{53} & a_{54} & a_{55} \end{vmatrix}$$

$$M_2=(a_{21}a_{43}-a_{23}a_{41}); \quad A_2=(-1)^{(2+4)+(1+3)} \begin{vmatrix} a_{12} & a_{14} & a_{15} \\ a_{32} & a_{34} & a_{35} \\ a_{52} & a_{54} & a_{55} \end{vmatrix}$$

$$M_3=(a_{21}a_{44}-a_{24}a_{41}); \quad A_3=(-1)^{(2+4)+(1+4)} \begin{vmatrix} a_{12} & a_{13} & a_{15} \\ a_{32} & a_{33} & a_{35} \\ a_{52} & a_{53} & a_{55} \end{vmatrix}$$

$$M_4=(a_{21}a_{45}-a_{25}a_{41}); \quad A_4=(-1)^{(2+4)+(1+5)} \begin{vmatrix} a_{12} & a_{13} & a_{14} \\ a_{32} & a_{33} & a_{34} \\ a_{52} & a_{53} & a_{54} \end{vmatrix}$$

$$M_5=(a_{22}a_{43}-a_{23}a_{42}); \quad A_5=(-1)^{(2+4)+(2+3)} \begin{vmatrix} a_{11} & a_{14} & a_{15} \\ a_{31} & a_{34} & a_{35} \\ a_{51} & a_{54} & a_{55} \end{vmatrix}$$

$$M_6=(a_{22}a_{44}-a_{24}a_{42}); \quad A_6=(-1)^{(2+4)+(2+4)} \begin{vmatrix} a_{11} & a_{13} & a_{15} \\ a_{31} & a_{33} & a_{35} \\ a_{51} & a_{53} & a_{55} \end{vmatrix}$$

$$M_7=(a_{22}a_{45}-a_{25}a_{42}); \quad A_7=(-1)^{(2+4)+(2+5)} \begin{vmatrix} a_{11} & a_{13} & a_{14} \\ a_{31} & a_{33} & a_{34} \\ a_{51} & a_{53} & a_{54} \end{vmatrix}$$

$$M_8=(a_{23}a_{44}-a_{24}a_{43}); \quad A_8=(-1)^{(2+4)+(3+4)} \begin{vmatrix} a_{11} & a_{12} & a_{15} \\ a_{31} & a_{32} & a_{35} \\ a_{51} & a_{52} & a_{55} \end{vmatrix}$$

$$M_9=(a_{23}a_{45}-a_{25}a_{43}); \quad A_9=(-1)^{(2+4)+(3+5)} \begin{vmatrix} a_{11} & a_{12} & a_{14} \\ a_{31} & a_{32} & a_{34} \\ a_{51} & a_{52} & a_{54} \end{vmatrix}$$

$$M_{10}=(a_{24}a_{45}-a_{25}a_{44}); \quad A_{10}=(-1)^{(2+4)+(4+5)} \begin{vmatrix} a_{11} & a_{12} & a_{13} \\ a_{31} & a_{32} & a_{33} \\ a_{51} & a_{52} & a_{53} \end{vmatrix}$$

So ,the value of determinant of the 5x5 matrix in accordance with Laplace's theorem for marked out jointly the second and fourth rows can be determined by the following formula in general view:

$$|A| = -M_1A_1 + M_2A_2 - M_3A_3 + M_4A_4 - M_5A_5 + M_6A_6 - M_7A_7 - M_8A_8 + M_9A_9 - M_{10}A_{10} \; (\blacktriangle)$$

In Figure 90 is shown the tied graph(tree) for elements of the determinant of the 5x5 matrix at marked out jointly the second and fourth rows applicably to Laplace's theorem.

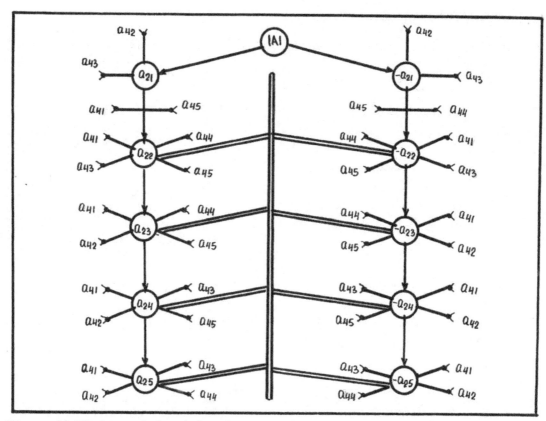

Figure 90 Tied graph (tree) for elements of the determinant of the 5x5 matrix at marked out jointly the second and fourth rows applicably to Laplace's theorem

After of some transformations the above-indicated formula ($\blacktriangle$) for evaluation of the determinant of the 5x5 matrix for marked out jointly the second and fourth rows has the following view in general form:

$$
\begin{aligned}
|A| =\; & a_{21}a_{42}[a_{13}(a_{35}a_{54}-a_{34}a_{55})+a_{14}(a_{33}a_{55}-a_{35}a_{53})+a_{15}(a_{34}a_{53}-a_{33}a_{54})]+ \\
& +a_{21}a_{43}[a_{12}(a_{34}a_{55}-a_{35}a_{54})+a_{14}(a_{35}a_{52}-a_{32}a_{55})+a_{15}(a_{32}a_{54}-a_{34}a_{52})]+ \\
& +a_{21}a_{44}[a_{12}(a_{35}a_{53}-a_{33}a_{55})+a_{13}(a_{32}a_{55}-a_{35}a_{52})+a_{15}(a_{33}a_{52}-a_{32}a_{53})]+ \\
& +a_{21}a_{45}[a_{12}(a_{33}a_{54}-a_{34}a_{53})+a_{13}(a_{34}a_{52}-a_{32}a_{54})+a_{14}(a_{32}a_{53}-a_{33}a_{52})]+ \\
\\
& +a_{22}a_{41}[a_{13}(a_{34}a_{55}-a_{35}a_{54})+a_{14}(a_{35}a_{53}-a_{33}a_{55})+a_{15}(a_{33}a_{54}-a_{34}a_{53})]+ \\
& +a_{22}a_{43}[a_{11}(a_{35}a_{54}-a_{34}a_{55})+a_{14}(a_{31}a_{55}-a_{35}a_{51})+a_{15}(a_{34}a_{51}-a_{31}a_{54})]+ \\
& +a_{22}a_{44}[a_{11}(a_{33}a_{55}-a_{35}a_{53})+a_{13}(a_{35}a_{51}-a_{31}a_{55})+a_{15}(a_{31}a_{53}-a_{33}a_{51})]+ \\
& +a_{22}a_{45}[a_{11}(a_{34}a_{53}-a_{33}a_{54})+a_{13}(a_{31}a_{54}-a_{34}a_{51})+a_{14}(a_{33}a_{51}-a_{31}a_{53})]+
\end{aligned}
$$

$+a_{23}a_{41}[a_{12}(a_{35}a_{54}-a_{34}a_{55})+a_{14}(a_{32}a_{55}-a_{35}a_{52})+a_{15}(a_{34}a_{52}-a_{32}a_{54})]+$
$+a_{23}a_{42}[a_{11}(a_{34}a_{55}-a_{35}a_{54})+a_{14}(a_{35}a_{51}-a_{31}a_{55})+a_{15}(a_{31}a_{54}-a_{34}a_{51})]+$
$+a_{23}a_{44}[a_{11}(a_{35}a_{52}-a_{32}a_{55})+a_{12}(a_{31}a_{55}-a_{35}a_{51})+a_{15}(a_{32}a_{51}-a_{31}a_{52})]+$
$+a_{23}a_{45}[a_{11}(a_{32}a_{54}-a_{34}a_{52})+a_{12}(a_{34}a_{51}-a_{31}a_{54})+a_{14}(a_{31}a_{52}-a_{32}a_{51})]+$

$+a_{24}a_{41}[a_{12}(a_{33}a_{55}-a_{35}a_{53})+a_{13}(a_{35}a_{52}-a_{32}a_{55})+a_{15}(a_{32}a_{53}-a_{33}a_{52})]+$
$+a_{24}a_{42}[a_{11}(a_{35}a_{53}-a_{33}a_{55})+a_{13}(a_{31}a_{55}-a_{35}a_{51})+a_{15}(a_{33}a_{51}-a_{31}a_{53})]+$
$+a_{24}a_{43}[a_{11}(a_{32}a_{55}-a_{35}a_{52})+a_{12}(a_{35}a_{51}-a_{31}a_{55})+a_{15}(a_{31}a_{52}-a_{32}a_{51})]+$
$+a_{24}a_{45}[a_{11}(a_{33}a_{52}-a_{32}a_{53})+a_{12}(a_{31}a_{53}-a_{33}a_{51})+a_{13}(a_{32}a_{51}-a_{31}a_{52})]+$

$+a_{25}a_{41}[a_{12}(a_{34}a_{53}-a_{33}a_{54})+a_{13}(a_{32}a_{54}-a_{34}a_{52})+a_{14}(a_{33}a_{52}-a_{32}a_{53})]+$
$+a_{25}a_{42}[a_{11}(a_{33}a_{54}-a_{34}a_{53})+a_{13}(a_{34}a_{51}-a_{31}a_{54})+a_{14}(a_{31}a_{53}-a_{33}a_{51})]+$
$+a_{25}a_{43}[a_{11}(a_{34}a_{52}-a_{32}a_{54})+a_{12}(a_{31}a_{54}-a_{34}a_{51})+a_{14}(a_{32}a_{51}-a_{31}a_{52})]+$
$+a_{25}a_{44}[a_{11}(a_{32}a_{53}-a_{33}a_{52})+a_{12}(a_{33}a_{51}-a_{31}a_{53})+a_{13}(a_{31}a_{52}-a_{32}a_{51})]$ (91)

In Appendix 8 is shown evaluation of the determinant of the 5x5 matrix for marked out jointly the second and fourth rows applicably to Laplace's theorem.

g) For the second and fifth rows

Referring to Laplace's theorem, we can now to write for marked out two jointly rows in view of the second and fifth rows of the determinant of the 5x5 matrix the following minors of the second order (M_{ij})and also the algebraic supplemental (A_{ij})in general view and then to express the determinant by the following formula:

$$|A| = M_1A_1 - M_2A_2 + M_3A_3 - M_4A_4 + M_5A_5 - M_6A_6 + M_7A_7 + M_8A_8 - M_9A_9 + M_{10}A_{10} \ (\Diamond)$$

where,
M_{ij}-minors of the second order;
A_{ij}-algebraic (cofactors) supplemental .
In detail the formula ($\Diamond$) for evaluation of the determinant of the 5x5 matrix after of some transformations, has the following view in general form:

$|A| = a_{21}a_{52}[a_{13}(a_{34}a_{45}-a_{35}a_{44})+a_{14}(a_{35}a_{43}-a_{33}a_{45})+a_{15}(a_{33}a_{44}-a_{34}a_{43})]+$
$+a_{21}a_{53}[a_{12}(a_{35}a_{44}-a_{34}a_{45})+a_{14}(a_{32}a_{45}-a_{35}a_{42})+a_{15}(a_{34}a_{42}-a_{32}a_{44})]+$
$+a_{21}a_{54}[a_{12}(a_{33}a_{45}-a_{35}a_{43})+a_{13}(a_{35}a_{42}-a_{32}a_{45})+a_{15}(a_{32}a_{43}-a_{33}a_{42})]+$
$+a_{21}a_{55}[a_{12}(a_{34}a_{43}-a_{33}a_{44})+a_{13}(a_{32}a_{44}-a_{34}a_{42})+a_{14}(a_{33}a_{42}-a_{32}a_{43})]+$

$+a_{22}a_{51}[a_{13}(a_{35}a_{44}-a_{34}a_{45})+a_{14}(a_{33}a_{45}-a_{35}a_{43})+a_{15}(a_{34}a_{43}-a_{33}a_{44})]+$
$+a_{22}a_{53}[a_{11}(a_{34}a_{45}-a_{35}a_{44})+a_{14}(a_{35}a_{41}-a_{31}a_{45})+a_{15}(a_{31}a_{44}-a_{34}a_{41})]+$
$+a_{22}a_{54}[a_{11}(a_{35}a_{43}-a_{33}a_{45})+a_{13}(a_{31}a_{45}-a_{35}a_{41})+a_{15}(a_{33}a_{41}-a_{31}a_{43})]+$
$+a_{22}a_{55}[a_{11}(a_{33}a_{44}-a_{34}a_{43})+a_{13}(a_{34}a_{41}-a_{31}a_{44})+a_{14}(a_{31}a_{43}-a_{33}a_{41})]+$

$+a_{23}a_{51}[a_{12}(a_{34}a_{45}-a_{35}a_{44})+a_{14}(a_{35}a_{42}-a_{32}a_{45})+a_{15}(a_{32}a_{44}-a_{34}a_{42})]+$
$+a_{23}a_{52}[a_{11}(a_{35}a_{44}-a_{34}a_{45})+a_{14}(a_{31}a_{45}-a_{35}a_{41})+a_{15}(a_{34}a_{41}-a_{31}a_{44})]+$
$+a_{23}a_{54}[a_{11}(a_{32}a_{45}-a_{35}a_{42})+a_{12}(a_{35}a_{41}-a_{31}a_{45})+a_{15}(a_{31}a_{42}-a_{32}a_{41})]+$
$+a_{23}a_{55}[a_{11}(a_{34}a_{42}-a_{32}a_{44})+a_{12}(a_{31}a_{44}-a_{34}a_{41})+a_{14}(a_{32}a_{41}-a_{31}a_{42})]+$

$+a_{24}a_{51}[a_{12}(a_{35}a_{43}-a_{33}a_{45})+a_{13}(a_{32}a_{45}-a_{35}a_{42})+a_{15}(a_{33}a_{42}-a_{32}a_{43})]+$
$+a_{24}a_{52}[a_{11}(a_{33}a_{45}-a_{35}a_{43})+a_{13}(a_{35}a_{41}-a_{31}a_{45})+a_{15}(a_{31}a_{43}-a_{33}a_{41})]+$
$+a_{24}a_{53}[a_{11}(a_{35}a_{42}-a_{32}a_{45})+a_{12}(a_{31}a_{45}-a_{35}a_{41})+a_{15}(a_{32}a_{41}-a_{31}a_{42})]+$
$+a_{24}a_{55}[a_{11}(a_{32}a_{43}-a_{33}a_{42})+a_{12}(a_{33}a_{41}-a_{31}a_{43})+a_{13}(a_{31}a_{42}-a_{32}a_{41})]+$

$+a_{25}a_{51}[a_{12}(a_{33}a_{44}-a_{34}a_{43})+a_{13}(a_{34}a_{42}-a_{32}a_{44})+a_{14}(a_{32}a_{43}-a_{33}a_{42})]+$

$+a_{25}a_{52}[a_{11}(a_{34}a_{43}-a_{33}a_{44})+a_{13}(a_{31}a_{44}-a_{34}a_{41})+a_{14}(a_{33}a_{41}-a_{31}a_{43})]+$

$+a_{25}a_{53}[a_{11}(a_{32}a_{44}-a_{34}a_{42})+a_{12}(a_{34}a_{41}-a_{31}a_{44})+a_{14}(a_{31}a_{42}-a_{32}a_{41})]+$

$+a_{25}a_{54}[a_{11}(a_{33}a_{42}-a_{32}a_{43})+a_{12}(a_{31}a_{43}-a_{33}a_{41})+a_{13}(a_{32}a_{41}-a_{31}a_{42})]\quad(92)$

$$M_1=(a_{21}a_{52}-a_{22}a_{51}); \quad A_1=(-1)^{(2+5)+(1+2)}\begin{vmatrix} a_{13} & a_{14} & a_{15} \\ a_{33} & a_{34} & a_{35} \\ a_{43} & a_{44} & a_{45} \end{vmatrix}$$

$$M_2=(a_{21}a_{53}-a_{23}a_{51}); \quad A_2=(-1)^{(2+5)+(1+3)}\begin{vmatrix} a_{12} & a_{14} & a_{15} \\ a_{32} & a_{34} & a_{35} \\ a_{42} & a_{44} & a_{45} \end{vmatrix}$$

$$M_3=(a_{21}a_{54}-a_{24}a_{51}); \quad A_3=(-1)^{(2+5)+(1+4)}\begin{vmatrix} a_{12} & a_{13} & a_{15} \\ a_{32} & a_{33} & a_{35} \\ a_{42} & a_{43} & a_{45} \end{vmatrix}$$

$$M_4=(a_{21}a_{55}-a_{25}a_{51}); \quad A_4=(-1)^{(2+5)+(1+5)}\begin{vmatrix} a_{12} & a_{13} & a_{14} \\ a_{32} & a_{33} & a_{34} \\ a_{42} & a_{43} & a_{44} \end{vmatrix}$$

$$M_5=(a_{22}a_{53}-a_{23}a_{52}); \quad A_5=(-1)^{(2+5)+(2+3)}\begin{vmatrix} a_{11} & a_{14} & a_{15} \\ a_{31} & a_{34} & a_{35} \\ a_{41} & a_{44} & a_{45} \end{vmatrix}$$

$$M_6=(a_{22}a_{54}-a_{24}a_{52}); \quad A_6=(-1)^{(2+5)+(2+4)}\begin{vmatrix} a_{11} & a_{13} & a_{15} \\ a_{31} & a_{33} & a_{35} \\ a_{41} & a_{43} & a_{45} \end{vmatrix}$$

$$M_7=(a_{22}a_{55}-a_{25}a_{52}); \quad A_7=(-1)^{(2+5)+(2+5)}\begin{vmatrix} a_{11} & a_{13} & a_{14} \\ a_{31} & a_{33} & a_{34} \\ a_{41} & a_{43} & a_{44} \end{vmatrix}$$

$$M_8=(a_{23}a_{54}-a_{24}a_{53}); \quad A_8=(-1)^{(2+5)+(3+4)}\begin{vmatrix} a_{11} & a_{12} & a_{15} \\ a_{31} & a_{32} & a_{35} \\ a_{41} & a_{42} & a_{45} \end{vmatrix}$$

$$M_9=(a_{23}a_{55}-a_{25}a_{53}); \quad A_9=(-1)^{(2+5)+(3+5)}\begin{vmatrix} a_{11} & a_{12} & a_{14} \\ a_{31} & a_{32} & a_{34} \\ a_{41} & a_{42} & a_{44} \end{vmatrix}$$

$$M_{10}=(a_{24}a_{55}-a_{25}a_{54}); \quad A_{10}=(-1)^{(2+5)+(4+5)}\begin{vmatrix} a_{11} & a_{12} & a_{13} \\ a_{31} & a_{32} & a_{33} \\ a_{41} & a_{42} & a_{43} \end{vmatrix}$$

And besides analysis of formula (92) indicates on fact that elements $(a_{21}, a_{22}, a_{23}, a_{24}, a_{25})$ of the second row present as the linking elements with other elements, participating in process of evaluation the determinant of the 5x5 matrix for marked out jointly the second and fifth rows. And these conclusions can be confirmed by the data shown in Figure 91.

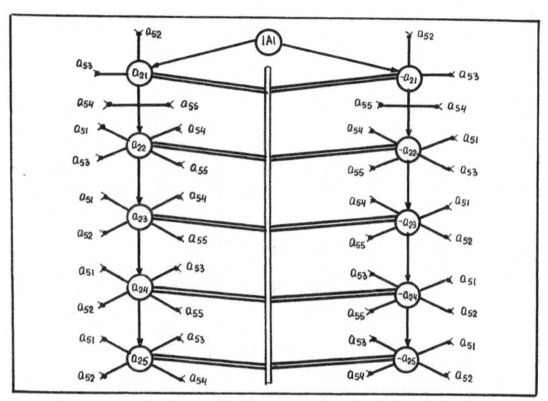

Figure 91 Tied graph (tree) for elements of the determinant of the 5x5 matrix at marked out jointly the second and fifth rows applicably to Laplace's theorem

In Appendix 9 is shown evaluation of the determinant of the 5x5 matrix for marked out jointly the second and fifth rows applicably to Laplace's theorem.

h) For the third and fourth rows

Considering the general features of Laplace's theorem applicably to evaluation of the determinant of the 5x5 matrix *for marked out jointly the third and fourth rows* ,we can now to write the following minors of the second order(M_{ij}) and also the algebraic(cofactors) supplemental (A_{ij}) and also to express the determinant by the following formula:

$$|A| = M_1A_1 - M_2A_2 + M_3A_3 - M_4A_4 + M_5A_5 - M_6A_6 + M_7A_7 + M_8A_8 - M_9A_9 + M_{10}A_{10} \quad (\square)$$

where,
M_{ij}-minors of the second order
A_{ij}-algebraic(cofactors) supplemental .
In detail the formula ($\square$) for evaluation of the determinant $|A|$ of the 5x5 matrix has the following view after of some transformations:

107

$$M_1 = (a_{31}a_{42} - a_{32}a_{41}); \quad A_1 = (-1)^{(3+4)+(1+2)} \begin{vmatrix} a_{13} & a_{14} & a_{15} \\ a_{23} & a_{24} & a_{25} \\ a_{53} & a_{54} & a_{55} \end{vmatrix}$$

$$M_2 = (a_{31}a_{43} - a_{33}a_{41}); \quad A_2 = (-1)^{(3+4)+(1+3)} \begin{vmatrix} a_{12} & a_{14} & a_{15} \\ a_{22} & a_{24} & a_{25} \\ a_{52} & a_{54} & a_{55} \end{vmatrix}$$

$$M_3 = (a_{31}a_{44} - a_{34}a_{41}); \quad A_3 = (-1)^{(3+4)+(1+4)} \begin{vmatrix} a_{12} & a_{13} & a_{15} \\ a_{22} & a_{23} & a_{25} \\ a_{52} & a_{53} & a_{55} \end{vmatrix}$$

$$M_4 = (a_{31}a_{45} - a_{35}a_{41}); \quad A_4 = (-1)^{(3+4)+(1+5)} \begin{vmatrix} a_{12} & a_{13} & a_{14} \\ a_{22} & a_{23} & a_{24} \\ a_{52} & a_{53} & a_{54} \end{vmatrix}$$

$$M_5 = (a_{32}a_{43} - a_{33}a_{42}); \quad A_5 = (-1)^{(3+4)+(2+3)} \begin{vmatrix} a_{11} & a_{14} & a_{15} \\ a_{21} & a_{24} & a_{25} \\ a_{51} & a_{54} & a_{55} \end{vmatrix}$$

$$M_6 = (a_{32}a_{44} - a_{34}a_{42}); \quad A_6 = (-1)^{(3+4)+(2+4)} \begin{vmatrix} a_{11} & a_{13} & a_{15} \\ a_{21} & a_{23} & a_{25} \\ a_{51} & a_{53} & a_{55} \end{vmatrix}$$

$$M_7 = (a_{32}a_{45} - a_{35}a_{42}); \quad A_7 = (-1)^{(3+4)+(2+5)} \begin{vmatrix} a_{11} & a_{13} & a_{14} \\ a_{21} & a_{23} & a_{24} \\ a_{51} & a_{53} & a_{54} \end{vmatrix}$$

$$M_8 = (a_{33}a_{44} - a_{34}a_{43}); \quad A_8 = (-1)^{(3+4)+(3+4)} \begin{vmatrix} a_{11} & a_{12} & a_{15} \\ a_{21} & a_{22} & a_{25} \\ a_{51} & a_{52} & a_{55} \end{vmatrix}$$

$$M_9 = (a_{33}a_{45} - a_{35}a_{43}); \quad A_9 = (-1)^{(3+4)+(3+5)} \begin{vmatrix} a_{11} & a_{12} & a_{14} \\ a_{21} & a_{22} & a_{24} \\ a_{51} & a_{52} & a_{54} \end{vmatrix}$$

$$M_{10} = (a_{34}a_{45} - a_{35}a_{44}); \quad A_{10} = (-1)^{(3+4)+(4+5)} \begin{vmatrix} a_{11} & a_{12} & a_{13} \\ a_{21} & a_{22} & a_{23} \\ a_{51} & a_{52} & a_{53} \end{vmatrix}$$

In Figure 92 is shown the tied graph (tree) for elements of the determinant of the 5x5 matrix at marked out the third and fourth rows applicably to Laplace's theorem. **In Appendix 10** is shown evaluation of the determinant of the 5x5 matrix for marked out jointly the third and fourth rows applicably to Laplace's theorem.

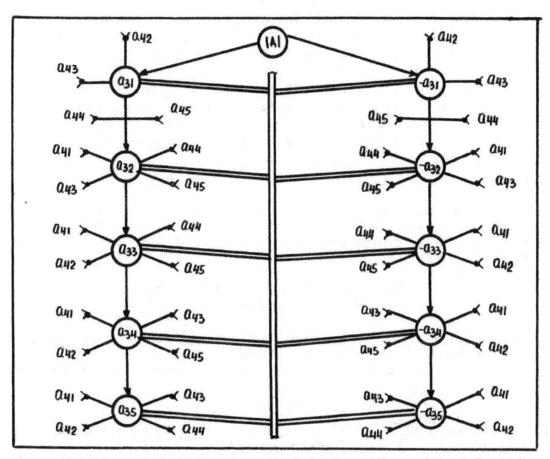

Figure 92 Tied graph (tree) for elements of the determinant of the 5x5 matrix at marked out jointly the third and fourth rows applicably to Laplace's theorem

$$
\begin{aligned}
\left|\,A\,\right| =\; &a_{31}a_{42}[a_{13}(a_{24}a_{55}-a_{25}a_{54})+a_{14}(a_{25}a_{53}-a_{23}a_{55})+a_{15}(a_{23}a_{54}-a_{24}a_{53})]+\\
&+a_{31}a_{43}[a_{12}(a_{25}a_{54}-a_{24}a_{55})+a_{14}(a_{22}a_{55}-a_{25}a_{52})+a_{15}(a_{24}a_{52}-a_{22}a_{54})]+\\
&+a_{31}a_{44}[a_{12}(a_{23}a_{55}-a_{25}a_{53})+a_{13}(a_{25}a_{52}-a_{22}a_{55})+a_{15}(a_{22}a_{53}-a_{23}a_{52})]+\\
&+a_{31}a_{45}[a_{12}(a_{24}a_{53}-a_{23}a_{54})+a_{13}(a_{22}a_{54}-a_{24}a_{52})+a_{14}(a_{23}a_{52}-a_{22}a_{53})]+\\[8pt]
&+a_{32}a_{41}[a_{13}(a_{25}a_{54}-a_{24}a_{55})+a_{14}(a_{23}a_{55}-a_{25}a_{53})+a_{15}(a_{24}a_{53}-a_{23}a_{54})]+\\
&+a_{32}a_{43}[a_{11}(a_{24}a_{55}-a_{25}a_{54})+a_{14}(a_{25}a_{51}-a_{21}a_{55})+a_{15}(a_{21}a_{54}-a_{24}a_{51})]+\\
&+a_{32}a_{44}[a_{11}(a_{25}a_{53}-a_{23}a_{55})+a_{13}(a_{21}a_{55}-a_{25}a_{51})+a_{15}(a_{23}a_{51}-a_{21}a_{53})]+\\
&+a_{32}a_{45}[a_{11}(a_{23}a_{54}-a_{24}a_{53})+a_{13}(a_{24}a_{51}-a_{21}a_{54})+a_{14}(a_{21}a_{53}-a_{23}a_{51})]+\\[8pt]
&+a_{33}a_{41}[a_{12}(a_{24}a_{55}-a_{25}a_{54})+a_{14}(a_{25}a_{52}-a_{22}a_{55})+a_{15}(a_{22}a_{54}-a_{24}a_{52})]+\\
&+a_{33}a_{42}[a_{11}(a_{25}a_{54}-a_{24}a_{55})+a_{14}(a_{21}a_{55}-a_{25}a_{51})+a_{15}(a_{24}a_{51}-a_{21}a_{54})]+\\
&+a_{33}a_{44}[a_{11}(a_{22}a_{55}-a_{25}a_{52})+a_{12}(a_{25}a_{51}-a_{21}a_{55})+a_{15}(a_{21}a_{52}-a_{22}a_{51})]+\\
&+a_{33}a_{45}[a_{11}(a_{24}a_{52}-a_{22}a_{54})+a_{12}(a_{21}a_{54}-a_{24}a_{51})+a_{14}(a_{22}a_{51}-a_{21}a_{52})]+
\end{aligned}
$$

$+a_{34}a_{41}[a_{12}(a_{25}a_{53}{-}a_{23}a_{55}){+}a_{13}(a_{22}a_{55}{-}a_{25}a_{52}){+}a_{15}(a_{23}a_{52}{-}a_{22}a_{53})]+$
$+a_{34}a_{42}[a_{11}(a_{23}a_{55}{-}a_{25}a_{53}){+}a_{13}(a_{25}a_{51}{-}a_{21}a_{55}){+}a_{15}(a_{21}a_{53}{-}a_{23}a_{51})]+$
$+a_{34}a_{43}[a_{11}(a_{25}a_{52}{-}a_{22}a_{55}){+}a_{12}(a_{21}a_{55}{-}a_{25}a_{51}){+}a_{15}(a_{22}a_{51}{-}a_{21}a_{52})]+$
$+a_{34}a_{45}[a_{11}(a_{22}a_{53}{-}a_{23}a_{52}){+}a_{12}(a_{23}a_{51}{-}a_{21}a_{53}){+}a_{13}(a_{21}a_{52}{-}a_{22}a_{51})]+$

$+a_{35}a_{41}[a_{12}(a_{23}a_{54}{-}a_{24}a_{53}){+}a_{13}(a_{24}a_{52}{-}a_{22}a_{54}){+}a_{14}(a_{22}a_{53}{-}a_{23}a_{52})]+$
$+a_{35}a_{42}[a_{11}(a_{24}a_{53}{-}a_{23}a_{54}){+}a_{13}(a_{21}a_{54}{-}a_{24}a_{51}){+}a_{14}(a_{23}a_{51}{-}a_{21}a_{53})]+$
$+a_{35}a_{43}[a_{11}(a_{22}a_{54}{-}a_{24}a_{52}){+}a_{12}(a_{24}a_{51}{-}a_{21}a_{54}){+}a_{14}(a_{21}a_{52}{-}a_{22}a_{51})]+$
$+a_{35}a_{44}[a_{11}(a_{23}a_{52}{-}a_{22}a_{53}){+}a_{12}(a_{21}a_{53}{-}a_{23}a_{51}){+}a_{13}(a_{22}a_{51}{-}a_{21}a_{52})]$ (93)

i)   For the third and fifth rows

Referring to general features of Laplace's theorem applicably to evaluation of the determinant of the 5x5 matrix *for marked out jointly the third and fifth rows,* we can now to write the following minors of the second order(M_{ij}) and also the algebraic(cofactors) supplemental(A_{ij}) and also to express the determinant by the following formula:

$$|A| = -M_1A_1+M_2A_2-M_2A_2-M_3A_3+M_4A_4-M_5A_5+$$
$$+ M_6A_6-M_7A_7-M_8A_8+M_9A_9-M_{10}A_{10}   (**)$$

In Figure 93 is shown the tied graph(tree) for elements of the determinant of the 5x5 matrix at marked out jointly the third and fifth rows applicably to Laplace's theorem.

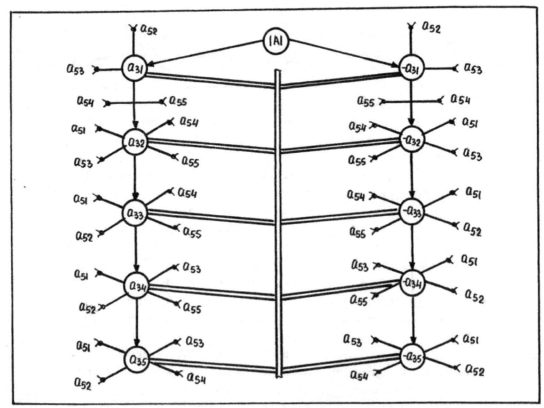

Figure 93 Tied graph (tree) for elements of the determinant of the 5x5 matrix at marked out jointly the third and fifth rows applicably to Laplace's theorem

$$M_1 = (a_{31}a_{52} - a_{32}a_{51}); \quad A_1 = (-1)^{(3+5)+(1+2)} \begin{vmatrix} a_{13} & a_{14} & a_{15} \\ a_{23} & a_{24} & a_{25} \\ a_{43} & a_{44} & a_{45} \end{vmatrix}$$

$$M_2 = (a_{31}a_{53} - a_{33}a_{51}); \quad A_2 = (-1)^{(3+5)+(1+3)} \begin{vmatrix} a_{12} & a_{14} & a_{15} \\ a_{22} & a_{24} & a_{25} \\ a_{42} & a_{44} & a_{45} \end{vmatrix}$$

$$M_3 = (a_{31}a_{54} - a_{34}a_{51}); \quad A_3 = (-1)^{(3+5)+(1+4)} \begin{vmatrix} a_{12} & a_{13} & a_{15} \\ a_{22} & a_{23} & a_{25} \\ a_{42} & a_{43} & a_{45} \end{vmatrix}$$

$$M_4 = (a_{31}a_{55} - a_{35}a_{51}); \quad A_4 = (-1)^{(3+5)+(1+5)} \begin{vmatrix} a_{12} & a_{13} & a_{14} \\ a_{22} & a_{23} & a_{24} \\ a_{42} & a_{43} & a_{44} \end{vmatrix}$$

$$M_5 = (a_{32}a_{53} - a_{33}a_{52}); \quad A_5 = (-1)^{(3+5)+(2+3)} \begin{vmatrix} a_{11} & a_{14} & a_{15} \\ a_{21} & a_{24} & a_{25} \\ a_{41} & a_{44} & a_{45} \end{vmatrix}$$

$$M_6 = (a_{32}a_{54} - a_{34}a_{52}); \quad A_6 = (-1)^{(3+5)+(2+4)} \begin{vmatrix} a_{11} & a_{13} & a_{15} \\ a_{21} & a_{23} & a_{25} \\ a_{41} & a_{43} & a_{45} \end{vmatrix}$$

$$M_7 = (a_{32}a_{55} - a_{35}a_{52}); \quad A_7 = (-1)^{(3+4)+(2+5)} \begin{vmatrix} a_{11} & a_{13} & a_{14} \\ a_{21} & a_{23} & a_{24} \\ a_{41} & a_{43} & a_{44} \end{vmatrix}$$

$$M_8 = (a_{33}a_{54} - a_{34}a_{53}); \quad A_8 = (-1)^{(3+5)+(3+4)} \begin{vmatrix} a_{11} & a_{12} & a_{15} \\ a_{21} & a_{22} & a_{25} \\ a_{41} & a_{42} & a_{45} \end{vmatrix}$$

$$M_9 = (a_{33}a_{55} - a_{35}a_{53}); \quad A_9 = (-1)^{(3+5)+(3+5)} \begin{vmatrix} a_{11} & a_{12} & a_{14} \\ a_{21} & a_{22} & a_{24} \\ a_{41} & a_{42} & a_{44} \end{vmatrix}$$

$$M_{10} = (a_{34}a_{55} - a_{35}a_{54}); \quad A_{10} = (-1)^{(3+5)+(4+5)} \begin{vmatrix} a_{11} & a_{12} & a_{13} \\ a_{21} & a_{22} & a_{23} \\ a_{41} & a_{42} & a_{43} \end{vmatrix}$$

In detail the formula (**) ,after of some transformations, has the following view in general form:

$$|A| = a_{31}a_{52}[a_{13}(a_{25}a_{44}-a_{24}a_{45})+a_{14}(a_{23}a_{45}-a_{25}a_{43})+a_{15}(a_{24}a_{43}-a_{23}a_{44})]+$$
$$+a_{31}a_{53}[a_{12}(a_{24}a_{45}-a_{25}a_{44})+a_{14}(a_{25}a_{42}-a_{22}a_{45})+a_{15}(a_{22}a_{44}-a_{24}a_{42})]+$$
$$+a_{31}a_{54}[a_{12}(a_{25}a_{43}-a_{23}a_{45})+a_{13}(a_{22}a_{45}-a_{25}a_{42})+a_{15}(a_{23}a_{42}-a_{22}a_{43})]+$$
$$+a_{31}a_{55}[a_{12}(a_{23}a_{44}-a_{24}a_{43})+a_{13}(a_{24}a_{42}-a_{22}a_{44})+a_{14}(a_{22}a_{43}-a_{23}a_{42})]+$$

$$+a_{32}a_{51}[a_{13}(a_{24}a_{45}-a_{25}a_{44})+a_{14}(a_{25}a_{43}-a_{23}a_{45})+a_{15}(a_{23}a_{44}-a_{24}a_{43})]+$$
$$+a_{32}a_{53}[a_{11}(a_{25}a_{44}-a_{24}a_{45})+a_{14}(a_{21}a_{45}-a_{25}a_{41})+a_{15}(a_{24}a_{41}-a_{21}a_{44})]+$$
$$+a_{32}a_{54}[a_{11}(a_{23}a_{45}-a_{25}a_{43})+a_{13}(a_{25}a_{41}-a_{21}a_{45})+a_{15}(a_{21}a_{43}-a_{23}a_{41})]+$$
$$+a_{32}a_{55}[a_{11}(a_{24}a_{43}-a_{23}a_{44})+a_{13}(a_{21}a_{44}-a_{24}a_{41})+a_{14}(a_{23}a_{41}-a_{21}a_{43})]+$$

$$+a_{33}a_{51}[a_{12}(a_{25}a_{44}-a_{24}a_{45})+a_{14}(a_{22}a_{45}-a_{25}a_{42})+a_{15}(a_{24}a_{42}-a_{22}a_{44})]+$$
$$+a_{33}a_{52}[a_{11}(a_{24}a_{45}-a_{25}a_{44})+a_{14}(a_{25}a_{41}-a_{21}a_{45})+a_{15}(a_{21}a_{44}-a_{24}a_{41})]+$$
$$+a_{33}a_{54}[a_{11}(a_{25}a_{42}-a_{22}a_{45})+a_{12}(a_{21}a_{45}-a_{25}a_{41})+a_{15}(a_{22}a_{41}-a_{21}a_{42})]+$$
$$+a_{33}a_{55}[a_{11}(a_{22}a_{44}-a_{24}a_{42})+a_{12}(a_{24}a_{41}-a_{21}a_{44})+a_{14}(a_{21}a_{42}-a_{22}a_{41})]+$$

$$+a_{34}a_{51}[a_{12}(a_{23}a_{45}-a_{25}a_{43})+a_{13}(a_{25}a_{42}-a_{22}a_{45})+a_{15}(a_{22}a_{43}-a_{23}a_{42})]+$$
$$+a_{34}a_{52}[a_{11}(a_{25}a_{43}-a_{23}a_{45})+a_{13}(a_{21}a_{45}-a_{25}a_{41})+a_{15}(a_{23}a_{41}-a_{21}a_{43})]+$$
$$+a_{34}a_{53}[a_{11}(a_{22}a_{45}-a_{25}a_{42})+a_{12}(a_{25}a_{41}-a_{21}a_{45})+a_{15}(a_{21}a_{42}-a_{22}a_{41})]+$$
$$+a_{34}a_{55}[a_{11}(a_{23}a_{42}-a_{22}a_{43})+a_{12}(a_{21}a_{43}-a_{23}a_{41})+a_{13}(a_{22}a_{41}-a_{21}a_{42})]+$$

$$+a_{35}a_{51}[a_{12}(a_{24}a_{43}-a_{23}a_{44})+a_{13}(a_{22}a_{44}-a_{24}a_{42})+a_{14}(a_{23}a_{42}-a_{22}a_{43})]+$$
$$+a_{35}a_{52}[a_{11}(a_{23}a_{44}-a_{24}a_{43})+a_{13}(a_{24}a_{41}-a_{21}a_{44})+a_{14}(a_{21}a_{43}-a_{23}a_{41})]+$$
$$+a_{35}a_{53}[a_{11}(a_{24}a_{42}-a_{22}a_{44})+a_{12}(a_{21}a_{44}-a_{24}a_{41})+a_{14}(a_{22}a_{41}-a_{21}a_{42})]+$$
$$+a_{35}a_{54}[a_{11}(a_{22}a_{43}-a_{23}a_{42})+a_{12}(a_{23}a_{41}-a_{21}a_{43})+a_{13}(a_{21}a_{42}-a_{22}a_{41})] \quad (94)$$

In Appendix 11 is shown evaluation of the determinant of the 5x5 matrix with using of Laplace's theorem for marked out jointly the third and fifth rows.

j) For the fourth and fifth rows

Considering the general features of Laplace's theorem applicably to evaluation of the determinant of the 5x5 matrix *for marked out jointly the fourth and fifth rows ,*we can now to write the following minors of the second order(M_{ij}) and also the algebraic (cofactors) supplemental (A_{ij}) and also to express the determinant by the following formula:

$$|A| = M_1A_1 - M_2A_2 + M_3A_3 - M_4A_4 + M_5A_5 - M_6A_6 + M_7A_7 + M_8A_8 - M_9A_9 + M_{10}A_{10} \quad (\blacktriangle)$$

where,
$M_{(ij)}$- minors of the second order of i-row and j-column
$A_{(ij)}$-algebraic(cofactors) supplemental of i-row and j-column.

In Figure 94 is shown the tied graph (tree) for elements of the determinant of the 5x5 matrix at marked out jointly the fourth and fifth rows applicably to Laplace's theorem. In detail the formula ($\blacktriangle$), after of some transformations, has the following view in general form:

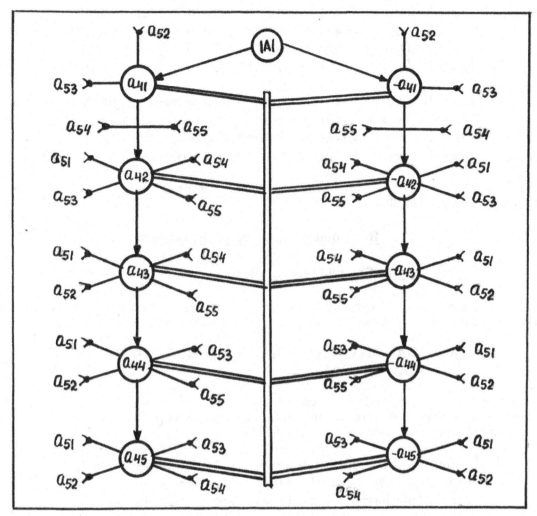

Figure 94 Tied graph (tree) for elements of the determinant of the 5x5 matrix at marked out jointly the fourth and fifth rows applicably to Laplace's theorem

$$|A| = a_{41}a_{52}[a_{13}(a_{24}a_{35}-a_{25}a_{34})+a_{14}(a_{25}a_{33}-a_{23}a_{35})+a_{15}(a_{23}a_{34}-a_{24}a_{33})]+$$
$$+a_{41}a_{53}[a_{12}(a_{25}a_{34}-a_{24}a_{35})+a_{14}(a_{22}a_{35}-a_{25}a_{32})+a_{15}(a_{24}a_{32}-a_{22}a_{34})]+$$
$$+a_{41}a_{54}[a_{12}(a_{23}a_{35}-a_{25}a_{33})+a_{13}(a_{25}a_{32}-a_{22}a_{35})+a_{15}(a_{22}a_{33}-a_{23}a_{32})]+$$
$$+a_{41}a_{55}[a_{12}(a_{24}a_{33}-a_{23}a_{34})+a_{13}(a_{22}a_{34}-a_{24}a_{32})+a_{14}(a_{23}a_{32}-a_{22}a_{33})]+$$

$$+a_{42}a_{51}[a_{13}(a_{25}a_{34}-a_{24}a_{35})+a_{14}(a_{23}a_{35}-a_{25}a_{33})+a_{15}(a_{24}a_{33}-a_{23}a_{34})]+$$
$$+a_{42}a_{53}[a_{11}(a_{24}a_{35}-a_{25}a_{34})+a_{14}(a_{25}a_{31}-a_{21}a_{35})+a_{15}(a_{21}a_{34}-a_{24}a_{31})]+$$
$$+a_{42}a_{54}[a_{11}(a_{25}a_{33}-a_{23}a_{35})+a_{13}(a_{21}a_{35}-a_{25}a_{31})+a_{15}(a_{23}a_{31}-a_{21}a_{33})]+$$
$$+a_{42}a_{55}[a_{11}(a_{23}a_{34}-a_{24}a_{33})+a_{13}(a_{24}a_{31}-a_{21}a_{34})+a_{14}(a_{21}a_{33}-a_{23}a_{31})]+$$

$$+a_{43}a_{51}[a_{12}(a_{24}a_{35}-a_{25}a_{34})+a_{14}(a_{25}a_{32}-a_{22}a_{35})+a_{15}(a_{22}a_{34}-a_{24}a_{32})]+$$
$$+a_{43}a_{52}[a_{11}(a_{25}a_{34}-a_{24}a_{35})+a_{14}(a_{21}a_{35}-a_{25}a_{31})+a_{15}(a_{24}a_{31}-a_{21}a_{34})]+$$
$$+a_{43}a_{54}[a_{11}(a_{22}a_{35}-a_{25}a_{32})+a_{12}(a_{25}a_{31}-a_{21}a_{35})+a_{15}(a_{21}a_{32}-a_{22}a_{31})]+$$
$$+a_{43}a_{55}[a_{11}(a_{24}a_{32}-a_{22}a_{34})+a_{12}(a_{21}a_{34}-a_{24}a_{31})+a_{14}(a_{22}a_{31}-a_{21}a_{32})]+$$

113

$$+a_{44}a_{51}[a_{12}(a_{25}a_{33}-a_{23}a_{35})+a_{13}(a_{22}a_{35}-a_{25}a_{32})+a_{15}(a_{23}a_{32}-a_{22}a_{33})]+$$
$$+a_{44}a_{52}[a_{11}(a_{23}a_{35}-a_{25}a_{33})+a_{13}(a_{25}a_{31}-a_{21}a_{35})+a_{15}(a_{21}a_{33}-a_{23}a_{31})]+$$
$$+a_{44}a_{53}[a_{11}(a_{25}a_{32}-a_{22}a_{35})+a_{12}(a_{21}a_{35}-a_{25}a_{31})+a_{15}(a_{22}a_{31}-a_{21}a_{32})]+$$
$$+a_{44}a_{55}[a_{11}(a_{22}a_{33}-a_{23}a_{32})+a_{12}(a_{23}a_{31}-a_{21}a_{33})+a_{13}(a_{21}a_{32}-a_{22}a_{31})]+$$

$$+a_{45}a_{51}[a_{12}(a_{23}a_{34}-a_{24}a_{33})+a_{13}(a_{24}a_{32}-a_{22}a_{34})+a_{14}(a_{22}a_{33}-a_{23}a_{32})]+$$
$$+a_{45}a_{52}[a_{11}(a_{24}a_{33}-a_{23}a_{34})+a_{13}(a_{21}a_{34}-a_{24}a_{31})+a_{14}(a_{23}a_{31}-a_{21}a_{33})]+$$
$$+a_{45}a_{53}[a_{11}(a_{22}a_{34}-a_{24}a_{32})+a_{12}(a_{24}a_{31}-a_{21}a_{34})+a_{14}(a_{21}a_{32}-a_{22}a_{31})]+$$
$$+a_{45}a_{54}[a_{11}(a_{23}a_{32}-a_{22}a_{33})+a_{12}(a_{21}a_{33}-a_{23}a_{31})+a_{13}(a_{22}a_{31}-a_{21}a_{32})] \qquad (95)$$

In **Appendix 12** is shown evaluation of the determinant of the 5x5 matrix with using the Laplace's theorem for marked out jointly the fourth and fifth rows.

❖ By Laplace's theorem for the columns

a) For the first and second columns

The value of determinant of the 5x5 matrix at conditions when its elements marked out jointly to the first and second columns can be expressed applicably to Laplace's theorem in such general view as:

$$|A| = M_1A_1 - M_2A_2 + M_3A_3 - M_4A_4 + M_5A_5 - M_6A_6 + M_7A_7 +$$
$$+ M_8A_8 - M_9A_9 + M_{10}A_{10} \qquad (*)$$

where,

M_{ij}- minor of the second order of i-rows and j-columns;

A_{ij}-algebraic supplemental (cofactor) of the third order of i-rows and j-columns .

In detail the formula (*) ,after of some transformations, has the following view:

$$|A| = a_{11}a_{22}[a_{33}(a_{44}a_{55}-a_{45}a_{54})+a_{34}(a_{45}a_{53}-a_{43}a_{55})+a_{35}(a_{43}a_{54}-a_{44}a_{53})]+$$
$$+a_{11}a_{32}[a_{23}(a_{45}a_{54}-a_{44}a_{55})+a_{24}(a_{43}a_{55}-a_{45}a_{53})+a_{25}(a_{44}a_{53}-a_{43}a_{54})]+$$
$$+a_{11}a_{42}[a_{23}(a_{34}a_{55}-a_{35}a_{54})+a_{24}(a_{35}a_{53}-a_{33}a_{55})+a_{25}(a_{33}a_{54}-a_{34}a_{53})]+$$
$$+a_{11}a_{52}[a_{23}(a_{35}a_{44}-a_{34}a_{45})+a_{24}(a_{33}a_{45}-a_{35}a_{43})+a_{25}(a_{34}a_{43}-a_{33}a_{44})]+$$

$$+a_{21}a_{12}[a_{33}(a_{45}a_{54}-a_{44}a_{55})+a_{34}(a_{43}a_{55}-a_{45}a_{53})+a_{35}(a_{44}a_{53}-a_{43}a_{54})]+$$
$$+a_{21}a_{32}[a_{13}(a_{44}a_{55}-a_{45}a_{54})+a_{14}(a_{45}a_{53}-a_{43}a_{55})+a_{15}(a_{43}a_{54}-a_{44}a_{53})]+$$
$$+a_{21}a_{42}[a_{13}(a_{35}a_{54}-a_{34}a_{55})+a_{14}(a_{33}a_{55}-a_{35}a_{53})+a_{15}(a_{34}a_{53}-a_{33}a_{54})]+$$
$$+a_{21}a_{52}[a_{13}(a_{34}a_{45}-a_{35}a_{44})+a_{14}(a_{35}a_{43}-a_{33}a_{45})+a_{15}(a_{33}a_{44}-a_{34}a_{43})]+$$

$$+a_{31}a_{12}[a_{23}(a_{44}a_{55}-a_{45}a_{54})+a_{24}(a_{45}a_{53}-a_{43}a_{55})+a_{25}(a_{43}a_{54}-a_{44}a_{53})]+$$
$$+a_{31}a_{22}[a_{13}(a_{45}a_{54}-a_{44}a_{55})+a_{14}(a_{43}a_{55}-a_{45}a_{53})+a_{15}(a_{44}a_{53}-a_{43}a_{54})]+$$
$$+a_{31}a_{42}[a_{13}(a_{24}a_{55}-a_{25}a_{54})+a_{14}(a_{25}a_{53}-a_{23}a_{55})+a_{15}(a_{23}a_{54}-a_{24}a_{53})]+$$
$$+a_{31}a_{52}[a_{13}(a_{25}a_{44}-a_{24}a_{45})+a_{14}(a_{23}a_{45}-a_{25}a_{43})+a_{15}(a_{24}a_{43}-a_{23}a_{44})]+$$

$$+a_{41}a_{12}[a_{23}(a_{35}a_{54}-a_{34}a_{55})+a_{24}(a_{33}a_{55}-a_{35}a_{53})+a_{25}(a_{34}a_{53}-a_{33}a_{54})]+$$
$$+a_{41}a_{22}[a_{13}(a_{34}a_{55}-a_{35}a_{54})+a_{14}(a_{35}a_{53}-a_{33}a_{55})+a_{15}(a_{33}a_{54}-a_{34}a_{53})]+$$
$$+a_{41}a_{32}[a_{13}(a_{25}a_{54}-a_{24}a_{55})+a_{14}(a_{23}a_{55}-a_{25}a_{53})+a_{15}(a_{24}a_{53}-a_{23}a_{54})]+$$
$$+a_{41}a_{52}[a_{13}(a_{24}a_{35}-a_{25}a_{34})+a_{14}(a_{25}a_{33}-a_{23}a_{35})+a_{15}(a_{23}a_{34}-a_{24}a_{33})]+$$

$$+a_{51}a_{12}[a_{23}(a_{34}a_{45}-a_{35}a_{44})+a_{24}(a_{35}a_{43}-a_{33}a_{45})+a_{25}(a_{33}a_{44}-a_{34}a_{43})]+$$
$$+a_{51}a_{22}[a_{13}(a_{35}a_{44}-a_{34}a_{45})+a_{14}(a_{33}a_{45}-a_{35}a_{43})+a_{15}(a_{34}a_{43}-a_{33}a_{44})]+$$
$$+a_{51}a_{32}[a_{13}(a_{24}a_{45}-a_{25}a_{44})+a_{14}(a_{25}a_{43}-a_{23}a_{45})+a_{15}(a_{23}a_{44}-a_{24}a_{43})]+$$
$$+a_{51}a_{42}[a_{13}(a_{25}a_{34}-a_{24}a_{35})+a_{14}(a_{23}a_{35}-a_{25}a_{33})+a_{15}(a_{24}a_{33}-a_{23}a_{34})] \quad (96)$$

$$M_1=(a_{11}a_{22}-a_{12}a_{21}) \qquad A_1=(-1)^{(1+2)+(1+2)}\begin{vmatrix} a_{33} & a_{34} & a_{35} \\ a_{43} & a_{44} & a_{45} \\ a_{53} & a_{54} & a_{55} \end{vmatrix}$$

$$M_2=(a_{11}a_{32}-a_{12}a_{31}) \qquad A_2=(-1)^{(1+2)+(1+3)}\begin{vmatrix} a_{23} & a_{24} & a_{25} \\ a_{43} & a_{44} & a_{45} \\ a_{53} & a_{54} & a_{55} \end{vmatrix}$$

$$M_3=(a_{11}a_{42}-a_{12}a_{41}) \qquad A_3=(-1)^{(1+2)+(1+4)}\begin{vmatrix} a_{23} & a_{24} & a_{25} \\ a_{33} & a_{34} & a_{35} \\ a_{53} & a_{54} & a_{55} \end{vmatrix}$$

$$M_4=(a_{11}a_{52}-a_{12}a_{51}) \qquad A_4=(-1)^{(1+2)+(1+5)}\begin{vmatrix} a_{23} & a_{24} & a_{25} \\ a_{33} & a_{34} & a_{35} \\ a_{43} & a_{44} & a_{45} \end{vmatrix}$$

$$M_5=(a_{21}a_{32}-a_{22}a_{31}) \qquad A_5=(-1)^{(1+2)+(2+3)}\begin{vmatrix} a_{13} & a_{14} & a_{15} \\ a_{43} & a_{44} & a_{45} \\ a_{53} & a_{54} & a_{55} \end{vmatrix}$$

$$M_6=(a_{21}a_{42}-a_{22}a_{41}) \qquad A_6=(-1)^{(1+2)+(2+4)}\begin{vmatrix} a_{13} & a_{14} & a_{15} \\ a_{33} & a_{34} & a_{35} \\ a_{53} & a_{54} & a_{55} \end{vmatrix}$$

$$M_7=(a_{21}a_{52}-a_{22}a_{51}) \qquad A_7=(-1)^{(1+2)+(2+5)}\begin{vmatrix} a_{13} & a_{14} & a_{15} \\ a_{33} & a_{34} & a_{35} \\ a_{43} & a_{44} & a_{45} \end{vmatrix}$$

$$M_8=(a_{31}a_{42}-a_{32}a_{41}) \qquad A_8=(-1)^{(1+2)+(3+4)}\begin{vmatrix} a_{13} & a_{14} & a_{15} \\ a_{23} & a_{24} & a_{25} \\ a_{53} & a_{54} & a_{55} \end{vmatrix}$$

$$M_9=(a_{31}a_{52}-a_{32}a_{51}) \qquad A_9=(-1)^{(1+2)+(3+5)}\begin{vmatrix} a_{13} & a_{14} & a_{15} \\ a_{23} & a_{24} & a_{25} \\ a_{43} & a_{44} & a_{45} \end{vmatrix}$$

$$M_{10}=(a_{41}a_{52}-a_{42}a_{51}) \qquad A_{10}=(-1)^{(1+2)+(4+5)}\begin{vmatrix} a_{13} & a_{14} & a_{15} \\ a_{23} & a_{24} & a_{25} \\ a_{33} & a_{34} & a_{35} \end{vmatrix}$$

In Figure 95 is shown the tied graph (tree) for elements of the determinant of the 5x5 matrix at marked out jointly the first and second columns applicably to Laplace's theorem .

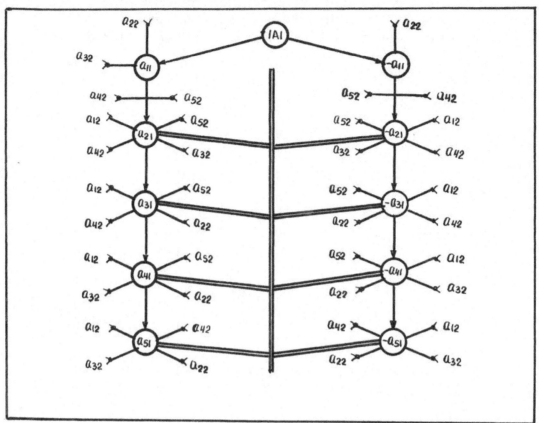

Figure 95 Tied graph (tree) for elements of the determinant of the 5x5 matrix at marked out jointly the first and second columns applicably to Laplace's theorem

In Appendix 13 is shown evaluation of the determinant of the 5x5 matrix with using of Laplace's theorem for marked out jointly the first and second columns.

b)   **For the first and third columns**

The value of determinant $|A|$ of the 5x5 matrix at conditions when marked out jointly two columns, such as the first and third columns applicably to Laplace's theorem, can be expressed by the following formula in general view:

$$|A| = -M_1A_1 + M_2A_2 - M_3A_3 + M_4A_4 - M_5A_5 + M_6A_6 - M_7A_7 - M_8A_8 + M_9A_9 - M_{10}A_{10} \ (x)$$

where,

M_1-minor and A_1-algebraic supplemental (cofactor) for marked out jointly the first and third columns and also the first and second rows;

M_2-minor and A_2-algebraic supplemental (cofactor) for marked out jointly the first and third columns and also the first and third rows;

M_3 –minor and A_3-algebraic supplemental (cofactor) for marked out jointly the first and third columns and also the first and fourth rows;

M_4-minor and A_4-algebraic supplemental (cofactor) for marked out jointly the first and third columns and also the first and fifth rows;

M_5-minor and A_5-algebraic supplemental (cofactor) for marked out jointly the first and third columns and also the second and third rows;

116

M_6-minor and A_6-algebraic supplemental (cofactor) for marked out jointly the first and third columns and also the second and fourth rows;

M_7-minor and A_7-algebraic supplemental (cofactor) for marked out jointly the first and third columns and also the second and fifth rows;

M_8-minor and A_8-algebraic supplemental (cofactor) for marked out jointly the first and third columns and also the third and fourth rows;

M_9-minor and A_9-algebraic supplemental (cofactor) for marked out jointly the first and third columns and also the third and fifth rows;

M_{10}-minor and A_{10}-algebraic supplemental (cofactor) for marked out jointly the first and third columns and also the fourth and fifth rows.

$$M_1 = (a_{11}a_{23} - a_{13}a_{21}) \qquad A_1 = (-1)^{(1+3)+(1+2)} \begin{vmatrix} a_{32} & a_{34} & a_{35} \\ a_{42} & a_{44} & a_{45} \\ a_{52} & a_{54} & a_{55} \end{vmatrix}$$

$$M_2 = (a_{11}a_{33} - a_{13}a_{31}) \qquad A_2 = (-1)^{(1+3)+(1+3)} \begin{vmatrix} a_{22} & a_{24} & a_{25} \\ a_{42} & a_{44} & a_{45} \\ a_{52} & a_{54} & a_{55} \end{vmatrix}$$

$$M_3 = (a_{11}a_{43} - a_{13}a_{41}) \qquad A_3 = (-1)^{(1+3)+(1+4)} \begin{vmatrix} a_{22} & a_{24} & a_{25} \\ a_{32} & a_{34} & a_{35} \\ a_{52} & a_{54} & a_{55} \end{vmatrix}$$

$$M_4 = (a_{11}a_{53} - a_{13}a_{51}) \qquad A_4 = (-1)^{(1+3)+(1+5)} \begin{vmatrix} a_{22} & a_{24} & a_{25} \\ a_{32} & a_{34} & a_{35} \\ a_{42} & a_{44} & a_{45} \end{vmatrix}$$

$$M_5 = (a_{21}a_{33} - a_{23}a_{31}) \qquad A_5 = (-1)^{(1+3)+(2+3)} \begin{vmatrix} a_{12} & a_{14} & a_{15} \\ a_{42} & a_{44} & a_{45} \\ a_{52} & a_{54} & a_{55} \end{vmatrix}$$

$$M_6 = (a_{21}a_{43} - a_{23}a_{41}) \qquad A_6 = (-1)^{(1+3)+(2+4)} \begin{vmatrix} a_{12} & a_{14} & a_{15} \\ a_{32} & a_{34} & a_{35} \\ a_{52} & a_{54} & a_{55} \end{vmatrix}$$

$$M_7 = (a_{21}a_{53} - a_{23}a_{51}) \qquad A_7 = (-1)^{(1+3)+(2+5)} \begin{vmatrix} a_{12} & a_{14} & a_{15} \\ a_{32} & a_{34} & a_{35} \\ a_{42} & a_{44} & a_{45} \end{vmatrix}$$

$$M_8 = (a_{31}a_{43} - a_{33}a_{41}) \qquad A_8 = (-1)^{(1+3)+(3+4)} \begin{vmatrix} a_{12} & a_{14} & a_{15} \\ a_{22} & a_{24} & a_{25} \\ a_{52} & a_{54} & a_{55} \end{vmatrix}$$

$$M_9 = (a_{31}a_{53} - a_{33}a_{51}) \qquad A_9 = (-1)^{(1+3)+(3+5)} \begin{vmatrix} a_{12} & a_{14} & a_{15} \\ a_{22} & a_{24} & a_{25} \\ a_{42} & a_{44} & a_{45} \end{vmatrix}$$

$$M_{10} = (a_{41}a_{53} - a_{43}a_{51}) \qquad A_{10} = (-1)^{(1+3)+(4+5)} \begin{vmatrix} a_{12} & a_{14} & a_{15} \\ a_{22} & a_{24} & a_{25} \\ a_{32} & a_{34} & a_{35} \end{vmatrix}$$

In Figure 96 is shown the tied graph (tree) for elements of the determinant of the 5x5 matrix at marked out jointly the first and third columns applicably to Laplace's theorem.

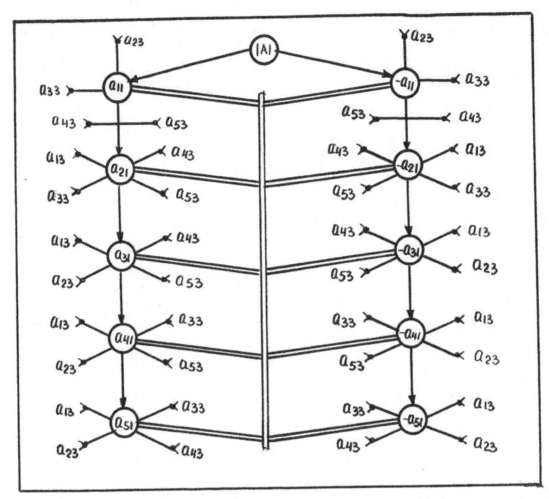

Figure 96 Tied graph (tree) for elements of the determinant of the 5x5 matrix at marked out jointly the first and third columns applicably to Laplace's theorem

After of some calculations and transformations the formula(x) has the following view for marked out jointly the first and third columns applicably to Laplace's theorem:

$$|A| = a_{11}a_{23}[a_{32}(a_{45}a_{54}-a_{44}a_{55})+a_{34}(a_{42}a_{55}-a_{45}a_{52})+a_{35}(a_{44}a_{52}-a_{42}a_{54})]+$$
$$+a_{11}a_{33}[a_{22}(a_{44}a_{55}-a_{45}a_{54})+a_{24}(a_{45}a_{52}-a_{42}a_{55})+a_{25}(a_{42}a_{54}-a_{44}a_{52})]+$$
$$+a_{11}a_{43}[a_{22}(a_{35}a_{54}-a_{34}a_{55})+a_{24}(a_{32}a_{55}-a_{35}a_{52})+a_{25}(a_{34}a_{52}-a_{32}a_{54})]+$$
$$+a_{11}a_{53}[a_{22}(a_{34}a_{45}-a_{35}a_{44})+a_{24}(a_{35}a_{42}-a_{32}a_{45})+a_{25}(a_{32}a_{44}-a_{34}a_{42})]+$$

$$+a_{21}a_{13}[a_{32}(a_{44}a_{55}-a_{45}a_{54})+a_{34}(a_{45}a_{52}-a_{42}a_{55})+a_{35}(a_{42}a_{54}-a_{44}a_{52})]+$$
$$+a_{21}a_{33}[a_{12}(a_{45}a_{54}-a_{44}a_{55})+a_{14}(a_{42}a_{55}-a_{45}a_{52})+a_{15}(a_{44}a_{52}-a_{42}a_{54})]+$$
$$+a_{21}a_{43}[a_{12}(a_{34}a_{55}-a_{35}a_{54})+a_{14}(a_{35}a_{52}-a_{32}a_{55})+a_{15}(a_{32}a_{54}-a_{34}a_{52})]+$$
$$+a_{21}a_{53}[a_{12}(a_{35}a_{44}-a_{34}a_{45})+a_{14}(a_{32}a_{45}-a_{35}a_{42})+a_{15}(a_{34}a_{42}-a_{32}a_{44})]+$$

118

$+a_{31}a_{13}[a_{22}(a_{45}a_{54}-a_{44}a_{55})+a_{24}(a_{42}a_{55}-a_{45}a_{52})+a_{25}(a_{44}a_{52}-a_{42}a_{54})]+$
$+a_{31}a_{23}[a_{12}(a_{44}a_{55}-a_{45}a_{54})+a_{14}(a_{45}a_{52}-a_{42}a_{55})+a_{15}(a_{42}a_{54}-a_{44}a_{52})]+$
$+a_{31}a_{43}[a_{12}(a_{25}a_{54}-a_{24}a_{55})+a_{14}(a_{22}a_{55}-a_{25}a_{52})+a_{15}(a_{24}a_{52}-a_{22}a_{54})]+$
$+a_{31}a_{53}[a_{12}(a_{24}a_{45}-a_{25}a_{44})+a_{14}(a_{25}a_{42}-a_{22}a_{45})+a_{15}(a_{22}a_{44}-a_{24}a_{42})]+$

$+a_{41}a_{13}[a_{22}(a_{34}a_{55}-a_{35}a_{54})+a_{24}(a_{35}a_{52}-a_{32}a_{55})+a_{25}(a_{32}a_{54}-a_{34}a_{52})]+$
$+a_{41}a_{23}[a_{12}(a_{35}a_{54}-a_{34}a_{55})+a_{14}(a_{32}a_{55}-a_{35}a_{52})+a_{15}(a_{34}a_{52}-a_{32}a_{54})]+$
$+a_{41}a_{33}[a_{12}(a_{24}a_{55}-a_{25}a_{54})+a_{14}(a_{25}a_{52}-a_{22}a_{55})+a_{15}(a_{22}a_{54}-a_{24}a_{52})]+$
$+a_{41}a_{53}[a_{12}(a_{25}a_{34}-a_{24}a_{35})+a_{14}(a_{22}a_{35}-a_{25}a_{32})+a_{15}(a_{24}a_{32}-a_{22}a_{34})]+$

$+a_{51}a_{13}[a_{22}(a_{35}a_{44}-a_{34}a_{45})+a_{24}(a_{32}a_{45}-a_{35}a_{42})+a_{25}(a_{34}a_{42}-a_{32}a_{44})]+$
$+a_{51}a_{23}[a_{12}(a_{34}a_{45}-a_{35}a_{44})+a_{14}(a_{35}a_{42}-a_{32}a_{45})+a_{15}(a_{32}a_{44}-a_{34}a_{42})]+$
$+a_{51}a_{33}[a_{12}(a_{25}a_{44}-a_{24}a_{45})+a_{14}(a_{22}a_{45}-a_{25}a_{42})+a_{15}(a_{24}a_{42}-a_{22}a_{44})]+$
$+a_{51}a_{43}[a_{12}(a_{24}a_{35}-a_{25}a_{34})+a_{14}(a_{25}a_{32}-a_{22}a_{35})+a_{15}(a_{22}a_{34}-a_{24}a_{32})]$ (97)

In Appendix 14 is shown evaluation of the determinant of the 5x5 matrix with using of Laplace's theorem for marked out jointly the first and third columns.

c) For the first and fourth columns

The value of determinant $|A|$ of the 5x5 matrix at conditions when marked out jointly two columns applicably to Laplace's theorem can be expressed by the following formula in general view:

$$|A|=M_1A_1-M_2A_2+M_3A_3-M_4A_4+M_5A_5-M_6A_6+M_7A_7+M_8A_8-M_9A_9+M_{10}A_{10} \ (\odot)$$

where,
M_1-minor and A_1-algebraic supplemental (cofactor) for marked out jointly the first and fourth columns and also the first and second rows;
M_2-minor and A_2-algebraic supplemental (cofactor) for marked out jointly the first and fourth columns and also the first and third rows;
M_3-minor and A_3-algebraic supplemental (cofactor) for marked out jointly the first and fourth columns and also the first and fourth rows;
M_4-minor and A_4-algebraic supplemental (cofactor) for marked out jointly the first and fourth columns and also the first and fifth rows;
M_5-minor and A_5-algebraic supplemental (cofactor) for marked out jointly the first and fourth columns and also the second and third rows;
M_6-minor and A_6-algebraic supplemental (cofactor) for marked out jointly the first and fourth columns and also the second and fourth rows;
M_7-minor and A_7-algebraic supplemental (cofactor) for marked out jointly the first and fourth columns and also the second and fifth rows;
M_8-minor and A_8-algebraic supplemental (cofactor) for marked out jointly the first and fourth columns and also the third and fourth rows;
M_9-minor and A_9-algebraic supplemental (cofactor) for marked out jointly the first and fourth columns and also the third and fifth rows;
M_{10}-minor and A_{10}-algebraic supplemental (cofactor) for marked out jointly the first and fourth columns and also the fourth and fifth rows.

In Figure 97 is shown the tied graph (tree) for elements of the determinant of the 5x5 matrix at marked out jointly the first and fourth columns applicably to Laplace's theorem.

$$M_1 = (a_{11}a_{24} - a_{14}a_{21}) \qquad A_1 = (-1)^{(1+4)+(1+2)} \begin{vmatrix} a_{32} & a_{33} & a_{35} \\ a_{42} & a_{43} & a_{45} \\ a_{52} & a_{53} & a_{55} \end{vmatrix}$$

$$M_2 = (a_{11}a_{34} - a_{14}a_{31}) \qquad A_2 = (-1)^{(1+4)+(1+3)} \begin{vmatrix} a_{22} & a_{23} & a_{25} \\ a_{42} & a_{43} & a_{45} \\ a_{52} & a_{53} & a_{55} \end{vmatrix}$$

$$M_3 = (a_{11}a_{44} - a_{14}a_{41}) \qquad A_3 = (-1)^{(1+4)+(1+4)} \begin{vmatrix} a_{22} & a_{23} & a_{25} \\ a_{32} & a_{33} & a_{35} \\ a_{52} & a_{53} & a_{55} \end{vmatrix}$$

$$M_4 = (a_{11}a_{54} - a_{14}a_{51}) \qquad A_4 = (-1)^{(1+4)+(1+5)} \begin{vmatrix} a_{22} & a_{23} & a_{25} \\ a_{32} & a_{33} & a_{35} \\ a_{42} & a_{43} & a_{45} \end{vmatrix}$$

$$M_5 = (a_{21}a_{34} - a_{24}a_{31}) \qquad A_5 = (-1)^{(1+4)+(2+3)} \begin{vmatrix} a_{12} & a_{13} & a_{15} \\ a_{42} & a_{43} & a_{45} \\ a_{52} & a_{53} & a_{55} \end{vmatrix}$$

$$M_6 = (a_{21}a_{44} - a_{24}a_{41}) \qquad A_6 = (-1)^{(1+4)+(2+4)} \begin{vmatrix} a_{12} & a_{13} & a_{15} \\ a_{32} & a_{33} & a_{35} \\ a_{52} & a_{53} & a_{55} \end{vmatrix}$$

$$M_7 = (a_{21}a_{54} - a_{24}a_{51}) \qquad A_7 = (-1)^{(1+4)+(2+5)} \begin{vmatrix} a_{12} & a_{13} & a_{15} \\ a_{32} & a_{33} & a_{35} \\ a_{42} & a_{43} & a_{45} \end{vmatrix}$$

$$M_8 = (a_{31}a_{44} - a_{34}a_{41}) \qquad A_8 = (-1)^{(1+4)+(3+4)} \begin{vmatrix} a_{12} & a_{13} & a_{15} \\ a_{22} & a_{23} & a_{25} \\ a_{52} & a_{53} & a_{55} \end{vmatrix}$$

$$M_9 = (a_{31}a_{54} - a_{34}a_{51}) \qquad A_9 = (-1)^{(1+4)+(3+5)} \begin{vmatrix} a_{12} & a_{13} & a_{15} \\ a_{22} & a_{23} & a_{25} \\ a_{42} & a_{43} & a_{45} \end{vmatrix}$$

$$M_{10} = (a_{41}a_{54} - a_{44}a_{51}) \qquad A_{10} = (-1)^{(1+4)+(4+5)} \begin{vmatrix} a_{12} & a_{13} & a_{15} \\ a_{22} & a_{23} & a_{25} \\ a_{32} & a_{33} & a_{35} \end{vmatrix}$$

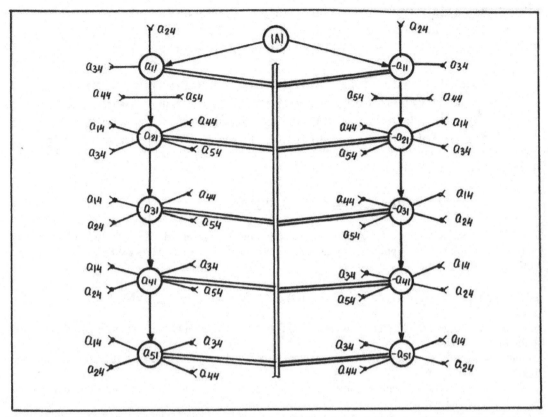

Figure 97 Tied graph (tree) for elements of the determinant of the 5x5 matrix at marked out jointly the first and fourth columns applicably to Laplace's theorem

After of some calculations and transformations, the formula (◉) has the following view for marked out jointly the first and fourth columns applicably to Laplace's theorem for the determinant of the 5x5 matrix in general view:

$$|\,A\,| = a_{11}a_{24}[a_{32}(a_{43}a_{55}-a_{45}a_{53})+a_{33}(a_{45}a_{52}-a_{42}a_{55})+a_{35}(a_{42}a_{53}-a_{43}a_{52})]+$$
$$+a_{11}a_{34}[a_{22}(a_{45}a_{53}-a_{43}a_{55})+a_{23}(a_{42}a_{55}-a_{45}a_{52})+a_{25}(a_{43}a_{52}-a_{42}a_{53})]+$$
$$+a_{11}a_{44}[a_{22}(a_{33}a_{55}-a_{35}a_{53})+a_{23}(a_{35}a_{52}-a_{32}a_{55})+a_{25}(a_{32}a_{53}-a_{33}a_{52})]+$$
$$+a_{11}a_{54}[a_{22}(a_{35}a_{43}-a_{33}a_{45})+a_{23}(a_{32}a_{45}-a_{35}a_{42})+a_{25}(a_{33}a_{42}-a_{32}a_{43})]+$$

$$+a_{21}a_{14}[a_{32}(a_{45}a_{53}-a_{43}a_{55})+a_{33}(a_{42}a_{55}-a_{45}a_{52})+a_{35}(a_{43}a_{52}-a_{42}a_{53})]+$$
$$+a_{21}a_{34}[a_{12}(a_{43}a_{55}-a_{45}a_{53})+a_{13}(a_{45}a_{52}-a_{42}a_{55})+a_{15}(a_{42}a_{53}-a_{43}a_{52})]+$$
$$+a_{21}a_{44}[a_{12}(a_{35}a_{53}-a_{33}a_{55})+a_{13}(a_{32}a_{55}-a_{35}a_{52})+a_{15}(a_{33}a_{52}-a_{32}a_{53})]+$$
$$+a_{21}a_{54}[a_{12}(a_{33}a_{45}-a_{35}a_{43})+a_{13}(a_{35}a_{42}-a_{32}a_{45})+a_{15}(a_{32}a_{43}-a_{33}a_{42})]+$$

$$+a_{31}a_{14}[a_{22}(a_{43}a_{55}-a_{45}a_{53})+a_{23}(a_{45}a_{52}-a_{42}a_{55})+a_{25}(a_{42}a_{53}-a_{43}a_{52})]+$$
$$+a_{31}a_{24}[a_{12}(a_{45}a_{53}-a_{43}a_{55})+a_{13}(a_{42}a_{55}-a_{45}a_{52})+a_{15}(a_{43}a_{52}-a_{42}a_{53})]+$$
$$+a_{31}a_{44}[a_{12}(a_{23}a_{55}-a_{25}a_{53})+a_{13}(a_{25}a_{52}-a_{22}a_{55})+a_{15}(a_{22}a_{53}-a_{23}a_{52})]+$$
$$+a_{31}a_{54}[a_{12}(a_{25}a_{43}-a_{23}a_{45})+a_{13}(a_{22}a_{45}-a_{25}a_{42})+a_{15}(a_{23}a_{42}-a_{22}a_{43})]+$$

121

$$+a_{41}a_{14}[a_{22}(a_{35}a_{53}-a_{33}a_{55})+a_{23}(a_{32}a_{55}-a_{35}a_{52})+a_{25}(a_{33}a_{52}-a_{32}a_{53})]+$$
$$+a_{41}a_{24}[a_{12}(a_{33}a_{55}-a_{35}a_{53})+a_{13}(a_{35}a_{52}-a_{32}a_{55})+a_{15}(a_{32}a_{53}-a_{33}a_{52})]+$$
$$+a_{41}a_{34}[a_{12}(a_{25}a_{53}-a_{23}a_{55})+a_{13}(a_{22}a_{55}-a_{25}a_{52})+a_{15}(a_{23}a_{52}-a_{22}a_{53})]+$$
$$+a_{41}a_{54}[a_{12}(a_{23}a_{35}-a_{25}a_{33})+a_{13}(a_{25}a_{32}-a_{22}a_{35})+a_{15}(a_{22}a_{33}-a_{23}a_{32})]+$$

$$+a_{51}a_{14}[a_{22}(a_{33}a_{45}-a_{35}a_{43})+a_{23}(a_{35}a_{42}-a_{32}a_{45})+a_{25}(a_{32}a_{43}-a_{33}a_{42})]+$$
$$+a_{51}a_{24}[a_{12}(a_{35}a_{43}-a_{33}a_{45})+a_{13}(a_{32}a_{45}-a_{35}a_{42})+a_{15}(a_{33}a_{42}-a_{32}a_{43})]+$$
$$+a_{51}a_{34}[a_{12}(a_{23}a_{45}-a_{25}a_{43})+a_{13}(a_{25}a_{42}-a_{22}a_{45})+a_{15}(a_{22}a_{43}-a_{23}a_{42})]+$$
$$+a_{51}a_{44}[a_{12}(a_{25}a_{33}-a_{23}a_{35})+a_{13}(a_{22}a_{35}-a_{25}a_{32})+a_{15}(a_{23}a_{32}-a_{22}a_{33})] \qquad (98)$$

In Appendix 15 is shown evaluation of the determinant of the 5x5 matrix with using of Laplace's theorem for marked out jointly the first and fourth columns.

d) **For the first and fifth columns**

The value of determinant $|A|$ of the 5x5 matrix at conditions when marked out jointly two columns ,such as the first and fifth columns applicably to Laplace's theorem can be expressed by the following formula in general view:

$$|A| = -M_1A_1+M_2A_2-M_3A_3+M_4A_4-M_5A_5+M_6A_6-M_7A_7-M_8A_8+M_9A_9-M_{10}A_{10} \quad (*)$$

where,

M_1-minor and A_1-algebraic supplemental (cofactor) for marked out jointly the first and fifth columns and also the first and second rows;

M_2-minor and A_2—algebraic supplemental (cofactor) for marked out jointly the first and fifth columns and also the first and third rows;

M_3-minor and A_3-algebraic supplemental (cofactor) for marked out jointly the first and fifth columns and also the first and fourth rows;

M_4-minor and A_4-algebraic supplemental (cofactor) for marked out jointly the first and fifth columns and also the first and fifth rows;

M_5-minor and A_5-algebraic supplemental (cofactor) for marked out jointly the first and fifth columns and also the second and third rows;

M_6—minor and A_6-algebraic supplemental (cofactor) for marked out jointly the first and fifth columns and also the second and fourth rows;

M_7-minor and A_7-algebraic supplemental (cofactor) for marked out jointly the first and fifth columns and also the second and fifth rows;

M_8—minor and A_8-algebraic supplemental (cofactor) for marked out jointly the first and fifth columns and also the third and fourth rows;

M_9-minor and A_9-algebraic supplemental (cofactor) for marked out jointly the first and fifth columns and also the third and fifth rows;

M_{10}-minor and A_{10}-algebraic supplemental (cofactor) for marked out jointly the first and fifth columns and also the fourth and fifth rows.

In Figure 98 is shown the functional graph of all cascade first block(a_{11}) elements of the first column applicably to Laplace's theorem for positive (———) and negative (-------) areas. In Figure 99 is shown the functional graph of all cascade second block (a_{21}) elements of the first column applicably to Laplace's theorem (———) and negative (---------) areas. In Figure 100 is shown the functional graph for all cascade third block (a_{31}) elements of the first column applicably to Laplace's theorem for positive (———) and negative (------) areas. In Figure 101 is shown the functional graph of all cascade fourth block (a_{41}) elements of the first column applicably to Laplace's theorem for positive(———) and negative (------) areas. In Figure 102 is shown the functional graph of all cascade fifth block(a_{51}) elements of the first column applicably to Laplace's theorem for positive (———) and negative areas.

After of some calculations and transformations , the formula (*) has the following view for marked out jointly the first and fifth columns applicably to Laplace's theorem for the determinant of the 5x5 matrix in general view:

$$M_1 = (a_{11}a_{25} - a_{15}a_{21})$$

$$A_1 = (-1)^{(1+5)+(1+2)} \begin{vmatrix} a_{32} & a_{33} & a_{34} \\ a_{42} & a_{43} & a_{44} \\ a_{52} & a_{53} & a_{54} \end{vmatrix}$$

$$M_2 = (a_{11}a_{35} - a_{15}a_{31})$$

$$A_2 = (-1)^{(1+5)+(1+3)} \begin{vmatrix} a_{22} & a_{23} & a_{24} \\ a_{42} & a_{43} & a_{44} \\ a_{52} & a_{53} & a_{54} \end{vmatrix}$$

$$M_3 = (a_{11}a_{45} - a_{15}a_{41})$$

$$A_3 = (-1)^{(1+5)+(1+4)} \begin{vmatrix} a_{22} & a_{23} & a_{24} \\ a_{32} & a_{33} & a_{34} \\ a_{52} & a_{53} & a_{54} \end{vmatrix}$$

$$M_4 = (a_{11}a_{55} - a_{15}a_{51})$$

$$A_4 = (-1)^{(1+5)+(1+5)} \begin{vmatrix} a_{22} & a_{23} & a_{24} \\ a_{32} & a_{33} & a_{34} \\ a_{42} & a_{43} & a_{44} \end{vmatrix}$$

$$M_5 = (a_{21}a_{35} - a_{25}a_{31})$$

$$A_5 = (-1)^{(1+5)+(2+3)} \begin{vmatrix} a_{12} & a_{13} & a_{14} \\ a_{42} & a_{43} & a_{44} \\ a_{52} & a_{53} & a_{54} \end{vmatrix}$$

$$M_6 = (a_{21}a_{45} - a_{25}a_{41})$$

$$A_6 = (-1)^{(1+5)+(2+4)} \begin{vmatrix} a_{12} & a_{13} & a_{14} \\ a_{32} & a_{33} & a_{34} \\ a_{52} & a_{53} & a_{54} \end{vmatrix}$$

$$M_7 = (a_{21}a_{55} - a_{25}a_{51})$$

$$A_7 = (-1)^{(1+5)+(2+5)} \begin{vmatrix} a_{12} & a_{13} & a_{14} \\ a_{32} & a_{33} & a_{34} \\ a_{42} & a_{43} & a_{44} \end{vmatrix}$$

$$M_8 = (a_{31}a_{45} - a_{35}a_{41})$$

$$A_8 = (-1)^{(1+5)+(3+4)} \begin{vmatrix} a_{12} & a_{13} & a_{14} \\ a_{22} & a_{23} & a_{24} \\ a_{52} & a_{53} & a_{54} \end{vmatrix}$$

$$M_9 = (a_{31}a_{55} - a_{35}a_{51})$$

$$A_9 = (-1)^{(1+5)+(3+5)} \begin{vmatrix} a_{12} & a_{13} & a_{14} \\ a_{22} & a_{23} & a_{24} \\ a_{42} & a_{43} & a_{44} \end{vmatrix}$$

$$M_{10} = (a_{41}a_{55} - a_{45}a_{51})$$

$$A_{10} = (-1)^{(1+5)+(4+5)} \begin{vmatrix} a_{12} & a_{13} & a_{14} \\ a_{22} & a_{23} & a_{24} \\ a_{32} & a_{33} & a_{34} \end{vmatrix}$$

$$|A| = a_{11}a_{25}[a_{32}(a_{44}a_{53}-a_{43}a_{54})+a_{33}(a_{42}a_{54}-a_{44}a_{52})+a_{34}(a_{43}a_{52}-a_{42}a_{53})]+$$
$$+a_{11}a_{35}[a_{22}(a_{43}a_{54}-a_{44}a_{53})+a_{23}(a_{44}a_{52}-a_{42}a_{54})+a_{24}(a_{42}a_{53}-a_{43}a_{52})]+$$
$$+a_{11}a_{45}[a_{22}(a_{34}a_{53}-a_{33}a_{54})+a_{23}(a_{32}a_{54}-a_{34}a_{52})+a_{24}(a_{33}a_{52}-a_{32}a_{53})]+$$
$$+a_{11}a_{55}[a_{22}(a_{33}a_{44}-a_{34}a_{43})+a_{23}(a_{34}a_{42}-a_{32}a_{44})+a_{24}(a_{32}a_{43}-a_{33}a_{42})]+$$

$$+a_{21}a_{15}[a_{32}(a_{43}a_{54}-a_{44}a_{53})+a_{33}(a_{44}a_{52}-a_{42}a_{54})+a_{34}(a_{42}a_{53}-a_{43}a_{52})]+$$
$$+a_{21}a_{35}[a_{12}(a_{44}a_{53}-a_{43}a_{54})+a_{13}(a_{42}a_{54}-a_{44}a_{52})+a_{14}(a_{43}a_{52}-a_{42}a_{53})]+$$
$$+a_{21}a_{45}[a_{12}(a_{33}a_{54}-a_{34}a_{53})+a_{13}(a_{34}a_{52}-a_{32}a_{54})+a_{14}(a_{32}a_{53}-a_{33}a_{52})]+$$
$$+a_{21}a_{55}[a_{12}(a_{34}a_{43}-a_{33}a_{44})+a_{13}(a_{32}a_{44}-a_{34}a_{42})+a_{14}(a_{33}a_{42}-a_{32}a_{43})]+$$

$$+a_{31}a_{15}[a_{22}(a_{44}a_{53}-a_{43}a_{54})+a_{23}(a_{42}a_{54}-a_{44}a_{52})+a_{24}(a_{43}a_{52}-a_{42}a_{53})]+$$
$$+a_{31}a_{25}[a_{12}(a_{43}a_{54}-a_{44}a_{53})+a_{13}(a_{44}a_{52}-a_{42}a_{54})+a_{14}(a_{42}a_{53}-a_{43}a_{52})]+$$
$$+a_{31}a_{45}[a_{12}(a_{24}a_{53}-a_{23}a_{54})+a_{13}(a_{22}a_{54}-a_{24}a_{52})+a_{14}(a_{23}a_{52}-a_{22}a_{53})]+$$
$$+a_{31}a_{55}[a_{12}(a_{23}a_{44}-a_{24}a_{43})+a_{13}(a_{24}a_{42}-a_{22}a_{44})+a_{14}(a_{22}a_{43}-a_{23}a_{42})]+$$

$$+a_{41}a_{15}[a_{22}(a_{33}a_{54}-a_{34}a_{53})+a_{23}(a_{34}a_{52}-a_{32}a_{54})+a_{24}(a_{32}a_{53}-a_{33}a_{52})]+$$
$$+a_{41}a_{25}[a_{12}(a_{34}a_{53}-a_{33}a_{54})+a_{13}(a_{32}a_{54}-a_{34}a_{52})+a_{14}(a_{33}a_{52}-a_{32}a_{53})]+$$
$$+a_{41}a_{35}[a_{12}(a_{23}a_{54}-a_{24}a_{53})+a_{13}(a_{24}a_{52}-a_{22}a_{54})+a_{14}(a_{22}a_{53}-a_{23}a_{52})]+$$
$$+a_{41}a_{55}[a_{12}(a_{24}a_{33}-a_{23}a_{34})+a_{13}(a_{22}a_{34}-a_{24}a_{32})+a_{14}(a_{23}a_{32}-a_{22}a_{33})]+$$

$$+a_{51}a_{15}[a_{22}(a_{34}a_{43}-a_{33}a_{44})+a_{23}(a_{32}a_{44}-a_{34}a_{42})+a_{24}(a_{33}a_{42}-a_{32}a_{43})]+$$
$$+a_{51}a_{25}[a_{12}(a_{33}a_{44}-a_{34}a_{43})+a_{13}(a_{34}a_{42}-a_{32}a_{44})+a_{14}(a_{32}a_{43}-a_{33}a_{42})]+$$
$$+a_{51}a_{35}[a_{12}(a_{24}a_{43}-a_{23}a_{44})+a_{13}(a_{22}a_{44}-a_{24}a_{42})+a_{14}(a_{23}a_{42}-a_{22}a_{43})]+$$
$$+a_{51}a_{45}[a_{12}(a_{23}a_{34}-a_{24}a_{33})+a_{13}(a_{24}a_{32}-a_{22}a_{34})+a_{14}(a_{22}a_{33}-a_{23}a_{32})] \quad (99)$$

Analysis of data shown in Figure 98 indicates on the following facts applicably to Laplace's theorem for marked out jointly the first and fifth columns of the 5x5 matrix *for all cascade first block (a_{11}):*

- Each element of the fifth column($a_{25},a_{35},a_{45},a_{55}$) has frequency of functional ties equal f=1 with element(a_{11}) of the first column;
- Each element (a_{22},a_{23},a_{24})of the second row, besides of element (a_{25}) has frequency of functional ties equal f=3 with elements of the fifth column;
- Each element(a_{32},a_{33},a_{34}) of the third row, besides of element (a_{35}) ,has frequency of functional ties equal f=5 with elements of the second row for positive and negative areas;
- Each element(a_{42},a_{43},a_{44})of the fourth row, besides of element (a_{45}) ,has frequency of functional ties equal f=6 with elements of the second and third rows for positive and negative areas;
- Each element (a_{52},a_{53},a_{54}) of the fifth row, besides of element (a_{55}) ,has frequency of functional ties equal f=6 with elements of the third and fourth rows for positive and negative areas.

So, in total we can conclude the following frequency of functional ties between elements of the 5x5 matrix for marked out jointly the first and fifth columns applicably to Laplace's theorem *for all cascade first block(a_{11}):*

f=1 for elements a_{35},a_{45},a_{55}
f=3 for elements a_{22},a_{23},a_{24}
f=5 for elements a_{32},a_{33},a_{34}
f=6 for elements a_{42},a_{43},a_{44} and a_{52},a_{53},a_{54}

Analysis of data shown in Figure 99 indicates on the following facts applicably to Laplace's theorem for marked out jointly the first and fifth columns of the 5x5 matrix *for all cascade second block(a_{21}):*

124

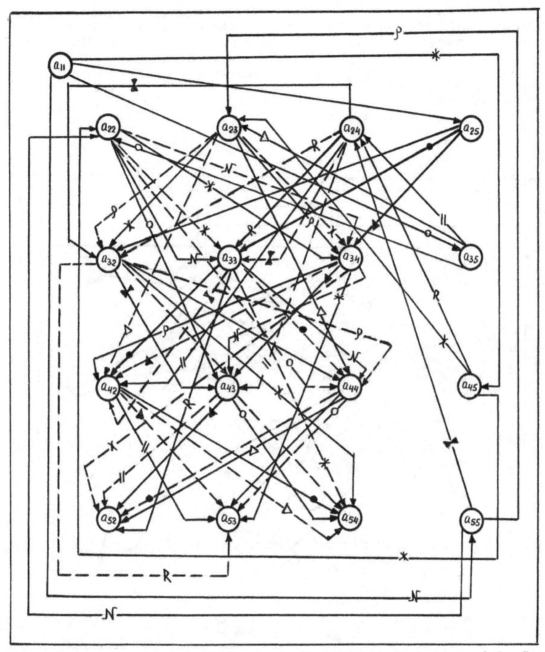

Figure 98 Functional graph of all cascade first block (a₁₁) elements of the first column applicably to Laplace's theorem for positive (——) and negative (------) areas

- Each element of the fifth column $(a_{15}, a_{35}, a_{45}, a_{55})$ has frequency of functional ties equal f=1 with the second element (a_{21}) of the first column;
- Each element (a_{12}, a_{13}, a_{14}) of the first row, besides of element (a_{15}) has frequency of functional ties equal f=3 with elements of the fifth column;
- Each element (a_{32}, a_{33}, a_{34}) of the third row, besides of element (a_{35}), has frequency of functional ties equal f=5 with elements of the first row for positive and negative areas;

- Each element (a_{42}, a_{43}, a_{44}) of the fourth row, besides of element(a_{45}),has frequency of functional ties equal f=6 with elements of the first and third rows for positive and negative areas;
- Each element (a_{52}, a_{53}, a_{54}) of the fifth row ,besides of element (a_{55}) ,has frequency of functional ties equal f=6 with elements of the third and fourth rows for positive and negative areas;

So, in total we can conclude the following frequency of functional ties between elements of the 5x5 matrix for marked out jointly elements of the first and fifth columns applicably to Laplace's theorem *for all cascade second block(a_{21}):*

f=1 for elements $a_{15}, a_{35}, a_{45}, a_{55}$

f=3 for elements a_{12}, a_{13}, a_{14}

f=5 for elements a_{32}, a_{33}, a_{34}

f=6 for elements a_{42}, a_{42}, a_{43} and a_{52}, a_{53}, a_{54}

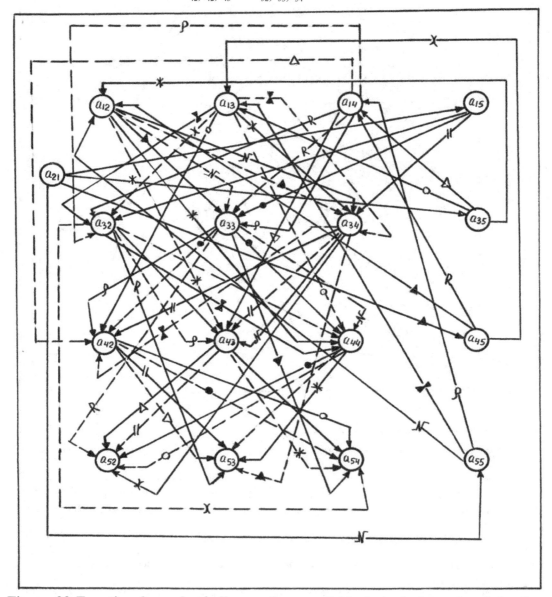

Figure 99 Functional graph of all cascade second block (a_{21}) elements of the first column applicably to Laplace's theorem for positive (———) and negative (-------) areas

126

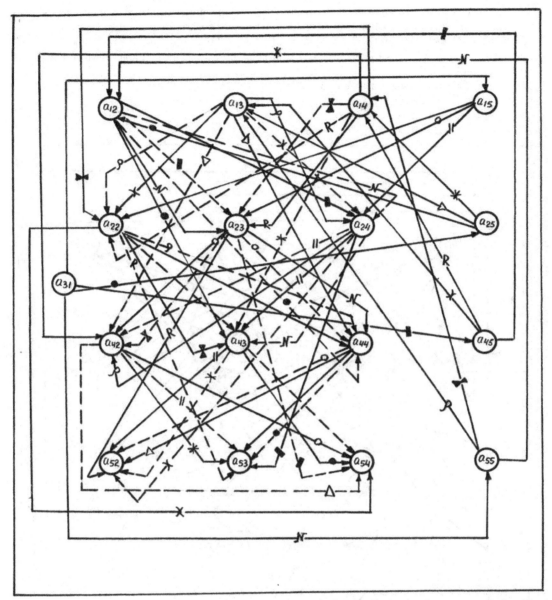

Figure 100 Functional graph for all cascade third block (a₃₁) elements of the first column applicably to Laplace's theorem for positive (————) and negative (-----) areas

Analysis of data shown in Figure 100 indicates on the following facts applicably to Laplace's theorem for marked out jointly the first and fifth columns of the 5x5 matrix for *all cascade third block(a₃₁):*

- Each element of the fifth column $(a_{15}, a_{25}, a_{45}, a_{55})$ has frequency of functional ties equal f=1 with the third element (a_{31}) of the first column;
- Each element (a_{12}, a_{13}, a_{14}) of the first row, besides of element (a_{15}), has frequency of functional ties equal f=3 with elements of the fifth column;
- Each element (a_{22}, a_{23}, a_{24}) of the second row, besides of element (a_{25}), has frequency of functional ties equal f=5 with elements of the first row for positive and negative areas;
- Each element (a_{42}, a_{43}, a_{44}) of the fourth row, besides of element (a_{45}), has frequency of functional ties equal f=6 with elements of the first and second rows for positive and negative areas;

127

- Each element (a_{52}, a_{53}, a_{54}) of the fifth row, besides of element(a_{55}) ,has frequency of functional ties equal f=6 with elements of the second and fourth rows for positive and negative areas.

So, in total we can conclude the following frequency of functional ties between elements of the 5x5 matrix for marked out jointly elements of the first and fifth columns applicably to Laplace's theorem for *all cascade third block(a_{31})*:

  f=1   for elements $a_{15}, a_{25}, a_{45}, a_{55}$
  f=3   for elements a_{12}, a_{13}, a_{14}
  f=5   for elements a_{22}, a_{23}, a_{24}
  f=6   for elements a_{42}, a_{43}, a_{44} and a_{52}, a_{53}, a_{54}

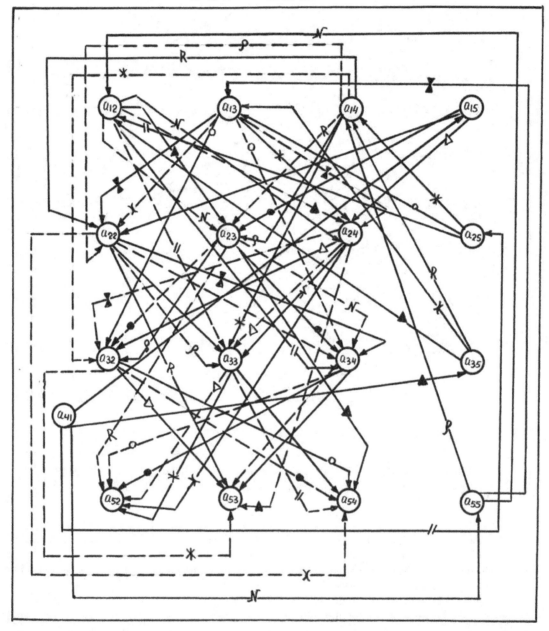

Figure 101 Functional graph of all cascade fourth block (a_{41}) elements of the first column applicably to Laplace's theorem for positive (——) and negative areas

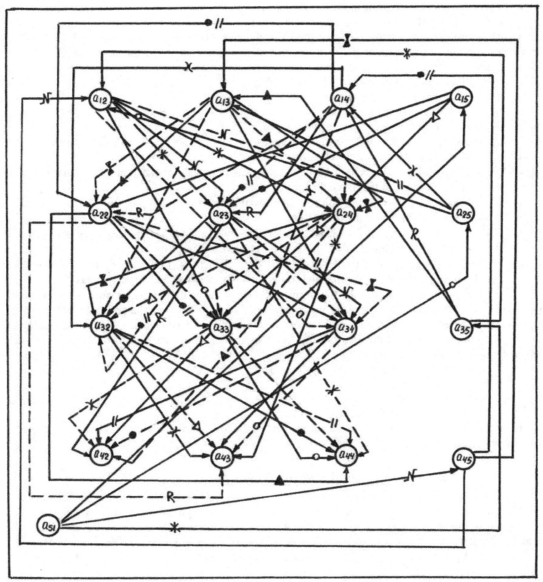

Figure 102 Functional graph of all cascade fifth block(a_{51}) elements of the first column applicably to Laplace's theorem for positive (———) and negative(-------) areas

Analysis of data shown in Figure 101 indicates on the following facts applicably to Laplace's theorem for marked out jointly the first and fifth columns of the 5x5 matrix *for all cascade fourth block(a_{41})*:

- Each element of the fifth column ($a_{15}, a_{25}, a_{35}, a_{55}$) has frequency of functional ties equal f=1 with the fourth element(a_{41}) of the first column;
- Each element (a_{12}, a_{13}, a_{14}) of the first row, besides of element (a_{15}),has frequency of functional ties equal f=3 with elements of the fifth column;
- Each element (a_{22}, a_{23}, a_{24}) of the second row, besides of element(a_{25}),has frequency of functional ties equal f=5 with elements of the first row for positive and negative areas;
- Each element(a_{32}, a_{33}, a_{34})of the third row, besides of element(a_{35}), has frequency of functional ties equal f=6 with elements of the first and second rows for positive and negative areas;

129

- Each element (a_{52}, a_{53}, a_{54}) of the fifth row ,besides of element(a_{55}),has frequency of functional ties equal f=6 with elements of the second and third rows for positive and negative areas.

So, in total we can conclude the following frequency of functional ties between elements of the 5x5 matrix for marked out jointly elements of the first and fifth columns applicably to Laplace's theorem *for all cascade fourth block(a_{41}):*

f=1 for elements $a_{15}, a_{25}, a_{35}, a_{55}$

f=3 for elements a_{12}, a_{13}, a_{14}

f=5 for elements a_{22}, a_{23}, a_{24}

f=6 for elements a_{32}, a_{33}, a_{34} and a_{52}, a_{53}, a_{54}

Analysis of data shown in Figure 102 indicates on the following facts applicably to Laplace's theorem for marked out jointly the first and fifth elements of the 5x5 matrix *for all cascade fifth block(a_{51}):*

- Each element of the fifth column ($a_{15}, a_{25}, a_{35}, a_{45}$) has frequency of functional ties equal f=1 with the fifth element (a_{51}) of the fifth column;
- Each element(a_{12}, a_{13}, a_{14}) of the fifth row ,besides of element (a_{15}) ,has frequency of functional ties equal f=3 with elements of the fifth column;
- Each element(a_{22}, a_{23}, a_{24})of the second row, besides of element(a_{25}) ,has frequency of functional ties equal f=5 with elements of the first row for positive and negative areas;
- Each element (a_{32}, a_{33}, a_{34}) of the third row, besides of element (a_{35}) has frequency of functional ties equal f=6 with elements of the first and second rows for positive and negative areas;
- Each element (a_{42}, a_{43}, a_{44}) of the fourth row, besides of element (a_{45}) ,has frequency of functional ties equal f=6 with elements of the second and third rows for positive and negative areas.

So, in total we can conclude the following frequency functional ties between elements of the 5x5 matrix for marked out jointly elements of the first and fifth columns applicably to Laplace's theorem *for all cascade fifth block (a_{51}):*

f=1 for elements $a_{15}, a_{25}, a_{35}, a_{45}$

f=3 for elements a_{12}, a_{13}, a_{14}

f=5 for elements a_{22}, a_{23}, a_{24}

f=6 for elements a_{32}, a_{33}, a_{34} and a_{42}, a_{43}, a_{44}

In Appendix 16 is shown evaluation of the determinant of the 5x5 matrix with using of Laplace's theorem for marked out jointly the first and fifth columns.

e) <u>For the second and third columns</u>

The value of determinant $|A|$ of the 5x5 matrix at conditions when marked out jointly two columns, such as the second and third columns applicably to Laplace's theorem can be expressed by the following formula in general view:

$$|A| = M_1A_1 - M_2A_2 + M_3A_3 - M_4A_4 + M_5A_5 - M_6A_6 + M_7A_7 - M_8A_8 - M_9A_9 + M_{10}A_{10} \quad (\bullet)$$

where,

M_1-minor and A_1-algebraic supplemental (cofactor) for marked out jointly the second and third columns and also the first and second rows;

M_2-minor and A_2-algebraic supplemental (cofactor) for marked out jointly the second and third columns and also the first and third rows;

M_3-minor and A_3-algebraic supplemental (cofactor) for marked out jointly the second and third columns and also the first and fourth rows;

M_4-minor and A_4-algebraic supplemental (cofactor) for marked out jointly the second and third columns and also the first and fifth rows;

M_5-minor and A_5-algebraic supplemental (cofactor) for marked out jointly the second and third columns and also the second and third rows;

M_6-minor and A_6-algebraic supplemental (cofactor) for marked out jointly the second and third columns and also the second and fourth rows;

M_7-minor and A_7- algebraic supplemental (cofactor) for marked out jointly the second and third columns and also the second and fifth rows;

M_8-minor and A_8-algebraic supplemental (cofactor) for marked out jointly the second and third columns and also the third and fourth rows;

130

M_9-minor and A_9-algebraic supplemental (cofactor) for marked out jointly the second and third columns and also the third and fifth rows;

M_{10}-minor and A_{10}-algebraic supplemental (cofactor) for marked out jointly the second and third columns and also the fourth and fifth rows.

$$M_1 = (a_{12}a_{23} - a_{13}a_{22}) \qquad A_1 = (-1)^{(2+3)+(1+2)} \begin{vmatrix} a_{31} & a_{34} & a_{35} \\ a_{41} & a_{44} & a_{45} \\ a_{51} & a_{54} & a_{55} \end{vmatrix}$$

$$M_2 = (a_{12}a_{33} - a_{13}a_{32}) \qquad A_2 = (-1)^{(2+3)+(1+3)} \begin{vmatrix} a_{21} & a_{24} & a_{25} \\ a_{41} & a_{44} & a_{45} \\ a_{51} & a_{54} & a_{55} \end{vmatrix}$$

$$M_3 = (a_{12}a_{43} - a_{13}a_{42}) \qquad A_3 = (-1)^{(2+3)+(1+4)} \begin{vmatrix} a_{21} & a_{24} & a_{25} \\ a_{31} & a_{34} & a_{35} \\ a_{51} & a_{54} & a_{55} \end{vmatrix}$$

$$M_4 = (a_{12}a_{53} - a_{13}a_{52}) \qquad A_4 = (-1)^{(2+3)+(1+5)} \begin{vmatrix} a_{21} & a_{24} & a_{25} \\ a_{31} & a_{34} & a_{35} \\ a_{41} & a_{44} & a_{45} \end{vmatrix}$$

$$M_5 = (a_{22}a_{33} - a_{23}a_{32}) \qquad A_5 = (-1)^{(2+3)+(2+3)} \begin{vmatrix} a_{11} & a_{14} & a_{15} \\ a_{41} & a_{44} & a_{45} \\ a_{51} & a_{54} & a_{55} \end{vmatrix}$$

$$M_6 = (a_{22}a_{43} - a_{23}a_{42}) \qquad A_6 = (-1)^{(2+3)+(2+4)} \begin{vmatrix} a_{11} & a_{14} & a_{15} \\ a_{31} & a_{34} & a_{35} \\ a_{51} & a_{54} & a_{55} \end{vmatrix}$$

$$M_7 = (a_{22}a_{53} - a_{23}a_{52}) \qquad A_7 = (-1)^{(2+3)+(2+5)} \begin{vmatrix} a_{11} & a_{14} & a_{15} \\ a_{31} & a_{34} & a_{35} \\ a_{41} & a_{44} & a_{45} \end{vmatrix}$$

$$M_8 = (a_{32}a_{43} - a_{33}a_{42}) \qquad A_8 = (-1)^{(2+3)+(3+4)} \begin{vmatrix} a_{11} & a_{14} & a_{15} \\ a_{21} & a_{24} & a_{25} \\ a_{51} & a_{54} & a_{55} \end{vmatrix}$$

$$M_9 = (a_{32}a_{53} - a_{33}a_{52}) \qquad A_9 = (-1)^{(2+3)+(3+5)} \begin{vmatrix} a_{11} & a_{14} & a_{15} \\ a_{21} & a_{24} & a_{25} \\ a_{41} & a_{44} & a_{45} \end{vmatrix}$$

$$M_{10} = (a_{42}a_{53} - a_{43}a_{52}) \qquad A_{10} = (-1)^{(2+3)+(4+5)} \begin{vmatrix} a_{11} & a_{14} & a_{15} \\ a_{21} & a_{24} & a_{25} \\ a_{31} & a_{34} & a_{35} \end{vmatrix}$$

In Figure 103 is shown the functional graph for all cascade first block(a_{12}) elements of the second column applicably to Laplace's theorem for positive (———) and negative (---------) areas .In Figure 104 is shown the functional graph for all cascade second block(a_{22}) elements of the second column applicably to Laplace's theorem for positive (———) and negative(-------) areas. In Figure 105 is shown the functional graph for all cascade third block (a_{32}) elements of the second column applicably to Laplace's theorem for positive(———) and negative(------) areas. In Figure 106 is shown the functional graph for all cascade fourth block (a_{42}) elements of the second column applicably to Laplace's theorem for positive (———) and negative(------) areas. In Figure 107 is shown the functional graph for all cascade fifth block (a_{52}) of the second column applicably to Laplace's theorem for positive (———) and negative (------) areas.

After of some calculations and transformations the formula (●) has the following view *for marked out jointly the second and third columns* applicably to Laplace's theorem for the determinant of the 5x5 matrix in general view:

$$|A| = a_{12}a_{23}[a_{31}(a_{44}a_{55}-a_{45}a_{54})+a_{34}(a_{45}a_{51}-a_{41}a_{55})+a_{35}(a_{41}a_{54}-a_{44}a_{51})]+$$
$$+a_{12}a_{33}[a_{21}(a_{45}a_{54}-a_{44}a_{55})+a_{24}(a_{41}a_{55}-a_{45}a_{51})+a_{25}(a_{44}a_{51}-a_{41}a_{54})]+$$
$$+a_{12}a_{43}[a_{21}(a_{34}a_{55}-a_{35}a_{54})+a_{24}(a_{35}a_{51}-a_{31}a_{55})+a_{25}(a_{31}a_{54}-a_{34}a_{51})]+$$
$$+a_{12}a_{53}[a_{21}(a_{35}a_{44}-a_{34}a_{45})+a_{24}(a_{31}a_{45}-a_{35}a_{41})+a_{25}(a_{34}a_{41}-a_{31}a_{44})]+$$

$$+a_{22}a_{13}[a_{31}(a_{45}a_{54}-a_{44}a_{55})+a_{34}(a_{41}a_{55}-a_{45}a_{51})+a_{35}(a_{44}a_{51}-a_{41}a_{54})]+$$
$$+a_{22}a_{33}[a_{11}(a_{44}a_{55}-a_{45}a_{54})+a_{14}(a_{45}a_{51}-a_{41}a_{55})+a_{15}(a_{41}a_{54}-a_{44}a_{51})]+$$
$$+a_{22}a_{43}[a_{11}(a_{35}a_{54}-a_{34}a_{55})+a_{14}(a_{31}a_{55}-a_{35}a_{51})+a_{15}(a_{34}a_{51}-a_{31}a_{54})]+$$
$$+a_{22}a_{53}[a_{11}(a_{34}a_{45}-a_{35}a_{44})+a_{14}(a_{35}a_{41}-a_{31}a_{45})+a_{15}(a_{31}a_{44}-a_{34}a_{41})]+$$

$$+a_{32}a_{13}[a_{21}(a_{44}a_{55}-a_{45}a_{54})+a_{24}(a_{45}a_{51}-a_{41}a_{55})+a_{25}(a_{41}a_{54}-a_{44}a_{51})]+$$
$$+a_{32}a_{23}[a_{11}(a_{45}a_{54}-a_{44}a_{55})+a_{14}(a_{41}a_{55}-a_{45}a_{51})+a_{15}(a_{44}a_{51}-a_{41}a_{54})]+$$
$$+a_{32}a_{43}[a_{11}(a_{24}a_{55}-a_{25}a_{54})+a_{14}(a_{25}a_{51}-a_{21}a_{55})+a_{15}(a_{21}a_{54}-a_{24}a_{51})]+$$
$$+a_{32}a_{53}[a_{11}(a_{25}a_{44}-a_{24}a_{45})+a_{14}(a_{21}a_{45}-a_{25}a_{41})+a_{15}(a_{24}a_{41}-a_{21}a_{44})]+$$

$$+a_{42}a_{13}[a_{21}(a_{35}a_{54}-a_{34}a_{55})+a_{24}(a_{31}a_{55}-a_{35}a_{51})+a_{25}(a_{34}a_{51}-a_{31}a_{54})]+$$
$$+a_{42}a_{23}[a_{11}(a_{34}a_{55}-a_{35}a_{54})+a_{14}(a_{35}a_{51}-a_{31}a_{55})+a_{15}(a_{31}a_{54}-a_{34}a_{51})]+$$
$$+a_{42}a_{33}[a_{11}(a_{25}a_{54}-a_{24}a_{55})+a_{14}(a_{21}a_{55}-a_{25}a_{51})+a_{15}(a_{24}a_{51}-a_{21}a_{54})]+$$
$$+a_{42}a_{53}[a_{11}(a_{24}a_{35}-a_{25}a_{34})+a_{14}(a_{25}a_{31}-a_{21}a_{35})+a_{15}(a_{21}a_{34}-a_{24}a_{31})]+$$

$$+a_{52}a_{13}[a_{21}(a_{34}a_{45}-a_{35}a_{44})+a_{24}(a_{35}a_{41}-a_{31}a_{45})+a_{25}(a_{31}a_{44}-a_{34}a_{41})]+$$
$$+a_{52}a_{23}[a_{11}(a_{35}a_{44}-a_{34}a_{45})+a_{14}(a_{31}a_{45}-a_{35}a_{41})+a_{15}(a_{34}a_{41}-a_{31}a_{44})]+$$
$$+a_{52}a_{33}[a_{11}(a_{24}a_{45}-a_{25}a_{44})+a_{14}(a_{25}a_{41}-a_{21}a_{45})+a_{15}(a_{21}a_{44}-a_{24}a_{41})]+$$
$$+a_{52}a_{43}[a_{11}(a_{25}a_{34}-a_{24}a_{35})+a_{14}(a_{21}a_{35}-a_{25}a_{31})+a_{15}(a_{24}a_{31}-a_{21}a_{34})]\qquad (100)$$

Analysis of data shown in Figure 103 indicates on the following facts applicably to Laplace's theorem for marked out jointly the second and third columns of the 5x5 matrix *for all cascade first block (a_{12})* elements of the second column:

- Each element of the third column($a_{23},a_{33},a_{43},a_{53}$) has frequency of functional ties equal f=1 with element (a_{12}) of the second column;
- Each element(a_{21},a_{24},a_{25}) of the second row, besides of element (a_{23}), has frequency of functional ties equal f=3 with elements of the third column;
- Each element (a_{31},a_{34},a_{35}) of the third row, besides of element (a_{32}) ,has frequency of functional ties equal f=5 with elements of the second row for positive and negative areas;
- Each element (a_{41},a_{44},a_{45}) of the fourth row ,besides of element (a_{43}) ,has frequency of functional ties equal f=6 with elements of the second and third rows for positive and negative areas;

- Each element (a_{51}, a_{54}, a_{55}) of the fifth row, besides of element (a_{53}), has frequency of functional ties equal f=6 with elements of the third and fourth rows for positive and negative areas.

So, in total we can conclude the following frequency functional ties between elements of the 5x5 matrix for marked out jointly elements of the second and third columns applicably to Laplace's theorem *for all cascade first block (a_{12}) elements of the second column:*

$$f=1 \text{ for elements } (a_{23}, a_{33}, a_{43}, a_{53})$$
$$f=3 \text{ for elements } (a_{21}, a_{24}, a_{25})$$
$$f=5 \text{ for elements } (a_{31}, a_{34}, a_{35})$$
$$f=6 \text{ for elements } (a_{41}, a_{44}, a_{45}) \text{ and } (a_{51}, a_{54}, a_{55}).$$

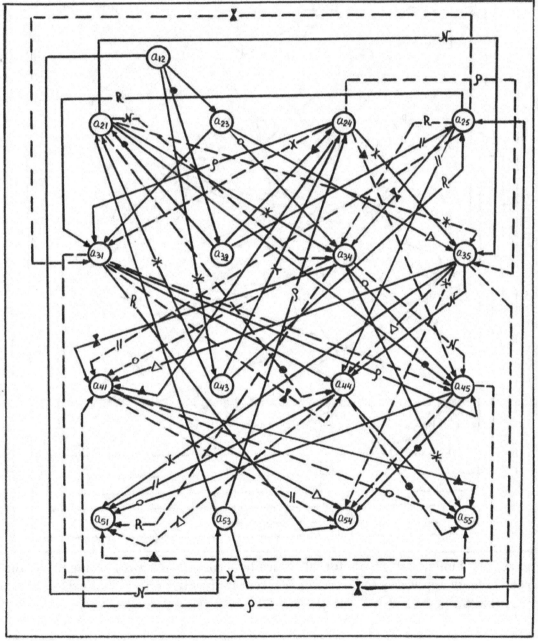

Figure 103 Functional graph of all cascade first block (a_{12}) elements of the second column applicably to Laplace's theorem for positive (———) and negative(— — —) areas

133

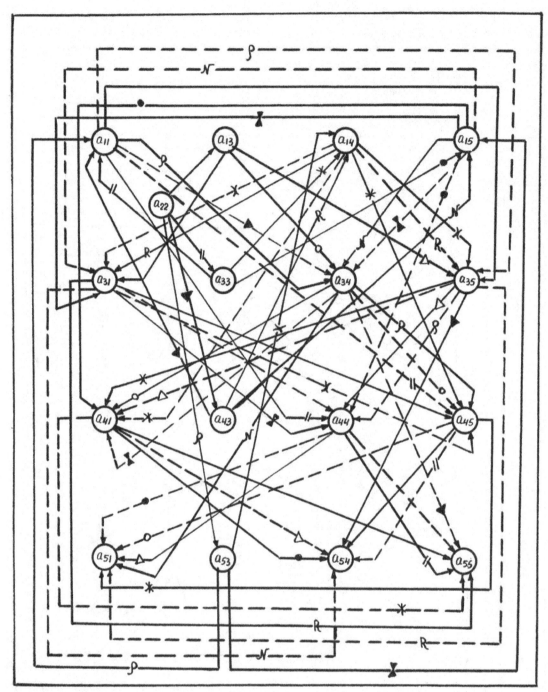

Figure 104 Functional graph for all cascade second block (a₂₂) elements of the second column applicably to Laplace's theorem for positive (———) and negative (— — —) areas

Analysis of data shown in Figure 104 indicates on the following facts applicably to Laplace's theorem for marked out jointly the second and third columns of the 5x5 matrix *for all cascade second block (a₂₂) elements of the second column:*

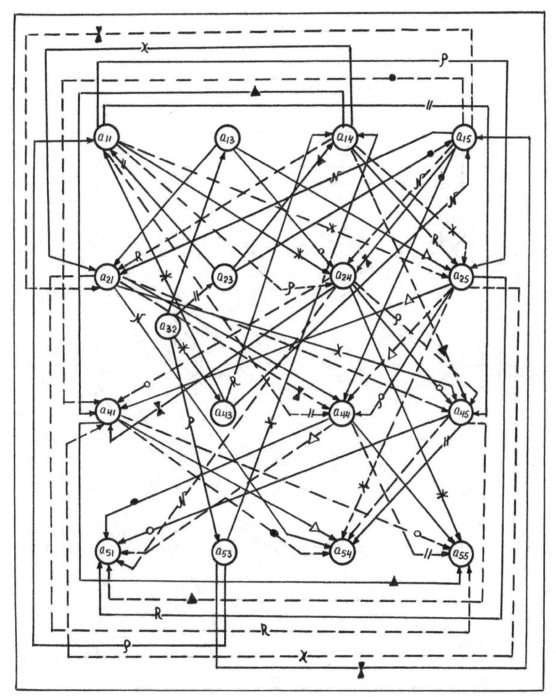

Figure 105 Functional graph for all cascade third block (a_{32}) elements of the second column applicably to Laplace's theorem for positive (———) and negative (— — —) areas

- Each element of the third column($a_{13}, a_{33}, a_{43}, a_{53}$) has frequency of functional ties equal f=1 with element(a_{22}) of the second column;
- Each element (a_{11}, a_{14}, a_{15}) of the first row ,besides of element(a_{13}), has frequency of functional ties equal f=3 with elements of the third column;

135

- Each element (a_{31}, a_{34}, a_{35}) of the third row, besides of element (a_{33}) ,has frequency of functional ties equal f=5 with elements of the first row for positive and negative areas;
- Each element (a_{41}, a_{44}, a_{45}) of the fourth row, besides of element (a_{43}),has frequency of functional ties equal f=6 with elements of the first and third rows for positive and negative areas;
- Each element (a_{51}, a_{54}, a_{55}) of the fifth row, besides of element (a_{53}) ,has frequency of functional ties equal f=6 with elements of the third and fourth rows for positive and negative areas.

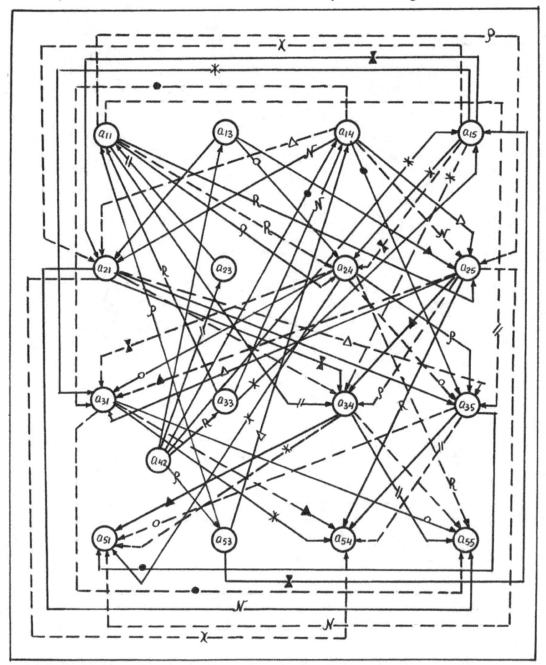

Figure 106 Functional graph for all cascade fourth block (a_{42}) elements of the second column applicably to Laplace's theorem for positive (——) and negative (— — —)areas

136

So, in total we can conclude the following frequency functional ties between elements of the 5x5 matrix for marked out jointly elements of the second and third columns applicably to Laplace's theorem *for all cascade second block(a_{22}) elements of the second column:*

f=1 for elements ($a_{13}, a_{33}, a_{43}, a_{53}$)

f=3 for elements (a_{11}, a_{14}, a_{15})

f=5 for elements (a_{31}, a_{34}, a_{35})

f=6 for elements (a_{41}, a_{44}, a_{45}) and (a_{51}, a_{54}, a_{55})

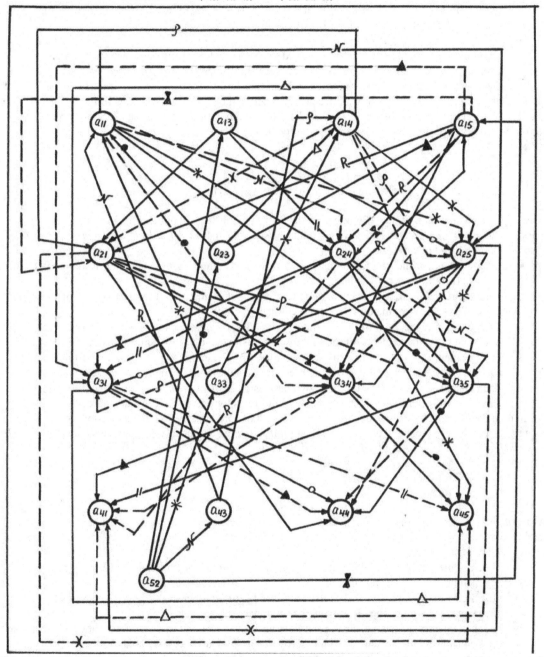

Figure 107 Functional graph for all cascade fifth block(a_{52}) elements of the second column applicably to Laplace's theorem for positive (———) and negative (— — —) areas

137

Analysis of data shown in Figure 105 indicates on the following facts applicably to Laplace's theorem for marked out jointly the second and third columns of the 5x 5 matrix *for all cascade third block (a_{32}) elements of the second column:*

- Each element of the third column ($a_{13}, a_{23}, a_{43}, a_{53}$) has frequency of functional ties equal f=1 with element (a_{32}) of the second column;
- Each element (a_{11}, a_{14}, a_{15}) of the first row, besides of element (a_{13}) ,has frequency of functional ties equal f=3 with elements of the third column;
- Each element (a_{21}, a_{24}, a_{25}) of the second row, besides of element (a_{23}), has frequency of functional ties equal f=5 with elements of the first row for positive and negative areas;
- Each element(a_{41}, a_{44}, a_{45}) of the fourth row ,besides of element (a_{43}) ,has frequency of functional ties equal f=6 with elements of the first and second rows for positive and negative areas;
- Each element (a_{51}, a_{54}, a_{55}) of the fifth row , besides of element (a_{53}), has frequency of functional ties equal f=6 with elements of the second and fourth rows for positive and negative areas.

So, in total we can conclude the following frequency functional ties between elements of the 5x5 matrix for marked out jointly elements of the second and third columns applicably to Lapalce's theorem *for all cascade third block (a_{32}) elements of the second column:*

f=1 for elements ($a_{13}, a_{23}, a_{43}, a_{53}$)
f=3 for elements (a_{11}, a_{14}, a_{15})
f=5 for elements (a_{21}, a_{24}, a_{25})
f=6 for elements (a_{41}, a_{44}, a_{45}) and (a_{51}, a_{54}, a_{55}).

Analysis of data shown in Figure 106 indicates on the following facts applicably to Laplace's theorem for marked out jointly the second and third columns of the 5x5 matrix *for all cascade fourth block (a_{42}) elements of the second column:*

- Each element of the third column ($a_{13}, a_{23}, a_{33}, a_{53}$)has frequency of functional ties equal f=1 with element (a_{42}) of the second column;
- Each element (a_{11}, a_{14}, a_{15}) of the first row, besides of element (a_{13}) ,has frequency of functional ties equal f=3 with elements of the third column;
- Each element (a_{21}, a_{24}, a_{25}) of the second row, besides of element(a_{23}) ,has frequency of functional ties equal f=5 with elements of the first row for positive and negative areas;
- Each element (a_{31}, a_{34}, a_{35}) of the third row ,besides of element(a_{33}),has frequency of functional ties equal f=6 with elements of the first and second rows for positive and negative areas;
- Each element (a_{51}, a_{54}, a_{55}) of the fifth row, besides of element (a_{53}), has frequency of functional ties equal f=6 with elements of the second and third rows for positive and negative area.

So, in total we can conclude the following frequency of functional ties between elements of the 5x5 matrix for marked out jointly elements of the second and third columns applicably to Laplace's theorem *for all cascade fourth block(a_{42}) elements of the second column:*

f=1 for elements ($a_{13}, a_{23}, a_{33}, a_{53}$)
f=3 for elements (a_{11}, a_{14}, a_{15})
f=5 for elements (a_{21}, a_{24}, a_{25})
f=6 for elements (a_{31}, a_{34}, a_{35}) and (a_{51}, a_{54}, a_{55})

Analysis of data shown in Figure 107 indicates on the following facts applicably to Laplace's theorem for marked out jointly the second and third columns of the 5x5 matrix *for all cascade fifth block (a_{52}) elements of the second column:*

- Each element of the third column ($a_{13}, a_{23}, a_{33}, a_{43}$) has frequency of functional ties equal f=1 with element(a_{52}) of the second column;
- Each element (a_{11}, a_{14}, a_{15}) of the first row, besides of element(a_{13}) ,has frequency of functional ties equal f=3 with elements of the third column;
- Each element(a_{21}, a_{24}, a_{25})of the second row, besides of element (a_{23}),has frequency of functional ties equal f=5 with elements of the first row for positive and negative areas;
- Each element(a_{31}, a_{34}, a_{35}) of the third row ,besides of element(a_{33}),has frequency of functional ties equal f=6 with elements of the first and second rows for positive and negative areas;
- Each element (a_{41}, a_{44}, a_{45}) of the fourth row, besides of element (a_{43}), has frequency of functional ties equal f=6 with elements of the second and third rows for positive and negative areas.

So, in total we can conclude the following frequency of functional ties between elements of the 5x5 matrix for marked out jointly elements of the second and third columns applicably to Laplace's theorem *for all cascade fifth block (a_{52}) elements of the second column:*

$f=1$ for elements ($a_{13},a_{23},a_{33},a_{43}$)

$f=3$ for elements (a_{11},a_{14},a_{15})

$f=5$ for elements (a_{21},a_{24},a_{25})

$f=6$ for elements (a_{31},a_{34},a_{35}) and (a_{41},a_{44},a_{45}).

In Appendix 17 is shown evaluation of the determinant of the 5x5 matrix with using of Laplace's theorem for marked out jointly the second and third columns.

f)   <u>For the second and fourth columns</u>

The value of determinant $|A|$ of the 5x5 matrix at conditions when marked out jointly two columns, such as the second and fourth columns applicably to Laplace's theorem ,can be expressed by the following formula in general view:

$$|A| = -M_1A_1+M_2A_2-M_3A_3+M_4A_4-M_5A_5+M_6A_6-M_7A_7-M_8A_8+M_9A_9-M_{10}A_{10} \text{ (x)}$$

where,

M_1-minor and A_1-algebraic supplemental (cofactor) for marked out jointly the second and fourth columns and also the first and second rows;

M_2-minor and A_2-algebraic supplemental (cofactor) for marked out jointly the second and fourth columns and also the first and third rows;

M_3-minor and A_3-algebraic supplemental (cofactor) for marked out jointly the second and fourth columns and also the first and fourth rows;

M_4-minor and A_4-algebraic supplemental (cofactor) for marked out jointly the second and fourth columns and also the first and fifth rows;

M_5-minor and A_5-algebraic supplemental (cofactor) for marked out jointly the second and fourth columns and also the second and third rows;

M_6-minor and A_6-algebraic supplemental (cofactor) for marked out jointly the second and fourth columns and also the second and fourth rows;

M_7-minor and A_7-algebraic supplemental (cofactor) for marked out jointly the second and fourth columns and also the second and fifth rows;

M_8-minor and A_8-algebraic supplemental (cofactor) for marked out jointly the second and fourth columns and also the third and fourth rows;

M_9-minor and A_9-algebraic supplemental (cofactor) for marked out jointly the second and fourth columns and also the third and fifth rows;

M_{10}-minor and A_{10}- algebraic supplemental (cofactor) for marked out jointly the second and fourth columns and also the fourth and fifth rows.

After of some calculations and transformations the formula(x) has the following view *for marked out jointly the second and fourth columns* applicably to Laplace's theorem for the determinant of the 5x5 matrix in general view:

$$|A| = a_{12}a_{24}[a_{31}(a_{45}a_{53}-a_{43}a_{55})+a_{33}(a_{41}a_{55}-a_{45}a_{51})+a_{35}(a_{43}a_{51}-a_{41}a_{53})]+$$
$$+a_{12}a_{34}[a_{21}(a_{43}a_{55}-a_{45}a_{53})+a_{23}(a_{45}a_{51}-a_{41}a_{55})+a_{25}(a_{41}a_{53}-a_{43}a_{51})]+$$
$$+a_{12}a_{44}[a_{21}(a_{35}a_{53}-a_{33}a_{55})+a_{23}(a_{31}a_{55}-a_{35}a_{51})+a_{25}(a_{33}a_{51}-a_{31}a_{53})]+$$
$$+a_{12}a_{54}[a_{21}(a_{33}a_{45}-a_{35}a_{43})+a_{23}(a_{35}a_{41}-a_{31}a_{45})+a_{25}(a_{31}a_{43}-a_{33}a_{41})]+$$

$$+a_{22}a_{14}[a_{31}(a_{43}a_{55}-a_{45}a_{53})+a_{33}(a_{45}a_{51}-a_{41}a_{55})+a_{35}(a_{41}a_{53}-a_{43}a_{51})]+$$
$$+a_{22}a_{34}[a_{11}(a_{45}a_{53}-a_{43}a_{55})+a_{13}(a_{41}a_{55}-a_{45}a_{51})+a_{15}(a_{43}a_{51}-a_{41}a_{53})]+$$
$$+a_{22}a_{44}[a_{11}(a_{33}a_{55}-a_{35}a_{53})+a_{13}(a_{35}a_{51}-a_{31}a_{55})+a_{15}(a_{31}a_{53}-a_{33}a_{51})]+$$
$$+a_{22}a_{54}[a_{11}(a_{35}a_{43}-a_{33}a_{45})+a_{13}(a_{31}a_{45}-a_{35}a_{41})+a_{15}(a_{33}a_{41}-a_{31}a_{43})]+$$

$+a_{32}a_{14}[a_{21}(a_{45}a_{53}-a_{43}a_{55})+a_{23}(a_{41}a_{55}-a_{45}a_{51})+a_{25}(a_{43}a_{51}-a_{41}a_{53})]+$
$+a_{32}a_{24}[a_{11}(a_{43}a_{55}-a_{45}a_{53})+a_{13}(a_{45}a_{51}-a_{41}a_{55})+a_{15}(a_{41}a_{53}-a_{43}a_{51})]+$
$+a_{32}a_{44}[a_{11}(a_{25}a_{53}-a_{23}a_{55})+a_{13}(a_{21}a_{55}-a_{25}a_{51})+a_{15}(a_{23}a_{51}-a_{21}a_{53})]+$
$+a_{32}a_{54}[a_{11}(a_{23}a_{45}-a_{25}a_{43})+a_{13}(a_{25}a_{41}-a_{21}a_{45})+a_{15}(a_{21}a_{43}-a_{23}a_{41})]+$

$+a_{42}a_{14}[a_{21}(a_{33}a_{55}-a_{35}a_{53})+a_{23}(a_{35}a_{51}-a_{31}a_{55})+a_{25}(a_{31}a_{53}-a_{33}a_{51})]+$
$+a_{42}a_{24}[a_{11}(a_{35}a_{53}-a_{33}a_{55})+a_{13}(a_{31}a_{55}-a_{35}a_{51})+a_{15}(a_{33}a_{51}-a_{31}a_{53})]+$
$+a_{42}a_{34}[a_{11}(a_{23}a_{55}-a_{25}a_{53})+a_{13}(a_{25}a_{51}-a_{21}a_{55})+a_{15}(a_{21}a_{53}-a_{23}a_{51})]+$
$+a_{42}a_{54}[a_{11}(a_{25}a_{33}-a_{23}a_{35})+a_{13}(a_{21}a_{35}-a_{25}a_{31})+a_{15}(a_{23}a_{31}-a_{21}a_{33})]+$

$+a_{52}a_{14}[a_{21}(a_{35}a_{43}-a_{33}a_{45})+a_{23}(a_{31}a_{45}-a_{35}a_{41})+a_{25}(a_{33}a_{41}-a_{31}a_{43})]+$
$+a_{52}a_{24}[a_{11}(a_{33}a_{45}-a_{35}a_{43})+a_{13}(a_{35}a_{41}-a_{31}a_{45})+a_{15}(a_{31}a_{43}-a_{33}a_{41})]+$
$+a_{52}a_{34}[a_{11}(a_{25}a_{43}-a_{23}a_{45})+a_{13}(a_{21}a_{45}-a_{25}a_{41})+a_{15}(a_{23}a_{41}-a_{21}a_{43})]+$
$+a_{52}a_{44}[a_{11}(a_{23}a_{35}-a_{25}a_{33})+a_{13}(a_{25}a_{31}-a_{21}a_{35})+a_{15}(a_{21}a_{33}-a_{23}a_{31})]$ (101)

In Figure 108 is shown the functional graph for all cascade first block (a_{12}) elements of the second column applicably to Laplace's theorem for positive (———) and negative (-------) areas. In Figure 109 is shown the functional graph for all cascade second block (a_{22}) elements of the second column applicably to Laplace's theorem for positive (———) and negative (-------) areas. In Figure 110 is shown the functional graph for all cascade third block(a_{32}) elements of the second column applicably to Laplace's theorem for positive(———) and negative (-------) areas. In Figure 111 is shown the functional graph for all cascade fourth block(a_{42}) elements of the second column applicably to Laplace's theorem for positive (———) and negative(------) areas. In Figure 112 is shown the functional graph for all cascade fifth block (a_{52}) elements of the second column applicably to Laplace's theorem for positive (———) and negative areas.

Analysis of data shown in Figure 108 indicates on the following facts applicably to Laplace's theorem for marked out jointly the second and fourth columns of the 5x5 matrix *for all cascade first block (a_{12}) elements of the second column:*

- Each element of the fourth column ($a_{24},a_{34},a_{44},a_{54}$)has frequency of functional ties equal f=1 with element (a_{12}) of the second column;
- Each element (a_{21},a_{23},a_{25}) of the second row, besides of element (a_{24}),has frequency of functional ties equal f=3 with elements of the fourth column;
- Each element (a_{31},a_{33},a_{35}) of the third row, besides of element(a_{34}) ,has frequency of functional ties equal f=5 with elements of the second row for positive and negative areas;
- Each element(a_{41},a_{43},a_{45}) of the fourth row, besides of element (a_{44}), has frequency of functional ties equal f=6 with elements of the second and third rows for positive and negative areas;
- Each element (a_{51},a_{53},a_{55}) of the fifth row, besides of element (a_{54}) ,has frequency of functional ties equal f=6 with elements of the third and fourth row for positive and negative areas.

So, in total we can conclude the following frequency of functional ties between elements of the 5x5 matrix for marked out jointly elements of the second and fourth columns applicably to Laplace's theorem *for all cascade first block (a_{12}) elements of the second column:*

         f=1 for elements ($a_{24},a_{34},a_{44},a_{54}$)
         f=3 for elements (a_{21},a_{23},a_{25})
         f=5 for elements (a_{31},a_{33},a_{35})
         f=6 for elements (a_{41},a_{43},a_{45}) and (a_{51},a_{53},a_{55}).

Analysis of data shown in Figure 109 indicates on the following facts applicably to Laplace's theorem for marked out joint the second and fourth columns of the 5x5 matrix *for all cascade second block(a_{22}) elements of the second column:*

$$M_1=(a_{12}a_{24}-a_{14}a_{22})$$

$$A_1=(-1)^{(2+4)+(1+2)}\begin{vmatrix} a_{31} & a_{33} & a_{35} \\ a_{41} & a_{43} & a_{45} \\ a_{51} & a_{53} & a_{55} \end{vmatrix}$$

$$M_2=(a_{12}a_{34}-a_{14}a_{32})$$

$$A_2=(-1)^{(2+4)+(1+3)}\begin{vmatrix} a_{21} & a_{23} & a_{25} \\ a_{41} & a_{43} & a_{45} \\ a_{51} & a_{53} & a_{55} \end{vmatrix}$$

$$M_3=(a_{12}a_{44}-a_{14}a_{42})$$

$$A_3=(-1)^{(2+4)+(1+4)}\begin{vmatrix} a_{21} & a_{23} & a_{25} \\ a_{31} & a_{33} & a_{35} \\ a_{51} & a_{53} & a_{55} \end{vmatrix}$$

$$M_4=(a_{12}a_{54}-a_{14}a_{52})$$

$$A_4=(-1)^{(2+4)+(1+5)}\begin{vmatrix} a_{21} & a_{23} & a_{25} \\ a_{31} & a_{33} & a_{35} \\ a_{41} & a_{43} & a_{45} \end{vmatrix}$$

$$M_5=(a_{22}a_{34}-a_{24}a_{32})$$

$$A_5=(-1)^{(2+4)+(2+3)}\begin{vmatrix} a_{11} & a_{13} & a_{15} \\ a_{41} & a_{43} & a_{45} \\ a_{51} & a_{53} & a_{55} \end{vmatrix}$$

$$M_6=(a_{22}a_{44}-a_{24}a_{42})$$

$$A_6=(-1)^{(2+4)+(2+4)}\begin{vmatrix} a_{11} & a_{13} & a_{15} \\ a_{31} & a_{33} & a_{35} \\ a_{51} & a_{53} & a_{55} \end{vmatrix}$$

$$M_7=(a_{22}a_{54}-a_{24}a_{52})$$

$$A_7=(-1)^{(2+4)+(2+5)}\begin{vmatrix} a_{11} & a_{13} & a_{15} \\ a_{31} & a_{33} & a_{35} \\ a_{41} & a_{43} & a_{45} \end{vmatrix}$$

$$M_8=(a_{32}a_{44}-a_{34}a_{42})$$

$$A_8=(-1)^{(2+4)+(3+4)}\begin{vmatrix} a_{11} & a_{13} & a_{15} \\ a_{21} & a_{23} & a_{25} \\ a_{51} & a_{53} & a_{55} \end{vmatrix}$$

$$M_9=(a_{32}a_{54}-a_{34}a_{52})$$

$$A_9=(-1)^{(2+4)+(3+5)}\begin{vmatrix} a_{11} & a_{13} & a_{15} \\ a_{21} & a_{23} & a_{25} \\ a_{41} & a_{43} & a_{45} \end{vmatrix}$$

$$M_{10}=(a_{42}a_{54}-a_{44}a_{52})$$

$$A_{10}=(-1)^{(2+4)+(4+5)}\begin{vmatrix} a_{11} & a_{13} & a_{15} \\ a_{21} & a_{23} & a_{25} \\ a_{31} & a_{33} & a_{35} \end{vmatrix}$$

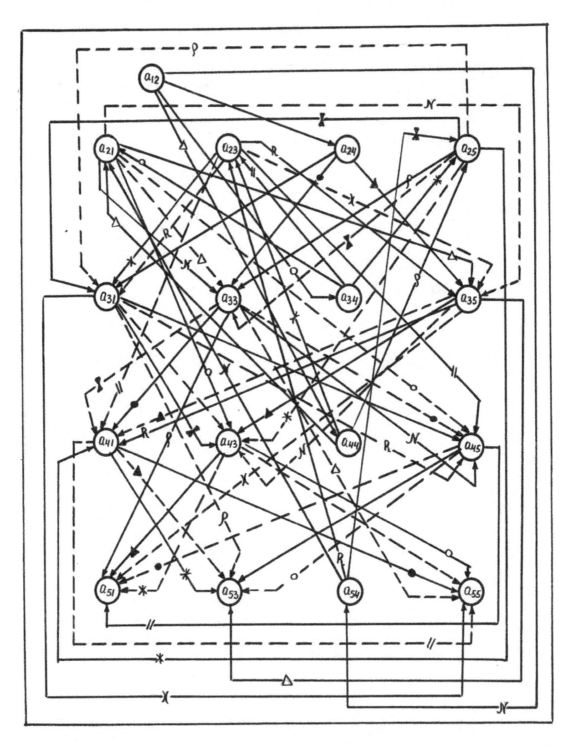

Figure 108 Functional graph for all cascade first block (a₁₂) elements of the second column applicably to Laplace's theorem for positive (——) and negative(— — —) areas

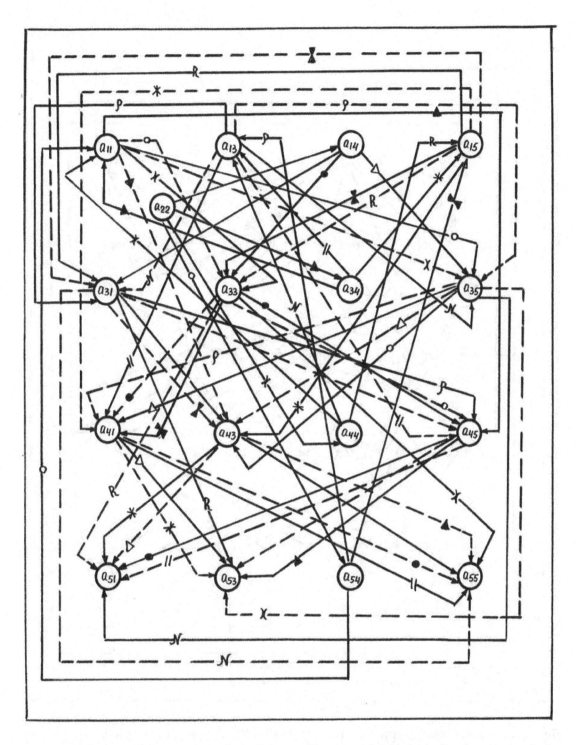

Figure 109 Functional graph for all cascade second block (a₂₂) elements of the second column applicably to Laplace's theorem for positive (——) and negative(— —) areas

143

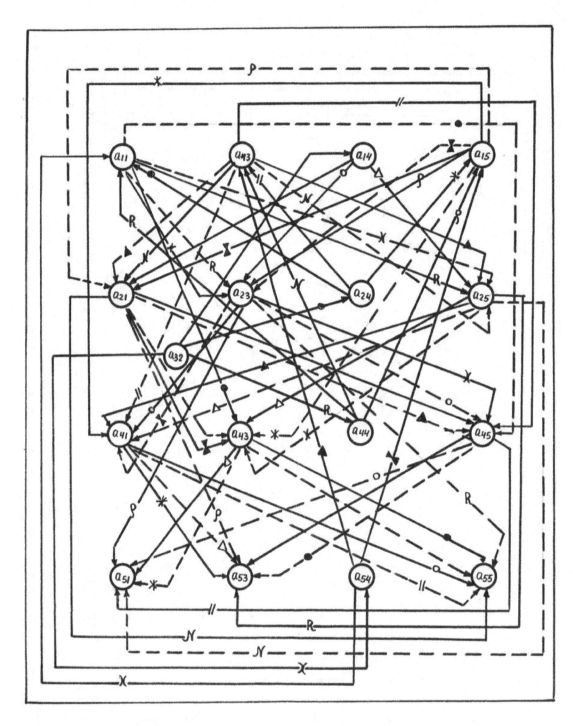

Figure 110 Functional graph for all cascade third block (a₃₂) elements of the second column applicably to Laplace's theorem for positive (——) and negative (— — —) areas

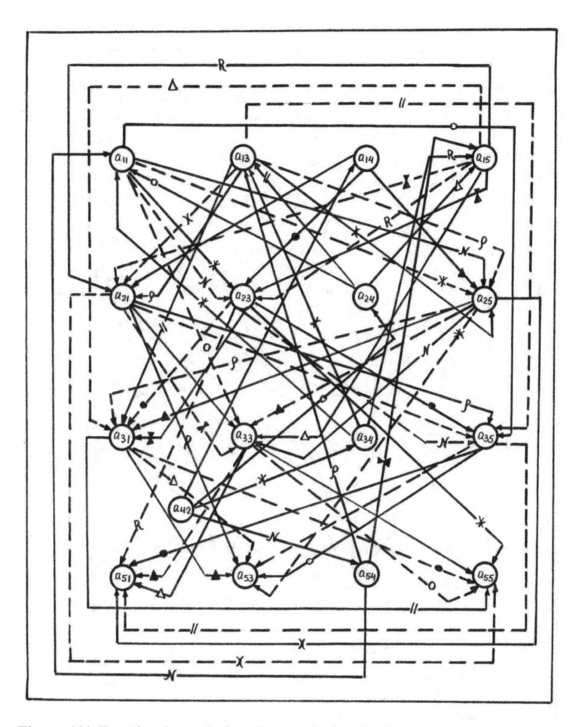

Figure 111 Functional graph for all cascade fourth block (a_{42}) elements of the second column applicably to Laplace's theorem for positive (———) and negative (———) areas

145

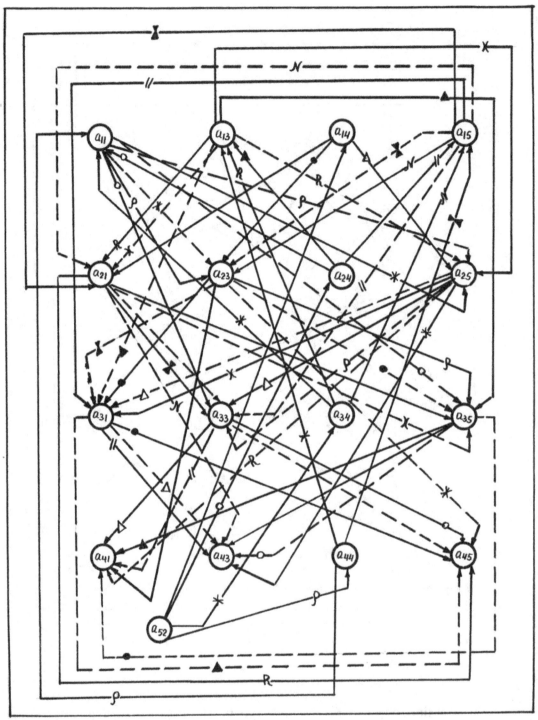

Figure 112 Functional graph for all cascade fifth block (a$_{52}$) elements of the second column applicably to Laplace's theorem for positive(———)and negative(— — —) areas

146

- Each element of the fourth column ($a_{14}, a_{34}, a_{44}, a_{54}$)has frequency of functional ties equal f=1 with element (a_{22}) of the second column;
- Each element (a_{11}, a_{13}, a_{14}) of the first row, besides of element (a_{14}),has frequency of functional ties equal f=3 with elements of the fourth column;
- Each element (a_{31}, a_{33}, a_{35})of the third row ,besides of element (a_{34}), has frequency of functional ties equal f=5 with elements of the first row for positive and negative areas;
- Each element (a_{41}, a_{43}, a_{45}) of the fourth row, besides of element (a_{44}), has frequency of functional ties equal f=6 with elements of the first and third rows for positive and negative areas;
- Each element(a_{51}, a_{53}, a_{55}) of the fifth row, besides of element (a_{54}) ,has frequency of functional ties equal f=6 with elements of the third and fourth rows for positive and negative areas.

So, in total we can conclude the following frequency of functional ties between elements of the 5x5 matrix for marked out jointly elements of the second and fourth columns applicably to Laplace's theorem *for all cascade second block (a_{22}) elements of the second column:*

> f=1 for elements ($a_{14}, a_{34}, a_{44}, a_{54}$)
> f=3 for elements (a_{11}, a_{13}, a_{14})
> f=5 for elements (a_{31}, a_{33}, a_{35})
> f=6 for elements (a_{41}, a_{43}, a_{45}) and (a_{51}, a_{53}, a_{55}).

Analysis of data shown in Figure 110 indicates on the following facts applicably to Laplace's theorem for marked out jointly the second and fourth columns of the 5x5 matrix *for all cascade third block (a_{32}) elements of the second column:*

- Each element of the fourth column ($a_{14}, a_{24}, a_{44}, a_{54}$) has frequency of functional ties equal f=1 with element (a_{32}) of the second column;
- Each element (a_{11}, a_{13}, a_{15})of the first row ,besides of element (a_{14}), has frequency of functional ties equal f=3 with elements of the fourth column;
- Each element(a_{21}, a_{23}, a_{25})of the second row ,besides of element (a_{24}), has frequency of functional ties equal f=5 with elements of the first row for positive and negative areas;
- Each element (a_{41}, a_{43}, a_{45}) of the fourth row, besides of element (a_{44}) has frequency of functional ties equal f=6 with elements of the first and second rows for positive and negative areas;
- Each element (a_{51}, a_{53}, a_{55}) of the fifth row, besides of element (a_{54}),has frequency of functional ties equal f=6 with elements of the second and fourth rows for positive and negative areas.

So, in total we can conclude the following frequency of functional ties between elements of the 5x5 matrix for marked out jointly elements of the second and fourth columns applicably to Laplace's theorem *for all cascade third block (a_{32}) elements of the second column:*

> f=1 for elements ($a_{14}, a_{24}, a_{44}, a_{54}$)
> f=3 for elements (a_{11}, a_{13}, a_{15})
> f=5 for elements (a_{21}, a_{23}, a_{25})
> f=6 for elements (a_{41}, a_{43}, a_{45}) and (a_{51}, a_{53}, a_{55}).

Analysis of data shown in Figure 111 indicates of the following facts applicably to Laplace's theorem for marked out jointly the second and fourth columns of the 5x5 matrix *for all cascade fourth block (a_{42}) elements of the second column:*

- Each element of the fourth column($a_{14}, a_{24}, a_{34}, a_{54}$) has frequency of functional ties equal f=1 with element (a_{42}) of the second column;
- Each element (a_{11}, a_{13}, a_{15}) of the first row, besides of element (a_{14}),has frequency of functional ties equal f=3 with elements of the fourth column;
- Each element (a_{21}, a_{23}, a_{25}) of the second row, besides of element (a_{24}) ,has frequency of functional ties equal f=5 with elements of the first row for positive and negative areas;
- Each element (a_{31}, a_{33}, a_{35}) of the third row, besides of element (a_{34}), has frequency of functional ties equal f=6 with elements of the first and second rows for positive and negative areas;
- Each element (a_{51}, a_{53}, a_{55}) of the fifth row, besides of element (a_{54}), has frequency of functional ties equal f=6 with elements of the second and third rows for positive and negative areas.

So, in total we can conclude the following frequency of functional ties between elements of the 5x5 matrix for marked out jointly elements of the second and fourth columns applicably to Laplace's theorem *for all cascade fourth block (a_{42}) elements of the second column:*

f=1 for elements ($a_{14}, a_{24}, a_{34}, a_{54}$)

f=3 for elements (a_{11}, a_{13}, a_{15})

f=5 for elements (a_{21}, a_{23}, a_{25})

f=6 for elements (a_{31}, a_{33}, a_{35}) and (a_{51}, a_{53}, a_{55}).

Analysis of data shown in Figure 112 indicates on the following facts applicably to Laplace's theorem for marked out jointly the second and fourth columns of the 5x5 matrix *for all cascade fifth block (a_{52}) elements of the second column:*

- Each element of the fourth column ($a_{14}, a_{24}, a_{34}, a_{44}$) has frequency of functional ties equal f=1;
- Each element (a_{11}, a_{13}, a_{15}) of the first row ,besides of element (a_{14}), has frequency of functional ties equal f=3 with elements of the fourth column;
- Each element (a_{21}, a_{23}, a_{25}) of the second row, besides of element (a_{24}) ,has frequency of functional ties equal f=5 with elements of the first row for positive and negative areas;
- Each element (a_{31}, a_{33}, a_{35}) of the third row, besides of element (a_{34}) ,has frequency of functional ties equal f=6 with elements of the first and second rows for positive and negative areas;
- Each element (a_{41}, a_{43}, a_{45}) of the fourth row, besides of element (a_{44}),has frequency of functional ties equal f=6 with elements of the second and third rows for positive and negative areas.

So, in total we can conclude the following frequency of functional ties between elements of the 5x5 matrix for marked out jointly elements of the second and fourth columns applicably to Laplace's theorem *for all cascade fifth block (a_{52}) elements of the second column:*

f=1for elements ($a_{14}, a_{24}, a_{34}, a_{44}$)

f=3 for elements (a_{11}, a_{13}, a_{15})

f=5 for elements (a_{21}, a_{23}, a_{25})

f=6 for elements (a_{31}, a_{33}, a_{35}) and (a_{41}, a_{43}, a_{45}).

In Appendix 18 is shown evaluation of the determinant of the 5x5 matrix with using of Laplace's theorem for marked out jointly the second and fourth columns.

g) For the second and fifth columns

The value of determinant $|A|$ of the 5x5 matrix at conditions when marked out jointly two columns, such as the second and fifth columns applicably to Laplace's theorem, can be expressed by the following formula in general view:

$$|A| = M_1A_1 - M_2A_2 + M_3A_3 - M_4A_4 + M_5A_5 - M_6A_6 + M_7A_7 + M_8A_8 - M_9A_9 + M_{10}A_{10} \quad (*)$$

where,

M_1-minor and A_1-algebraic supplemental (cofactor) for marked out jointly the second and fifth columns and also the first and second rows;

M_2- minor and A_2-algebraic supplemental (cofactor) for marked out jointly the second and fifth columns and also the first and third rows;

M_3-minor and A_3-algebraic supplemental (cofactor) for marked out jointly the second and fifth columns and also the first and fourth rows;

M_4-minor and A_4-algebraic supplemental (cofactor) for marked out jointly the second and fifth columns and also the first and fifth rows;

M_5-minor and A_5-algebraic supplemental (cofactor) for marked out jointly the second and fifth columns and also the second and third rows;

M_6-minor and A_6-algebraic supplemental (cofactor) for marked out jointly the second and fifth columns and also the second and fourth rows;

M_7-minor and A_7-algebraic supplemental (cofactor) for marked out jointly the second and fifth columns and also the second and fifth rows;

M_8-minor and A_8-algebraic supplemental (cofactor) for marked out jointly the second and fifth columns and also the third and fourth rows;

M_9-minor and A_9-algebraic supplemental (cofactor) for marked out jointly the second and fifth columns and also the third and fifth rows;

M_{10}-minor and A_{10}-algebraic supplemental (cofactor) for marked out jointly the second and fifth columns and also the fourth and fifth rows.

$$M_1=(a_{12}a_{25}-a_{15}a_{22}) \qquad A_1=(-1)^{(2+5)+(1+2)}\begin{vmatrix} a_{31} & a_{33} & a_{34} \\ a_{41} & a_{43} & a_{44} \\ a_{51} & a_{53} & a_{54} \end{vmatrix}$$

$$M_2=(a_{12}a_{35}-a_{15}a_{32}) \qquad A_2=(-1)^{(2+5)+(1+3)}\begin{vmatrix} a_{21} & a_{23} & a_{24} \\ a_{41} & a_{43} & a_{44} \\ a_{51} & a_{53} & a_{54} \end{vmatrix}$$

$$M_3=(a_{12}a_{45}-a_{15}a_{42}) \qquad A_3=(-1)^{(2+5)+(1+4)}\begin{vmatrix} a_{21} & a_{23} & a_{24} \\ a_{31} & a_{33} & a_{34} \\ a_{51} & a_{53} & a_{54} \end{vmatrix}$$

$$M_4=(a_{12}a_{55}-a_{15}a_{52}) \qquad A_4=(-1)^{(2+5)+(1+5)}\begin{vmatrix} a_{21} & a_{23} & a_{24} \\ a_{31} & a_{33} & a_{34} \\ a_{41} & a_{43} & a_{44} \end{vmatrix}$$

$$M_5=(a_{22}a_{35}-a_{25}a_{32}) \qquad A_5=(-1)^{(2+5)+(2+3)}\begin{vmatrix} a_{11} & a_{13} & a_{14} \\ a_{41} & a_{43} & a_{44} \\ a_{51} & a_{53} & a_{54} \end{vmatrix}$$

$$M_6=(a_{22}a_{45}-a_{25}a_{42}) \qquad A_6=(-1)^{(2+5)+(2+4)}\begin{vmatrix} a_{11} & a_{13} & a_{14} \\ a_{31} & a_{33} & a_{34} \\ a_{51} & a_{53} & a_{54} \end{vmatrix}$$

$$M_7=(a_{22}a_{55}-a_{25}a_{52}) \qquad A_7=(-1)^{(2+5)+(2+5)}\begin{vmatrix} a_{11} & a_{13} & a_{14} \\ a_{31} & a_{33} & a_{34} \\ a_{41} & a_{43} & a_{44} \end{vmatrix}$$

$$M_8=(a_{32}a_{45}-a_{35}a_{42}) \qquad A_8=(-1)^{(2+5)+(3+4)}\begin{vmatrix} a_{11} & a_{13} & a_{14} \\ a_{21} & a_{23} & a_{24} \\ a_{51} & a_{53} & a_{54} \end{vmatrix}$$

$$M_9=(a_{32}a_{55}-a_{35}a_{52}) \qquad A_9=(-1)^{(2+5)+(3+5)}\begin{vmatrix} a_{11} & a_{13} & a_{14} \\ a_{21} & a_{23} & a_{24} \\ a_{41} & a_{43} & a_{44} \end{vmatrix}$$

$$M_{10}=(a_{42}a_{55}-a_{45}a_{52}) \qquad A_{10}=(-1)^{(2+5)+(4+5)}\begin{vmatrix} a_{11} & a_{13} & a_{14} \\ a_{21} & a_{23} & a_{24} \\ a_{31} & a_{33} & a_{34} \end{vmatrix}$$

After of some calculations and transformations, the formula (*) has the following view *for marked out jointly the second and fifth columns* applicably to Laplace;s theorem for the determinant of the 5x5 matrix in general view:

$$
\begin{aligned}
|A| =\; & a_{12}a_{25}[a_{31}(a_{43}a_{54}-a_{44}a_{53})+a_{33}(a_{44}a_{51}-a_{41}a_{54})+a_{34}(a_{41}a_{53}-a_{43}a_{51})]+ \\
& +a_{12}a_{35}[a_{21}(a_{44}a_{53}-a_{43}a_{54})+a_{23}(a_{41}a_{54}-a_{44}a_{51})+a_{24}(a_{43}a_{51}-a_{41}a_{53})]+ \\
& +a_{12}a_{45}[a_{21}(a_{33}a_{54}-a_{34}a_{53})+a_{23}(a_{34}a_{51}-a_{31}a_{54})+a_{24}(a_{31}a_{53}-a_{33}a_{51})]+ \\
& +a_{12}a_{55}[a_{21}(a_{34}a_{43}-a_{33}a_{44})+a_{23}(a_{31}a_{44}-a_{34}a_{41})+a_{24}(a_{33}a_{41}-a_{31}a_{43})]+ \\[4pt]
& +a_{22}a_{15}[a_{31}(a_{44}a_{53}-a_{43}a_{54})+a_{33}(a_{41}a_{54}-a_{44}a_{51})+a_{34}(a_{43}a_{51}-a_{41}a_{53})]+ \\
& +a_{22}a_{35}[a_{11}(a_{43}a_{54}-a_{44}a_{53})+a_{13}(a_{44}a_{51}-a_{41}a_{54})+a_{14}(a_{41}a_{53}-a_{43}a_{51})]+ \\
& +a_{22}a_{45}[a_{11}(a_{34}a_{53}-a_{33}a_{54})+a_{13}(a_{31}a_{54}-a_{34}a_{51})+a_{14}(a_{33}a_{51}-a_{31}a_{53})]+ \\
& +a_{22}a_{55}[a_{11}(a_{33}a_{44}-a_{34}a_{43})+a_{13}(a_{34}a_{41}-a_{31}a_{44})+a_{14}(a_{31}a_{43}-a_{33}a_{41})]+ \\[4pt]
& +a_{32}a_{15}[a_{21}(a_{43}a_{54}-a_{44}a_{53})+a_{23}(a_{44}a_{51}-a_{41}a_{54})+a_{24}(a_{41}a_{53}-a_{43}a_{51})]+ \\
& +a_{32}a_{25}[a_{11}(a_{44}a_{53}-a_{43}a_{54})+a_{13}(a_{41}a_{54}-a_{44}a_{51})+a_{14}(a_{43}a_{51}-a_{41}a_{53})]+ \\
& +a_{32}a_{45}[a_{11}(a_{23}a_{54}-a_{24}a_{53})+a_{13}(a_{24}a_{51}-a_{21}a_{54})+a_{14}(a_{21}a_{53}-a_{23}a_{51})]+ \\
& +a_{32}a_{55}[a_{11}(a_{24}a_{43}-a_{23}a_{44})+a_{13}(a_{21}a_{44}-a_{24}a_{41})+a_{14}(a_{23}a_{41}-a_{21}a_{43})]+ \\[4pt]
& +a_{42}a_{15}[a_{21}(a_{34}a_{53}-a_{33}a_{54})+a_{23}(a_{31}a_{54}-a_{34}a_{51})+a_{24}(a_{33}a_{51}-a_{31}a_{53})]+ \\
& +a_{42}a_{25}[a_{11}(a_{33}a_{54}-a_{34}a_{53})+a_{13}(a_{34}a_{51}-a_{31}a_{54})+a_{14}(a_{31}a_{53}-a_{33}a_{51})]+ \\
& +a_{42}a_{35}[a_{11}(a_{24}a_{53}-a_{23}a_{54})+a_{13}(a_{21}a_{54}-a_{24}a_{51})+a_{14}(a_{23}a_{51}-a_{21}a_{53})]+ \\
& +a_{42}a_{55}[a_{11}(a_{23}a_{34}-a_{24}a_{33})+a_{13}(a_{24}a_{31}-a_{21}a_{34})+a_{14}(a_{21}a_{33}-a_{23}a_{31})]+ \\[4pt]
& +a_{52}a_{15}[a_{21}(a_{33}a_{44}-a_{34}a_{43})+a_{23}(a_{34}a_{41}-a_{31}a_{44})+a_{24}(a_{31}a_{43}-a_{33}a_{41})]+ \\
& +a_{52}a_{25}[a_{11}(a_{34}a_{43}-a_{33}a_{44})+a_{13}(a_{31}a_{44}-a_{34}a_{41})+a_{14}(a_{33}a_{41}-a_{31}a_{43})]+ \\
& +a_{52}a_{35}[a_{11}(a_{23}a_{44}-a_{24}a_{43})+a_{13}(a_{24}a_{41}-a_{21}a_{44})+a_{14}(a_{21}a_{43}-a_{23}a_{41})]+ \\
& +a_{52}a_{45}[a_{11}(a_{24}a_{33}-a_{23}a_{34})+a_{13}(a_{21}a_{34}-a_{24}a_{31})+a_{14}(a_{23}a_{31}-a_{21}a_{33})]
\end{aligned} \quad (102)
$$

In Figure 113 is shown the functional graph for all cascade first block(a_{12}) elements of the second column applicably to Laplaces' theorem for positive (———) and negative (------) areas. In Figure 114 is shown the functional graph for all cascade second block (a_{22}) elements to the second column applicably to Laplace's theorem for positive(———) and negative (-------) areas.

In Figure 115 is shown the functional graph for all cascade third block(a_{32}) elements of the second column applicably to Laplace's theorem for positive(———) and negative(-------) areas. In Figure 116 is shown the functional graph for all cascade fourth block(a_{42}) elements of the second column applicably to Laplace's theorem for positive(———) and negative (------) areas.

In Figure 117 is shown the functional graph for all cascade fifth block (a_{52}) elements of the second column applicably to Laplace's theorem for positive(———) and negative (------) areas.

In Appendix 19 is shown evaluation of the determinant of the 5x5 matrix with using of Laplace's theorem for marked out jointly the second and fifth columns.

Analysis of data shown in Figure 113 indicates on the following facts applicably to Laplace's theorem for marked out jointly the second and fifth columns of the 5x5 matrix *for all cascade the first block(a_{12}) elements of the second column:*

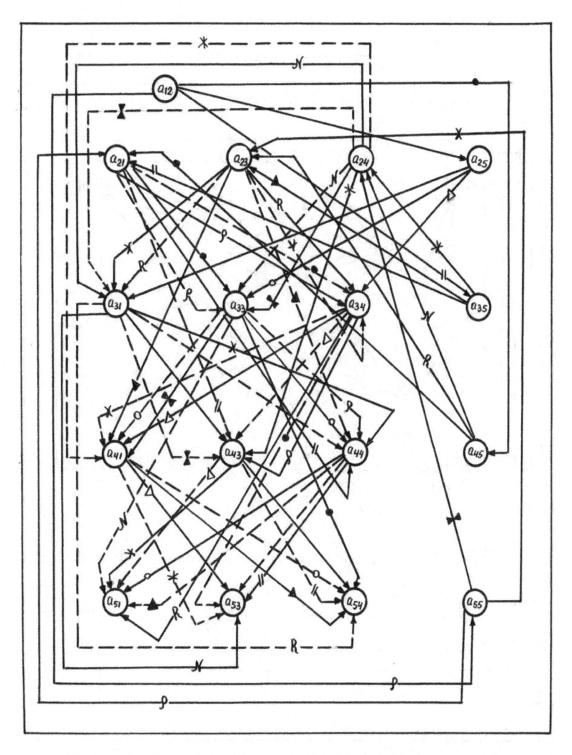

Figure 113 Functional graph for all cascade first block (a_{12}) elements of the second column applicably to Laplace's theorem for positive (———) and negative (-------) areas

151

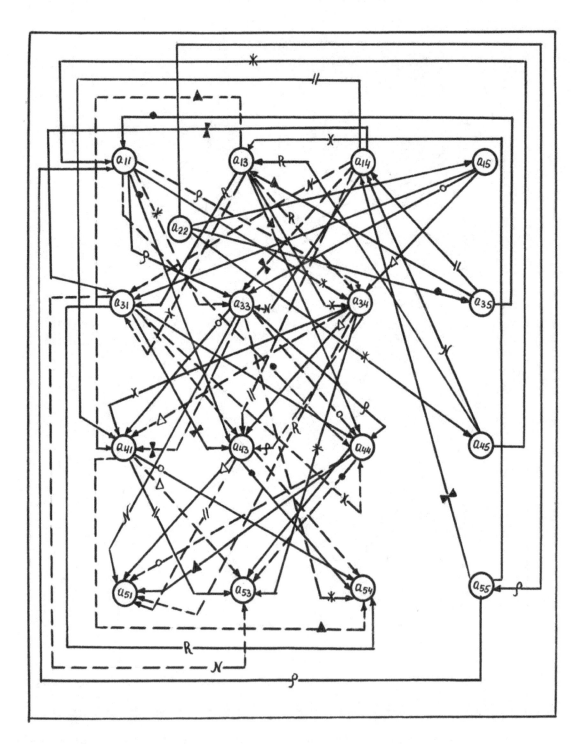

Figure 114 Functional graph for all cascade second block (a₂₂) elements of the second column applicably to Laplace's theorem for positive (———) and negative (-------) areas

152

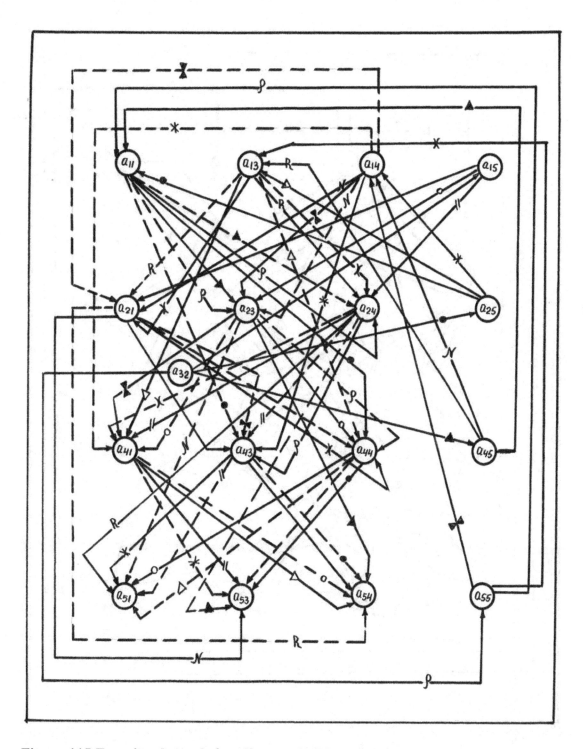

Figure 115 Functional graph for all cascade third block(a₃₂) elements of the second column applicably to Lapalce's theorem for positive(————) and negative(-------) areas

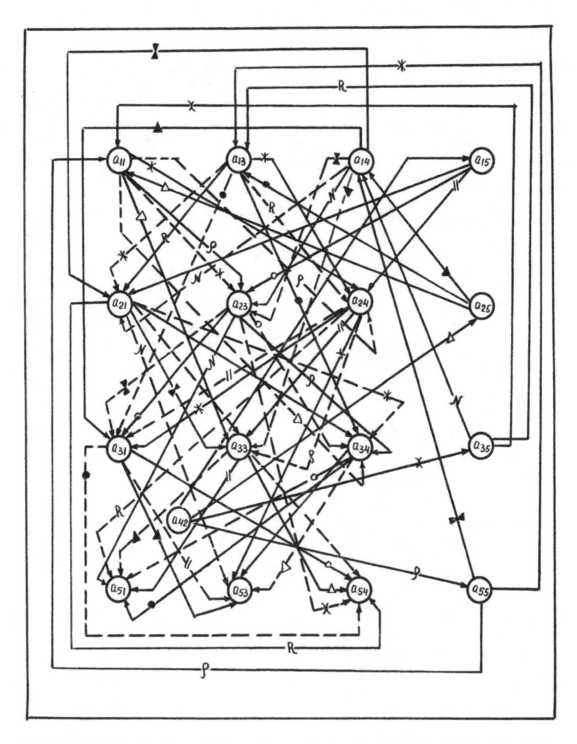

Figure 116 Functional graph for all cascade fourth block (a₄₂) elements of the second column applicably to Laplace's theorem for positive (———) and negative (-------) areas

154

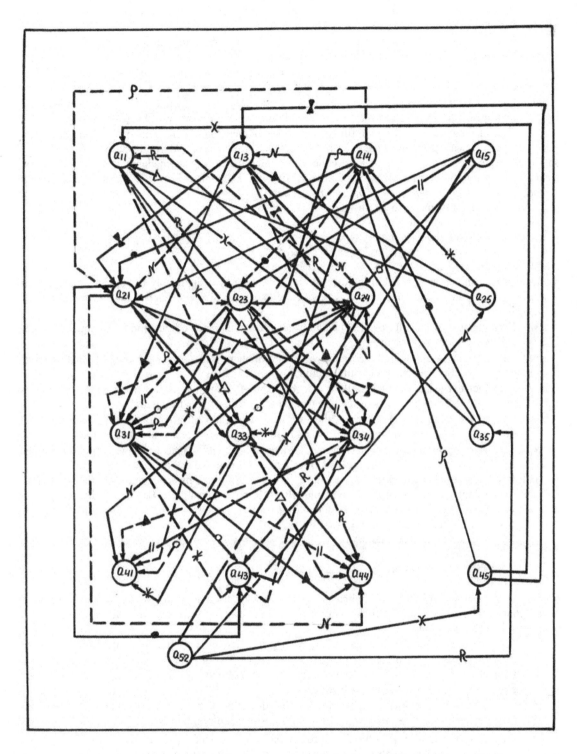

Figure 117 Functional graph for all cascade fifth block (a₅₂) elements of the second column applicably to Laplace's theorem for positive(———) and negative (-------) areas

155

- Each element of the fifth column(a_{25},a_{35},a_{45},a_{55})has frequency of functional ties equal f=1 with the first block (a_{12}) element;
- Each element (a_{21},a_{23},a_{24}) of the second row, besides of element (a_{25}) ,has frequency of functional ties equal f=3 with elements of the fifth column;
- Each element (a_{31},a_{33},a_{34})of the third row, besides of element (a_{35})has frequency of functional ties equal f=5 with elements of the second row for positive and negative areas;
- Each element (a_{41},a_{43},a_{44}) of the fourth row, besides of element (a_{45}) ,has frequency of functional ties equal f=6 with elements of the second and third rows for positive and negative areas;
- Each element (a_{51},a_{53},a_{54}) of the fifth row ,besides of element(a_{55}) ,has frequency of functional ties equal f=6 with elements of the third and fourth rows for positive and negative areas.

So, in total we can conclude the following frequency of functional ties between elements of the 5x5 matrix for marked out jointly elements of the second and fifth columns applicably to Laplace's theorem *for all cascade the first block(a_{12}) elements of the second column:*

f=1 for elements (a_{25},a_{35},a_{45},a_{55})

f=3 for elements (a_{21},a_{23},a_{24})

f=5 for elements (a_{31},a_{33},a_{34})

f=6 for elements (a_{41},a_{43},a_{44}) and (a_{51},a_{53},a_{54}).

Analysis of data shown in Figure 114 indicates on the following facts applicably to Laplace's theorem for marked out jointly the second and fifth columns of the 5x5 matrix *for all cascade the second block(a_{22}) elements of the second column:*
- Each element of the fifth column (a_{25},a_{35},a_{45},a_{55}) has frequency of functional ties equal f=1 with the second block (a_{22}) element;
- Each element (a_{11},a_{13},a_{14}) of the first row, besides of element (a_{15}) ,has frequency of functional ties equal f=3 with elements of the fifth column;
- Each element (a_{31},a_{33},a_{34}) of the third row ,besides of element (a_{35}) ,has frequency of functional ties equal f=5 with elements of the first row for positive and negative areas;
- Each element (a_{41},a_{43},a_{44}) of the fourth row, besides of element (a_{45}) has frequency of functional ties equal f=6 with elements of the first and third rows for positive and negative areas;
- Each element (a_{51},a_{53},a_{54}) of the fifth row, besides of element (a_{55}) has frequency of functional ties equal f=6 with elements of the third and fourth rows for positive and negative areas.

So , in total we can conclude the following frequency of functional ties between elements of the 5x5 matrix for marked out jointly elements of the second and fifth columns applicably to Laplace's theorem *for all cascade the second block(a_{22}) elements of the second column:*

f=1 for elements (a_{15},a_{35},a_{45},a_{55})

f=3 for elements (a_{11},a_{13},a_{14})

f=5 for elements (a_{31},a_{33},a_{34})

f=6 for elements (a_{41},a_{43},a_{44}) and (a_{51},a_{53},a_{54}).

Analysis of data shown in Figure 115 indicates on the following facts applicably to Laplace's theorem for marked out jointly the second and fifth columns of the 5x5 matrix *for all cascade the third block (a_{32}) elements of the second column:*
- Each element of the fifth column (a_{15},a_{25},a_{45},a_{55}) has frequency of functional ties equal f=1 with the third block(a_{32}) element ;
- Each element (a_{11},a_{13},a_{14}) of the first row ,besides of element(a_{15}), has frequency of functional ties equal f=3 with elements of the fifth column;
- Each element (a_{21},a_{23},a_{24}) of the second row, besides of element (a_{25}), has frequency of functional ties equal f=5 with elements of the first row for positive and negative areas;
- Each element (a_{41},a_{43},a_{44}) of the fourth row, besides of element (a_{45}) ,has frequency of functional ties equal f=6 with elements of the first and second rows for positive and negative areas;
- Each element (a_{51},a_{53},a_{54}) of the fifth row, besides of element (a_{55}), has frequency of functional ties equal f=6 with elements of the second and fourth rows for positive and negative areas.

So, in total we can conclude the following frequency of functional ties between elements of the 5x5 matrix for marked out jointly elements of the second and fifth columns applicably to Laplace's theorem *for all cascade the third block (a_{32}) elements of the second column:*

$f=1$ for elements ($a_{15}, a_{25}, a_{45}, a_{55}$)
$f=3$ for elements (a_{11}, a_{13}, a_{14})
$f=5$ for elements (a_{21}, a_{23}, a_{24})
$f=6$ for elements (a_{41}, a_{43}, a_{44}) and (a_{51}, a_{53}, a_{54}) .

Analysis of data shown in Figure 116 indicates on the following facts applicably to Laplace's theorem for marked out jointly the second and fifth columns of the 5x5 matrix *for all cascade the fourth block(a_{42}) elements of the second column:*

- Each element of the fifth column($a_{15}, a_{25}, a_{35}, a_{55}$) has frequency of functional ties equal $f=1$ with the fourth cascade element (a_{42});
- Each element (a_{11}, a_{13}, a_{14}) of the first row, besides of element (a_{15}), has frequency of functional ties equal $f=3$ with elements of the fifth column;
- Each element(a_{21}, a_{23}, a_{24}) of the second row, besides of element(a_{25}) has frequency of functional ties equal $f=5$ with elements of the first row for positive and negative areas;
- Each element(a_{31}, a_{33}, a_{34}) of the third row, besides of element(a_{35}),has frequency of functional ties equal $f=6$ with elements of the first and second rows for positive and negative areas;
- Each element (a_{51}, a_{53}, a_{54}) of the fifth row ,besides of element (a_{55}), has frequency of functional ties equal $f=6$ with elements of the second and third rows for positive and negative areas.

So, in total we can conclude the following frequency of functional ties between elements of the 5x5 matrix for marked out jointly elements of the second and fifth columns applicably to Laplace's theorem *for all cascade the fourth block (a_{42}) elements of the second column:*

$f=1$ for elements ($a_{15}, a_{25}, a_{35}, a_{55}$)
$f=3$ for elements (a_{11}, a_{13}, a_{14})
$f=5$ for elements (a_{21}, a_{23}, a_{24})
$f=6$ for elements (a_{31}, a_{33}, a_{34}) and (a_{51}, a_{53}, a_{54}).

Analysis of data shown in Figure 117 indicates on the following facts applicably to Laplace's theorem for marked out jointly the second and fifth columns of the 5x5 matrix *for all cascade the fifth block(a_{52}) elements of the second column:*

- Each element of the fifth column ($a_{15}, a_{25}, a_{35}, a_{45}$) has frequency of functional ties equal $f=1$ with the fifth cascade block(a_{52}) of the second column;
- Each element (a_{11}, a_{13}, a_{14}) of the first row, besides of element (a_{15}), has frequency of functional ties equal $f=3$ with elements of the fifth column;
- Each element (a_{21}, a_{23}, a_{24})of the second row, besides of element (a_{25}) ,has frequency of functional ties equal $f=5$ with elements of the first row for positive and negative areas;
- Each element (a_{31}, a_{33}, a_{34}) of the third row ,besides of element (a_{35}), has frequency of functional ties equal $f=6$ with elements of the first and second rows for positive and negative areas;
- Each element (a_{41}, a_{43}, a_{44}) of the fourth row, besides of element (a_{45}), has frequency of functional ties equal $f=6$ with elements of the second and third rows for positive and negative areas.

So, in total we can conclude the following frequency of functional ties between elements of the 5x5 matrix for marked out jointly elements of the second and fifth columns applicably to Laplace's theorem *for all cascade the fifth block(a_{52}) elements f the second column:*

$f=1$ for elements ($a_{15}, a_{25}, a_{35}, a_{45}$)
$f=3$ for elements (a_{11}, a_{13}, a_{14})
$f=5$ for elements (a_{21}, a_{23}, a_{24})
$f=6$ for elements (a_{31}, a_{33}, a_{34}) and (a_{41}, a_{43}, a_{44}).

h) **For the third and fourth columns**

The value of determinant $|A|$ of the 5x5 matrix at conditions when marked out jointly two columns , such as the *third and fourth columns* applicably to Laplace's theorem ,can be expressed by the following formula in general view:

$$|\mathbf{A}|=M_1A_1-M_2A_2+M_3A_3-M_4A_4+M_5A_5-M_6A_6+M_7A_7+M_8A_8-M_9A_9+M_{10}A_{10} \quad (**)$$

where,

M_1-minor and A_1-algebraic supplemental (cofactor) for marked out jointly the third and fourth columns and also the first and second rows;

M_2- minor and A_2-algebraic supplemental (cofactor) for marked out jointly the third and fourth columns and also the first and third rows;

M_3-minor and A_3-algebraic supplemental (cofactor) for marked out jointly the third and fourth columns and also the first and fourth rows;

M_4-minor and A_4-algebraic supplemental (cofactor) for marked out jointly the third and fourth columns and also the first and fifth rows;

M_5-minor and A_5-algebraic supplemental (cofactor) for marked out jointly the third and fourth columns and also the second and third rows;

M_6-minor and A_6-algebraic supplemental (cofactor) for marked out jointly the third and fourth columns and also the second and fourth rows;

M_7-minor and A_7-algebraic supplemental (cofactor) for marked out jointly the third and fourth columns and also the second and fifth rows;

M_8-minor and A_8-algebraic supplemental (cofactor) for marked out jointly the third and fourth columns and also the third and fourth rows;

M_9-minor and A_9-algebraic supplemental (cofactor) for marked out jointly the third and fourth columns and also the third and fifth rows;

M_{10}-minor and A_{10}-algebraic supplemental (cofactor) for marked out jointly the third and fourth columns and also the fourth and fifth rows.

After of some calculations and transformations, the formula (**) has the following view *for marked out jointly the third and fourth columns* applicably to Laplace's theorem for the determinant of the 5x5 matrix in general view:

$$|\mathbf{A}|=a_{13}a_{24}[a_{31}(a_{42}a_{55}-a_{45}a_{52})+a_{32}(a_{45}a_{51}-a_{41}a_{55})+a_{35}(a_{41}a_{52}-a_{42}a_{51})]+$$
$$+a_{13}a_{34}[a_{21}(a_{45}a_{52}-a_{42}a_{55})+a_{22}(a_{41}a_{55}-a_{45}a_{51})+a_{25}(a_{42}a_{51}-a_{41}a_{52})]+$$
$$+a_{13}a_{44}[a_{21}(a_{32}a_{55}-a_{35}a_{52})+a_{22}(a_{35}a_{51}-a_{31}a_{55})+a_{25}(a_{31}a_{52}-a_{32}a_{51})]+$$
$$+a_{13}a_{54}[a_{21}(a_{35}a_{42}-a_{32}a_{45})+a_{22}(a_{31}a_{45}-a_{35}a_{41})+a_{25}(a_{32}a_{41}-a_{31}a_{42})]+$$

$$+a_{23}a_{14}[a_{31}(a_{45}a_{52}-a_{42}a_{55})+a_{32}(a_{41}a_{55}-a_{45}a_{51})+a_{35}(a_{42}a_{51}-a_{41}a_{52})]+$$
$$+a_{23}a_{34}[a_{11}(a_{42}a_{55}-a_{45}a_{52})+a_{12}(a_{45}a_{51}-a_{41}a_{55})+a_{15}(a_{41}a_{52}-a_{42}a_{51})]+$$
$$+a_{23}a_{44}[a_{11}(a_{35}a_{52}-a_{32}a_{55})+a_{12}(a_{31}a_{55}-a_{35}a_{51})+a_{15}(a_{32}a_{51}-a_{31}a_{52})]+$$
$$+a_{23}a_{54}[a_{11}(a_{32}a_{45}-a_{35}a_{42})+a_{12}(a_{35}a_{41}-a_{31}a_{45})+a_{15}(a_{31}a_{42}-a_{32}a_{41})]+$$

$$+a_{33}a_{14}[a_{21}(a_{42}a_{55}-a_{45}a_{52})+a_{22}(a_{45}a_{51}-a_{41}a_{55})+a_{25}(a_{41}a_{52}-a_{42}a_{51})]+$$
$$+a_{33}a_{24}[a_{11}(a_{45}a_{52}-a_{42}a_{55})+a_{12}(a_{41}a_{55}-a_{45}a_{51})+a_{15}(a_{42}a_{51}-a_{41}a_{52})]+$$
$$+a_{33}a_{44}[a_{11}(a_{22}a_{55}-a_{25}a_{52})+a_{12}(a_{25}a_{51}-a_{21}a_{55})+a_{15}(a_{21}a_{52}-a_{22}a_{51})]+$$
$$+a_{33}a_{54}[a_{11}(a_{25}a_{42}-a_{22}a_{45})+a_{12}(a_{21}a_{45}-a_{25}a_{41})+a_{15}(a_{22}a_{41}-a_{21}a_{42})]+$$

$$+a_{43}a_{14}[a_{21}(a_{35}a_{52}-a_{32}a_{55})+a_{22}(a_{31}a_{55}-a_{35}a_{51})+a_{25}(a_{32}a_{51}-a_{31}a_{52})]+$$
$$+a_{43}a_{24}[a_{11}(a_{32}a_{55}-a_{35}a_{52})+a_{12}(a_{35}a_{51}-a_{31}a_{55})+a_{15}(a_{31}a_{52}-a_{32}a_{51})]+$$
$$+a_{43}a_{34}[a_{11}(a_{25}a_{52}-a_{22}a_{55})+a_{12}(a_{21}a_{55}-a_{25}a_{51})+a_{15}(a_{22}a_{51}-a_{21}a_{52})]+$$
$$+a_{43}a_{54}[a_{11}(a_{22}a_{35}-a_{25}a_{32})+a_{12}(a_{25}a_{31}-a_{21}a_{35})+a_{15}(a_{21}a_{32}-a_{22}a_{31})]+$$

$$+a_{53}a_{14}[a_{21}(a_{32}a_{45}-a_{35}a_{42})+a_{22}(a_{35}a_{41}-a_{31}a_{45})+a_{25}(a_{31}a_{42}-a_{32}a_{41})]+$$
$$+a_{53}a_{24}[a_{11}(a_{35}a_{42}-a_{32}a_{45})+a_{12}(a_{31}a_{45}-a_{35}a_{41})+a_{15}(a_{32}a_{41}-a_{31}a_{42})]+$$
$$+a_{53}a_{34}[a_{11}(a_{22}a_{45}-a_{25}a_{42})+a_{12}(a_{25}a_{41}-a_{21}a_{45})+a_{15}(a_{21}a_{42}-a_{22}a_{41})]+$$
$$+a_{53}a_{44}[a_{11}(a_{25}a_{32}-a_{22}a_{35})+a_{12}(a_{21}a_{35}-a_{25}a_{31})+a_{15}(a_{22}a_{31}-a_{21}a_{32})] \quad (103)$$

$M_1 = (a_{13}a_{24} - a_{14}a_{23})$

$A_1 = (-1)^{(3+4)+(1+2)} \begin{vmatrix} a_{31} & a_{32} & a_{35} \\ a_{41} & a_{42} & a_{45} \\ a_{51} & a_{52} & a_{55} \end{vmatrix}$

$M_2 = (a_{13}a_{34} - a_{14}a_{33})$

$A_2 = (-1)^{(3+4)+(1+3)} \begin{vmatrix} a_{21} & a_{22} & a_{25} \\ a_{41} & a_{42} & a_{45} \\ a_{51} & a_{52} & a_{55} \end{vmatrix}$

$M_3 = (a_{13}a_{44} - a_{14}a_{43})$

$A_3 = (-1)^{(3+4)+(1+4)} \begin{vmatrix} a_{21} & a_{22} & a_{25} \\ a_{31} & a_{32} & a_{35} \\ a_{51} & a_{52} & a_{55} \end{vmatrix}$

$M_4 = (a_{13}a_{54} - a_{14}a_{53})$

$A_4 = (-1)^{(3+4)+(1+5)} \begin{vmatrix} a_{21} & a_{22} & a_{25} \\ a_{31} & a_{32} & a_{35} \\ a_{41} & a_{42} & a_{45} \end{vmatrix}$

$M_5 = (a_{23}a_{34} - a_{24}a_{33})$

$A_5 = (-1)^{(3+4)+(2+3)} \begin{vmatrix} a_{11} & a_{12} & a_{15} \\ a_{41} & a_{42} & a_{45} \\ a_{51} & a_{52} & a_{55} \end{vmatrix}$

$M_6 = (a_{23}a_{44} - a_{24}a_{43})$

$A_6 = (-1)^{(3+4)+(2+4)} \begin{vmatrix} a_{11} & a_{12} & a_{15} \\ a_{31} & a_{32} & a_{35} \\ a_{51} & a_{52} & a_{55} \end{vmatrix}$

$M_7 = (a_{23}a_{54} - a_{24}a_{53})$

$A_7 = (-1)^{(3+4)+(2+5)} \begin{vmatrix} a_{11} & a_{12} & a_{15} \\ a_{31} & a_{32} & a_{35} \\ a_{41} & a_{42} & a_{45} \end{vmatrix}$

$M_8 = (a_{33}a_{44} - a_{34}a_{43})$

$A_8 = (-1)^{(3+4)+(3+4)} \begin{vmatrix} a_{11} & a_{12} & a_{15} \\ a_{21} & a_{22} & a_{25} \\ a_{51} & a_{52} & a_{55} \end{vmatrix}$

$M_9 = (a_{33}a_{54} - a_{34}a_{53})$

$A_9 = (-1)^{(3+4)+(3+5)} \begin{vmatrix} a_{11} & a_{12} & a_{15} \\ a_{21} & a_{22} & a_{25} \\ a_{41} & a_{42} & a_{45} \end{vmatrix}$

$M_{10} = (a_{43}a_{54} - a_{44}a_{53})$

$A_{10} = (-1)^{(3+4)+(4+5)} \begin{vmatrix} a_{11} & a_{12} & a_{15} \\ a_{21} & a_{22} & a_{25} \\ a_{31} & a_{32} & a_{35} \end{vmatrix}$

In Figure 118 is shown the functional graph of all cascade first block (a_{13})elements of the third column applicably to Lapalce's theorem for positive(———) and negative (--------) areas. In Figure 119 is shown the functional graph of all cascade second block(a_{23}) elements of the third column applicably to Laplace's theorem for positive(———) and negative (------) areas. In Figure 120 is shown the functional graph of all third block(a_{33}) elements of the third column applicably to Laplace's theorem for positive(———) and negative (-------) areas. In Figure 121 is shown the functional graph of all fourth block(a_{43})elements of the third column applicably to Laplace's theorem for positive (———) and negative(--------) areas. In Figure 122 is shown the functional graph of all cascade fifth block (a_{53}) elements of the third column applicably to Laplace's theorem for positive (———) and negative (-------) areas.

Analysis of data shown in Figure 118 indicates on the following facts applicably to Laplace's theorem for marked out jointly the third and fourth columns of the 5x5 matrix *for all cascade the first block (a_{13}) element of the third column:*
- Each element of the fourth column (a_{24},a_{34},a_{44},a_{54}) has frequency of functional ties equal f=1 with the first block(a_{13}) element of the third column;
- Each element (a_{21},a_{22},a_{25}) of the second row, besides of element (a_{24}) has frequency of functional ties equal f=3 with elements of the fourth column;
- Each element (a_{31},a_{32},a_{35}) of the third row, besides of element (a_{34}) ,has frequency of functional ties equal f=5 with elements of the second row for positive and negative areas;
- Each element (a_{41},a_{42},a_{45}),besides of element (a_{44}),has frequency of functional ties equal f=6 with elements of the second and third rows for positive and negative areas;
- Each element (a_{51},a_{52},a_{55}),besides of element (a_{54}),has frequency of functional ties equal f=6 with elements of the third and fourth rows for positive and negative areas.

So, in total we can conclude the following frequency of functional ties between elements of the 5x5 matrix for marked out jointly elements of the third and fourth columns applicably to Laplace's theorem *for all cascade the first block (a_{13}) element of the third column:*

f=1 for elements (a_{24},a_{34},a_{44},a_{54})
f=3 for elements (a_{21},a_{22},a_{25})
f=5 for elements (a_{31},a_{32},a_{35})
f=6 for elements (a_{41},a_{42},a_{45}) and (a_{51},a_{52},a_{55}).

Analysis of data shown in Figure 119 indicates on the following facts applicably to Laplace's theorem for marked out jointly the third and fourth columns of the 5x5 matrix *for all cascade the second block (a_{23}) element of the third column:*
- Each element of the fourth column (a_{14},a_{34},a_{44},a_{54}) has frequency of functional ties equal f=1 with the second block (a_{23}) element of the third column;
- Each element (a_{11},a_{12},a_{15})of the first row, besides of element (a_{14}) ,has frequency of functional ties equal f=3 with elements of the fourth column;
- Each element (a_{31},a_{32},a_{35}) of the third row, besides of element (a_{34}), has frequency of functional ties equal f=5 with elements of the first row for positive and negative areas;
- Each element (a_{41},a_{42},a_{45}) of the fourth row, besides of element (a_{44}) ,has frequency of functional ties equal f=6 with elements of the first and third rows for positive and negative areas;
- Each element (a_{51},a_{52},a_{55}) of the fifth row ,besides of element (a_{54}),has frequency of functional ties equal f=6 with elements of the third and fourth rows for positive and negative areas.

So, in total we can conclude the following frequency of functional ties between elements of the 5x5 matrix for marked out jointly elements of the third and fourth columns applicably to Laplace's theorem *for al cascade the second block (a_{23}) element of the third column:*

f=1 for elements (a_{14},a_{34},a_{44},a_{54})
f=3 for elements (a_{11},a_{12},a_{15})
f=5 for elements (a_{31},a_{32},a_{35})
f=6 for elements (a_{41},a_{42},a_{45}) and (a_{51},a_{52},a_{55}).

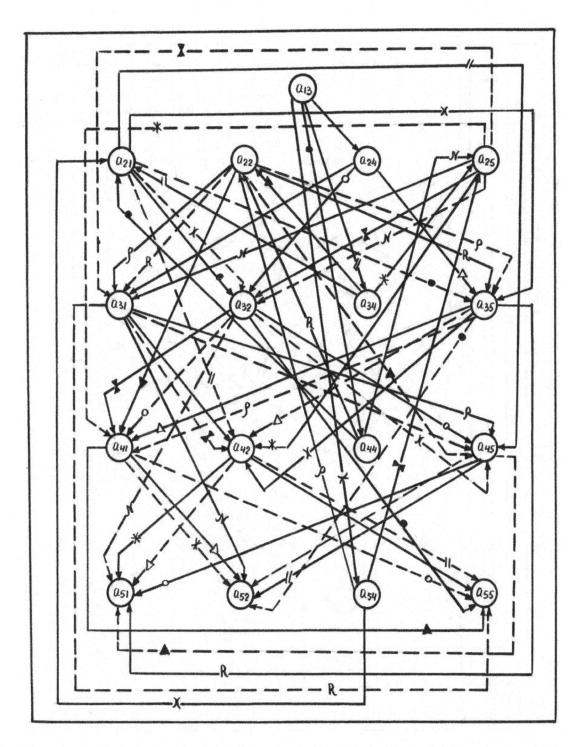

Figure 118 Functional graph of all cascade first block (a_{13}) elements of the third column applicably to Laplace's theorem for positive (———) and negative (-------) areas

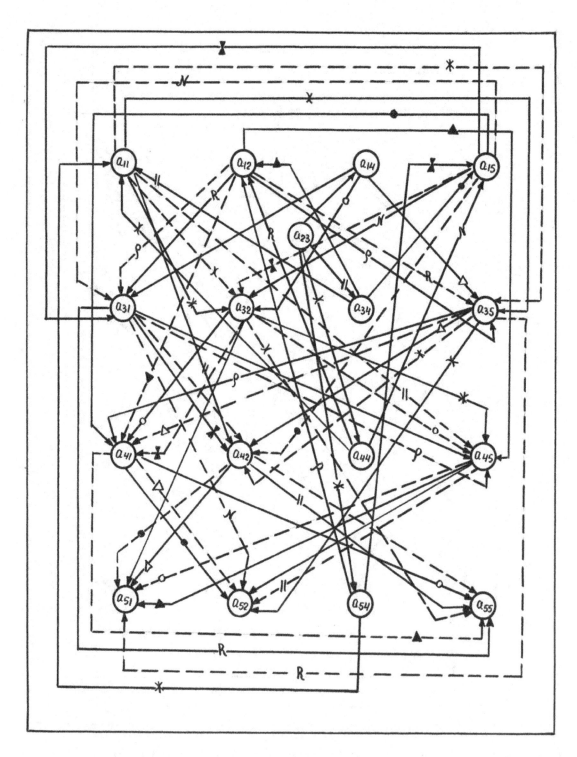

Figure 119 Functional graph of all cascade second block (a₂₃) elements of the third column applicably to Laplace's theorem for positive (———) and negative(-------) areas

162

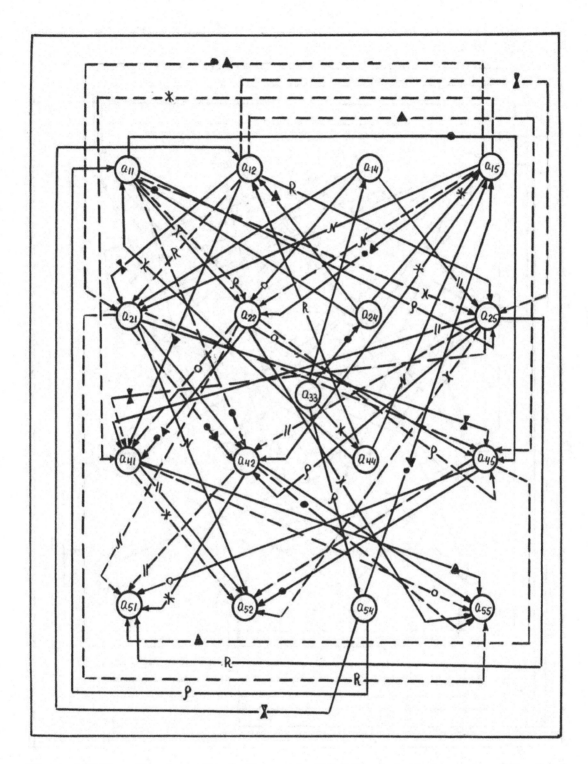

Figure 120 Functional graph of all cascade third block (a₃₃) elements of the third column applicably to Laplace's theorem for positive (——) and negative (------) areas

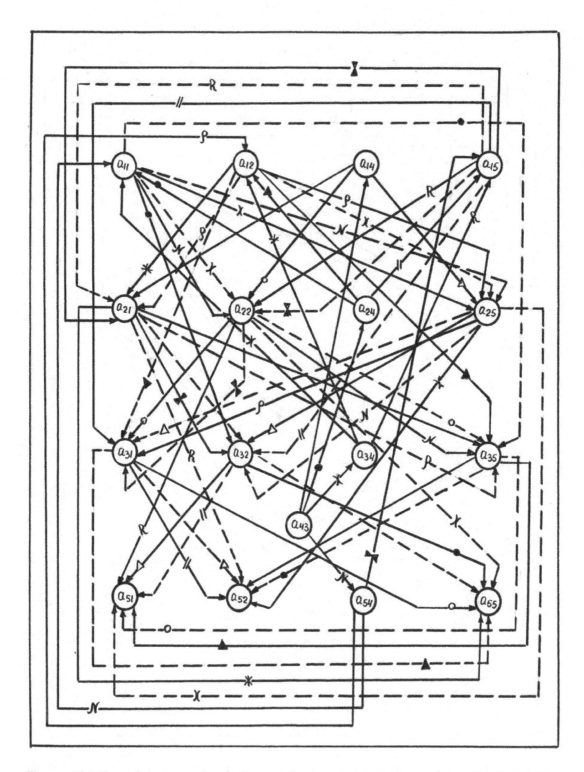

Figure 121 Functional graph of all cascade fourth block (a₄₃) elements of the third column applicably to Laplace's theorem for positive(————) and negative (-------) areas

164

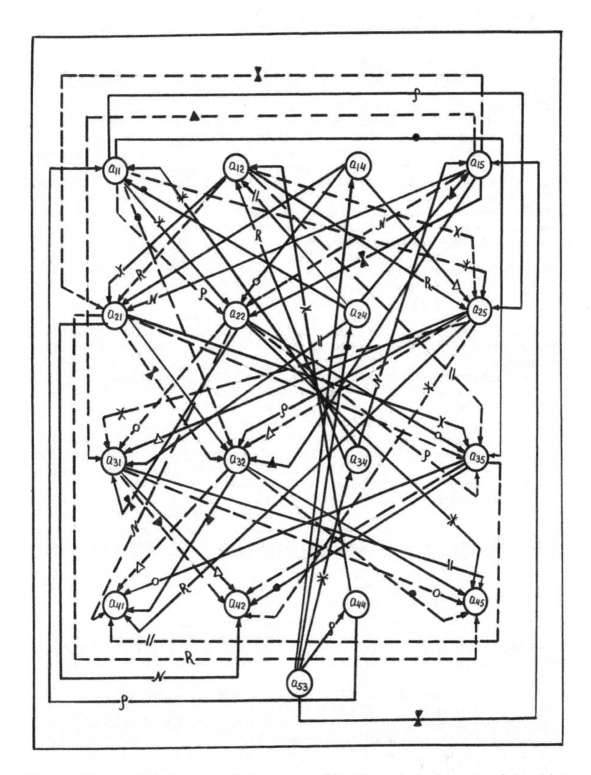

Figure 122 Functional graph of all cascade fifth block (a₅₃) elements of the third column applicably to Laplace's theorem for positive(———) and negative(-------) areas

Analysis of data shown in Figure 120 indicates on the following facts applicably to Laplace's theorem for marked out jointly the third and fourth columns of the 5x5 matrix *for all cascade the third block (a_{33}) element of the third column:*

- Each element of the fourth column ($a_{14}, a_{24}, a_{44}, a_{54}$)has frequency of functional ties equal f=1 with the third block (a_{33}) element of the third column;
- Each element (a_{11}, a_{12}, a_{15}) of the first row, besides of element(a_{14}) ,has frequency of functional ties equal f=3 with elements of the fourth column;
- Each element (a_{21}, a_{22}, a_{25}) of the second row, besides of element (a_{24}), has frequency of functional ties equal f=5 with elements of the first row for positive and negative areas;
- Each element (a_{41}, a_{42}, a_{45}) of the fourth row, besides of element (a_{44}), has frequency of functional ties equal f=6 with elements of the first and second rows for positive and negative areas;
- Each element (a_{51}, a_{52}, a_{55}) of the fifth row, besides of element (a_{54}) ,has frequency of functional ties equal f=6 with elements of the second and fourth rows for positive and negative areas.

So, in total we can conclude the following frequency of functional ties between elements of the 5x5 matrix for marked out jointly elements of the third and fourth columns applicably to Laplace's theorem *for all cascade the third block (a_{33}) element of the third column:*

$$f=1 \text{ for elements } (a_{14}, a_{24}, a_{44}, a_{54})$$
$$f=3 \text{ for elements } (a_{11}, a_{12}, a_{15})$$
$$f=5 \text{ for elements } (a_{21}, a_{22}, a_{25})$$
$$f=6 \text{ for elements } (a_{41}, a_{42}, a_{45}) \text{ and } (a_{51}, a_{52}, a_{55}).$$

Analysis of data shown in Figure 121 indicates on the following facts applicably to Laplace's theorem for marked out the third and fourth columns of the 5x5 matrix *for all cascade the fourth block(a_{43}) element of the third column:*

- Each element of the fourth column ($a_{14}, a_{24}, a_{34}, a_{54}$) has frequency of functional ties f=1 with the fourth block (a_{43}) element of the third column;
- Each element (a_{11}, a_{12}, a_{15}) of the first row, besides of element (a_{14}) ,has frequency of functional ties equal f=3 with elements of the fourth column;
- Each element(a_{21}, a_{22}, a_{25}) of the second row, besides of element (a_{24}) ,has frequency of functional ties equal f=5 with elements of the first row for positive and negative areas;
- Each element (a_{31}, a_{32}, a_{35}) of the third row, besides of element (a_{34}) ,has frequency of functional ties equal f=6 with elements of the first and second rows for positive and negative areas;
- Each element (a_{51}, a_{52}, a_{55}) of the fifth row, besides of element (a_{54}), has frequency of functional ties equal f=6 with elements of the second and third rows for positive and negative areas.

So, in total we can conclude the following frequency of functional ties between elements of the 5x5 matrix for marked out jointly elements of the third and fourth columns applicably to Laplace's theorem *for all cascade the fourth block (a_{43}) element of the third column:*

$$f=1 \text{ for elements } (a_{14}, a_{24}, a_{34}, a_{54})$$
$$f=3 \text{ for elements } (a_{11}, a_{12}, a_{15})$$
$$f=5 \text{ for elements } (a_{21}, a_{22}, a_{25})$$
$$f=6 \text{ for elements } (a_{31}, a_{32}, a_{35}) \text{ and } (a_{51}, a_{52}, a_{55}).$$

Analysis of data shown in Figure 122 indicates on the following facts applicably to Laplace's theorem for marked out jointly the third and fourth columns of the 5x5 matrix *for all cascade the fifth block (a_{53}) element of the third column:*

- Each element of the fourth column ($a_{14}, a_{24}, a_{34}, a_{44}$) has frequency of functional ties equal f=1 with the fifth block (a_{53}) elements of the third column;
- Each element (a_{11}, a_{12}, a_{15}) of the first row ,besides of element (a_{14}) ,has frequency of functional ties equal f=3 with elements of the fourth column;
- Each element (a_{21}, a_{22}, a_{25}) of the second row, besides of element (a_{24}) ,has frequency of functional ties equal f=5 with elements of the first row for positive and negative areas;
- Each element (a_{31}, a_{32}, a_{35}) of the third row, besides of element (a_{34}) ,has frequency of functional ties equal f=6 with elements of the first and second rows for positive and negative areas;

166

- Each element (a_{41}, a_{42}, a_{45}) of the fourth row, besides of element (a_{44}), has frequency of functional ties equal $f=6$ with elements of the second and third rows for positive and negative areas.

So, in total we can conclude the following frequency of functional ties between elements of the 5x5 matrix for marked out jointly elements of the third and fourth columns applicably to Laplace's theorem *for all cascade the fifth block (a_{53}) element of the third column:*

$f=1$ for elements $(a_{14}, a_{24}, a_{34}, a_{44})$
$f=3$ for elements (a_{11}, a_{12}, a_{15})
$f=5$ for elements (a_{21}, a_{22}, a_{25})
$f=6$ for elements (a_{31}, a_{32}, a_{35}) and (a_{41}, a_{42}, a_{45}).

In Appendix 20 is shown evaluation of the determinant of the 5x5 matrix with using of Laplace's theorem for marked out jointly the third and fourth columns.

i) **For the third and fifth columns**

The value of determinant $|A|$ of the 5x5 matrix at conditions when marked out jointly two columns, such as the third and fifth columns applicably to Laplace's theorem, can be expressed by the following formula in general view:

$$|A| = -M_1A_1 + M_2A_2 - M_3A_3 + M_4A_4 - M_5A_5 + M_6A_6 - M_7A_7 - M_8A_8 + M_9A_9 - M_{10}A_{10} \; (\square)$$

where,

M_1-minor and A_1-algebraic supplemental (cofactor) for marked out jointly the third and fifth columns and also the first and second rows;

M_2- minor and A_2-algebraic supplemental (cofactor) for marked out jointly the third and fifth columns and also the first and third rows;

M_3-minor and A_3-algebraic supplemental (cofactor) for marked out jointly the third and fifth columns and also the first and fourth rows;

M_4-minor and A_4-algebraic supplemental (cofactor) for marked out the third and fifth columns and also the first and fifth rows;

M_5-minor and A_5- algebraic supplemental (cofactor) for marked out the third and fifth columns and also the second and third rows;

M_6-minor and A_6-algebraic supplemental (cofactor) for marked out and third and fifth columns and also the second and fourth rows;

M_7-minor and A_7-algebraic supplemental (cofactor) for marked out the third and fifth columns and also the second and fifth rows;

M_8-minor and A_8- algebraic supplemental (cofactor) for marked out the third and fifth columns and also the third and fourth rows;

M_9-minor and A_9- algebraic supplemental (cofactor) for marked out the third and fifth columns and also the third and fifth rows;

M_{10}-minor and A_{10}-algebraic supplemental (cofactor) for marked out the third and fifth columns and also the fourth and fifth rows.

In Figure123 is shown the functional graph for all cascade first block (a_{13}) elements of the third column applicably to Laplace's theorem for positive (——) and negative (-----) areas. In Figure 124 is shown the functional graph for all cascade second block (a_{23}) elements of the third column applicably to Laplace's theorem for positive (——) and negative (-------) areas. In Figure 125 is shown the functional graphs for all cascade third block (a_{33}) elements of the third column applicably to Laplace's theorem for positive (——) and negative (-----) areas. In Figure 126 is shown the functional graphs for all cascade fourth block (a_{43}) elements of the fourth column applicably to Laplace's theorem for positive (——) and negative areas. In Figure 127 is shown the functional graph for all cascade fifth block (a_{53}) elements f the third column applicably to Laplace's theorem for positive (——) and negative (------) areas.

$M_1 = (a_{13}a_{25} - a_{15}a_{23})$

$$A_1 = (-1)^{(3+5)+(1+2)} \begin{vmatrix} a_{31} & a_{32} & a_{34} \\ a_{41} & a_{42} & a_{44} \\ a_{51} & a_{52} & a_{54} \end{vmatrix}$$

$M_2 = (a_{13}a_{35} - a_{15}a_{33})$

$$A_2 = (-1)^{(3+5)+(1+3)} \begin{vmatrix} a_{21} & a_{22} & a_{24} \\ a_{41} & a_{42} & a_{44} \\ a_{51} & a_{52} & a_{54} \end{vmatrix}$$

$M_3 = (a_{13}a_{45} - a_{15}a_{43})$

$$A_3 = (-1)^{(3+5)+(1+4)} \begin{vmatrix} a_{21} & a_{22} & a_{24} \\ a_{31} & a_{32} & a_{34} \\ a_{51} & a_{52} & a_{54} \end{vmatrix}$$

$M_4 = (a_{13}a_{55} - a_{15}a_{53})$

$$A_4 = (-1)^{(3+5)+(1+5)} \begin{vmatrix} a_{21} & a_{22} & a_{24} \\ a_{31} & a_{32} & a_{34} \\ a_{41} & a_{42} & a_{44} \end{vmatrix}$$

$M_5 = (a_{23}a_{35} - a_{25}a_{33})$

$$A_5 = (-1)^{(3+5)+(2+3)} \begin{vmatrix} a_{11} & a_{12} & a_{14} \\ a_{41} & a_{42} & a_{44} \\ a_{51} & a_{52} & a_{54} \end{vmatrix}$$

$M_6 = (a_{23}a_{45} - a_{25}a_{43})$

$$A_6 = (-1)^{(3+5)+(2+4)} \begin{vmatrix} a_{11} & a_{12} & a_{14} \\ a_{31} & a_{32} & a_{34} \\ a_{51} & a_{52} & a_{54} \end{vmatrix}$$

$M_7 = (a_{23}a_{55} - a_{25}a_{53})$

$$A_7 = (-1)^{(3+5)+(2+5)} \begin{vmatrix} a_{11} & a_{12} & a_{14} \\ a_{31} & a_{32} & a_{34} \\ a_{41} & a_{42} & a_{44} \end{vmatrix}$$

$M_8 = (a_{33}a_{45} - a_{35}a_{43})$

$$A_8 = (-1)^{(3+5)+(3+4)} \begin{vmatrix} a_{11} & a_{12} & a_{14} \\ a_{21} & a_{22} & a_{24} \\ a_{51} & a_{52} & a_{54} \end{vmatrix}$$

$M_9 = (a_{33}a_{55} - a_{35}a_{53})$

$$A_9 = (-1)^{(3+5)+(3+5)} \begin{vmatrix} a_{11} & a_{12} & a_{14} \\ a_{21} & a_{22} & a_{24} \\ a_{41} & a_{42} & a_{44} \end{vmatrix}$$

$M_{10} = (a_{43}a_{55} - a_{45}a_{53})$

$$A_{10} = (-1)^{(3+5)+(4+5)} \begin{vmatrix} a_{11} & a_{12} & a_{14} \\ a_{21} & a_{22} & a_{24} \\ a_{31} & a_{32} & a_{34} \end{vmatrix}$$

After of some calculations and transformations, the formula (□) has the following view *for marked out jointly the third and fifth columns* applicably to Laplace's theorem for the determinant of the 5x5 matrix in general view:

$$|A| = a_{13}a_{25}[a_{31}(a_{44}a_{52}-a_{42}a_{54})+a_{32}(a_{41}a_{54}-a_{44}a_{51})+a_{34}(a_{42}a_{51}-a_{41}a_{52})]+$$
$$+a_{13}a_{35}[a_{21}(a_{42}a_{54}-a_{44}a_{52})+a_{22}(a_{44}a_{51}-a_{41}a_{54})+a_{24}(a_{41}a_{52}-a_{42}a_{51})]+$$
$$+a_{13}a_{45}[a_{21}(a_{34}a_{52}-a_{32}a_{54})+a_{22}(a_{31}a_{54}-a_{34}a_{51})+a_{24}(a_{32}a_{51}-a_{31}a_{52})]+$$
$$+a_{13}a_{55}[a_{21}(a_{32}a_{44}-a_{34}a_{42})+a_{22}(a_{34}a_{41}-a_{31}a_{44})+a_{24}(a_{31}a_{42}-a_{32}a_{41})]+$$

$$+a_{23}a_{15}[a_{31}(a_{42}a_{54}-a_{44}a_{52})+a_{32}(a_{44}a_{51}-a_{41}a_{54})+a_{34}(a_{41}a_{52}-a_{42}a_{51})]+$$
$$+a_{23}a_{35}[a_{11}(a_{44}a_{52}-a_{42}a_{54})+a_{12}(a_{41}a_{54}-a_{44}a_{51})+a_{14}(a_{42}a_{51}-a_{41}a_{52})]+$$
$$+a_{23}a_{45}[a_{11}(a_{32}a_{54}-a_{34}a_{52})+a_{12}(a_{34}a_{51}-a_{31}a_{54})+a_{14}(a_{31}a_{52}-a_{32}a_{51})]+$$
$$+a_{23}a_{55}[a_{11}(a_{34}a_{42}-a_{32}a_{44})+a_{12}(a_{31}a_{44}-a_{34}a_{41})+a_{14}(a_{32}a_{41}-a_{31}a_{42})]+$$

$$+a_{33}a_{15}[a_{21}(a_{44}a_{52}-a_{42}a_{54})+a_{22}(a_{41}a_{54}-a_{44}a_{51})+a_{24}(a_{42}a_{51}-a_{41}a_{52})+$$
$$+a_{33}a_{25}[a_{11}(a_{42}a_{54}-a_{44}a_{52})+a_{12}(a_{44}a_{51}-a_{41}a_{54})+a_{14}(a_{41}a_{52}-a_{42}a_{51})]+$$
$$+a_{33}a_{45}[a_{11}(a_{24}a_{52}-a_{22}a_{54})+a_{12}(a_{21}a_{54}-a_{24}a_{51})+a_{14}(a_{22}a_{51}-a_{21}a_{52})]+$$
$$+a_{33}a_{55}[a_{11}(a_{22}a_{44}-a_{24}a_{42})+a_{12}(a_{24}a_{41}-a_{21}a_{44})+a_{14}(a_{21}a_{42}-a_{22}a_{41})]+$$

$$+a_{43}a_{15}[a_{21}(a_{32}a_{54}-a_{34}a_{52})+a_{22}(a_{34}a_{51}-a_{31}a_{54})+a_{24}(a_{31}a_{52}-a_{32}a_{51})]+$$
$$+a_{43}a_{25}[a_{11}(a_{34}a_{52}-a_{32}a_{54})+a_{12}(a_{31}a_{54}-a_{34}a_{51})+a_{14}(a_{32}a_{51}-a_{31}a_{52})]+$$
$$+a_{43}a_{35}[a_{11}(a_{22}a_{54}-a_{24}a_{52})+a_{12}(a_{24}a_{51}-a_{21}a_{54})+a_{14}(a_{21}a_{52}-a_{22}a_{51})]+$$
$$+a_{43}a_{55}[a_{11}(a_{24}a_{32}-a_{22}a_{34})+a_{12}(a_{21}a_{34}-a_{24}a_{31})+a_{14}(a_{22}a_{31}-a_{21}a_{32})]+$$

$$+a_{53}a_{15}[a_{21}(a_{34}a_{42}-a_{32}a_{44})+a_{22}(a_{31}a_{44}-a_{34}a_{41})+a_{24}(a_{32}a_{41}-a_{31}a_{42})]+$$
$$+a_{53}a_{25}[a_{11}(a_{32}a_{44}-a_{34}a_{42})+a_{12}(a_{34}a_{41}-a_{31}a_{44})+a_{14}(a_{31}a_{42}-a_{32}a_{41})]+$$
$$+a_{53}a_{35}[a_{11}(a_{24}a_{42}-a_{22}a_{44})+a_{12}(a_{21}a_{44}-a_{24}a_{41})+a_{14}(a_{22}a_{41}-a_{21}a_{42})]+$$
$$+a_{53}a_{45}[a_{11}(a_{22}a_{34}-a_{24}a_{32})+a_{12}(a_{24}a_{31}-a_{21}a_{34})+a_{14}(a_{21}a_{32}-a_{22}a_{31})] \quad (104)$$

Analysis of data shown in Figure 123 indicates on the following facts applicably to Laplace's theorem for marked out jointly the third and fifth columns of the 5x5 matrix *for all cascade the first block (a_{13})* element of the third column:

- Each element of the fifth column ($a_{25}, a_{35}, a_{45}, a_{55}$) has frequency of functional ties equal f=1 with the first block(a_{13}) of the third column;
- Each element (a_{21}, a_{22}, a_{24})of the second row, besides of element (a_{25}),has frequency of functional ties equal f=3 with elements of the fifth column;
- Each element (a_{31}, a_{32}, a_{34}) of the third row, besides of element (a_{35}) ,has frequency of functional ties equal f=5 with elements of the second row for positive and negative areas;
- Each element (a_{41}, a_{42}, a_{44}) of the fourth row, besides of element (a_{45}),has frequency of functional ties equal f=6 with elements of the second and third rows for positive and negative areas;
- Each element (a_{51}, a_{52}, a_{54}) of the fifth row, besides of element (a_{55}) ,has frequency of functional ties equal f=6 with elements of the third and fourth rows for positive and negative areas.

So, in total we can conclude the following frequency of functional ties between elements of the 5x5 matrix for marked out jointly elements of the third and fifth columns applicably to Laplace's theorem *for all cascade the first block(a_{13})* element of the third column:

- f=1 for elements ($a_{25}, a_{35}, a_{45}, a_{55}$)
- f=3 for elements (a_{21}, a_{22}, a_{24})
- f=5 for elements (a_{31}, a_{32}, a_{34})
- f=6 for elements (a_{41}, a_{42}, a_{44}) and (a_{51}, a_{52}, a_{54})

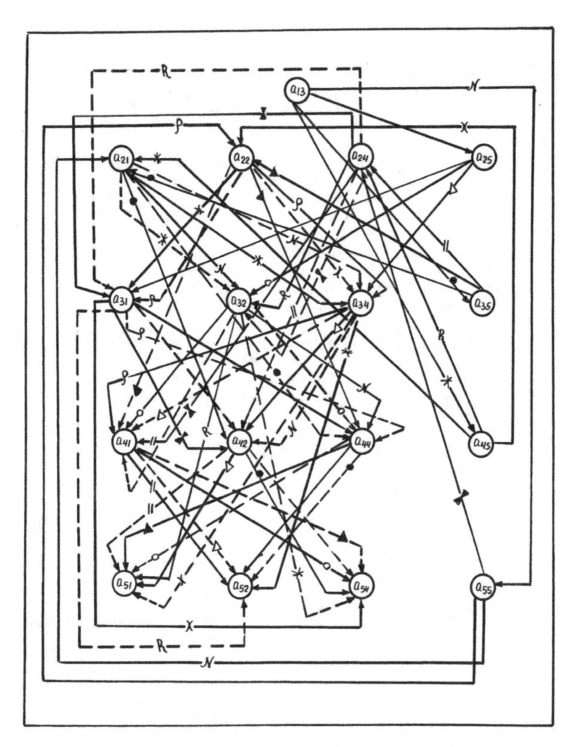

Figure 123 Functional graph for all cascade first block (a₁₃) elements of the third column applicably to Laplace's theorem for positive (———) and negative (------) areas

170

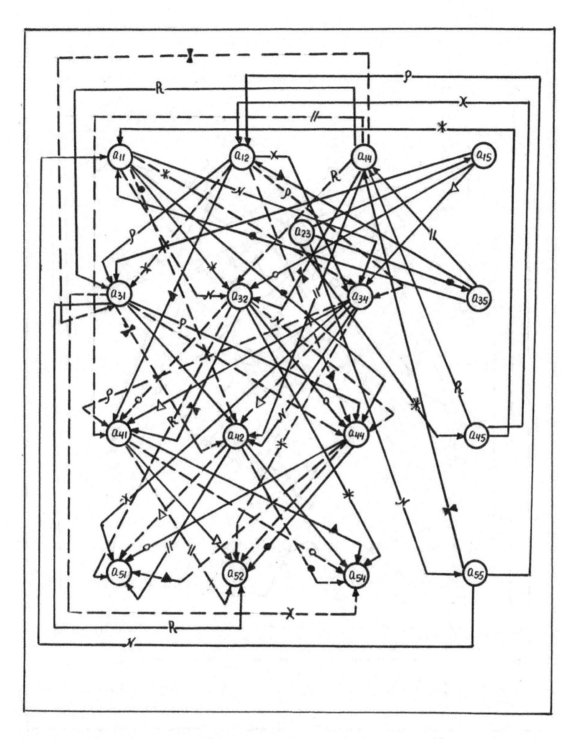

Figure 124 Functional graph for all cascade second block (a₂₃) elements of the third column applicably to Laplace's theorem for positive (——) and negative (------) areas

171

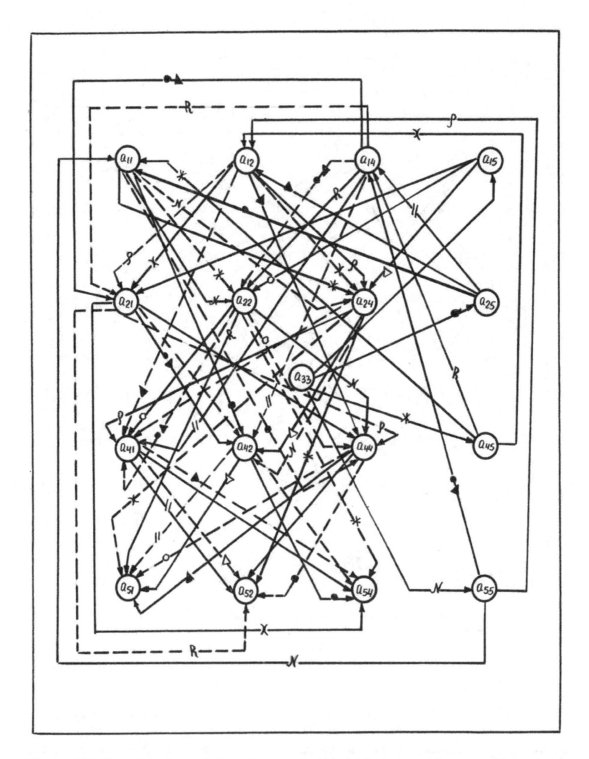

Figure 125 Functional graph for all cascade third block (a$_{33}$) elements of the third column applicably to Laplace's theorem for positive (———) and negative (-------) areas

172

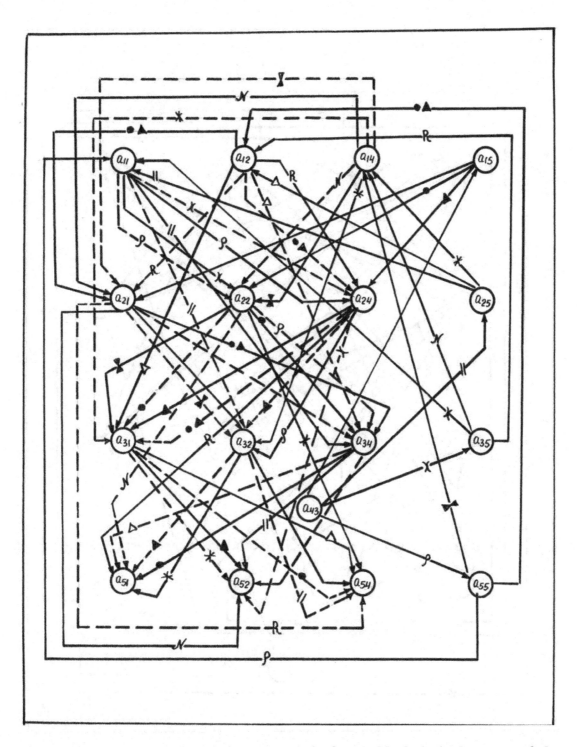

Figure 126 Functional graph for all cascade fourth block (a₄₃) elements   of the fourth column applicably to Laplace's theorem for positive (——) and negative (----) areas

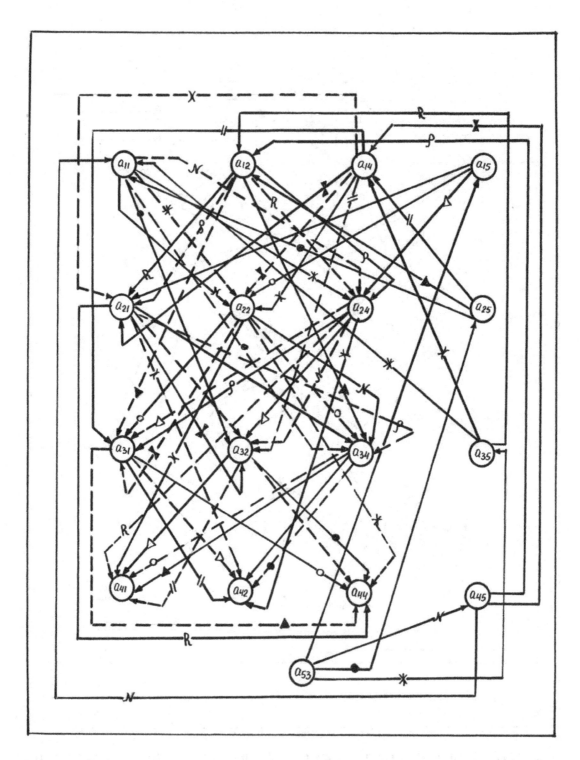

Figure 127 Functional graph for all cascade fifth block (a₅₃) elements of the third column applicably to Laplace's theorem for positive (———) and negative(------) areas

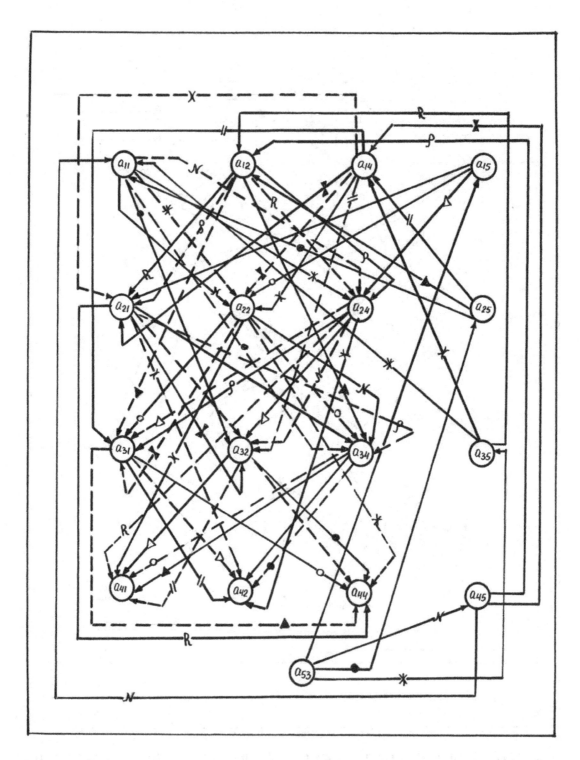

174

Analysis of data shown in Figure 124 indicates on the following facts applicably to Laplace's theorem for marked out jointly the third and fifth columns of the 5x5 matrix *for all cascade the second block (a_{23})* element of the third column:

- Each element of the (a_{15}, a_{35}, a_{45}, a_{55}) has frequency of functional ties equal f=1 with the second block (a_{23}) of the third column;
- Each element(a_{11}, a_{12}, a_{14}) of the first row, besides of element (a_{15}) ,has frequency of functional ties equal f=3 with elements of the fifth column;
- Each element (a_{31}, a_{32}, a_{34}) of the third row ,besides of element(a_{35}), has frequency of functional ties equal f=5 with elements of the first row for positive and negative areas;
- Each element (a_{41}, a_{42}, a_{44}) of the fourth row, besides of element (a_{45}),has frequency of functional ties equal f=6 with elements of the first and third rows for positive and negative areas;
- Each element (a_{51}, a_{52}, a_{54}) of the fifth row, besides of element (a_{55}) ,has frequency of functional ties equal f=6 with elements of the third and fourth rows for positive and negative areas.

So, in total we can conclude the following frequency of functional ties between elements of the 5x5 matrix for marked out jointly elements of the third and fifth columns applicably to Laplace's theorem *for all cascade the second block(a_{23}) element of the third column:*

f=1 for elements (a_{15}, a_{35}, a_{45}, a_{55})
f=3 for elements (a_{11}, a_{12}, a_{14})
f=5 for elements (a_{31}, a_{32}, a_{34})
f=6 for elements (a_{41}, a_{42}, a_{44}) and (a_{51}, a_{52}, a_{54}).

Analysis of data shown in Figure 125 indicates on the following facts applicably to Laplace's theorem for marked out jointly the third and fifth columns of the 5x5 matrix *for all cascade the third block (a_{33}) element of the third column:*

- Each element of the first column (a_{15}, a_{25}, a_{45}, a_{55})has frequency of functional ties equal f=1 with the third block (a_{33}) of the third column;
- Each element (a_{11}, a_{12}, a_{14}) of the first row , besides of element (a_{15}) ,has frequency of functional ties equal f=3 with elements of the fifth column;
- Each element (a_{21}, a_{22}, a_{24}) of the second row, besides of element (a_{25}) ,has frequency of functional ties equal f=5 with elements of the first row for positive and negative areas;
- Each element (a_{41}, a_{42}, a_{44})of the fourth row, besides of element (a_{45}),has frequency of functional ties equal f=6 with elements of the first and second rows for positive and negative areas;
- Each element (a_{51}, a_{52}, a_{54}) of the fifth row , besides of element (a_{55}) ,has frequency of functional ties equal f=6 with elements of the second and fourth rows for positive and negative areas.

So, in total we can conclude the following frequency of functional ties between elements of the third and fifth columns applicably to Laplace's theorem *for all cascade the third block (a_{33}) element of the third column:*

f=1 for elements (a_{15}, a_{25}, a_{45}, a_{55})
f=3 for elements (a_{11}, a_{12}, a_{14})
f=5 for elements (a_{21}, a_{22}, a_{24})
f=6 for elements (a_{41}, a_{42}, a_{44}) and (a_{51}, a_{52}, a_{54}).

Analysis of data shown in Figure 126 indicates on the following facts applicably to Laplace's theorem for marked out jointly the third and fifth columns of the 5x5 matrix *for all cascade the fourth block(a_{43})* element of the third column:

- Each element of the fifth column (a_{15}, a_{25}, a_{35}, a_{55}) has frequency of functional ties equal f=1 with the fourth block(a_{43}) of the third column;
- Each element (a_{11}, a_{12}, a_{14}) of the first row ,besides of element (a_{15}) ,has frequency of functional ties equal f=3 with elements of the fifth column;
- Each element(a_{21}, a_{22}, a_{24})of the second row, besides of element(a_{25}) ,has frequency of functional ties equal f=5 with elements of the first row for positive and negative areas;
- Each element (a_{31}, a_{32}, a_{34}) of the third row, besides of element(a_{35}) ,has frequency of functional ties equal f=6 with elements of the first and second rows for positive and negative areas;

- Each element (a_{51},a_{52},a_{54}) of the fifth row, besides of element (a_{55}) ,has frequency of functional ties equal f=6 with elements of the second and third rows for positive and negative areas.

So, in total we can conclude the following frequency of functional ties between elements of the third and fifth columns applicably to Laplace's theorem *for all cascade the fourth block(a_{43}) element of the third column:*

          f=1 for elements (a_{15},a_{25},a_{35},a_{55})

          f=3 for elements (a_{11},a_{12},a_{14})

          f=5 for elements (a_{21},a_{22},a_{24})

          f=6 for elements (a_{31},a_{32},a_{34}) and (a_{51},a_{52},a_{54}).

Analysis of data shown in Figure 127 indicates on the following facts applicably to Laplace's theorem for marked out jointly the third and fifth columns of the 5x5 matrix *for all cascade the fifth block (a_{53}) elements of the third column:*

- Each element of the fifth column (a_{15},a_{25},a_{35},a_{45}) has frequency of functional ties equal f=1 with the fifth block (a_{53}) of the third column;
- Each element (a_{11},a_{12},a_{14}) of the first row, besides of element(a_{15}) ,has frequency of functional ties equal f=3 with elements of the fifth column;
- Each element (a_{21},a_{22},a_{24}) of the second row, besides of element (a_{25}) ,has frequency of functional ties equal f=5 with elements of the first row for positive and negative areas;
- Each element (a_{31},a_{32},a_{34})of the third row, besides of element (a_{35}) ,has frequency of functional ties equal f=6 with elements of the first and second rows for positive and negative areas;
- Each element (a_{41},a_{42},a_{44}) of the fourth row, besides of element (a_{45}) ,has frequency of functional ties equal f=6 with elements of the second and third rows for positive and negative areas.

So, in total we can conclude the following frequency of functional ties between elements of the third and fifth columns applicably to Laplace's theorem *for all cascade the fifth block(a_{53}) element of the third column:*

          f=1 for elements (a_{15},a_{25},a_{35},a_{45})

          f=3 for elements (a_{11},a_{12},a_{14})

          f=5 for elements (a_{21},a_{22},a_{24}

          f=6 for elements (a_{31},a_{32},a_{34}) and (a_{41},a_{42},a_{44}).

In Appendix 21 is shown evaluation of the determinant of the 5x5 matrix with using of Laplace's theorem for marked out jointly the third and fifth columns.

j)   **For the fourth and fifth columns**

The value of determinant $|A|$ of the 5x5 matrix at conditions when marked out jointly two columns ,such as the *fourth and fifth columns* applicably to Laplace's theorem ,can be expressed by the following formula in general view:

$$|A| = M_1A_1 - M_2A_2 + M_3A_3 - M_4A_4 + M_5A_5 - M_6A_6 + M_7A_7 + M_8A_8 - M_9A_9 + M_{10}A_{10} \ (\blacktriangle)$$

where,

M_1-minor and A_1- algebraic supplemental (cofactor) for marked out jointly the fourth and fifth columns and also the first and second rows;

M_2- minor and A_2-algebraic supplemental (cofactor) for marked out jointly the fourth and fifth columns and also the first and third rows;

M_3-minor and A_3-algebraic supplemental (cofactor) for marked out jointly the fourth and fifth columns and also the first and fourth rows;

M_4- minor and A_4-algebraic supplemental (cofactor) for marked out jointly the fourth and fifth columns and also the first and fifth rows;

M_5- minor and A_5-algebraic supplemental (cofactor) for marked out jointly the fourth and fifth columns and also the second and third rows;

M_6-minor and A_6-algebraic supplemental (cofactor) for marked out jointly the fourth and fifth columns and also the second and fourth rows;

M_7-minor and A_7- algebraic supplemental (cofactor) for marked out jointly the fourth and fifth columns and also the second and fifth rows;

176

M_8- minor and A_8-algebraic supplemental (cofactor) for marked out jointly the fourth and fifth columns and also the third and fourth rows;

M_9-minor and A_9-algebraic supplemental (cofactor) for marked out jointly the fourth and fifth columns and also the third and fifth rows;

M_{10}-minor and A_{10}- algebraic supplemental (cofactor) for marked out jointly the fourth and fifth columns and also the fourth and fifth rows.

$$M_1 = (a_{14}a_{25} - a_{15}a_{24}) \qquad A_1 = (-1)^{(4+5)+(1+2)} \begin{vmatrix} a_{31} & a_{32} & a_{33} \\ a_{41} & a_{42} & a_{43} \\ a_{51} & a_{52} & a_{53} \end{vmatrix}$$

$$M_2 = (a_{14}a_{35} - a_{15}a_{34}) \qquad A_2 = (-1)^{(4+5)+(1+3)} \begin{vmatrix} a_{21} & a_{22} & a_{23} \\ a_{41} & a_{42} & a_{43} \\ a_{51} & a_{52} & a_{53} \end{vmatrix}$$

$$M_3 = (a_{14}a_{45} - a_{15}a_{44}) \qquad A_3 = (-1)^{(4+5)+(1+4)} \begin{vmatrix} a_{21} & a_{22} & a_{23} \\ a_{31} & a_{32} & a_{33} \\ a_{51} & a_{52} & a_{53} \end{vmatrix}$$

$$M_4 = (a_{14}a_{55} - a_{15}a_{54}) \qquad A_4 = (-1)^{(4+5)+(1+5)} \begin{vmatrix} a_{21} & a_{22} & a_{23} \\ a_{31} & a_{32} & a_{33} \\ a_{41} & a_{42} & a_{43} \end{vmatrix}$$

$$M_5 = (a_{24}a_{35} - a_{25}a_{34}) \qquad A_5 = (-1)^{(4+5)+(2+3)} \begin{vmatrix} a_{11} & a_{12} & a_{13} \\ a_{41} & a_{42} & a_{43} \\ a_{51} & a_{52} & a_{53} \end{vmatrix}$$

$$M_6 = (a_{243}a_{45} - a_{25}a_{44}) \qquad A_6 = (-1)^{(4+5)+(2+4)} \begin{vmatrix} a_{11} & a_{12} & a_{13} \\ a_{31} & a_{32} & a_{33} \\ a_{51} & a_{52} & a_{53} \end{vmatrix}$$

$$M_7 = (a_{24}a_{55} - a_{25}a_{54}) \qquad A_7 = (-1)^{(4+5)+(2+5)} \begin{vmatrix} a_{11} & a_{12} & a_{13} \\ a_{31} & a_{32} & a_{33} \\ a_{41} & a_{42} & a_{43} \end{vmatrix}$$

$$M_8 = (a_{34}a_{45} - a_{35}a_{44}) \qquad A_8 = (-1)^{(4+5)+(3+4)} \begin{vmatrix} a_{11} & a_{12} & a_{13} \\ a_{21} & a_{22} & a_{23} \\ a_{51} & a_{52} & a_{53} \end{vmatrix}$$

$$M_9 = (a_{34}a_{55} - a_{35}a_{54}) \qquad A_9 = (-1)^{(4+5)+(3+5)} \begin{vmatrix} a_{11} & a_{12} & a_{13} \\ a_{21} & a_{22} & a_{23} \\ a_{41} & a_{42} & a_{43} \end{vmatrix}$$

$$M_{10} = (a_{44}a_{55} - a_{45}a_{54}) \qquad A_{10} = (-1)^{(4+5)+(4+5)} \begin{vmatrix} a_{11} & a_{12} & a_{13} \\ a_{21} & a_{22} & a_{23} \\ a_{31} & a_{32} & a_{33} \end{vmatrix}$$

After of some calculations and transformations, the formula (▲) has the following view *for marked out jointly the fourth and fifth columns* applicably to Laplace's theorem for the determinant of the 5x5 matrix in general view:

$$|\,A\,| = a_{14}a_{25}[a_{31}(a_{42}a_{53}-a_{43}a_{52})+a_{32}(a_{43}a_{51}-a_{41}a_{53})+a_{33}(a_{41}a_{52}-a_{42}a_{51})]+$$
$$+a_{14}a_{35}[a_{21}(a_{43}a_{52}-a_{42}a_{53})+a_{22}(a_{41}a_{53}-a_{43}a_{51})+a_{23}(a_{42}a_{51}-a_{41}a_{52})]+$$
$$+a_{14}a_{45}[a_{21}(a_{32}a_{53}-a_{33}a_{52})+a_{22}(a_{33}a_{51}-a_{31}a_{53})+a_{23}(a_{31}a_{52}-a_{32}a_{51})]+$$
$$+a_{14}a_{55}[a_{21}(a_{33}a_{42}-a_{32}a_{43})+a_{22}(a_{31}a_{43}-a_{33}a_{41})+a_{23}(a_{32}a_{41}-a_{31}a_{42})]+$$

$$+a_{24}a_{15}[a_{31}(a_{43}a_{52}-a_{42}a_{53})+a_{32}(a_{41}a_{53}-a_{43}a_{51})+a_{33}(a_{42}a_{51}-a_{41}a_{52})]+$$
$$+a_{24}a_{35}[a_{11}(a_{42}a_{53}-a_{43}a_{52})+a_{12}(a_{43}a_{51}-a_{41}a_{53})+a_{13}(a_{41}a_{52}-a_{42}a_{51})]+$$
$$+a_{24}a_{45}[a_{11}(a_{33}a_{52}-a_{32}a_{53})+a_{12}(a_{31}a_{53}-a_{33}a_{51})+a_{13}(a_{32}a_{51}-a_{31}a_{52})]+$$
$$+a_{24}a_{55}[a_{11}(a_{32}a_{43}-a_{33}a_{42})+a_{12}(a_{33}a_{41}-a_{31}a_{43})+a_{13}(a_{31}a_{42}-a_{32}a_{41})]+$$

$$+a_{34}a_{15}[a_{21}(a_{42}a_{53}-a_{43}a_{52})+a_{22}(a_{43}a_{51}-a_{41}a_{53})+a_{23}(a_{41}a_{52}-a_{42}a_{51})]+$$
$$+a_{34}a_{25}[a_{11}(a_{43}a_{52}-a_{42}a_{53})+a_{12}(a_{41}a_{53}-a_{43}a_{51})+a_{13}(a_{42}a_{51}-a_{41}a_{52})]+$$
$$+a_{34}a_{45}[a_{11}(a_{22}a_{53}-a_{23}a_{52})+a_{12}(a_{23}a_{51}-a_{21}a_{53})+a_{13}(a_{21}a_{52}-a_{22}a_{51})]+$$
$$+a_{34}a_{55}[a_{11}(a_{23}a_{42}-a_{22}a_{43})+a_{12}(a_{21}a_{43}-a_{23}a_{41})+a_{13}(a_{22}a_{41}-a_{21}a_{42})]+$$

$$+a_{44}a_{15}[a_{21}(a_{33}a_{52}-a_{32}a_{53})+a_{22}(a_{31}a_{53}-a_{33}a_{51})+a_{23}(a_{32}a_{51}-a_{31}a_{52})]+$$
$$+a_{44}a_{25}[a_{11}(a_{32}a_{53}-a_{33}a_{52})+a_{12}(a_{33}a_{51}-a_{31}a_{53})+a_{13}(a_{31}a_{52}-a_{32}a_{51})]+$$
$$+a_{44}a_{35}[a_{11}(a_{23}a_{52}-a_{22}a_{53})+a_{12}(a_{21}a_{53}-a_{23}a_{51})+a_{13}(a_{22}a_{51}-a_{21}a_{52})]+$$
$$+a_{44}a_{55}[a_{11}(a_{22}a_{33}-a_{23}a_{32})+a_{12}(a_{23}a_{31}-a_{21}a_{33})+a_{13}(a_{21}a_{32}-a_{22}a_{31})]+$$

$$+a_{54}a_{15}[a_{21}(a_{32}a_{43}-a_{33}a_{42})+a_{22}(a_{33}a_{41}-a_{31}a_{43})+a_{23}(a_{31}a_{42}-a_{32}a_{41})]+$$
$$+a_{54}a_{25}[a_{11}(a_{33}a_{42}-a_{32}a_{43})+a_{12}(a_{31}a_{43}-a_{33}a_{41})+a_{13}(a_{32}a_{41}-a_{31}a_{42})]+$$
$$+a_{54}a_{35}[a_{11}(a_{22}a_{43}-a_{23}a_{42})+a_{12}(a_{23}a_{41}-a_{21}a_{43})+a_{13}(a_{21}a_{42}-a_{22}a_{41})]+$$
$$+a_{54}a_{45}[a_{11}(a_{23}a_{32}-a_{22}a_{33})+a_{12}(a_{21}a_{33}-a_{23}a_{31})+a_{13}(a_{22}a_{31}-a_{21}a_{32})] \qquad (105)$$

In Figure 128 is shown the functional graph of all cascade the first block (a_{14}) element of the fourth column applicably to Laplace's theorem for positive (——) and negative (-------) areas. In Figure 129 is shown the functional graph of all cascade the second block (a_{24}) element of the fourth column applicably to Laplace's theorem for positive (——) and negative(------) areas. In Figure 130 is shown the functional graph of all cascade the third block (a_{34}) element of the fourth column applicably to Laplace's theorem for positive (——) and negative (------) areas. In Figure 131 is shown the functional graph of all cascade the fourth block(a_{44}) element of the fourth column applicably to Lapalce's theorem for positive (——) and negative (-------) areas. In Figure 132 is shown the functional graph of all cascade fifth block (a_{54}) element of the fourth column applicably to Lapalce's theorem for positive (——) and negative (-------) areas.

Analysis of data shown in Figure 128 indicates on the following facts applicably to Laplace's theorem for marked out jointly the fourth and fifth columns of the 5x5 matrix *for all cascade the first block (a_{14}) element of the fourth column:*

- Each element of the fifth column($a_{25},a_{35},a_{45},a_{55}$) has frequency of functional ties equal f=1 with the first block (a_{14}) of the fourth column;
- Each element (a_{21},a_{22},a_{23}) of the second row, besides of element (a_{25}) ,has frequency of functional ties equal f=3 with elements of the fifth column;
- Each element (a_{31},a_{32},a_{33}) of the third row, besides of element(a_{35}) ,has frequency of functional ties equal f=5 with elements of the second row for positive and negative areas;

178

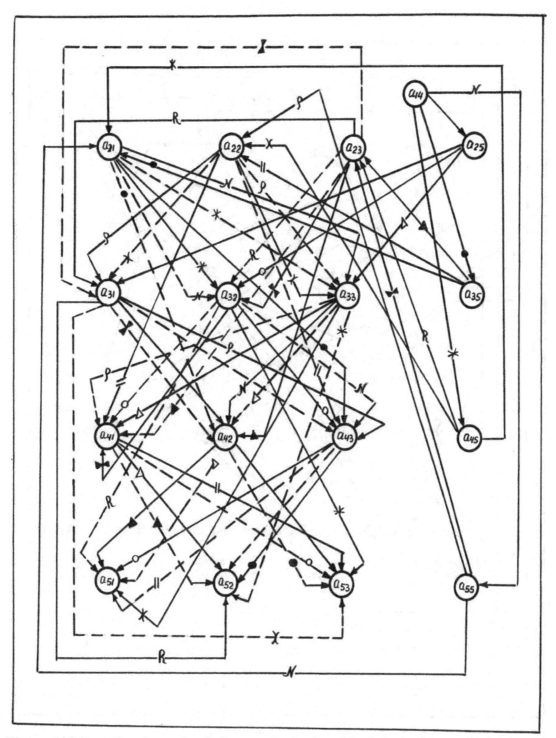

Figure 128 Functional graph of all cascade the first block (a₁₄) element of the fourth column applicably to Laplace's theorem for positive (————) and negative (------) areas

179

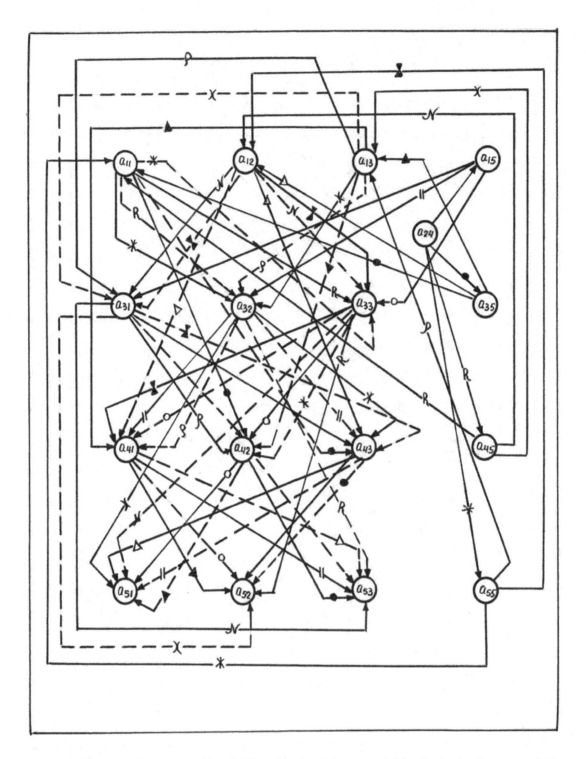

Figure 129 Functional graph of all cascade the second block (a₂₄) element of the fourth column applicably to Laplace's theorem for positive (———) and negative (-------) areas

180

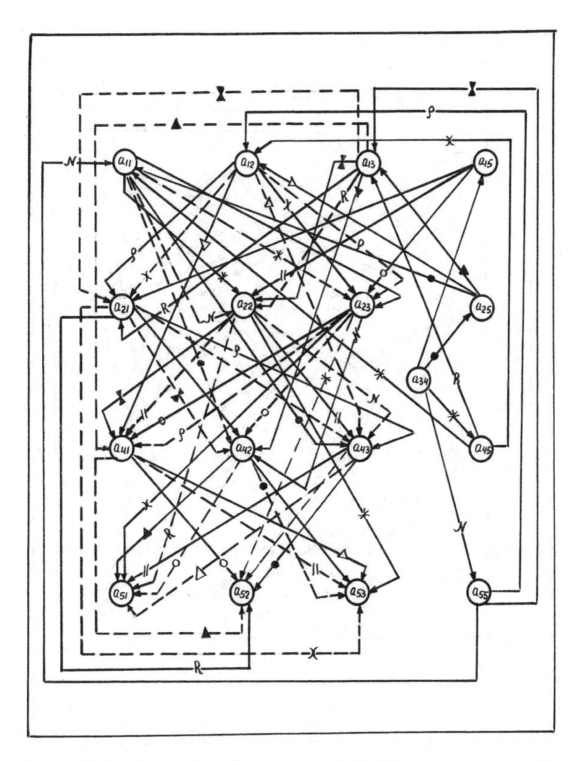

Figure 130 Functional graph of all cascade the third block (a_{34}) element of the fourth column applicably to Laplace's theorem for positive (———) and negative (------) areas

181

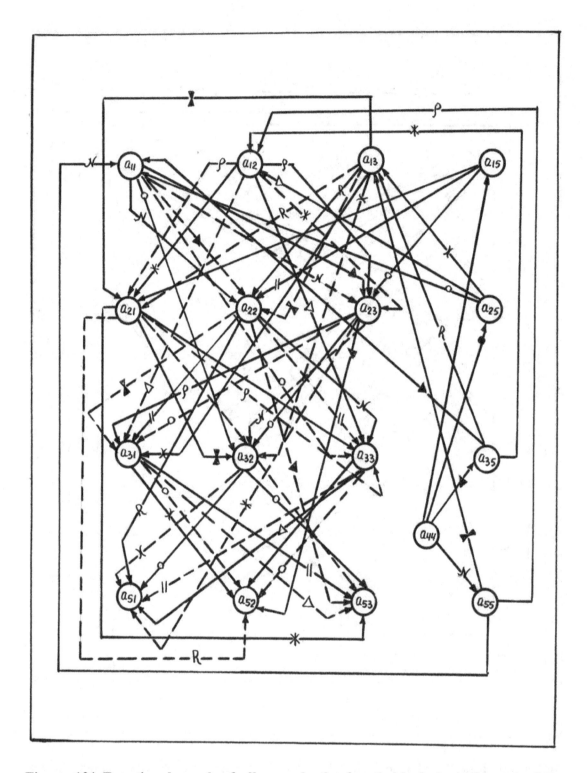

Figure 131 Functional graph of all cascade the fourth block (a₄₄) element of the fourth column applicably to Laplace's theorem for positive(——) and negative(-----) areas

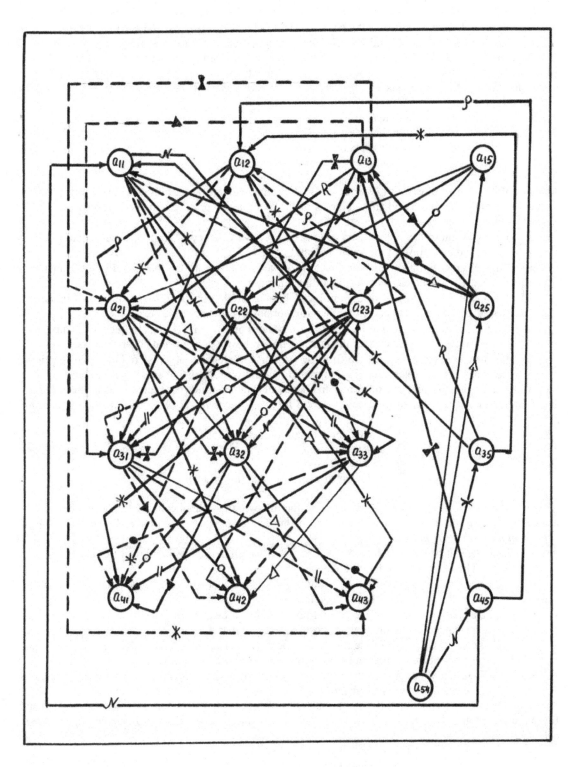

Figure 132 Functional graph of all cascade the fifth block (a₅₄) element of the fourth column applicably to Lapalce's theorem for positive(———) and negative (------) areas

- Each element (a_{41},a_{42},a_{43}) of the fourth row, besides of element (a_{45}) ,has frequency of functional ties equal f=6 with elements of the second and third rows for positive and negative areas;
- Each element (a_{51},a_{52},a_{53}) of the fifth row ,besides of element (a_{55}) ,has frequency of functional ties equal f=6 with elements of the third and fourth rows for positive and negative areas.

So, in total we can conclude the following frequency of functional ties between elements of the fourth and fifth columns applicably to Laplace's theorem *for all cascade the first block(a_{14}) element of the fourth column:*

$$f=1 \text{ for elements } (a_{25},a_{35},a_{45},a_{55})$$
$$f=3 \text{ for elements } (a_{21},a_{22},a_{23})$$
$$f=5 \text{ for elements } (a_{31},a_{32},a_{33})$$
$$f=6 \text{ for elements } (a_{41},a_{42},a_{43)} \text{ and } (a_{51},a_{52},a_{53}).$$

Analysis of data shown in Figure 129 indicates on the following facts applicably to Laplace's theorem for marked out jointly the fourth and fifth columns of the 5x5 matrix for *all cascade the second block (a_{24}) element of the fourth column:*

- Each element of the fifth column (a_{15},a_{35},a_{45},a_{55}) has frequency of functional ties equal f=1 with the second block (a_{24}) of the fourth column;
- Each element (a_{11},a_{12},a_{13}) of the first row, besides of element (a_{15}) ,has frequency of functional ties equal f=3 with elements of the fifth column;
- Each element (a_{31},a_{32},a_{33}) of the third row, besides of element (a_{35}) ,has frequency of functional ties equal f=5 with elements of the first row for positive and negative areas;
- Each element (a_{41},a_{42},a_{43}) of the fourth row, besides of element(a_{45}) ,has frequency of functional ties equal f=6 with elements of the first and third rows for positive and negative areas;
- Each element (a_{51},a_{52},a_{53}) of the fifth row ,besides of element (a_{55}) ,has frequency of functional ties equal f=6 with elements of the third and fourth rows for positive and negative areas.

So , in total we can conclude the following frequency of functional ties between elements of the fourth and fifth columns applicably to Laplace's theorem *for all cascade the second block(a_{24}) element o fthe fourth column:*

$$f=1 \text{ for elements } (a_{15},a_{35},a_{45},a_{55})$$
$$f=3 \text{ for elements } (a_{11},a_{12},a_{13})$$
$$f=5 \text{ for elements } (a_{31},a_{32},a_{33})$$
$$f=6 \text{ for elements } (a_{41},a_{42},a_{43}) \text{ and } (a_{51},a_{52},a_{53}).$$

Analysis of data shown in Figure 130 indicates on the following facts applicably to Laplace's theorem for marked out jointly the fourth and fifth columns of the 5x5 matrix *for all cascade the third block (a_{34}) element of the fourth column:*

- Each element of the fifth column (a_{15},a_{25},a_{45},a_{55}) has frequency of functional ties equal f=1 with the third block (a_{34}) of the fourth column;
- Each element (a_{11},a_{12},a_{13}) of the first row, besides of element (a_{15}),has frequency of functional ties equal f=3 with elements of the fifth column;
- Each element (a_{21},a_{22},a_{23}) of the second row ,besides of element (a_{25}),has frequency of ties equal f=5 with elements of the first row for positive and negative areas;
- Each element (a_{41},a_{42},a_{43}) of the fourth row, besides of element (a_{45}), has frequency of ties equal f=6 with elements of the first and second rows for positive and negative areas;
- Each element (a_{51},a_{52},a_{53}) of the fifth row, besides of element (a_{55}) ,has frequency of ties equal f=6 with elements of the second and fourth rows for positive and negative areas.

So, in total we can conclude the following frequency of functional ties between elements of the fourth and fifth columns applicably to Laplace's theorem *for all cascade the third block(a_{34}) element of the fourth column:*

$$f=1 \text{ for elements } (a_{15},a_{25},a_{45},a_{55})$$
$$f=3 \text{ for elements } (a_{11},a_{12},a_{13})$$
$$f=5 \text{ for elements } (a_{21},a_{22,}a_{23})$$
$$f=6 \text{ for elements } (a_{41},a_{42},a_{43}) \text{ and } (a_{51},a_{52},a_{53}).$$

Analysis of data shown in Figure 131indicates on the following facts applicably to Laplace's theorem for marked out jointly the fourth and fifth columns of the 5x5 matrix *for all cascade the fourth block (a_{44}) element of the fourth column:*

- Each element of the fifth column ($a_{15},a_{25},a_{35},a_{55}$) has frequency of functional ties equal f=1 with the fourth block (a_{44}) of the fourth column;
- Each element (a_{11},a_{12},a_{13}) of the first row, besides of element (a_{15}),has frequency of functional ties equal f=3 with elements of the fifth column;
- Each element (a_{21},a_{22},a_{23})of the second row, besides of element (a_{25}),has frequency of functional ties equal f=5 with elements of the first row for positive and negative areas;
- Each element (a_{31},a_{32},a_{33}) of the third row , besides of element (a_{35}) ,has frequency of functional ties equal f=6 with elements of the first and second rows for positive and negative areas;
- Each element(a_{51},a_{52},a_{53}) of the fifth row),besides of element (a_{55}),has frequency of functional ties equal f=6 with elements of the second and third rows for positive and negative areas.

So, in total we can conclude the following frequency of functional ties between elements of the fourth and fifth columns applicably to Lapalce's theorem *for all cascade the fourth block (a_{44}) element of the fourth column:*

f=1 for elements ($a_{15},a_{25},a_{35},a_{55}$)
f=3 for elements (a_{11},a_{12},a_{13})
f=5 for elements (a_{21},a_{22},a_{23})
f=6 for elements (a_{31},a_{32},a_{33}) and (a_{51},a_{52},a_{53}).

Analysis of data shown in Figure 132 indicates on the following facts applicably to Laplace's theorem for marked out jointly the fourth and fifth columns of the 5x5 matrix *for all cascade the fifth block (a_{54}) element of the fourth column:*

- Each element of the fifth column ($a_{15},a_{25},a_{35},a_{45}$) has frequency of functional ties equal f=1 with the fifth block (a_{54}) of the fourth column;
- Each element (a_{11},a_{12},a_{13}) of the first row, besides of element (a_{15}) ,has frequency of functional ties equal f=3 with elements of the fifth column;
- Each element (a_{21},a_{22},a_{23}) of the second row, besides of element (a_{25}),has frequency of ties equal f=5 with elements of the first row for positive and negative areas;
- Each element (a_{31},a_{32},a_{33}) of the third row, besides of element(a_{35}) ,has frequency of ties equal f=6 with elements of the first and second rows for positive and negative areas;
- Each element (a_{41},a_{42},a_{43}) of the fourth row, besides of element (a_{45}),has frequency of ties equal f=6 with elements of the second and third rows for positive and negative areas.

So ,in total we can conclude the following frequency of functional ties between elements of the fourth and fifth columns applicably to Laplace's theorem *for all cascade the fifth block (a_{54}) element of the fourth column:*

f=1 for elements ($a_{15},a_{25},a_{35},a_{45}$)
f=3 for elements (a_{11},a_{12},a_{13})
f=5 for elements (a_{21},a_{22},a_{23})
f=6 for elements (a_{31},a_{32},a_{33}) and (a_{41},a_{42},a_{43}).

In Appendix 22 is shown evaluation of the determinant of the 5x5 matrix with using of Laplace's theorem *for marked out jointly the fourth and fifth columns.*

CHAPTER FOUR ALGORITHMS AND GRAPHS FOR SOLVING DETERMINANTS WITH 6X6 MATRICES

1. The general algorithms and functional graphs for solving of the determinant of the 6x6 matrix

1.1 By expansion (cofactor) method for some rows

Theory of determinants advantageously is joined with studying of a system linear equations which can be introduced in general view as:

$$a_{11}X_1+a_{12}X_2+a_{13}X_3+\ldots\ldots a_{1n}X_n=b_1$$
$$a_{21}X_1+a_{22}X_2+a_{23}X_3+\ldots\ldots a_{2n}X_n=b_2 \qquad (*)$$
$$a_{31}X_1+a_{32}X_2+a_{33}X_3+\ldots\ldots a_{3n}X_n=b_3$$
$$\ldots\ldots\ldots\ldots\ldots\ldots\ldots\ldots\ldots$$
$$a_{n1}X_1+a_{n2}X_2+a_{n3}X_3+\ldots\ldots a_{nn}X_n=b_n$$

where,

$a_{11};a_{12};a_{13};\ldots.a_{1n}$-elements of the first row

$a_{21};a_{22};a_{23};\ldots.a_{2n}$-elemenets of the second row

$a_{31};a_{32};a_{33};\ldots.a_{3n}$-elemeents of the third row

$\ldots\ldots\ldots\ldots\ldots\ldots\ldots\ldots\ldots\ldots\ldots$

$a_{n1};a_{n2};a_{n3}$-elemeents of n-row

$b_1;b_2;b_3;\ldots\ldots b_n$-free elements ($b_n\neq0$ for not uniform system)

$X_1;X_2;X_3\ldots.X_n$-unknown parameters of system.

The above-named system (*) can be expressed in general view as nxn matrix:

$$A=\begin{vmatrix} a_{11} & a_{12} & a_{13}\ldots\ldots a_{1n} \\ a_{21} & a_{22} & a_{23}\ldots\ldots a_{2n} \\ a_{31} & a_{32} & a_{33}\ldots\ldots a_{3n} \\ \ldots & \ldots & \ldots\ldots\ldots\ldots \\ a_{n1} & a_{n2} & a_{n3}\ldots\ldots a_{nn} \end{vmatrix}$$

Referring to well-known references **[1]** ,**[2]** ,**[8]** ,we can now to conclude that determinant of nxn matrix evaluated by expansion (cofactor) method in general view across *to any row* can be defined as:

$$|A|=a_{11}A_{11}+a_{12}A_{12}+a_{13}A_{13}+\ldots.a_{1n}A_{1n}=$$
$$=a_{21}A_{21}+a_{22}A_{22}+a_{23}A_{23}+\ldots.a_{2n}A_{2n}= \qquad (106)$$
$$=a_{31}A_{31}+a_{32}A_{32}+a_{33}A_{33}+\ldots.a_{3n}A_{3n}=$$
$$\bullet$$
$$=a_{n1}A_{n1}+a_{n2}A_{n2}+a_{n3}A_{n3}+\ldots.a_{nn}A_{nn}=\sum_{i=1}^{n}a_{in}A_{in}\ (i=1,2,3\ldots n)$$

or *to any column:*

$$|A|=a_{11}A_{11}+a_{21}a_{21}+a_{31}A_{31}+\ldots\ldots a_{n1}A_{n1}=$$
$$=a_{12}A_{12}+a_{22}A_{22}+a_{32}A_{32}+\ldots\ldots a_{n2}A_{n2}= \qquad (107)$$
$$=a_{13}A_{13}+a_{23}A_{23}+a_{33}A_{33}+\ldots\ldots a_{n3}A_{n3}=$$
$$\bullet$$
$$=a_{1j}A_{1j}+a_{2j}A_{2j}+a_{3j}A_{3j}+\ldots.a_{nj}A_{nj}=\sum_{j=1}^{n}a_{nj}A_{nj}(j=1,2,3..n)$$

Analysis of formulas (105) and (106) shows the essential fact that evaluation of the determinant $|A|$ of any nxn matrix by Expansion (cofactor) method *always equal of sum product of its elements on their algebraic supplemental (cofactor) expressed in view of* $A_{ij}=(-1)^{i+j} M_{ij}$,*i.e* we can finally to conclude that determinant $|A|$ of any matrix at expansion of its elements *to any row has value equal:*

$$|A|=\sum_{i=1}^{n}(-1)^{i+j} a_{in}A_{in}\ (i=1,2,3\ldots n;j=1,2,3\ldots.n)$$

or *to any column* the value of determinant is equal:

$$|A|=\sum_{j=1}^{n}(-1)^{j+i} a_{nj}A_{nj}\ (i=1,2,3..n;\ j=1,2,3..n)$$

So, we see that in both versions the determinant $|A|$ of any nxn matrix has the same value, i.e we have

$$|A|=\sum_{i=1}^{N}(-1)^{i+j} a_{in}A_{in}=\sum_{j=1}^{n}(-1)^{j+I} a_{nj}A_{nj} \qquad (108)$$

The above-indicated conclusions can be confirmed by some evaluation of the determinant $|A|$ of the 6x6 matrix *by Expansion method* at conditions when its elements have expansion, for a instance , *to the first row:*

a) **To the first row**

$$A=\begin{vmatrix} a_{11} & a_{12} & a_{13} & a_{14} & a_{15} & a_{16} \\ a_{21} & a_{22} & a_{23} & a_{24} & a_{25} & a_{26} \\ a_{31} & a_{32} & a_{33} & a_{34} & a_{35} & a_{36} \\ a_{41} & a_{42} & a_{43} & a_{44} & a_{45} & a_{46} \\ a_{51} & a_{52} & a_{53} & a_{54} & a_{55} & a_{56} \\ a_{61} & a_{62} & a_{63} & a_{64} & a_{65} & a_{66} \end{vmatrix}$$

In this case, we have the value of the determinant of the 6x6 matrix in view of:

$$|A|=a_{11}A_{11}-a_{12}A_{12}+a_{13}A_{13}-a_{14}A_{14}+a_{15}A_{15}-a_{16}A_{16}$$

where,

- In the 6x6 matrix primary mark out jointly the first and then the first column and write the minor M_{11} of element a_{11} and algebraic supplemental (cofactor) A_{11}:

$$A_{11}=(-1)^{1+1}\begin{vmatrix} a_{22} & a_{23} & a_{24} & a_{25} & a_{26} \\ a_{32} & a_{33} & a_{34} & a_{35} & a_{36} \\ a_{42} & a_{43} & a_{44} & a_{45} & a_{46} \\ a_{52} & a_{53} & a_{54} & a_{55} & a_{56} \\ a_{62} & a_{63} & a_{64} & a_{65} & a_{66} \end{vmatrix}$$

- And then mark out the first row and the second column and write the minor M_{12} of element a_{12} and algebraic supplemental (cofactor)A_{12}:

$$A_{12}=(-1)^{1+2}\begin{vmatrix} a_{21} & a_{23} & a_{24} & a_{25} & a_{26} \\ a_{31} & a_{33} & a_{34} & a_{35} & a_{36} \\ a_{41} & a_{43} & a_{44} & a_{45} & a_{46} \\ a_{51} & a_{53} & a_{54} & a_{55} & a_{56} \\ a_{61} & a_{63} & a_{64} & a_{65} & a_{66} \end{vmatrix}$$

- And then for the third step mark out the first row and the third column and write the minor M_{13} of element a_{13} of the determinant and algebraic supplemental (cofactor)A_{13}:

$$A_{13}=(-1)^{1+3}\begin{vmatrix} a_{21} & a_{22} & a_{24} & a_{25} & a_{26} \\ a_{31} & a_{32} & a_{34} & a_{35} & a_{36} \\ a_{41} & a_{42} & a_{44} & a_{45} & a_{46} \\ a_{51} & a_{52} & a_{54} & a_{55} & a_{56} \\ a_{61} & a_{62} & a_{64} & a_{65} & a_{66} \end{vmatrix}$$

- And then mark the first row and the fourth column and write the minor M_{14} of element a_{14} of the determinant and algebraic supplemental (cofactor) A_{14}:

$$A_{14}=(-1)^{1+4}\begin{vmatrix} a_{21} & a_{22} & a_{23} & a_{25} & a_{26} \\ a_{31} & a_{32} & a_{33} & a_{35} & a_{36} \\ a_{41} & a_{42} & a_{43} & a_{45} & a_{46} \\ a_{51} & a_{52} & a_{53} & a_{55} & a_{56} \\ a_{61} & a_{62} & a_{63} & a_{65} & a_{66} \end{vmatrix}$$

- On the fifth step we mark out the first row and the fifth column and write the minor M_{15} of element a_{15} of the determinant and algebraic supplemental (cofactor) A_{15}:

$$A_{15}=(-1)^{1+5}\begin{vmatrix} a_{21} & a_{22} & a_{23} & a_{24} & a_{26} \\ a_{31} & a_{32} & a_{33} & a_{34} & a_{36} \\ a_{41} & a_{42} & a_{43} & a_{44} & a_{46} \\ a_{51} & a_{52} & a_{53} & a_{54} & a_{56} \\ a_{61} & a_{62} & a_{63} & a_{64} & a_{66} \end{vmatrix}$$

- And finally we mark out the first row and the six column and write the minor M_{16} of element a_{16} of the determinant and algebraic supplemental (cofactor) A_{16}:

$$A_{16}=(-1)^{1+6}\begin{vmatrix} a_{21} & a_{22} & a_{23} & a_{24} & a_{25} \\ a_{31} & a_{32} & a_{33} & a_{34} & a_{35} \\ a_{41} & a_{42} & a_{43} & a_{44} & a_{45} \\ a_{51} & a_{52} & a_{53} & a_{54} & a_{55} \\ a_{61} & a_{62} & a_{63} & a_{64} & a_{65} \end{vmatrix}$$

So, after *of changing all sizes of above-indicated 5x5 matrices for algebraic supplemental (A_{11}, A_{12}, A_{13}, A_{14}, A_{15}, A_{16}) to the sizes of 3x3 matrices* , we have after of some calculations and transformations ,the following formula for evaluation of the combined summary determinant $\sum |A|$ of the 6x6 matrix in general view is equal:

$$\sum |A| = a_{11}A_{11} - a_{12}A_{12} + a_{13}A_{13} - a_{14}A_{14} + a_{15}A_{15} - a_{16}A_{16} =$$
$$= |A'_{11}| + |A'_{12}| + |A'_{13}| + |A'_{14}| + |A'_{15}| + |A'_{16}| \qquad (109)$$

where,
- combined partial determinant: $|A'_{11}| = |A_1| + |A_2| + |A_3| + |A_4| + |A_5|$;
- combined partial determinant: $|A'_{12}| = |A_6| + |A_7| + |A_8| + |A_9| + |A_{10}|$;
- combined partial determinant: $|A'_{13}| = |A_{11}| + |A_{12}| + |A_{13}| + |A_{14}| + |A_{15}|$;
- combined partial determinant: $|A'_{14}| = |A_{16}| + |A_{17}| + |A_{18}| + |A_{19}| + |A_{20}|$;
- combined partial determinant $|A'_{15}| = |A_{21}| + |A_{22}| + |A_{23}| + |A_{24}| + |A_{25}|$;
- combined partial determinant $|A'_{16}| = |A_{26}| + |A_{27}| + |A_{28}| + |A_{29}| + |A_{30}|$

$$\begin{aligned}|A'_{11}| =\;& a_{11}a_{22}a_{33}[a_{44}(a_{55}a_{66}-a_{56}a_{65})+a_{45}(a_{56}a_{64}-a_{54}a_{66})+a_{46}(a_{54}a_{65}-a_{55}a_{64})]+ \\ &+a_{11}a_{22}a_{34}[a_{43}(a_{56}a_{65}-a_{55}a_{66})+a_{45}(a_{53}a_{66}-a_{56}a_{63})+a_{46}(a_{55}a_{63}-a_{53}a_{65})]+ \quad |A_1| \\ &+a_{11}a_{22}a_{35}[a_{43}(a_{54}a_{66}-a_{56}a_{64})+a_{44}(a_{56}a_{63}-a_{53}a_{66})+a_{46}(a_{53}a_{64}-a_{54}a_{63})]+ \\ &+a_{11}a_{22}a_{36}[a_{43}(a_{55}a_{64}-a_{54}a_{65})+a_{44}(a_{53}a_{65}-a_{55}a_{63})+a_{45}(a_{54}a_{63}-a_{53}a_{64})]+ \\[6pt] &+a_{11}a_{23}a_{32}[a_{44}(a_{56}a_{65}-a_{55}a_{66})+a_{45}(a_{54}a_{66}-a_{56}a_{64})+a_{46}(a_{55}a_{64}-a_{54}a_{65})]+ \\ &+a_{11}a_{23}a_{34}[a_{42}(a_{55}a_{66}-a_{56}a_{65})+a_{45}(a_{56}a_{62}-a_{52}a_{66})+a_{46}(a_{52}a_{65}-a_{55}a_{62})]+ \quad |A_2| \\ &+a_{11}a_{23}a_{35}[a_{42}(a_{56}a_{64}-a_{54}a_{66})+a_{44}(a_{52}a_{66}-a_{56}a_{62})+a_{46}(a_{54}a_{62}-a_{52}a_{64})]+ \\ &+a_{11}a_{23}a_{36}[a_{42}(a_{54}a_{65}-a_{55}a_{64})+a_{44}(a_{55}a_{62}-a_{52}a_{65})+a_{45}(a_{52}a_{64}-a_{54}a_{62})]+ \\[6pt] &+a_{11}a_{24}a_{32}[a_{43}(a_{55}a_{66}-a_{56}a_{65})+a_{45}(a_{56}a_{63}-a_{53}a_{66})+a_{46}(a_{53}a_{65}-a_{55}a_{63})]+ \\ &+a_{11}a_{24}a_{33}[a_{42}(a_{56}a_{65}-a_{55}a_{66})+a_{45}(a_{52}a_{66}-a_{56}a_{62})+a_{46}(a_{55}a_{62}-a_{52}a_{65})]+ \quad |A_3| \\ &+a_{11}a_{24}a_{35}[a_{42}(a_{53}a_{66}-a_{56}a_{63})+a_{43}(a_{56}a_{62}-a_{52}a_{66})+a_{46}(a_{52}a_{63}-a_{53}a_{62})]+ \\ &+a_{11}a_{24}a_{36}[a_{42}(a_{55}a_{63}-a_{53}a_{65})+a_{43}(a_{52}a_{65}-a_{55}a_{62})+a_{45}(a_{53}a_{62}-a_{52}a_{63})]+ \end{aligned}$$

188

$$+a_{11}a_{25}a_{32}[a_{43}(a_{56}a_{64}-a_{54}a_{66})+a_{44}(a_{53}a_{66}-a_{56}a_{63})+a_{46}(a_{54}a_{63}-a_{53}a_{64})]+$$
$$+a_{11}a_{25}a_{33}[a_{42}(a_{54}a_{66}-a_{56}a_{64})+a_{44}(a_{56}a_{62}-a_{52}a_{66})+a_{46}(a_{52}a_{64}-a_{54}a_{62})]+ \quad |A_4|$$
$$+a_{11}a_{25}a_{34}[a_{42}(a_{56}a_{63}-a_{53}a_{66})+a_{43}(a_{52}a_{66}-a_{56}a_{62})+a_{46}(a_{53}a_{62}-a_{52}a_{63})]+$$
$$+a_{11}a_{25}a_{36}[a_{42}(a_{53}a_{64}-a_{54}a_{63})+a_{43}(a_{54}a_{62}-a_{52}a_{64})+a_{44}(a_{52}a_{63}-a_{53}a_{62})]+$$

$$+a_{11}a_{26}a_{32}[a_{43}(a_{54}a_{65}-a_{55}a_{64})+a_{44}(a_{55}a_{63}-a_{53}a_{65})+a_{45}(a_{53}a_{64}-a_{54}a_{63})]+$$
$$+a_{11}a_{26}a_{33}[a_{42}(a_{55}a_{64}-a_{54}a_{65})+a_{44}(a_{52}a_{65}-a_{55}a_{62})+a_{45}(a_{54}a_{62}-a_{52}a_{64})]+ \quad |A_5|$$
$$+a_{11}a_{26}a_{34}[a_{42}(a_{53}a_{65}-a_{55}a_{63})+a_{43}(a_{55}a_{62}-a_{52}a_{65})+a_{45}(a_{52}a_{63}-a_{53}a_{62})]+$$
$$+a_{11}a_{26}a_{35}[a_{42}(a_{54}a_{63}-a_{53}a_{64})+a_{43}(a_{52}a_{64}-a_{54}a_{62})+a_{44}(a_{53}a_{62}-a_{52}a_{63})] \quad (110)$$

A. Analysis of combined partial determinant $|A'_{11}|$ of the 6x6 matrix

Combined partial determinant $|A'_{11}|$ is expressed by formula (110) and consists from five separate parts $(|A_1|, |A_2|, |A_3|, |A_4|, |A_5|)$ which are shown in Figures 133 to 137 for positive (——) and negative(-------) areas:

❖ For Figure 133 of the combined partial determinant $|A_1|$:

Analysis of data shown in Figure 133 for combined partial determinant $|A_1|$ of the 6x6 matrix indicates on the following result in general view:

- The first element (a_{11}) of the first row is joined with the second element (a_{22}) of the second row;
- The second element (a_{22}) of the second row is joined with elements ($a_{33}, a_{34}, a_{35}, a_{36}$) of the third row;

a) For elements of the third row

- The third element (a_{33}) of the third row is joined with elements (a_{44}, a_{45}, a_{46}) of the fourth row;
- The fourth element (a_{34})of the third row is joined with elements (a_{43}, a_{45}, a_{46})of the fourth row;
- The fifth element (a_{35}) of the third row is joined with elements (a_{43}, a_{44}, a_{46})of the fourth row;
- The sixth element (a_{36}) of the third row is joined with elements (a_{43}, a_{44}, a_{45})of the fourth row-in total we can conclude that sequence of elements connection for the third row with elements of the fourth row has the following view, forming the frequency of ties equal f=3:

$$a_{33} \rightarrow a_{44}, a_{45}, a_{46}$$
$$a_{34} \rightarrow a_{43}, a_{45}, a_{46}$$
$$a_{35} \rightarrow a_{43}, a_{44}, a_{46}$$
$$a_{36} \rightarrow a_{43}, a_{44}, a_{45}$$

b) For elements of the fourth row

- The third element (a_{43}) of the fourth row is joined with elements ($a_{54}, a_{54}, a_{55}, a_{55}, a_{56}, a_{56}$) of the fifth row for positive and negative areas;
- The fourth element (a_{44}) of the fourth row is joined with elements ($a_{53}, a_{53}, a_{55}, a_{55}, a_{56}, a_{56}$) of the fifth row for positive and negative areas;
- The fifth element (a_{45})of the fourth row is joined with elements ($a_{53}, a_{53}, a_{54}, a_{54}, a_{56}, a_{56}$) of the fifth row for positive and negative areas;
- The sixth element (a_{46}) of the fourth row is joined with elements ($a_{53}, a_{53}, a_{54}, a_{54}, a_{55}, a_{55}$)for positive and negative areas-in total we can conclude that sequence of elements connection for the fourth row with elements of the fifth row has the following view, forming the frequency of ties equal f=6:

$$a_{43} \rightarrow a_{54}, a_{54}, a_{55}, a_{55}, a_{56}, a_{56}$$
$$a_{44} \rightarrow a_{53}, a_{53}, a_{55}, a_{55}, a_{56}, a_{56}$$
$$a_{45} \rightarrow a_{53}, a_{53}, a_{54}, a_{54}, a_{56}, a_{56}$$
$$a_{46} \rightarrow a_{53}, a_{53}, a_{54}, a_{54}, a_{55}, a_{55}$$

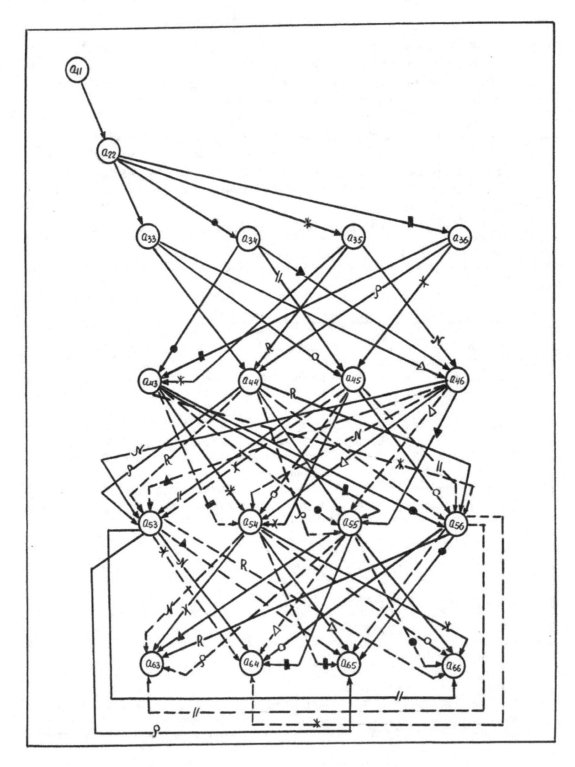

Figure 133 Functional graphs between of elements for the combined partial 6x6 matrix determinant $|A_1|$

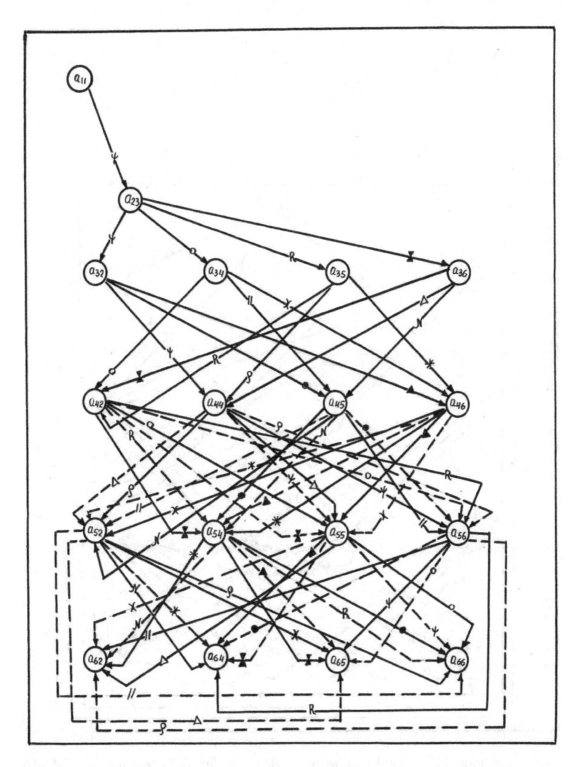

Figure 134 Functional graphs between of elements for the combined partial 6x6 matrix determinant $|A_2|$

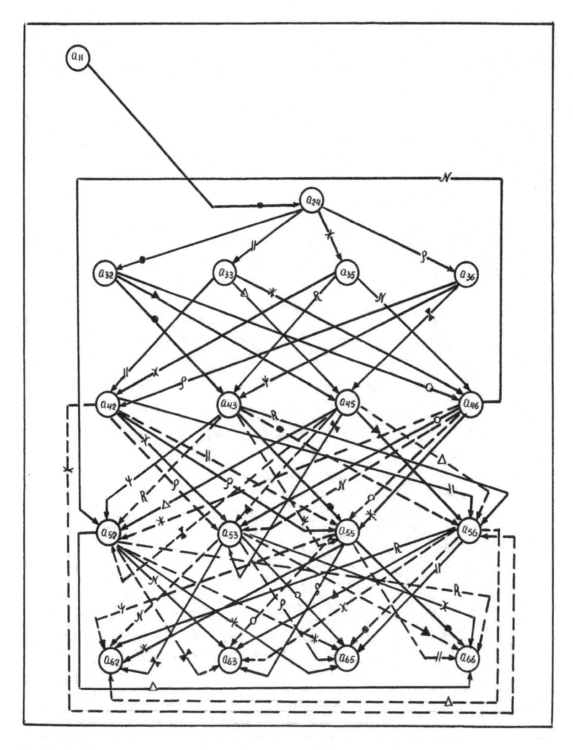

Figure 135 Functional graphs between of elements for the combined partial 6x6 matrix determinant $\left| A_3 \right|$

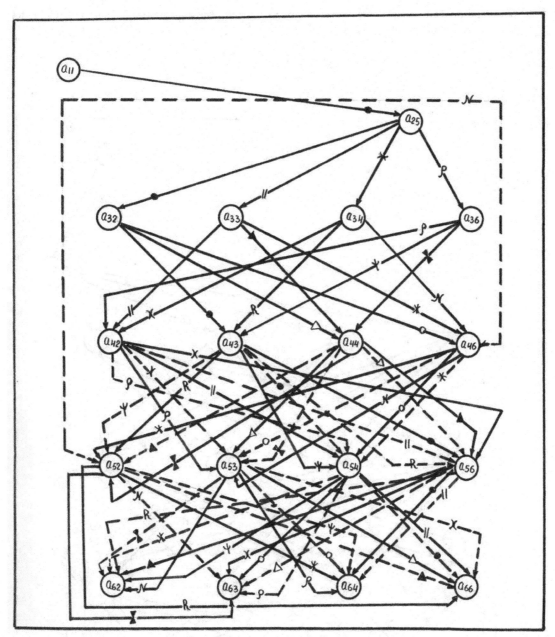

Figure 136 Functional graphs between of elements for the combined partial 6x6 matrix determinant $\left| A_4 \right|$

    c) For elements of the fifth row
- The third element (a_{53}) of the fifth row is joined with elements ($a_{64}, a_{64}, a_{65}, a_{65}, a_{66}, a_{66}$) of the six row for positive and negative areas;
- The fourth element (a_{54}) of the fifth row is joined with elements ($a_{63}, a_{63}, a_{65}, a_{65}, a_{66}, a_{66}$) of the six row for positive and negative areas;
- The fifth element (a_{55}) of the fifth row is joined with elements ($a_{63}, a_{63}, a_{64}, a_{64}, a_{66}, a_{66}$) of the six row for positive and negative areas;
- The sixth element (a_{56}) of the fifth row is joined with elements ($a_{63}, a_{63}, a_{64}, a_{64}, a_{65}, a_{65}$) of the sixth

row for positive and negative areas-in total we can conclude that sequence of elements connection for the fifth row with elements of the sixth row has the following view, forming the frequency of ties equal f=6:

$$a_{53} \rightarrow a_{64}, a_{64}, a_{65}, a_{65}, a_{66}, a_{66}$$
$$a_{54} \rightarrow a_{63}, a_{63}, a_{65}, a_{65}, a_{66}, a_{66}$$
$$a_{55} \rightarrow a_{63}, a_{63}, a_{64}, a_{64}, a_{66}, a_{66}$$
$$a_{56} \rightarrow a_{63}, a_{63}, a_{64}, a_{64}, a_{65}, a_{65}$$

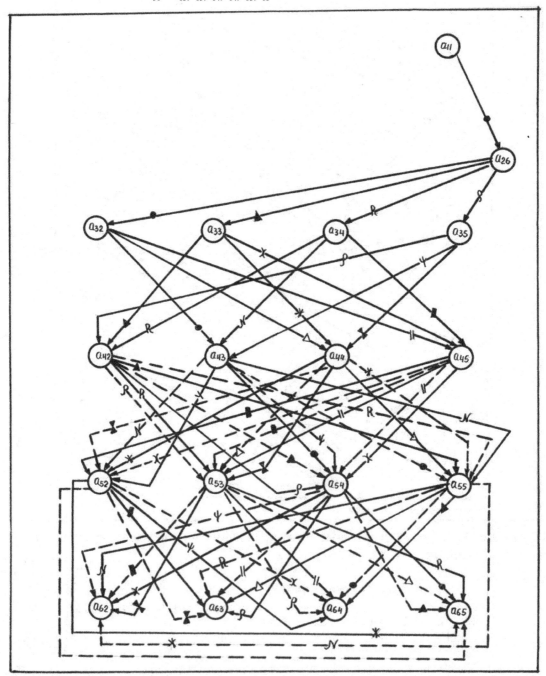

Figure 137 Functional graphs between of elements for the combined partial 6x6 matrix determinant $|A_5|$

194

❖ For Figure 134 of the combined partial determinant $|A_2|$:

Analysis of data shown in Figure 134 in accordance with formula (110) for the determinant $|A_2|$ of the 6x6 matrix indicates on the following results in general view for positive and negative areas:
- The first element (a_{11}) of the first row is joined with the third element (a_{23}) of the second row;
- The third element (a_{23}) of the second row is joined with elements $(a_{32}, a_{34}, a_{35}, a_{36})$ of the third row;

a) *For elements of the third row*
- The second element (a_{32}) of the third row is joined with elements (a_{44}, a_{45}, a_{46}) of the fourth row;
- The fourth element (a_{34}) of the third row is joined with elements (a_{42}, a_{45}, a_{46}) of the fourth row;
- The fifth element (a_{35}) of the third row is joined with elements (a_{42}, a_{44}, a_{46}) of the fourth row;
- The sixth element (a_{36}) of the third row is joined with elements (a_{42}, a_{44}, a_{45}) of the fourth row- in total we can conclude that sequence of elements connection for the third row with elements of the fourth row has the following view, forming the frequency of ties equal f=3:

$$a_{32} \rightarrow a_{44}, a_{45}, a_{46}$$
$$a_{34} \rightarrow a_{42}, a_{45}, a_{46}$$
$$a_{35} \rightarrow a_{42}, a_{44}, a_{46}$$
$$a_{36} \rightarrow a_{42}, a_{44}, a_{45}$$

b) *For elements of the fourth row*
- The second element (a_{42}) of the fourth row is joined with elements $(a_{54}, a_{54}, a_{55}, a_{55}, a_{56}, a_{56})$ of the fifth row for positive and negative areas;
- The fourth element (a_{44}) of the fourth row is joined with elements $(a_{52}, a_{52}, a_{55}, a_{55}, a_{56}, a_{56})$ of the fifth row for positive and negative areas;
- The fifth element (a_{45}) of the fourth row is joined with elements $(a_{52}, a_{52}, a_{54}, a_{54}, a_{56}, a_{56})$ of the fifth row for positive and negative areas;
- The sixth element (a_{46}) of the fourth row is joined with elements $(a_{52}, a_{52}, a_{54}, a_{54}, a_{55}, a_{55})$ of the fifth row for positive and negative areas- in total we can conclude that sequence of elements connection for the fourth row with elements of the fifth row has the following view, forming the frequency of ties equal f=6:

$$a_{42} \rightarrow a_{54}, a_{54}, a_{55}, a_{55}, a_{56}, a_{56}$$
$$a_{44} \rightarrow a_{52}, a_{52}, a_{55}, a_{55}, a_{56}, a_{56}$$
$$a_{45} \rightarrow a_{52}, a_{52}, a_{54}, a_{54}, a_{56}, a_{56}$$
$$a_{46} \rightarrow a_{52}, a_{52}, a_{54}, a_{54}, a_{55}, a_{55}$$

c) *For elements of the fifth row*
- The second element (a_{52}) of the fifth row is joined with elements $(a_{64}, a_{64}, a_{65}, a_{65}, a_{66}, a_{66})$ of the six row;
- The fifth element (a_{55}) of the fifth row is joined with elements $(a_{62}, a_{62}, a_{64}, a_{64}, a_{66}, a_{66})$ of the sixth row;
- The sixth element (a_{56}) of the fifth row is joined with elements $(a_{62}, a_{62}, a_{64}, a_{64}, a_{65}, a_{65})$ of the sixth row-in total we can conclude that sequence of elements connection for the fifth row with elements of the sixth row has the following view, forming the frequency of ties equal f=6:

$$a_{52} \rightarrow a_{64}, a_{64}, a_{65}, a_{65}, a_{66}, a_{66}$$
$$a_{54} \rightarrow a_{62}, a_{62}, a_{65}, a_{65}, a_{66}, a_{66}$$
$$a_{55} \rightarrow a_{62}, a_{62}, a_{64}, a_{64}, a_{66}, a_{66}$$
$$a_{56} \rightarrow a_{62}, a_{62}, a_{64}, a_{64}, a_{65}, a_{65}$$

❖ For Figure 135 of the combined partial determinant $|A_3|$:

Analysis of data shown in Figure 135 in accordance with formula (110) of the combined partial determinant $|A_3|$ of the 6x6 matrix indicates on the following results in general view for positive and negative areas:

- The first element (a_{11}) of the first row is joined with the fourth element (a_{24}) of the second row;
- The fourth element (a_{24}) of the second row is joined with elements ($a_{32}, a_{33}, a_{35}, a_{36}$) of the third row;

a) *For elements of the third row*

- The second element (a_{32}) of the third row is joined with elements (a_{43}, a_{45}, a_{46}) of the fourth row;
- The third element (a_{33}) of the third row is joined with elements (a_{42}, a_{45}, a_{46}) of the fourth row;
- The fifth element (a_{35}) of the third row is joined with elements (a_{42}, a_{43}, a_{46}) of the fourth row;
- The sixth element (a_{36}) of the third row is joined with elements (a_{42}, a_{43}, a_{45}) of the fourth row- in total we can conclude that sequence of elements connection for the third row with elements of the fourth row has the following view, forming the frequency of ties equal f=3:

$$a_{32} \rightarrow a_{43}, a_{45}, a_{46}$$
$$a_{33} \rightarrow a_{42}, a_{45}, a_{46}$$
$$a_{35} \rightarrow a_{42}, a_{43}, a_{46}$$
$$a_{36} \rightarrow a_{42}, a_{43}, a_{45}$$

b) *For elements of the fourth row*

- The second element (a_{42}) of the fourth row is joined with elements ($a_{53}, a_{53}, a_{55}, a_{55}, a_{56}, a_{56}$) of the fifth row for positive and negative areas;
- The third element (a_{43}) of the fourth row is joined with elements ($a_{52}, a_{52}, a_{55}, a_{55}, a_{56}, a_{56}$) of the fifth row for positive and negative areas;
- The fifth element (a_{45}) of the fourth row is joined with elements ($a_{52}, a_{52}, a_{53}, a_{53}, a_{56}, a_{56}$) of the fifth row for positive and negative areas;
- The sixth element (a_{46}) of the fourth row is joined with elements ($a_{52}, a_{52}, a_{53}, a_{53}, a_{55}, a_{55}$) of the fifth row for positive and negative areas- in total we can conclude that sequence of elements connection for the fourth row with elements of the fifth row has the following view, forming the frequency of ties equal f=6:

$$a_{42} \rightarrow a_{53}, a_{53}, a_{55}, a_{55}, a_{56}, a_{56}$$
$$a_{43} \rightarrow a_{52}, a_{52}, a_{55}, a_{55}, a_{56}, a_{56}$$
$$a_{45} \rightarrow a_{52}, a_{52}, a_{53}, a_{53}, a_{56}, a_{56}$$
$$a_{46} \rightarrow a_{52}, a_{52}, a_{53}, a_{53}, a_{55}, a_{55}$$

c) *For elements of the fifth row*

- The second element (a_{52}) of the fifth row is joined with elements ($a_{63}, a_{63}, a_{65}, a_{65}, a_{66}, a_{66}$) of the sixth row for positive and negative areas;
- The third element (a_{53}) of the fifth row is joined with elements ($a_{62}, a_{62}, a_{65}, a_{65}, a_{66}, a_{66}$) of the sixth row for positive and negative areas;
- The fifth element (a_{55}) of the fifth row is joined with elements ($a_{62}, a_{62}, a_{63}, a_{63}, a_{66}, a_{66}$) of the sixth row for positive and negative areas;
- The sixth element (a_{56}) of the fifth row is joined with elements ($a_{62}, a_{62}, a_{63}, a_{63}, a_{65}, a_{65}$) of the sixth row for positive and negative areas-in total we can conclude that sequence of elements connection for the fifth row with elements of the six row has the following view, forming the frequency of ties equal f=6:

$$a_{52} \rightarrow a_{63}, a_{63}, a_{65}, a_{65}, a_{66}, a_{66}$$
$$a_{53} \rightarrow a_{62}, a_{62}, a_{65}, a_{65}, a_{66}, a_{66}$$
$$a_{55} \rightarrow a_{62}, a_{62}, a_{63}, a_{63}, a_{66}, a_{66}$$
$$a_{56} \rightarrow a_{62}, a_{62}, a_{63}, a_{63}, a_{65}, a_{65}$$

❖ For Figure 136 of the combined partial determinant $|A_4|$:

Analysis of data shown in Figure 136 applicably to formula (110) of the combined partial determinant $|A_4|$ of the 6x6 matrix for positive and negative areas indicates on the following results in general view:

- The first element (a_{11}) of the first row is joined with the fifth element (a_{25}) of the second row;

- The fifth element (a_{25}) of the second row is joined with elements ($a_{32},a_{33},a_{34},a_{36}$) of the third row;

a) *For elements of the third row*

- The second element (a_{32}) of the third row is joined with elements (a_{43},a_{44},a_{46}) of the fourth row;
- The third element (a_{33}) of the third row is joined with elements (a_{42},a_{44},a_{46}) of the fourth row;
- The fourth element (a_{34}) of the third row is joined with elements (a_{42},a_{43},a_{46}) of the fourth row;
- The sixth element(a_{36}) of the third row is joined with elements (a_{42},a_{43},a_{44}) of the fourth row-in total we can conclude that sequence of elements connection for the third row with elements of the fourth row has the following view, forming the frequency of ties equal f=3:

$$a_{32}{\rightarrow}a_{43},a_{44},a_{46}$$
$$a_{33}{\rightarrow}a_{42},a_{44},a_{46}$$
$$a_{34}{\rightarrow}a_{42},a_{43},a_{46}$$
$$a_{36}{\rightarrow}a_{42},a_{43},a_{44}$$

b) *For elements of the fourth row*

- The second element(a_{42}) of the fourth row is joined with elements ($a_{53},a_{53},a_{54},a_{54},a_{56},a_{56}$)of the fifth row for positive and negative areas;
- The third element (a_{43}) of the fourth row is joined with elements ($a_{52},a_{52},a_{54},a_{54},a_{56},a_{56}$) of the fifth row for positive and negative areas;
- The fourth element (a_{44}) of the fourth row is joined with elements ($a_{52},a_{52},a_{53},a_{53},a_{56},a_{56}$) of the fifth row for positive and negative areas;
- The sixth element(a_{46}) of the fourth row is joined with elements ($a_{52},a_{52},a_{53},a_{53},a_{54},a_{54}$) of the fifth row for positive and negative areas-in total we can conclude that sequence of elements connection for the fourth row with elements of the fifth row has the following view, forming the frequency of ties equal f=6:

$$a_{42}{\rightarrow}a_{53},a_{53},a_{54},a_{54},a_{56},a_{56}$$
$$a_{43}{\rightarrow}a_{52},a_{52},a_{54},a_{54},a_{56},a_{56}$$
$$a_{44}{\rightarrow}a_{52},a_{52},a_{53},a_{53},a_{56},a_{56}$$
$$a_{46}{\rightarrow}a_{52},a_{52},a_{53},a_{53},a_{54},a_{54}$$

c) *For elements of the fifth row*

- The second element (a_{52}) of the fifth row is joined with elements ($a_{63},a_{63},a_{64},a_{64},a_{66},a_{66}$)of the sixth row for positive and negative areas;
- The third element(a_{53}) of the fifth row is joined with elements ($a_{62},a_{62},a_{64},a_{64},a_{66},a_{66}$) of the sixth row for positive and negative areas;
- The fourth element (a_{54}) of the fifth row is joined with elements ($a_{62},a_{62},a_{63},a_{63},a_{66},a_{66}$) of the sixth row for positive and negative areas;
- The sixth element(a_{56}) of the fifth row is joined with elements ($a_{62},a_{62},a_{63},a_{63},a_{64},a_{64}$) of the sixth row for positive and negative areas-in total we can conclude that sequence of elements connection for the fifth row with elements of the sixth row has the following view, forming the frequency of ties equal f=6:

$$a_{52}{\rightarrow}a_{63},a_{63},a_{64},a_{64},a_{66},a_{66}$$
$$a_{53}{\rightarrow}a_{62},a_{62},a_{64},a_{64},a_{66},a_{66}$$
$$a_{54}{\rightarrow}a_{62},a_{62},a_{63},a_{63},a_{66},a_{66}$$
$$a_{56}{\rightarrow}a_{62},a_{62},a_{63},a_{63},a_{64},a_{64}$$

❖ For Figure 137 of the combined partial determinant $|A_5|$:

Analysis of data shown in Figure 137 and applicably to formula (110) of the combined partial determinant $|A_5|$ of the 6x6 matrix for positive and negative areas indicates on the following results in general view:
- The first element (a_{11}) of the first row is joined with the sixth element(a_{26}) of the second row;
- The sixth element (a_{26})of the second row is joined with elements ($a_{32},a_{33},a_{34},a_{35}$) of the third row;

197

a) *For elements of the third row*

- The second element(a_{32})of the third row is joined with elements (a_{43}, a_{44}, a_{45}) of the fourth row;
- The third element (a_{33}) of the third row is joined with elements (a_{42}, a_{44}, a_{45}) of the fourth row;
- The fourth element (a_{34}) of the third row is joined with elements (a_{42}, a_{43}, a_{45}) of the fourth row;
- The fifth element (a_{35}) of the third row is joined with elements (a_{42}, a_{43}, a_{44}) of the fourth row-in total we can conclude that sequence of elements connection for the third row with elements of the fourth row has the following view, forming the frequency of ties equal f=3:

$$a_{32} \longrightarrow a_{43}, a_{44}, a_{45}$$
$$a_{33} \longrightarrow a_{42}, a_{44}, a_{45}$$
$$a_{34} \longrightarrow a_{42}, a_{43}, a_{45}$$
$$a_{35} \longrightarrow a_{42}, a_{43}, a_{44}$$

b) *For elements of the fourth row*

- The second element (a_{42})of the fourth row is joined with elements ($a_{53}, a_{53}, a_{54}, a_{54}, a_{55}, a_{55}$) of the fifth row for positive and negative areas;
- The third element (a_{43}) of the fourth row is joined with elements ($a_{52}, a_{52}, a_{54}, a_{54}, a_{55}, a_{55}$) of the fifth row for positive and negative areas;
- The fourth element(a_{44}) of the fourth row is joined with elements ($a_{52}, a_{52}, a_{53}, a_{53}, a_{55}, a_{55}$) of the fifth row for positive and negative areas;
- The fifth element(a_{45}) of the fourth row is joined with elements ($a_{52}, a_{52}, a_{53}, a_{53}, a_{54}, a_{54}$) of the fifth row for positive and negative areas-in total we can conclude that sequence of elements connection for the fourth row with elements of the fifth row has the following view, forming the frequency of ties equal f=6:

$$a_{42} \longrightarrow a_{53}, a_{53}, a_{54}, a_{54}, a_{55}, a_{55}$$
$$a_{43} \longrightarrow a_{52}, a_{52}, a_{54}, a_{54}, a_{55}, a_{55}$$
$$a_{44} \longrightarrow a_{52}, a_{52}, a_{53}, a_{53}, a_{55}, a_{55}$$
$$a_{45} \longrightarrow a_{52}, a_{52}, a_{53}, a_{53}, a_{54}, a_{54}$$

c) *For elements of the fifth row*

- The second element (a_{52}) of the fifth row is joined with elements ($a_{63}, a_{63}, a_{64}, a_{64}, a_{65}, a_{65}$) of the sixth row for positive and negative areas;
- The third element(a_{53}) of the fifth row is joined with elements ($a_{62}, a_{62}, a_{64}, a_{64}, a_{65}, a_{65}$) of the sixth row for positive and negative areas;
- The fourth element(a_{54}) of the fifth row is joined with elements ($a_{62}, a_{62}, a_{63}, a_{63}, a_{65}, a_{65}$) of the sixth row for positive and negative areas;
- The fifth element(a_{55}) of the fifth row is joined with elements ($a_{62}, a_{62}, a_{63}, a_{63}, a_{64}, a_{64}$) of the sixth row for positive and negative areas-in total we can conclude that sequence of elements connection for the fifth row with elements of the sixth row has the following view, forming the frequency of ties equal f=6:

$$a_{52} \longrightarrow a_{63}, a_{63}, a_{64}, a_{64}, a_{65}, a_{65}$$
$$a_{53} \longrightarrow a_{62}, a_{62}, a_{64}, a_{64}, a_{65}, a_{65}$$
$$a_{54} \longrightarrow a_{62}, a_{62}, a_{63}, a_{63}, a_{65}, a_{65}$$
$$a_{55} \longrightarrow a_{62}, a_{62}, a_{63}, a_{63}, a_{64}, a_{64}$$

B. **Analysis of combined partial determinant $\left| A'_{12} \right|$ of the 6x6 matrix:**

The combined partial determinant $\left| A'_{12} \right|$ consists from five separate parts and is expressed by the following primary formula in view of

$$\left| A'_{12} \right| = \left| A_6 \right| + \left| A_7 \right| + \left| A_8 \right| + \left| A_9 \right| + \left| A_{10} \right|$$

And also the above-said determinant introduced by the following finally formula (111) ,after of some transformations ,elements of which are shown in view of functional graphs in Figures 138 to 142 :

198

$$|A'_{12}| = a_{12}a_{21}a_{33}[a_{44}(a_{56}a_{65}-a_{55}a_{66})+a_{45}(a_{54}a_{66}-a_{56}a_{64})+a_{46}(a_{55}a_{64}-a_{54}a_{65})]+$$
$$+a_{12}a_{21}a_{34}[a_{43}(a_{55}a_{66}-a_{56}a_{65})+a_{45}(a_{56}a_{63}-a_{53}a_{66})+a_{46}(a_{53}a_{65}-a_{55}a_{63})]+ \quad |A_6|$$
$$+a_{12}a_{21}a_{35}[a_{43}(a_{56}a_{64}-a_{54}a_{66})+a_{44}(a_{53}a_{66}-a_{56}a_{63})+a_{46}(a_{54}a_{63}-a_{53}a_{64})]+$$
$$+a_{12}a_{21}a_{36}[a_{43}(a_{54}a_{65}-a_{55}a_{64})+a_{44}(a_{55}a_{63}-a_{53}a_{65})+a_{45}(a_{53}a_{64}-a_{54}a_{63})]+$$

$$+a_{12}a_{23}a_{31}[a_{44}(a_{55}a_{66}-a_{56}a_{65})+a_{45}(a_{56}a_{64}-a_{54}a_{66})+a_{46}(a_{54}a_{65}-a_{55}a_{64})]+$$
$$+a_{12}a_{23}a_{34}[a_{41}(a_{56}a_{65}-a_{55}a_{66})+a_{45}(a_{51}a_{66}-a_{56}a_{61})+a_{46}(a_{55}a_{61}-a_{51}a_{65})]+ \quad |A_7|$$
$$+a_{12}a_{23}a_{35}[a_{41}(a_{54}a_{66}-a_{56}a_{64})+a_{44}(a_{56}a_{61}-a_{51}a_{66})+a_{46}(a_{51}a_{64}-a_{54}a_{61})]+$$
$$+a_{12}a_{23}a_{36}[a_{41}(a_{55}a_{64}-a_{54}a_{65})+a_{44}(a_{51}a_{65}-a_{55}a_{61})+a_{45}(a_{54}a_{61}-a_{51}a_{64})]+$$

$$+a_{12}a_{24}a_{31}[a_{43}(a_{56}a_{65}-a_{55}a_{66})+a_{45}(a_{53}a_{66}-a_{56}a_{63})+a_{46}(a_{55}a_{63}-a_{53}a_{65})]+$$
$$+a_{12}a_{24}a_{33}[a_{41}(a_{55}a_{66}-a_{56}a_{65})+a_{45}(a_{56}a_{61}-a_{51}a_{66})+a_{46}(a_{51}a_{65}-a_{55}a_{61})]+ \quad |A_8|$$
$$+a_{12}a_{24}a_{35}[a_{41}(a_{56}a_{63}-a_{53}a_{66})+a_{43}(a_{51}a_{66}-a_{56}a_{61})+a_{46}(a_{53}a_{61}-a_{51}a_{63})]+$$
$$+a_{12}a_{24}a_{36}[a_{41}(a_{53}a_{65}-a_{55}a_{63})+a_{43}(a_{55}a_{61}-a_{51}a_{65})+a_{45}(a_{51}a_{63}-a_{53}a_{61})]+$$

$$+a_{12}a_{25}a_{31}[a_{43}(a_{54}a_{66}-a_{56}a_{64})+a_{44}(a_{56}a_{63}-a_{53}a_{66})+a_{46}(a_{53}a_{64}-a_{54}a_{63})]+$$
$$+a_{12}a_{25}a_{33}[a_{41}(a_{56}a_{64}-a_{54}a_{66})+a_{44}(a_{51}a_{66}-a_{56}a_{61})+a_{46}(a_{54}a_{61}-a_{51}a_{64})]+ \quad |A_9|$$
$$+a_{12}a_{25}a_{34}[a_{41}(a_{53}a_{66}-a_{56}a_{63})+a_{43}(a_{56}a_{61}-a_{51}a_{66})+a_{46}(a_{51}a_{63}-a_{53}a_{61})]+$$
$$+a_{12}a_{25}a_{36}[a_{41}(a_{54}a_{63}-a_{53}a_{64})+a_{43}(a_{51}a_{64}-a_{54}a_{61})+a_{44}(a_{53}a_{61}-a_{51}a_{63})]+$$

$$+a_{12}a_{26}a_{31}[a_{43}(a_{55}a_{64}-a_{54}a_{65})+a_{44}(a_{53}a_{65}-a_{55}a_{63})+a_{45}(a_{54}a_{63}-a_{53}a_{64})]+$$
$$+a_{12}a_{26}a_{33}[a_{41}(a_{54}a_{65}-a_{55}a_{64})+a_{44}(a_{55}a_{61}-a_{51}a_{65})+a_{45}(a_{51}a_{64}-a_{54}a_{61})]+ \quad |A_{10}|$$
$$+a_{12}a_{26}a_{34}[a_{41}(a_{55}a_{63}-a_{53}a_{65})+a_{43}(a_{51}a_{65}-a_{55}a_{61})+a_{45}(a_{53}a_{61}-a_{51}a_{63})]+$$
$$+a_{12}a_{26}a_{35}[a_{41}(a_{53}a_{64}-a_{54}a_{63})+a_{43}(a_{54}a_{61}-a_{51}a_{64})+a_{44}(a_{51}a_{63}-a_{53}a_{61})] \quad (111)$$

In **Appendix 23** is shown evaluation of the determinant of the 6x6 matrix in numerical data with using of Expansion (cofactor) method *elements of which have expansion to the first row*. And also are shown the functional ties between of the considered 6x6 matrix elements.

❖ **For Figure 138 of the combined partial determinant** $|A_6|$:

Analysis of data shown in Figure 138 applicably to formula (111) of the combined partial determinant $|A_6|$ of the 6x6 matrix for positive and negative areas indicates on the following results in general view:
- The second element (a_{12}) of the first row is joined with the first element (a_{21}) of the second row;
- The first element (a_{21}) of the second row is joined with elements ($a_{33},a_{34},a_{35},a_{36}$) of the third row;
a) *For elements of the third row*
- The third element(a_{33}) of the third row is joined with elements (a_{44},a_{45},a_{46}) of the fourth row;
- The fourth element (a_{34}) of the third row is joined with elements (a_{43},a_{45},a_{46}) of the fourth row;
- The fifth element (a_{35}) of the third row is joined with elements (a_{43},a_{44},a_{46}) of the fourth row;
- The sixth element (a_{36}) of the third row is joined with elements (a_{43},a_{44},a_{45}) of the fourth row-in total we can conclude that sequence of elements connection for the third row with elements of the fourth row has the following view, forming the frequency of ties equal f=3:

$$a_{33}\rightarrow a_{44},a_{45},a_{46}$$
$$a_{34}\rightarrow a_{43},a_{45},a_{46}$$
$$a_{35}\rightarrow a_{43},a_{44},a_{46}$$
$$a_{36}\rightarrow a_{43},a_{44},a_{45}$$

b) *For elements of the fourth row*
- The third element(a_{43}) of the fourth row is joined with elements ($a_{54},a_{54},a_{55},a_{55},a_{56},a_{56}$) of the fifth row for positive and negative areas;

199

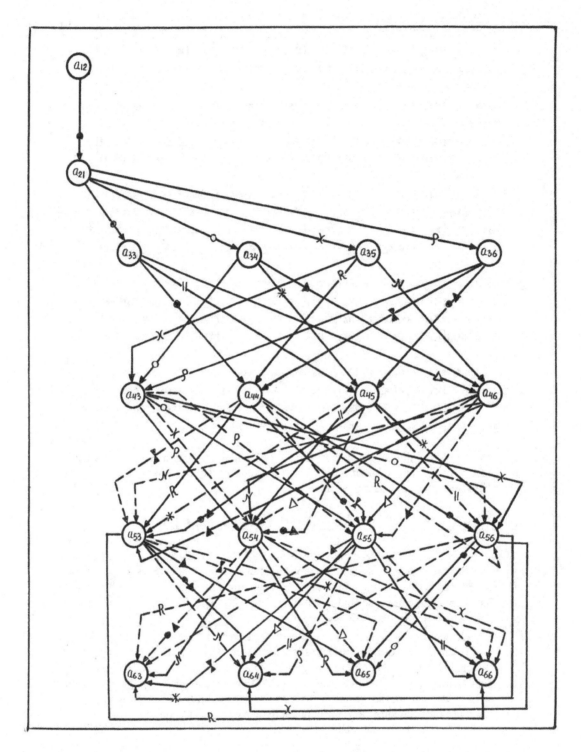

Figure 138 Functional graphs between of elements for the combined partial 6x6 matrix determinant $\left| A_6 \right|$

200

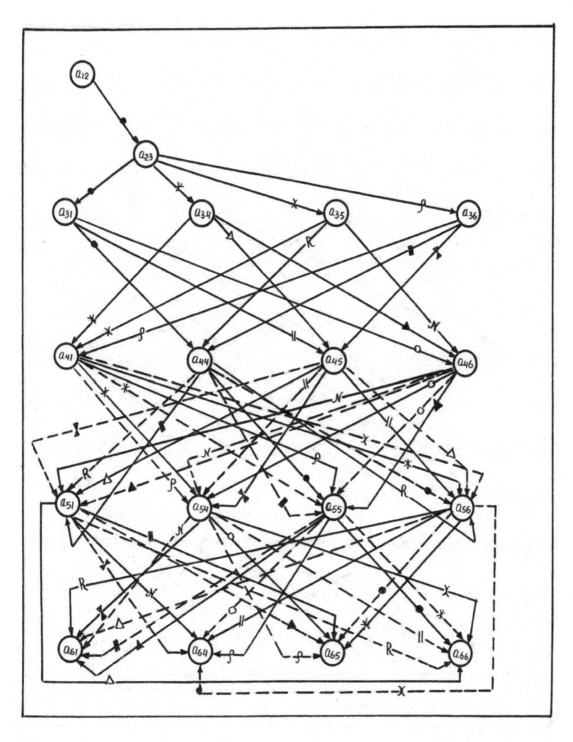

Figure 139 Functional graphs between of elements for the combined partial 6x6 matrix determinant $\left| A_7 \right|$

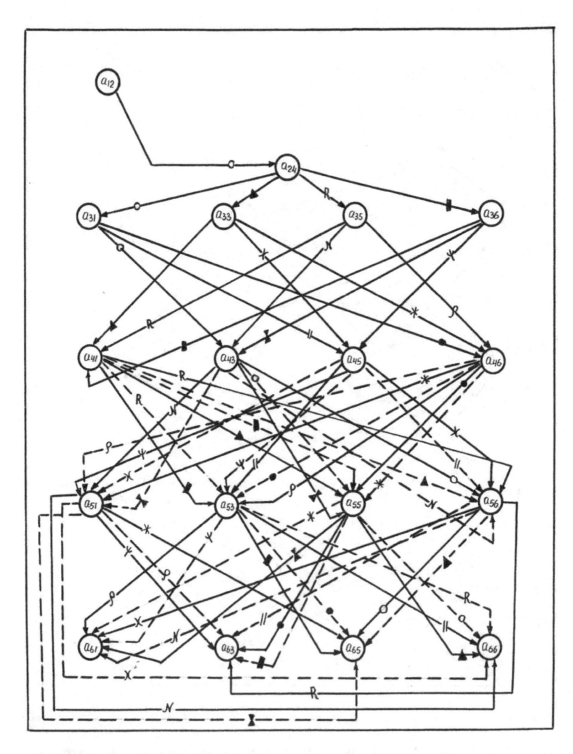

Figure 140 Functional graphs between of elements for the combined partial 6x6 matrix determinant $\left| A_8 \right|$

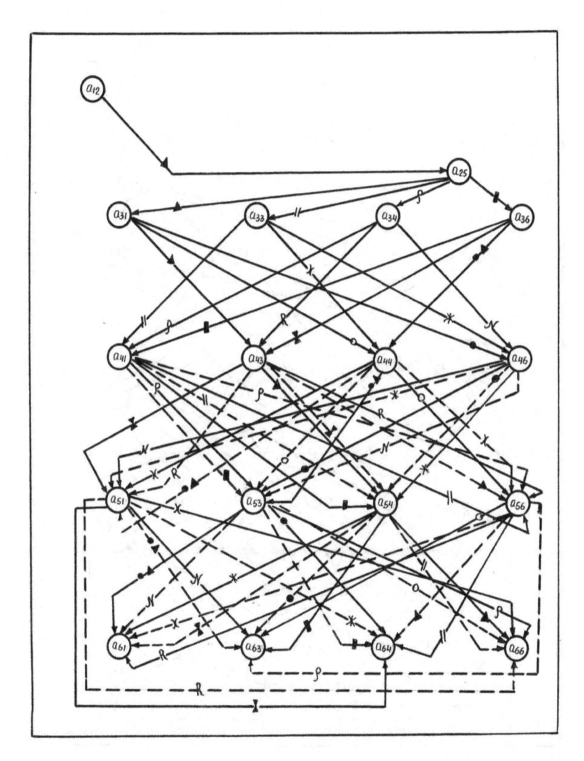

Figure 141 Functional graphs between of elements for the combined partial 6x6 determinant $\left| A_9 \right|$

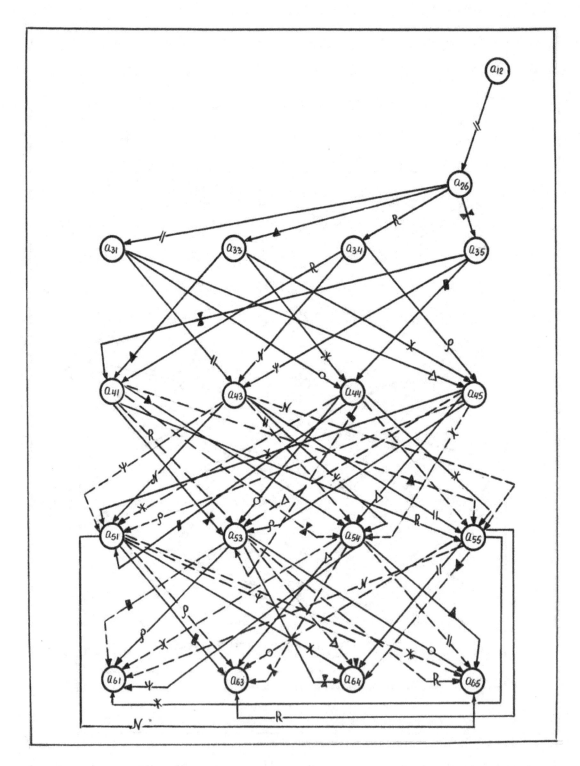

Figure 142 Functional graphs between of elements for the combined partial 6x6 matrix determinant $\left| A_{10} \right|$

- The fourth element(a_{44}) of the fourth row is joined with elements ($a_{53},a_{53},a_{55},a_{55},a_{56},a_{56}$) of the fifth row for positive and negative areas;
- The fifth element (a_{45}) of the fourth row is joined with elements ($a_{53},a_{53},a_{54},a_{54},a_{56},a_{56}$) of the fifth row for positive and negative areas;
- The sixth element(a_{46})of the fourth row is joined with elements ($a_{53},a_{53},a_{54},a_{54},a_{55},a_{55}$) for positive and negative areas- in total we can conclude that sequence of elements connection for the fourth row with elements of the fifth row has the following view, forming the frequency of ties equal f=6:

$$a_{43}\rightarrow a_{54},a_{54},a_{55},a_{55},a_{56},a_{56}$$
$$a_{44}\rightarrow a_{53},a_{53},a_{55},a_{55},a_{56},a_{56}$$
$$a_{45}\rightarrow a_{53},a_{53},a_{54},a_{54},a_{56},a_{56}$$
$$a_{46}\rightarrow a_{53},a_{53},a_{54},a_{54},a_{55},a_{55}$$

c) *For elements of the fifth row*

- The third element(a_{53}) of the fifth row is joined with elements ($a_{64},a_{64},a_{65},a_{65},a_{66},a_{66}$) of the sixth row for positive and negative areas;
- The fourth element(a_{54}) of the fifth row is joined with elements ($a_{63},a_{63},a_{65},a_{65},a_{66},a_{66}$) of the sixth row for positive and negative areas;
- The fifth element (a_{55}) of the fifth row is joined with elements ($a_{63},a_{63},a_{64},a_{64},a_{66},a_{66}$) of the sixth row for positive and negative areas;
- The sixth element(a_{56}) of the fifth row is joined with elements ($a_{63},a_{63},a_{64},a_{64},a_{65},a_{65}$) of the sixth row for positive and negative areas- in total we can conclude that sequence of elements connection for the fifth row with elements of the sixth row has the following view ,forming the frequency of ties equal f=6:

$$a_{53}\rightarrow a_{64},a_{64},a_{65},a_{65},a_{66},a_{66}$$
$$a_{54}\rightarrow a_{63},a_{63},a_{65},a_{65},a_{66},a_{66}$$
$$a_{55}\rightarrow a_{63},a_{63},a_{64},a_{64},a_{66},a_{66}$$
$$a_{56}\rightarrow a_{63},a_{63},a_{64},a_{64},a_{65},a_{65}$$

❖ For Figure 139 of the combined partial determinant $\lvert A_7 \rvert$:

Analysis of data shown in Figure 139 applicably to formula (111) of the combined partial determinant $\lvert A_7 \rvert$ of the 6x6 matrix for positive and negative areas indicates on the following results in general view:
- The second element (a_{12}) of the first row is joined with the third element (a_{23}) of the second row;
- The third element (a_{23}) of the second row is joined with elements ($a_{31},a_{34},a_{35},a_{36}$) of the third row;

a) *For elements of the third row*

- The first element(a_{31}) of the third row is joined with elements (a_{44},a_{45},a_{46}) of the fourth row;
- The fourth element(a_{34}) of the third row is joined with elements (a_{41},a_{45},a_{46}) of the fourth row;
- The fifth element(a_{35}) of the third row is joined with elements (a_{41},a_{44},a_{46}) of the fourth row;
- The sixth element (a_{36}) of the third row is joined with elements (a_{41},a_{44},a_{45}) of the fourth row-in total we can conclude that sequence of elements connection for the third row with elements of the fourth row has the following view, forming the frequency of ties equal f=3:

$$a_{31}\rightarrow a_{44},a_{45},a_{46}$$
$$a_{34}\rightarrow a_{41},a_{45},a_{46}$$
$$a_{35}\rightarrow a_{41},a_{44},a_{46}$$
$$a_{36}\rightarrow a_{41},a_{44},a_{45}$$

b) *For elements of the fourth row*

- The first element(a_{41}) of the fourth row is joined with elements ($a_{54},a_{54},a_{55},a_{55},a_{56},a_{56}$) of the fifth row for positive and negative areas;
- The fourth element(a_{44})of the fourth row is joined with elements ($a_{51},a_{51},a_{55},a_{55},a_{56},a_{56}$) of the fifth row for positive and negative areas;

- The fifth element (a_{45}) of the fourth row is joined with elements ($a_{51},a_{51},a_{54},a_{54},a_{56},a_{56}$) of the fifth row for positive and negative areas;
- The sixth element(a_{46}) of the fourth row is joined with elements ($a_{51},a_{51},a_{54},a_{54},a_{55},a_{55}$) of the fifth row for positive and negative areas-in total we can conclude that sequence of elements connection for the fourth row with elements of the fifth row has the following view, forming the frequency of ties equal f=6:

$$a_{41} \rightarrow a_{54},a_{54},a_{55},a_{55},a_{56},a_{56}$$
$$a_{44} \rightarrow a_{51},a_{51},a_{55},a_{55},a_{56},a_{56}$$
$$a_{45} \rightarrow a_{51},a_{51},a_{54},a_{54},a_{56},a_{56}$$
$$a_{46} \rightarrow a_{51},a_{51},a_{54},a_{54},a_{55},a_{55}$$

c) For elements of the fifth row

- The first element(a_{51}) of the fifth row is joined with elements ($a_{64},a_{64},a_{65},a_{65},a_{66},a_{66}$) of the sixth row for positive and negative areas;
- The fourth element(a_{54}) of the fifth row is joined with elements ($a_{61},a_{61},a_{65},a_{65},a_{66},a_{66}$)of the sixth row for positive and negative areas;
- The fifth element(a_{55}) of the fifth row is joined with elements ($a_{61},a_{61},a_{64},a_{64},a_{66},a_{66}$) of the sixth row for positive and negative areas;
- The sixth element(a_{56}) of the fifth row is joined with elements ($a_{61},a_{61},a_{64},a_{64},a_{65},a_{65}$) of the sixth row for positive and negative areas-in total we can conclude that sequence of elements connection for the fifth row with elements of the sixth row has the following view ,forming the frequency of ties equal f=6:

$$a_{51} \rightarrow a_{64},a_{64},a_{65},a_{65},a_{66},a_{66}$$
$$a_{54} \rightarrow a_{61},a_{61},a_{65},a_{65},a_{66},a_{66}$$
$$a_{55} \rightarrow a_{61},a_{61},a_{64},a_{64},a_{66},a_{66}$$
$$a_{56} \rightarrow a_{61},a_{61},a_{64},a_{64},a_{65},a_{65}$$

❖ For Figure 140 of the combined partial determinant $\left| A_8 \right|$:

Analysis of data shown in Figure 140 applicably to formula (111) of the combined partial determinant $\left| A_8 \right|$ of the 6x6 matrix for positive and negative areas indicates on the following results in general view:
- The second element(a_{12}) of the first row is joined with the fourth element(a_{24}) of the second row;
- The fourth element(a_{24}) of the second row is joined with elements ($a_{31},a_{33},a_{35},a_{36}$) of the third row;

a) For elements of the third row

- The first element(a_{31})of the third row is joined with elements (a_{43},a_{45},a_{46}) of the fourth row;
- The third element(a_{33})of the third row is joined with elements (a_{41},a_{45},a_{46}) of the fourth row;
- The fifth element(a_{35}) of the third row is joined with elements (a_{41},a_{43},a_{46}) of the fourth row;
- The sixth element(a_{36})of the third row is joined with elements (a_{41},a_{43},a_{45}) of the fourth row-in total we can conclude that sequence of elements connection for the third row with elements of the fourth row has the following view, forming the frequency of ties equal f=3:

$$a_{31} \rightarrow a_{43},a_{45},a_{46}$$
$$a_{33} \rightarrow a_{41},a_{45},a_{46}$$
$$a_{35} \rightarrow a_{41},a_{43},a_{46}$$
$$a_{36} \rightarrow a_{41},a_{43},a_{45}$$

b) For elements of the fourth row

- The first element (a_{41}) of the fourth row is joined with elements ($a_{53},a_{53},a_{55},a_{55},a_{56},a_{56}$)of the fifth row for positive and negative areas;
- The third element (a_{43}) of the fourth row is joined with elements ($a_{51},a_{51},a_{55},a_{55},a_{56},a_{56}$) of the fifth row for positive and negative areas;
- The fifth element(a_{45})of the fourth row is joined with elements ($a_{51},a_{51},a_{53},a_{53},a_{56},a_{56}$) of the fifth row for positive and negative areas;

- The sixth element(a_{46}) of the fourth row is joined with elements ($a_{51},a_{51},a_{53},a_{53},a_{55},a_{55}$) of the fifth row for positive and negative areas-in total we can conclude that sequence of elements connection for the fourth row with elements of the fifth row has the following view, forming the frequency of ties equal f=6:

$$a_{41}\rightarrow a_{53},a_{53},a_{55},a_{55},a_{56},a_{56}$$
$$a_{43}\rightarrow a_{51},a_{51},a_{55},a_{55},a_{56},a_{56}$$
$$a_{45}\rightarrow a_{51},a_{51},a_{53},a_{53},a_{56},a_{56}$$
$$a_{46}\rightarrow a_{51},a_{51},a_{53},a_{53},a_{55},a_{55}$$

c) *For elements of the fifth row*

- The first element (a_{51}) of the fifth row is joined with elements ($a_{63},a_{63},a_{65},a_{65},a_{66},a_{66}$) of the sixth row for positive and negative areas;
- The third element (a_{53}) of the fifth row is joined with elements ($a_{61},a_{61},a_{65},a_{65},a_{66},a_{66}$) of the sixth row for positive and negative areas;
- The fifth element (a_{55}) of the fifth row is joined with elements ($a_{61},a_{61},a_{63},a_{63},a_{66},a_{66}$) for positive and negative areas;
- The sixth element (a_{56}) of the fifth row is joined with elements ($a_{61},a_{61},a_{63},a_{63},a_{65},a_{65}$)for positive and negative areas-in total we can conclude that sequence of elements connection for the fifth row with elements of the sixth row has the following view, forming the frequency of ties equal f=6:

$$a_{51}\rightarrow a_{63},a_{63},a_{65},a_{65},a_{66},a_{66}$$
$$a_{53}\rightarrow a_{61},a_{61},a_{65},a_{65},a_{66},a_{66}$$
$$a_{55}\rightarrow a_{61},a_{61},a_{63},a_{63},a_{66},a_{66}$$
$$a_{56}\rightarrow a_{61},a_{61},a_{63},a_{63},a_{65},a_{65}$$

❖ For Figure 141 of the combined partial determinant $|A_9|$:

Analysis of data shown in Figure 141 applicably to formula (111) of the combined partial determinant $|A_9|$ of the 6x6 matrix for positive and negative areas indicates on the following results in general view:
- The second element(a_{12}) of the first row is joined with the fifth element(a_{25}) of the second row;
- The fifth element(a_{25}) of the second row is joined with elements ($a_{31},a_{33},a_{34},a_{36}$) of the third row;

a) *For elements of the third row*

- The first element(a_{31}) of the third row is joined with elements (a_{43},a_{44},a_{46}) of the fifth row;
- The third element(a_{33}) of the third row is joined with elements (a_{41},a_{44},a_{46}) of the fourth row;
- The fourth element(a_{34}) of the third row is joined with elements (a_{41},a_{43},a_{46}) of the fourth row;
- The sixth element(a_{36}) of the third row is joined with elements (a_{41},a_{43},a_{44}) of the fourth row –in total we can conclude that sequence of elements connection for the third row with elements of the fourth row has the following view, forming the frequency of ties equal f=3:

$$a_{31}\rightarrow a_{43},a_{44},a_{46}$$
$$a_{33}\rightarrow a_{41},a_{44},a_{46}$$
$$a_{34}\rightarrow a_{41},a_{43},a_{46}$$
$$a_{36}\rightarrow a_{41},a_{43},a_{44}$$

b) *For elements of the fourth row*

- The first element(a_{41}) of the fourth row is joined with elements ($a_{53},a_{53},a_{54},a_{54},a_{56},a_{56}$) of the fifth row for positive and negative areas;
- The third element(a_{43}) of the fourth row is joined with elements ($a_{51},a_{51},a_{54},a_{54},a_{56},a_{56}$) of the fifth row for positive and negative areas;
- The fourth element(a_{44}) of the fourth row is joined with elements ($a_{51},a_{51},a_{53},a_{53},a_{56},a_{56}$) of the fifth row for positive and negative areas;
- The sixth element(a_{46}) of the fourth row is joined with elements ($a_{51},a_{51},a_{53},a_{53},a_{54},a_{54}$) of the fifth row for positive and negative areas- in total we can conclude that sequence of elements connection

for the fourth row with elements of the fifth row has the following view, forming the frequency of ties equal f=6:

$$a_{41} \rightarrow a_{53}, a_{53}, a_{54}, a_{54}, a_{56}, a_{56}$$
$$a_{43} \rightarrow a_{51}, a_{51}, a_{54}, a_{54}, a_{56}, a_{56}$$
$$a_{44} \rightarrow a_{51}, a_{51}, a_{53}, a_{53}, a_{56}, a_{56}$$
$$a_{46} \rightarrow a_{51}, a_{51}, a_{53}, a_{53}, a_{54}, a_{54}$$

c) For elements of the fifth row

- The first element (a_{51}) of the fifth row is joined with elements ($a_{63}, a_{63}, a_{64}, a_{64}, a_{66}, a_{66}$) of the sixth row for positive and negative areas;
- The third element (a_{53}) of the fifth row is joined with elements ($a_{61}, a_{61}, a_{64}, a_{64}, a_{66}, a_{66}$) of the sixth row for positive and negative areas;
- The fourth element(a_{54}) of the fifth row is joined with elements ($a_{61}, a_{61}, a_{63}, a_{63}, a_{66}, a_{66}$) of the sixth row for positive and negative areas;
- The sixth element (a_{56}) of the fifth row is joined with elements ($a_{61}, a_{61}, a_{63}, a_{63}, a_{64}, a_{64}$)-in total we can conclude that sequence of elements connection for the fifth row with elements of the sixth row has the following view, forming the frequency of ties equal f=6:

$$a_{51} \rightarrow a_{63}, a_{63}, a_{64}, a_{64}, a_{66}, a_{66}$$
$$a_{53} \rightarrow a_{61}, a_{61}, a_{64}, a_{64}, a_{66}, a_{66}$$
$$a_{54} \rightarrow a_{61}, a_{61}, a_{63}, a_{63}, a_{66}, a_{66}$$
$$a_{56} \rightarrow a_{61}, a_{61}, a_{63}, a_{63}, a_{64}, a_{64}$$

❖ For Figure 142 of the combined partial determinant $|A_{10}|$:

Analysis of data shown in Figure 142 applicably to formula (111) of the combined partial determinant $|A_{10}|$ of the 6x6 matrix for positive and negative areas indicates on the following results in general view:
- The second element(a_{12}) of the first row is joined with the sixth element (a_{26}) of the second row;
- The sixth element (a_{26}) of the second row is joined with elements ($a_{31}, a_{33}, a_{34}, a_{35}$) of the third row;

a) For elements of the third row

- The first element(a_{31}) of the third row is joined with elements (a_{43}, a_{44}, a_{45}) of the fourth row;
- The third element(a_{33}) of the third row is joined with elements (a_{41}, a_{44}, a_{45}) of the fourth row;
- The fourth element(a_{34}) of the third row is joined with elements (a_{41}, a_{43}, a_{45}) of the fourth row;
- The fifth element(a_{35}) of the third row is joined with elements (a_{41}, a_{43}, a_{44}) of the fourth row-in total we can conclude that sequence of elements connection for the third row with elements of the fourth row has the following view, forming the frequency of ties equal f=3:

$$a_{31} \rightarrow a_{43}, a_{44}, a_{45}$$
$$a_{33} \rightarrow a_{41}, a_{44}, a_{45}$$
$$a_{34} \rightarrow a_{41}, a_{43}, a_{45}$$
$$a_{35} \rightarrow a_{41}, a_{43}, a_{44}$$

b) For elements of the fourth row

- The first element(a_{41}) of the fourth row is joined with elements ($a_{53}, a_{53}, a_{54}, a_{54}, a_{55}, a_{55}$) of the fifth row for positive and negative areas;
- The third element(a_{43}) of the fourth row is joined with elements ($a_{51}, a_{51}, a_{54}, a_{54}, a_{55}, a_{55}$)of the fifth row for positive and negative areas;
- The fourth element(a_{44}) of the fourth row is joined with elements ($a_{51}, a_{51}, a_{53}, a_{53}, a_{55}, a_{55}$) of the fifth row for positive and negative areas;
- The fifth element(a_{45}) of the fourth row is joined with elements ($a_{51}, a_{51}, a_{53}, a_{53}, a_{54}, a_{54}$) of the fifth row for positive and negative areas- in total we can conclude that sequence of elements connection for the fourth row with elements of the fifth row has the following view, forming the frequency of ties equal f=6:

$$a_{41} \rightarrow a_{53}, a_{53}, a_{54}, a_{54}, a_{55}, a_{55}$$
$$a_{43} \rightarrow a_{51}, a_{51}, a_{54}, a_{54}, a_{55}, a_{55}$$
$$a_{44} \rightarrow a_{51}, a_{51}, a_{53}, a_{53}, a_{55}, a_{55}$$
$$a_{45} \rightarrow a_{51}, a_{51}, a_{53}, a_{53}, a_{54}, a_{54}$$

c) *For elements of the fifth row*

- The first element (a_{51}) of the fifth row is joined with elements ($a_{63}, a_{63}, a_{64}, a_{64}, a_{65}, a_{65}$) of the sixth row for positive and negative areas;
- The third element(a_{53}) of the fifth row is joined with elements ($a_{61}, a_{61}, a_{64}, a_{64}, a_{65}, a_{65}$) of the sixth row for positive and negative areas;
- The fourth element(a_{54}) of the fifth row is joined with elements ($a_{61}, a_{61}, a_{63}, a_{63}, a_{65}, a_{65}$) of the sixth row for positive and negative areas;
- The fifth element(a_{55}) of the fifth row is joined with elements ($a_{61}, a_{61}, a_{63}, a_{63}, a_{64}, a_{64}$) of the sixth row for positive and negative areas-in total we can conclude that sequence of elements connection for the fifth row with elements of the sixth row has the following view, forming the frequency of ties equal f=6:

$$a_{51} \rightarrow a_{63}, a_{63}, a_{64}, a_{64}, a_{65}, a_{65}$$
$$a_{53} \rightarrow a_{61}, a_{61}, a_{64}, a_{64}, a_{65}, a_{65}$$
$$a_{54} \rightarrow a_{61}, a_{61}, a_{63}, a_{63}, a_{65}, a_{65}$$
$$a_{55} \rightarrow a_{61}, a_{61}, a_{63}, a_{63}, a_{64}, a_{64}$$

C. Analysis of combined partial determinant $\left| A'_{13} \right|$ of the 6x6 matrix:

The combined partial determinant $\left| A'_{13} \right|$ consists from five separate parts and is expressed by the following formula in view of:

$$\left| A'_{13} \right| = \left| A_{11} \right| + \left| A_{12} \right| + \left| A_{13} \right| + \left| A_{14} \right| + \left| A_{15} \right|$$

And also the above-said determinant $\left| A'_{13} \right|$ is introduced by the following finally formula (112),after of some transformations, elements of which are shown in view of functional graphs in Figures 143 to 147:

$$\left| A'_{13} \right| = a_{13}a_{21}a_{32}[a_{44}(a_{55}a_{66} - a_{56}a_{65}) + a_{45}(a_{56}a_{64} - a_{54}a_{66}) + a_{46}(a_{54}a_{65} - a_{55}a_{64})] +$$
$$+ a_{13}a_{21}a_{34}[a_{42}(a_{56}a_{65} - a_{55}a_{66}) + a_{45}(a_{52}a_{66} - a_{56}a_{62}) + a_{46}(a_{55}a_{62} - a_{52}a_{65})] + \quad \left| A_{11} \right|$$
$$+ a_{13}a_{21}a_{35}[a_{42}(a_{54}a_{66} - a_{56}a_{64}) + a_{44}(a_{56}a_{62} - a_{52}a_{66}) + a_{46}(a_{52}a_{64} - a_{54}a_{62})] +$$
$$+ a_{13}a_{21}a_{36}[a_{42}(a_{55}a_{64} - a_{54}a_{65}) + a_{44}(a_{52}a_{65} - a_{55}a_{62}) + a_{45}(a_{54}a_{62} - a_{52}a_{64})] +$$

$$+ a_{13}a_{22}a_{31}[a_{44}(a_{56}a_{65} - a_{55}a_{66}) + a_{45}(a_{54}a_{66} - a_{56}a_{64}) + a_{46}(a_{55}a_{64} - a_{54}a_{65})] +$$
$$+ a_{13}a_{22}a_{34}[a_{41}(a_{55}a_{66} - a_{56}a_{65}) + a_{45}(a_{56}a_{61} - a_{51}a_{66}) + a_{46}(a_{51}a_{65} - a_{55}a_{61})] + \quad \left| A_{12} \right|$$
$$+ a_{13}a_{22}a_{35}[a_{41}(a_{56}a_{64} - a_{54}a_{66}) + a_{44}(a_{51}a_{66} - a_{56}a_{61}) + a_{46}(a_{54}a_{61} - a_{51}a_{64})] +$$
$$+ a_{13}a_{22}a_{36}[a_{41}(a_{54}a_{65} - a_{55}a_{64}) + a_{44}(a_{55}a_{61} - a_{51}a_{65}) + a_{45}(a_{51}a_{64} - a_{54}a_{61})] +$$

$$+ a_{13}a_{24}a_{31}[a_{42}(a_{55}a_{66} - a_{56}a_{65}) + a_{45}(a_{56}a_{62} - a_{52}a_{66}) + a_{46}(a_{52}a_{65} - a_{55}a_{62})] +$$
$$+ a_{13}a_{24}a_{32}[a_{41}(a_{56}a_{65} - a_{55}a_{66}) + a_{45}(a_{51}a_{66} - a_{56}a_{61}) + a_{46}(a_{55}a_{61} - a_{51}a_{65})] + \quad \left| A_{13} \right|$$
$$+ a_{13}a_{24}a_{35}[a_{41}(a_{52}a_{66} - a_{56}a_{62}) + a_{42}(a_{56}a_{61} - a_{51}a_{66}) + a_{46}(a_{51}a_{62} - a_{52}a_{61})] +$$
$$+ a_{13}a_{24}a_{36}[a_{41}(a_{55}a_{62} - a_{52}a_{65}) + a_{42}(a_{51}a_{65} - a_{55}a_{61}) + a_{45}(a_{52}a_{61} - a_{51}a_{62})] +$$

$$+ a_{13}a_{25}a_{31}[a_{42}(a_{56}a_{64} - a_{54}a_{66}) + a_{44}(a_{52}a_{66} - a_{56}a_{62}) + a_{46}(a_{54}a_{62} - a_{52}a_{64})] +$$
$$+ a_{13}a_{25}a_{32}[a_{41}(a_{54}a_{66} - a_{56}a_{64}) + a_{44}(a_{56}a_{61} - a_{51}a_{66}) + a_{46}(a_{51}a_{64} - a_{54}a_{61})] + \quad \left| A_{14} \right|$$
$$+ a_{13}a_{25}a_{34}[a_{41}(a_{56}a_{62} - a_{52}a_{66}) + a_{42}(a_{51}a_{66} - a_{56}a_{61}) + a_{46}(a_{52}a_{61} - a_{51}a_{62})] +$$
$$+ a_{13}a_{25}a_{36}[a_{41}(a_{52}a_{64} - a_{54}a_{62}) + a_{42}(a_{54}a_{61} - a_{51}a_{64}) + a_{44}(a_{51}a_{62} - a_{52}a_{61})] +$$

$$+a_{13}a_{26}a_{31}[a_{42}(a_{54}a_{65}-a_{55}a_{64})+a_{44}(a_{55}a_{62}-a_{52}a_{65})+a_{45}(a_{52}a_{64}-a_{54}a_{62})]+$$
$$+a_{13}a_{26}a_{32}[a_{41}(a_{55}a_{64}-a_{54}a_{65})+a_{44}(a_{51}a_{65}-a_{55}a_{61})+a_{45}(a_{54}a_{61}-a_{51}a_{64})]+ \quad |A_{15}|$$
$$+a_{13}a_{26}a_{34}[a_{41}(a_{52}a_{65}-a_{55}a_{62})+a_{42}(a_{55}a_{61}-a_{51}a_{65})+a_{45}(a_{51}a_{62}-a_{52}a_{61})]+$$
$$+a_{13}a_{26}a_{35}[a_{41}(a_{54}a_{62}-a_{52}a_{64})+a_{42}(a_{51}a_{64}-a_{54}a_{61})+a_{44}(a_{52}a_{61}-a_{51}a_{62})] \quad (112)$$

❖ For Figure 143 of the combined partial determinant $|A_{11}|$:

Analysis of data shown in Figure 143 applicably to formula (112) of the combined partial determinant $|A_{11}|$ of the 6x6 matrix for positive and negative areas indicates on the following results in general view:
- The third element (a_{13}) of the first row is joined with the first element (a_{21}) of the second row;
- The first element(a_{21}) of the second row is joined with elements ($a_{32},a_{34},a_{35},a_{36}$);

a) *For elements of the third row*
- The second element(a_{32}) of the third row is joined with elements (a_{44},a_{45},a_{46}) of the fourth row;
- The fourth element(a_{34}) of the third row is joined with elements (a_{42},a_{45},a_{46}) of the fourth row;
- The fifth element (a_{35}) of the third row is joined with elements (a_{42},a_{44},a_{46}) of the fourth row;
- The sixth element(a_{36}) of the third row is joined with elements (a_{42},a_{44},a_{45}) of the fourth row –in total we can conclude that sequence of elements connection for the third row with elements of the fourth row has the following view ,forming the frequency of ties equal f=3:

$$a_{32} \rightarrow a_{44},a_{45},a_{46}$$
$$a_{34} \rightarrow a_{42},a_{45},a_{46}$$
$$a_{35} \rightarrow a_{42},a_{44},a_{46}$$
$$a_{36} \rightarrow a_{42},a_{44},a_{45}$$

b) *For elements of the fourth row*
- The second element(a_{42}) of the fourth row is joined with elements ($a_{54},a_{54},a_{55},a_{55},a_{56},a_{56}$) of the fifth row for positive and negative areas;
- The fourth element (a_{44}) of the fourth row is joined with elements ($a_{52},a_{52},a_{55},a_{55},a_{56},a_{56}$) of the fifth row for positive and negative areas ;
- The fifth element (a_{45}) of the fourth row is joined with elements ($a_{52},a_{52},a_{54},a_{54},a_{56},a_{56}$) of the fifth row for positive and negative areas;
- The sixth element (a_{46}) of the fourth row is joined with elements ($a_{52},a_{52},a_{54},a_{54},a_{55},a_{55}$) of the fifth row for positive and negative areas-in total we can conclude that sequence of elements connection for the fourth row with elements of the fifth row has the following view, forming the frequency of ties equal f=6:

$$a_{42} \rightarrow a_{54},a_{54},a_{55},a_{55},a_{56},a_{56}$$
$$a_{44} \rightarrow a_{52},a_{52},a_{55},a_{55},a_{56},a_{56}$$
$$a_{45} \rightarrow a_{52},a_{52},a_{54},a_{54},a_{56},a_{56}$$
$$a_{46} \rightarrow a_{52},a_{52},a_{54},a_{54},a_{55},a_{55}$$

c) *For elements of the fifth row*
- The second element(a_{52}) of the fifth row is joined with elements ($a_{64},a_{64},a_{65},a_{65},a_{66},a_{66}$) of the sixth row for positive and negative areas;
- The fourth element(a_{54}) of the fifth row is joined with elements ($a_{62},a_{62},a_{65},a_{65},a_{66},a_{66}$) of the sixth row for positive and negative areas;
- The fifth element(a_{55}) of the fifth row is joined with elements ($a_{62},a_{62},a_{64},a_{64},a_{66},a_{66}$)of the sixth row for positive and negative areas;
- The sixth element (a_{56}) of the fifth row is joined with elements ($a_{62},a_{62},a_{64},a_{64},a_{65},a_{65}$) of the sixth row for positive and negative areas-in total we can conclude that sequence of elements connection for the fifth row with elements of the sixth row has the following view, forming the frequency of ties equal f=6:

$$a_{52} \rightarrow a_{64}, a_{64}, a_{65}, a_{65}, a_{66}, a_{66}$$
$$a_{54} \rightarrow a_{62}, a_{62}, a_{65}, a_{65}, a_{66}, a_{66}$$
$$a_{55} \rightarrow a_{62}, a_{62}, a_{64}, a_{64}, a_{66}, a_{66}$$
$$a_{56} \rightarrow a_{62}, a_{62}, a_{64}, a_{64}, a_{65}, a_{65}$$

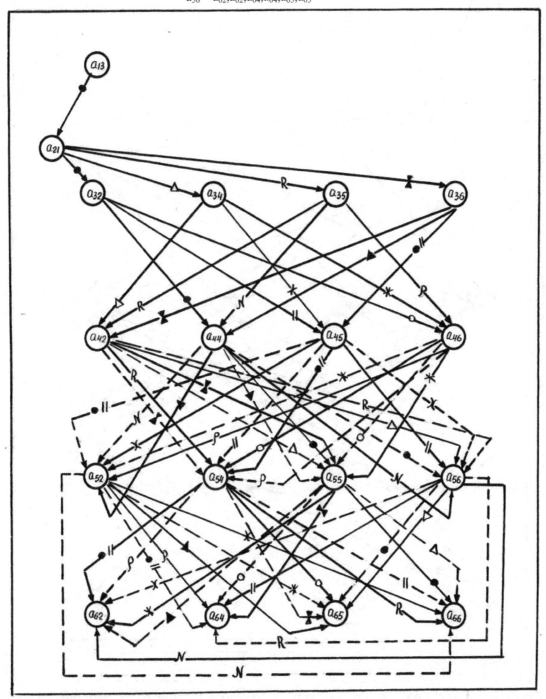

Figure 143 Functional graphs between of elements for the combined partial 6x6 matrix determinant $|A_{11}|$

211

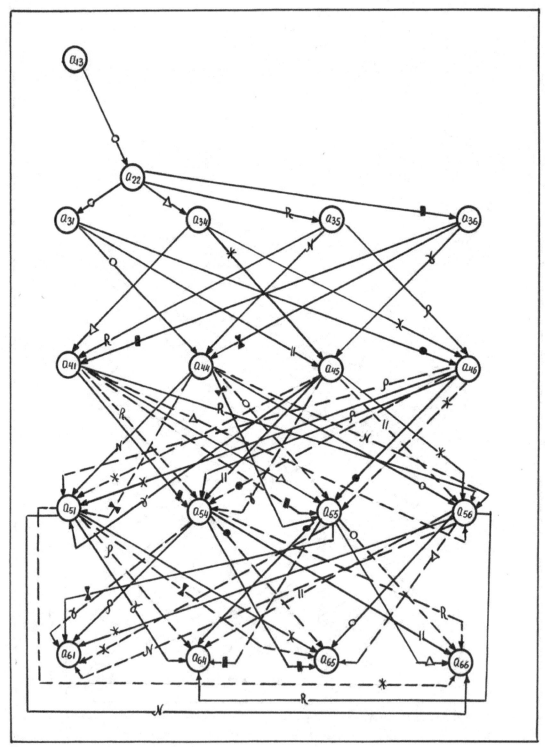

Figure 144 Functional graphs between of elements for the combined partial 6x6 matrix determinant $|A_{12}|$

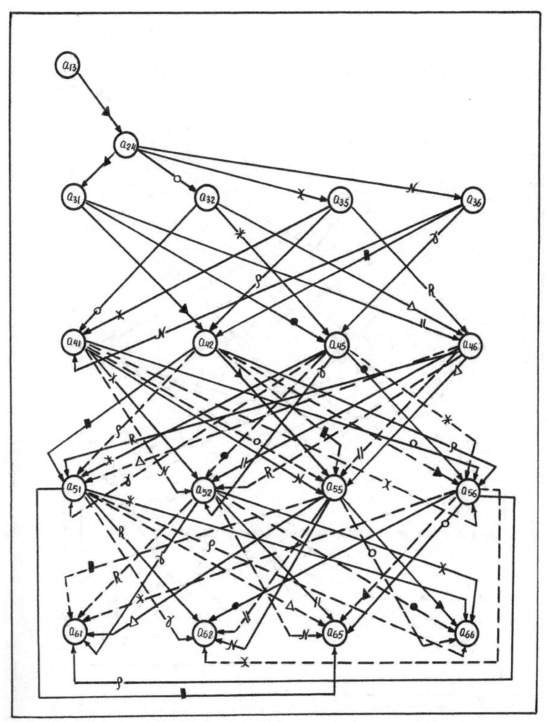

Figure 145 Functional graphs between of elements for the combined partial 6x6 matrix determinant $\left| A_{13} \right|$

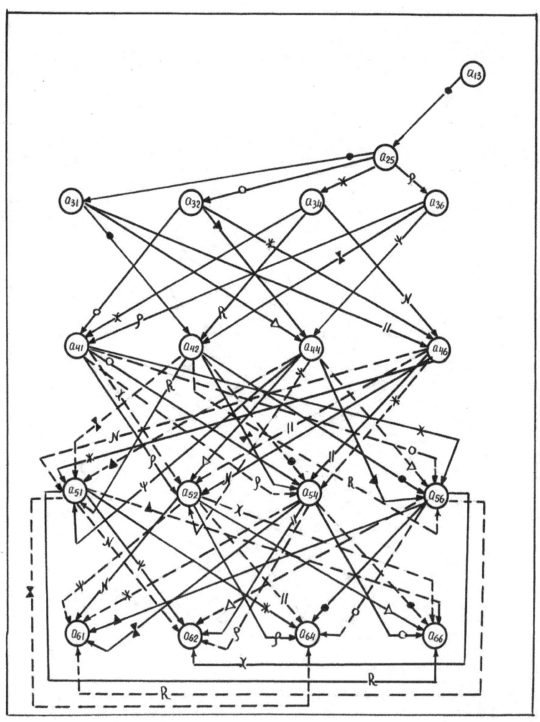

Figure 146 Functional graphs between of elements for the combined partial 6x6 matrix determinant $\left| A_{14} \right|$

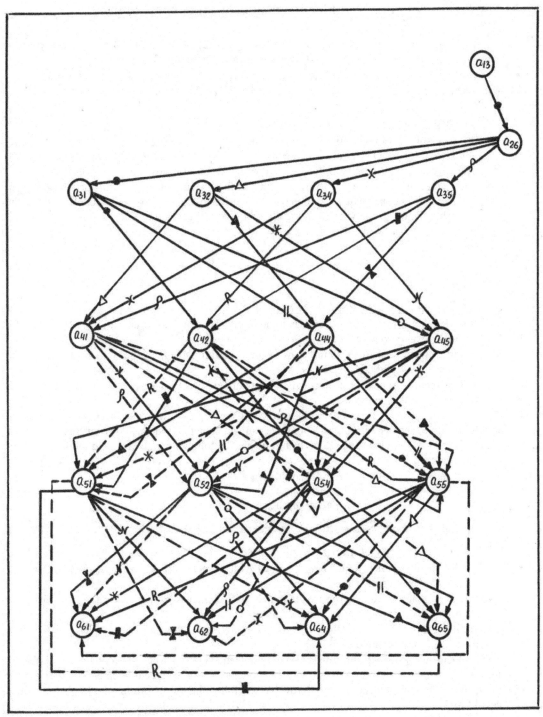

Figure 147 Functional graphs between of elements for the combined partial 6x6 matrix determinant $\left| A_{15} \right|$

❖ For Figure 144 of the combined partial determinant $|A_{12}|$:

Analysis of data shown in Figure 144 applicably to formula (112) of the combined partial determinant $|A_{12}|$ for positive and negative areas of the 6x6 matrix indicates on the following results in general view :
- The third element(a_{13}) of the first row is joined with the second element (a_{22}) of the second row;
- The second element(a_{22}) of the second row is joined with elements ($a_{31},a_{34},a_{35},a_{36}$);

a) *For elements of the third row*
- The first element(a_{31}) of the third row is joined with elements (a_{44},a_{45},a_{46}) of the fourth row;
- The fourth element(a_{34}) of the third row is joined with elements (a_{41},a_{45},a_{46}) of the fourth row;
- The fifth element(a_{35}) of the third row is joined with elements (a_{41},a_{44},a_{46}) of the fourth row;
- The sixth element(a_{36}) of the third row is joined with elements (a_{41},a_{44},a_{45}) of the fourth row-in total we can conclude that sequence of elements connection for the third row with elements of the fourth row has the following view ,forming the frequency of ties equal f=3:

$$a_{31} \rightarrow a_{44},a_{45},a_{46}$$
$$a_{34} \rightarrow a_{41},a_{45},a_{46}$$
$$a_{35} \rightarrow a_{41},a_{44},a_{46}$$
$$a_{36} \rightarrow a_{41},a_{44},a_{45}$$

b) *For elements of the fourth row*
- The first element(a_{41}) of the fourth row is joined with elements ($a_{54},a_{54},a_{55},a_{55},a_{56},a_{56}$) of the fifth row for positive and negative areas;
- The fourth element (a_{44}) of the fourth row is joined with elements ($a_{51},a_{51},a_{55},a_{55},a_{56},a_{56}$) of the fifth row for positive and negative areas;
- The fifth element (a_{45}) of the fourth row is joined with elements ($a_{51},a_{51},a_{54},a_{54},a_{56},a_{56}$) of the fifth row for positive and negative areas;
- The sixth element (a_{46}) of the fourth row is joined with elements ($a_{51},a_{51},a_{54},a_{54},a_{55},a_{55}$) of the fifth row for positive and negative areas-in total we can conclude that sequence of elements connection for the fourth row with elements of the fifth row has the following view ,forming the frequency of ties equal f=6:

$$a_{41} \rightarrow a_{54},a_{54},a_{55},a_{55},a_{56},a_{56}$$
$$a_{44} \rightarrow a_{51},a_{51},a_{55},a_{55},a_{56},a_{56}$$
$$a_{45} \rightarrow a_{51},a_{51},a_{54},a_{54},a_{56},a_{56}$$
$$a_{46} \rightarrow a_{51},a_{51},a_{54},a_{54},a_{55},a_{55}$$

c) *For elements of the fifth row*
- The first element(a_{51}) of the fifth row is joined with elements ($a_{64},a_{64},a_{65},a_{65},a_{66},a_{66}$) of the sixth row for positive and negative areas;
- The fourth element(a_{54}) of the fifth row is joined with elements ($a_{61},a_{61},a_{65},a_{65},a_{66},a_{66}$) of the sixth row for positive and negative areas;
- The fifth element(a_{55}) of the fifth row is joined with elements ($a_{61},a_{61},a_{64},a_{64},a_{66},a_{66}$) of the sixth row for positive and negative areas;
- The six element(a_{56}) of the fifth row is joined with elements ($a_{61},a_{61},a_{64},a_{64},a_{65},a_{65}$) of the sixth row for positive and negative areas- in total we can conclude that sequence of elements connection for the fifth row with elements of the sixth row has the following view, forming the frequency of ties equal f=6:

$$a_{51} \rightarrow a_{64},a_{64},a_{65},a_{65},a_{66},a_{66}$$
$$a_{54} \rightarrow a_{61},a_{61},a_{65},a_{65},a_{66},a_{66}$$
$$a_{55} \rightarrow a_{61},a_{61},a_{64},a_{64},a_{66},a_{66}$$
$$a_{56} \rightarrow a_{61},a_{61},a_{64},a_{64},a_{65},a_{65}$$

❖ **For Figure 145 of the combined partial determinant $|A_{13}|$:**

Analysis of data shown in Figure 145 applicably to formula (112) of the combined partial determinant $|A_{13}|$ for positive (——) and negative(--------) areas of the 6x6 matrix indicates on the following results in general view:

- The third element(a_{13}) of the first row is joined with the fourth element(a_{24}) of the second row;
- The fourth element (a_{24}) of the second row is joined with elements ($a_{31}, a_{32}, a_{35}, a_{36}$);

a) *For elements of the third row*

- The first element (a_{31}) of the third row is joined with elements (a_{42}, a_{45}, a_{46}) of the fourth row;
- The second element(a_{32}) of the third row is joined with elements (a_{41}, a_{45}, a_{46}) of the fourth row;
- The fifth element (a_{35}) of the third row is joined with elements (a_{41}, a_{42}, a_{46}) of the fourth row;
- The sixth element(a_{36}) of the third row is joined with elements (a_{41}, a_{42}, a_{45}) of the fourth row-in total we can conclude that sequence of elements connection for the third row with elements of the fourth row has the following view, forming the frequency of ties equal f=3:

$$a_{31} \rightarrow a_{42}, a_{45}, a_{46}$$
$$a_{32} \rightarrow a_{41}, a_{45}, a_{46}$$
$$a_{35} \rightarrow a_{41}, a_{42}, a_{46}$$
$$a_{36} \rightarrow a_{41}, a_{42}, a_{46}$$

b) *For elements of the fourth row*

- The first element(a_{41}) of the fourth row is joined with elements ($a_{52}, a_{52}, a_{55}, a_{55}, a_{56}, a_{56}$) of the fifth row for positive and negative areas;
- The second element(a_{42}) of the fourth row is joined with elements ($a_{51}, a_{51}, a_{55}, a_{55}, a_{56}, a_{56}$) of the fifth row for positive and negative areas;
- The fifth element(a_{45}) of the fourth row is joined with elements ($a_{51}, a_{51}, a_{52}, a_{52}, a_{56}, a_{56}$) of the fifth row for positive and negative areas;
- The sixth element(a_{46}) of the fourth row is joined with elements ($a_{51}, a_{51}, a_{52}, a_{52}, a_{55}, a_{55}$) of the fifth row for positive and negative areas-in total we can conclude that sequence of elements connection for the fourth row with elements of the fifth row has the following view, forming the frequency of ties equal f=6:

$$a_{41} \rightarrow a_{52}, a_{52}, a_{55}, a_{55}, a_{56}, a_{56}$$
$$a_{42} \rightarrow a_{51}, a_{51}, a_{55}, a_{55}, a_{56}, a_{56}$$
$$a_{45} \rightarrow a_{51}, a_{51}, a_{52}, a_{52}, a_{56}, a_{56}$$
$$a_{46} \rightarrow a_{51}, a_{51}, a_{52}, a_{52}, a_{55}, a_{55}$$

c) *For elements of the fifth row*

- The first element(a_{51}) of the fifth row is joined with elements ($a_{62}, a_{62}, a_{65}, a_{65}, a_{66}, a_{66}$) of the sixth row for positive and negative areas;
- The second element(a_{52}) of the fifth row is joined with elements ($a_{61}, a_{61}, a_{65}, a_{65}, a_{66}, a_{66}$) of the sixth row for positive and negative areas;
- The fifth element(a_{55}) of the fifth row is joined with elements ($a_{61}, a_{61}, a_{62}, a_{62}, a_{66}, a_{66}$) of the sixth row for positive and negative areas;
- The sixth element(a_{56}) of the fifth row is joined with elements ($a_{61}, a_{61}, a_{62}, a_{62}, a_{65}, a_{65}$) of the sixth row for positive and negative areas-in total we can conclude that sequence of elements connection for the fifth row with elements of the sixth row has the following view, forming the frequency of ties equal f=6:

$$a_{51} \rightarrow a_{62}, a_{62}, a_{65}, a_{65}, a_{66}, a_{66}$$
$$a_{52} \rightarrow a_{61}, a_{61}, a_{65}, a_{65}, a_{66}, a_{66}$$
$$a_{55} \rightarrow a_{61}, a_{61}, a_{62}, a_{62}, a_{66}, a_{66}$$
$$a_{56} \rightarrow a_{61}, a_{61}, a_{62}, a_{62}, a_{65}, a_{65}$$

217

❖ For Figure 146 of the combined partial determinant $|A_{14}|$:

Analysis of data shown in Figure 146 applicably to formula (112) of the combined partial determinant $|A_{14}|$ of the 6x6 matrix for positive and negative areas indicates on the following results in general view:
- The third element (a_{13}) of the first row is joined with the fifth element (a_{25}) of the second row;
- The fifth element (a_{25}) of the second row is joined with elements ($a_{31},a_{32},a_{34},a_{36}$) of the third row;

a) *For elements of the third row*
- The first element(a_{31}) of the third row is joined with elements (a_{42},a_{44},a_{46}) of the fourth row;
- The second element(a_{32}) of the third row is joined with elements (a_{41},a_{44},a_{46}) of the fourth row;
- The fourth element (a_{34}) of the third row is joined with elements (a_{41},a_{42},a_{46}) of the fourth row;
- The sixth element(a_{36}) of the third row is joined with elements (a_{41},a_{42},a_{44}) of the fourth row-in total we can conclude that sequence of elements connection for the third row with elements of the fourth row has the following view, forming the frequency of ties equal f=3:

$$a_{31}{\rightarrow}a_{42},a_{44},a_{46}$$
$$a_{32}{\rightarrow}a_{41},a_{44},a_{46}$$
$$a_{34}{\rightarrow}a_{41},a_{42},a_{4}$$
$$a_{36}{\rightarrow}a_{41},a_{42},a_{44}$$

b) *For elements of the fourth row*
- The first element (a_{41}) of the fourth row is joined with elements ($a_{52},a_{52},a_{54},a_{54},a_{56},a_{56}$) of the fifth row for positive and negative areas;
- The second element(a_{42}) of the fourth row is joined with elements ($a_{51},a_{51},a_{54},a_{54},a_{56},a_{56}$) of the fifth row for positive and negative areas;
- The fourth element (a_{44}) of the fourth row is joined with elements ($a_{51},a_{51},a_{52},a_{52},a_{56},a_{56}$) of the fifth row for positive and negative areas;
- The sixth element(a_{46}) of the fourth row is joined with elements ($a_{51},a_{51},a_{52},a_{52},a_{54},a_{54}$) of the fifth row for positive and negative areas-in total we can conclude that sequence of elements connection fot he fourth row with elements of the fifth row has the following view, forming the frequency of ties equal f=6:

$$a_{41}{\rightarrow}a_{52},a_{52},a_{54},a_{54},a_{56},a_{56}$$
$$a_{42}{\rightarrow}a_{51},a_{51},a_{54},a_{54},a_{56},a_{56}$$
$$a_{44}{\rightarrow}a_{51},a_{51},a_{52},a_{52},a_{56},a_{56}$$
$$a_{46}{\rightarrow}a_{51},a_{51},a_{52},a_{52},a_{54},a_{54}$$

c) *For elements of the fifth row*
- The first element (a_{51}) of the fifth row is joined with elements ($a_{62},a_{62},a_{64},a_{64},a_{66},a_{66}$) of the sixth row for positive and negative areas;
- The second element(a_{52}) of the fifth row is joined with elements ($a_{61},a_{61},a_{64},a_{64},a_{66},a_{66}$) of the sixth row for positive and negative areas;
- The fourth element(a_{54}) of the fifth row is joined with elements ($a_{61},a_{61},a_{62},a_{62},a_{66},a_{66}$) of the sixth row for positive and negative areas;
- The sixth element (a_{56})of the fifth row is joined with elements ($a_{61},a_{61},a_{62},a_{62},a_{64},a_{64}$) of the sixth row for positive and negative areas –in total we can conclude that sequence of elements connection for the fifth row with elements of the sixth row has the following view, forming the frequency of ties equal f=6:

$$a_{51}{\rightarrow}a_{62},a_{62},a_{64},a_{64},a_{66},a_{66}$$
$$a_{52}{\rightarrow}a_{61},a_{61},a_{64},a_{64},a_{66},a_{66}$$
$$a_{54}{\rightarrow}a_{61},a_{61},a_{62},a_{62},a_{66},a_{66}$$
$$a_{56}{\rightarrow}a_{61},a_{61},a_{62},a_{62},a_{64},a_{64}$$

❖ For Figure 147 of the combined partial determinant $\left| A_{15} \right|$:

Analysis of data shown in Figure 147 applicably to formula (112) of the combined partial determinant $\left| A_{15} \right|$ of the 6x6 matrix for positive and negative areas indicates on the following results in general view:

- The third element(a_{13}) of the first row is joined with the sixth element (a_{26})of the second row;
- The sixth element (a_{26}) of the second row is joined with elements ($a_{31},a_{32},a_{34},a_{35}$);

a) For elements of the third row

- The first element(a_{31}) of the third row is joined with elements (a_{42},a_{44},a_{45}) of the fourth row;
- The second element(a_{32}) of the third row is joined with elements (a_{41},a_{44},a_{45}) of the fourth row;
- The fourth element(a_{34}) of the third row is joined with elements (a_{41},a_{42},a_{45}) of the fourth row;
- The fifth element(a_{35}) of the third row is joined with elements (a_{41},a_{42},a_{44}) of the fourth row-in total we can conclude that sequence of elements connection for the third row with elements of the fourth row has the following view, forming the frequency of ties equal f=3:

$$a_{31} \rightarrow a_{42},a_{44},a_{45}$$
$$a_{32} \rightarrow a_{41},a_{44},a_{45}$$
$$a_{34} \rightarrow a_{41},a_{42},a_{45}$$
$$a_{35} \rightarrow a_{41},a_{42},a_{44}$$

b) For elements of the fourth row

- The first element(a_{41}) of the fourth row is joined with elements ($a_{52},a_{52},a_{54},a_{54},a_{55},a_{55}$) of the fifth row for positive and negative areas;
- The second element(a_{42}) of the fourth row is joined with elements ($a_{51},a_{51},a_{54},a_{54},a_{55},a_{55}$) of the fifth row for positive and negative areas;
- The fourth element(a_{44}) of the fourth row is joined with elements ($a_{51},a_{51},a_{52},a_{52},a_{55},a_{55}$)of the fifth row for positive and negative areas;
- The fifth element(a_{45}) of the fourth row is joined with elements ($a_{51},a_{51},a_{52},a_{52},a_{54},a_{54}$)of the fifth row for positive and negative areas-in total we can conclude that sequence of elements connection for the fourth row with elements of the fifth row has the following view, forming the frequency of ties equal f=6:

$$a_{41} \rightarrow a_{52},a_{52},a_{54},a_{54},a_{55},a_{55}$$
$$a_{42} \rightarrow a_{51},a_{51},a_{54},a_{54},a_{55},a_{55}$$
$$a_{44} \rightarrow a_{51},a_{51},a_{52},a_{52},a_{55},a_{55}$$
$$a_{45} \rightarrow a_{51},a_{51},a_{52},a_{52},a_{54},a_{54}$$

c) For elements of the fifth row

- The first element(a_{51}) of the fifth row is joined with elements ($a_{62},a_{62},a_{64},a_{64},a_{65},a_{65}$) of the sixth row for positive and negative areas;
- The second element(a_{52}) of the fifth row is joined with elements ($a_{61},a_{61},a_{64},a_{64},a_{65},a_{65}$) of the sixth row for positive and negative areas;
- The fourth element(a_{54}) of the fifth row is joined with elements ($a_{61},a_{61},a_{62},a_{62},a_{65},a_{65}$) of the sixth row for positive and negative areas;
- The fifth element (a_{55}) of the fifth row is joined with elements ($a_{61},a_{61},a_{62},a_{62},a_{64},a_{64}$) of the sixth row for positive and negative areas-in total we can conclude that sequence of elements connection for the fifth row with elements of the sixth row has the following view , forming the frequency of ties equal f=6:

$$a_{51} \rightarrow a_{62},a_{62},a_{64},a_{64},a_{65},a_{65}$$
$$a_{52} \rightarrow a_{61},a_{61},a_{64},a_{64},a_{65},a_{65}$$
$$a_{54} \rightarrow a_{61},a_{61},a_{62},a_{62},a_{65},a_{65}$$
$$a_{55} \rightarrow a_{61},a_{61},a_{62},a_{62},a_{64},a_{64}$$

D. Analysis of combined partial determinant $|A'_{14}|$ of the combined matrix:

The combined partial determinant $|A'_{14}|$ consists from five separate parts and is expressed by the following form in view of

$$|A'_{14}| = |A_{16}| + |A_{17}| + |A_{18}| + |A_{19}| + |A_{20}|$$

And also the above-indicated determinant $|A'_{14}|$ is introduced by the following finally formula (113),after of some transformations, elements of which are shown in view of functional graphs in Figures 148 to 152:

$$
\begin{aligned}
|A'_{14}| = {} & a_{14}a_{21}a_{32}[a_{43}(a_{56}a_{65}-a_{55}a_{66})+a_{45}(a_{53}a_{66}-a_{56}a_{63})+a_{46}(a_{55}a_{63}-a_{53}a_{65})]+ \\
& +a_{14}a_{21}a_{33}[a_{42}(a_{55}a_{66}-a_{56}a_{65})+a_{45}(a_{56}a_{62}-a_{52}a_{66})+a_{46}(a_{52}a_{65}-a_{55}a_{62})]+ \quad |A_{16}| \\
& +a_{14}a_{21}a_{35}[a_{42}(a_{56}a_{63}-a_{53}a_{66})+a_{43}(a_{52}a_{66}-a_{56}a_{62})+a_{46}(a_{53}a_{62}-a_{52}a_{63})]+ \\
& +a_{14}a_{21}a_{36}[a_{42}(a_{53}a_{65}-a_{55}a_{63})+a_{43}(a_{55}a_{62}-a_{52}a_{65})+a_{45}(a_{52}a_{63}-a_{53}a_{62})]+ \\[6pt]
& +a_{14}a_{22}a_{31}[a_{43}(a_{55}a_{66}-a_{56}a_{65})+a_{45}(a_{56}a_{63}-a_{53}a_{66})+a_{46}(a_{53}a_{65}-a_{55}a_{63})]+ \\
& +a_{14}a_{22}a_{33}[a_{41}(a_{56}a_{65}-a_{55}a_{66})+a_{45}(a_{51}a_{66}-a_{56}a_{61})+a_{46}(a_{55}a_{61}-a_{51}a_{65})]+ \quad |A_{17}| \\
& +a_{14}a_{22}a_{35}[a_{41}(a_{53}a_{66}-a_{56}a_{63})+a_{43}(a_{56}a_{61}-a_{51}a_{66})+a_{46}(a_{51}a_{63}-a_{53}a_{61})]+ \\
& +a_{14}a_{22}a_{36}[a_{41}(a_{55}a_{63}-a_{53}a_{65})+a_{43}(a_{51}a_{65}-a_{55}a_{61})+a_{45}(a_{53}a_{61}-a_{51}a_{63})]+ \\[6pt]
& +a_{14}a_{23}a_{31}[a_{42}(a_{56}a_{65}-a_{55}a_{66})+a_{45}(a_{52}a_{66}-a_{56}a_{62})+a_{46}(a_{55}a_{62}-a_{52}a_{65})]+ \\
& +a_{14}a_{23}a_{32}[a_{41}(a_{55}a_{66}-a_{56}a_{65})+a_{45}(a_{56}a_{61}-a_{51}a_{66})+a_{46}(a_{51}a_{65}-a_{55}a_{61})]+ \quad |A_{18}| \\
& +a_{14}a_{23}a_{35}[a_{41}(a_{56}a_{62}-a_{52}a_{66})+a_{42}(a_{51}a_{66}-a_{56}a_{61})+a_{46}(a_{52}a_{61}-a_{51}a_{62})]+ \\
& +a_{14}a_{23}a_{36}[a_{41}(a_{52}a_{65}-a_{55}a_{62})+a_{42}(a_{55}a_{61}-a_{51}a_{65})+a_{45}(a_{51}a_{62}-a_{52}a_{61})]+ \\[6pt]
& +a_{14}a_{25}a_{31}[a_{42}(a_{53}a_{66}-a_{56}a_{63})+a_{43}(a_{56}a_{62}-a_{52}a_{66})+a_{46}(a_{52}a_{63}-a_{53}a_{62})]+ \\
& +a_{14}a_{25}a_{32}[a_{41}(a_{56}a_{63}-a_{53}a_{66})+a_{43}(a_{51}a_{66}-a_{56}a_{61})+a_{46}(a_{53}a_{61}-a_{51}a_{63})]+ \quad |A_{19}| \\
& +a_{14}a_{25}a_{33}[a_{41}(a_{52}a_{66}-a_{56}a_{62})+a_{42}(a_{56}a_{61}-a_{51}a_{66})+a_{46}(a_{51}a_{62}-a_{52}a_{61})]+ \\
& +a_{14}a_{25}a_{36}[a_{41}(a_{53}a_{62}-a_{52}a_{63})+a_{42}(a_{51}a_{63}-a_{53}a_{61})+a_{43}(a_{52}a_{61}-a_{51}a_{62})]+ \\[6pt]
& +a_{14}a_{26}a_{31}[a_{42}(a_{55}a_{63}-a_{53}a_{65})+a_{43}(a_{52}a_{65}-a_{55}a_{62})+a_{45}(a_{53}a_{62}-a_{52}a_{63})]+ \\
& +a_{14}a_{26}a_{32}[a_{41}(a_{53}a_{65}-a_{55}a_{63})+a_{43}(a_{55}a_{61}-a_{51}a_{65})+a_{45}(a_{51}a_{63}-a_{53}a_{61})]+ \quad |A_{20}| \\
& +a_{14}a_{26}a_{33}[a_{41}(a_{55}a_{62}-a_{52}a_{65})+a_{42}(a_{51}a_{65}-a_{55}a_{61})+a_{45}(a_{52}a_{61}-a_{51}a_{62})]+ \\
& +a_{14}a_{26}a_{35}[a_{41}(a_{52}a_{63}-a_{53}a_{62})+a_{42}(a_{53}a_{61}-a_{51}a_{63})+a_{43}(a_{51}a_{62}-a_{52}a_{61})] \quad (113)
\end{aligned}
$$

❖ For Figure 148 of the combined partial determinant $|A_{16}|$:

Analysis of data shown in Figure 148 applicably to formula (113) of the combined partial determinant $|A_{16}|$ of the 6x6 matrix for positive and negative areas indicates on the following results in general view:
- The fourth element (a_{14}) of the first row is joined with the first element(a_{21}) of the second row;
- The first element(a_{21}) of the second row is joined with elements ($a_{32},a_{33},a_{35},a_{36}$);

a) For elements of the third row
- The second element(a_{32}) of the third row is joined with elements (a_{43},a_{45},a_{46}) of the fourth row;
- The third element(a_{33}) of the third row is joined with elements (a_{42},a_{45},a_{46}) of the fourth row;
- The fifth element(a_{35}) of the third row is joined with elements (a_{42},a_{43},a_{46}) of the fourth row;
- The sixth element(a_{36}) of the third row is joined with elements (a_{42},a_{43},a_{45}) of the fourth row-in total we can conclude that sequence of elements connection for the third row with elements of the fourth row has the following view, forming the frequency of ties equal f=3:

220

$$a_{32} \rightarrow a_{43}, a_{45}, a_{46}$$
$$a_{33} \rightarrow a_{42}, a_{45}, a_{46}$$
$$a_{35} \rightarrow a_{42}, a_{43}, a_{46}$$
$$a_{36} \rightarrow a_{42}, a_{43}, a_{45}$$

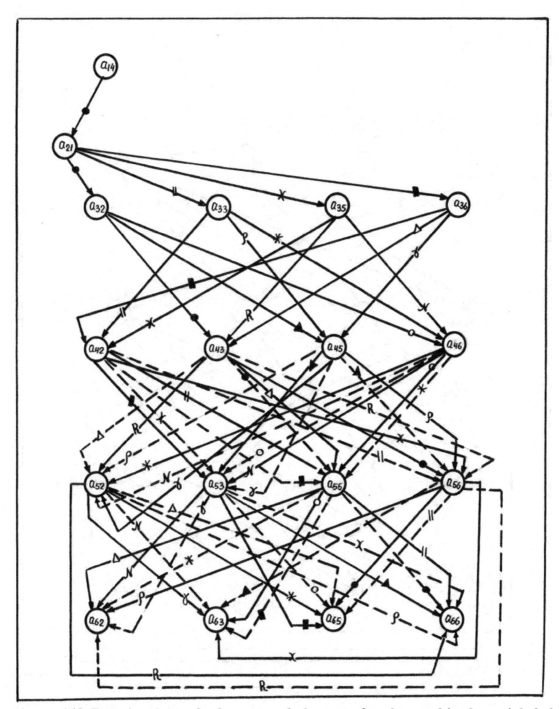

Figure 148 Functional graphs between of elements for the combined partial 6x6 matrix determinant $\left| A_{16} \right|$

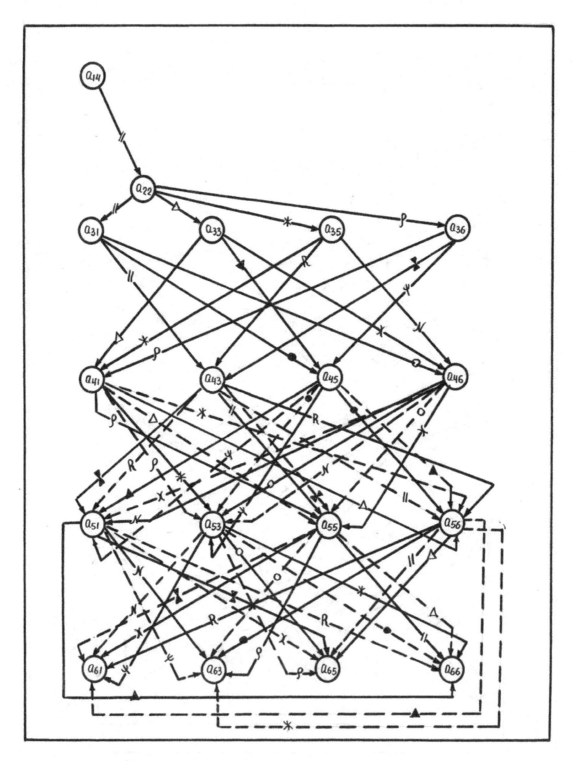

Figure 149 Functional graphs between of elements for the combined partial 6x6 matrix determinant $|A_{17}|$

222

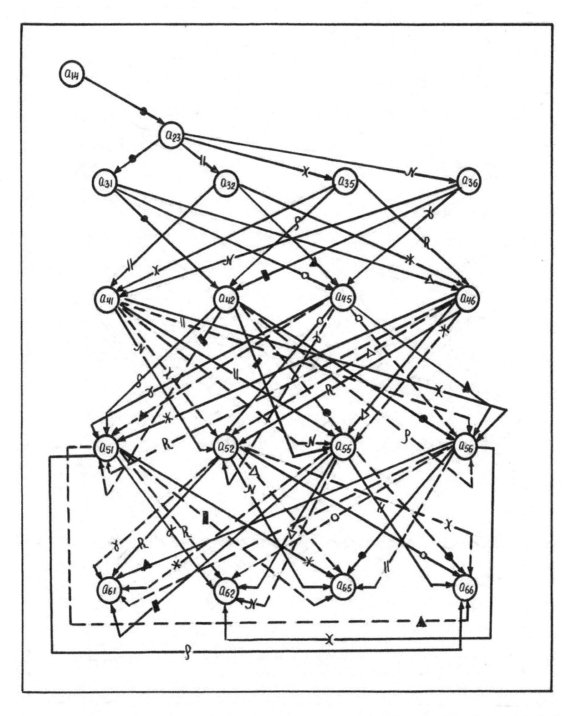

Figure 150 Functional graphs between of elements for the combined partial 6x6 matrix determinant $|A_{18}|$

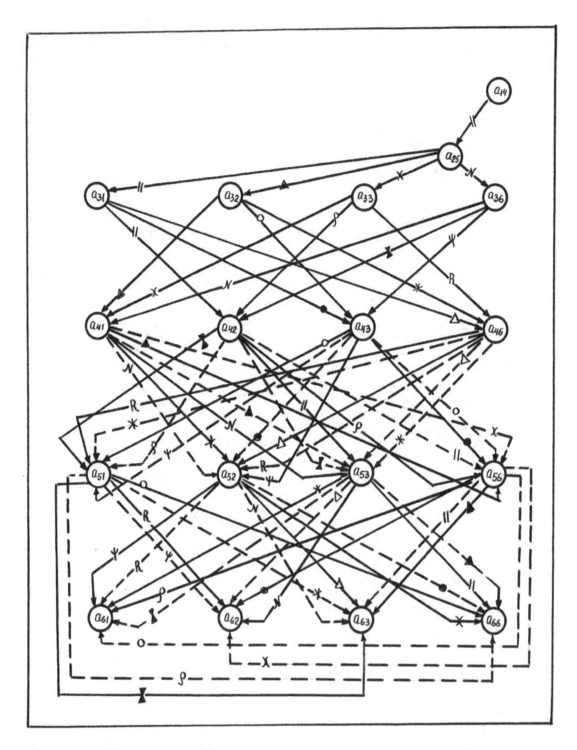

Figure 151 Functional graphs between of elements for the combined partial 6x6 matrix determinant $\left| A_{19} \right|$

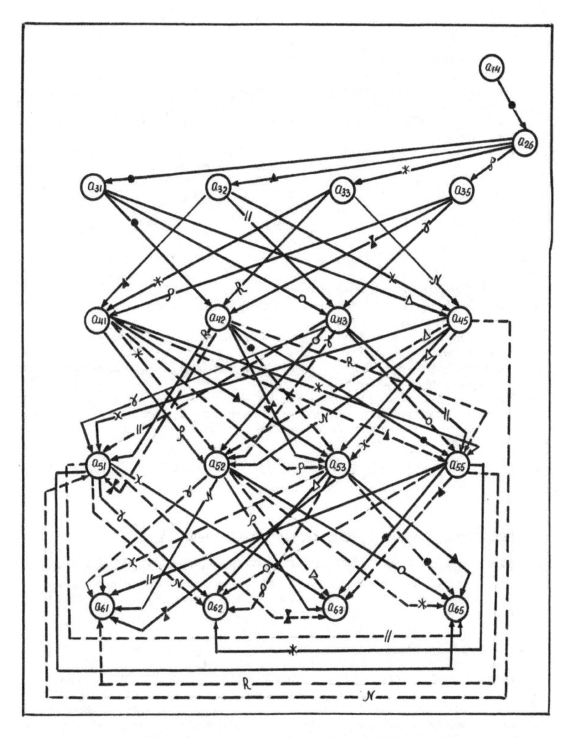

Figure 152 Functional graphs between of elements for the combined partial 6x6 matrix determinant $\left| A_{20} \right|$

b) *For elements of the fourth row*

- The second element (a_{42}) of the fourth row is joined with elements ($a_{53},a_{53},a_{55},a_{55},a_{56},a_{56}$) of the fifth row for positive and negative areas;
- The third element(a_{43}) of the fourth row is joined with elements ($a_{52},a_{52},a_{55},a_{55},a_{56},a_{56}$) of the fifth row for positive and negative areas;
- The fifth element(a_{45})of the fourth row is joined with elements ($a_{52},a_{52},a_{53},a_{53},a_{56},a_{56}$) of the fifth row for positive and negative areas;
- The sixth element (a_{46})of the fourth row is joined with elements ($a_{52},a_{52},a_{53},a_{53},a_{55},a_{55}$) of the fifth row for positive and negative areas-in total we can conclude that sequence of elements connection fourth row with elements of the fifth row has the following view, forming the frequency of ties equal f=6:

$$a_{42} \rightarrow a_{53},a_{53},a_{55},a_{55},a_{56},a_{56}$$
$$a_{43} \rightarrow a_{52},a_{52},a_{55},a_{55},a_{56},a_{56}$$
$$a_{45} \rightarrow a_{52},a_{52},a_{53},a_{53},a_{56},a_{56}$$
$$a_{46} \rightarrow a_{52},a_{52},a_{53},a_{53},a_{55},a_{55}$$

c) *For elements of the fifth row*

- The second element (a_{52}) of the fifth row is joined with elements ($a_{63},a_{63},a_{65},a_{65},a_{66},a_{66}$) of the sixth row for positive and negative areas;
- The third element(a_{53}) of the fifth row is joined with elements ($a_{62},a_{62},a_{65},a_{65},a_{66},a_{66}$) of the sixth row for positive and negative areas;
- The fifth element(a_{55}) of the fifth row is joined with elements ($a_{62},a_{62},a_{63},a_{63},a_{66},a_{66}$)of the sixth row for positive and negative areas;
- The sixth element(a_{56}) of the fifth row is joined with elements ($a_{62},a_{62},a_{63},a_{63},a_{65},a_{65}$) of the sixth row for positive and negative areas-in total we can conclude that sequence of elements connection for the fifth row with elements of the sixth row has the following view, forming the frequency of ties equal f=6:

$$a_{52} \rightarrow a_{63},a_{63},a_{65},a_{65},a_{66},a_{66}$$
$$a_{53} \rightarrow a_{62},a_{62},a_{65},a_{65},a_{66},a_{66}$$
$$a_{55} \rightarrow a_{62},a_{62},a_{63},a_{63},a_{66},a_{66}$$
$$a_{56} \rightarrow a_{62},a_{62},a_{63},a_{63},a_{65},a_{65}$$

❖ For Figure 149 of the combined partial determinant $\left| A_{17} \right|$:

Analysis of data shown in Figure 149 applicably to formula (113) of the combined partial determinant $\left| A_{17} \right|$ of the 6x6 matrix for positive and negative areas indicates on the following results in general view:
- The fourth element (a_{14}) of the first row is joined with the second element (a_{22}) of the second row;
- The second element(a_{22}) of the second row is joined with elements ($a_{31},a_{33},a_{35},a_{36}$);

a) *For elements of the third row*

- The first element(a_{31}) of the third row is joined with elements (a_{43},a_{45},a_{46}) of the fourth row;
- The third element (a_{33}) of the third row is joined with elements (a_{41},a_{45},a_{46}) of the fourth row;
- The fifth element(a_{35}) of the third row is joined with elements (a_{41},a_{43},a_{46}) of the fourth row;
- The sixth element(a_{36}) of the third row is joined with elements (a_{41},a_{43},a_{45}) of the fourth row- in total we can conclude that sequence of elements connection for the third row with elements of the fourth row has the following view, forming the frequency of ties equal f=3:

$$a_{31} \rightarrow a_{43},a_{45},a_{46}$$
$$a_{33} \rightarrow a_{41},a_{45},a_{46}$$
$$a_{35} \rightarrow a_{41},a_{43},a_{46}$$
$$a_{36} \rightarrow a_{41},a_{43},a_{45}$$

b) *For elements of the fourth row*

- The first element(a_{41}) of the fourth row is joined with elements ($a_{53},a_{53},a_{55},a_{55},a_{56},a_{56}$) of the fifth row for positive and negative areas;
- The third element(a_{43}) of the fourth row is joined with elements ($a_{51},a_{51},a_{55},a_{55},a_{56},a_{56}$) of the fifth row for positive and negative areas;
- The fifth element(a_{45}) of the fourth row is joined with elements ($a_{51},a_{51},a_{53},a_{53},a_{56},a_{56}$) of the fifth row for positive and negative areas;
- The sixth element(a_{46}) of the fourth row is joined with elements ($a_{51},a_{51},a_{53},a_{53},a_{55},a_{55}$) of the fifth row for positive and negative areas-in total we can conclude that sequence of elements connection fourth row with elements of the fifth row has the following view, forming the frequency of ties equal f=6:

$$a_{41} \rightarrow a_{53},a_{53},a_{55},a_{55},a_{56},a_{56}$$
$$a_{43} \rightarrow a_{51},a_{51},a_{55},a_{55},a_{56},a_{56}$$
$$a_{45} \rightarrow a_{51},a_{51},a_{53},a_{53},a_{56},a_{56}$$
$$a_{46} \rightarrow a_{51},a_{51},a_{53},a_{53},a_{55},a_{55}$$

c) *For elements of the fifth row*

- The first element(a_{51}) of the fifth row is joined with elements ($a_{63},a_{63},a_{65},a_{65},a_{66},a_{66}$) of the sixth row for positive and negative areas;
- The third element(a_{53}) of the fifth row is joined with elements ($a_{61},a_{61},a_{65},a_{65},a_{66},a_{66}$) of the sixth row for positive and negative areas;
- The fifth element(a_{55}) of the fifth row is joined with elements ($a_{61},a_{61},a_{63},a_{63},a_{66},a_{66}$) of the sixth row for positive and negative areas;
- The sixth element (a_{56}) of the fifth row is joined with elements ($a_{61},a_{61},a_{63},a_{63},a_{65},a_{65}$) of the sixth row for positive and negative areas-in total we can conclude that sequence of elements connection for the fifth row with elements of the sixth row has the following view ,forming the frequency of ties equal f=6:

$$a_{51} \rightarrow a_{63},a_{63},a_{65},a_{65},a_{66},a_{66}$$
$$a_{53} \rightarrow a_{61},a_{61},a_{65},a_{65},a_{66},a_{66}$$
$$a_{55} \rightarrow a_{61},a_{61},a_{63},a_{63},a_{66},a_{66}$$
$$a_{56} \rightarrow a_{61},a_{61},a_{63},a_{63},a_{65},a_{65}$$

❖ For Figure 150 of the combined partial determinant $\lfloor A_{18} \rfloor$:

Analysis of data shown in Figure 150 applicably to formula (113) of the combined partial determinant $\lfloor A_{18} \rfloor$ of the 6x6 matrix for positive and negative areas indicates on the following results in general view:

- The fourth element(a_{14}) of the first row is joined with the third element(a_{23}) of the second row;
- The third element(a_{23}) of the second row is joined with elements ($a_{31},a_{32},a_{35},a_{36}$);

a) *For elements of the third row*

- The first element (a_{31}) of the third row is joined with elements (a_{42},a_{45},a_{46}) of the fourth row;
- The second element(a_{32}) of the third row is joined with elements (a_{41},a_{45},a_{46}) of the fourth row;
- The fifth element(a_{35}) of the third row is joined with elements (a_{41},a_{42},a_{46}) of the fourth row;
- The sixth element(a_{36}) of the third row is joined with elements (a_{41},a_{42},a_{45}) of the fourth row-in total we can conclude that sequence of elements connection for the third row with elements of the fourth row has the following view, forming the frequency of ties equal f=3:

$$a_{31} \rightarrow a_{42},a_{45},a_{46}$$
$$a_{32} \rightarrow a_{41},a_{45},a_{46}$$
$$a_{35} \rightarrow a_{41},a_{42},a_{46}$$
$$a_{36} \rightarrow a_{41},a_{42},a_{45}$$

b) For elements of the fourth row

- The first element(a_{41}) of the fourth row is joined with elements ($a_{52}, a_{52}, a_{55}, a_{55}, a_{56}, a_{56}$) of the fifth row for positive and negative areas;
- The second element(a_{42}) of the fourth row is joined with elements ($a_{51}, a_{51}, a_{55}, a_{55}, a_{56}, a_{56}$) of the fifth row for positive and negative areas;
- The fifth element(a_{45}) of the fourth row is joined with elements($a_{51}, a_{51}, a_{52}, a_{52}, a_{56}, a_{56}$) of the fifth row for positive and negative areas;
- The six element(a_{46}) of the fourth row is joined with elements ($a_{51}, a_{51}, a_{52}, a_{52}, a_{55}, a_{55}$) of the fifth row for positive and negative areas- in total we can conclude that sequence of elements connection fourth row with elements of the fifth row has the following view ,forming the frequency of ties equal f=6:

$$a_{41} \rightarrow a_{52}, a_{52}, a_{55}, a_{55}, a_{56}, a_{56}$$
$$a_{42} \rightarrow a_{51}, a_{51}, a_{55}, a_{55}, a_{56}, a_{56}$$
$$a_{45} \rightarrow a_{51}, a_{51}, a_{52}, a_{52}, a_{56}, a_{56}$$
$$a_{46} \rightarrow a_{51}, a_{51}, a_{52}, a_{52}, a_{55}, a_{55}$$

c) For elements of the fifth row

- The first element(a_{51}) of the fifth row is joined with elements ($a_{62}, a_{62}, a_{65}, a_{65}, a_{66}, a_{66}$) of the sixth row for positive and negative areas;
- The second element(a_{52}) of the fifth row is joined with elements ($a_{61}, a_{61}, a_{65}, a_{65}, a_{66}, a_{66}$) of the sixth row for positive and negative areas;
- The fifth element(a_{55}) of the fifth row is joined with elements ($a_{61}, a_{61}, a_{62}, a_{62}, a_{66}, a_{66}$)of the sixth row for positive and negative areas;
- The sixth element(a_{56})of the fifth row is joined with elements ($a_{61}, a_{61}, a_{62}, a_{62}, a_{65}, a_{65}$) of the sixth row for positive and negative areas-in total we can conclude that sequence of elements connection for the fifth row with elements of the sixth row has the following view ,forming the frequency of ties equal f=6:

$$a_{51} \rightarrow a_{62}, a_{62}, a_{65}, a_{65}, a_{66}, a_{66}$$
$$a_{52} \rightarrow a_{61}, a_{61}, a_{65}, a_{65}, a_{66}, a_{66}$$
$$a_{55} \rightarrow a_{61}, a_{61}, a_{62}, a_{62}, a_{66}, a_{66}$$
$$a_{56} \rightarrow a_{61}, a_{61}, a_{62}, a_{62}, a_{65}, a_{65}$$

❖ For Figure 151 of the combined partial determinant $|A_{19}|$:

Analysis of data shown in Figure 151 applicably to formula (113) of the combined partial determinant $|A_{19}|$ of the 6x6 matrix for positive and negative areas indicates on the following results in general view:
- The fourth element(a_{14}) of the first row is joined with the fifth element(a_{25}) of the second row;
- The fifth element(a_{25}) of the second row is joined with elements ($a_{31}, a_{32}, a_{33}, a_{36}$);

a) For elements of the third row

- The first element(a_{31}) of the third row is joined with elements (a_{42}, a_{43}, a_{46}) of the fourth row;
- The second element(a_{32}) of the third row is joined with elements (a_{41}, a_{43}, a_{46}) of the fourth row;
- The third element(a_{33}) of the third row is joined with elements (a_{41}, a_{42}, a_{46}) of the fourth row;
- The sixth element(a_{36}) of the third row is joined with elements (a_{41}, a_{42}, a_{43}) of the fourth row-in total we can conclude that sequence of elements connection for the third row with elements of the fourth row has the following view, forming the frequency of ties equal f=3:

$$a_{31} \rightarrow a_{42}, a_{43}, a_{46}$$
$$a_{32} \rightarrow a_{41}, a_{43}, a_{46}$$
$$a_{33} \rightarrow a_{41}, a_{42}, a_{46}$$
$$a_{36} \rightarrow a_{41}, a_{42}, a_{43}$$

b) *For elements of the fourth row*

- The first element(a_{41}) of the fourth row is joined with elements ($a_{52},a_{52},a_{53},a_{53},a_{56},a_{56}$) of the fifth row for positive and negative areas;
- The second element(a_{42}) of the fourth row is joined with elements ($a_{51},a_{51},a_{53},a_{53},a_{56},a_{56}$) of the fifth row for positive and negative areas;
- The third element(a_{43}) of the fourth row is joined with elements ($a_{51},a_{51},a_{52},a_{52},a_{56},a_{56}$) of the fifth row for positive and negative areas;
- The sixth element(a_{46}) of the fourth row is joined with elements ($a_{51},a_{51},a_{52},a_{52},a_{53},a_{53}$) of the fifth row for positive and negative areas- in total we can conclude that sequence of elements connection fourth row with elements of the fifth row has the following view, forming the frequency of ties equal f=6:

$$a_{41} \rightarrow a_{52},a_{52},a_{53},a_{53},a_{56},a_{56}$$
$$a_{42} \rightarrow a_{51},a_{51},a_{53},a_{53},a_{56},a_{56}$$
$$a_{43} \rightarrow a_{51},a_{51},a_{52},a_{52},a_{56},a_{56}$$
$$a_{46} \rightarrow a_{51},a_{51},a_{52},a_{52},a_{53},a_{53}$$

c) *For elements of the fifth row*

- The first element(a_{51}) of the fifth row is joined with elements ($a_{62},a_{62},a_{63},a_{63},a_{66},a_{66}$) of the six row for positive and negative areas;
- The second element(a_{52}) of the fifth row is joined with elements ($a_{61},a_{61},a_{63},a_{63},a_{66},a_{66}$) of the sixth row for positive and negative areas;
- The third element(a_{53}) of the fifth row is joined with elements ($a_{61},a_{61},a_{62},a_{62},a_{66},a_{66}$) of the sixth row for positive and negative areas;
- The sixth element(a_{56}) of the fifth row is joined with elements ($a_{61},a_{61},a_{62},a_{62},a_{63},a_{63}$) of the sixth row for positive and negative areas- in total we can conclude that sequence of elements connection for the fifth row with elements of the sixth row has the following view, forming the frequency of ties equal f=6:

$$a_{51} \rightarrow a_{62},a_{62},a_{63},a_{63},a_{66},a_{66}$$
$$a_{52} \rightarrow a_{61},a_{61},a_{63},a_{63},a_{66},a_{66}$$
$$a_{53} \rightarrow a_{61},a_{61},a_{62},a_{62},a_{66},a_{66}$$
$$a_{56} \rightarrow a_{61},a_{61},a_{62},a_{62},a_{63},a_{63}$$

❖ For Figure 152 of the combined partial determinant $\lvert A_{20} \rvert$:

Analysis of data shown in Figure 152 applicably to formula (113) of the combined partial determinant $\lvert A_{20} \rvert$ for positive and negative areas of the 6x6 matrix indicates on the following results in general view:
- The fourth element(a_{14}) of the first row is joined with the sixth element(a_{26}) of the second row;
- The sixth element(a_{26}) of the second row is joined with elements ($a_{31},a_{32},a_{33},a_{35}$);

a) *For elements of the third row*

- The first element (a_{31}) of the third row is joined with elements (a_{42},a_{43},a_{45}) of the fourth row;
- The second element(a_{32})of the third row is joined with elements (a_{41},a_{43},a_{45}) of the fourth row;
- The third element (a_{33}) of the third row is joined with elements (a_{41},a_{42},a_{45}) of the fourth row;
- The fifth element(a_{35}) of the third row is joined with elements (a_{41},a_{42},a_{43}) of the fourth row-in total we can conclude that sequence of elements connection for the third row with elements of the fourth row has the following view, forming the frequency of ties equal f=3:

$$a_{31} \rightarrow a_{42},a_{43},a_{45}$$
$$a_{32} \rightarrow a_{41},a_{43},a_{45}$$
$$a_{33} \rightarrow a_{41},a_{42},a_{45}$$
$$a_{35} \rightarrow a_{41},a_{42},a_{43}$$

b) *For elements of the fourth row*

- The first element(a_{41}) of the fourth row is joined with elements ($a_{52}, a_{52}, a_{53}, a_{53}, a_{55}, a_{55}$) of the fifth row for positive and negative areas;
- The second element (a_{42}) of the fourth row is joined with elements ($a_{51}, a_{51}, a_{53}, a_{53}, a_{55}, a_{55}$) of the fifth row for positive and negative areas;
- The third element (a_{43}) of the fourth row is joined with elements ($a_{51}, a_{51}, a_{52}, a_{52}, a_{55}, a_{55}$) of the fifth row for positive and negative areas
- The fifth element (a_{45}) of the fourth row is joined with elements ($a_{51}, a_{51}, a_{52}, a_{52}, a_{53}, a_{53}$) of the fifth row for positive and negative areas-in total we can conclude that sequence of elements connection fourth row with elements of the fifth row has the following view, forming the frequency of ties equal f=6:

$$a_{41} \rightarrow a_{52}, a_{52}, a_{53}, a_{53}, a_{55}, a_{55}$$
$$a_{42} \rightarrow a_{51}, a_{51}, a_{53}, a_{53}, a_{55}, a_{55}$$
$$a_{43} \rightarrow a_{51}, a_{51}, a_{52}, a_{52}, a_{55}, a_{55}$$
$$a_{45} \rightarrow a_{51}, a_{51}, a_{52}, a_{52}, a_{53}, a_{53}$$

c) *For elements of the fifth row*

- The first element(a_{51}) of the fifth row is joined with elements ($a_{62}, a_{62}, a_{63}, a_{63}, a_{65}, a_{65}$) of the sixth row for positive and negative areas;
- The second element(a_{52}) of the fifth row is joined with elements ($a_{61}, a_{61}, a_{63}, a_{63}, a_{65}, a_{65}$) of the sixth row for positive and negative areas;
- The third element(a_{53}) of the fifth row is joined with elements ($a_{61}, a_{61}, a_{62}, a_{62}, a_{65}, a_{65}$) of the sixth row for positive and negative areas;
- The fifth element(a_{55}) of the fifth row is joined with elements ($a_{61}, a_{61}, a_{62}, a_{62}, a_{63}, a_{63}$) of the sixth row for positive and negative areas-in total we can conclude that sequence of elements connection for the fifth row with elements of the sixth row has the following view, forming the frequency of ties equal f=6:

$$a_{51} \rightarrow a_{62}, a_{62}, a_{63}, a_{63}, a_{65}, a_{65}$$
$$a_{51} \rightarrow a_{61}, a_{61}, a_{63}, a_{63}, a_{65}, a_{65}$$
$$a_{53} \rightarrow a_{61}, a_{61}, a_{62}, a_{62}, a_{65}, a_{65}$$
$$a_{55} \rightarrow a_{61}, a_{61}, a_{62}, a_{62}, a_{63}, a_{63}$$

E. __Analysis of combined partial determinant $|A'_{15}|$ of the 6x6 matrix:__

The combined partial determinant $|A'_{15}|$ consists from five separate parts and is expressed by the following form in view of

$$\left|A'_{15}\right| = \left|A_{21}\right| + \left|A_{22}\right| + \left|A_{23}\right| + \left|A_{24}\right| + \left|A_{25}\right|$$

And also the above-said determinant $|A'_{15}|$ introduced by the following finally formula (114),after of some transformations, elements of which are shown in view of functional graphs in Figures 153 to 157:

$$
\begin{aligned}
\left|A'_{15}\right| = & a_{15}a_{21}a_{32}[a_{43}(a_{54}a_{66}-a_{56}a_{64})+a_{44}(a_{56}a_{63}-a_{53}a_{66})+a_{46}(a_{53}a_{64}-a_{54}a_{63})]+ \\
& +a_{15}a_{21}a_{33}[a_{42}(a_{56}a_{64}-a_{54}a_{66})+a_{44}(a_{52}a_{66}-a_{56}a_{62})+a_{46}(a_{54}a_{62}-a_{52}a_{64})]+ \quad \left|A_{21}\right| \\
& +a_{15}a_{21}a_{34}[a_{42}(a_{53}a_{66}-a_{56}a_{63})+a_{43}(a_{56}a_{62}-a_{52}a_{66})+a_{46}(a_{52}a_{63}-a_{53}a_{62})]+ \\
& +a_{15}a_{21}a_{36}[a_{42}(a_{54}a_{63}-a_{53}a_{64})+a_{43}(a_{52}a_{64}-a_{54}a_{62})+a_{44}(a_{53}a_{62}-a_{52}a_{63})]+
\end{aligned}
$$

$+a_{15}a_{22}a_{31}[a_{43}(a_{56}a_{64}-a_{54}a_{66})+a_{44}(a_{53}a_{66}-a_{56}a_{63})+a_{46}(a_{54}a_{63}-a_{53}a_{64})]+$
$+a_{15}a_{22}a_{33}[a_{41}(a_{54}a_{66}-a_{56}a_{64})+a_{44}(a_{56}a_{61}-a_{51}a_{66})+a_{46}(a_{51}a_{64}-a_{54}a_{61})]+$ $\quad |A_{22}|$
$+a_{15}a_{22}a_{34}[a_{41}(a_{56}a_{63}-a_{53}a_{66})+a_{43}(a_{51}a_{66}-a_{56}a_{61})+a_{46}(a_{53}a_{61}-a_{51}a_{63})]+$
$+a_{15}a_{22}a_{36}[a_{41}(a_{53}a_{64}-a_{54}a_{63})+a_{43}(a_{54}a_{61}-a_{51}a_{64})+a_{44}(a_{51}a_{63}-a_{53}a_{61})]+$

$+a_{15}a_{23}a_{31}[a_{42}(a_{54}a_{66}-a_{56}a_{64})+a_{44}(a_{56}a_{62}-a_{52}a_{66})+a_{46}(a_{52}a_{64}-a_{54}a_{62})]+$
$+a_{15}a_{23}a_{32}[a_{41}(a_{56}a_{64}-a_{54}a_{66})+a_{44}(a_{51}a_{66}-a_{56}a_{61})+a_{46}(a_{54}a_{61}-a_{51}a_{64})]+$ $\quad |A_{23}|$
$+a_{15}a_{23}a_{34}[a_{41}(a_{52}a_{66}-a_{56}a_{62})+a_{42}(a_{56}a_{61}-a_{51}a_{66})+a_{46}(a_{51}a_{62}-a_{52}a_{61})]+$
$+a_{15}a_{23}a_{36}[a_{41}(a_{54}a_{62}-a_{52}a_{64})+a_{42}(a_{51}a_{64}-a_{54}a_{61})+a_{44}(a_{52}a_{61}-a_{51}a_{62})]+$

$+a_{15}a_{24}a_{31}[a_{42}(a_{56}a_{63}-a_{53}a_{66})+a_{43}(a_{52}a_{66}-a_{56}a_{62})+a_{46}(a_{53}a_{62}-a_{52}a_{63})]+$
$+a_{15}a_{24}a_{32}[a_{41}(a_{53}a_{66}-a_{56}a_{63})+a_{43}(a_{56}a_{61}-a_{51}a_{66})+a_{46}(a_{51}a_{63}-a_{53}a_{61})]+$ $\quad |A_{24}|$
$+a_{15}a_{24}a_{33}[a_{41}(a_{56}a_{62}-a_{52}a_{66})+a_{42}(a_{51}a_{66}-a_{56}a_{61})+a_{46}(a_{52}a_{61}-a_{51}a_{62})]+$
$+a_{15}a_{24}a_{36}[a_{41}(a_{52}a_{63}-a_{53}a_{62})+a_{42}(a_{53}a_{61}-a_{51}a_{63})+a_{43}(a_{51}a_{62}-a_{52}a_{61})]+$

$+a_{15}a_{26}a_{31}[a_{42}(a_{53}a_{64}-a_{54}a_{63})+a_{43}(a_{54}a_{62}-a_{52}a_{64})+a_{44}(a_{52}a_{63}-a_{53}a_{62})]+$
$+a_{15}a_{26}a_{32}[a_{41}(a_{54}a_{63}-a_{53}a_{64})+a_{43}(a_{51}a_{64}-a_{54}a_{61})+a_{44}(a_{53}a_{61}-a_{51}a_{63})]+$ $\quad |A_{25}|$
$+a_{15}a_{26}a_{33}[a_{41}(a_{52}a_{64}-a_{54}a_{62})+a_{42}(a_{54}a_{61}-a_{51}a_{64})+a_{44}(a_{51}a_{62}-a_{52}a_{61})]+$
$+a_{15}a_{26}a_{34}[a_{41}(a_{53}a_{62}-a_{52}a_{63})+a_{42}(a_{51}a_{63}-a_{53}a_{61})+a_{43}(a_{52}a_{61}-a_{51}a_{62})]$ $\quad (114)$

❖ For Figure 153 of the combined partial determinant $|A_{21}|$:

Analysis of data shown in Figure 153 applicably to formula (114) of the combined partial determinant $|A_{21}|$ of the 6x6 matrix for positive and negative areas indicates on the following results in general view:

- The fifth element(a_{15}) of the first row is joined with the first element (a_{21}) of the second row;
- The first element(a_{21}) of the second row is joined with elements ($a_{32},a_{33},a_{34},a_{36}$);

a) *For elements of the third row*

- The second element (a_{32}) of the third row is joined with elements (a_{43},a_{44},a_{46}) of the fourth row;
- The third element(a_{33}) of the third row is joined with elements (a_{42},a_{44},a_{46}) of the fourth row;
- The fourth element(a_{34}) of the third row is joined with elements (a_{42},a_{43},a_{46}) of the fourth row;
- The sixth element(a_{36}) of the third row is joined with elements (a_{42},a_{43},a_{44}) of the fourth row- in total we can conclude that sequence of elements connection for the third row with elements of the fourth row has the following view ,forming the frequency of ties equal f=3:

$$a_{32}\rightarrow a_{43},a_{44},a_{46}$$
$$a_{33}\rightarrow a_{42},a_{44},a_{46}$$
$$a_{34}\rightarrow a_{42},a_{43},a_{46}$$
$$a_{36}\rightarrow a_{42},a_{43},a_{44}$$

b) *For elements of the fourth row*

- The second element(a_{42})of the fourth row is joined with elements ($a_{53},a_{53},a_{54},a_{54},a_{56},a_{56}$) of the fifth row for positive and negative areas;
- The third element (a_{43}) of the fourth row is joined with elements ($a_{52},a_{52},a_{54},a_{54},a_{56},a_{56}$) of the fifth row for positive and negative areas;
- The fourth element(a_{44}) of the fourth row is joined with elements ($a_{52},a_{52},a_{53},a_{53},a_{56},a_{56}$) of the fifth row for positive and negative areas;
- The sixth element(a_{46}) of the fourth row is joined with elements ($a_{52},a_{52},a_{53},a_{53},a_{54},a_{54}$) of the fifth row for positive and negative areas- in total we can conclude that sequence of elements connection fourth row with elements of the fifth row has the following view, forming the frequency of ties equal f=6:

231

$a_{42} \rightarrow a_{53}, a_{53}, a_{54}, a_{54}, a_{56}, a_{56}$
$a_{43} \rightarrow a_{52}, a_{52}, a_{54}, a_{54}, a_{56}, a_{56}$
$a_{44} \rightarrow a_{52}, a_{52}, a_{53}, a_{53}, a_{56}, a_{56}$
$a_{46} \rightarrow a_{52}, a_{52}, a_{53}, a_{53}, a_{54}, a_{54}$

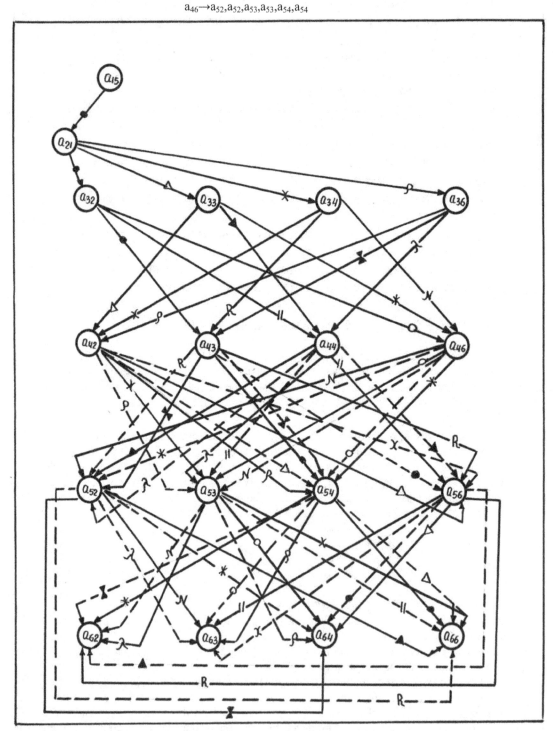

Figure 153 Functional graphs between of elements for the combined partial 6x6 matrix determinant $|A_{21}|$ of positive (————) and negative (-------) areas

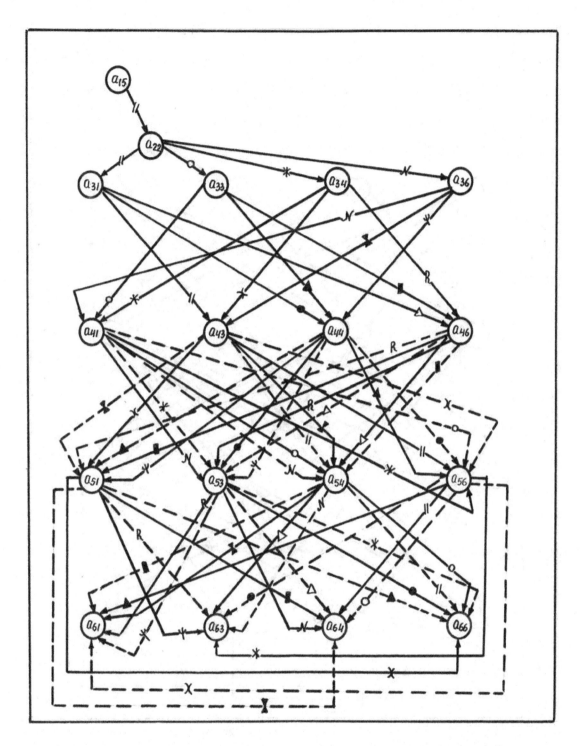

Figure 154 Functional graphs between of elements for the combined partial 6x6 matrix determinant $\left|A_{22}\right|$ **of positive (———) and negative (--------)areas**

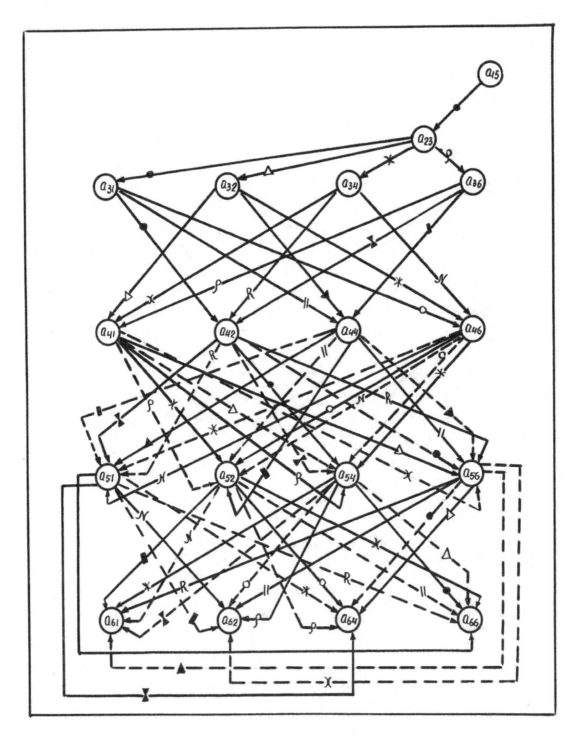

Figure 155 Functional graphs between of elements for the combined partial 6x6 matrix determinant $\left| A_{23} \right|$ of positive (———) and negative (--------) areas

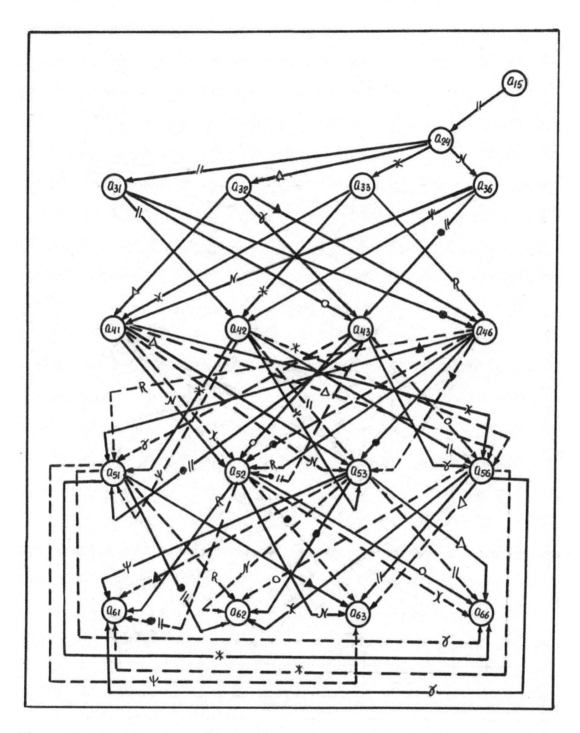

Figure 156 Functional graphs between of elements for the combined partial 6x6 matrix determinant $\left| A_{24} \right|$ of positive (————) and negative (--------)areas

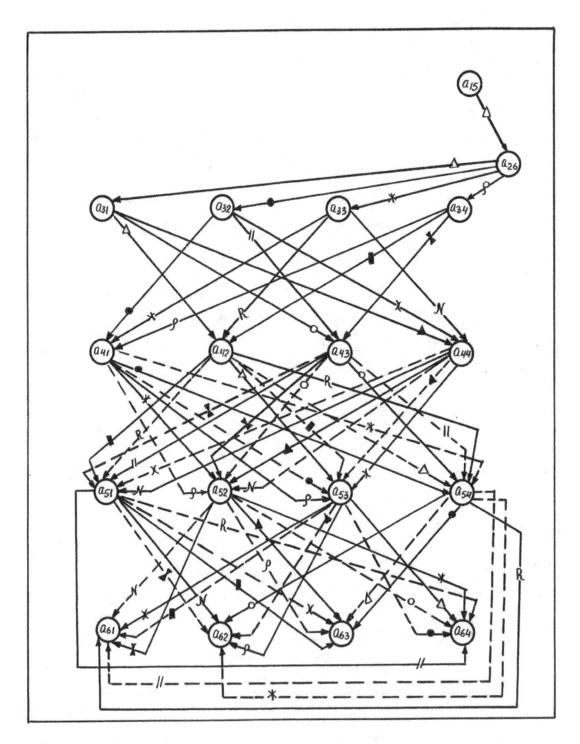

Figure 157 Functional graphs between of elements for the combined partial 6x6 matrix determinant $\left| A_{25} \right|$ **of the positive (———) and negative (--------) areas**

c) *For elements of the fifth row*

- The second element(a_{52}) of the fifth row is joined with elements ($a_{63}, a_{63}, a_{64}, a_{64}, a_{66}, a_{66}$) of the sixth row for positive and negative areas;
- The third element(a_{53}) of the fifth row is joined with elements ($a_{62}, a_{62}, a_{64}, a_{64}, a_{66}, a_{66}$) of the sixth row for positive and negative areas;
- The fourth element(a_{54}) of the fifth row is joined with elements ($a_{62}, a_{62}, a_{63}, a_{63}, a_{66}, a_{66}$) of the sixth row for positive and negative areas;
- The sixth element(a_{56}) of the fifth row is joined with elements ($a_{62}, a_{62}, a_{63}, a_{63}, a_{64}, a_{64}$) of the sixth row for positive and negative areas-in total we can conclude that sequence of elements connection for the fifth row with elements of the sixth row has the following view, forming the frequency of ties equal f=6:

$$a_{52} \rightarrow a_{63}, a_{63}, a_{64}, a_{64}, a_{66}, a_{66}$$
$$a_{53} \rightarrow a_{62}, a_{62}, a_{64}, a_{64}, a_{66}, a_{66}$$
$$a_{54} \rightarrow a_{62}, a_{62}, a_{63}, a_{63}, a_{66}, a_{66}$$
$$a_{56} \rightarrow a_{62}, a_{62}, a_{63}, a_{63}, a_{66}, a_{66}$$

❖ For Figure 154 of the combined partial determinant $|A_{22}|$:

Analysis of data shown in Figure 154 applicably to formula (114)of the combined partial determinant $|A_{22}|$ of the 6x6 matrix indicates on the following results in general view:

- The fifth element(a_{15}) of the first row is joined with the second element(a_{22}) of the second row;
- The second element(a_{22}) of the second row is joined with elements ($a_{31}, a_{33}, a_{34}, a_{36}$) of the third row;

a) *For elements of the third row*

- The first element(a_{31}) of the third row is joined with elements (a_{43}, a_{44}, a_{46}) of the fourth row;
- The third element(a_{33}) of the third row is joined with elements (a_{41}, a_{44}, a_{46}) of the fourth row;
- The fourth element(a_{34}) of the third row is joined with elements (a_{41}, a_{43}, a_{46}) of the fourth row;
- The sixth element(a_{36}) of the third row is joined with elements (a_{41}, a_{43}, a_{44}) of the fourth row –in total we can conclude that sequence of elements connection for the third row with elements of the fourth row has the following view, forming the frequency of ties equal f=3:

$$a_{31} \rightarrow a_{43}, a_{44}, a_{46}$$
$$a_{33} \rightarrow a_{41}, a_{44}, a_{46}$$
$$a_{34} \rightarrow a_{41}, a_{43}, a_{46}$$
$$a_{36} \rightarrow a_{41}, a_{43}, a_{44}$$

b) *For elements of the fourth row*

- The first element(a_{41}) of the fourth row is joined with elements ($a_{53}, a_{53}, a_{54}, a_{54}, a_{56}, a_{56}$) of the fifth row for positive and negative areas;
- The third element(a_{43}) of the fourth row is joined with elements ($a_{51}, a_{51}, a_{54}, a_{54}, a_{56}, a_{56}$) of the fifth row for positive and negative areas;
- The fourth element(a_{44}) of the fourth row is joined with elements ($a_{51}, a_{51}, a_{53}, a_{53}, a_{56}, a_{56}$) of the fifth row for positive and negative areas;
- The sixth element (a_{46}) of the fourth row is joined with elements ($a_{51}, a_{51}, a_{53}, a_{53}, a_{54}, a_{54}$) of the fifth row for positive and negative areas-in total we can conclude that sequence of elements connection fourth row with elements of the fifth row has the following view, forming the frequency of ties equal f=6:

$$a_{41} \rightarrow a_{53}, a_{53}, a_{54}, a_{54}, a_{56}, a_{56}$$
$$a_{43} \rightarrow a_{51}, a_{51}, a_{54}, a_{54}, a_{56}, a_{56}$$
$$a_{44} \rightarrow a_{51}, a_{51}, a_{53}, a_{53}, a_{56}, a_{56}$$
$$a_{46} \rightarrow a_{51}, a_{51}, a_{53}, a_{53}, a_{54}, a_{54}$$

c) *For elements of the fifth row*

- The first element (a_{51}) of the fifth row is joined with elements ($a_{63}, a_{63}, a_{64}, a_{64}, a_{66}, a_{66}$) of the sixth row for positive and negative areas;
- The third element(a_{53}) of the fifth row is joined with elements ($a_{61}, a_{61}, a_{64}, a_{64}, a_{66}, a_{66}$) of the sixth row for positive and negative areas;
- The fourth element(a_{54}) of the fifth row is joined with elements ($a_{61}, a_{61}, a_{63}, a_{63}, a_{66}, a_{66}$) of the sixth row for positive and negative areas;
- The sixth element(a_{56}) of the fifth row is joined with elements ($a_{61}, a_{61}, a_{63}, a_{63}, a_{64}, a_{64}$) of the sixth row for positive and negative areas- in total we can conclude that sequence of elements connection for the fifth row with elements of the sixth row has the following view, forming the frequency of ties equal f=6:

$$a_{51} \rightarrow a_{63}, a_{63}, a_{64}, a_{64}, a_{66}, a_{66}$$
$$a_{53} \rightarrow a_{61}, a_{61}, a_{64}, a_{64}, a_{66}, a_{66}$$
$$a_{54} \rightarrow a_{61}, a_{61}, a_{63}, a_{63}, a_{66}, a_{66}$$
$$a_{56} \rightarrow a_{61}, a_{61}, a_{63}, a_{63}, a_{64}, a_{64}$$

❖ **For Figure 155 of the combined partial determinant $|A_{23}|$:**

Analysis of data shown in Figure 155 applicably to formula (114) of the combined partial determinant $|A_{23}|$ of the 6x6 matrix for positive and negative areas indicates on the following results in general view:
- The fifth element(a_{15})of the first row is joined with the third element(a_{23}) of the second row;
- The third element(a_{23}) of the second row is joined with elements($a_{31}, a_{32}, a_{34}, a_{36}$) of the third row;

a) *For elements of the third row*

- The first element(a_{31}) of the third row is joined with elements (a_{42}, a_{44}, a_{46}) of the fourth row;
- The second element(a_{32}) of the third row is joined with elements (a_{41}, a_{44}, a_{46}) of the fourth row;
- The fourth element(a_{34}) of the third row is joined with elements (a_{41}, a_{42}, a_{46}) of the fourth row;
- The sixth element(a_{36}) of the third row is joined with elements (a_{41}, a_{42}, a_{44}) of the fourth row-in total we can conclude that sequence of elements connection for the third row with elements of the fourth row has the following view ,forming the frequency of ties equal f=3:

$$a_{31} \rightarrow a_{42}, a_{44}, a_{46}$$
$$a_{32} \rightarrow a_{41}, a_{44}, a_{46}$$
$$a_{34} \rightarrow a_{41}, a_{42}, a_{46}$$
$$a_{36} \rightarrow a_{41}, a_{42}, a_{44}$$

b) *For elements of the fourth row*

- The first element (a_{41}) of the fourth row is joined with elements ($a_{52}, a_{52}, a_{54}, a_{54}, a_{56}, a_{56}$) of the fifth row for positive and negative areas;
- The second element(a_{42}) of the fourth row is joined with elements ($a_{51}, a_{51}, a_{54}, a_{54}, a_{56}, a_{56}$) of the fifth row for positive and negative areas;
- The fourth element(a_{44}) of the fourth row is joined with elements ($a_{51}, a_{51}, a_{52}, a_{52}, a_{56}, a_{56}$) of the fifth row for positive and negative areas;
- The sixth element(a_{46}) of the fourth row is joined with elements ($a_{51}, a_{51}, a_{52}, a_{52}, a_{54}, a_{54}$) of the fifth row for positive and negative areas- in total we can conclude that sequence of elements connection fourth row with elements of the fifth row has the following view, forming the frequency of ties equal f=6:

$$a_{41} \rightarrow a_{52}, a_{52}, a_{54}, a_{54}, a_{56}, a_{56}$$
$$a_{42} \rightarrow a_{51}, a_{51}, a_{54}, a_{54}, a_{56}, a_{56}$$
$$a_{44} \rightarrow a_{51}, a_{51}, a_{52}, a_{52}, a_{56}, a_{56}$$
$$a_{46} \rightarrow a_{51}, a_{51}, a_{52}, a_{52}, a_{54}, a_{54}$$

c) *For elements of the fifth row*

- The first element(a_{51}) of the fifth row is joined with elements ($a_{62}, a_{62}, a_{64}, a_{64}, a_{66}, a_{66}$) of the sixth row for positive and negative areas;
- The second element(a_{52}) of the fifth row is joined with elements($a_{61}, a_{61}, a_{64}, a_{64}, a_{66}, a_{66}$) of the sixth row for positive and negative areas;
- The fourth element(a_{54}) of the fifth row is joined with elements ($a_{61}, a_{61}, a_{62}, a_{62}, a_{66}, a_{66}$)of the sixth row for positive and negative areas;
- The sixth element(a_{56}) of the fifth row is joined with elements ($a_{61}, a_{61}, a_{62}, a_{62}, a_{64}, a_{64}$) of the sixth row for positive and negative areas-in total we can conclude that sequence of elements connection for the fifth row with elements of the sixth row has the following view, forming the frequency of ties equal f=6:

$$a_{51} \rightarrow a_{62}, a_{62}, a_{64}, a_{64}, a_{66}, a_{66}$$
$$a_{52} \rightarrow a_{61}, a_{61}, a_{64}, a_{64}, a_{66}, a_{66}$$
$$a_{54} \rightarrow a_{61}, a_{61}, a_{62}, a_{62}, a_{66}, a_{66}$$
$$a_{56} \rightarrow a_{61}, a_{61}, a_{62}, a_{62}, a_{64}, a_{64}$$

❖ For Figure 156 of the combined partial determinant $|A_{24}|$:

Analysis of data shown in Figure 156 applicably to formula (114) of the combined partial determinant $|A_{24}|$ of the 6x6 matrix for positive and negative areas indicates on the following results in general view:
- The fifth element(a_{15}) of the first row is joined with the fourth element (a_{24}) of the second row;
- The fourth element (a_{24}) of the second row is joined with elements ($a_{31}, a_{32}, a_{33}, a_{36}$) of the third row;

a) *For elements of the third row*

- The first element(a_{31}) of the third row is joined with elements (a_{42}, a_{43}, a_{46}) of the fourth row;
- The second element(a_{32}) of the third row is joined with elements (a_{41}, a_{43}, a_{46}) of the fourth row;
- The third element(a_{33}) of the third row is joined with elements (a_{41}, a_{42}, a_{46}) of the fourth row;
- The sixth element(a_{36}) of the third row is joined with elements (a_{41}, a_{42}, a_{43}) of the fourth row- in total we can conclude that sequence of elements connection for the third row with elements of the fourth row has the following view, forming the frequency of ties equal f=3:

$$a_{31} \rightarrow a_{42}, a_{43}, a_{46}$$
$$a_{32} \rightarrow a_{41}, a_{43}, a_{46}$$
$$a_{33} \rightarrow a_{41}, a_{42}, a_{46}$$
$$a_{36} \rightarrow a_{41}, a_{42}, a_{43}$$

b) *For elements of the fourth row*

- The first element(a_{41}) of the fourth row is joined with elements ($a_{52}, a_{52}, a_{53}, a_{53}, a_{56}, a_{56}$) of the fifth row for positive and negative areas;
- The second element(a_{42}) of the fourth row is joined with elements($a_{51}, a_{51}, a_{53}, a_{53}, a_{56}, a_{56}$) of the fifth row for positive and negative areas;
- The third element (a_{43}) of the fourth row is joined with elements ($a_{51}, a_{51}, a_{52}, a_{52}, a_{56}, a_{56}$) of the fifth row for positive and negative areas;
- The sixth element (a_{46}) of the fourth row is joined with elements ($a_{51}, a_{51}, a_{52}, a_{52}, a_{53}, a_{53}$) of the fifth row for positive and negative areas-in total we can conclude that sequence of elements connection fourth row with elements of the fifth row has the following view ,forming the frequency of ties equal f=6:

$$a_{41} \rightarrow a_{52}, a_{52}, a_{53}, a_{53}, a_{56}, a_{56}$$
$$a_{42} \rightarrow a_{51}, a_{51}, a_{53}, a_{53}, a_{56}, a_{56}$$
$$a_{43} \rightarrow a_{51}, a_{51}, a_{52}, a_{52}, a_{56}, a_{56}$$
$$a_{46} \rightarrow a_{51}, a_{51}, a_{52}, a_{52}, a_{53}, a_{53}$$

c) For elements of the fifth row

- The first element (a_{51}) of the fifth row is joined with elements ($a_{62}, a_{62}, a_{63}, a_{63}, a_{66}, a_{66}$) of the sixth row for positive and negative areas;
- The second element(a_{52}) of the fifth row is joined with elements ($a_{61}, a_{61}, a_{63}, a_{63}, a_{66}, a_{66}$) of the sixth row for positive and negative areas;
- The third element(a_{53}) of the fifth row is joined with elements ($a_{61}, a_{61}, a_{62}, a_{62}, a_{66}, a_{66}$) of the sixth row for positive and negative areas;
- The sixth element (a_{56}) of the fifth row is joined with elements ($a_{61}, a_{61}, a_{62}, a_{62}, a_{63}, a_{63}$) of the sixth row for positive and negative areas-in total we can conclude that sequence of elements connection for the fifth row with elements of the sixth row has the following view, forming the frequency of ties equal f=6:

$$a_{51} \rightarrow a_{62}, a_{62}, a_{63}, a_{63}, a_{66}, a_{66}$$
$$a_{52} \rightarrow a_{61}, a_{61}, a_{63}, a_{63}, a_{66}, a_{66}$$
$$a_{53} \rightarrow a_{61}, a_{61}, a_{62}, a_{62}, a_{66}, a_{66}$$
$$a_{56} \rightarrow a_{61}, a_{61}, a_{62}, a_{62}, a_{63}, a_{63}$$

❖ For Figure 157 of the combined partial determinant $|A_{25}|$:

Analysis of data shown in Figure 157 applicably to formula (114) of the combined partial determinant $|A_{25}|$ of the 6x6 matrix for positive and negative areas indicates on the following results in general view:

- The fifth element(a_{15}) of the first row is joined with the sixth element(a_{26}) of the second row;
- The sixth element(a_{26}) of the second row is joined with elements ($a_{31}, a_{32}, a_{33}, a_{34}$);

a) For elements of the third row

- The first element(a_{31}) of the third row is joined with elements (a_{42}, a_{43}, a_{44}) of the fourth row;
- The second element(a_{32}) of the third row is joined with elements (a_{41}, a_{43}, a_{44}) of the fourth row;
- The third element(a_{33})of the third row is joined with elements (a_{41}, a_{42}, a_{44}) of the fourth row;
- The fourth element(a_{34}) of the third row is joined with elements (a_{41}, a_{42}, a_{43}) of the fourth row –in total we can conclude that sequence of elements connection for the third row with elements of the fourth row has the following view, forming the frequency of ties equal f=3:

$$a_{31} \rightarrow a_{42}, a_{43}, a_{44}$$
$$a_{32} \rightarrow a_{41}, a_{43}, a_{44}$$
$$a_{33} \rightarrow a_{41}, a_{42}, a_{44}$$
$$a_{34} \rightarrow a_{41}, a_{42}, a_{43}$$

b) For elements of the fourth row

- The first element(a_{41}) of the fourth row is joined with elements ($a_{52}, a_{52}, a_{53}, a_{53}, a_{54}, a_{54}$) of the fifth row for positive and negative areas;
- The second element (a_{42}) of the fourth row is joined with elements ($a_{51}, a_{51}, a_{53}, a_{53}, a_{54}, a_{54}$) of the fifth row for positive and negative areas;
- The third element(a_{43}) of the fourth row is joined with elements ($a_{51}, a_{51}, a_{52}, a_{52}, a_{54}, a_{54}$)of the fifth row for positive and negative areas;
- The fourth element(a_{44}) of the fourth row is joined with elements ($a_{51}, a_{51}, a_{52}, a_{52}, a_{53}, a_{53}$) of the fifth row for positive and negative areas-in total we can conclude that sequence of elements connection fourth row with elements of the fifth row has the following view , forming the frequency of ties equal f=6:

$$a_{41} \rightarrow a_{52}, a_{52}, a_{53}, a_{53}, a_{54}, a_{54}$$
$$a_{42} \rightarrow a_{51}, a_{51}, a_{53}, a_{53}, a_{54}, a_{54}$$
$$a_{43} \rightarrow a_{51}, a_{51}, a_{52}, a_{52}, a_{54}, a_{54}$$
$$a_{44} \rightarrow a_{51}, a_{51}, a_{52}, a_{52}, a_{53}, a_{53}$$

c) *For elements of the fifth row*

- The first element(a_{51}) of the fifth row is joined with elements ($a_{62},a_{62},a_{63},a_{63},a_{64},a_{64}$) of the sixth row for positive and negative areas;
- The second element(a_{52}) of the fifth row is joined with elements ($a_{61},a_{61},a_{63},a_{63},a_{64},a_{64}$) of the sixth row for positive and negative areas;
- The third element(a_{53}) of the fifth row is joined with elements ($a_{61},a_{61},a_{62},a_{62},a_{64},a_{64}$) of the sixth row for positive and negative areas;
- The fourth element(a_{54}) of the fifth row is joined with elements ($a_{61},a_{61},a_{62},a_{62},a_{63},a_{63}$) of the sixth row for positive and negative areas-in total we can conclude that sequence of elements connection for the fifth row with elements of the sixth row has the following view, forming the frequency of ties equal f=6:

$$a_{51} \rightarrow a_{62},a_{62},a_{63},a_{63},a_{64},a_{64}$$
$$a_{52} \rightarrow a_{61},a_{61},a_{63},a_{63},a_{64},a_{64}$$
$$a_{53} \rightarrow a_{61},a_{61},a_{62},a_{62},a_{64},a_{64}$$
$$a_{54} \rightarrow a_{61},a_{61},a_{62},a_{62},a_{63},a_{63}$$

F. Analysis of combined partial determinant $\left|A'_{16}\right|$ of the 6x6 matrix:

The combined partial determinant $\left|A'_{16}\right|$ consists from five separate parts and is expressed by the following for in view of

$$\left|A'_{16}\right| = \left|A_{26}\right| + \left|A_{27}\right| + \left|A_{28}\right| + \left|A_{29}\right| + \left|A_{30}\right|$$

And also the above-indicated determinant $\left|A'_{16}\right|$ is expressed by the following finally formula (115) ,after of some transformations ,elements of which are shown in view of functional graphs in Figures 158 to 162:

$$
\begin{aligned}
\left|A'_{16}\right| = &\; a_{16}a_{21}a_{32}[a_{43}(a_{55}a_{64}-a_{54}a_{65})+a_{44}(a_{53}a_{65}-a_{55}a_{63})+a_{45}(a_{54}a_{63}-a_{53}a_{64})]+ \\
&+ a_{16}a_{21}a_{33}[a_{42}(a_{54}a_{65}-a_{55}a_{64})+a_{44}(a_{55}a_{62}-a_{52}a_{65})+a_{45}(a_{52}a_{64}-a_{54}a_{62})]+ \quad \left|A_{26}\right| \\
&+ a_{16}a_{21}a_{34}[a_{42}(a_{55}a_{63}-a_{53}a_{65})+a_{43}(a_{52}a_{65}-a_{55}a_{62})+a_{45}(a_{53}a_{62}-a_{52}a_{63})]+ \\
&+ a_{16}a_{21}a_{35}[a_{42}(a_{53}a_{64}-a_{54}a_{63})+a_{43}(a_{54}a_{62}-a_{52}a_{64})+a_{44}(a_{52}a_{63}-a_{53}a_{62})]+ \\
\\
&+ a_{16}a_{22}a_{31}[a_{43}(a_{54}a_{65}-a_{55}a_{64})+a_{44}(a_{55}a_{63}-a_{53}a_{65})+a_{45}(a_{53}a_{64}-a_{54}a_{63})]+ \\
&+ a_{16}a_{22}a_{33}[a_{41}(a_{55}a_{64}-a_{54}a_{65})+a_{44}(a_{51}a_{65}-a_{55}a_{61})+a_{45}(a_{54}a_{61}-a_{51}a_{64})]+ \quad \left|A_{27}\right| \\
&+ a_{16}a_{22}a_{34}[a_{41}(a_{53}a_{65}-a_{55}a_{63})+a_{43}(a_{55}a_{61}-a_{51}a_{65})+a_{45}(a_{51}a_{63}-a_{53}a_{61})]+ \\
&+ a_{16}a_{22}a_{35}[a_{41}(a_{54}a_{63}-a_{53}a_{64})+a_{43}(a_{51}a_{64}-a_{54}a_{61})+a_{44}(a_{53}a_{61}-a_{51}a_{63})]+ \\
\\
&+ a_{16}a_{23}a_{31}[a_{42}(a_{55}a_{64}-a_{54}a_{65})+a_{44}(a_{52}a_{65}-a_{55}a_{62})+a_{45}(a_{54}a_{62}-a_{52}a_{64})]+ \\
&+ a_{16}a_{23}a_{32}[a_{41}(a_{54}a_{65}-a_{55}a_{64})+a_{44}(a_{55}a_{61}-a_{51}a_{65})+a_{45}(a_{51}a_{64}-a_{54}a_{61})]+ \quad \left|A_{28}\right| \\
&+ a_{16}a_{23}a_{34}[a_{41}(a_{55}a_{62}-a_{52}a_{65})+a_{42}(a_{51}a_{65}-a_{55}a_{61})+a_{45}(a_{52}a_{61}-a_{51}a_{62})]+ \\
&+ a_{16}a_{23}a_{35}[a_{41}(a_{52}a_{64}-a_{54}a_{62})+a_{42}(a_{54}a_{61}-a_{51}a_{64})+a_{44}(a_{51}a_{62}-a_{52}a_{61})]+ \\
\\
&+ a_{16}a_{24}a_{31}[a_{42}(a_{53}a_{65}-a_{55}a_{63})+a_{43}(a_{55}a_{62}-a_{52}a_{65})+a_{45}(a_{52}a_{63}-a_{53}a_{62})]+ \\
&+ a_{16}a_{24}a_{32}[a_{41}(a_{55}a_{63}-a_{53}a_{65})+a_{43}(a_{51}a_{65}-a_{55}a_{61})+a_{45}(a_{53}a_{61}-a_{51}a_{63})]+ \quad \left|A_{29}\right| \\
&+ a_{16}a_{24}a_{33}[a_{41}(a_{52}a_{65}-a_{55}a_{62})+a_{42}(a_{55}a_{61}-a_{51}a_{65})+a_{45}(a_{51}a_{62}-a_{52}a_{61})]+ \\
&+ a_{16}a_{24}a_{35}[a_{41}(a_{53}a_{62}-a_{52}a_{63})+a_{42}(a_{51}a_{63}-a_{53}a_{61})+a_{43}(a_{52}a_{61}-a_{51}a_{62})]+ \\
\\
&+ a_{16}a_{25}a_{31}[a_{42}(a_{54}a_{63}-a_{53}a_{64})+a_{43}(a_{52}a_{64}-a_{54}a_{62})+a_{44}(a_{53}a_{62}-a_{52}a_{63})]+ \\
&+ a_{16}a_{25}a_{32}[a_{41}(a_{53}a_{64}-a_{54}a_{63})+a_{43}(a_{54}a_{61}-a_{51}a_{64})+a_{44}(a_{51}a_{63}-a_{53}a_{61})]+ \quad \left|A_{30}\right| \\
&+ a_{16}a_{25}a_{33}[a_{41}(a_{54}a_{62}-a_{52}a_{64})+a_{42}(a_{51}a_{64}-a_{54}a_{61})+a_{44}(a_{52}a_{61}-a_{51}a_{62})]+ \\
&+ a_{16}a_{25}a_{34}[a_{41}(a_{52}a_{63}-a_{53}a_{62})+a_{42}(a_{53}a_{61}-a_{51}a_{63})+a_{43}(a_{51}a_{62}-a_{52}a_{61})] \quad (115)
\end{aligned}
$$

241

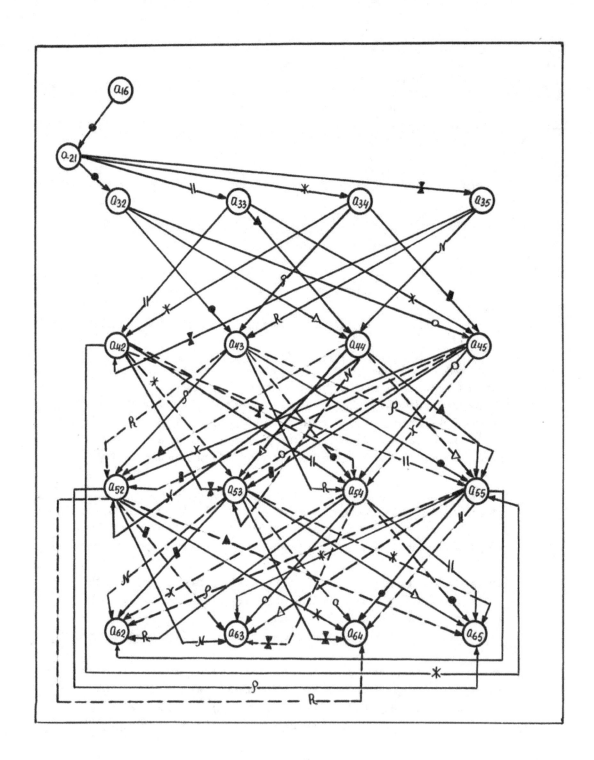

Figure 158 Functional graphs between elements for the combined partial 6x6 matrix determinant $\left| A_{26} \right|$ for positive (————) and negative (--------) areas

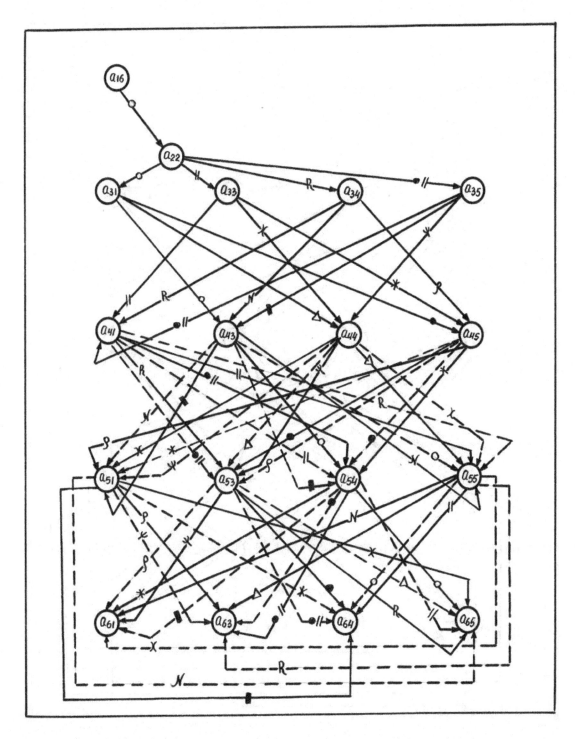

Figure 159 Functional graphs between of elements for the combined partial 6x6 matrix determinant │A₂₇│ for positive (————) and negative (--------)areas

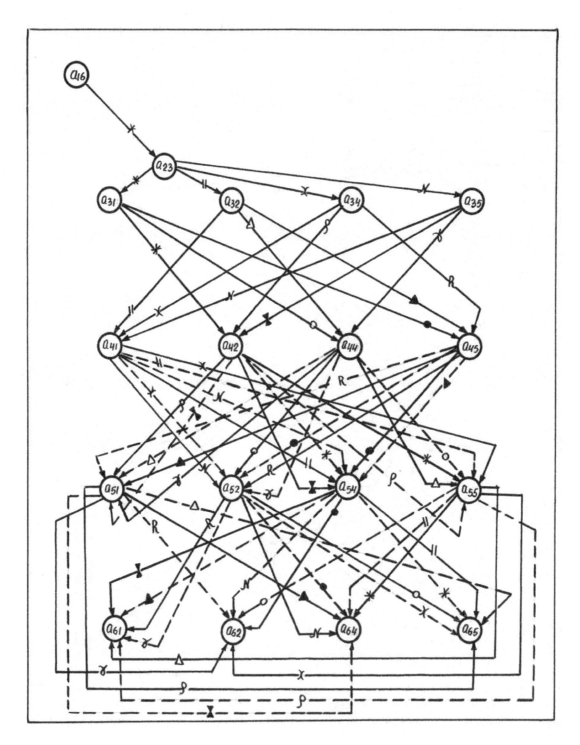

Figure 160 Functional graphs between of elements for the combined partial 6x6 matrix determinant $|A_{28}|$ for positive (——) and negative (-------) areas

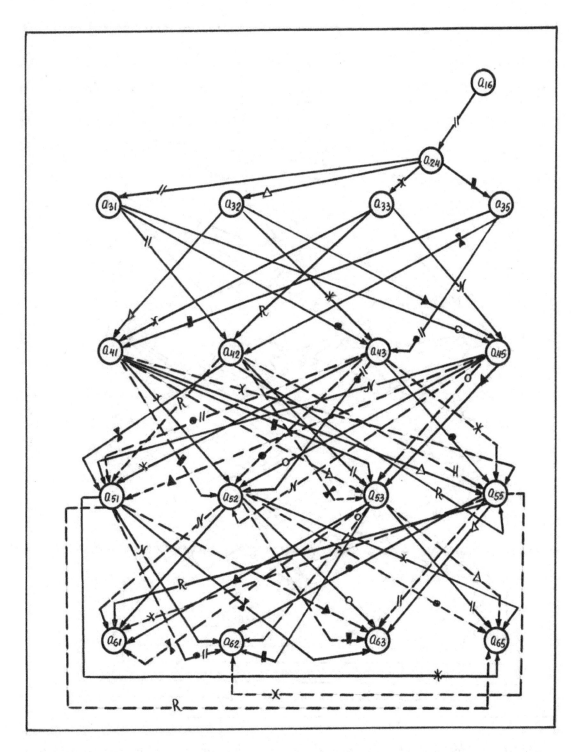

Figure 161 Functional graphs between of elements for the combined partial 6x6 matrix determinant $\left| A_{29} \right|$ for positive (——) and negative (------) areas

245

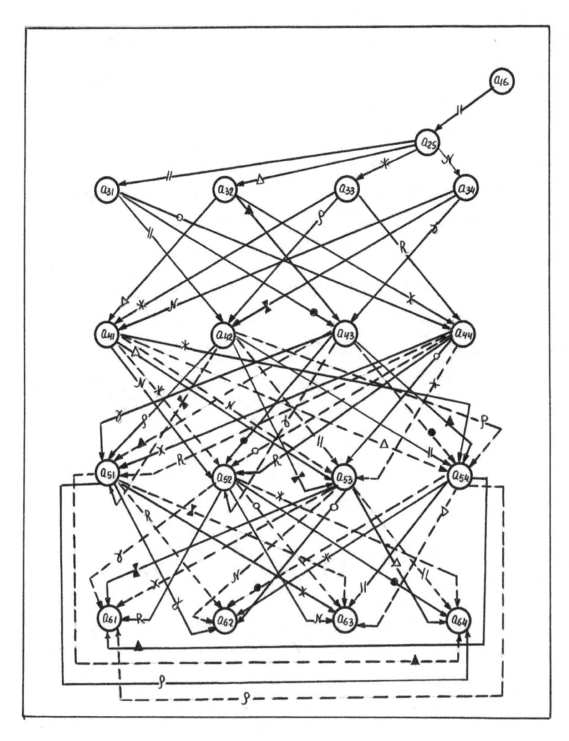

Figure 162 Functional graphs between of elements for the combined partial 6x6 matrix determinant $|A_{30}|$ for positive (——) and negative (-------) areas

246

❖ For Figure 158 of the combined partial determinant $\left|A_{26}\right|$:

Analysis of data shown in Figure 158 applicably to formula (115) of the combined partial determinant $\left|A_{26}\right|$ of the 6x6 matrix for positive and negative areas indicates on the following results in general view:
- The sixth element(a_{16}) of the first row is joined with the first element (a_{21}) of the second row;
- The first element (a_{21}) of the second row is joined with elements ($a_{32}, a_{33}, a_{34}, a_{35}$);

a) *For elements of the third row*
- The second element(a_{32}) of the third row is joined with elements (a_{43}, a_{44}, a_{45}) of the fourth row;
- The third element(a_{33}) of the third row is joined with elements (a_{42}, a_{44}, a_{45}) of the fourth row;
- The fourth element(a_{34}) of the third row is joined with elements (a_{42}, a_{43}, a_{45}) of the fourth row;
- The fifth element(a_{35}) of the third row is joined with elements (a_{42}, a_{43}, a_{44}) of the fourth row –in total we can conclude that sequence of elements connection for the third row with elements of the fourth row has the following view, forming the frequency of ties equal f=3:

$$a_{32} \longrightarrow a_{43}, a_{44}, a_{45}$$
$$a_{33} \longrightarrow a_{42}, a_{44}, a_{45}$$
$$a_{34} \longrightarrow a_{42}, a_{43}, a_{45}$$
$$a_{35} \longrightarrow a_{42}, a_{43}, a_{44}$$

b) *For elements of the fourth row*
- The second element(a_{42}) of the fourth row is joined with elements ($a_{53}, a_{53}, a_{54}, a_{54}, a_{55}, a_{55}$) of the fifth row for positive and negative areas;
- The third element (a_{43}) of the fourth row is joined with elements ($a_{52}, a_{52}, a_{54}, a_{54}, a_{55}, a_{55}$) of the fifth row for positive and negative areas;
- The fourth element(a_{44}) of the fourth row is joined with elements ($a_{52}, a_{52}, a_{53}, a_{53}, a_{55}, a_{55}$) of the fifth row for positive and negative areas;
- The fifth element(a_{45})of the fourth row is joined with elements ($a_{52}, a_{52}, a_{53}, a_{53}, a_{54}, a_{54}$) of the fifth row for positive and negative areas-in total we can conclude that sequence of elements connection fourth row with elements of the fifth row has the following view, forming the frequency of ties equal f=6:

$$a_{42} \longrightarrow a_{53}, a_{53}, a_{54}, a_{54}, a_{55}, a_{55}$$
$$a_{43} \longrightarrow a_{52}, a_{52}, a_{54}, a_{54}, a_{55}, a_{55}$$
$$a_{44} \longrightarrow a_{52}, a_{52}, a_{53}, a_{53}, a_{55}, a_{55}$$
$$a_{45} \longrightarrow a_{52}, a_{52}, a_{53}, a_{53}, a_{54}, a_{54}$$

c) *For elements of the fifth row*
- The second element (a_{52})of the fifth row is joined with elements ($a_{63}, a_{63}, a_{64}, a_{64}, a_{65}, a_{65}$) of the sixth row for positive and negative areas;
- The third element(a_{53}) of the fifth row is joined with elements ($a_{62}, a_{62}, a_{64}, a_{64}, a_{65}, a_{65}$) of the sixth row for positive and negative areas;
- The fourth element(a_{54}) of the fifth row is joined with elements ($a_{62}, a_{62}, a_{63}, a_{63}, a_{65}, a_{65}$) of the sixth row for positive and negative areas;
- The fifth element(a_{55}) of the fifth row is joined with elements ($a_{62}, a_{62}, a_{63}, a_{63}, a_{64}, a_{64}$) of the sixth row for positive and negative areas- in total we can conclude that sequence of elements connection for the fifth row with elements of the sixth row has the following view, forming the frequency of ties equal f=6:

$$a_{52} \longrightarrow a_{63}, a_{63}, a_{64}, a_{64}, a_{65}, a_{65}$$
$$a_{53} \longrightarrow a_{62}, a_{62}, a_{64}, a_{64}, a_{65}, a_{65}$$
$$a_{54} \longrightarrow a_{62}, a_{62}, a_{63}, a_{63}, a_{65}, a_{65}$$
$$a_{55} \longrightarrow a_{62}, a_{62}, a_{63}, a_{63}, a_{64}, a_{64}$$

❖ For Figure 159 of the combined partial determinant $|A_{27}|$:

Analysis of data shown in Figure 159 applicably to formula (115) of the combined partial determinant $|A_{27}|$ of the 6x6 matrix for positive and negative areas indicates on the following results in general view:
- The sixth element(a_{16}) of the first row is joined with the second element(a_{22}) of the second row;
- The second element (a_{22}) of the second row is joined with elements ($a_{31},a_{33},a_{34},a_{35}$) of the third row;

a) *For elements of the third row*
- The first element(a_{31}) of the third row is joined with elements (a_{43},a_{44},a_{45}) of the fourth row;
- The third element(a_{33}) of the third row is joined with elements (a_{41},a_{44},a_{45}) of the fourth row;
- The fourth element(a_{34}) of the third row is joined with elements (a_{41},a_{43},a_{45}) of the fourth row;
- The fifth element(a_{35}) of the third row is joined with elements(a_{41},a_{43},a_{44}) of the fourth row-in total we can conclude that sequence of elements connection for the third row with elements of the fourth row has the following view, forming the frequency of ties equal f=3:

$$a_{31} \rightarrow a_{43},a_{44},a_{45}$$
$$a_{33} \rightarrow a_{41},a_{44},a_{45}$$
$$a_{34} \rightarrow a_{41},a_{43},a_{45}$$
$$a_{35} \rightarrow a_{41},a_{43},a_{44}$$

b) *For elements of the fourth row*
- The first element (a_{41}) of the fourth row is joined with elements ($a_{53},a_{53},a_{54},a_{54},a_{55},a_{55}$) of the fifth row for positive and negative areas;
- The third element (a_{43}) of the fourth row is joined with elements ($a_{51},a_{51},a_{54},a_{54},a_{55},a_{55}$) of the fifth row for positive and negative areas;
- The fourth element (a_{44}) of the fourth row is joined with elements ($a_{51},a_{51},a_{53},a_{53},a_{55},a_{55}$) of the fifth row for positive and negative areas;
- The fifth element(a_{45}) of the fourth row is joined with elements ($a_{51},a_{51},a_{53},a_{53},a_{54},a_{54}$) of the fifth row for positive and negative areas- in total we can conclude that sequence of elements connection fourth row with elements of the fifth row has the following view, forming the frequency of ties equal f=6:

$$a_{41} \rightarrow a_{53},a_{53},a_{54},a_{54},a_{55},a_{55}$$
$$a_{43} \rightarrow a_{51},a_{51},a_{54},a_{54},a_{55},a_{55}$$
$$a_{44} \rightarrow a_{51},a_{51},a_{53},a_{53},a_{55},a_{55}$$
$$a_{45} \rightarrow a_{51},a_{51},a_{53},a_{53},a_{54},a_{54}$$

c) *For elements of the fifth row*
- The first element(a_{51}) of the fifth row is joined with elements ($a_{63},a_{63},a_{64},a_{64},a_{65},a_{65}$) of the sixth row for positive and negative areas;
- The third element(a_{53}) of the fifth row is joined with elements ($a_{61},a_{61},a_{64},a_{64},a_{65},a_{65}$) of the sixth row for positive and negative areas;
- The fourth element (a_{54}) of the fifth row is joined with elements ($a_{61},a_{61},a_{63},a_{63},a_{65},a_{65}$) of the sixth row for positive and negative areas;
- The fifth element(a_{55}) of the fifth row is joined with elements ($a_{61},a_{61},a_{63},a_{63},a_{64},a_{64}$) of the sixth row for positive and negative areas-in total we can conclude that sequence of elements connection for the fifth row with elements of the sixth row has the following view, forming the frequency of ties equal f=6:

$$a_{51} \rightarrow a_{63},a_{63},a_{64},a_{64},a_{65},a_{65}$$
$$a_{53} \rightarrow a_{61},a_{61},a_{64},a_{64},a_{65},a_{65}$$
$$a_{54} \rightarrow a_{61},a_{61},a_{63},a_{63},a_{65},a_{65}$$
$$a_{55} \rightarrow a_{61},a_{61},a_{63},a_{63},a_{64},a_{64}$$

248

❖ For Figure 160 of the combined partial determinant $|A_{28}|$:

Analysis of data shown in Figure 160 applicably to formula (115) of the combined partial determinant $|A_{28}|$ of the 6x6 matrix for positive and negative areas indicates on the following results in general view:

- The six element(a_{16}) of the first row is joined with the third element (a_{23}) of the second row;
- The third element (a_{23}) of the second row is joined with elements ($a_{31}, a_{32}, a_{34}, a_{35}$) of the third row;

a) For elements of the third row

- The first element(a_{31}) of the third row is joined with elements (a_{42}, a_{44}, a_{45}) of the fourth row;
- The second element(a_{32}) of the third row is joined with elements (a_{41}, a_{44}, a_{45}) of the fourth row;
- The fourth element(a_{34}) of the third row is joined with elements (a_{41}, a_{42}, a_{45}) of the fourth row;
- The fifth element (a_{35}) of the third row is joined with elements (a_{41}, a_{42}, a_{44}) of the fourth row-in total we can conclude that sequence of elements connection for the third row with elements of the fourth row has the following view, forming the frequency of ties equal f=3:

$$a_{31} \rightarrow a_{42}, a_{44}, a_{45}$$
$$a_{32} \rightarrow a_{41}, a_{44}, a_{45}$$
$$a_{34} \rightarrow a_{41}, a_{42}, a_{45}$$
$$a_{35} \rightarrow a_{41}, a_{42}, a_{44}$$

b) For elements of the fourth row

- The first element(a_{41}) of the fourth row is joined with elements ($a_{52}, a_{52}, a_{54}, a_{54}, a_{55}, a_{55}$) of the fifth row for positive and negative areas;
- The second element(a_{42}) of the fourth row is joined with elements ($a_{51}, a_{51}, a_{54}, a_{54}, a_{55}, a_{55}$) of the fifth row for positive and negative areas;
- The fourth element(a_{44}) of the fourth row is joined with elements ($a_{51}, a_{51}, a_{52}, a_{52}, a_{55}, a_{55}$) of the fifth row for positive and negative areas;
- The fifth element(a_{45}) of the fourth row is joined with elements($a_{51}, a_{51}, a_{52}, a_{52}, a_{54}, a_{54}$) of the fifth row for positive and negative areas-in total we can conclude that sequence of elements connection fourth row with elements of the fifth row has the following view, forming the frequency of ties equal f=6:

$$a_{41} \rightarrow a_{52}, a_{52}, a_{54}, a_{54}, a_{55}, a_{55}$$
$$a_{42} \rightarrow a_{51}, a_{51}, a_{54}, a_{54}, a_{55}, a_{55}$$
$$a_{44} \rightarrow a_{51}, a_{51}, a_{52}, a_{52}, a_{55}, a_{55}$$
$$a_{45} \rightarrow a_{51}, a_{51}, a_{52}, a_{52}, a_{54}, a_{54}$$

c) For elements of the fifth row

- The first element(a_{51}) of the fifth row is joined with elements ($a_{62}, a_{62}, a_{64}, a_{64}, a_{65}, a_{65}$) of the sixth row for positive and negative areas;
- The second element(a_{52}) of the fifth row is joined with elements ($a_{61}, a_{61}, a_{64}, a_{64}, a_{65}, a_{65}$) of the sixth row for positive and negative areas;
- The fourth element(a_{54}) of the fifth row is joined with elements ($a_{61}, a_{61}, a_{62}, a_{62}, a_{65}, a_{65}$) of the sixth row for positive and negative areas;
- The fifth element(a_{55}) of the fifth row is joined with elements($a_{61}, a_{61}, a_{62}, a_{62}, a_{64}, a_{64}$) of the sixth row for positive and negative areas-in total we can conclude that sequence of elements connection for the fifth row with elements of the sixth row has the following view ,forming the frequency of ties equal f=6:

$$a_{51} \rightarrow a_{62}, a_{62}, a_{64}, a_{64}, a_{65}, a_{65}$$
$$a_{52} \rightarrow a_{61}, a_{61}, a_{64}, a_{64}, a_{65}, a_{65}$$
$$a_{54} \rightarrow a_{61}, a_{61}, a_{62}, a_{62}, a_{65}, a_{65}$$
$$a_{55} \rightarrow a_{61}, a_{61}, a_{62}, a_{62}, a_{64}, a_{64}$$

249

❖ **For Figure 161 of the combined partial determinant $|A_{29}|$:**

Analysis of data shown in Figure 161 applicably to formula (115) of the combined partial determinant $|A_{29}|$ of the 6x6 matrix for positive and negative areas indicates on the following results in general view:

- The sixth element(a_{16}) of the first row is joined with the fourth element(a_{24}) of the second row;
- The fourth element (a_{24}) of the second row is joined with elements ($a_{31},a_{32},a_{33},a_{35}$) of the third row;

a) *For elements of the third row*

- The first element(a_{31}) of the third row is joined with elements (a_{42},a_{43},a_{45}) of the fourth row;
- The second element(a_{32}) of the third row is joined with elements (a_{41},a_{43},a_{45}) of the fourth row;
- The third element(a_{33}) of the third row is joined with elements (a_{41},a_{42},a_{45}) of the fourth row;
- The fifth element(a_{35}) of the third row is joined with elements (a_{41},a_{42},a_{43})of the fourth row-in total we can conclude that sequence of elements connection for the third row with elements of the fourth row has the following view, forming the frequency of ties equal f=3:

$$a_{31} \rightarrow a_{42},a_{43},a_{45}$$
$$a_{32} \rightarrow a_{41},a_{43},a_{45}$$
$$a_{33} \rightarrow a_{41},a_{42},a_{45}$$
$$a_{35} \rightarrow a_{41},a_{42},a_{43}$$

b) *For elements of the fourth row*

- The first element (a_{41}) of the fourth row is joined with elements ($a_{52},a_{52},a_{53},a_{53},a_{55},a_{55}$) of the fifth row for positive and negative areas;
- The second element (a_{42}) of the fourth row is joined with elements ($a_{51},a_{51},a_{53},a_{53},a_{55},a_{55}$) of the fifth row for positive and negative areas;
- The third element(a_{43}) of the fourth row is joined with elements ($a_{51},a_{51},a_{52},a_{52},a_{55},a_{55}$) of the fifth row for positive and negative areas;
- The fifth element (a_{45}) of the fourth row is joined with elements ($a_{51},a_{51},a_{52},a_{52},a_{53},a_{53}$) of the fifth row for positive and negative areas- in total we can conclude that sequence of elements connection fourth row with elements of the fifth row has the following view, forming the frequency of ties equal f=6:

$$a_{41} \rightarrow a_{52},a_{52},a_{53},a_{53},a_{55},a_{55}$$
$$a_{42} \rightarrow a_{51},a_{51},a_{53},a_{53},a_{55},a_{55}$$
$$a_{43} \rightarrow a_{51},a_{51},a_{52},a_{52},a_{55},a_{55}$$
$$a_{45} \rightarrow a_{51},a_{51},a_{52},a_{52},a_{53},a_{53}$$

c) *For elements of the fifth row*

- The first element(a_{51}) of the fifth row is joined with elements($a_{62},a_{62},a_{63},a_{63},a_{65},a_{65}$) of the sixth row for positive and negative areas;
- The second element(a_{52}) of the fifth row is joined with elements ($a_{61},a_{61},a_{63},a_{63},a_{65},a_{65}$) of the sixth row for positive and negative areas;
- The third element(a_{53}) of the fifth row is joined with elements ($a_{61},a_{61},a_{62},a_{62},a_{65},a_{65}$) of the sixth row for positive and negative areas;
- The fifth element(a_{55}) of the fifth row is joined with elements($a_{61},a_{61},a_{62},a_{62},a_{63},a_{63}$) of the sixth row for positive and negative areas- in total we can conclude that sequence of elements connection for the fifth row with elements of the sixth row has the following view, forming the frequency of ties equal f=6:

$$a_{51} \rightarrow a_{62},a_{62},a_{63},a_{63},a_{65},a_{65}$$
$$a_{52} \rightarrow a_{61},a_{61},a_{63},a_{63},a_{65},a_{65}$$
$$a_{53} \rightarrow a_{61},a_{61},a_{62},a_{62},a_{65},a_{65}$$
$$a_{55} \rightarrow a_{61},a_{61},a_{62},a_{62},a_{63},a_{63}$$

❖ For Figure 162 of the combined partial determinant $\left|A_{30}\right|$:

Analysis of data shown in Figure 162 applicably to formula (115) of the combined partial determinant $\left|A_{30}\right|$ for positive and negative areas indicates on the following results in general view:
- The sixth element(a_{16})of the first row is joined with the fifth element(a_{25}) of the second row;
- The fifth element (a_{25}) of the second row is joined with elements ($a_{31},a_{32},a_{33},a_{34}$) of the third row;

a) *For elements of the third row*
- The first element(a_{31}) of the third row is joined with elements (a_{42},a_{43},a_{44}) of the fourth row;
- The second element(a_{32}) of the third row is joined with elements (a_{41},a_{43},a_{44}) of the fourth row;
- The third element(a_{33}) of the third row is joined with elements (a_{41},a_{42},a_{44}) of the fourth row;
- The fourth element(a_{34}) of the third row is joined with elements (a_{41},a_{42},a_{43}) of the fourth row-in total we can conclude that sequence of elements connection for the third row with elements of the fourth row has the following view, forming the frequency of ties equal f=3:

$$a_{31}\rightarrow a_{42},a_{43},a_{44}$$
$$a_{32}\rightarrow a_{41},a_{43},a_{44}$$
$$a_{33}\rightarrow a_{41},a_{42},a_{44}$$
$$a_{34}\rightarrow a_{41},a_{42},a_{43}$$

b) *For elements of the fourth row*
- The first element(a_{41}) of the fourth row is joined with elements($a_{52},a_{52},a_{53},a_{53},a_{54},a_{54}$) of the fifth row for positive and negative areas;
- The second element(a_{42}) of the fourth row is joined with elements ($a_{51},a_{51},a_{53},a_{53},a_{54},a_{54}$) of the fifth row for positive and negative areas;
- The third element (a_{43}) of the fourth row is joined with elements ($a_{51},a_{51},a_{52},a_{52},a_{54},a_{54}$) of the fifth row for positive and negative areas;
- The fourth element (a_{44}) of the fourth row is joined with elements ($a_{51},a_{51},a_{52},a_{52},a_{53},a_{53}$) of the fifth row for positive and negative areas- in total we can conclude that sequence of elements connection fourth row with elements of the fifth row has the following view, forming the frequency of ties equal f=6:

$$a_{41}\rightarrow a_{52},a_{52},a_{53},a_{53},a_{54},a_{54}$$
$$a_{42}\rightarrow a_{51},a_{51},a_{53},a_{53},a_{54},a_{54}$$
$$a_{43}\rightarrow a_{51},a_{51},a_{52},a_{52},a_{54},a_{54}$$
$$a_{44}\rightarrow a_{51},a_{51},a_{52},a_{52},a_{53},a_{53}$$

c) *For elements of the fifth row*
- The first element(a_{51}) of the fifth row is joined with elements ($a_{62},a_{62},a_{63},a_{63},a_{64},a_{64}$)of the sixth row for positive and negative areas;
- The second element (a_{52}) of the fifth row is joined with elements ($a_{61},a_{61},a_{63},a_{63},a_{64},a_{64}$) of the sixth row for positive and negative areas;
- The third element(a_{53}) of the fifth row is joined with elements ($a_{61},a_{61},a_{62},a_{62},a_{64},a_{64}$) of the sixth row for positive and negative areas;
- The fourth element(a_{54}) of the fifth row is joined with elements($a_{61},a_{61},a_{62},a_{62},a_{63},a_{63}$) of the sixth row for positive and negative areas- in total we can conclude that sequence of elements connection for the fifth row with elements of the sixth row has the following view, forming the frequency of ties equal f=6:

$$a_{51}\rightarrow a_{62},a_{62},a_{63},a_{63},a_{64},a_{64}$$
$$a_{52}\rightarrow a_{61},a_{61},a_{63},a_{63},a_{64},a_{64}$$
$$a_{53}\rightarrow a_{61},a_{61},a_{62},a_{62},a_{64},a_{64}$$
$$a_{54}\rightarrow a_{61},a_{61},a_{62},a_{62},a_{63},a_{63}$$

In Appendix 24 is shown solving of the system normal six equations with using of the Expansion (cofactor) method and Cramer's rule of the determinants of the 6x 6 matrices *applicably to the first row*. And also are shown the functional ties between elements of the 6x6 matrices in graphs.

In **Appendix 25** is shown solving of system normal six linear equations with using of Expansion (cofactor) and Laplace's methods for evaluation of the determinants of the 6x6 matrices. **In Appendix 26** is shown the evaluation of determinant $|A|$ of the 6x6 matrix by Expansion and Laplace's methods with using of the functional graphs presenting between elements.

In Appendix 27 is shown determination of system normal six linear equations with using of Cramer's rule, Expansion and Laplace's methods with objective of comparative analysis.

1.2 <u>By Laplace's theorem for some columns</u>

❖ *For marked out jointly the third, fourth and fifth columns*

Applicably to Laplace's theorem [2] we consider the conditions in general view, for a instance, when in the determinant $|A|$ of the 6x6 matrix marked out jointly the third, fourth and fifth columns. The combinations of used rows and other parameters of the determinant are shown in Table 8.

Table 8 Combination of used rows and other parameters of the determinant of the 6x6 matrix

n/n	Combination of used rows	Parameters of the determinant of the 6x6 matrix	
		Minors of the third order (M_{ij})	Algebraic(A_{ij})supplemental (cofactors) of the third order
1	$i_1+i_2+i_3$	$M_1=\begin{vmatrix} a_{13} & a_{14} & a_{15} \\ a_{23} & a_{24} & a_{25} \\ a_{33} & a_{34} & a_{35} \end{vmatrix}$	$A_1=\begin{vmatrix} a_{41} & a_{42} & a_{46} \\ a_{51} & a_{52} & a_{56} \\ a_{61} & a_{62} & a_{66} \end{vmatrix}$
2	$i_1+i_2+i_4$	$M_2=\begin{vmatrix} a_{13} & a_{14} & a_{15} \\ a_{23} & a_{24} & a_{25} \\ a_{43} & a_{44} & a_{45} \end{vmatrix}$	$A_2=\begin{vmatrix} a_{31} & a_{32} & a_{36} \\ a_{51} & a_{52} & a_{56} \\ a_{61} & a_{62} & a_{66} \end{vmatrix}$
3	$i_1+i_2+i_5$	$M_3=\begin{vmatrix} a_{13} & a_{14} & a_{15} \\ a_{23} & a_{24} & a_{25} \\ a_{53} & a_{54} & a_{55} \end{vmatrix}$	$A_3=\begin{vmatrix} a_{31} & a_{32} & a_{36} \\ a_{41} & a_{42} & a_{46} \\ a_{61} & a_{62} & a_{66} \end{vmatrix}$
4	$i_1+i_2+i_6$	$M_4=\begin{vmatrix} a_{13} & a_{14} & a_{15} \\ a_{23} & a_{24} & a_{25} \\ a_{63} & a_{64} & a_{65} \end{vmatrix}$	$A_4=\begin{vmatrix} a_{31} & a_{32} & a_{36} \\ a_{41} & a_{42} & a_{46} \\ a_{51} & a_{52} & a_{56} \end{vmatrix}$
5	$i_1+i_3+i_4$	$M_5=\begin{vmatrix} a_{13} & a_{14} & a_{15} \\ a_{33} & a_{34} & a_{35} \\ a_{43} & a_{44} & a_{45} \end{vmatrix}$	$A_5=\begin{vmatrix} a_{21} & a_{22} & a_{26} \\ a_{51} & a_{52} & a_{56} \\ a_{61} & a_{62} & a_{66} \end{vmatrix}$
6	$i_1+i_3+i_5$	$M_6=\begin{vmatrix} a_{13} & a_{14} & a_{15} \\ a_{33} & a_{34} & a_{35} \\ a_{53} & a_{54} & a_{55} \end{vmatrix}$	$A_6=\begin{vmatrix} a_{21} & a_{22} & a_{26} \\ a_{41} & a_{42} & a_{46} \\ a_{61} & a_{62} & a_{66} \end{vmatrix}$
7	$i_1+i_3+i_6$	$M_7=\begin{vmatrix} a_{13} & a_{14} & a_{15} \\ a_{33} & a_{34} & a_{35} \\ a_{63} & a_{64} & a_{65} \end{vmatrix}$	$A_7=\begin{vmatrix} a_{21} & a_{22} & a_{26} \\ a_{41} & a_{42} & a_{46} \\ a_{51} & a_{52} & a_{56} \end{vmatrix}$
8	$i_1+i_4+i_5$	$M_8=\begin{vmatrix} a_{13} & a_{14} & a_{15} \\ a_{43} & a_{44} & a_{45} \\ a_{53} & a_{54} & a_{55} \end{vmatrix}$	$A_8=\begin{vmatrix} a_{21} & a_{22} & a_{26} \\ a_{31} & a_{32} & a_{36} \\ a_{61} & a_{62} & a_{66} \end{vmatrix}$
9	$i_1+i_4+i_6$	$M_9=\begin{vmatrix} a_{13} & a_{14} & a_{15} \\ a_{43} & a_{44} & a_{45} \\ a_{63} & a_{64} & a_{65} \end{vmatrix}$	$A_9=\begin{vmatrix} a_{21} & a_{22} & a_{26} \\ a_{31} & a_{32} & a_{36} \\ a_{51} & a_{52} & a_{56} \end{vmatrix}$
10	$i_2+i_3+i_4$	$M_{10}=\begin{vmatrix} a_{23} & a_{24} & a_{25} \\ a_{33} & a_{34} & a_{35} \\ a_{43} & a_{44} & a_{45} \end{vmatrix}$	$A_{10}=\begin{vmatrix} a_{11} & a_{12} & a_{16} \\ a_{51} & a_{52} & a_{56} \\ a_{61} & a_{62} & a_{66} \end{vmatrix}$

№	index	M	A
11	$i_2+i_3+i_5$	$M_{11}=\begin{vmatrix} a_{23} & a_{24} & a_{25} \\ a_{33} & a_{34} & a_{35} \\ a_{53} & a_{54} & a_{55} \end{vmatrix}$	$A_{11}=\begin{vmatrix} a_{11} & a_{12} & a_{16} \\ a_{41} & a_{42} & a_{46} \\ a_{61} & a_{62} & a_{66} \end{vmatrix}$
12	$i_2+i_3+i_6$	$M_{12}=\begin{vmatrix} a_{23} & a_{24} & a_{25} \\ a_{33} & a_{34} & a_{35} \\ a_{63} & a_{64} & a_{65} \end{vmatrix}$	$A_{12}=\begin{vmatrix} a_{11} & a_{12} & a_{16} \\ a_{41} & a_{42} & a_{46} \\ a_{51} & a_{52} & a_{56} \end{vmatrix}$
13	$i_3+i_4+i_5$	$M_{13}=\begin{vmatrix} a_{33} & a_{34} & a_{35} \\ a_{43} & a_{44} & a_{45} \\ a_{53} & a_{54} & a_{55} \end{vmatrix}$	$A_{13}=\begin{vmatrix} a_{11} & a_{12} & a_{16} \\ a_{21} & a_{22} & a_{26} \\ a_{61} & a_{62} & a_{66} \end{vmatrix}$
14	$i_3+i_4+i_6$	$M_{14}=\begin{vmatrix} a_{33} & a_{34} & a_{35} \\ a_{43} & a_{44} & a_{45} \\ a_{63} & a_{64} & a_{65} \end{vmatrix}$	$A_{14}=\begin{vmatrix} a_{11} & a_{12} & a_{16} \\ a_{21} & a_{22} & a_{26} \\ a_{51} & a_{52} & a_{56} \end{vmatrix}$
15	$i_3+i_5+i_6$	$M_{15}=\begin{vmatrix} a_{33} & a_{34} & a_{35} \\ a_{53} & a_{54} & a_{55} \\ a_{63} & a_{64} & a_{65} \end{vmatrix}$	$A_{15}=\begin{vmatrix} a_{11} & a_{12} & a_{16} \\ a_{21} & a_{22} & a_{26} \\ a_{41} & a_{42} & a_{46} \end{vmatrix}$
16	$i_4+i_5+i_6$	$M_{16}=\begin{vmatrix} a_{43} & a_{44} & a_{45} \\ a_{53} & a_{54} & a_{55} \\ a_{63} & a_{64} & a_{65} \end{vmatrix}$	$A_{16}=\begin{vmatrix} a_{11} & a_{12} & a_{16} \\ a_{21} & a_{22} & a_{26} \\ a_{31} & a_{32} & a_{36} \end{vmatrix}$

So, from Table 8 we can conclude that value of the determinant $|A|$ of the 6x6 matrix applicably to Lapalce's theorem at marked out jointly the third , fourth and fifth columns expresses by the following formula in general view as:

$$|A| = M_1A_1-M_2A_2+M_3A_3-M_4A_4+M_5A_5-M_6A_6+M_7A_7+M_8A_8-$$
$$-M_9A_9-M_{10}A_{10}+M_{11}A_{11}-M_{12}A_{12}+M_{13}A_{13}-M_{14}A_{14}+M_{15}A_{15}-M_{16}A_{16} \quad (116)$$

After of some calculations and transformations, the value of determinant $|A|$ of the 6x6 matrix for the above-considered conditions has the following view in general form as:

$$|A|=\{a_{13}a_{24}a_{35}[a_{41}(a_{52}a_{66}-a_{56}a_{62})+a_{42}(a_{56}a_{61}-a_{51}a_{66})+a_{46}(a_{51}a_{62}-a_{52}a_{61})]+$$
$$+a_{13}a_{24}a_{45}[a_{31}(a_{56}a_{62}-a_{52}a_{66})+a_{32}(a_{51}a_{66}-a_{56}a_{61})+a_{36}(a_{52}a_{61}-a_{51}a_{62})]+$$
$$+a_{13}a_{24}a_{55}[a_{31}(a_{42}a_{66}-a_{46}a_{62})+a_{32}(a_{46}a_{61}-a_{41}a_{66})+a_{36}(a_{41}a_{62}-a_{42}a_{61})]+ \quad |A_1|$$
$$+a_{13}a_{24}a_{65}[a_{31}(a_{46}a_{52}-a_{42}a_{56})+a_{32}(a_{41}a_{56}-a_{46}a_{51})+a_{36}(a_{42}a_{51}-a_{41}a_{52})]+$$

$$+a_{13}a_{25}a_{34}[a_{41}(a_{56}a_{62}-a_{52}a_{66})+a_{42}(a_{51}a_{66}-a_{56}a_{61})+a_{46}(a_{52}a_{61}-a_{51}a_{62})]+$$
$$+a_{13}a_{25}a_{44}[a_{31}(a_{52}a_{66}-a_{56}a_{62})+a_{32}(a_{56}a_{61}-a_{51}a_{66})+a_{36}(a_{51}a_{62}-a_{52}a_{61})]+$$
$$+a_{13}a_{25}a_{54}[a_{31}(a_{46}a_{62}-a_{42}a_{66})+a_{32}(a_{41}a_{66}-a_{46}a_{61})+a_{36}(a_{42}a_{61}-a_{41}a_{62})]+ \quad |A_2|$$
$$+a_{13}a_{25}a_{64}[a_{31}(a_{42}a_{56}-a_{46}a_{52})+a_{32}(a_{46}a_{51}-a_{41}a_{56})+a_{36}(a_{41}a_{52}-a_{42}a_{51})]+$$

$$+a_{13}a_{34}a_{45}[a_{21}(a_{52}a_{66}-a_{56}a_{62})+a_{22}(a_{56}a_{61}-a_{51}a_{66})+a_{26}(a_{51}a_{62}-a_{52}a_{61})]+$$
$$+a_{13}a_{34}a_{55}[a_{21}(a_{46}a_{62}-a_{42}a_{66})+a_{22}(a_{41}a_{66}-a_{46}a_{61})+a_{26}(a_{42}a_{61}-a_{41}a_{62})]+ \quad |A_3|$$
$$+a_{13}a_{34}a_{65}[a_{21}(a_{42}a_{56}-a_{46}a_{52})+a_{22}(a_{46}a_{51}-a_{41}a_{56})+a_{26}(a_{41}a_{52}-a_{42}a_{51})]+$$

$$+a_{13}a_{35}a_{44}[a_{21}(a_{56}a_{62}-a_{52}a_{66})+a_{22}(a_{51}a_{66}-a_{56}a_{61})+a_{26}(a_{52}a_{61}-a_{51}a_{62})]+$$
$$+a_{13}a_{35}a_{54}[a_{21}(a_{42}a_{66}-a_{46}a_{62})+a_{22}(a_{46}a_{61}-a_{41}a_{66})+a_{26}(a_{41}a_{62}-a_{42}a_{61})]+ \quad |A_4|$$
$$+a_{13}a_{35}a_{64}[a_{21}(a_{46}a_{52}-a_{42}a_{56})+a_{22}(a_{41}a_{56}-a_{46}a_{51})+a_{26}(a_{42}a_{51}-a_{41}a_{52})]+$$

$$+a_{13}a_{44}a_{55}[a_{21}(a_{32}a_{66}-a_{36}a_{62})+a_{22}(a_{36}a_{61}-a_{31}a_{66})+a_{26}(a_{31}a_{62}-a_{32}a_{61})]+$$
$$+a_{13}a_{44}a_{65}[a_{21}(a_{36}a_{52}-a_{32}a_{56})+a_{22}(a_{31}a_{56}-a_{36}a_{51})+a_{26}(a_{32}a_{51}-a_{31}a_{52})]+$$
$$+a_{13}a_{45}a_{54}[a_{21}(a_{36}a_{62}-a_{32}a_{66})+a_{22}(a_{31}a_{66}-a_{36}a_{61})+a_{26}(a_{32}a_{61}-a_{31}a_{62})]+ \quad |A_5|$$
$$+a_{13}a_{45}a_{64}[a_{21}(a_{32}a_{56}-a_{36}a_{52})+a_{22}(a_{36}a_{51}-a_{31}a_{56})+a_{26}(a_{31}a_{52}-a_{32}a_{51})] \}+$$

$$+\{a_{14}a_{23}a_{35}[a_{41}(a_{56}a_{62}-a_{52}a_{66})+a_{42}(a_{51}a_{66}-a_{56}a_{61})+a_{46}(a_{52}a_{61}-a_{51}a_{62})]+$$
$$+a_{14}a_{23}a_{45}[a_{31}(a_{52}a_{66}-a_{56}a_{62})+a_{32}(a_{56}a_{61}-a_{51}a_{66})+a_{36}(a_{51}a_{62}-a_{52}a_{61})]+$$
$$+a_{14}a_{23}a_{55}[a_{31}(a_{46}a_{62}-a_{42}a_{66})+a_{32}(a_{41}a_{66}-a_{46}a_{61})+a_{36}(a_{42}a_{61}-a_{41}a_{62})]+ \quad |A_6|$$
$$+a_{14}a_{23}a_{65}[a_{31}(a_{42}a_{56}-a_{46}a_{52})+a_{32}(a_{46}a_{51}-a_{41}a_{56})+a_{36}(a_{41}a_{52}-a_{42}a_{51})]+$$

$$+a_{14}a_{25}a_{33}[a_{41}(a_{52}a_{66}-a_{56}a_{62})+a_{42}(a_{56}a_{61}-a_{51}a_{66})+a_{46}(a_{51}a_{62}-a_{52}a_{61})]+$$
$$+a_{14}a_{25}a_{43}[a_{31}(a_{56}a_{62}-a_{52}a_{66})+a_{32}(a_{51}a_{66}-a_{56}a_{61})+a_{36}(a_{52}a_{61}-a_{51}a_{62})]+$$
$$+a_{14}a_{25}a_{53}[a_{31}(a_{42}a_{66}-a_{46}a_{62})+a_{32}(a_{46}a_{61}-a_{41}a_{66})+a_{36}(a_{41}a_{62}-a_{42}a_{61})]+ \quad |A_7|$$
$$+a_{14}a_{25}a_{63}[a_{31}(a_{46}a_{52}-a_{42}a_{56})+a_{32}(a_{41}a_{56}-a_{46}a_{51})+a_{36}(a_{42}a_{51}-a_{41}a_{52})]+$$

$$+a_{14}a_{33}a_{45}[a_{21}(a_{56}a_{62}-a_{52}a_{66})+a_{22}(a_{51}a_{66}-a_{56}a_{61})+a_{26}(a_{52}a_{61}-a_{51}a_{62})]+$$
$$+a_{14}a_{33}a_{55}[a_{21}(a_{42}a_{66}-a_{46}a_{62})+a_{22}(a_{46}a_{61}-a_{41}a_{66})+a_{26}(a_{41}a_{62}-a_{42}a_{61})]+ \quad |A_8|$$
$$+a_{14}a_{33}a_{65}[a_{21}(a_{46}a_{52}-a_{42}a_{56})+a_{22}(a_{41}a_{56}-a_{46}a_{51})+a_{26}(a_{42}a_{51}-a_{41}a_{52})]+$$

$$+a_{14}a_{35}a_{43}[a_{21}(a_{52}a_{66}-a_{56}a_{62})+a_{22}(a_{56}a_{61}-a_{51}a_{66})+a_{26}(a_{51}a_{62}-a_{52}a_{61})]+$$
$$+a_{14}a_{35}a_{53}[a_{21}(a_{46}a_{62}-a_{42}a_{66})+a_{22}(a_{41}a_{66}-a_{46}a_{61})+a_{26}(a_{42}a_{61}-a_{41}a_{62})]+ \quad |A_9|$$
$$+a_{14}a_{35}a_{63}[a_{21}(a_{42}a_{56}-a_{46}a_{52})+a_{22}(a_{46}a_{51}-a_{41}a_{56})+a_{26}(a_{41}a_{52}-a_{42}a_{51})]+$$

$$+a_{14}a_{43}a_{55}[a_{21}(a_{36}a_{62}-a_{32}a_{66})+a_{22}(a_{31}a_{66}-a_{36}a_{61})+a_{26}(a_{32}a_{61}-a_{31}a_{62})]+$$
$$+a_{14}a_{43}a_{65}[a_{21}(a_{32}a_{56}-a_{36}a_{52})+a_{22}(a_{36}a_{51}-a_{31}a_{56})+a_{26}(a_{31}a_{52}-a_{32}a_{51})]+$$
$$+a_{14}a_{45}a_{53}[a_{21}(a_{32}a_{66}-a_{36}a_{62})+a_{22}(a_{36}a_{61}-a_{31}a_{66})+a_{26}(a_{31}a_{62}-a_{32}a_{61})]+ \quad |A_{10}|$$
$$+a_{14}a_{45}a_{63}[a_{21}(a_{36}a_{52}-a_{32}a_{56})+a_{22}(a_{31}a_{56}-a_{36}a_{51})+a_{26}(a_{32}a_{51}-a_{31}a_{52})]\}+$$

••

$$+\{a_{15}a_{23}a_{34}[a_{41}(a_{52}a_{66}-a_{56}a_{62})+a_{42}(a_{56}a_{61}-a_{51}a_{66})+a_{46}(a_{51}a_{62}-a_{52}a_{61})]+$$
$$+a_{15}a_{23}a_{44}[a_{31}(a_{56}a_{62}-a_{52}a_{66})+a_{32}(a_{51}a_{66}-a_{56}a_{61})+a_{36}(a_{52}a_{61}-a_{51}a_{62})]+$$
$$+a_{15}a_{23}a_{54}[a_{31}(a_{32}a_{66}-a_{36}a_{62})+a_{32}(a_{46}a_{61}-a_{41}a_{66})+a_{36}(a_{41}a_{62}-a_{42}a_{61})]+ \quad |A_{11}|$$
$$+a_{15}a_{23}a_{64}[a_{31}(a_{46}a_{52}-a_{42}a_{56})+a_{32}(a_{41}a_{56}-a_{46}a_{51})+a_{36}(a_{42}a_{51}-a_{41}a_{52})]+$$

$$+a_{15}a_{24}a_{33}[a_{41}(a_{56}a_{62}-a_{52}a_{66})+a_{42}(a_{51}a_{66}-a_{56}a_{61})+a_{46}(a_{52}a_{61}-a_{51}a_{62})]+$$
$$+a_{15}a_{24}a_{43}[a_{31}(a_{52}a_{66}-a_{56}a_{62})+a_{32}(a_{56}a_{61}-a_{51}a_{66})+a_{36}(a_{51}a_{62}-a_{52}a_{61})]+$$
$$+a_{15}a_{24}a_{53}[a_{31}(a_{46}a_{62}-a_{42}a_{66})+a_{32}(a_{41}a_{66}-a_{46}a_{61})+a_{36}(a_{42}a_{61}-a_{41}a_{62})]+ \quad |A_{12}|$$
$$+a_{15}a_{24}a_{63}[a_{31}(a_{42}a_{56}-a_{46}a_{52})+a_{32}(a_{46}a_{51}-a_{41}a_{56})+a_{36}(a_{41}a_{52}-a_{42}a_{51})]+$$

$$+a_{15}a_{33}a_{44}[a_{21}(a_{52}a_{66}-a_{56}a_{62})+a_{22}(a_{56}a_{61}-a_{51}a_{66})+a_{26}(a_{51}a_{62}-a_{52}a_{61})]+$$
$$+a_{15}a_{33}a_{54}[a_{21}(a_{46}a_{62}-a_{42}a_{66})+a_{22}(a_{41}a_{66}-a_{46}a_{61})+a_{26}(a_{42}a_{61}-a_{41}a_{62})]+ \quad |A_{13}|$$
$$+a_{15}a_{33}a_{64}[a_{21}(a_{42}a_{56}-a_{46}a_{52})+a_{22}(a_{46}a_{51}-a_{41}a_{56})+a_{26}(a_{41}a_{52}-a_{42}a_{51})]+$$

$$+a_{15}a_{34}a_{43}[a_{21}(a_{56}a_{62}-a_{52}a_{66})+a_{22}(a_{51}a_{66}-a_{56}a_{61})+a_{26}(a_{52}a_{61}-a_{51}a_{62})]+$$
$$+a_{15}a_{34}a_{53}[a_{21}(a_{42}a_{66}-a_{46}a_{62})+a_{22}(a_{46}a_{61}-a_{41}a_{66})+a_{26}(a_{41}a_{62}-a_{42}a_{61})]+ \quad |A_{14}|$$
$$+a_{15}a_{34}a_{63}[a_{21}(a_{46}a_{52}-a_{42}a_{56})+a_{22}(a_{41}a_{56}-a_{46}a_{51})+a_{26}(a_{42}a_{51}-a_{41}a_{52})]+$$

$$+a_{15}a_{43}a_{54}[a_{21}(a_{32}a_{66}-a_{36}a_{62})+a_{22}(a_{36}a_{61}-a_{31}a_{66})+a_{26}(a_{31}a_{62}-a_{32}a_{61})]+$$
$$+a_{15}a_{43}a_{64}[a_{21}(a_{36}a_{52}-a_{32}a_{56})+a_{22}(a_{31}a_{56}-a_{36}a_{51})+a_{26}(a_{32}a_{51}-a_{31}a_{52})]+$$
$$+a_{15}a_{44}a_{53}[a_{21}(a_{36}a_{62}-a_{32}a_{66})+a_{22}(a_{31}a_{66}-a_{36}a_{61})+a_{26}(a_{32}a_{61}-a_{31}a_{62})]+ \quad |A_{15}|$$
$$+a_{15}a_{44}a_{63}[a_{21}(a_{32}a_{56}-a_{36}a_{52})+a_{22}(a_{36}a_{51}-a_{31}a_{56})+a_{26}(a_{31}a_{52}-a_{32}a_{51})]\}+$$

254

$\bullet\bullet$ $+\{a_{23}a_{34}a_{45}[a_{11}(a_{56}a_{62}-a_{52}a_{66})+a_{12}(a_{51}a_{66}-a_{56}a_{61})+a_{16}(a_{52}a_{61}-a_{51}a_{62})]+$
$+a_{23}a_{34}a_{55}[a_{11}(a_{42}a_{66}-a_{46}a_{62})+a_{12}(a_{46}a_{51}-a_{41}a_{66})+a_{16}(a_{41}a_{62}-a_{42}a_{61})]+$ $\quad |A_{16}|$
$+a_{23}a_{34}a_{65}[a_{11}(a_{46}a_{52}-a_{42}a_{56})+a_{12}(a_{41}a_{56}-a_{46}a_{51})+a_{16}(a_{42}a_{61}-a_{41}a_{52})]+$

$+a_{23}a_{35}a_{44}[a_{11}(a_{52}a_{66}-a_{56}a_{62})+a_{12}(a_{56}a_{61}-a_{51}a_{66})+a_{16}(a_{51}a_{62}-a_{52}a_{61})]+$
$+a_{23}a_{35}a_{54}[a_{11}(a_{46}a_{62}-a_{42}a_{66})+a_{12}(a_{41}a_{66}-a_{46}a_{51})+a_{16}(a_{42}a_{61}-a_{41}a_{62})]+$ $\quad |A_{17}|$
$+a_{23}a_{35}a_{64}[a_{11}(a_{42}a_{56}-a_{46}a_{52})+a_{12}(a_{46}a_{51}-a_{41}a_{56})+a_{16}(a_{41}a_{52}-a_{42}a_{61})]+$

$+a_{24}a_{33}a_{45}[a_{11}(a_{52}a_{66}-a_{56}a_{62})+a_{12}(a_{56}a_{61}-a_{51}a_{66})+a_{16}(a_{51}a_{62}-a_{52}a_{61})]+$
$+a_{24}a_{33}a_{55}[a_{11}(a_{46}a_{62}-a_{42}a_{66})+a_{12}(a_{41}a_{66}-a_{46}a_{51})+a_{16}(a_{42}a_{61}-a_{41}a_{62})]+$
$+a_{24}a_{33}a_{65}[a_{11}(a_{42}a_{56}-a_{46}a_{52})+a_{12}(a_{46}a_{51}-a_{41}a_{56})+a_{16}(a_{41}a_{52}-a_{42}a_{61})]+$ $\quad |A_{18}|$

$+a_{24}a_{35}a_{43}[a_{11}(a_{56}a_{62}-a_{52}a_{66})+a_{12}(a_{51}a_{66}-a_{56}a_{61})+a_{16}(a_{52}a_{61}-a_{51}a_{62})]+$
$+a_{24}a_{35}a_{53}[a_{11}(a_{42}a_{66}-a_{46}a_{62})+a_{12}(a_{46}a_{51}-a_{41}a_{66})+a_{16}(a_{41}a_{62}-a_{42}a_{61})]+$ $\quad |A_{19}|$
$+a_{24}a_{35}a_{63}[a_{11}(a_{46}a_{52}-a_{42}a_{56})+a_{12}(a_{41}a_{56}-a_{46}a_{51})+a_{16}(a_{42}a_{61}-a_{41}a_{52})]+$

$+a_{25}a_{33}a_{44}[a_{11}(a_{56}a_{62}-a_{52}a_{66})+a_{12}(a_{51}a_{66}-a_{56}a_{61})+a_{16}(a_{52}a_{61}-a_{51}a_{62})]+$
$+a_{25}a_{33}a_{54}[a_{11}(a_{42}a_{66}-a_{46}a_{62})+a_{12}(a_{46}a_{51}-a_{41}a_{66})+a_{16}(a_{41}a_{62}-a_{42}a_{61})]+$ $\quad |A_{20}|$
$+a_{25}a_{33}a_{64}[a_{11}(a_{46}a_{52}-a_{42}a_{56})+a_{12}(a_{41}a_{56}-a_{46}a_{51})+a_{16}(a_{42}a_{61}-a_{41}a_{52})]+$

$+a_{25}a_{34}a_{43}[a_{11}(a_{52}a_{66}-a_{56}a_{62})+a_{12}(a_{56}a_{61}-a_{51}a_{66})+a_{16}(a_{51}a_{62}-a_{52}a_{61})]+$
$+a_{25}a_{34}a_{53}[a_{11}(a_{46}a_{62}-a_{42}a_{66})+a_{12}(a_{41}a_{66}-a_{46}a_{51})+a_{16}(a_{42}a_{61}-a_{41}a_{62})]+$ $\quad |A_{21}|$
$+a_{25}a_{34}a_{63}[a_{11}(a_{42}a_{56}-a_{46}a_{52})+a_{12}(a_{46}a_{51}-a_{41}a_{56})+a_{16}(a_{41}a_{52}-a_{42}a_{61})]\}+$

$\bullet\bullet$ $+\{a_{33}a_{44}a_{55}[a_{11}(a_{22}a_{66}-a_{26}a_{62})+a_{12}(a_{26}a_{61}-a_{21}a_{66})+a_{16}(a_{21}a_{62}-a_{22}a_{61})]+$
$+a_{33}a_{44}a_{65}[a_{11}(a_{26}a_{52}-a_{22}a_{56})+a_{12}(a_{21}a_{56}-a_{26}a_{51})+a_{16}(a_{22}a_{51}-a_{21}a_{52})]+$ $\quad |A_{22}|$
$+a_{33}a_{45}a_{54}[a_{11}(a_{26}a_{62}-a_{22}a_{66})+a_{12}(a_{21}a_{66}-a_{26}a_{61})+a_{16}(a_{22}a_{61}-a_{21}a_{62})]+$

$+a_{33}a_{45}a_{64}[a_{11}(a_{22}a_{56}-a_{26}a_{52})+a_{12}(a_{26}a_{51}-a_{21}a_{56})+a_{16}(a_{21}a_{52}-a_{22}a_{51})]+$
$+a_{33}a_{54}a_{65}[a_{11}(a_{22}a_{46}-a_{26}a_{42})+a_{12}(a_{26}a_{41}-a_{21}a_{46})+a_{16}(a_{21}a_{42}-a_{22}a_{41})]+$ $\quad |A_{23}|$
$+a_{33}a_{55}a_{64}[a_{11}(a_{26}a_{42}-a_{22}a_{46})+a_{12}(a_{21}a_{46}-a_{26}a_{41})+a_{16}(a_{22}a_{41}-a_{21}a_{42})]+$

$+a_{34}a_{43}a_{55}[a_{11}(a_{26}a_{62}-a_{22}a_{66})+a_{12}(a_{21}a_{66}-a_{26}a_{61})+a_{16}(a_{22}a_{61}-a_{21}a_{62})]+$
$+a_{34}a_{43}a_{65}[a_{11}(a_{22}a_{56}-a_{26}a_{52})+a_{12}(a_{26}a_{51}-a_{21}a_{56})+a_{16}(a_{21}a_{52}-a_{22}a_{51})]+$ $\quad |A_{24}|$
$+a_{34}a_{45}a_{53}[a_{11}(a_{22}a_{66}-a_{26}a_{62})+a_{12}(a_{26}a_{61}-a_{21}a_{66})+a_{16}(a_{21}a_{62}-a_{22}a_{61})]+$

$+a_{34}a_{45}a_{63}[a_{11}(a_{26}a_{52}-a_{22}a_{56})+a_{12}(a_{21}a_{56}-a_{26}a_{51})+a_{16}(a_{22}a_{51}-a_{21}a_{52})]+$
$+a_{34}a_{53}a_{65}[a_{11}(a_{26}a_{42}-a_{22}a_{46})+a_{12}(a_{21}a_{46}-a_{26}a_{41})+a_{16}(a_{22}a_{41}-a_{21}a_{42})]+$ $\quad |A_{25}|$
$+a_{34}a_{55}a_{63}[a_{11}(a_{22}a_{46}-a_{26}a_{42})+a_{12}(a_{26}a_{41}-a_{21}a_{46})+a_{16}(a_{21}a_{42}-a_{22}a_{41})]+$

$+a_{35}a_{43}a_{54}[a_{11}(a_{22}a_{66}-a_{26}a_{62})+a_{12}(a_{26}a_{61}-a_{21}a_{66})+a_{16}(a_{21}a_{62}-a_{22}a_{61})]+$
$+a_{35}a_{43}a_{64}[a_{11}(a_{26}a_{52}-a_{22}a_{56})+a_{12}(a_{21}a_{56}-a_{26}a_{51})+a_{16}(a_{22}a_{51}-a_{21}a_{52})]+$ $\quad |A_{26}|$
$+a_{35}a_{44}a_{53}[a_{11}(a_{26}a_{62}-a_{22}a_{66})+a_{12}(a_{21}a_{66}-a_{26}a_{61})+a_{16}(a_{22}a_{61}-a_{21}a_{62})]+$

$$+a_{35}a_{44}a_{63}[a_{11}(a_{22}a_{56}-a_{26}a_{52})+a_{12}(a_{26}a_{51}-a_{21}a_{56})+a_{16}(a_{21}a_{52}-a_{22}a_{51})]+$$
$$+a_{35}a_{53}a_{64}[a_{11}(a_{22}a_{46}-a_{26}a_{42})+a_{12}(a_{26}a_{41}-a_{21}a_{46})+a_{16}(a_{21}a_{42}-a_{22}a_{41})]+ \qquad |A_{27}|$$
$$+a_{35}a_{54}a_{63}[a_{11}(a_{26}a_{42}-a_{22}a_{46})+a_{12}(a_{21}a_{46}-a_{26}a_{41})+a_{16}(a_{22}a_{41}-a_{21}a_{42})]+$$

$$+a_{43}a_{54}a_{65}[a_{11}(a_{26}a_{32}-a_{22}a_{36})+a_{12}(a_{21}a_{36}-a_{26}a_{31})+a_{16}(a_{22}a_{31}-a_{21}a_{32})]+$$
$$+a_{43}a_{55}a_{64}[a_{11}(a_{22}a_{36}-a_{26}a_{32})+a_{12}(a_{26}a_{31}-a_{21}a_{36})+a_{16}(a_{21}a_{32}-a_{22}a_{31})]+ \qquad |A_{28}|$$
$$+a_{44}a_{53}a_{65}[a_{11}(a_{22}a_{36}-a_{26}a_{32})+a_{12}(a_{26}a_{31}-a_{21}a_{36})+a_{16}(a_{21}a_{32}-a_{22}a_{31})]+$$

$$+a_{44}a_{55}a_{63}[a_{11}(a_{26}a_{32}-a_{22}a_{36})+a_{12}(a_{21}a_{36}-a_{26}a_{31})+a_{16}(a_{22}a_{31}-a_{21}a_{32})]+$$
$$+a_{45}a_{53}a_{64}[a_{11}(a_{26}a_{32}-a_{22}a_{36})+a_{12}(a_{21}a_{36}-a_{26}a_{31})+a_{16}(a_{22}a_{31}-a_{21}a_{32})]+ \qquad |A_{29}|$$
$$+a_{45}a_{54}a_{63}[a_{11}(a_{22}a_{36}-a_{26}a_{32})+a_{12}(a_{26}a_{31}-a_{21}a_{36})+a_{16}(a_{21}a_{32}-a_{22}a_{31})]\}$$

$$(117)$$

In Appendix 28 is shown evaluation of the determinants of the 6x6 matrices and also solving of the system six normal equations with using of Laplace's theorem for the above-indicated conditions and Cramer's rule.

So, we can conclude that combined summary determinant $\sum|A|$ of the 6x6 matrix applicably to Laplace's theorem, for a instance , *at marked out jointly the third, fourth and fifth columns,* in general view is equal:

$$\sum|A|=|A_1|+|A_2|+|A_3|+|A_4|+\ldots\ldots|A_{29}| \qquad (117a)$$

A. Analysis of combined partial determinant $|A_1|$ of the 6x6 matrix:

❖ *For Figure 163 of the combined partial determinant $|A_1|$*

- The combined partial determinant $|A_1|$ in general view includes the third element (a_{13}) of the first row and the fourth element (a_{24}) of the second row which are joined between itself with frequency of tie equal f=1;
- The fourth element (a_{24}) of the second row has the functional tie with each element ($a_{35},a_{45},a_{55},a_{65}$)of the fifth column ,forming the frequency of tie equal f=1 for above-indicated elements ($a_{35},a_{45},a_{55},a_{65}$) for positive area.

For elements of the fifth column

- The third element(a_{35}) of the fifth column has the functional ties with elements (a_{41},a_{42},a_{46}) of the fourth row;
- The fourth element (a_{45}) of the fifth column has the functional ties with elements (a_{31},a_{32},a_{36}) of the third row;
- The fifth element (a_{55}) of the fifth column has the functional ties with elements(a_{31},a_{32},a_{36})of the third row;
- The sixth element (a_{65}) of the fifth column has the functional ties with elements (a_{31},a_{32},a_{36})of the third row, forming the frequency of ties equal f=1 for above-indicated elements ($a_{35},a_{45},a_{55},a_{65}$)in such sequence as for positive area:

$$a_{35}\rightarrow a_{41},a_{42},a_{46}$$
$$a_{45}\rightarrow a_{31},a_{32},a_{36}$$
$$a_{55}\rightarrow a_{31},a_{32},a_{36}$$
$$a_{65}\rightarrow a_{31},a_{32},a_{36}$$

For elements of the third row

- The first element(a_{31}) of the third row has the functional ties with the second and sixth elements (a_{42}, a_{42},a_{46},a_{46}) of the fourth row and sixth element (a_{56}) of the fifth row for positive (———) and negative(-------) areas;
- The second element (a_{32}) of the third row has the functional ties with the first and sixth elements (a_{41},a_{41},a_{46},a_{46}) of the fourth row and the first element (a_{51}) of the fifth row for positive and negative areas;
- The sixth element (a_{36}) has the functional ties with the first and second elements (a_{41},a_{41},a_{42},a_{42})of the fourth row and the second element(a_{52}) of the fifth row for positive and negative areas, forming for above-indicated elements (a_{31},a_{32},a_{36})the frequency of functional ties equal f=3,besides of element(a_{35}) in such sequence as:

$$a_{31} \rightarrow a_{42},a_{42},a_{46},a_{46},a_{56}$$
$$a_{32} \rightarrow a_{41},a_{41},a_{46},a_{46},a_{51}$$
$$a_{36} \rightarrow a_{41},a_{41},a_{42},a_{42},a_{52}$$

For elements of the fourth row

- The first element(a_{41})of the fourth row has the functional ties with the second(a_{52},a_{52}) and sixth(a_{56},a_{56}) elements of the fifth row and the second(a_{62}) and sixth(a_{66})elements of the sixth row for positive and negative areas;
- The second element(a_{42}) of the fourth row has the functional ties with the first (a_{51},a_{51})and sixth(a_{56},a_{56}) elements of the fifth row and the first (a_{61}) and sixth(a_{66}) elements of the sixth row for positive and negative areas;
- The sixth element(a_{46}) of the fourth row has the functional ties with the first(a_{51},a_{51}) and second(a_{52},a_{52}) elements of the fifth row and the first(a_{61}) and second(a_{62}) elements of the sixth row for positive and negative areas, forming the frequency of functional ties equal f=5 for above-indicated elements (a_{41},a_{42},a_{46}),besides of element (a_{45}),in such sequence as:

$$a_{41} \rightarrow a_{52},a_{52},a_{56},a_{56},a_{62},a_{66}$$
$$a_{42} \rightarrow a_{51},a_{51},a_{56},a_{56},a_{61},a_{66}$$
$$a_{46} \rightarrow a_{51},a_{51},a_{52},a_{52},a_{61},a_{62}$$

For elements of the fifth row

- The first element(a_{51}) of the fifth row has the functional ties with the second(a_{62},a_{62}) and sixth(a_{66},a_{66}) elements of the sixth row for positive and negative areas;
- The second element(a_{52}) of the fifth row has the functional ties with the first(a_{61},a_{61}) and sixth(a_{66},a_{66}) elements of the sixth row for positive and negative areas;
- The sixth element (a_{56}) of the fifth row has the functional ties with the first (a_{61},a_{61}) and second (a_{62},a_{62}) elements of the sixth row for positive and negative areas, forming the frequency of functional ties equal f=6 for above-indicated elements (a_{51},a_{52},a_{56}) of the fifth row and elements (a_{61},a_{62},a_{66}), besides of elements(a_{55},a_{65}) ,in such sequence as:

$$a_{51} \rightarrow a_{62},a_{62},a_{66},a_{66}$$
$$a_{52} \rightarrow a_{61},a_{61},a_{66},a_{66}$$
$$a_{56} \rightarrow a_{61},a_{61},a_{62},a_{62}$$

In Figure 163 is shown the functional ties between elements for combined partial determinant $|A_1|$ of the 6x6 matrix applicably to Laplace's theorem *at marked out jointly the third, fourth and fifth columns* for positive and negative areas.

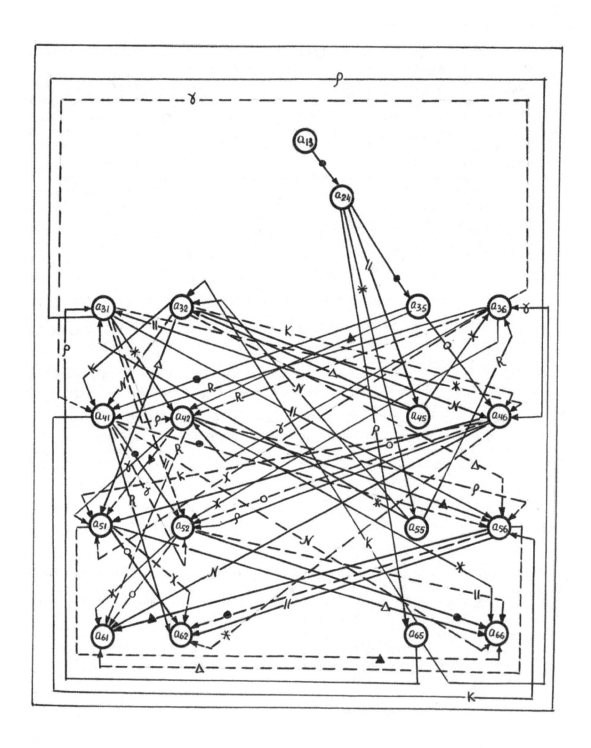

Figure 163 The functional ties between elements for combined partial determinant $|A_1|$ of the 6x6 matrix applicably to Laplace's theorem at marked out jointly the third, fourth and fifth columns for positive(——) and negative (--------) areas

258

B. Analysis of combined partial determinant $\left|A_2\right|$ of the 6x6 matrix:

❖ *For Figure 164 of the combined partial determinant $\left|A_2\right|$*

- Combined partial determinant $\left|A_2\right|$ in general view includes the third element (a_{13}) of the first row and the fifth element (a_{25}) of the second row which are joined with frequency of tie equal f=1;
- The fifth element (a_{25}) of the second row has the functional ties with elements $(a_{34},a_{44},a_{54},a_{64})$ of the fourth column, forming the frequency of tie equal f=1 for above-indicated elements.

For elements of the fourth column

- The third element (a_{34}) of the fourth column has the functional ties with elements (a_{41},a_{42},a_{46}) of the fourth row;
- The fourth element (a_{44}) of the fourth column has the functional ties with elements (a_{31},a_{32},a_{36}) of the third row;
- The fifth element (a_{54}) of the fourth column has the functional ties with elements (a_{31},a_{32},a_{36}) of the third row;
- The sixth element (a_{64}) of the fourth column has the functional ties with elements (a_{31},a_{32},a_{36}) of the third row in the following sequence, forming the frequency of tie equal f=1 for elements $(a_{34},a_{44},a_{54},a_{64})$ in such sequence as:

$$a_{34} \rightarrow a_{41},a_{42},a_{46}$$
$$a_{44} \rightarrow a_{31},a_{32},a_{36}$$
$$a_{54} \rightarrow a_{31},a_{32},a_{36}$$
$$a_{64} \rightarrow a_{31},a_{32},a_{36}$$

For elements of the third row

- The first element (a_{31}) of the third row has the functional ties with the second and sixth elements $(a_{42},a_{42},a_{46},a_{46})$ of the fourth row and sixth element (a_{56}) of the fifth row for positive and negative areas;
- The second element (a_{32}) has the functional ties with the first and sixth elements $(a_{41},a_{41},a_{46},a_{46})$ of the fourth row and the first element (a_{51}) of the fifth row for positive and negative areas;
- The sixth element (a_{36}) has the functional ties with the first and second $(a_{41},a_{41},a_{42},a_{42})$ elements of the fourth row and the second element (a_{52}) of the fifth row for positive and negative areas, forming for elements (a_{31},a_{32},a_{36}) the frequency of functional ties equal f=3, besides of element (a_{34}), in such sequence as:

$$a_{31} \rightarrow a_{42},a_{42},a_{46},a_{46},a_{56}$$
$$a_{32} \rightarrow a_{41},a_{41},a_{46},a_{46},a_{51}$$
$$a_{36} \rightarrow a_{41},a_{41},a_{42},a_{42},a_{52}$$

For elements of the fourth row

- The first element (a_{41}) of the fourth row has the functional ties with the second and sixth $(a_{52},a_{52},a_{56},a_{56})$ of the fifth row and the sixth (a_{66}) element of the sixth row for positive and negative areas;
- The second element (a_{42}) of the fourth row has the functional ties with the first and sixth $(a_{51},a_{51},a_{56},a_{56})$ elements of the fifth row and the first (a_{61}) and sixth (a_{66}) elements of the sixth row for positive and negative areas;
- The sixth element (a_{46}) of the fourth row has the functional ties with the first and second $(a_{51},a_{51},a_{52},a_{52})$ elements of the fifth row and the first (a_{61}) and second (a_{62}) elements of the sixth row for positive and negative areas, forming the frequency of functional ties equal f=5 for

for above-indicated elements (a_{41}, a_{42}, a_{46}),besides of element (a_{45}) in such sequence as:

$$a_{41} \rightarrow a_{52}, a_{52}, a_{56}, a_{56}, a_{66}$$
$$a_{42} \rightarrow a_{51}, a_{51}, a_{56}, a_{56}, a_{66}$$
$$a_{46} \rightarrow a_{51}, a_{51}, a_{52}, a_{52}, a_{62}$$

For elements of the fifth row

- The first element(a_{51}) of the fifth row has the functional ties with the second(a_{62}, a_{62}) and sixth(a_{66}, a_{66}) elements of the sixth row for positive and negative areas;
- The second element(a_{52})of the fifth row has the functional ties with the first and sixth($a_{61}, a_{61}, a_{66}, a_{66}$) elements of the sixth row for positive and negative areas;
- The sixth element(a_{56}) of the fifth row has the functional ties with the first and second($a_{61}, a_{61}, a_{62}, a_{62}$) elements of the sixth row for positive and negative areas, forming the frequency of functional ties equal f=6 for above-indicated elements(a_{51}, a_{52}, a_{56})of the fifth row and elements(a_{61}, a_{62}, a_{66}) of the sixth row ,besides of elements (a_{54}, a_{64}) in such sequence as:

$$a_{51} \rightarrow a_{62}, a_{62}, a_{66}, a_{66}$$
$$a_{52} \rightarrow a_{61}, a_{61}, a_{66}, a_{66}$$
$$a_{56} \rightarrow a_{61}, a_{61}, a_{62}, a_{62}$$

In Figure 164 is shown the functional ties for combined partial determinant $\left| A_2 \right|$ of the 6x6 matrix.

C. Analysis of combined partial determinant $\left| A_3 \right|$ of the 6x6 matrix

❖ *For Figure 165 of the combined partial determinant $\left| A_3 \right|$:*

- Combined partial determinant $\left| A_3 \right|$ in general view includes the third element(a_{13}) of the first row and the fourth element(a_{34}) of the third row which are joined each other ,forming the frequency of tie equal f=1;
- The fourth element(a_{34}) of the third row has functional tie with each element(a_{45}, a_{55}, a_{65}) of the fifth column, forming the frequency of tie equal f=1 for above-indicated elements.

For elements of the fifth column

- The fourth element (a_{45}) of the fifth column has the functional ties with elements(a_{21}, a_{22}, a_{26})of the second row;
- The fifth element(a_{55})of the fifth column has the functional ties with elements (a_{21}, a_{22}, a_{26})of the second row;
- The sixth element (a_{65}) of the fifth column has the functional ties with elements (a_{21}, a_{22}, a_{26}) of the second row in the following sequence, forming the frequency of tie equal f=1 for elements (a_{45}, a_{55}, a_{65}):

$$a_{45} \rightarrow a_{21}, a_{22}, a_{26}$$
$$a_{55} \rightarrow a_{21}, a_{22}, a_{26}$$
$$a_{65} \rightarrow a_{21}, a_{22}, a_{26}$$

For elements of the second row

- The first element(a_{21}) of the second row has the functional ties with the second and sixth ($a_{42}, a_{42}, a_{46}, a_{46}$) of the fourth row and sixth element(a_{56}) of the fifth row for positive and negative areas;
- The second element(a_{22}) of the second row has the functional ties with the first and sixth($a_{41}, a_{41}, a_{46}, a_{46}$) elements of the fourth row and the first element (a_{51}) of the fifth row for positive and negative areas;

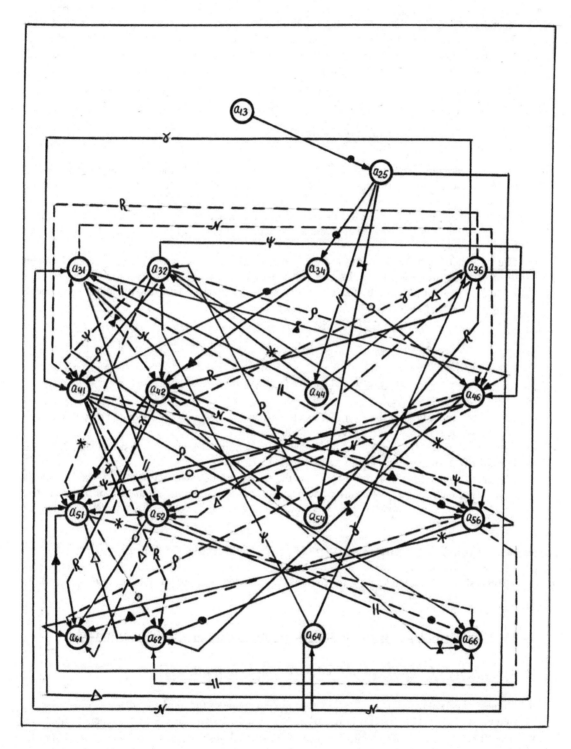

Figure 164 The functional ties between elements for combined partial determinant $|A_2|$ of the 6x6 matrix applicably to Laplace's theorem at marked out jointly the third, fourth and fifth columns for positive(———) and negative (-------) areas

261

- The sixth element(a_{26}) of the second row has the functional ties with the first and second($a_{41},a_{41},a_{42},a_{42}$) elements of the fourth row and the second element (a_{52})of the fifth row for positive and negative areas, forming for elements (a_{21},a_{22},a_{26}) the frequency of functional ties equal f=3 in such sequence as:

$$a_{21} \rightarrow a_{42},a_{42},a_{46},a_{46},a_{56}$$
$$a_{22} \rightarrow a_{41},a_{41},a_{46},a_{46},a_{51}$$
$$a_{26} \rightarrow a_{41},a_{41},a_{42},a_{42},a_{52}$$

For elements of the fourth row

- The first element(a_{41}) of the fourth row has the functional ties with the second and sixth elements (a_{52},a_{56})of the fifth row and the second and sixth elements (a_{62},a_{66}) of the sixth row for positive and negative areas;
- The second element(a_{42}) of the fourth row has the functional ties with the first and sixth elements (a_{51},a_{56}) of the fifth row and the first and sixth elements (a_{61},a_{66}) of the sixth row for positive and negative areas;
- The sixth element(a_{46}) of the fourth row has the functional ties with the first and second elements (a_{51},a_{52}) of the fifth row and the first and second elements (a_{61},a_{62}) of the sixth row for positive and negative areas, forming the frequency of functional ties equal f=4 for above- indicated elements (a_{41},a_{42},a_{46}) in such sequence as:

$$a_{41} \rightarrow a_{52},a_{56},a_{62},a_{66}$$
$$a_{42} \rightarrow a_{51},a_{56},a_{61},a_{66}$$
$$a_{46} \rightarrow a_{51},a_{52},a_{61},a_{62}$$

For elements of the fifth row

- The first element (a_{51}) of the fifth row has the functional ties with the second and sixth elements (a_{62},a_{66}) of the sixth row for positive and negative areas;
- The second element(a_{52}) of the fifth row has the functional ties with the first and sixth elements (a_{61},a_{66}) of the sixth row for positive and negative areas;
- The sixth element(a_{56}) of the fifth row has the functional ties with the first(a_{61}) and second(a_{62}) elements of the sixth row for positive and negative areas, forming the frequency of functional ties equal f=4 for above-indicated elements (a_{51},a_{52},a_{56})of the fifth row and elements (a_{61},a_{62},a_{66}) in such sequence as:

$$a_{51} \rightarrow a_{62},a_{66}$$
$$a_{52} \rightarrow a_{61},a_{66}$$
$$a_{56} \rightarrow a_{61},a_{62}$$

In Figure 165 is shown the functional ties for combined partial determinant $\left| A_3 \right|$ of the 6x6 matrix.

D. Analysis of combined partial determinant $\left| A_4 \right|$ of the 6x6 matrix :

❖ *For Figure 166 of the combined partial determinant $\left| A_4 \right|$:*

- Combined partial determinant $\left| A_4 \right|$ in general view includes the third element(a_{13}) of the first row and the fifth element(a_{35}) of the third row which have the functional ties with frequency equal f=1;
- The fifth element(a_{35}) of the third row has the functional ties with each elements(a_{44},a_{54},a_{64}) of the fourth column, forming the frequency of tie equal f=1 for above-indicated elements.

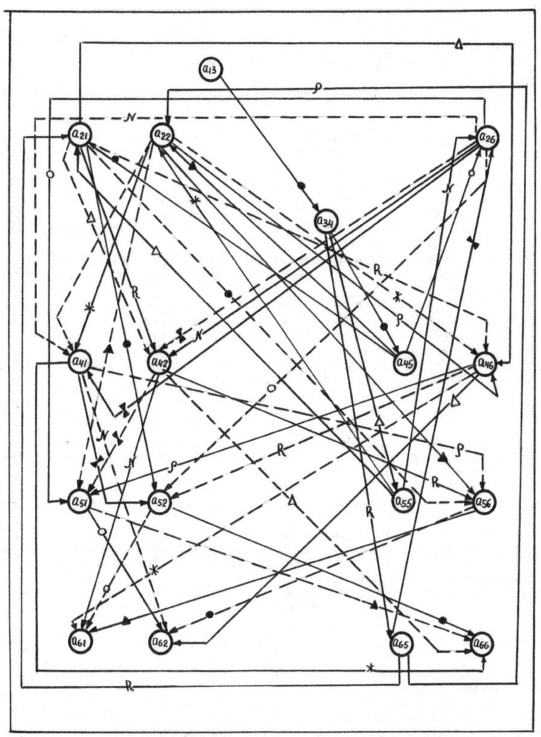

Figure 165 The functional ties between elements for combined partial determinant $|A_3|$ **of the 6x6 matrix applicably to Laplace's theorem at marked out jointly the third, fourth and fifth columns for positive(———) and negative(------) areas**

263

For elements of the fourth column

- The fourth element(a_{44})of the fourth column has the functional ties with elements (a_{21}, a_{22}, a_{26}) of the second row;
- The fifth element(a_{54}) of the fourth column has the functional ties with elements (a_{21}, a_{22}, a_{26})of the second row;
- The sixth element(a_{64}) of the fourth column has the functional ties with elements (a_{21}, a_{22}, a_{26}) of the second row in the following sequence, forming the frequency of tie equal f=1 for elements (a_{44}, a_{54}, a_{64}):

$$a_{44} \rightarrow a_{21}, a_{22}, a_{26}$$
$$a_{54} \rightarrow a_{21}, a_{22}, a_{26}$$
$$a_{64} \rightarrow a_{21}, a_{22}, a_{26}$$

For elements of the second row

- The first element(a_{21}) of the second row has the functional ties with the second(a_{42}, a_{42}) and sixth(a_{46}, a_{46}) elements of the fourth row and sixth(a_{56})element of the fifth row for positive and negative areas;
- The second element(a_{22}) of the second row has the functional ties with the first(a_{41}, a_{41}) and sixth(a_{46}, a_{46}) elements of the fourth row and the first element(a_{51}) of the fifth row for positive and negative areas;
- The sixth element(a_{26})of the second row has the functional ties with the first(a_{41}, a_{41}) and second(a_{42}, a_{42}) elements of the fourth row and the second element (a_{52}) of the fifth row for positive and negative areas, forming for elements (a_{21}, a_{22}, a_{26}) the frequency of functional ties equal f=3 in such sequence as:

$$a_{21} \rightarrow a_{42}, a_{42}, a_{46}, a_{46}, a_{56}$$
$$a_{22} \rightarrow a_{41}, a_{41}, a_{46}, a_{46}, a_{51}$$
$$a_{26} \rightarrow a_{41}, a_{41}, a_{42}, a_{42}, a_{52}$$

For elements of the fourth row

- The first element(a_{41})of the fourth row has the functional ties with the second(a_{52}) and sixth(a_{56}) elements of the fifth row and the second(a_{62}) and sixth(a_{66}) elements of the sixth row for positive and negative areas;
- The second element(a_{42}) of the fourth row has the functional ties with the first(a_{51}) and sixth(a_{56}) elements of the fifth row and the first(a_{61}) and sixth(a_{66}) elements of the sixth row for positive and negative areas;
- The sixth element(a_{46})of the fourth row has the functional ties with the first(a_{51}) and second(a_{52}) elements of the fifth row and the first(a_{61}) and second(a_{62}) elements of the sixth row for positive and negative areas, forming the frequency of functional ties equal f=4 for above-indicated elements (a_{41}, a_{42}, a_{46}) in such sequence as:

$$a_{41} \rightarrow a_{52}, a_{56}, a_{62}, a_{66}$$
$$a_{42} \rightarrow a_{51}, a_{56}, a_{61}, a_{66}$$
$$a_{46} \rightarrow a_{51}, a_{52}, a_{61}, a_{62}$$

For elements of the fifth row

- The first element(a_{51}) of the fifth row has the functional ties with the second(a_{62}) and sixth(a_{66}) elements of the sixth row for positive and negative areas;
- The second element(a_{52}) of the fifth row has the functional ties with the first(a_{61}) and sixth(a_{66}) elements of the sixth row for positive and negative areas;
- The sixth element(a_{56}) of the fifth row has the functional ties with the first(a_{61}) and second (a_{62}) elements of the sixth row for positive and negative areas, forming the frequency of functional ties equal f=4 for above-indicated elements(a_{51}, a_{52}, a_{56}) of the fifth row and elements (a_{61}, a_{62}, a_{66}) in such sequence as :

$$a_{51} \rightarrow a_{62}, a_{66}$$
$$a_{52} \rightarrow a_{61}, a_{66}$$
$$a_{56} \rightarrow a_{61}, a_{62}$$

In **Figure 166** is shown the functional ties for combined partial determinant $\left|A_4\right|$ of the 6x6 matrix.

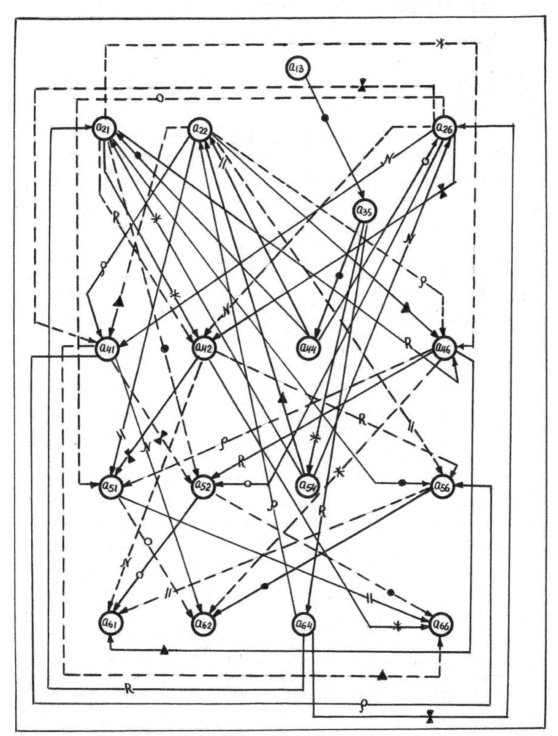

Figure 166 The functional ties between elements for combined partial determinant $\left|A_4\right|$ **of the 6x6 matrix applicably to Laplace's theorem at marked out jointly the third, fourth and fifth columns for positive(———) and negative(------) areas**

E. Analysis of combined partial determinant $|A_5|$ of the 6x6 matrix:

❖ *For Figure 167 of the combined partial determinant $|A_5|$*

- Combined partial determinant $|A_5|$ in general view includes the third element(a_{13}) of the first row and the fourth (a_{44}) and fifth (a_{45}) elements of the fourth row which is joined with these elements, forming the functional tie equal f=1;

For elements of the fourth and fifth columns

- The fourth element(a_{44}) of the fourth row is joined with the fifth(a_{55}) and sixth(a_{65}) elements of the fifth column ,forming the functional ties for elements(a_{55},a_{65}) the frequency equal f=1;
- The fifth element(a_{45})of the fourth row is joined with the fifth (a_{54}) and sixth(a_{64}) elements of the fourth column, forming the functional ties for elements (a_{54},a_{64}) the frequency equal f=1 in such sequence as:

$$a_{44} \rightarrow a_{55}, a_{65}$$
$$a_{45} \rightarrow a_{54}, a_{64}$$

- The fifth (a_{54}) element of the fourth column is joined with elements(a_{21},a_{22},a_{26}) of the second row;
- The sixth(a_{64}) element of the fourth column is joined with elements (a_{21},a_{22},a_{26}) of the second row;
- The fifth(a_{55}) element of the fifth column is joined with elements (a_{21},a_{22},a_{26}) of the second row, forming the functional ties for elements (a_{21},a_{22},a_{26}) with frequency equal f=3 in such sequence of elements as:

$$a_{54} \rightarrow a_{21}, a_{22}, a_{26}$$
$$a_{64} \rightarrow a_{21}, a_{22}, a_{26}$$
$$a_{55} \rightarrow a_{21}, a_{22}, a_{26}$$

For elements of the second row

- The first element(a_{21}) of the second row is joined with elements (a_{32},a_{32},a_{32},a_{32},a_{36},a_{36},a_{36},a_{36}) of the third row for positive and negative areas;
- The second element(a_{22}) of the second row is joined with elements (a_{31},a_{31},a_{31},a_{31},a_{36},a_{36},a_{36},a_{36}) of the third row for positive and negative areas;
- The sixth element(a_{26})of the second row is joined with elements(a_{31},a_{31},a_{31},a_{31},a_{32},a_{32},a_{32},a_{32}) of the third row for positive and negative areas, forming the frequency of functional ties equal f=4 for above-indicated elements (a_{21},a_{22},a_{26}) in such sequence as:

$$a_{21} \rightarrow a_{32}, a_{32}, a_{32}, a_{32}, a_{36}, a_{36}, a_{36}, a_{36}$$
$$a_{22} \rightarrow a_{31}, a_{31}, a_{31}, a_{31}, a_{36}, a_{36}, a_{36}, a_{36}$$
$$a_{26} \rightarrow a_{31}, a_{31}, a_{31}, a_{31}, a_{32}, a_{32}, a_{32}, a_{32}$$

For elements of the third row

- The first element(a_{31}) of the third row is joined with elements (a_{52},a_{52},a_{56},a_{56}) of the fifth row and elements (a_{62},a_{62},a_{66},a_{66}) of the sixth row for positive and negative areas;
- The second element (a_{32}) of the third row is joined with elements (a_{51},a_{51},a_{56},a_{56}) of the fifth row and elements (a_{61},a_{61},a_{66},a_{66}) of the sixth row for positive and negative areas;
- The sixth element (a_{36}) of the third row is joined with elements (a_{51},a_{51},a_{52},a_{52}) of the fifth row and elements (a_{61},a_{61},a_{62},a_{62}) of the sixth row for positive and negative areas , forming the frequency of ties equal f=8 for above-indicated elements (a_{31},a_{32},a_{36}) of the third row in such sequence as:

$$a_{31} \rightarrow a_{52}, a_{52}, a_{56}, a_{56}, a_{62}, a_{62}, a_{66}, a_{66}$$
$$a_{32} \rightarrow a_{51}, a_{51}, a_{56}, a_{56}, a_{61}, a_{61}, a_{66}, a_{66}$$
$$a_{36} \rightarrow a_{51}, a_{51}, a_{52}, a_{52}, a_{61}, a_{61}, a_{62}, a_{62}$$

and frequency of functional ties for elements (a_{51},a_{52},a_{56}) of the fifth row and elements (a_{61},a_{62},a_{66}) of the fifth row equal f=4.

In Figure 167 is shown the functional ties for combined partial determinant $|A_5|$ of the 6x6 matrix.

266

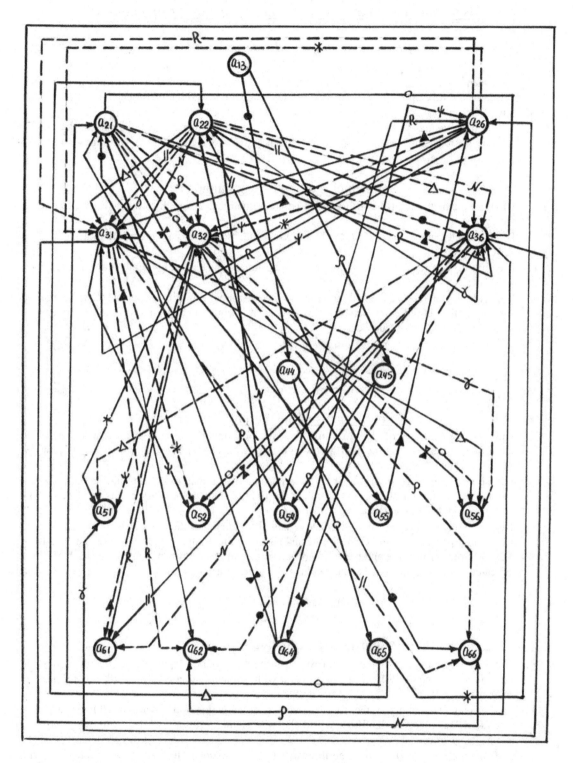

Figure 167 The functional ties between elements for combined partial determinant $|A_5|$ of the 6x6 matrix applicably to Laplace's theorem at marked out jointly the third, fourth and fifth columns for positive(———) and negative(--------) areas

267

F. **Analysis of combined partial determinant $|A_6|$ of the 6x6 matrix:**

❖ *For Figure 168 of the combined partial determinant $|A_6|$*

- Combined partial determinant $|A_6|$ in general view includes the fourth element(a_{14})of the first row and the third element(a_{23}) of the second row which are joined with frequency of functional tie equal f=1;
- The third element(a_{23}) of the second row has the functional ties with each element($a_{35},a_{45},a_{55},a_{65}$) of the fifth column, forming the frequency of tie equal f=1 for above-indicated elements($a_{35},a_{45},a_{55},a_{65}$).

For elements of the fifth column

- The third element(a_{35}) of the fifth column has the functional ties with elements(a_{41},a_{42},a_{46}) of the fourth row;
- The fourth element(a_{45}) of the fifth column has the functional ties with elements(a_{31},a_{32},a_{36}) of the third row;
- The fifth element(a_{55})of the fifth column has the functional ties with elements(a_{31},a_{32},a_{36}) of the third row;
- The sixth element(a_{65}) of the fifth column has the functional ties with elements (a_{31},a_{32},a_{36}) of the third row, forming the frequency of functional ties equal f=1 for above-indicated elements ($a_{35},a_{45},a_{55},a_{65}$) in such sequence as:

$$a_{35} \rightarrow a_{41},a_{42},a_{46}$$
$$a_{45} \rightarrow a_{31},a_{32},a_{36}$$
$$a_{55} \rightarrow a_{31},a_{32},a_{36}$$
$$a_{65} \rightarrow a_{31},a_{32},a_{36}$$

For elements of the third row

- The first element(a_{31}) of the third row has the functional ties with the second(a_{42},a_{42}) and sixth(a_{46},a_{46}) elements of the fourth row and also with elements (a_{52},a_{56}) of the fifth row for positive and negative areas;
- The second element(a_{32}) of the third row has the functional ties with the first(a_{41},a_{41}) and sixth(a_{46},a_{46}) of the fourth row and also with the first(a_{51}) and sixth(a_{56}) elements of the fifth row for positive and negative areas;
- The sixth element(a_{36}) of the third row has the functional ties with the first(a_{41},a_{41}) and second(a_{42},a_{42}) elements of the fourth row and also with the first(a_{51}) and second (a_{52}) elements of the fifth row for positive and negative areas, forming the frequency of functional ties for above-indicated elements (a_{31},a_{32},a_{36}) equal f=3 in such sequence as:

$$a_{31} \rightarrow a_{42},a_{42},a_{46},a_{46},a_{52},a_{56}$$
$$a_{32} \rightarrow a_{41},a_{41},a_{46},a_{46},a_{51},a_{56}$$
$$a_{36} \rightarrow a_{41},a_{41},a_{42},a_{42},a_{51},a_{52}$$

For elements of the fourth row

- The first element (a_{41}) of the fourth row has the functional ties with the second(a_{52},a_{52}) and sixth(a_{56},a_{56}) elements of the fifth row and also with elements (a_{62},a_{66}) of the sixth row for positive and negative areas;
- The second element(a_{42})of the fourth row has the functional ties with the first(a_{51},a_{51}) and sixth(a_{56},a_{56}) elements of the fifth row and also with elements (a_{61},a_{66}) of the sixth row for positive and negative areas;
- The sixth element(a_{46}) of the fourth row has the functional ties with the first(a_{51},a_{51}) and second(a_{52},a_{52}) elements of the fifth row and also with elements (a_{61},a_{62}) of the sixth row for positive and negative areas, forming the frequency of functional ties for above-indicated elements (a_{41},a_{42},a_{46})equal f=5 in such sequence as:

$$a_{41} \rightarrow a_{52}, a_{52}, a_{56}, a_{56}, a_{62}, a_{66}$$
$$a_{42} \rightarrow a_{51}, a_{51}, a_{56}, a_{56}, a_{61}, a_{66}$$
$$a_{46} \rightarrow a_{51}, a_{51}, a_{52}, a_{52}, a_{61}, a_{62}$$

For elements of the fifth row

- The first element(a_{51})of the fifth row has the functional ties with the second(a_{62}, a_{62}) and sixth(a_{66}, a_{66}) elements of the sixth row for positive and negative areas;
- The second element(a_{52}) of the fifth row has the functional ties with the first(a_{61}, a_{61}) and sixth(a_{66}, a_{66}) elements of the sixth row for positive and negative areas;
- The sixth element(a_{56}) of the fifth row has the functional ties with the first(a_{61}, a_{61}) and second(a_{62}, a_{62}) elements of the sixth row for positive and negative areas, forming the frequency of functional ties for above-indicated elements (a_{51}, a_{52}, a_{56}) of the fifth row and elements (a_{61}, a_{62}, a_{66}) of the sixth row equal f=6 in such sequence as:

$$a_{51} \rightarrow a_{62}, a_{62}, a_{66}, a_{66}$$
$$a_{52} \rightarrow a_{61}, a_{61}, a_{66}, a_{66}$$
$$a_{56} \rightarrow a_{61}, a_{61}, a_{62}, a_{62}$$

In Figure 168 is shown the functional ties for combined partial determinant $\left| A_6 \right|$ of the 6x6 matrix.

G. Analysis of combined partial determinant $\left| A_7 \right|$ of the 6x6 matrix:

❖ *For Figure 169 of the combined partial determinant $\left| A_7 \right|$*

- The combined partial determinant $\left| A_7 \right|$ of the 6x6 matrix in general view includes the fourth element(a_{14}) of the first row and the fifth element(a_{25}) of the second row which are joined with frequency of tie equal f=1;
- The fifth element(a_{25}) of the second row has the functional ties with each element($a_{33}, a_{43}, a_{53}, a_{63}$) of the third column, forming the frequency of tie equal f=1 for above-indicated elements($a_{33}, a_{43}, a_{53}, a_{63}$).

For elements of the third column

- The third element(a_{33}) of the third column has the functional ties with elements(a_{41}, a_{42}, a_{46})of the fourth row;
- The fourth element(a_{43}) of the third column has the functional ties with elements (a_{31}, a_{32}, a_{36})of the third row;
- The fifth element(a_{53})of the third column has the functional ties with elements (a_{31}, a_{32}, a_{36})of the third row;
- The sixth element(a_{63}) of the third column has the functional ties with elements (a_{31}, a_{32}, a_{36}) of the third row, forming the frequency of tie equal f=1 for above-named elements($a_{33}, a_{43}, a_{53}, a_{63}$) in such sequence as:

$$a_{33} \rightarrow a_{41}, a_{42}, a_{46}$$
$$a_{43} \rightarrow a_{31}, a_{32}, a_{36}$$
$$a_{53} \rightarrow a_{31}, a_{32}, a_{36}$$
$$a_{63} \rightarrow a_{31}, a_{32}, a_{36}$$

For elements of third row

- The first element(a_{31}) of the third row has the functional ties with the second(a_{42}, a_{42}) and sixth(a_{46}, a_{46}) elements of the fourth row and also with the second(a_{52}) and sixth(a_{56}) elements of the fifth row for positive and negative areas;
- The second element(a_{32})of the third row has the functional ties with the first(a_{41}, a_{41}) and sixth(a_{46}, a_{46}) elements of the fourth row and also with the first(a_{51}) and sixth(a_{56}) elements of the fifth row for positive and negative areas;

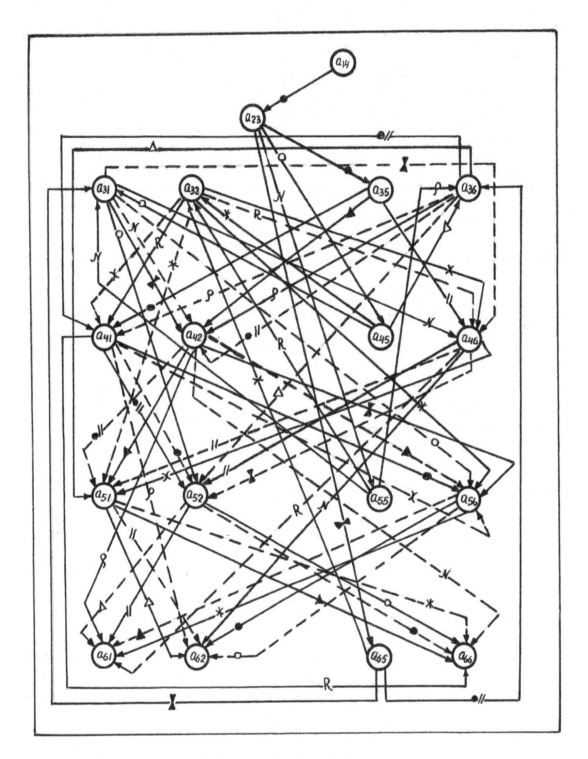

Figure 168 The functional ties between elements for combined partial determinant $|A_6|$ of the 6x6 matrix applicably to Laplace's theorem at marked out jointly the third ,fourth and fifth columns for positive (———) and negative (-------) areas

270

- The sixth element(a_{36}) of the third row has the functional ties with the first(a_{41},a_{41}) and second (a_{42},a_{42}) elements of the fourth row and also with the first (a_{51}) and second(a_{52}) elements of the fifth row for positive and negative areas, forming of functional ties for above-indicated elements (a_{31},a_{32},a_{36}) equal f=3 in such sequence as:

$$a_{31} \rightarrow a_{42},a_{42},a_{46},a_{46},a_{52},a_{56}$$
$$a_{32} \rightarrow a_{41},a_{41},a_{46},a_{46},a_{51},a_{56}$$
$$a_{36} \rightarrow a_{41},a_{41},a_{42},a_{42},a_{51},a_{52}$$

For elements of the fourth row

- The first element(a_{41})of the fourth row has the functional ties with the second(a_{52},a_{52}) and sixth(a_{56},a_{56}) elements of the fifth row and also has ties with the second(a_{62}) and sixth(a_{66}) elements of the sixth row for positive and negative areas;
- The second element(a_{42}) of the fourth row has the functional ties with the first(a_{51},a_{51}) and sixth(a_{56},a_{56}) elements of the fifth row and also has ties with the first(a_{61}) and sixth(a_{66}) elements of the sixth row for positive and negative areas;
- The sixth element(a_{46}) of the fourth row has the functional ties with the first(a_{51},a_{51}) and second(a_{52},a_{52}) elements of the fifth row and also ties with the first (a_{61}) and second (a_{62}) elements of the sixth row for positive and negative areas, forming the frequency of functional ties for above-indicated elements (a_{41},a_{42},a_{46})equal f=5 in such sequence as:

$$a_{41} \rightarrow a_{52},a_{52},a_{56},a_{56},a_{62},a_{66}$$
$$a_{42} \rightarrow a_{51},a_{51},a_{56},a_{56},a_{61},a_{66}$$
$$a_{46} \rightarrow a_{51},a_{51},a_{52},a_{52},a_{61},a_{62}$$

For elements of the fifth row

- The first element(a_{51})of the fifth row has the functional ties with the second(a_{62},a_{62}) and sixth(a_{66},a_{66}) elements of the sixth row for positive and negative areas;
- The second element(a_{52}) of the fifth row has the functional ties with the first(a_{61},a_{61}) and sixth(a_{66},a_{66}) elements of the sixth row for positive and negative areas;
- The sixth element(a_{56}) of the fifth row has the functional ties with the first(a_{61},a_{61}) and second(a_{62},a_{62}) elements of the sixth row for positive and negative areas, forming the frequency of functional ties for above-indicated elements (a_{51},a_{52},a_{56}) of the fifth row and elements (a_{61},a_{62},a_{66}) of the sixth row equal f=6 in such sequence as:

$$a_{51} \rightarrow a_{62},a_{62},a_{66},a_{66}$$
$$a_{52} \rightarrow a_{61},a_{61},a_{66},a_{66}$$
$$a_{56} \rightarrow a_{61},a_{61},a_{62},a_{62}$$

In **Figure 169** is shown the functional ties for combined partial determinant $\left| A_7 \right|$ of the 6x6 matrix.

H. Analysis of combined partial determinant $\left| A_8 \right|$ of the 6x6 matrix:

❖ *For Figure 170 of the combined partial determinant $\left| A_8 \right|$*

- The combined partial determinant $\left| A_8 \right|$ of the 6x6 matrix in general view includes the the fourth element(a_{14}) of the first row and the third element(a_{33}) of the third row which are joined with frequency of tie equal f=1;
- The third element(a_{33}) of the third row has the functional ties with each element (a_{45},a_{55},a_{65}) of the fifth column, forming the frequency of tie equal f=1 for above-indicated elements (a_{45},a_{55},a_{65}).

For elements of the fifth column

- The fourth element(a_{45}) of the fifth column has the functional ties with elements (a_{21},a_{22},a_{26}) of the second row;
- The fifth element(a_{55}) of the fifth column has the functional ties with elements(a_{21},a_{22},a_{26}) of the second row;

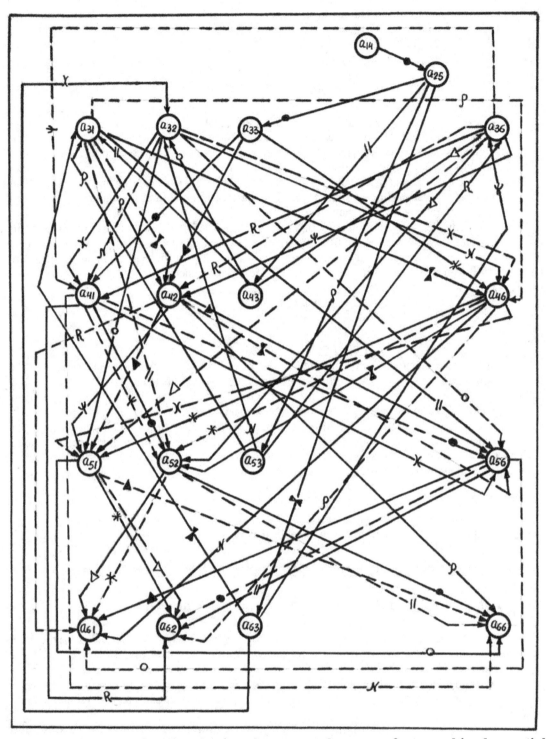

Figure 169 The functional ties between elements for combined partial determinant $\lvert A_7 \rvert$ of the 6x6 matrix applicably to Laplace's theorem at marked out jointly the third, fourth and fifth columns for positive(——) and negative(-----) areas

- The sixth element(a_{65}) of the fifth column has the functional ties with elements (a_{21}, a_{22}, a_{26}) of the second row, forming the frequency of tie equal f=1 for above-named-elements (a_{45}, a_{55}, a_{65}) of the fifth column in such sequence as:

$$a_{45} \rightarrow a_{21}, a_{22}, a_{26}$$
$$a_{55} \rightarrow a_{21}, a_{22}, a_{26}$$
$$a_{65} \rightarrow a_{21}, a_{22}, a_{26}$$

For elements of the second row

- The first element (a_{21}) of the second row has the functional ties with the second(a_{42}, a_{42}) and sixth(a_{46}, a_{46})elements of the fourth row also with the second(a_{52}) and sixth(a_{56}) elements of the fifth row for positive and negative areas;
- The second element(a_{22}) of the second row has the functional ties with the first(a_{41}, a_{41}) and sixth(a_{46}, a_{46}) elements of the fourth row and also with the first(a_{51}) and sixth(a_{56}) elements of the fifth row for positive and negative areas;
- The sixth element (a_{26}) of the second row has the functional ties with the first(a_{41}, a_{41}) and second(a_{42}, a_{42}) elements of the fourth row and also with the first(a_{51})and second(a_{52})elements of the fifth row for positive and negative areas, forming the frequency of functional ties for above-indicated elements(a_{21}, a_{22}, a_{26}) equal f=3 in such sequence as:

$$a_{21} \rightarrow a_{42}, a_{42}, a_{46}, a_{46}, a_{52}, a_{56}$$
$$a_{22} \rightarrow a_{41}, a_{41}, a_{46}, a_{46}, a_{51}, a_{56}$$
$$a_{26} \rightarrow a_{41}, a_{41}, a_{42}, a_{42}, a_{51}, a_{52}$$

For elements of the fourth row

- The first element(a_{41}) of the fourth row has the functional ties with the second(a_{52}) and sixth(a_{56})elements of the fifth row and also ties with the second(a_{62}) and sixth(a_{66}) elements of the sixth row for positive and negative areas;
- The second element(a_{42}) of the fourth row has the functional ties with the first(a_{51}) and sixth(a_{56}) elements of the fifth row and also ties with the first(a_{61}) and sixth(a_{66}) elements of the sixth row for positive and negative areas;
- The sixth element(a_{46})of the fourth row has the functional ties with the first(a_{51}) and second(a_{52}) elements of the fifth row and also ties with the first(a_{61}) and second(a_{62}) elements of the sixth row for positive and negative areas, forming the frequency of functional ties for above-indicated elements (a_{41}, a_{42}, a_{46}) equal f=4 in such sequence as:

$$a_{41} \rightarrow a_{52}, a_{56}, a_{62}, a_{66}$$
$$a_{42} \rightarrow a_{51}, a_{56}, a_{61}, a_{66}$$
$$a_{46} \rightarrow a_{51}, a_{52}, a_{61}, a_{62}$$

For elements of the fifth row

- The first element(a_{51}) of the fifth row has the functional ties with the second(a_{62}) and sixth(a_{66}) elements of the sixth row for positive and negative areas;
- The second element(a_{52}) of the fifth row has the functional ties with the first(a_{61}) and sixth(a_{66}) elements of the sixth row for positive and negative areas;
- The sixth element(a_{56})of the fifth row has the functional ties with the first(a_{61}) and second (a_{62}) elements of the sixth row for positive and negative areas, forming the frequency of functional ties for above-indicated elements (a_{51}, a_{52}, a_{56}) of the fifth row and elements (a_{61}, a_{62}, a_{66}) of the sixth row equal f=4 in such sequence as:

$$a_{51} \rightarrow a_{62}, a_{66}$$
$$a_{52} \rightarrow a_{61}, a_{66}$$
$$a_{56} \rightarrow a_{61}, a_{62}$$

In Figure 170 is shown the functional ties for combined partial determinant $\left| A_8 \right|$ of the 6x6 matrix.

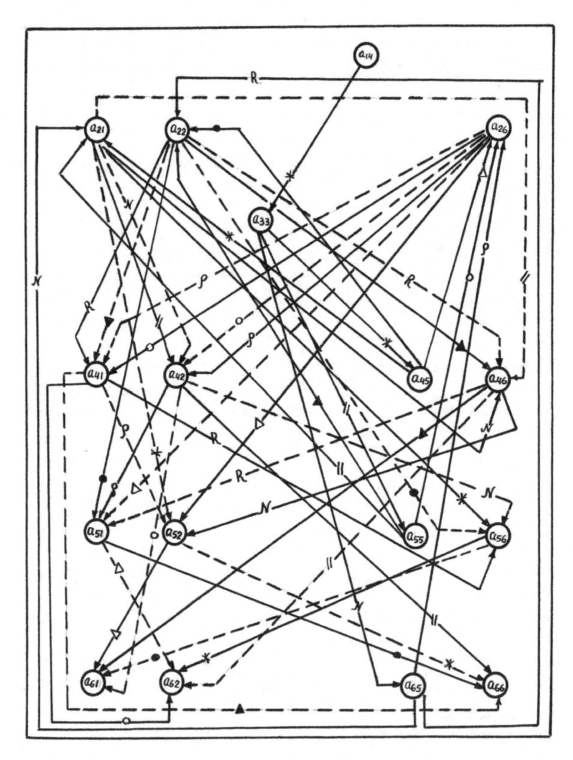

Figure 170 The functional ties between elements for combined partial determinant $|A_8|$ of the 6x6 matrix applicably to Laplace's theorem at marked out jointly the third, fourth and fifth columns for positive(——) and negative (------) areas

I. Analysis of combined partial determinant $\left|A_9\right|$ of the 6x6 matrix:

❖ *For Figure 171 of the combined partial determinant $\left|A_9\right|$*

- The combined partial determinant $\left|A_9\right|$ of the 6x6 matrix in general view includes the fourth element(a_{14}) of the first row and the fifth(a_{35}) element of the third row which are joined with frequency of tie equal f=1;
- The fifth element (a_{35}) of the third row has the functional ties with each element(a_{43},a_{53},a_{63}) of the third column, forming the frequency of tie equal f=1 for above-indicated elements (a_{43},a_{53},a_{63}).

For elements of the third column
- The fourth element(a_{43}) of the third column has the functional ties with elements(a_{21},a_{22},a_{26}) of the second row;
- The fifth element(a_{53}) of the third column has the functional ties with elements (a_{21},a_{22},a_{26})of the second row;
- The sixth element(a_{63}) of the third column has the functional ties with elements (a_{21},a_{22},a_{26}) of the second row ,forming the frequency of tie equal f=1 for above-named elements (a_{43},a_{53},a_{63})of the third column in such sequence as:

$$a_{43} \rightarrow a_{21},a_{22},a_{26}$$
$$a_{53} \rightarrow a_{21},a_{22},a_{26}$$
$$a_{63} \rightarrow a_{21},a_{22},a_{26}$$

For elements of the second row
- The first element(a_{21})of the second row is joined with the second(a_{42}) and sixth(a_{46}) elements of the fourth row and the second(a_{52}) and sixth(a_{56})elements of the fifth row for positive and negative areas;
- The second element (a_{22}) of the second row is joined with the first(a_{41}) and sixth(a_{46}) elements of the fourth row and the first(a_{51}) and sixth(a_{56}) elements of the fifth row for positive and negative areas;
- The sixth element(a_{26}) of the second row is joined with the first(a_{41}) and the second(a_{42})elements of the fourth row and the first(a_{51}) and second(a_{52}) elements of the fifth row for positive and negative areas, forming the frequency of functional ties for above-named elements (a_{21},a_{22},a_{26}) equal f=3 of the second row in such sequence as:

$$a_{21} \rightarrow a_{42},a_{46},a_{52},a_{56}$$
$$a_{22} \rightarrow a_{41},a_{46},a_{51},a_{56}$$
$$a_{26} \rightarrow a_{41},a_{42},a_{51},a_{52}$$

For elements of the fourth row
- The first element (a_{41}) of the fourth row has the functional ties with the second(a_{52}) and sixth(a_{56}) elements of the fifth row and the second(a_{62})and sixth(a_{66}) elements of the sixth row for positive and negative areas;
- The second element(a_{42}) of the fourth row has the functional ties with the first(a_{51}) and sixth(a_{56}) elements of the fifth row and the first (a_{61}) and sixth(a_{66}) elements of the sixth row for positive and negative areas;
- The sixth element(a_{46}) of the fourth row has the functional ties with the first (a_{51}) and second(a_{52}) elements of the fifth row and the first(a_{61}) and second(a_{62}) elements of the sixth row, forming the frequency of functional ties for above-indicated elements (a_{41},a_{42},a_{46}) equal f=4 of the fourth row in such sequence as:

$$a_{41} \rightarrow a_{52},a_{56},a_{62},a_{66}$$
$$a_{42} \rightarrow a_{51},a_{56},a_{61},a_{66}$$
$$a_{46} \rightarrow a_{51},a_{52},a_{61},a_{62}$$

For elements of the fifth row

- The first element(a_{51}) of the fifth row has the functional ties with the second(a_{62}) and sixth(a_{66}) elements of the sixth row for positive and negative areas;
- The second element(a_{52}) of the fifth row has the functional ties with the first(a_{61}) and sixth(a_{66}) elements of the sixth row for positive and negative areas;
- The sixth element(a_{56}) of the fifth row has the functional ties with the first (a_{61}) and second(a_{62}) elements of the sixth row for positive and negative areas, forming the frequency of functional ties for above-indicated elements (a_{51}, a_{52}, a_{56}) of the fifth row and elements (a_{61}, a_{62}, a_{66}) equal f=4 in such sequence as:

$$a_{51} \rightarrow a_{62}, a_{66}$$
$$a_{52} \rightarrow a_{61}, a_{66}$$
$$a_{56} \rightarrow a_{61}, a_{62}$$

In Figure 171 is shown the functional ties for combined partial determinant $\mid A_9 \mid$ of the 6x6 matrix.

K. Analysis of combined partial determinant $\mid A_{10} \mid$ of the 6x6 matrix:

❖ *For Figure 172 of the combined partial determinant $\mid A_{10} \mid$*

- The combined partial determinant $\mid A_{10} \mid$ of the 6x6 matrix in general view includes the fourth element(a_{14})of the first row and the third(a_{43}) and fifth(a_{45}) elements of the fourth row. And besides each element is joined with above-indicated fourth element(a_{14}), forming the frequency of functional tie equal f=1 for elements (a_{43}, a_{45}) of the fourth row;
- The third element(a_{43}) of the fourth row has functional ties with the fifth(a_{55}) and sixth(a_{65}) elements of the fifth column, forming the frequency of functional tie equal f=1 for elements (a_{55}, a_{65}) of the fifth column;
- The fifth(a_{45}) element of the fourth row has the functional ties with the fifth(a_{53})and sixth(a_{63}) elements of the third column, forming the frequency of functional tie equal f= for elements (a_{53}, a_{63}) of the third column.

For elements of the third and fifth columns

- The fourth element(a_{43})of the third column is joined with elements(a_{55}, a_{65}) of the fifth column;
- The fifth element(a_{53}) f the third column has the functional ties with elements (a_{21}, a_{22}, a_{26})of the second row;
- The sixth element(a_{63}) of the third column has the functional ties with elements (a_{21}, a_{22}, a_{26}) of the second row;
- The fourth element(a_{45}) of the fifth column is joined with elements (a_{53}, a_{63}) of the third column;
- The fifth element(a_{55}) of the fifth column has the functional ties with elements (a_{21}, a_{22}, a_{26}) of the second row;
- The sixth element(a_{65}) of the fifth column has the functional ties with elements (a_{21}, a_{22}, a_{26}) of the second row, forming for all above-indicated elements ($a_{43}, a_{53}, a_{63}, a_{45}, a_{55}, a_{65}$)the frequency of functional ties equal f=1 in such sequence of ties as:

$$a_{43} \rightarrow a_{55}, a_{65}$$
$$a_{53} \rightarrow a_{21}, a_{22}, a_{26}$$
$$a_{63} \rightarrow a_{21}, a_{22}, a_{26}$$
$$a_{45} \rightarrow a_{53}, a_{63}$$
$$a_{55} \rightarrow a_{21}, a_{22}, a_{26}$$
$$a_{65} \rightarrow a_{21}, a_{22}, a_{26}$$

For elements of the second row

- The first element(a_{21}) of the second row has the functional ties with the second($a_{32}, a_{32}, a_{32}, a_{32}$) and sixth($a_{36}, a_{36}, a_{36}, a_{36}$) elements of the third row for positive and negative areas;

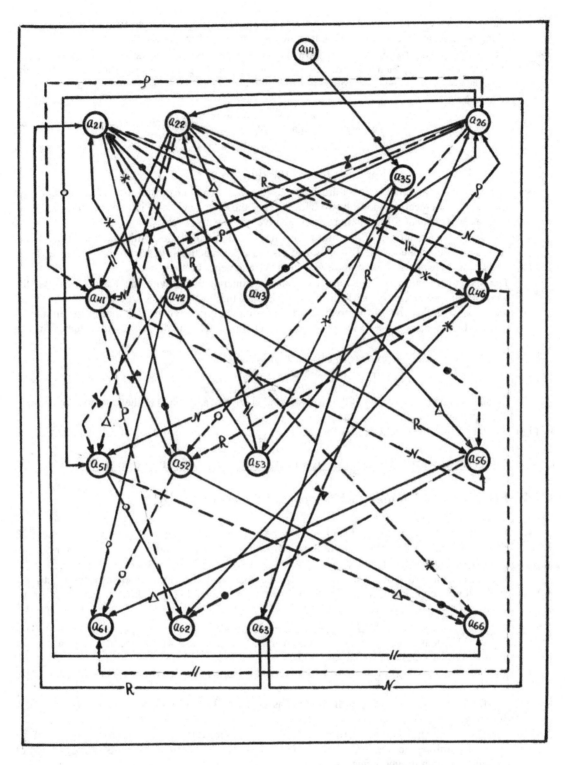

Figure 171 The functional ties between elements for combined partial determinant
$\left| A_9 \right|$ **of the 6x6 matrix applicably to Laplace's theorem at marked out jointly the**
third, fourth and fifth columns for positive (———) and negative (------) areas

- The second element(a_{22}) of the second row has the functional ties with the first($a_{31},a_{31},a_{31},a_{31}$) and sixth($a_{36},a_{36},a_{36},a_{36}$) elements of the third row for positive and negative areas;
- The sixth element (a_{26}) of the second row has the functional ties with the first($a_{31},a_{31},a_{31},a_{31}$) and second($a_{32},a_{32},a_{32},a_{32}$) elements of the third row for positive and negative areas, forming the frequency of functional ties equal f=4 for all above-indicated elements(a_{21},a_{22},a_{26}) in such sequence as:

$$a_{21} \rightarrow a_{32},a_{32},a_{32},a_{32},a_{36},a_{36},a_{36},a_{36}$$
$$a_{22} \rightarrow a_{31},a_{31},a_{31},a_{31},a_{36},a_{36},a_{36},a_{36}$$
$$a_{26} \rightarrow a_{31},a_{31},a_{31},a_{31},a_{32},a_{32},a_{32},a_{32}$$

For elements of the third row

- The first element (a_{31}) of the third row has the functional ties with the second(a_{52},a_{52}) and sixth(a_{56},a_{56}) elements of the fifth row and also has the functional ties with the second(a_{62},a_{62}) and sixth(a_{66},a_{66}) elements of the sixth row for positive and negative areas;
- The second element(a_{32}) of the third row has the functional ties with the first(a_{51},a_{51}) and sixth(a_{56},a_{56}) elements of the fifth row and also has the functional ties with the first(a_{61},a_{61}) and sixth(a_{66},a_{66}) elements of the sixth row for positive and negative areas;
- The sixth element(a_{36}) of the third row has the functional ties with the first(a_{51},a_{51}) and the second(a_{52},a_{52}) elements of the fifth row and also has the functional ties with the first(a_{61},a_{61}) and second(a_{62},a_{62}) elements of the sixth row for positive and negative areas, forming for all above-indicated elements(a_{31},a_{32},a_{36}) the frequency of functional ties equal f=8 in such sequence of ties as:

$$a_{31} \rightarrow a_{52},a_{52},a_{56},a_{56},a_{62},a_{62},a_{66},a_{66}$$
$$a_{32} \rightarrow a_{51},a_{51},a_{56},a_{56},a_{61},a_{61},a_{66},a_{66}$$
$$a_{36} \rightarrow a_{51},a_{51},a_{52},a_{52},a_{61},a_{61},a_{62},a_{62}$$

and other elements(a_{51},a_{52},a_{56}) of the fifth row and elements (a_{61},a_{62},a_{66}) of the sixth row, forming the frequency of functional ties equal f=4.

In Figure 172 is shown the functional ties for combined partial determinant $\mid A_{10} \mid$ of the 6x6 matrix.

L. Analysis of combined partial determinant $\mid A_{11} \mid$ of the 6x6 matrix:

❖ For Figure 173 of the combined partial determinant $\mid A_{11} \mid$

- The combined partial determinant $\mid A_{11} \mid$ of the 6x6 matrix in general view includes the fifth element(a_{15}) of the first row and the third element(a_{23}) of the second row which are joined with frequency of tie equal f=1;
- The third element (a_{23}) of the second row has the functional ties with each element ($a_{34},a_{44},a_{54},a_{64}$) of the fourth column, forming the frequency of tie equal f=1 for above-indicated elements ($a_{34},a_{44},a_{54},a_{64}$).

For elements of the fourth column

- The third element(a_{34}) of the fourth column has the functional ties with elements ($a_{41},a_{42},a_{42},a_{46}$) of the fourth row for positive and negative areas;
- The fourth element(a_{44}) of the fourth column has the functional ties with elements (a_{31},a_{32},a_{36}) of the third row;
- The fifth element(a_{54}) of the fourth column has the functional ties with elements(a_{31},a_{32},a_{36}) of the third row;
- The sixth element(a_{64}) of the fourth column has the functional ties with elements(a_{31},a_{32},a_{36}) of the third row, forming the frequency of tie equal f=1 for above-indicated elements ($a_{34},a_{44},a_{54},a_{64}$) of the fourth column in such sequence as:

$$a_{34} \rightarrow a_{41},a_{42},a_{42},a_{46}$$
$$a_{44} \rightarrow a_{31},a_{32},a_{36}$$
$$a_{54} \rightarrow a_{31},a_{32},a_{36}$$
$$a_{64} \rightarrow a_{31},a_{32},a_{36}$$

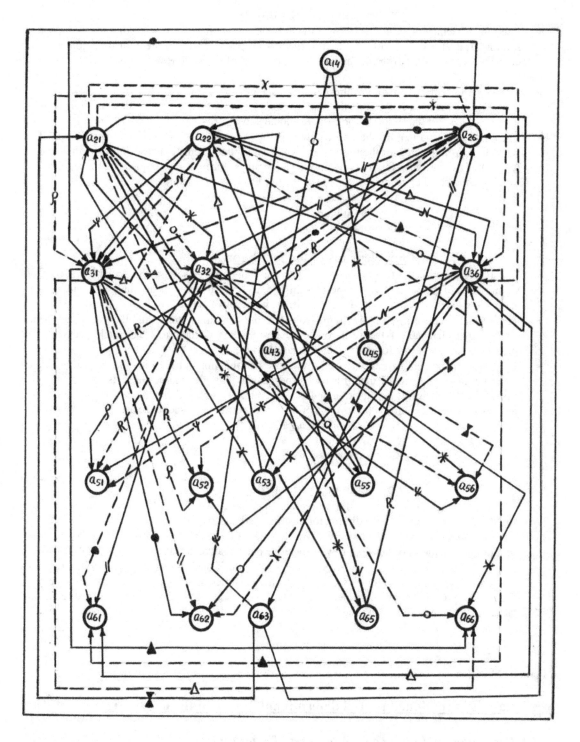

Figure 172 The functional ties between elements for combined partial determinant $|A_{10}|$ of the 6x6 matrix applicably to Laplace's theorem at marked out jointly the third, fourth and fifth columns for positive(——) and negative(-----) areas

279

For elements of the third row

- The first element(a_{31}) of the third row has the functional ties with the second(a_{42}) and sixth(a_{46},a_{46},a_{46}) elements of the fourth row. And besides this element also is joined with elements(a_{52},a_{56}) of the fifth row and sixth(a_{66}) element of the sixth row for positive and negative areas;
- The second element(a_{32}) of the third row has the functional ties with the first(a_{41},a_{41}) and sixth(a_{46}) elements of the fourth row . And besides this element also is joined with elements(a_{51},a_{56})of the fifth row for positive and negative areas;
- The sixth element(a_{36}) of the third row has the functional ties with the first (a_{41},a_{41})and second(a_{42},a_{42}) elements of the fourth row. And besides this element also is joined with elements (a_{51},a_{52}) of the fifth row for positive and negative areas, forming the frequency of ties equal f=3 for above-indicated elements (a_{31},a_{32},a_{36}) of the third row in such sequence as:

$$a_{31} \rightarrow a_{42},a_{46},a_{46},a_{46},a_{52},a_{56},a_{66}$$
$$a_{32} \rightarrow a_{41},a_{41},a_{46},a_{51},a_{56}$$
$$a_{36} \rightarrow a_{41},a_{41},a_{42},a_{42},a_{51},a_{52}$$

For elements of the fourth row

- The first element(a_{41}) of the fourth row has the functional ties with the second (a_{52},a_{52}) and sixth(a_{56},a_{56}) of the fifth row. And besides this element(a_{41}) also is joined with the second(a_{62}) and sixth(a_{66}) elements of the sixth row for positive and negative areas;
- The second element(a_{42}) of the fourth row has the functional ties with the first(a_{51},a_{51}) and sixth(a_{56},a_{56}) elements of the fifth row. And besides this element(a_{42}) also is joined with the first(a_{61}) of the sixth row for positive and negative areas;
- The sixth element(a_{46}) of the fourth row has the functional ties with the first(a_{51},a_{51}) and second(a_{52},a_{52}) elements of the fifth row. And besides this element(a_{46}) also is joined with the first(a_{61})of the sixth row for positive and negative areas, forming the frequency of ties equal f=5 for above-indicated elements (a_{41},a_{42},a_{46}) of the fourth row in such sequence as:

$$a_{41} \rightarrow a_{52},a_{52},a_{56},a_{56},a_{62},a_{66}$$
$$a_{42} \rightarrow a_{51},a_{51},a_{56},a_{56},a_{61}$$
$$a_{46} \rightarrow a_{51},a_{51},a_{52},a_{52},a_{61},a_{62}$$

For elements of the fifth row

- The first element(a_{51}) of the fifth row has the functional ties with the second(a_{62},a_{62})and sixth(a_{66},a_{66}) elements of the sixth row for positive and negative areas;
- The second element(a_{52}) of the fifth row has the functional ties with the first(a_{61},a_{61}) and sixth(a_{66},a_{66}) elements of the sixth row for positive and negative areas;
- The sixth element (a_{56}) of the fifth row has the functional ties with the first(a_{61},a_{61}) and second(a_{62},a_{62}) elements of the sixth row for positive and negative areas, forming the frequency of ties equal f=6 for above-indicated elements (a_{51},a_{52},a_{56}) and also for considered elements (a_{61},a_{62},a_{66}) of the sixth row in such sequence as:

$$a_{51} \rightarrow a_{62},a_{62},a_{66},a_{66}$$
$$a_{52} \rightarrow a_{61},a_{61},a_{66},a_{66}$$
$$a_{56} \rightarrow a_{61},a_{61},a_{62},a_{62}$$

In **Figure 173** is shown the functional ties for combined partial determinant $|A_{11}|$ of the 6x6 matrix.

M. Analysis of combined partial determinant $|A_{12}|$ of the 6x6 matrix:

❖ *For Figure 174 of the combined partial determinant $|A_{12}|$*

- The combined partial determinant $|A_{12}|$ of the 6x6 matrix in general view includes the fifth element(a_{15}) of the first row and the fourth element(a_{24}) of the second row which are joined with frequency of tie equal f=1;

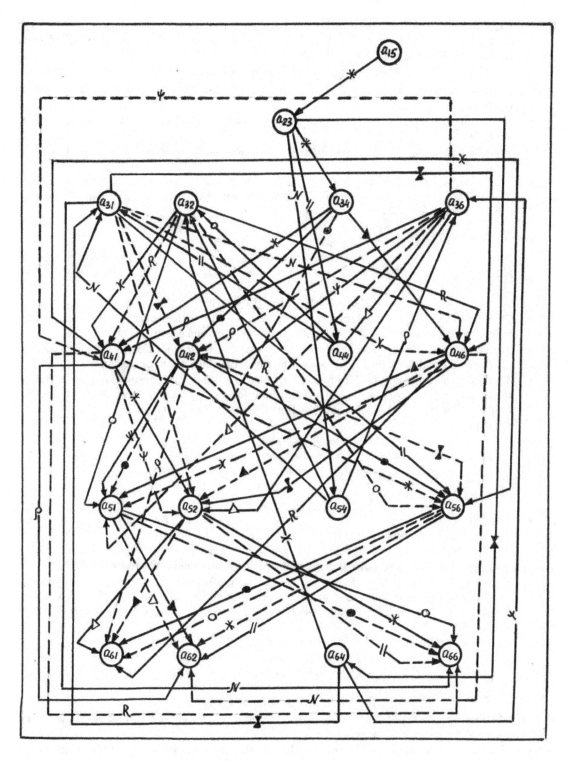

Figure 173 The functional ties between elements for combined partial determinant $|A_{11}|$ of the 6x6 matrix applicably to Laplace's theorem at marked out jointly the third ,fourth and fifth columns for positive (———) and negative(--------) areas

281

- The fourth element(a_{24})of the second row has the functional ties with each element(a_{33},a_{43},a_{53},a_{63}) of the third column, forming the frequency of tie equal f=1 for above- indicated elements(a_{33},a_{43},a_{53},a_{63}).

For elements of the third column

- The third element(a_{33}) of the third column has the functional ties with elements (a_{41},a_{42},a_{46}) of the fourth row;
- The fourth element(a_{43}) of the third column has the functional ties with elements (a_{31},a_{32},a_{36}) of the third row;
- The fifth element(a_{53}) of the third column has the functional ties with elements (a_{31},a_{32},a_{36}) of the third row;
- The sixth element(a_{63}) of the third column has the functional ties with elements (a_{31},a_{32},a_{36}), forming the frequency of tie equal f=1 for above-indicated elements (a_{33},a_{43},a_{53},a_{63}) of the third column in such sequence as:

$$a_{33} \rightarrow a_{41}, a_{42}, a_{46}$$
$$a_{43} \rightarrow a_{31}, a_{32}, a_{36}$$
$$a_{53} \rightarrow a_{31}, a_{32}, a_{36}$$
$$a_{63} \rightarrow a_{31}, a_{32}, a_{36}$$

For elements of the third row

- The first element (a_{31}) of the third row has the functional ties with elements (a_{42},a_{42},a_{46},a_{46}) of the fourth row and elements (a_{52},a_{56}) of the fifth row for positive and negative areas;
- The second element(a_{32}) of the third row has the functional ties with elements (a_{41},a_{41},a_{46},a_{46}) of the fourth row and elements (a_{51},a_{56}) of the fifth row for positive and negative areas;
- The sixth element(a_{36}) of the third row has the functional ties with elements (a_{41},a_{41},a_{42},a_{42})of the fourth row and elements (a_{51},a_{52}) of the fifth row for positive and negative areas, forming the frequency of ties equal f=3 for above-indicated elements (a_{31},a_{32},a_{36}) of the third row in such sequence as:

$$a_{31} \rightarrow a_{42}, a_{42}, a_{46}, a_{46}, a_{52}, a_{56}$$
$$a_{32} \rightarrow a_{41}, a_{41}, a_{46}, a_{46}, a_{51}, a_{56}$$
$$a_{36} \rightarrow a_{41}, a_{41}, a_{42}, a_{42}, a_{51}, a_{52}$$

For elements of the fourth row

- The first element (a_{41}) of the fourth row has the functional ties with elements (a_{52},a_{52},a_{56},a_{56}) of the fifth row and elements (a_{62},a_{66}) of the sixth row for positive and negative areas;
- The second element (a_{42}) of the fourth row has the functional ties with elements (a_{51},a_{51},a_{56},a_{56}) of the fifth row and elements (a_{61},a_{66}) of the sixth row for positive and negative areas;
- The sixth element (a_{46}) of the fourth row has the functional ties with elements (a_{51},a_{51},a_{52},a_{52}) of the fifth row and elements (a_{61},a_{62}) of the sixth row for positive and negative areas, forming the frequency of ties equal f=5 for above-indicated elements (a_{41},a_{42},a_{46}) of the fourth row in such sequence as:

$$a_{41} \rightarrow a_{52}, a_{52}, a_{56}, a_{56}, a_{62}, a_{66}$$
$$a_{42} \rightarrow a_{51}, a_{51}, a_{56}, a_{56}, a_{61}, a_{66}$$
$$a_{46} \rightarrow a_{51}, a_{51}, a_{52}, a_{52}, a_{61}, a_{62}$$

For elements of the fifth row

- The first element(a_{51}) of the fifth row has the functional ties with elements (a_{62},a_{62},a_{66},a_{66}) of the sixth row for positive and negative areas;
- The second element (a_{52}) of the fifth row has the functional ties with elements (a_{61},a_{61},a_{66},a_{66}) of the sixth row for positive and negative areas;
- The sixth element(a_{56}) of the fifth row has the functional ties with elements (a_{61},a_{61},a_{62},a_{62}) of the sixth row for positive and negative areas, forming the frequency of ties equal f=6 for above-indicated elements (a_{51},a_{52},a_{56}) of the fifth row and also elements (a_{61},a_{62},a_{66}) of the sixth row in such sequence as:

$$a_{51} \rightarrow a_{62}, a_{62}, a_{66}, a_{66}$$
$$a_{52} \rightarrow a_{61}, a_{61}, a_{66}, a_{66}$$
$$a_{56} \rightarrow a_{61}, a_{61}, a_{62}, a_{62}$$

In **Figure 174** is shown the functional ties for combined partial determinant $\left| A_{12} \right|$ of the 6x6 matrix.

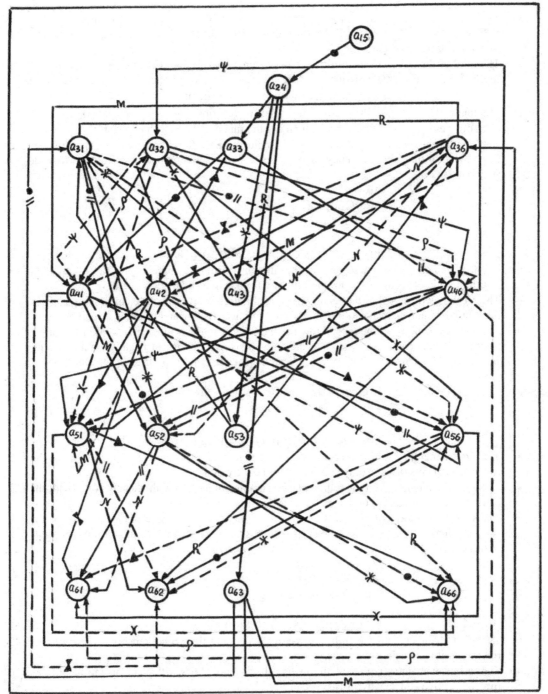

Figure 174 The functional ties between elements for combined partial determinant $\left| A_{12} \right|$ **of the 6x6 matrix applicably to Laplace's theorem at marked out jointly the third, fourth and fifth columns for positive(——) and negative (-------) areas**

283

N. Analysis of combined partial determinant $|A_{13}|$ of the 6x6 matrix:

❖ For Figure 175 of the combined partial determinant $|A_{13}|$

- The combined determinant $|A_{13}|$ of the 6x6 matrix in general view includes the fifth element(a_{15})of the first row and the third element(a_{33}) of the third row which are joined with frequency of tie equal f=1;
- The third element(a_{33}) of the third row has the functional ties with each element(a_{44},a_{54},a_{64})of the fourth column ,forming the frequency of tie equal f=1 for above-indicated elements(a_{44},a_{54},a_{64}).

For elements of the fourth column
- The fourth element(a_{44})of the fourth column has the functional ties with elements (a_{21},a_{22},a_{26});
- The fifth element(a_{54}) of the fourth column has the functional ties with elements (a_{21},a_{22},a_{26});
- The sixth element (a_{64}) of the fourth column has the functional ties with elements (a_{21},a_{22},a_{26}),forming the frequency of tie equal f=1 for above-indicated elements (a_{44},a_{54},a_{64}) in such sequence as:

$$a_{44} \rightarrow a_{21},a_{22},a_{26}$$
$$a_{54} \rightarrow a_{21},a_{22},a_{26}$$
$$a_{64} \rightarrow a_{21},a_{22},a_{26}$$

For elements of the second row
- The first element(a_{21}) of the second row has the functional ties with elements (a_{42},a_{42},a_{46},a_{46}) of the fourth row and elements(a_{52},a_{56})of the fifth row for positive and negative areas;
- The second element(a_{22}) of the second row has the functional ties with elements (a_{41},a_{41},a_{46},a_{46}) of the fourth row and elements(a_{51},a_{56}) of the fifth row for positive and negative areas;
- The sixth element(a_{26})of the second row has the functional ties with elements (a_{41},a_{41},a_{42},a_{42}) of the fourth row and elements (a_{51},a_{52}) of the fifth row for positive and negative areas, forming the frequency of ties equal f=3 for above- indicated elements (a_{21},a_{22},a_{26}) in such sequence as:

$$a_{21} \rightarrow a_{42},a_{42},a_{46},a_{46},a_{52},a_{56}$$
$$a_{22} \rightarrow a_{41},a_{41},a_{46},a_{46},a_{51},a_{56}$$
$$a_{26} \rightarrow a_{41},a_{41},a_{42},a_{42},a_{51},a_{52}$$

For elements of the fourth row
- The first element(a_{41}) of the fourth row has the functional ties with elements (a_{52},a_{56}) of the fifth row and elements (a_{62},a_{66}) of the sixth row for positive and negative areas;
- The second element(a_{42}) of the fourth row has the functional ties with elements (a_{51},a_{56})of the fifth row and elements (a_{61},a_{66}) of the sixth row for positive and negative areas;
- The sixth element(a_{46}) of the fourth row has the functional ties with elements (a_{51},a_{52}) of the fifth row and elements (a_{61},a_{62}) of the sixth row for positive and negative areas, forming the frequency of ties equal f=4 for above—indicated elements (a_{41},a_{42},a_{46}) in such sequence as:

$$a_{41} \rightarrow a_{52},a_{56},a_{62},a_{66}$$
$$a_{42} \rightarrow a_{51},a_{56},a_{61},a_{66}$$
$$a_{46} \rightarrow a_{51},a_{52},a_{61},a_{62}$$

For elements of the fifth row
- The first element(a_{51})of the fifth row has the functional ties with elements (a_{62},a_{66}) of the sixth row for positive and negative areas;
- The second element(a_{52}) of the fifth row has the functional ties with elements (a_{61},a_{66}) of the sixth row for positive and negative areas;
- The sixth element(a_{56}) of the fifth row has the functional ties with elements (a_{61},a_{62}) of the sixth row for positive and negative areas, forming the frequency of ties equal f=4 for above-indicated elements (a_{51},a_{52},a_{56}) and also for elements (a_{61},a_{62},a_{66}) of the sixth row in such sequence as:

$$a_{51} \rightarrow a_{62}, a_{66}$$
$$a_{52} \rightarrow a_{61}, a_{66}$$
$$a_{56} \rightarrow a_{61}, a_{62}$$

In Figure 175 is shown the functional ties for combined partial determinant $\left| A_{13} \right|$ of the 6x6 matrix

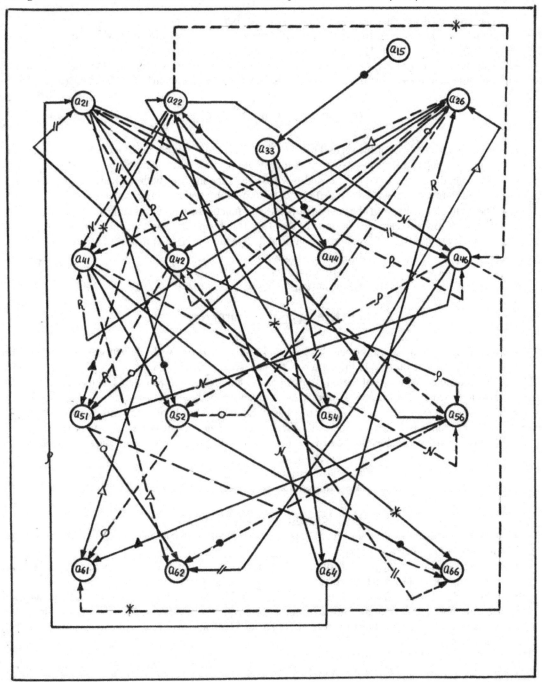

Figure 175 The functional ties between elements for combined partial determinant $\left| A_{13} \right|$ **of the 6x6 matrix applicably to Laplace's theorem at marked out jointly the third, fourth and fifth columns for positive(———) and negative(------) areas**

O. Analysis of combined partial determinant $|A_{14}|$ of the 6x 6 matrix:

❖ *For Figure 176 of the combined partial determinant $|A_{14}|$*

- The combined partial determinant $|A_{14}|$ of the 6x6 matrix in general view includes the fifth element (a_{15})of the first row and the fourth element(a_{34}) of the third row which are joined with frequency of tie equal f=1;
- The fourth element(a_{34}) of the third row has the functional ties with each element(a_{43},a_{53},a_{63}) of the third column, forming the frequency of tie equal f=1 for above-indicated elements(a_{43},a_{53},a_{63}).

For elements of the third column

- The first element(a_{21}) of the second row has the functional ties with elements ($a_{42},a_{42},a_{46},a_{46}$) of the fourth row and elements (a_{52},a_{56}) of the fifth row for positive and negative areas;
- The second element(a_{22}) of the second row has the functional ties with elements ($a_{41},a_{41},a_{46},a_{46}$) of the fourth row and elements (a_{51},a_{56}) of the fifth row for positive and negative areas;
- The sixth element(a_{26}) of the second row has the functional ties with elements ($a_{41},a_{41},a_{42},a_{42}$) of the fourth row and elements (a_{51},a_{52}) of the fifth row for positive and negative areas, forming the frequency of ties equal f=3 for above-indicated elements (a_{21},a_{22},a_{26}) in such sequence as:

$$a_{21} \rightarrow a_{42},a_{42},a_{46},a_{46},a_{52},a_{56}$$
$$a_{22} \rightarrow a_{41},a_{41},a_{46},a_{46},a_{51},a_{56}$$
$$a_{26} \rightarrow a_{41},a_{41},a_{42},a_{42},a_{51},a_{52}$$

For elements of the fourth row

- The first element(a_{41}) of the fourth row has the functional ties with elements(a_{52},a_{56}) of the fifth row and elements(a_{62},a_{66}) of the sixth row for positive and negative areas;
- The second element(a_{42}) of the fourth row has the functional ties with elements (a_{51},a_{56})of the fifth row and elements (a_{61},a_{66}) of the sixth row for positive and negative areas;
- The sixth element (a_{46}) of the fourth row has the functional ties with elements (a_{51},a_{52}) of the fifth row and elements (a_{61},a_{62}) of the sixth row for positive and negative areas, forming the frequency of ties equal f=4 for above-indicated elements (a_{41},a_{42},a_{46}) f the fourth row in such sequence as:

$$a_{41} \rightarrow a_{52},a_{56},a_{62},a_{66}$$
$$a_{42} \rightarrow a_{51},a_{56},a_{61},a_{66}$$
$$a_{46} \rightarrow a_{51},a_{52},a_{61},a_{62}$$

For elements of the fifth row

- The first element(a_{51}) of the fifth row has the functional ties with elements (a_{62},a_{66})of the sixth row for positive and negative areas;
- The second element(a_{52}) of the fifth row has the functional ties with elements (a_{61},a_{66})of the sixth row for positive and negative areas;
- The sixth element(a_{56}) of the fifth row has the functional ties with elements (a_{61},a_{62}) of the sixth row for positive and negative areas, forming the frequency of ties equal f=4 for above—indicated elements (a_{51},a_{52},a_{56})of the fifth row and also for elements (a_{61},a_{62},a_{66}) of the sixth row in such sequence as:

$$a_{51} \rightarrow a_{62},a_{66}$$
$$a_{52} \rightarrow a_{61},a_{66}$$
$$a_{56} \rightarrow a_{61},a_{62}$$

In Figure 176 is shown the functional ties for combined partial determinant $|A_{14}|$ of the 6x6 matrix.

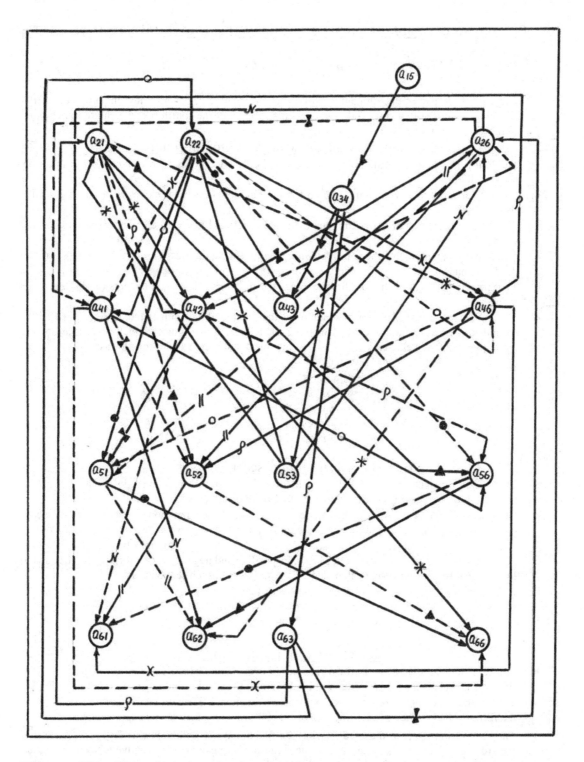

Figure 176 The functional ties between elements for combined partial determinant $\left| A_{14} \right|$ **of the 6x6 matrix applicably to Laplace's theorem at marked out jointly the third, fourth and fifth columns for positive(——) and negative (-----) areas**

287

P. Analysis of combined partial determinant $\left|A_{15}\right|$ of the 6x 6 matrix:

❖ *For Figure 177 of the combined partial determinant* $\left|A_{15}\right|$

- The combined partial determinant $\left|A_{15}\right|$ of the 6x6 matrix in general view includes the fifth element (a_{15}) of the first row and both elements (a_{43},a_{44}) of the fourth row which are joined with element (a_{15}), forming the frequency of tie equal f=1.

For elements of the third and fourth columns

- The fourth element(a_{43}) of the third column has the functional ties with elements (a_{54},a_{64}) of the fourth column;
- The fifth element(a_{53}) of the third column has the functional ties with elements (a_{21},a_{22},a_{26}) of the second row;
- The sixth element(a_{63}) of the third column has the functional ties with elements(a_{21},a_{22},a_{26}) of the second row;
- The fourth element(a_{44}) of the fourth column has the functional ties with elements(a_{53},a_{63}) of the third column;
- The fifth element(a_{54}) of the fourth column has the functional ties with elements (a_{21},a_{22},a_{26}) ft he second row;
- The sixth element(a_{64}) of the fourth column has the functional ties with elements (a_{21},a_{22},a_{26}) of the second row, forming the frequency of tie equal f=1 for above-indicated elements $(a_{43},a_{53},a_{63})$of the third column and elements (a_{44},a_{54},a_{64}) of the fourth column in such sequence as:

$$a_{43}{\rightarrow}a_{54},a_{64}$$
$$a_{53}{\rightarrow}a_{21},a_{22},a_{26}$$
$$a_{63}{\rightarrow}a_{21},a_{22},a_{26}$$
$$a_{44}{\rightarrow}a_{53},a_{63}$$
$$a_{54}{\rightarrow}a_{21},a_{22},a_{26}$$
$$a_{64}{\rightarrow}a_{21},a_{22},a_{26}$$

For elements of the second row

- The first element(a_{21}) of the second row has the functional ties with elements $(a_{32},a_{32},a_{32},a_{32},a_{36},a_{36},a_{36},a_{36})$ of the third row for positive and negative areas;
- The second element(a_{22}) of the second row has the functional ties with elements $(a_{31},a_{31},a_{31},a_{31},a_{36},a_{36},a_{36},a_{36})$ of the third row for positive and negative areas;
- The sixth element(a_{26}) of the second row has the functional ties with elements $(a_{31},a_{31},a_{31},a_{31},a_{32},a_{32},a_{32},a_{32})$ of the third row for positive and negative areas, forming the frequency of ties equal f=4 for above-indicated elements (a_{21},a_{22},a_{26}) in such sequence as:

$$a_{21}{\rightarrow}a_{32},a_{32},a_{32},a_{32},a_{36},a_{36},a_{36},a_{36}$$
$$a_{22}{\rightarrow}a_{31},a_{31},a_{31},a_{31},a_{36},a_{36},a_{36},a_{36}$$
$$a_{26}{\rightarrow}a_{31},a_{31},a_{31},a_{31},a_{32},a_{32},a_{32},a_{32}$$

For elements of the third row

- The first element(a_{31}) of the third row has the functional ties with elements $(a_{52},a_{52},a_{56},a_{56})$ of the fifth row and elements $(a_{62},a_{62},a_{66},a_{66})$ of the sixth row for positive and negative areas;
- The second element(a_{32}) of the third row has the functional ties with elements $(a_{51},a_{51},a_{56},a_{56})$ of the fifth row and elements $(a_{61},a_{61},a_{66},a_{66})$ of the sixth row for positive and negative areas;
- The sixth element(a_{36}) of the third row has the functional ties with elements $(a_{51},a_{51},a_{52},a_{52})$of the fifth row and elements $(a_{61},a_{61},a_{62},a_{62})$ of the sixth row for positive and negative areas, forming the frequency of ties equal f=8 for above-indicated elements (a_{31},a_{32},a_{36}) in such sequence as:

$$a_{31}{\rightarrow}a_{52},a_{52},a_{56},a_{56},a_{62},a_{62},a_{66},a_{66}$$
$$a_{32}{\rightarrow}a_{51},a_{51},a_{56},a_{56},a_{61},a_{61},a_{66},a_{66}$$
$$a_{36}{\rightarrow}a_{51},a_{51},a_{52},a_{52},a_{61},a_{61},a_{62},a_{62}$$

forming the frequency of functional ties equal f=4 also for above-considered elements(a_{51},a_{52},a_{56})

of the fifth row and elements (a_{61}, a_{62}, a_{66}) of the sixth row for positive and negative areas of this combined partial determinant $|A_{15}|$.

In **Figure 177** is shown the functional ties for combined partial determinant $|A_{15}|$ of the 6x6 matrix.

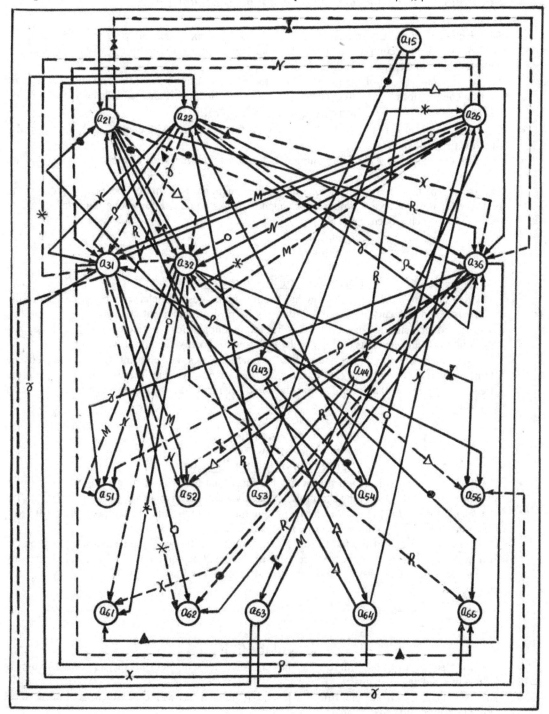

Figure 177 The functional ties between elements for combined partial determinant $|A_{15}|$ **of the 6x6 matrix applicably to Laplace's theorem at marked out jointly the third, fourth and fifth columns for positive (———) and negative(------) areas**

Q. Analysis of combined partial determinant $\left|A_{16}\right|$ of the 6x6 matrix:

❖ *For Figure 178 of the combined partial determinant* $\left|A_{16}\right|$

- The combined partial determinant $\left|A_{16}\right|$ of the 6x6 matrix in general view includes the third element (a_{23}) of the second row and also the fourth element(a_{34}) of the third row which are joined with frequency of tie equal f=1;
- The fourth element (a_{34}) of the third row has the functional ties with each element(a_{45}, a_{55}, a_{65}) of the fifth column, forming the frequency of tie equal f=1 for above-indicated elements (a_{45}, a_{55}, a_{65}).

For elements of the fifth column

- The fourth element(a_{45}) of the fifth column has the functional ties with elements (a_{11}, a_{12}, a_{16}) of the first row for positive area;
- The fifth element (a_{55})of the fifth column has the functional ties with elements (a_{11}, a_{12}, a_{16})of the first row for positive area;
- The sixth element(a_{65}) of the fifth column has the functional ties with elements(a_{11}, a_{12}, a_{16}) of the first row for positive area, forming the frequency of ties equal f=1 for above-indicated elements (a_{11}, a_{12}, a_{16}) in such sequence as:

$$a_{45} \rightarrow a_{11}, a_{12}, a_{16}$$
$$a_{55} \rightarrow a_{11}, a_{12}, a_{16}$$
$$a_{65} \rightarrow a_{11}, a_{12}, a_{16}$$

For elements of the first row

- The first element(a_{11}) of the first row has the functional ties with elements ($a_{42}, a_{42}, a_{46}, a_{46}$)of the fourth row and elements (a_{52}, a_{56})of the fifth row for positive and negative areas;
- The second element (a_{12}) of the first row has the functional ties with elements ($a_{41}, a_{41}, a_{46}, a_{46}$) of the fourth row and elements (a_{51}, a_{56}) of the fifth row for positive and negative areas;
- The sixth element (a_{16}) of the first row has the functional ties with elements ($a_{41}, a_{41}, a_{42}, a_{42}$) of the fourth row and elements (a_{51}, a_{52}) of the fifth row for positive and negative areas, forming the frequency of ties equal f=3 for above-indicated elements (a_{11}, a_{12}, a_{16}) of the first row in such sequence as:

$$a_{11} \rightarrow a_{42}, a_{42}, a_{46}, a_{46}, a_{52}, a_{56}$$
$$a_{12} \rightarrow a_{41}, a_{41}, a_{46}, a_{46}, a_{51}, a_{56}$$
$$a_{16} \rightarrow a_{41}, a_{41}, a_{42}, a_{42}, a_{51}, a_{52}$$

For elements of the fourth row

- The first element(a_{41}) of the fourth row has the functional ties with elements (a_{52}, a_{56}) of the fifth row and elements (a_{62}, a_{66}) of the sixth row for positive and negative areas;
- The second element(a_{42}) of the fourth row has the functional ties with element(a_{56}) of the fifth row and elements (a_{61}, a_{61}, a_{66})of the sixth row for positive and negative areas;
- The sixth element(a_{46}) of the fourth row has the functional ties with elements (a_{51}, a_{51}, a_{52}) of the fifth row and element(a_{62})of the sixth row for positive and negative areas, forming the frequency of functional ties equal f=4 for above-indicated elements (a_{41}, a_{42}, a_{46}) of the fourth row in such sequence as:

$$a_{41} \rightarrow a_{52}, a_{56}, a_{62}, a_{66}$$
$$a_{42} \rightarrow a_{56}, a_{61}, a_{61}, a_{66}$$
$$a_{46} \rightarrow a_{51}, a_{51}, a_{52}, a_{62}$$

For elements of the fifth row

- The first element(a_{51})of the fifth row has the functional ties with elements (a_{62}, a_{66}) of the sixth row for positive and negative areas;
- The second element(a_{52}) of the fifth row has the functional ties with elements(a_{61}, a_{66}) of the sixth row for positive and negative areas;
- The sixth element(a_{56})of the fifth row has the functional ties with elements(a_{61}, a_{62}) of the sixth row for positive and negative areas, forming the frequency of functional ties equal f=4 for above-indicated elements (a_{51}, a_{52}, a_{56}) of the fifth row and also for elements (a_{61}, a_{62}, a_{66}) of the sixth row in such sequence as:

290

$a_{51} \rightarrow a_{62}, a_{66}$

$a_{52} \rightarrow a_{61}, a_{66}$

$a_{56} \rightarrow a_{61}, a_{62}$

In Figure 178 is shown the functional ties for combined partial determinant $\left| A_{16} \right|$ of the 6x6 matrix.

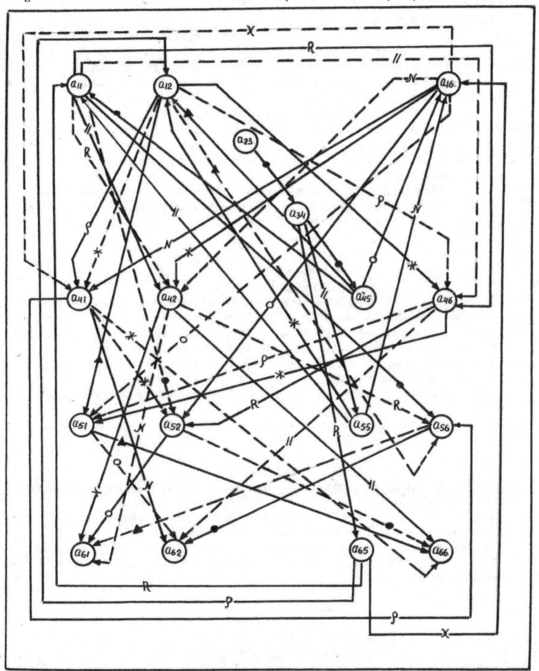

Figure 178 The functional ties between elements for combined partial determinant $\left| A_{16} \right|$ **of the 6x6 matrix applicably to Laplace's theorem at marked out jointly the third, fourth and fifth columns for positive(———) and negative (------) areas**

R . Analysis of combined partial determinant $|A_{17}|$ of the 6x6 matrix:

❖ *For Figure 179 of the combined partial determinant $|A_{17}|$*

- The combined partial determinant $|A_{17}|$ of the 6x6 matrix in general view includes the third element (a_{23}) of the second row and the fifth element(a_{35}) of the third row which are joined with frequency of tie equal f=1;
- The fifth element(a_{35}) of the third row has the functional ties with each element (a_{44}, a_{54}, a_{64}) of the fourth column, forming the frequency of tie equal f=1 for above-indicated elements (a_{44}, a_{54}, a_{64}) of the fourth column.

For elements of the fourth column

- The fourth element(a_{44})of the fourth column has the functional ties with elements (a_{11}, a_{12}, a_{16}) of the first row for positive area;
- The fifth element(a_{54}) of the fourth column has the functional ties with elements (a_{11}, a_{12}, a_{16}) of the first row for positive area;
- The sixth element(a_{64}) of the fourth column has the functional ties with elements (a_{11}, a_{12}, a_{16}) of the first row for positive area, forming the frequency of tie equal f=1 for above-indicated elements (a_{44}, a_{54}, a_{64}) of the fourth column in such sequence as:

$$a_{44} \rightarrow a_{11}, a_{12}, a_{16}$$
$$a_{54} \rightarrow a_{11}, a_{12}, a_{16}$$
$$a_{64} \rightarrow a_{11}, a_{12}, a_{16}$$

For elements of the first row

- The first element(a_{11}) of the first row has the functional ties with elements($a_{42}, a_{42}, a_{46}, a_{46}$)of the fourth row and elements (a_{52}, a_{56}) of the fifth row for positive and negative areas;
- The second element(a_{12}) of the first row has the functional ties with elements ($a_{41}, a_{41}, a_{46}, a_{46}$)of the fourth row and elements (a_{51}, a_{56}) of the fifth row for positive and negative areas;
- The sixth element(a_{16}) of the first row has the functional ties with elements ($a_{41}, a_{41}, a_{42}, a_{42}$) of the fourth row and elements (a_{51}, a_{52}) of the fifth row for positive and negative areas, forming the frequency of functional ties equal f=3 for above-indicated elements (a_{11}, a_{12}, a_{16}) in such sequence as:

$$a_{11} \rightarrow a_{42}, a_{42}, a_{46}, a_{46}, a_{52}, a_{56}$$
$$a_{12} \rightarrow a_{41}, a_{41}, a_{46}, a_{46}, a_{51}, a_{56}$$
$$a_{16} \rightarrow a_{41}, a_{41}, a_{42}, a_{42}, a_{51}, a_{52}$$

For elements of the fourth row

- The first element(a_{41}) of the fourth row has the functional ties with elements (a_{52}, a_{56}) of the fifth row and elements (a_{62}, a_{66}) of the sixth row for positive and negative areas;
- The second element (a_{42}) of the fourth row has the functional ties with element(a_{56}) of the fifth row and elements (a_{61}, a_{61}, a_{66}) of the sixth row for positive and negative areas;
- The sixth element(a_{46}) of the fourth row has the functional ties with elements (a_{51}, a_{51}, a_{52}) of the fifth row and element(a_{62}) of the sixth row for positive and negative areas, forming the frequency of functional ties equal f=4 for above-indicated elements(a_{41}, a_{42}, a_{46}) of the fourth row in such sequence as:

$$a_{41} \rightarrow a_{52}, a_{56}, a_{62}, a_{66}$$
$$a_{42} \rightarrow a_{56}, a_{61}, a_{61}, a_{66}$$
$$a_{46} \rightarrow a_{51}, a_{51}, a_{52}, a_{62}$$

For elements of the fifth row

- The first element(a_{51}) of the fifth row has the functional ties with elements (a_{62}, a_{66}) of the sixth row for positive and negative areas;
- The second element(a_{52}) of the fifth row has the functional ties with elements (a_{61}, a_{66}) of the sixth row for positive and negative areas;
- The sixth element(a_{56}) of the fifth row has the functional ties with elements(a_{61}, a_{62}) of the sixth row for positive and negative areas, forming the frequency of functional ties equal f=4 for above-

indicated elements(a_{51}, a_{52}, a_{56}) of the fifth row and also for elements (a_{61}, a_{62}, a_{66}) of the sixth row in such sequence as:

$$a_{51} \rightarrow a_{62}, a_{66}$$
$$a_{52} \rightarrow a_{61}, a_{66}$$
$$a_{56} \rightarrow a_{61}, a_{62}$$

In Figure 179 is shown the functional ties for combined partial determinant $\left| A_{17} \right|$ of the 6x6 matrix.

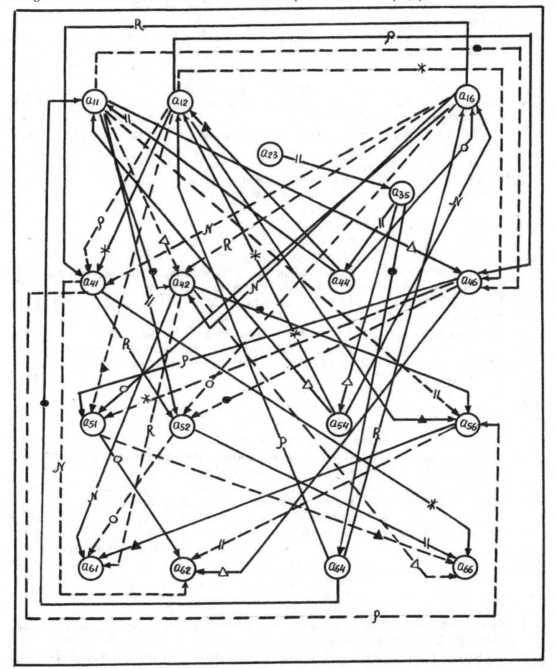

Figure 179 The functional ties between elements for combined partial determinant $\left| A_{17} \right|$ of the 6x6 matrix applicably to Laplace's theorem at marked out jointly the third, fourth and fifth columns for positive (——) and negative(------) areas

S. Analysis of combined partial determinant $\left|A_{18}\right|$ of the 6x6 matrix:

❖ For Figure 180 of the combined partial determinant $\left|A_{18}\right|$

- The combined partial determinant $\left|A_{18}\right|$ of the 6x6 matrix in general view includes the fourth element(a_{24}) of the second row and third element(a_{33}) of the third row which are joined with frequency of tie equal f=1;
- The third element(a_{33}) of the third row has the functional ties with each element(a_{45}, a_{55}, a_{65}) of the fifth column, forming the frequency of tie equal f=1 for above-indicated elements (a_{45}, a_{55}, a_{65})of the fifth column.

For elements of the fifth column
- The fourth element(a_{45}) of the fifth column has the functional ties with elements(a_{11}, a_{12}, a_{16})of the first row for positive area;
- The fifth element(a_{55}) of the fifth column has the functional ties with elements(a_{11}, a_{12}, a_{16}) of the first row for positive area;
- The sixth element(a_{65}) of the fifth column has the functional ties with elements(a_{11}, a_{12}, a_{16}) of the first row for positive area, forming the frequency of functional ties equal f=1 for above-indicated elements (a_{45}, a_{55}, a_{65}) of the fifth column in such sequence as:

$$a_{45} \rightarrow a_{11}, a_{12}, a_{16}$$
$$a_{55} \rightarrow a_{11}, a_{12}, a_{16}$$
$$a_{65} \rightarrow a_{11}, a_{12}, a_{16}$$

For elements of the first row
- The first element(a_{11})of the first row has the functional ties with elements ($a_{42}, a_{42}, a_{46}, a_{46}$) of the fourth row and elements (a_{52}, a_{56}) of the fifth row for positive and negative areas;
- The second element(a_{12}) of the first row has the functional ties with elements($a_{41}, a_{41}, a_{46}, a_{46}$)of the fourth row and elements (a_{51}, a_{56}) of the fifth row for positive and negative areas;
- The sixth element(a_{16}) of the first row has the functional ties with elements ($a_{41}, a_{41}, a_{42}, a_{42}$)of the fourth row and elements(a_{51}, a_{52}) of the fifth row for positive and negative areas, forming the frequency of ties equal f=3 for above-indicated elements (a_{11}, a_{12}, a_{16}) of the first row in such sequence as:

$$a_{11} \rightarrow a_{42}, a_{42}, a_{46}, a_{46}, a_{52}, a_{56}$$
$$a_{12} \rightarrow a_{41}, a_{41}, a_{46}, a_{46}, a_{51}, a_{56}$$
$$a_{16} \rightarrow a_{41}, a_{41}, a_{42}, a_{42}, a_{51}, a_{52}$$

For elements of the fourth row
- The first element(a_{41}) of the fourth row has the functional ties with elements(a_{52}, a_{56}) of the fifth row and elements(a_{62}, a_{66}) of the sixth row for positive and negative areas;
- The second element(a_{42}) of the fourth row has the functional ties with elements (a_{61}, a_{61}, a_{66}) of the sixth row and element (a_{56}) of the fifth row for positive and negative areas;
- The sixth element(a_{46}) of the fourth row has the functional ties with elements (a_{51}, a_{51}, a_{52}) of the fifth row and element(a_{62}) of the sixth row for positive and negative areas, forming the frequency of ties equal f=4 for above-indicated elements (a_{41}, a_{42}, a_{46}) of the fourth row in such sequence as:

$$a_{41} \rightarrow a_{52}, a_{56}, a_{62}, a_{66}$$
$$a_{42} \rightarrow a_{56}, a_{61}, a_{61}, a_{66}$$
$$a_{46} \rightarrow a_{51}, a_{51}, a_{52}, a_{62}$$

For elements of the fifth row
- The first element(a_{51}) of the fifth row has the functional ties with elements (a_{62}, a_{66}) of the sixth row for positive and negative areas;
- The second element (a_{52}) of the fifth row has the functional ties with elements (a_{61}, a_{66}) of the sixth row for positive and negative areas;
- The sixth element(a_{56}) of the fifth row has the functional ties with elements (a_{61}, a_{62}) of the sixth row for positive and negative areas, forming the frequency of ties equal f=4 for above-indicated elements(a_{51}, a_{52}, a_{56}) of the fifth row and also for elements (a_{61}, a_{62}, a_{66})in such sequence as:

$a_{51} \rightarrow a_{62}, a_{66}$
$a_{52} \rightarrow a_{61}, a_{66}$
$a_{56} \rightarrow a_{61}, a_{62}$

In Figure 180 is shown the functional ties for combined partial determinant $\left| A_{18} \right|$ of the 6x6 matrix.

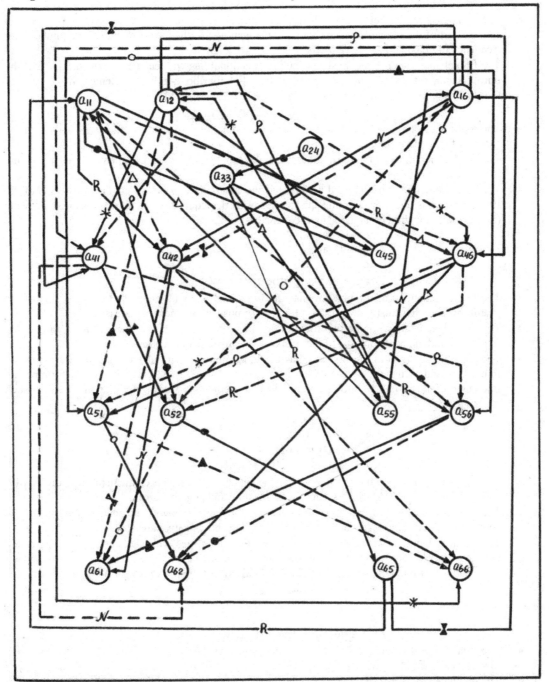

Figure 180 The functional ties between elements for combined partial determinant $\left| A_{18} \right|$ of the 6x 6 matrix applicably to Laplace's theorem at marked out jointly the third ,fourth and fifth columns for positive(——) and negative(-----) areas

T. Analysis of combined partial determinant $|A_{19}|$ of the 6x6 matrix:

❖ *For Figure 181 of the combined partial determinant* $|A_{19}|$

- The combined partial determinant $|A_{19}|$ of the 6x6 matrix in general vie includes the fourth element(a_{24}) of the second row and the fifth element(a_{35}) of the third row which are joined with frequency of tie equal f=1;
- The fifth element (a_{35}) of the third row has the functional ties with each element(a_{43},a_{53},a_{63}) of the third column, forming the frequency of tie equal f=1 for above-indicated elements(a_{43},a_{53},a_{63}) of the third column.

For elements of the third column

- The fourth element(a_{43}) of the third column has the functional ties with elements(a_{11},a_{12},a_{16}) of the first row for positive area;
- The fifth element(a_{53})of the third column has the functional ties with elements(a_{11},a_{12},a_{16}) of the first row for positive area;
- The sixth element(a_{63}) of the third column has the functional ties with elements(a_{11},a_{12},a_{16}) of the first row for positive area, forming the frequency of ties equal f=1 for above-indicated elements (a_{43},a_{53},a_{63}) of the third column in such sequence as:

$$a_{43} \rightarrow a_{11},a_{12},a_{16}$$
$$a_{53} \rightarrow a_{11},a_{12},a_{16}$$
$$a_{63} \rightarrow a_{11},a_{12},a_{16}$$

For elements of the first row

- The first element(a_{11}) of the first row has the functional ties with elements (a_{42},a_{42},a_{46},a_{46}) of the fourth row and elements(a_{52},a_{56}) of the fifth row for positive and negative areas;
- The second element(a_{12}) of the first row has the functional ties with elements (a_{41},a_{41},a_{46},a_{46}) of the fourth row and elements (a_{51},a_{56}) of the fifth row for positive and negative areas;
- The sixth element(a_{16}) of the first row has the functional ties with elements(a_{41},a_{41},a_{42},a_{42}) of the fourth row and elements (a_{51},a_{52}) of the fifth row for positive and negative areas, forming the frequency of ties equal f=3 for above-indicated elements(a_{11},a_{12},a_{16}) of the first row in such sequence as:

$$a_{11} \rightarrow a_{42},a_{42},a_{46},a_{46},a_{52},a_{56}$$
$$a_{12} \rightarrow a_{41},a_{41},a_{46},a_{46},a_{51},a_{56}$$
$$a_{16} \rightarrow a_{41},a_{41},a_{42},a_{42},a_{51},a_{52}$$

For elements of the fourth row

- The first element (a_{41})of the fourth row has the functional ties with elements(a_{52},a_{56}) of the fifth row and elements (a_{62},a_{66}) of the sixth row for positive and negative areas;
- The second element(a_{42}) of the fourth row has the functional ties with elements (a_{61},a_{61},a_{66}) of the sixth row and element(a_{56}) of the fifth row for positive and negative areas;
- The sixth element(a_{46}) of the fourth row has the functional ties with elements (a_{51},a_{51},a_{52}) of the fifth row and element(a_{62}) of the sixth row for positive and negative areas, forming the frequency of ties equal f=4 for above-indicated elements(a_{41},a_{42},a_{46}) of the fourth row in such sequence as:

$$a_{41} \rightarrow a_{52},a_{56},a_{62},a_{66}$$
$$a_{42} \rightarrow a_{56},a_{61},a_{61},a_{66}$$
$$a_{46} \rightarrow a_{51},a_{51},a_{52},a_{62}$$

For elements of the fifth row

- The first element(a_{51}) of the firth row has the functional ties with elements (a_{62},a_{66}) of the sixth row for positive and negative areas;
- The second element(a_{52}) of the fifth row has the functional ties with elements(a_{61},a_{66}) of the sixth row for positive and negative areas;
- The sixth element(a_{56}) of the fifth row has the functional ties with elements (a_{61},a_{62}) of the sixth row for positive and negative areas, forming the frequency of ties equal f=4 for above-indicated elements (a_{51},a_{52},a_{56}) of the fifth row and also for elements (a_{61},a_{62},a_{66}) of the sixth row in such

sequence as: $a_{51} \rightarrow a_{62}, a_{66}$

$a_{52} \rightarrow a_{61}, a_{66}$

$a_{56} \rightarrow a_{61}, a_{62}$

In Figure 181 is shown the functional ties for combined partial determinant $|A_{19}|$ of the 6x6 matrix.

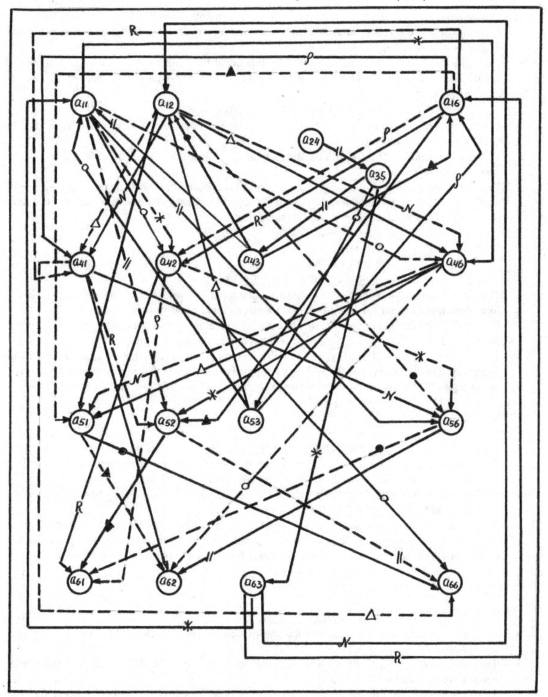

Figure 181 The functional ties between elements for combined partial determinant $|A_{19}|$ of the 6x6 matrix applicably to Laplace's theorem at marked out jointly the third ,fourth and fifth columns for positive(——) and negative (------) areas

297

U. Analysis of combined partial determinant $|A_{20}|$ of the 6x6 matrix:

❖ *For Figure 182 of the combined partial determinant* $|A_{20}|$

- The combined partial determinant $|A_{20}|$ of the 6x6 matrix in general view includes the fifth element(a_{25}) of the second row and the third element(a_{33}) of the third row which are joined with frequency of tie equal f=1;
- The third element(a_{33})f the third row has the functional ties with each element(a_{44}, a_{54}, a_{64})of the fourth column, forming the frequency of tie equal f=1 for above-indicated elements(a_{44}, a_{54}, a_{64}) of the fourth column.

For elements of the fourth column

- The fourth element(a_{44}) of the fourth column has the functional ties with elements(a_{11}, a_{12}, a_{16}) of the first row for positive area;
- The fifth element(a_{54}) of the fourth column has the functional ties with elements(a_{11}, a_{12}, a_{16}) of the first row for positive are;
- The sixth element (a_{64}) of the fourth column has the functional ties with elements (a_{11}, a_{12}, a_{16}) of the first row for positive area, forming the frequency of tie equal f=1 for above-indicated elements (a_{44}, a_{54}, a_{64})of the fourth column in such sequence as:

$$a_{44} \rightarrow a_{11}, a_{12}, a_{16}$$
$$a_{54} \rightarrow a_{11}, a_{12}, a_{16}$$
$$a_{64} \rightarrow a_{11}, a_{12}, a_{16}$$

For elements of the first row

- The first element(a_{11}) of the first row has the functional ties with elements($a_{42}, a_{42}, a_{46}, a_{46}$) of the fourth row and element(a_{56}) of the fifth row for positive and negative areas;
- The second element(a_{12}) of the first row has the functional ties with elements($a_{41}, a_{41}, a_{46}, a_{46}$) of the fourth row and element (a_{51}) of the fifth row for positive and negative areas;
- The sixth element(a_{16}) of the first row has the functional ties with elements($a_{41}, a_{41}, a_{42}, a_{42}$)of the fourth row and element(a_{52}) of the fifth row for positive and negative areas, forming the frequency of ties equal f=3 for above-indicated elements (a_{11}, a_{12}, a_{16}) of the first row in such sequence as:

$$a_{11} \rightarrow a_{42}, a_{42}, a_{46}, a_{46}, a_{56}$$
$$a_{12} \rightarrow a_{41}, a_{41}, a_{46}, a_{46}, a_{51}$$
$$a_{16} \rightarrow a_{41}, a_{41}, a_{42}, a_{42}, a_{52}$$

For elements of the fourth row

- The first element(a_{41}) of the fourth row has the functional ties with elements(a_{52}, a_{56}) of the fifth row and elements(a_{62}, a_{66}) of the sixth row for positive and negative areas;
- The second element(a_{42})of the fourth row has the functional ties with elements(a_{61}, a_{61}, a_{66}) of the sixth row and element(a_{56}) of the fifth row for positive and negative areas;
- The sixth element (a_{46}) of the fourth row has the functional ties with elements (a_{51}, a_{51}, a_{52}) of the fifth row and element(a_{62}) of the sixth row for positive and negative areas, forming the frequency of ties equal f=4 for above-indicated elements (a_{41}, a_{42}, a_{46}) of the fourth row in such sequence as:

$$a_{41} \rightarrow a_{52}, a_{56}, a_{62}, a_{66}$$
$$a_{42} \rightarrow a_{56}, a_{61}, a_{61}, a_{66}$$
$$a_{46} \rightarrow a_{51}, a_{51}, a_{52}, a_{62}$$

For elements of the fifth row

- The first element (a_{51})of the fifth row has the functional ties with elements (a_{62}, a_{66}) of the sixth row for positive and negative areas;
- The second element (a_{52})of the fifth row has the functional ties with elements(a_{61}, a_{66}) of the sixth row for positive and negative areas;
- The sixth element (a_{56}) of the fifth row has the functional ties with elements (a_{61}, a_{62}) of the sixth row for positive and negative areas, forming the frequency of ties equal f=4 for above-indicated elements(a_{51}, a_{52}, a_{56}) of the fifth row and elements(a_{61}, a_{62}, a_{66})of the sixth row in such sequence as:

$$a_{51} \rightarrow a_{62}, a_{66}$$
$$a_{52} \rightarrow a_{61}, a_{66}$$
$$a_{56} \rightarrow a_{61}, a_{62}$$

In Figure 182 is shown the functional ties for combined partial determinant $\left| A_{20} \right|$ of the 6x6 matrix.

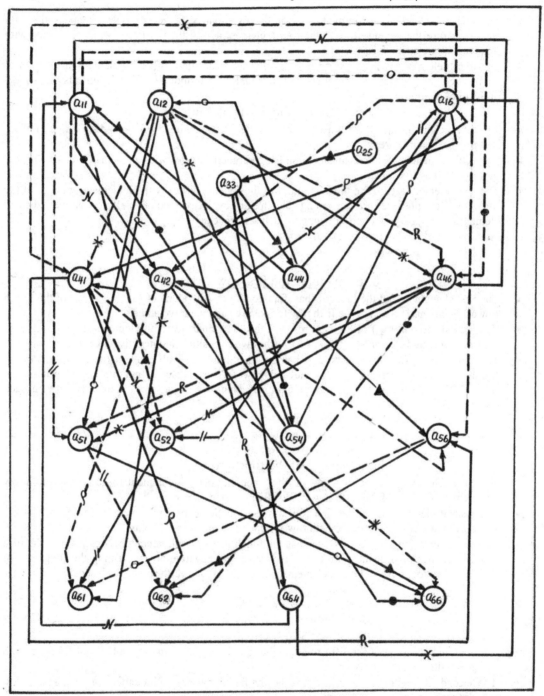

Figure 182 The functional ties between elements for combined partial determinant $\left| A_{20} \right|$ **of the 6x 6 matrix applicably to Laplace's theorem at marked out jointly the third, fourth and fifth columns for positive (———) and negative(------) areas**

299

V. Analysis of combined partial determinant $|A_{21}|$ of the 6x6 matrix:

❖ *For Figure 183 of the combined partial deteminant $|A_{21}|$*

- The combined partial determinant $|A_{21}|$ of the 6x6 matrix in general view includes the fifth element(a_{25}) of the second row and the fourth element(a_{34}) of the third row which are joined with frequency of tie equal f=1;
- The fourth element(a_{34}) of the third row has the functional ties with each element (a_{43},a_{53},a_{63}) of the third column, forming the frequency of tie equal f=1 for above-indicated elements (a_{43},a_{53},a_{63}) of the third column.

For elements of the third column

- The fourth element (a_{43}) of the third column has the functional ties with elements(a_{11},a_{12},a_{16}) of the first row for positive area;
- The fifth element(a_{53}) of the third column has the functional ties with elements (a_{11},a_{12},a_{16}) of the first row for positive area;
- The sixth element(a_{63}) of the third column has the functional ties with elements(a_{11},a_{12},a_{16}) of the first row for positive area, forming the frequency of tie equal f=1 for above-indicated elements(a_{43},a_{53},a_{63}) of the third column in such sequence as:

$$a_{43}\rightarrow a_{11},a_{12},a_{16}$$
$$a_{53}\rightarrow a_{11},a_{12},a_{16}$$
$$a_{63}\rightarrow a_{11},a_{12},a_{16}$$

For elements of the first row

- The first element (a_{11})of the first row has functional ties with elements(a_{42},a_{42},a_{46},a_{46}) of the fourth row and elements (a_{52},a_{56}) of the fifth row for positive and negative areas;
- The second element (a_{12}) of the first row has the functional ties with elements (a_{41},a_{41},a_{46},a_{46}) of the fourth row and elements(a_{51},a_{56}) of the fifth row for positive and negative areas;
- The sixth element(a_{16}) of the first row has the functional ties with elements(a_{41},a_{41},a_{42},a_{42}) of the fourth row and elements(a_{51},a_{52}) of the fifth row for positive and negative areas, forming the frequency of ties equal f=3 for above-indicated elements (a_{11},a_{12},a_{16}) of the first row in such sequence as:

$$a_{11}\rightarrow a_{42},a_{42},a_{46},a_{46},a_{52},a_{56}$$
$$a_{12}\rightarrow a_{41},a_{41},a_{46},a_{46},a_{51},a_{56}$$
$$a_{16}\rightarrow a_{41},a_{41},a_{42},a_{42},a_{51},a_{52}$$

For elements of the fourth row

- The first element(a_{41}) of the first row has the functional ties with elements(a_{52},a_{56}) of the fifth row and elements(a_{62},a_{66}) of the sixth row for positive and negative areas;
- The second element (a_{42})of the first row has the functional ties with elements(a_{61},a_{61},a_{66}) of the sixth row and element(a_{56}) of the fifth row for positive and negative areas;
- The sixth element(a_{46}) of the first row has the functional ties with elements(a_{51},a_{51},a_{52}) of the fifth row and element(a_{62}) of the sixth row for positive and negative areas, forming the frequency of ties equal f=4 for above-indicated elements(a_{41},a_{42},a_{46}) of the fourth row in such sequence as:

$$a_{41}\rightarrow a_{52},a_{56},a_{62},a_{66}$$
$$a_{42}\rightarrow a_{56},a_{61},a_{61},a_{66}$$
$$a_{46}\rightarrow a_{51},a_{51},a_{52},a_{62}$$

For elements of the fifth row

- The first element (a_{51}) of the fifth row has the functional ties with elements (a_{62},a_{66}) of the sixth row for positive and negative areas;
- The second element (a_{52}) of the fifth row has the functional ties with elements(a_{61},a_{66}) of the sixth row for positive and negative areas;
- The sixth element(a_{56}) of the fifth row has the functional ties with elements(a_{61},a_{62}) of the sixth row for positive and negative areas, forming the frequency of ties equal f=4 for above-indicated elements(a_{51},a_{52},a_{56}) and elements(a_{61},a_{62},a_{66}) in such sequence as:

$$a_{51} \rightarrow a_{62}, a_{66}$$
$$a_{52} \rightarrow a_{61}, a_{66}$$
$$a_{56} \rightarrow a_{61}, a_{62}$$

In Figure 183 is shown the functional ties for combined partial determinant $\left| A_{21} \right|$ of the 6x6 matrix.

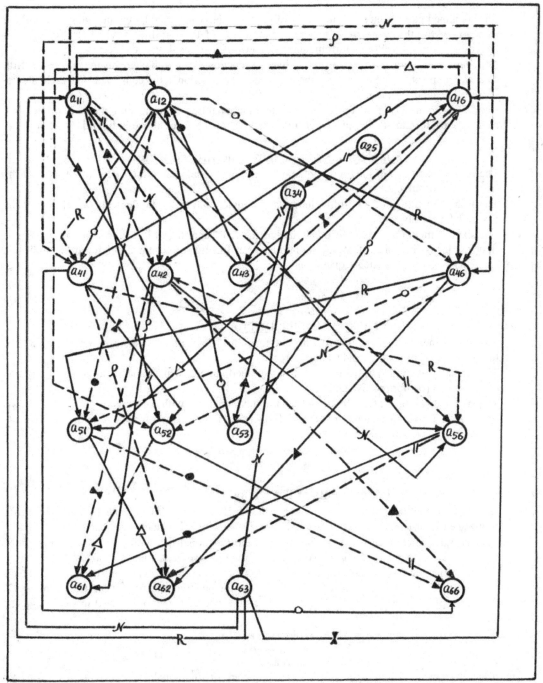

Figure 183 The functional ties between elements for combined partial determinant $\left| A_{21} \right|$ **of the 6x6 matrix applicably to Laplace's theorem at marked out jointly the third, fourth and fifth columns for positive(——) and negative(-----) areas**

W. Analysis of combined partial determinant $|A_{22}|$ of the 6x6 matrix:

❖ *For Figure 184 of the combined partial determinant $|A_{22}|$*

- The combined partial determinant $|A_{22}|$ of the 6x6 matrix in general view includes the third(a_{33}) element of the third row and both elements(a_{44},a_{45}) of the fourth row which are joined with above-indicated element(a_{33}), forming the frequency of functional tie equal f=1;
- The fourth element(a_{44}) of the fourth row has the functional ties with elements (a_{55},a_{65}) of the fifth column for positive area, forming the frequency of functional tie equal f=1 for each above-indicated element (a_{55},a_{65});
- The fifth element(a_{45}) of the fourth row has the functional tie with element(a_{54}) of the fifth row for positive area, forming the frequency of functional tie equal f=1 for above-indicated element(a_{54}).

For elements of the fourth and fifth columns
- The fifth element(a_{54}) of the fourth column has the functional ties with elements(a_{11},a_{12},a_{16}) of the first row for positive area;
- The fifth element(a_{55}) of the fifth column has the functional ties with elements (a_{11},a_{12},a_{16}) of the first row for positive area;
- The sixth element (a_{65}) of the fifth column has the functional ties with elements(a_{11},a_{12},a_{16}) of the first row for positive area, forming the frequency of functional ties equal f=1 for above-indicated elements (a_{54},a_{55},a_{65}) in such sequence as:

$$a_{54} \rightarrow a_{11},a_{12},a_{16}$$
$$a_{55} \rightarrow a_{11},a_{12},a_{16}$$
$$a_{65} \rightarrow a_{11},a_{12},a_{16}$$

For elements of the first row
- The first element(a_{11}) of the first row has the functional ties with elements(a_{22},a_{22},a_{22},a_{26},a_{26},a_{26}) of the second row for positive and negative areas;
- The second element(a_{12}) of the first row has the functional ties with elements (a_{21},a_{21},a_{21},a_{26},a_{26},a_{26}) of the second row for positive and negative areas;
- The sixth element(a_{16}) of the first row has the functional ties with elements (a_{21},a_{21},a_{21},a_{22},a_{22},a_{22}) of the second row, forming the frequency of functional ties equal f=3 for above-indicated elements (a_{11},a_{12},a_{16}) in such sequence as:

$$a_{11} \rightarrow a_{22},a_{22},a_{22},a_{26},a_{26},a_{26}$$
$$a_{12} \rightarrow a_{21},a_{21},a_{21},a_{26},a_{26},a_{26}$$
$$a_{16} \rightarrow a_{21},a_{21},a_{21},a_{22},a_{22},a_{22}$$

For elements of the second row
- The first element (a_{21}) of the second row has the functional ties with elements (a_{52},a_{56}) of the fifth row and elements(a_{62},a_{62},a_{66},a_{66}) of the sixth row for positive and negative areas;
- The second element(a_{22}) of the second row has the functional ties with elements (a_{51},a_{56}) of the fifth row and elements(a_{61},a_{61},a_{66},a_{66}) of the sixth row for positive and negative areas;
- The sixth element (a_{26}) of the second row has the functional ties with elements (a_{51},a_{52}) of the fifth row and elements(a_{61},a_{61},a_{62},a_{62}) of the sixth row for positive and negative areas, forming the frequency of functional ties equal f=6 for above-indicated elements (a_{21},a_{22},a_{26}) in such sequence as:

$$a_{21} \rightarrow a_{52},a_{56},a_{62},a_{62},a_{66},a_{66}$$
$$a_{22} \rightarrow a_{51},a_{56},a_{61},a_{61},a_{66},a_{66}$$
$$a_{26} \rightarrow a_{51},a_{52},a_{61},a_{61},a_{62},a_{62}$$

and also for elements (a_{51},a_{52},a_{56}) ,forming the frequency of functional ties equal f=2 and elements (a_{61},a_{62},a_{66}), forming the frequency of functional ties equal f=4 for positive and negative areas of the combined partial determinant $|A_{22}|$ of the 6x6 matrix.

In Figure 184 is shown the functional ties for combined partial determinant $|A_{22}|$ of the 6x6 matrix.

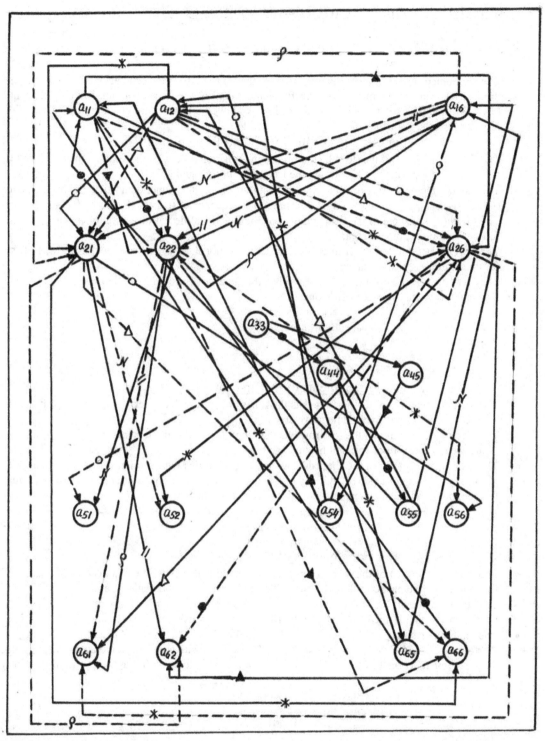

Figure 184 The functional ties between elements for combined partial determinant │A₂₂│ of the 6x6 matrix applicably to Laplace's theorem at marked out jointly the third, fourth and fifth columns for positive (──) and negative(-----) areas

X. Analysis of combined partial determinant $|A_{23}|$ of the 6x6 matrix:

❖ *For Figure 185 of the combined partial determinant $|A_{23}|$*

- The combined partial determinant $|A_{23}|$ of the 6x6 matrix in general view includes the third(a_{33})element of the third row and fifth element (a_{54}) of the fourth column, and also elements (a_{45},a_{55}) of the fifth column which are joined with above-indicated third element (a_{33}) of the third row ,forming the frequency of functional tie equal f=1.

For elements of the fourth and fifth columns

- The fifth element(a_{54})of the fourth column has the functional tie with the sixth element(a_{65}) of the fifth column for positive area, forming the frequency of tie equal f=1;
- The sixth element (a_{64})of the fourth column has the functional ties with elements (a_{11},a_{11},a_{12},a_{12},a_{16},a_{16}) for positive area, forming the frequency of ties equal f=2;
- The fourth(a_{45}) element of the fifth column has the functional tie with the sixth element(a_{64}) of the fourth column for positive area;
- The fifth element(a_{55}) of the fifth column has the functional tie with the sixth element(a_{64}) of the fourth column for positive area;
- The sixth element(a_{65}) of the fifth column has the functional ties with elements(a_{11},a_{12},a_{16}) of the first row for positive area in such sequence as:

$$a_{54} \rightarrow a_{65}$$
$$a_{64} \rightarrow a_{11},a_{11},a_{12},a_{12},a_{16},a_{16}$$
$$a_{45} \rightarrow a_{64}$$
$$a_{55} \rightarrow a_{64}$$
$$a_{65} \rightarrow a_{11},a_{12},a_{16}$$

For elements of the first row

- The first element(a_{11}) of the first row has the functional ties with elements (a_{22},a_{22},a_{22},a_{26},a_{26},a_{26})of the second row for positive and negative areas;
- The second element(a_{12}) of the first row has the functional ties with elements (a_{21},a_{21},a_{21},a_{26},a_{26},a_{26}) of the second row for positive and negative areas;
- The sixth element (a_{16}) of the first row has the functional ties with elements(a_{21},a_{21},a_{21},a_{22},a_{22},a_{22}) of the second row for positive and negative areas, forming the frequency of functional ties equal f=3 for above-indicated elements(a_{11},a_{12},a_{16}) in such sequence as:

$$a_{11} \rightarrow a_{22},a_{22},a_{22},a_{26},a_{26},a_{26}$$
$$a_{12} \rightarrow a_{21},a_{21},a_{21},a_{26},a_{26},a_{26}$$
$$a_{16} \rightarrow a_{21},a_{21},a_{21},a_{22},a_{22},a_{22}$$

For elements of the second row

- The first element(a_{21}) of the second row has the functional; ties with elements (a_{42},a_{42},a_{46},a_{46}) of the fourth row and elements(a_{52},a_{56}) of the fifth row for positive and negative areas;
- The second element(a_{22}) of the second row has the functional ties with elements(a_{41},a_{41},a_{46},a_{46}) of the fourth row and elements (a_{51},a_{56}) of the fifth row for positive and negative areas;
- The sixth element(a_{26}) of the second row has the functional ties with elements (a_{41},a_{41},a_{42},a_{42}) of the fourth row and elements(a_{51},a_{52}) of the fifth row for positive and negative areas, forming the frequency of functional ties equal f=6 for above-indicated elements (a_{21},a_{22},a_{26}) in such sequence as:

$$a_{21} \rightarrow a_{42},a_{42},a_{46},a_{46},a_{52},a_{56}$$
$$a_{22} \rightarrow a_{41},a_{41},a_{46},a_{46},a_{51},a_{56}$$
$$a_{26} \rightarrow a_{41},a_{41},a_{42},a_{42},a_{51},a_{52}$$

And besides analysis of functional ties between elements of the fourth and fifth rows indicates on fact that above-indicated elements (a_{41},a_{42},a_{46}) of the fourth row have the frequency of functional ties equal f=4 and elements (a_{51},a_{52},a_{56}) of the fifth row have the frequency of functional ties equal f=2.

In Figure 185 is shown the functional ties for combined partial determinant $|A_{23}|$ of the 6x6 matrix.

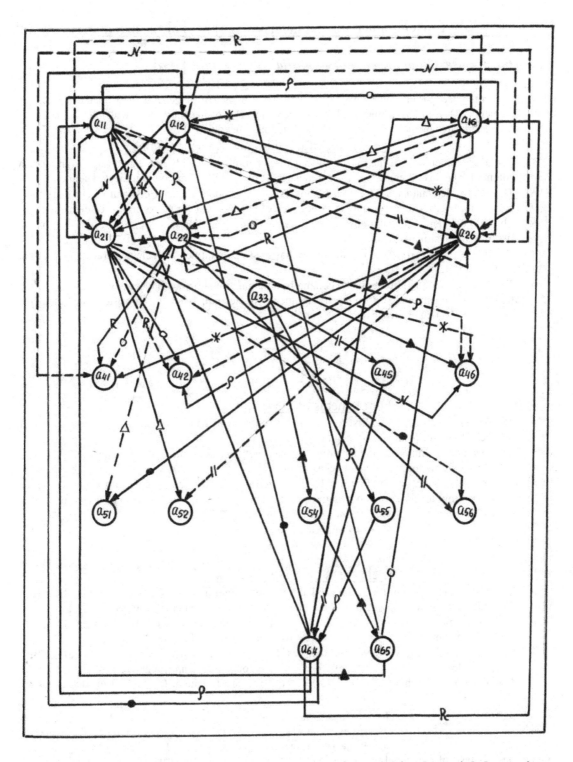

Figure 185 The functional ties between elements for combined partial determinant $|A_{23}|$ **of the 6x6 matrix applicably to Laplace's theorem at marked out jointly the third, fourth and fifth columns for positive(——) and negative (------)areas**

305

Y. Analysis of combined partial determinant $|A_{24}|$ of the 6x6 matrix:

❖ For Figure 186 of the combined partial determinant $|A_{24}|$

- The combined partial determinant $|A_{24}|$ of the 6x6 matrix in general view includes the fifth element (a_{35}) of the third row and both elements (a_{43},a_{44}) of the fourth row which are joined with above-indicated element(a_{35}). Forming the functional tie equal f=1.

For elements of the third and fourth columns

- The third element(a_{43}) of the fourth row has the functional ties with elements(a_{54},a_{64}) of the fourth column for positive area, forming the frequency of tie equal f=1 for above-indicated elements;
- The fourth element(a_{44}) of the fourth row has the functional tie with element (a_{53}) of the third column for positive area, forming the frequency of tie equal f=1 for above-indicated elements;
- The fifth element(a_{53}) of the third column has the functional ties with elements(a_{11},a_{12},a_{16}) of the first row for positive area;
- The fifth element (a_{54}) of the fourth column has the functional ties with elements(a_{11},a_{12},a_{16}) of the first row for positive area;
- The sixth element(a_{64}) of the fourth column has the functional ties with elements (a_{11},a_{12},a_{16}) of the first row for positive area, forming the frequency of ties equal f=3 for above-indicated elements (a_{11},a_{12},a_{16}) in such sequence as:

$$a_{43} \rightarrow a_{54}$$
$$a_{44} \rightarrow a_{53}$$
$$a_{53} \rightarrow a_{11},a_{12},a_{16}$$
$$a_{54} \rightarrow a_{11},a_{12},a_{16}$$
$$a_{64} \rightarrow a_{11},a_{12},a_{16}$$

For elements of the first row

- The first element (a_{11}) of the first row has the functional ties with elements(a_{22},a_{22},a_{22},a_{26},a_{26},a_{26}) of the second row for positive and negative areas;
- The second element(a_{12}) of the first row has the functional ties with elements(a_{21},a_{21},a_{21},a_{26},a_{26},a_{26}) of the second row for positive and negative areas;
- The sixth element(a_{16})of the first row has the functional ties with elements(a_{21},a_{21},a_{21},a_{22},a_{22},a_{22}) of the second row for positive and negative areas, forming the frequency of ties equal f=3 for above-indicated elements(a_{11},a_{12},a_{16}) in such sequence as:

$$a_{11} \rightarrow a_{22},a_{22},a_{22},a_{26},a_{26},a_{26}$$
$$a_{12} \rightarrow a_{21},a_{21},a_{21},a_{26},a_{26},a_{26}$$
$$a_{16} \rightarrow a_{21},a_{21},a_{21},a_{22},a_{22},a_{22}$$

For elements of the second row

- The first element (a_{21}) of the second row has the functional ties with elements (a_{52},a_{56}) of the fifth row and elements(a_{62},a_{62},a_{66},a_{66}) of the sixth row for positive and negative areas;
- The second element(a_{22})of the second row has the functional ties with elements(a_{51},a_{56}) of the fifth row and elements(a_{61},a_{61},a_{66},a_{66}) of the sixth row for positive and negative areas;
- The sixth element (a_{26}) of the second row has the functional ties with elements(a_{51},a_{52}) of the fifth row and elements(a_{61},a_{61},a_{62},a_{62}) of the sixth row for positive and negative areas, forming the frequency of functional ties equal f=6 for above-indicated elements(a_{21},a_{22},a_{26}) in such sequence as:

$$a_{21} \rightarrow a_{52},a_{56},a_{62},a_{62},a_{66},a_{66}$$
$$a_{22} \rightarrow a_{51},a_{56},a_{61},a_{61},a_{66},a_{66}$$
$$a_{26} \rightarrow a_{51},a_{52},a_{61},a_{61},a_{62},a_{62}$$

And besides analysis of functional ties between elements of the combined partial determinant $|A_{24}|$ of the 6x6 matrix show the following fact that frequency of ties for elements (a_{51},a_{52},a_{56}) of the fifth row is equal f=2 and for above-indicated elements(a_{61},a_{62},a_{66}) of the sixth row is equal f=4 for positive and negative areas.

In Figure 186 is shown the functional ties for combined partial determinant $|A_{24}|$ of the 6x6 matrix.

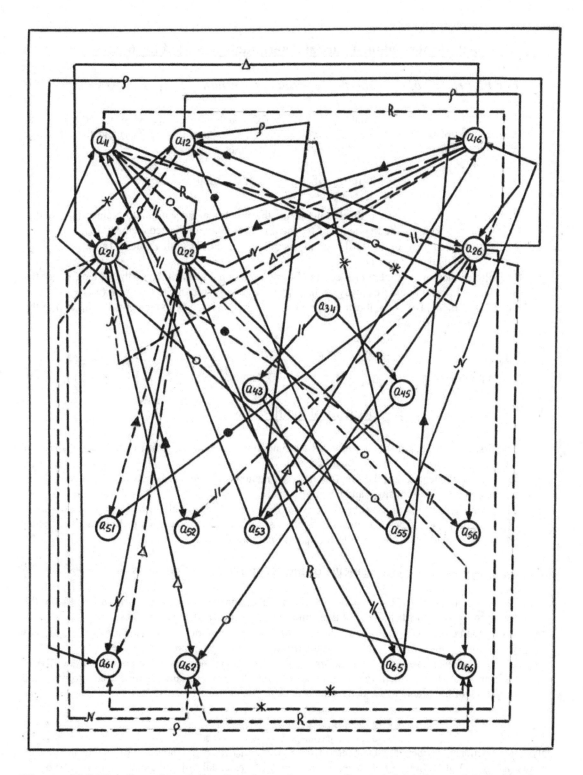

Figure 186 The functional ties between elements for combined partial determinant $|A_{24}|$ of the 6x6 matrix applicably to Lapalce's theorem at marked out jointly the third, fourth and fifth columns for positive(———) and negative(------) areas

307

Z. Analysis of combined partial determinant $|A_{25}|$ of the 6x6 matrix:

❖ For Figure 187 of the combined partial determinant $|A_{25}|$

- The combined partial determinant $|A_{25}|$ of the 6x6 matrix in general view includes the fifth element (a_{35}) of the third row which are joined with elements of the third(a_{53}) and fourth(a_{44},a_{54}) columns, forming the frequency of tie equal f=1 for elements (a_{44},a_{53},a_{54}).

For elements of the third and fourth columns

- The fifth element (a_{53}) of the third column has the functional tie with sixth element(a_{64}) of the fourth column, forming the frequency of tie equal f=1 for above-indicated element(a_{64});
- The sixth element (a_{63}) of the third column has the functional ties with elements (a_{11},a_{12},a_{16}) of the first row for positive area;
- The fifth element(a_{54}) of the fourth column has the functional tie with the sixth element(a_{63}) of the third column ,forming the frequency of ties equal f=2 for above-indicated element (a_{63});
- The fifth element (a_{64}) of the fourth column has the functional ties with elements (a_{11},a_{12},a_{16}) of the first row in such sequence as:

$$a_{53} \rightarrow a_{64}$$
$$a_{63} \rightarrow a_{11},a_{12},a_{16}$$
$$a_{54} \rightarrow a_{63}$$
$$a_{64} \rightarrow a_{11},a_{12},a_{16}$$

For elements of the first row

- The first element(a_{11}) of the first row has the functional ties with elements$(a_{22},a_{22},a_{22},a_{26},a_{26},a_{26})$ of the second row for positive and negative areas;
- The second element(a_{12}) of the first row has the functional ties with elements $(a_{21},a_{21},a_{21},a_{26},a_{26},a_{26})$ of the second row for positive and negative areas;
- The sixth element(a_{16}) of the first row has the functional ties with elements $(a_{21},a_{21},a_{21},a_{22},a_{22},a_{22})$ of the second row for positive and negative areas, forming the frequency of functional ties equal f=3 for above-indicated elements(a_{11},a_{12},a_{16}) of the first row.

For elements of the second row

- The first element(a_{21}) of the second row has the functional ties with elements$(a_{42},a_{42},a_{46},a_{46})$ of the fourth row and elements (a_{52},a_{56}) of the fifth row for positive and negative areas;
- The second element $(a_{22})$of the second row has the functional ties with elements$(a_{41},a_{41},a_{46},a_{46})$ of the fourth row and elements (a_{51},a_{56}) of the fifth row for positive and negative areas;
- The sixth element(a_{26}) of the second row has the functional ties with elements $(a_{41},a_{41},a_{42},a_{42})$ of the fourth row and elements(a_{51},a_{52}) of the fifth row for positive and negative areas, forming the frequency of functional ties equal f=6 in such sequence as:

$$a_{21} \rightarrow a_{42},a_{42},a_{46},a_{46},a_{52},a_{56}$$
$$a_{22} \rightarrow a_{41},a_{41},a_{46},a_{46},a_{51},a_{56}$$
$$a_{26} \rightarrow a_{41},a_{41},a_{42},a_{42},a_{51},a_{52}$$

Analysis of functional ties between elements of the fourth and fifth rows show the following facts when elements (a_{41},a_{42},a_{46}) of the fourth row have the frequency of ties equal f=4 and elements(a_{51},a_{52},a_{56}) of the fifth row have the frequency of ties equal f=2 for positive and negative areas.

In Figure 187 is shown the functional ties for combined partial determinant $|A_{25}|$ of the 6x6 matrix.

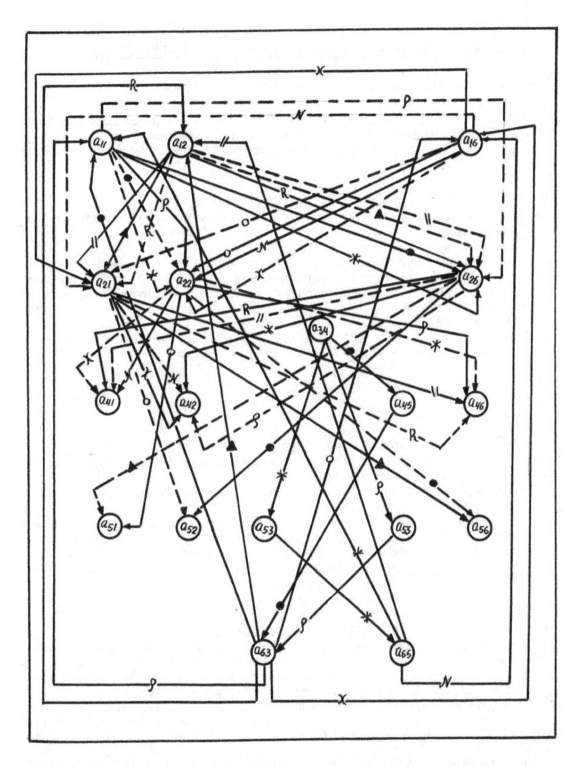

Figure 187 The functional ties between elements for combined partial determinant $|A_{25}|$ of the 6x6 matrix applicably to Laplace's theorem at marked out jointly the third, fourth and fifth columns for positive(———) and negative(------) areas

309

A.A. Analysis of combined partial determinant $|A_{26}|$ of the 6x6 matrix:

❖ For Figure 188 of the combined partial determinant $|A_{26}|$

- The combined partial determinant $|A_{26}|$ of the 6x6 matrix in general view includes the fourth element(a_{34}) of the third row which has the functional ties with element(a_{43}) of the third column and element (a_{45}) of the fifth column ,forming the frequency equal f=1 for above-indicated elements (a_{43},a_{45}).

For elements of the third and fifth columns
- The fourth element(a_{43}) of the third column has the functional ties with elements(a_{55},a_{65}) of the fifth column for positive area, forming the frequency of tie equal f=1 for above-indicated elements (a_{55},a_{65});
- The fourth element(a_{45}) of the fifth column has the functional tie with element(a_{53}) of the third column for positive area, forming the frequency of tie equal f=1 for above-indicated element(a_{53});
- The fifth element(a_{55}) of the fifth column has the functional ties with elements(a_{11},a_{12},a_{16}) of the first row for positive area;
- The sixth element (a_{65}) of the fifth column has the functional ties with elements(a_{11},a_{12},a_{16}) of the first row for positive area, forming the frequency of functional ties equal f=1 for above-indicated elements(a_{43},a_{45},a_{55},a_{65}) of the third and fifth columns in such sequence as:

$$a_{43} \rightarrow a_{55},a_{65}$$
$$a_{45} \rightarrow a_{53}$$
$$a_{55} \rightarrow a_{11},a_{12},a_{16}$$
$$a_{65} \rightarrow a_{11},a_{12},a_{16}$$

For elements of the first row
- The first element(a_{11}) of the first row has the functional ties with elements(a_{22},a_{22},a_{22},a_{26},a_{26},a_{26}) of the second row for positive and negative areas;
- The second element(a_{12}) of the first row has the functional ties with elements(a_{21},a_{21},a_{21},a_{26},a_{26},a_{26}) of the second row for positive and negative areas;
- The sixth element (a_{16}) of the first row has the functional ties with elements (a_{21},a_{21},a_{21},a_{22},a_{22},a_{22}) of the second row for positive and negative areas, forming the frequency of functional ties equal f=3 for above-indicated elements (a_{11},a_{12},a_{16}) of the first row in such sequence as:

$$a_{11} \rightarrow a_{22},a_{22},a_{22},a_{26},a_{26},a_{26}$$
$$a_{12} \rightarrow a_{21},a_{21},a_{21},a_{26},a_{26},a_{26}$$
$$a_{16} \rightarrow a_{21},a_{21},a_{21},a_{22},a_{22},a_{22}$$

For elements of the second row
- The first element(a_{21}) of the second row has the functional ties with elements (a_{52},a_{56}) of the fifth row and elements(a_{62},a_{62},a_{66},a_{66}) of the sixth row for positive and negative areas;
- The second element(a_{22}) of the second row has the functional ties with elements(a_{51},a_{56}) of the fifth row and elements(a_{61},a_{61},a_{66},a_{66}) of the sixth row for positive and negative areas;
- The sixth element(a_{26}) of the second row has the functional ties with elements(a_{51},a_{52}) of the fifth row and elements(a_{61},a_{61},a_{62},a_{62}) of the sixth row for positive and negative areas, forming the frequency of functional ties equal f=6 for above-indicated elements(a_{21},a_{22},a_{26}) of the second row.

Analysis of functional ties between elements of the fifth and sixth rows show the following fact that frequency of ties for elements (a_{51},a_{52},a_{56}) of the fifth row is equal f=2 and the frequency of ties for elements (a_{61},a_{62},a_{66}) of the sixth row is equal f=4 for positive and negative areas.
In Figure 188 is shown the functional ties for combined partial determinant $|A_{26}|$ of the 6x6 matrix.

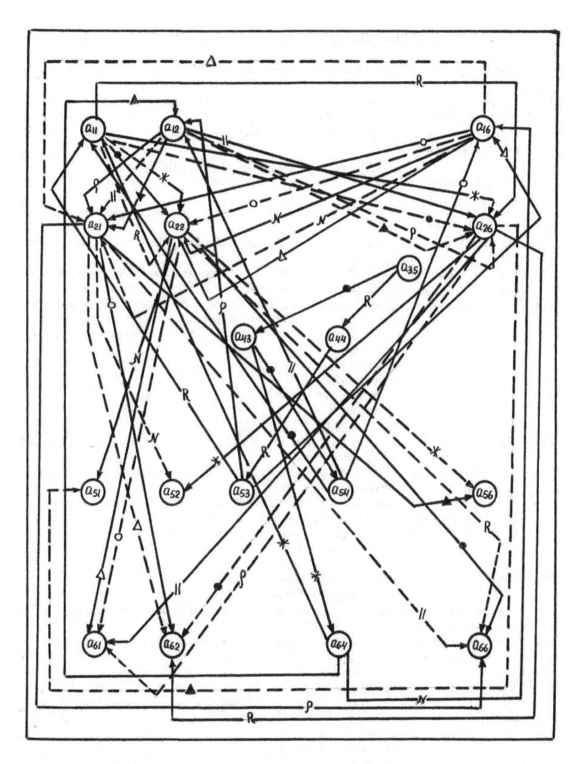

Figure 188 The functional ties between elements for combined partial determinant $\left|A_{26}\right|$ of the 6x6 matrix applicably to Laplace's theorem at marked out jointly the third, fourth and fifth columns for positive (——) and negative (------) areas

B.B Analysis of combined partial determinant $|A_{27}|$ of the 6x6 matrix:

❖ *For Figure 189 of the combined partial determinant* $|A_{27}|$

- The combined partial determinant $|A_{27}|$ of the 6x6 matrix in general view includes the fourth element(a_{34}) of the third row which has the functional ties with elements (a_{45},a_{53},a_{65}) of the third and fifth columns accordingly , forming the frequency of tie equal f=1 for above-indicated elements (a_{45},a_{53},a_{65}).

For elements of the third and fifth column

- The fifth element(a_{53}) of the third column has the functional tie with element (a_{65}) for positive area;
- The fourth element(a_{45}) of the fifth column has the functional tie with the sixth element(a_{63}) of the third column for positive area;
- The fifth element(a_{55}) of the fifth column has the functional tie with the sixth element(a_{63}) of the third column for positive area;
- The sixth element(a_{63}) of the third column has the functional ties with elements (a_{11},a_{11},a_{12},a_{12},a_{16},a_{16}) of the first row for positive area;
- The sixth element(a_{65}) of the fifth column has the functional ties with elements(a_{11},a_{12},a_{16}) of the first row for positive area, forming the frequency of functional ties equal f=2 for above-indicated element(a_{63}) and frequency equal f=1 for elements (a_{45},a_{53},a_{55},a_{65}) in such sequence as:

$$a_{53} \to a_{65}$$
$$a_{45} \to a_{63}$$
$$a_{55} \to a_{63}$$
$$a_{63} \to a_{11},a_{11},a_{12},a_{12},a_{16},a_{16}$$
$$a_{65} \to a_{11},a_{12},a_{16}$$

For elements of the first row

- The first element (a_{11}) of the first row has the functional ties with elements(a_{22},a_{22},a_{22},a_{26},a_{26},a_{26})of the second row for positive and negative areas;
- The second element(a_{12}) of the first row has the functional ties with elements(a_{21},a_{21},a_{21},a_{26},a_{26},a_{26}) of the second row for positive and negative areas;
- The sixth element(a_{16}) of the first row has the functional ties with elements(a_{21},a_{21},a_{21},a_{22},a_{22},a_{22}) of the second row for positive and negative areas, forming the frequency of functional ties equal f=3 for above-indicated elements(a_{11},a_{12},a_{16}) in such sequence as:

$$a_{11} \to a_{22},a_{22},a_{22},a_{26},a_{26},a_{26}$$
$$a_{12} \to a_{21},a_{21},a_{21},a_{26},a_{26},a_{26}$$
$$a_{16} \to a_{21},a_{21},a_{21},a_{22},a_{22},a_{22}$$

For elements of the second row

- The first element(a_{21}) of the second row has the functional ties with elements(a_{42},a_{42},a_{46},a_{46}) of the fourth row and elements(a_{52},a_{56}) of the fifth row for positive and negative areas;
- The second element (a_{22}) of the second row has the functional ties with elements(a_{41},a_{41},a_{46},a_{46}) of the fourth row and elements(a_{51},a_{56}) of the fifth row for positive and negative areas;
- The sixth element(a_{26}) of the second row has the functional ties with elements(a_{41},a_{41},a_{42},a_{42}) of the fourth row and elements(a_{51},a_{52}) of the fifth row for positive and negative areas , forming the frequency of functional ties equal f=6 for above-indicated elements (a_{21},a_{22},a_{26}) in such sequence as:

$$a_{21} \to a_{42},a_{42},a_{46},a_{46},a_{52},a_{56}$$
$$a_{22} \to a_{41},a_{41},a_{46},a_{46},a_{51},a_{56}$$
$$a_{26} \to a_{41},a_{41},a_{42},a_{42},a_{51},a_{52}$$

Analysis of functional ties show the following facts about of frequency for elements of the fourth and fifth rows, such as: the frequency of ties is equal f=4 for elements (a_{41},a_{42},a_{46}) of the fourth row and frequency of ties is equal f=2 for elements (a_{51},a_{52},a_{56}) of the fifth row.

In Figure 189 is shown the functional ties for combined partial determinant $|A_{27}|$ of the 6x6 matrix.

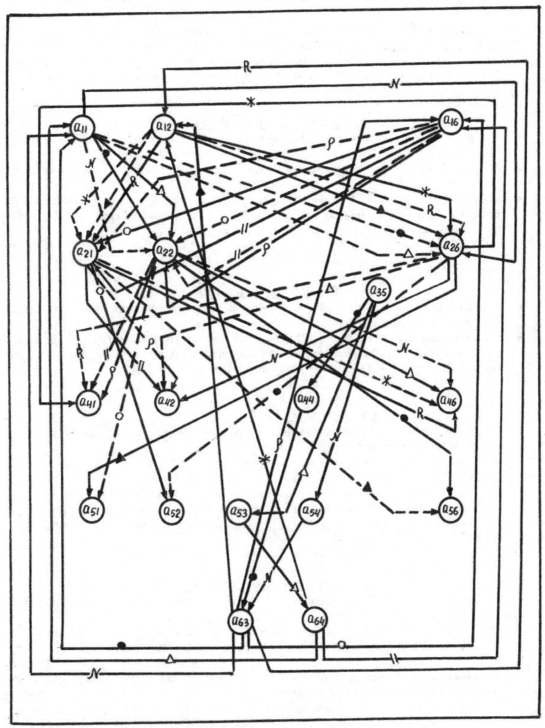

Figure 189 The functional ties between elements for combined partial determinant $|A_{27}|$ of the 6x6 matrix applicably to Laplace's theorem at marked out jointly the third, fourth and fifth columns for positive(———) and negative (------) areas

313

C.C. Analysis of combined partial determinant $\left| A_{28} \right|$ of the 6x6 matrix:

❖ *For Figure 190 of the combined partial determinant* $\left| A_{28} \right|$

- The combined partial determinant $\left| A_{28} \right|$ of the 6x6 matrix in general view includes both elements(a_{43},a_{44}) of the fourth row which are joined with elements(a_{53},a_{54},a_{55}) of the fifth row, forming the frequency equal f=1 with such functional ties for above-indicated elements (a_{43},a_{44}) as:

$$a_{43} \rightarrow a_{54}, a_{55}$$
$$a_{44} \rightarrow a_{53}$$

For elements of the fifth row

- The third element(a_{53}) of the fifth row has the functional tie with element(a_{65}) of the sixth row for positive area;
- The fourth element(a_{54}) of the fifth row has the functional tie with element(a_{65}) of the sixth row for positive area;
- The fifth element(a_{55}) of the fifth row has the functional tie with element(a_{64}) of the sixth row for positive area, forming the frequency of tie equal f=1 for above-indicated elements(a_{53},a_{54},a_{55}) of the fifth row in such sequence as:

$$a_{53} \rightarrow a_{65}$$
$$a_{54} \rightarrow a_{65}$$
$$a_{55} \rightarrow a_{64}$$

For elements of the sixth row

- The fourth element (a_{64})of the sixth row has the functional ties with elements(a_{11},a_{12},a_{16}) of the first row for positive area;
- The fifth element(a_{65}) of the sixth row has the functional ties with elements(a_{11},a_{11},a_{12},a_{12},a_{16},a_{16}) of the first row for positive area, forming the frequency of ties equal f=1 for element(a_{64}) and equal f=2 for element (a_{65}) of the sixth row in such sequence as:

$$a_{64} \rightarrow a_{11}, a_{12}, a_{16}$$
$$a_{65} \rightarrow a_{11}, a_{11}, a_{12}, a_{12}, a_{16}, a_{16}$$

For elements of the first row

- The first element(a_{11}) of the first row has the functional ties with elements(a_{22},a_{22},a_{22},a_{26},a_{26},a_{26}) of the second row for positive and negative areas;
- The second element (a_{12}) of the first row has the functional ties with elements (a_{21},a_{21},a_{21},a_{26},a_{26},a_{26}) of the second row for positive and negative areas;
- The sixth element(a_{16}) of the first row has the functional ties with elements (a_{21},a_{21},a_{21},a_{22},a_{22},a_{22}) of the second row for positive and negative areas, forming the frequency of ties equal f=3 for above-indicated elements (a_{11},a_{12},a_{16}) of the first row in such sequence as:

$$a_{11} \rightarrow a_{22}, a_{22}, a_{22}, a_{26}, a_{26}, a_{26}$$
$$a_{12} \rightarrow a_{21}, a_{21}, a_{21}, a_{26}, a_{26}, a_{26}$$
$$a_{16} \rightarrow a_{21}, a_{21}, a_{21}, a_{22}, a_{22}, a_{22}$$

For elements of the second row

- The first element (a_{21}) of the second row has the functional ties with elements(a_{32},a_{32},a_{32},a_{36},a_{36},a_{36}) of the third row for positive and negative areas;
- The second element(a_{22}) of the second row has the functional ties with elements(a_{31},a_{31},a_{31},a_{36},a_{36},a_{36}) of the third row for positive and negative areas;
- The sixth element(a_{26}) of the second row has the functional ties with elements(a_{31},a_{31},a_{31},a_{32},a_{32},a_{32}) of the third row for positive and negative areas, forming the frequency of ties equal f=6 in such sequence as:

$$a_{21} \rightarrow a_{32}, a_{32}, a_{32}, a_{36}, a_{36}, a_{36}$$
$$a_{22} \rightarrow a_{31}, a_{31}, a_{31}, a_{36}, a_{36}, a_{36}$$
$$a_{26} \rightarrow a_{31}, a_{31}, a_{31}, a_{32}, a_{32}, a_{32}$$

Analysis of functional ties between of the third row elements show the following fact that each element(a_{31},a_{32},a_{36}) has the frequency of ties equal f=6 for positive and negative areas.

In Figure 190 is shown the functional ties for combined partial determinant $\left| A_{28} \right|$ of the 6x6 matrix.

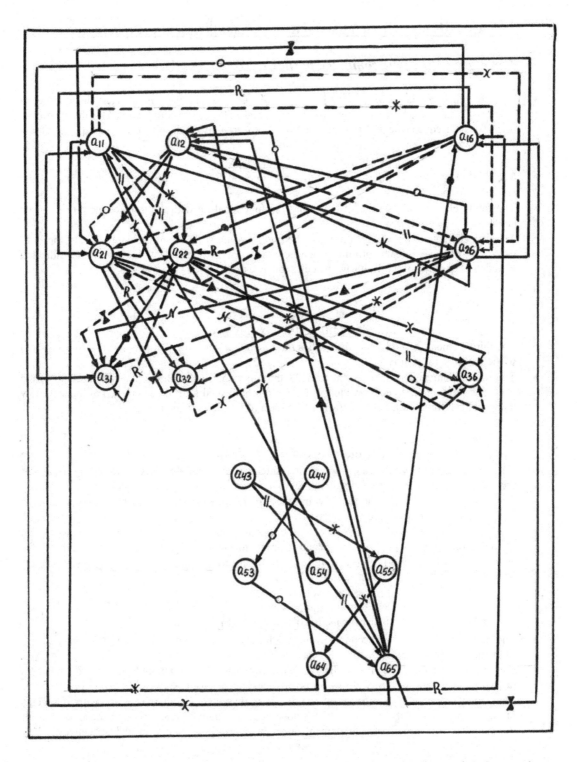

Figure 190 The functional ties between elements for combined partial determinant $|A_{28}|$ of the 6x6 matrix applicably to Laplace's theorem at marked out jointly the third, fourth and fifth columns for positive(———) and negative (-------) areas

315

D.D. Analysis of combined partial determinant $|A_{29}|$ of the 6x6 matrix:

❖ *For Figure 191 of the combined partial determinant* $|A_{29}|$

- The combined partial determinant $|A_{29}|$ of the 6x6 matrix in general view includes both elements (a_{44},a_{45}) of the fourth row which are joined with elements (a_{53},a_{54},a_{55}) of the fifth row, forming the frequency equal f=1 of functional tie for above-indicated elements(a_{44},a_{45}) of the fourth row: $\quad a_{44}{\rightarrow}a_{55}$; $\quad a_{45}{\rightarrow}a_{53},a_{54}$

For elements of the fifth row

- The third element(a_{53}) of the fifth row has the functional tie with the fourth element(a_{64}) of the sixth row for positive area;
- The fourth element (a_{54}) of the fifth row has the functional tie with the third element (a_{63}) of the sixth row for positive area;
- The fifth element(a_{55}) of the fifth row has the functional tie with the third element(a_{63}) of the sixth row for positive area, forming the frequency of tie equal f=1 for above-indicated elements(a_{53},a_{54},a_{55}) of the fifth row in such sequence as:

$$a_{53}{\rightarrow}a_{64}$$
$$a_{54}{\rightarrow}a_{63}$$
$$a_{55}{\rightarrow}a_{63}$$

For elements of the sixth row

- The third element (a_{63}) of the sixth row has the functional ties with elements $(a_{11},a_{11},a_{12},a_{12},a_{16},a_{16})$ of the first row for positive area;
- The fourth element(a_{64}) of the sixth row has the functional ties with elements (a_{11},a_{12},a_{16}) of the first row, forming the frequency equal f=1 for above-indicated fourth element(a_{64}) of the sixth row and equal f=2 for the third element(a_{63}) of the sixth row in such sequence as:

$$a_{63}{\rightarrow}a_{11},a_{11},a_{12},a_{12},a_{16},a_{16}$$
$$a_{64}{\rightarrow}a_{11},a_{12},a_{16}$$

For elements of the first row

- The first element $(a_{11})$of the first row has the functional ties with elements$(a_{22},a_{22},a_{22},a_{26},a_{26},a_{26})$ of the second row for positive and negative areas;
- The second element(a_{12}) of the first row has the functional ties with elements$(a_{21},a_{21},a_{21},a_{26},a_{26},a_{26})$ of the second row for positive and negative areas;
- The sixth element (a_{16}) of the first row has the functional ties with elements $(a_{21},a_{21},a_{21},a_{22},a_{22},a_{22})$ of the second row for positive and negative areas, forming the frequency of ties equal f=3 for above-indicated elements (a_{11},a_{12},a_{16}) of the first row in such sequence as:

$$a_{11}{\rightarrow}a_{22},a_{22},a_{22},a_{26},a_{26},a_{26}$$
$$a_{12}{\rightarrow}a_{21},a_{21},a_{21},a_{26},a_{26},a_{26}$$
$$a_{16}{\rightarrow}a_{21},a_{21},a_{21},a_{22},a_{22},a_{22}$$

For elements of the second row

- The first element $(a_{21})$of the second row has the functional ties with elements$(a_{32},a_{32},a_{32},a_{36},a_{36},a_{36})$ of the third row for positive and negative areas;
- The second element (a_{22}) of the second row has the functional ties with elements $(a_{31},a_{31},a_{31},a_{36},a_{36},a_{36})$ of the third row for positive and negative areas;
- The sixth element (a_{26}) of the second row has the functional ties with elements $(a_{31},a_{31},a_{31},a_{32},a_{32},a_{32})$ of the third row for positive and negative areas, forming the frequency of ties equal f=6 for above-indicated elements(a_{21},a_{22},a_{26}) of the second row in such sequence as:

$$a_{21}{\rightarrow}a_{32},a_{32},a_{32},a_{36},a_{36},a_{36}$$
$$a_{22}{\rightarrow}a_{31},a_{31},a_{31},a_{36},a_{36},a_{36}$$
$$a_{26}{\rightarrow}a_{31},a_{31},a_{31},a_{32},a_{32},a_{32}$$

Analysis of functional ties between of the third row elements show the following fact that each element(a_{31},a_{32},a_{36}) has the frequency of ties equal f=6 for positive and negative areas.

In Figure 191 is shown the functional ties for combined partial determinant $|A_{29}|$ of the 6x6 matrix.

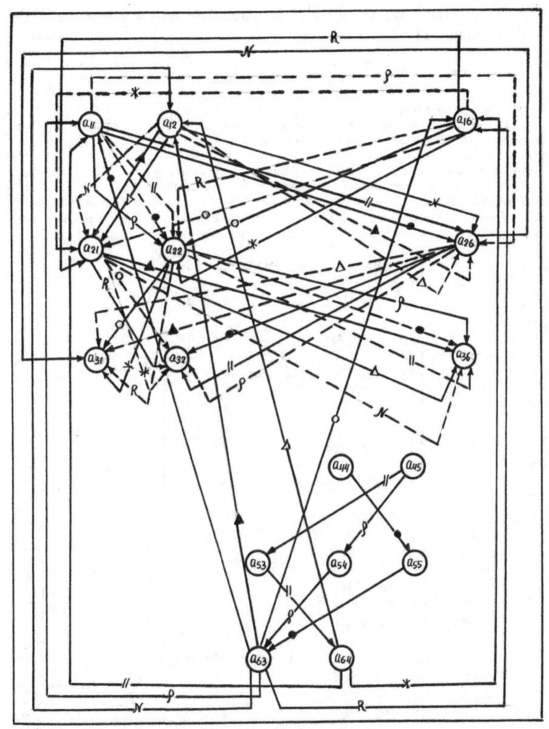

Figure 191 The functional ties between elements for combined partial determinant $\left|A_{29}\right|$ of the 6x6 matrix applicably to Laplace's theorem at marked out jointly the third, fourth and fifth columns for positive(——) and negative (----) areas

317

2. The general algorithms and functional graphs for solving of the determinant of the 7x7 matrix

2.1 By Laplace's theorem for some rows

The above- shown analysis of evaluation determinants for matrices ,having interval of sizes equal $5 \leq n \leq 10$ and more indicated on the essential fact that these determinants have very complicated algorithms, functional graphs and also the calculation procedures particularly for solving of the multiple system of linear equations in general view.

And for this reason with goal of reducing the calculation processes in evaluation of value determinants for above-named matrices advantageously in numerical data should use the designed algorithms,formulas and functional graphs. And as the prove of this conclusion we now consider the well-known Laplace's method [2] for solving determinant $|A|$ of 7x7 matrix($\blacktriangle$) ,for a instance ,when we randomly in this matrix mark out four rows in view of the first,second ,third and fourth rows($i_1+i_2+i_3+i_4$):

$$A = \begin{vmatrix} a_{11} & a_{12} & a_{13} & a_{14} & a_{15} & a_{16} & a_{17} \\ a_{21} & a_{22} & a_{23} & a_{24} & a_{25} & a_{26} & a_{27} \\ a_{31} & a_{32} & a_{33} & a_{34} & a_{35} & a_{36} & a_{37} \\ a_{41} & a_{42} & a_{43} & a_{44} & a_{45} & a_{46} & a_{47} \\ a_{51} & a_{52} & a_{53} & a_{54} & a_{55} & a_{56} & a_{57} \\ a_{61} & a_{62} & a_{63} & a_{64} & a_{65} & a_{66} & a_{67} \\ a_{71} & a_{72} & a_{73} & a_{74} & a_{75} & a_{76} & a_{77} \end{vmatrix} \qquad (\blacktriangle)$$

And then we write the different minors(M_{ij}) of the fourth order,forming in these rows and corresponding columns the algebraic supplemental of the third order,as this shown in the following view:

$$M_1 = \begin{vmatrix} a_{11} & a_{12} & a_{13} & a_{14} \\ a_{21} & a_{22} & a_{23} & a_{24} \\ a_{31} & a_{32} & a_{33} & a_{34} \\ a_{41} & a_{42} & a_{43} & a_{44} \end{vmatrix} \qquad A_1 = (-1)^{(1+2+3+4)+(1+2+3+4)} \begin{vmatrix} a_{55} & a_{56} & a_{57} \\ a_{65} & a_{66} & a_{67} \\ a_{75} & a_{76} & a_{77} \end{vmatrix}$$

$$M_2 = \begin{vmatrix} a_{11} & a_{12} & a_{13} & a_{15} \\ a_{21} & a_{22} & a_{23} & a_{25} \\ a_{31} & a_{32} & a_{33} & a_{35} \\ a_{41} & a_{42} & a_{43} & a_{45} \end{vmatrix} \qquad A_2 = (-1)^{(1+2+3+4)+(1+2+3+5)} \begin{vmatrix} a_{54} & a_{56} & a_{57} \\ a_{64} & a_{66} & a_{67} \\ a_{74} & a_{76} & a_{77} \end{vmatrix}$$

$$M_3 = \begin{vmatrix} a_{11} & a_{12} & a_{13} & a_{16} \\ a_{21} & a_{22} & a_{23} & a_{26} \\ a_{31} & a_{32} & a_{33} & a_{36} \\ a_{41} & a_{42} & a_{43} & a_{46} \end{vmatrix} \qquad A_3 = (-1)^{(1+2+3+4)+(1+2+3+6)} \begin{vmatrix} a_{54} & a_{55} & a_{57} \\ a_{64} & a_{65} & a_{67} \\ a_{74} & a_{75} & a_{77} \end{vmatrix}$$

318

$$M_4=\begin{vmatrix} a_{11} & a_{12} & a_{13} & a_{17} \\ a_{21} & a_{22} & a_{23} & a_{27} \\ a_{31} & a_{32} & a_{33} & a_{37} \\ a_{41} & a_{42} & a_{43} & a_{47} \end{vmatrix} \qquad A_4=(-1)^{(1+2+3+4)+(1+2+3+7)}\begin{vmatrix} a_{54} & a_{55} & a_{56} \\ a_{64} & a_{65} & a_{66} \\ a_{74} & a_{75} & a_{76} \end{vmatrix}$$

$$M_5=\begin{vmatrix} a_{12} & a_{13} & a_{14} & a_{15} \\ a_{22} & a_{23} & a_{24} & a_{25} \\ a_{32} & a_{33} & a_{34} & a_{35} \\ a_{42} & a_{43} & a_{44} & a_{45} \end{vmatrix} \qquad A_5=(-1)^{(1+2+3+4)+(2+3+4+5)}\begin{vmatrix} a_{51} & a_{56} & a_{57} \\ a_{61} & a_{66} & a_{67} \\ a_{71} & a_{76} & a_{77} \end{vmatrix}$$

and more:

$$M_6=\begin{vmatrix} a_{12} & a_{13} & a_{14} & a_{16} \\ a_{22} & a_{23} & a_{24} & a_{26} \\ a_{32} & a_{33} & a_{34} & a_{36} \\ a_{42} & a_{43} & a_{44} & a_{46} \end{vmatrix} \qquad A_6=(-1)^{(1+2+3+4)+(2+3+4+6)}\begin{vmatrix} a_{51} & a_{55} & a_{57} \\ a_{61} & a_{65} & a_{67} \\ a_{71} & a_{75} & a_{77} \end{vmatrix}$$

$$M_7=\begin{vmatrix} a_{12} & a_{13} & a_{14} & a_{17} \\ a_{22} & a_{23} & a_{24} & a_{27} \\ a_{32} & a_{33} & a_{34} & a_{37} \\ a_{42} & a_{43} & a_{44} & a_{47} \end{vmatrix} \qquad A_7=(-1)^{(1+2+3+4)+(2+3+4+7)}\begin{vmatrix} a_{51} & a_{55} & a_{56} \\ a_{61} & a_{65} & a_{66} \\ a_{71} & a_{75} & a_{76} \end{vmatrix}$$

$$M_8=\begin{vmatrix} a_{13} & a_{14} & a_{15} & a_{16} \\ a_{23} & a_{24} & a_{25} & a_{26} \\ a_{33} & a_{34} & a_{35} & a_{36} \\ a_{43} & a_{44} & a_{45} & a_{46} \end{vmatrix} \qquad A_8=(-1)^{(1+2+3+4)+(3+4+5+6)}\begin{vmatrix} a_{51} & a_{52} & a_{57} \\ a_{61} & a_{62} & a_{67} \\ a_{71} & a_{72} & a_{77} \end{vmatrix}$$

$$M_9=\begin{vmatrix} a_{13} & a_{14} & a_{15} & a_{17} \\ a_{23} & a_{24} & a_{25} & a_{27} \\ a_{33} & a_{34} & a_{35} & a_{37} \\ a_{43} & a_{44} & a_{45} & a_{47} \end{vmatrix} \qquad A_9=(-1)^{(1+2+3+4)+(3+4+5+7)}\begin{vmatrix} a_{51} & a_{52} & a_{56} \\ a_{61} & a_{62} & a_{66} \\ a_{71} & a_{72} & a_{76} \end{vmatrix}$$

$$M_{10}=\begin{vmatrix} a_{14} & a_{15} & a_{16} & a_{17} \\ a_{24} & a_{25} & a_{26} & a_{27} \\ a_{34} & a_{35} & a_{36} & a_{37} \\ a_{44} & a_{45} & a_{46} & a_{47} \end{vmatrix} \qquad A_{10}=(-1)^{(1+2+3+4)+(4+5+6+7)}\begin{vmatrix} a_{51} & a_{52} & a_{53} \\ a_{61} & a_{62} & a_{63} \\ a_{71} & a_{72} & a_{73} \end{vmatrix}$$

So, after of some transformations and computations we can define the value of determinant $|A|$ of 7x7 matrix in accordance with Laplace's method:

$$|A| = M_1A_1 - M_2A_2 + M_3A_3 - M_4A_4 + M_5A_5 - M_6A_6 + M_7A_7 + M_8A_8 - M_9A_9 + M_{10}A_{10} =$$
$$= |D_1|A_1 - |D_2|A_2 + |D_3|A_3 - |D_4|A_4 + |D_5|A_5 - |D_6|A_6 + |D_7|A_7 +$$
$$+ |D_8|A_8 - |D_9|A_9 + |D_{10}|A_{10} \qquad (**)$$

where,

$M_1 \ldots\ldots\ldots M_{10}$ –minors of the fourth order for matrix determinant 7x7

$A_1 \ldots\ldots\ldots A_{10}$ -algebraic supplemental

$|D_1| \ldots\ldots |D_{10}|$ -partial determinants for the above-indicated matrix determinant 7x7 which are equal:

$$M_1 = |D_1| = a_{11}(a_{22}a_{33}a_{44} + a_{23}a_{34}a_{42} + a_{24}a_{32}a_{43} - a_{24}a_{33}a_{42} - a_{22}a_{34}a_{43} - a_{23}a_{32}a_{44}) -$$
$$- a_{12}(a_{21}a_{33}a_{44} + a_{23}a_{34}a_{41} + a_{24}a_{31}a_{43} - a_{24}a_{33}a_{41} - a_{21}a_{34}a_{43} - a_{23}a_{31}a_{44}) +$$
$$+ a_{13}(a_{21}a_{32}a_{44} + a_{22}a_{34}a_{41} + a_{24}a_{31}a_{42} - a_{24}a_{32}a_{41} - a_{21}a_{34}a_{42} - a_{22}a_{31}a_{44}) -$$
$$- a_{14}(a_{21}a_{32}a_{43} + a_{22}a_{33}a_{41} + a_{23}a_{31}a_{42} - a_{23}a_{32}a_{41} - a_{21}a_{33}a_{42} - a_{22}a_{31}a_{43})$$

$$M_2 = |D_2| = a_{11}(a_{22}a_{33}a_{45} + a_{23}a_{35}a_{42} + a_{25}a_{32}a_{43} - a_{25}a_{33}a_{42} - a_{22}a_{35}a_{43} - a_{23}a_{32}a_{45}) -$$
$$- a_{12}(a_{21}a_{33}a_{45} + a_{23}a_{35}a_{41} + a_{25}a_{31}a_{43} - a_{25}a_{33}a_{41} - a_{21}a_{35}a_{43} - a_{23}a_{31}a_{45}) +$$
$$+ a_{13}(a_{21}a_{32}a_{45} + a_{22}a_{35}a_{41} + a_{25}a_{31}a_{42} - a_{25}a_{32}a_{41} - a_{21}a_{35}a_{42} - a_{22}a_{31}a_{45}) -$$
$$- a_{15}(a_{21}a_{32}a_{43} + a_{22}a_{33}a_{41} + a_{23}a_{31}a_{42} - a_{23}a_{32}a_{41} - a_{21}a_{33}a_{42} - a_{22}a_{31}a_{43})$$

$$M_3 = |D_3| = a_{11}(a_{22}a_{33}a_{46} + a_{23}a_{36}a_{42} + a_{26}a_{32}a_{43} - a_{26}a_{33}a_{42} - a_{22}a_{36}a_{43} - a_{23}a_{32}a_{46}) -$$
$$- a_{12}(a_{21}a_{33}a_{46} + a_{23}a_{36}a_{41} + a_{26}a_{31}a_{43} - a_{26}a_{33}a_{41} - a_{21}a_{36}a_{43} - a_{23}a_{31}a_{46}) +$$
$$+ a_{13}(a_{21}a_{32}a_{46} + a_{22}a_{36}a_{41} + a_{26}a_{31}a_{42} - a_{26}a_{32}a_{41} - a_{21}a_{36}a_{42} - a_{22}a_{31}a_{46}) -$$
$$- a_{16}(a_{21}a_{32}a_{43} + a_{22}a_{33}a_{41} + a_{23}a_{31}a_{42} - a_{23}a_{32}a_{41} - a_{21}a_{33}a_{42} - a_{22}a_{31}a_{43})$$

$$M_4 = |D_4| = a_{11}(a_{22}a_{33}a_{47} + a_{23}a_{37}a_{42} + a_{27}a_{32}a_{43} - a_{27}a_{33}a_{42} - a_{22}a_{37}a_{43} - a_{23}a_{32}a_{47}) -$$
$$- a_{12}(a_{21}a_{33}a_{47} + a_{23}a_{37}a_{41} + a_{27}a_{31}a_{43} - a_{27}a_{33}a_{41} - a_{21}a_{37}a_{43} - a_{23}a_{31}a_{47}) +$$
$$+ a_{13}(a_{21}a_{32}a_{47} + a_{22}a_{37}a_{41} + a_{27}a_{31}a_{42} - a_{27}a_{32}a_{41} - a_{21}a_{37}a_{42} - a_{22}a_{31}a_{47}) -$$
$$- a_{17}(a_{21}a_{32}a_{43} + a_{22}a_{33}a_{41} + a_{23}a_{31}a_{42} - a_{23}a_{32}a_{41} - a_{21}a_{33}a_{42} - a_{22}a_{31}a_{43})$$

$$M_5 = |D_5| = a_{12}(a_{23}a_{34}a_{45} + a_{24}a_{35}a_{43} + a_{25}a_{33}a_{44} - a_{25}a_{34}a_{43} - a_{23}a_{35}a_{44} - a_{24}a_{33}a_{45}) -$$
$$- a_{13}(a_{22}a_{34}a_{45} + a_{24}a_{35}a_{42} + a_{25}a_{32}a_{44} - a_{25}a_{34}a_{42} - a_{22}a_{35}a_{44} - a_{24}a_{32}a_{45}) +$$
$$+ a_{14}(a_{22}a_{33}a_{45} + a_{23}a_{35}a_{42} + a_{25}a_{32}a_{43} - a_{25}a_{33}a_{42} - a_{22}a_{35}a_{43} - a_{23}a_{32}a_{45}) -$$
$$- a_{15}(a_{22}a_{33}a_{44} + a_{23}a_{34}a_{42} + a_{24}a_{32}a_{43} - a_{24}a_{33}a_{42} - a_{22}a_{34}a_{43} - a_{23}a_{32}a_{44})$$

$$M_6 = |D_6| = a_{12}(a_{23}a_{34}a_{46} + a_{24}a_{36}a_{43} + a_{26}a_{33}a_{44} - a_{26}a_{34}a_{43} - a_{23}a_{36}a_{44} - a_{24}a_{33}a_{46}) -$$
$$- a_{13}(a_{22}a_{34}a_{46} + a_{24}a_{36}a_{42} + a_{26}a_{32}a_{44} - a_{26}a_{34}a_{42} - a_{22}a_{36}a_{44} - a_{24}a_{32}a_{46}) +$$
$$+ a_{14}(a_{22}a_{33}a_{46} + a_{23}a_{36}a_{42} + a_{26}a_{32}a_{43} - a_{26}a_{33}a_{42} - a_{22}a_{36}a_{43} - a_{23}a_{32}a_{46}) -$$
$$- a_{16}(a_{22}a_{33}a_{44} + a_{23}a_{34}a_{42} + a_{24}a_{32}a_{43} - a_{24}a_{33}a_{42} - a_{22}a_{34}a_{43} - a_{23}a_{32}a_{44})$$

$$M_7 = |D_7| = a_{12}(a_{23}a_{34}a_{47} + a_{24}a_{37}a_{43} + a_{27}a_{33}a_{44} - a_{27}a_{34}a_{43} - a_{23}a_{37}a_{44} - a_{24}a_{33}a_{47}) -$$
$$- a_{13}(a_{22}a_{34}a_{47} + a_{24}a_{37}a_{42} + a_{27}a_{32}a_{44} - a_{27}a_{34}a_{42} - a_{22}a_{37}a_{44} - a_{24}a_{32}a_{47}) +$$
$$+ a_{14}(a_{22}a_{33}a_{47} + a_{23}a_{37}a_{42} + a_{27}a_{32}a_{43} - a_{27}a_{33}a_{42} - a_{22}a_{37}a_{43} - a_{23}a_{32}a_{47}) -$$
$$- a_{17}(a_{22}a_{33}a_{44} + a_{23}a_{34}a_{42} + a_{24}a_{32}a_{43} - a_{24}a_{33}a_{42} - a_{22}a_{34}a_{43} - a_{23}a_{32}a_{44})$$

$M_8 = |D_8| = a_{13}(a_{24}a_{35}a_{46}+a_{25}a_{36}a_{44}+a_{26}a_{34}a_{45}-a_{26}a_{35}a_{44}-a_{24}a_{36}a_{45}-a_{25}a_{34}a_{46})-$

$-a_{14}(a_{23}a_{35}a_{46}+a_{25}a_{36}a_{43}+a_{26}a_{33}a_{45}-a_{26}a_{35}a_{43}-a_{23}a_{36}a_{45}-a_{25}a_{33}a_{46})+$

$+a_{15}(a_{23}a_{34}a_{46}+a_{24}a_{36}a_{43}+a_{26}a_{33}a_{44}-a_{26}a_{34}a_{43}-a_{23}a_{36}a_{44}-a_{24}a_{33}a_{46})-$

$-a_{16}(a_{23}a_{34}a_{45}+a_{24}a_{35}a_{43}+a_{25}a_{33}a_{44}-a_{25}a_{34}a_{43}-a_{23}a_{35}a_{44}-a_{24}a_{33}a_{45})$

$M_9 = |D_9| = a_{13}(a_{24}a_{35}a_{47}+a_{25}a_{37}a_{44}+a_{27}a_{34}a_{45}-a_{27}a_{35}a_{44}-a_{24}a_{37}a_{45}-a_{25}a_{34}a_{47})-$

$-a_{14}(a_{23}a_{35}a_{47}+a_{25}a_{37}a_{43}+a_{27}a_{33}a_{45}-a_{27}a_{35}a_{43}-a_{23}a_{37}a_{45}-a_{25}a_{33}a_{47})+$

$+a_{15}(a_{23}a_{34}a_{47}+a_{24}a_{37}a_{43}+a_{27}a_{33}a_{44}-a_{27}a_{34}a_{43}-a_{23}a_{37}a_{44}-a_{24}a_{33}a_{47})-$

$-a_{17}(a_{23}a_{34}a_{45}+a_{24}a_{35}a_{43}+a_{25}a_{33}a_{44}-a_{25}a_{34}a_{43}-a_{23}a_{35}a_{44}-a_{24}a_{33}a_{45})$

$M_{10} = |D_{10}| = a_{14}(a_{25}a_{36}a_{47}+a_{26}a_{37}a_{45}+a_{27}a_{35}a_{46}-a_{27}a_{36}a_{45}-a_{25}a_{37}a_{46}-a_{26}a_{35}a_{47})-$

$-a_{15}(a_{24}a_{36}a_{47}+a_{26}a_{37}a_{44}+a_{27}a_{34}a_{46}-a_{27}a_{36}a_{44}-a_{24}a_{37}a_{46}-a_{26}a_{34}a_{47})+$

$+a_{16}(a_{24}a_{35}a_{47}+a_{25}a_{37}a_{44}+a_{27}a_{34}a_{45}-a_{27}a_{35}a_{44}-a_{24}a_{37}a_{45}-a_{25}a_{34}a_{47})-$

$-a_{17}(a_{24}a_{35}a_{46}+a_{25}a_{36}a_{44}+a_{26}a_{34}a_{45}-a_{26}a_{35}a_{44}-a_{24}a_{36}a_{45}-a_{25}a_{34}a_{46})$

And finally the formula for evaluation of the matrix determinant 7x7 at Laplace's method has the following view:

$|A| = a_{11}a_{23}a_{32}a_{44}[a_{55}(a_{67}a_{76}-a_{66}a_{77})+a_{56}(a_{65}a_{77}-a_{67}a_{75})+a_{57}(a_{66}a_{75}-a_{65}a_{76})]+$

$+a_{11}a_{23}a_{32}a_{45}[a_{54}(a_{66}a_{77}-a_{67}a_{76})+a_{56}(a_{67}a_{74}-a_{64}a_{77})+a_{57}(a_{64}a_{76}-a_{66}a_{74})]+$

$+a_{11}a_{23}a_{32}a_{46}[a_{54}(a_{67}a_{75}-a_{65}a_{77})+a_{55}(a_{64}a_{77}-a_{67}a_{74})+a_{57}(a_{65}a_{74}-a_{64}a_{75})]+$

$+a_{11}a_{23}a_{32}a_{47}[a_{54}(a_{65}a_{76}-a_{66}a_{75})+a_{55}(a_{66}a_{74}-a_{64}a_{76})+a_{56}(a_{64}a_{75}-a_{65}a_{74})]+$

$+a_{11}a_{23}a_{34}a_{42}[a_{55}(a_{66}a_{77}-a_{67}a_{76})+a_{56}(a_{67}a_{75}-a_{65}a_{77})+a_{57}(a_{65}a_{76}-a_{66}a_{75})]+$

$+a_{11}a_{23}a_{36}a_{42}[a_{54}(a_{65}a_{77}-a_{67}a_{75})+a_{55}(a_{67}a_{74}-a_{64}a_{77})+a_{57}(a_{64}a_{75}-a_{65}a_{74})]+$

$+a_{11}a_{23}a_{37}a_{42}[a_{54}(a_{66}a_{75}-a_{65}a_{76})+a_{55}(a_{64}a_{76}-a_{66}a_{74})+a_{56}(a_{65}a_{74}-a_{64}a_{75})]+$

$+a_{11}a_{23}a_{35}a_{42}[a_{54}(a_{67}a_{76}-a_{66}a_{77})+a_{56}(a_{64}a_{77}-a_{67}a_{74})+a_{57}(a_{66}a_{74}-a_{64}a_{76})]+$

$+a_{11}a_{22}a_{33}a_{46}[a_{54}(a_{65}a_{77}-a_{67}a_{75})+a_{55}(a_{67}a_{74}-a_{64}a_{77})+a_{57}(a_{64}a_{75}-a_{65}a_{74})]+$

$+a_{11}a_{22}a_{35}a_{43}[a_{54}(a_{66}a_{77}-a_{67}a_{76})+a_{56}(a_{67}a_{74}-a_{64}a_{77})+a_{57}(a_{64}a_{76}-a_{66}a_{74})]+$

$+a_{11}a_{22}a_{36}a_{43}[a_{54}(a_{67}a_{75}-a_{65}a_{77})+a_{55}(a_{64}a_{77}-a_{67}a_{74})+a_{57}(a_{65}a_{74}-a_{64}a_{75})]+$

$+a_{11}a_{22}a_{33}a_{47}[a_{54}(a_{66}a_{75}-a_{65}a_{76})+a_{55}(a_{64}a_{76}-a_{66}a_{74})+a_{56}(a_{65}a_{74}-a_{64}a_{75})]+$

$+a_{11}a_{22}a_{37}a_{43}[a_{54}(a_{65}a_{76}-a_{66}a_{75})+a_{55}(a_{66}a_{74}-a_{64}a_{76})+a_{56}(a_{64}a_{75}-a_{65}a_{74})]+$

$+a_{11}a_{22}a_{33}a_{44}[a_{55}(a_{66}a_{77}-a_{67}a_{76})+a_{56}(a_{67}a_{75}-a_{65}a_{77})+a_{57}(a_{65}a_{76}-a_{66}a_{75})]+$

$+a_{11}a_{22}a_{34}a_{43}[a_{55}(a_{67}a_{76}-a_{66}a_{77})+a_{56}(a_{65}a_{77}-a_{67}a_{75})+a_{57}(a_{66}a_{75}-a_{65}a_{76})]+$

$+a_{11}a_{22}a_{33}a_{45}[a_{54}(a_{67}a_{76}-a_{66}a_{77})+a_{56}(a_{64}a_{77}-a_{67}a_{74})+a_{57}(a_{66}a_{74}-a_{64}a_{76})]+$

$+a_{11}a_{22}a_{33}a_{45}[a_{54}(a_{67}a_{76}-a_{66}a_{77})+a_{56}(a_{64}a_{77}-a_{67}a_{74})+a_{57}(a_{66}a_{74}-a_{64}a_{76})]+$

$+a_{11}a_{24}a_{32}a_{43}[a_{55}(a_{66}a_{77}-a_{67}a_{76})+a_{56}(a_{67}a_{75}-a_{65}a_{77})+a_{57}(a_{65}a_{76}-a_{66}a_{75})]+$

$+a_{11}a_{24}a_{33}a_{42}[a_{55}(a_{67}a_{76}-a_{66}a_{77})+a_{56}(a_{65}a_{77}-a_{67}a_{75})+a_{57}(a_{66}a_{75}-a_{65}a_{76})]+$

$+a_{11}a_{25}a_{32}a_{43}[a_{54}(a_{67}a_{76}-a_{66}a_{77})+a_{56}(a_{64}a_{77}-a_{67}a_{74})+a_{57}(a_{66}a_{74}-a_{64}a_{76})]+$

$+a_{11}a_{25}a_{33}a_{42}[a_{54}(a_{66}a_{77}-a_{67}a_{76})+a_{56}(a_{67}a_{74}-a_{64}a_{77})+a_{57}(a_{64}a_{76}-a_{66}a_{74})]+$

$+a_{11}a_{26}a_{32}a_{43}[a_{54}(a_{65}a_{77}-a_{67}a_{75})+a_{55}(a_{67}a_{74}-a_{64}a_{77})+a_{57}(a_{64}a_{75}-a_{65}a_{74})]+$

$+a_{11}a_{26}a_{33}a_{42}[a_{54}(a_{67}a_{75}-a_{65}a_{77})+a_{55}(a_{64}a_{77}-a_{67}a_{74})+a_{57}(a_{65}a_{74}-a_{64}a_{75})]+$

$+a_{11}a_{27}a_{32}a_{43}[a_{54}(a_{66}a_{75}-a_{65}a_{76})+a_{55}(a_{64}a_{76}-a_{66}a_{74})+a_{56}(a_{65}a_{74}-a_{64}a_{75})]+$

$+a_{11}a_{27}a_{33}a_{42}[a_{54}(a_{65}a_{76}-a_{66}a_{75})+a_{55}(a_{66}a_{74}-a_{64}a_{76})+a_{56}(a_{64}a_{75}-a_{65}a_{74})]+$

$$+a_{12}a_{21}a_{33}a_{44}[a_{55}(a_{67}a_{76}-a_{66}a_{77})+a_{56}(a_{65}a_{77}-a_{67}a_{75})+a_{57}(a_{66}a_{75}-a_{65}a_{76})]+$$
$$+a_{12}a_{21}a_{33}a_{45}[a_{54}(a_{66}a_{77}-a_{67}a_{76})+a_{56}(a_{67}a_{74}-a_{64}a_{77})+a_{57}(a_{64}a_{76}-a_{66}a_{74})]+$$
$$+a_{12}a_{21}a_{33}a_{46}[a_{54}(a_{67}a_{75}-a_{65}a_{77})+a_{55}(a_{64}a_{77}-a_{67}a_{74})+a_{57}(a_{65}a_{74}-a_{64}a_{75})]+$$
$$+a_{12}a_{21}a_{33}a_{47}[a_{54}(a_{65}a_{76}-a_{66}a_{75})+a_{55}(a_{66}a_{74}-a_{64}a_{76})+a_{56}(a_{64}a_{75}-a_{65}a_{74})]+$$
$$+a_{12}a_{21}a_{34}a_{43}[a_{55}(a_{66}a_{77}-a_{67}a_{76})+a_{56}(a_{67}a_{75}-a_{65}a_{77})+a_{57}(a_{65}a_{76}-a_{66}a_{75})]+$$
$$+a_{12}a_{21}a_{35}a_{43}[a_{54}(a_{67}a_{76}-a_{66}a_{77})+a_{56}(a_{64}a_{77}-a_{67}a_{74})+a_{57}(a_{66}a_{74}-a_{64}a_{76})]+$$
$$+a_{12}a_{21}a_{36}a_{43}[a_{54}(a_{65}a_{77}-a_{67}a_{75})+a_{55}(a_{67}a_{74}-a_{64}a_{75})+a_{57}(a_{64}a_{75}-a_{65}a_{74})]+$$
$$+a_{12}a_{21}a_{37}a_{43}[a_{54}(a_{66}a_{75}-a_{65}a_{76})+a_{55}(a_{64}a_{76}-a_{66}a_{74})+a_{56}(a_{65}a_{74}-a_{64}a_{75})]+$$
$$+a_{12}a_{23}a_{31}a_{44}[a_{55}(a_{66}a_{77}-a_{67}a_{76})+a_{56}(a_{67}a_{75}-a_{65}a_{77})+a_{57}(a_{65}a_{76}-a_{66}a_{75})]+$$
$$+a_{12}a_{23}a_{31}a_{45}[a_{54}(a_{67}a_{76}-a_{66}a_{77})+a_{56}(a_{64}a_{77}-a_{67}a_{74})+a_{57}(a_{66}a_{74}-a_{64}a_{76})]+$$
$$+a_{12}a_{23}a_{31}a_{46}[a_{54}(a_{65}a_{77}-a_{67}a_{75})+a_{55}(a_{67}a_{74}-a_{64}a_{77})+a_{57}(a_{64}a_{75}-a_{65}a_{74})]+$$
$$+a_{12}a_{23}a_{31}a_{47}[a_{54}(a_{66}a_{75}-a_{65}a_{76})+a_{55}(a_{64}a_{76}-a_{66}a_{74})+a_{56}(a_{65}a_{74}-a_{64}a_{75})]+$$
$$+a_{12}a_{23}a_{34}a_{41}[a_{55}(a_{67}a_{76}-a_{66}a_{77})+a_{56}(a_{65}a_{77}-a_{67}a_{75})+a_{57}(a_{66}a_{75}-a_{65}a_{76})]+$$
$$+a_{12}a_{23}a_{34}a_{45}[a_{51}(a_{66}a_{77}-a_{67}a_{76})+a_{56}(a_{67}a_{71}-a_{61}a_{77})+a_{57}(a_{61}a_{76}-a_{66}a_{71})]+$$
$$+a_{12}a_{23}a_{34}a_{46}[a_{51}(a_{67}a_{75}-a_{65}a_{77})+a_{55}(a_{61}a_{77}-a_{67}a_{71})+a_{57}(a_{65}a_{71}-a_{61}a_{75})]+$$
$$+a_{12}a_{23}a_{34}a_{47}[a_{51}(a_{65}a_{76}-a_{66}a_{75})+a_{55}(a_{66}a_{71}-a_{61}a_{76})+a_{56}(a_{61}a_{75}-a_{65}a_{71})]+$$
$$+a_{12}a_{23}a_{35}a_{41}[a_{54}(a_{66}a_{77}-a_{67}a_{76})+a_{56}(a_{67}a_{74}-a_{64}a_{77})+a_{57}(a_{64}a_{76}-a_{66}a_{74})]+$$
$$+a_{12}a_{23}a_{35}a_{44}[a_{51}(a_{67}a_{76}-a_{66}a_{77})+a_{56}(a_{61}a_{77}-a_{67}a_{71})+a_{57}(a_{66}a_{71}-a_{61}a_{76})]+$$
$$+a_{12}a_{23}a_{36}a_{41}[a_{54}(a_{67}a_{75}-a_{65}a_{77})+a_{55}(a_{64}a_{77}-a_{67}a_{74})+a_{57}(a_{65}a_{74}-a_{64}a_{75})]+$$
$$+a_{12}a_{23}a_{36}a_{44}[a_{51}(a_{65}a_{77}-a_{67}a_{75})+a_{55}(a_{67}a_{71}-a_{61}a_{77})+a_{57}(a_{61}a_{75}-a_{65}a_{71})]+$$
$$+a_{12}a_{23}a_{37}a_{41}[a_{54}(a_{65}a_{76}-a_{66}a_{75})+a_{55}(a_{66}a_{74}-a_{64}a_{76})+a_{56}(a_{64}a_{75}-a_{65}a_{74})]+$$
$$+a_{12}a_{23}a_{37}a_{44}[a_{51}(a_{66}a_{75}-a_{65}a_{76})+a_{55}(a_{61}a_{76}-a_{66}a_{71})+a_{56}(a_{65}a_{71}-a_{61}a_{75})]+$$
$$+a_{12}a_{24}a_{31}a_{43}[a_{55}(a_{67}a_{76}-a_{66}a_{77})+a_{56}(a_{65}a_{77}-a_{67}a_{75})+a_{57}(a_{66}a_{75}-a_{65}a_{76})]+$$
$$+a_{12}a_{24}a_{33}a_{41}[a_{55}(a_{66}a_{77}-a_{67}a_{76})+a_{56}(a_{67}a_{75}-a_{65}a_{77})+a_{57}(a_{65}a_{76}-a_{66}a_{75})]+$$
$$+a_{12}a_{24}a_{33}a_{45}[a_{51}(a_{67}a_{76}-a_{66}a_{77})+a_{56}(a_{61}a_{77}-a_{67}a_{71})+a_{57}(a_{66}a_{71}-a_{61}a_{76})]+$$
$$+a_{12}a_{24}a_{33}a_{46}[a_{51}(a_{65}a_{77}-a_{67}a_{75})+a_{55}(a_{67}a_{71}-a_{61}a_{77})+a_{57}(a_{61}a_{75}-a_{65}a_{71})]+$$
$$+a_{12}a_{24}a_{33}a_{47}[a_{51}(a_{66}a_{75}-a_{65}a_{76})+a_{55}(a_{61}a_{76}-a_{66}a_{71})+a_{56}(a_{65}a_{71}-a_{61}a_{75})]+$$
$$+a_{12}a_{24}a_{35}a_{43}[a_{51}(a_{66}a_{77}-a_{67}a_{76})+a_{56}(a_{67}a_{71}-a_{61}a_{77})+a_{57}(a_{61}a_{76}-a_{66}a_{71})]+$$
$$+a_{12}a_{24}a_{36}a_{43}[a_{51}(a_{67}a_{75}-a_{65}a_{77})+a_{55}(a_{61}a_{77}-a_{67}a_{71})+a_{57}(a_{65}a_{71}-a_{61}a_{75})]+$$
$$+a_{12}a_{24}a_{37}a_{43}[a_{51}(a_{65}a_{76}-a_{66}a_{75})+a_{55}(a_{66}a_{71}-a_{61}a_{76})+a_{56}(a_{61}a_{75}-a_{65}a_{71})]+$$
$$+a_{12}a_{25}a_{31}a_{43}[a_{54}(a_{66}a_{77}-a_{67}a_{76})+a_{56}(a_{67}a_{74}-a_{64}a_{77})+a_{57}(a_{64}a_{76}-a_{66}a_{74})]+$$
$$+a_{12}a_{25}a_{33}a_{41}[a_{54}(a_{67}a_{76}-a_{66}a_{77})+a_{56}(a_{64}a_{77}-a_{67}a_{74})+a_{57}(a_{66}a_{74}-a_{64}a_{76})]+$$
$$+a_{12}a_{25}a_{33}a_{44}[a_{51}(a_{66}a_{77}-a_{67}a_{76})+a_{56}(a_{67}a_{71}-a_{61}a_{77})+a_{57}(a_{61}a_{76}-a_{66}a_{71})]+$$
$$+a_{12}a_{25}a_{34}a_{43}[a_{51}(a_{67}a_{76}-a_{66}a_{77})+a_{56}(a_{61}a_{77}-a_{67}a_{71})+a_{57}(a_{66}a_{71}-a_{61}a_{76})]+$$
$$+a_{12}a_{26}a_{31}a_{43}[a_{54}(a_{67}a_{75}-a_{65}a_{77})+a_{55}(a_{64}a_{77}-a_{67}a_{74})+a_{57}(a_{65}a_{74}-a_{64}a_{75})]+$$
$$+a_{12}a_{26}a_{33}a_{41}[a_{54}(a_{65}a_{77}-a_{67}a_{75})+a_{55}(a_{67}a_{74}-a_{64}a_{77})+a_{57}(a_{64}a_{75}-a_{65}a_{74})]+$$
$$+a_{12}a_{26}a_{33}a_{44}[a_{51}(a_{67}a_{75}-a_{65}a_{77})+a_{55}(a_{61}a_{77}-a_{67}a_{71})+a_{57}(a_{65}a_{71}-a_{61}a_{75})]+$$
$$+a_{12}a_{26}a_{34}a_{43}[a_{51}(a_{65}a_{77}-a_{67}a_{75})+a_{55}(a_{67}a_{71}-a_{61}a_{77})+a_{57}(a_{61}a_{75}-a_{65}a_{71})]+$$
$$+a_{12}a_{27}a_{31}a_{43}[a_{54}(a_{65}a_{76}-a_{66}a_{75})+a_{55}(a_{66}a_{74}-a_{64}a_{76})+a_{56}(a_{64}a_{75}-a_{65}a_{74})]+$$
$$+a_{12}a_{27}a_{33}a_{41}[a_{54}(a_{66}a_{75}-a_{65}a_{76})+a_{55}(a_{64}a_{76}-a_{66}a_{74})+a_{56}(a_{65}a_{74}-a_{64}a_{75})]+$$
$$+a_{12}a_{27}a_{33}a_{44}[a_{51}(a_{65}a_{76}-a_{66}a_{75})+a_{55}(a_{66}a_{71}-a_{61}a_{76})+a_{56}(a_{61}a_{75}-a_{65}a_{71})]+$$
$$+a_{12}a_{27}a_{34}a_{43}[a_{51}(a_{66}a_{75}-a_{65}a_{76})+a_{55}(a_{61}a_{76}-a_{66}a_{71})+a_{56}(a_{65}a_{71}-a_{61}a_{75})]+$$

$$+a_{13}a_{21}a_{32}a_{44}[a_{55}(a_{66}a_{77}-a_{67}a_{76})+a_{56}(a_{67}a_{75}-a_{65}a_{77})+a_{57}(a_{65}a_{76}-a_{66}a_{75})]+$$
$$+a_{13}a_{21}a_{32}a_{45}[a_{54}(a_{67}a_{76}-a_{66}a_{77})+a_{56}(a_{64}a_{77}-a_{67}a_{74})+a_{57}(a_{66}a_{74}-a_{64}a_{76})]+$$
$$+a_{13}a_{21}a_{32}a_{46}[a_{54}(a_{65}a_{77}-a_{67}a_{75})+a_{55}(a_{67}a_{74}-a_{64}a_{77})+a_{57}(a_{64}a_{75}-a_{65}a_{74})]+$$
$$+a_{13}a_{21}a_{32}a_{47}[a_{54}(a_{66}a_{75}-a_{65}a_{76})+a_{55}(a_{64}a_{76}-a_{66}a_{74})+a_{56}(a_{65}a_{74}-a_{64}a_{75})]+$$
$$+a_{13}a_{21}a_{34}a_{42}[a_{55}(a_{67}a_{76}-a_{66}a_{77})+a_{56}(a_{65}a_{77}-a_{67}a_{75})+a_{57}(a_{66}a_{75}-a_{65}a_{76})]+$$
$$+a_{13}a_{21}a_{35}a_{42}[a_{54}(a_{66}a_{77}-a_{67}a_{76})+a_{56}(a_{67}a_{74}-a_{64}a_{77})+a_{57}(a_{64}a_{76}-a_{66}a_{74})]+$$
$$+a_{13}a_{21}a_{36}a_{42}[a_{54}(a_{67}a_{75}-a_{65}a_{77})+a_{55}(a_{64}a_{77}-a_{67}a_{74})+a_{57}(a_{65}a_{74}-a_{64}a_{75})]+$$
$$+a_{13}a_{21}a_{37}a_{42}[a_{54}(a_{65}a_{76}-a_{66}a_{75})+a_{55}(a_{66}a_{74}-a_{64}a_{76})+a_{56}(a_{64}a_{75}-a_{65}a_{74})]+$$
$$+a_{13}a_{22}a_{31}a_{44}[a_{55}(a_{67}a_{76}-a_{66}a_{77})+a_{56}(a_{65}a_{77}-a_{67}a_{75})+a_{57}(a_{66}a_{75}-a_{65}a_{76})]+$$
$$+a_{13}a_{22}a_{31}a_{45}[a_{54}(a_{66}a_{77}-a_{67}a_{76})+a_{56}(a_{67}a_{74}-a_{64}a_{77})+a_{57}(a_{64}a_{76}-a_{66}a_{74})]+$$
$$+a_{13}a_{22}a_{31}a_{46}[a_{54}(a_{67}a_{75}-a_{65}a_{77})+a_{55}(a_{64}a_{77}-a_{67}a_{74})+a_{57}(a_{65}a_{74}-a_{64}a_{75})]+$$
$$+a_{13}a_{22}a_{31}a_{47}[a_{54}(a_{65}a_{76}-a_{66}a_{75})+a_{55}(a_{66}a_{74}-a_{64}a_{76})+a_{56}(a_{64}a_{75}-a_{65}a_{74})]+$$
$$+a_{13}a_{22}a_{34}a_{41}[a_{55}(a_{66}a_{77}-a_{67}a_{76})+a_{56}(a_{67}a_{75}-a_{65}a_{77})+a_{57}(a_{65}a_{76}-a_{66}a_{75})]+$$
$$+a_{13}a_{22}a_{34}a_{45}[a_{51}(a_{67}a_{76}-a_{66}a_{77})+a_{56}(a_{61}a_{77}-a_{67}a_{71})+a_{57}(a_{66}a_{71}-a_{61}a_{76})]+$$
$$+a_{13}a_{22}a_{34}a_{46}[a_{51}(a_{65}a_{77}-a_{67}a_{75})+a_{55}(a_{67}a_{71}-a_{61}a_{77})+a_{57}(a_{61}a_{75}-a_{65}a_{71})]+$$
$$+a_{13}a_{22}a_{34}a_{47}[a_{51}(a_{66}a_{75}-a_{65}a_{76})+a_{55}(a_{61}a_{76}-a_{66}a_{71})+a_{56}(a_{65}a_{71}-a_{61}a_{75})]+$$
$$+a_{13}a_{22}a_{35}a_{41}[a_{54}(a_{67}a_{76}-a_{66}a_{77})+a_{56}(a_{64}a_{77}-a_{67}a_{74})+a_{57}(a_{66}a_{74}-a_{64}a_{76})]+$$
$$+a_{13}a_{22}a_{35}a_{44}[a_{51}(a_{66}a_{77}-a_{67}a_{76})+a_{56}(a_{67}a_{71}-a_{61}a_{77})+a_{57}(a_{61}a_{76}-a_{66}a_{71})]+$$
$$+a_{13}a_{22}a_{36}a_{41}[a_{54}(a_{65}a_{77}-a_{67}a_{75})+a_{55}(a_{67}a_{74}-a_{64}a_{77})+a_{57}(a_{64}a_{75}-a_{65}a_{74})]+$$
$$+a_{13}a_{22}a_{36}a_{44}[a_{51}(a_{67}a_{75}-a_{65}a_{77})+a_{55}(a_{61}a_{77}-a_{67}a_{71})+a_{57}(a_{65}a_{71}-a_{61}a_{75})]+$$
$$+a_{13}a_{22}a_{37}a_{41}[a_{54}(a_{66}a_{75}-a_{65}a_{76})+a_{55}(a_{64}a_{76}-a_{66}a_{74})+a_{56}(a_{65}a_{74}-a_{64}a_{75})]+$$
$$+a_{13}a_{22}a_{37}a_{44}[a_{51}(a_{65}a_{76}-a_{66}a_{75})+a_{55}(a_{66}a_{71}-a_{61}a_{76})+a_{56}(a_{61}a_{75}-a_{65}a_{71})]+$$
$$+a_{13}a_{24}a_{31}a_{42}[a_{55}(a_{66}a_{77}-a_{67}a_{76})+a_{56}(a_{67}a_{75}-a_{65}a_{77})+a_{57}(a_{65}a_{76}-a_{66}a_{75})]+$$
$$+a_{13}a_{24}a_{32}a_{41}[a_{55}(a_{67}a_{76}-a_{66}a_{77})+a_{56}(a_{65}a_{77}-a_{67}a_{75})+a_{57}(a_{66}a_{75}-a_{65}a_{76})]+$$
$$+a_{13}a_{24}a_{32}a_{45}[a_{51}(a_{66}a_{77}-a_{67}a_{76})+a_{56}(a_{67}a_{71}-a_{61}a_{77})+a_{57}(a_{61}a_{76}-a_{66}a_{71})]+$$
$$+a_{13}a_{24}a_{32}a_{46}[a_{51}(a_{67}a_{75}-a_{65}a_{77})+a_{55}(a_{61}a_{77}-a_{67}a_{71})+a_{57}(a_{65}a_{71}-a_{61}a_{75})]+$$
$$+a_{13}a_{24}a_{32}a_{47}[a_{51}(a_{65}a_{76}-a_{66}a_{75})+a_{55}(a_{66}a_{71}-a_{61}a_{76})+a_{56}(a_{61}a_{75}-a_{65}a_{71})]+$$
$$+a_{13}a_{24}a_{35}a_{42}[a_{51}(a_{67}a_{76}-a_{66}a_{77})+a_{56}(a_{61}a_{77}-a_{67}a_{71})+a_{57}(a_{66}a_{71}-a_{61}a_{76})]+$$
$$+a_{13}a_{24}a_{35}a_{46}[a_{51}(a_{62}a_{77}-a_{67}a_{72})+a_{52}(a_{67}a_{71}-a_{61}a_{77})+a_{57}(a_{61}a_{72}-a_{62}a_{71})]+$$
$$+a_{13}a_{24}a_{35}a_{47}[a_{51}(a_{66}a_{72}-a_{62}a_{76})+a_{52}(a_{61}a_{76}-a_{66}a_{71})+a_{56}(a_{62}a_{71}-a_{61}a_{72})]+$$
$$+a_{13}a_{24}a_{36}a_{42}[a_{51}(a_{65}a_{77}-a_{67}a_{75})+a_{55}(a_{67}a_{71}-a_{61}a_{77})+a_{57}(a_{61}a_{75}-a_{65}a_{71})]+$$
$$+a_{13}a_{24}a_{36}a_{45}[a_{51}(a_{67}a_{72}-a_{62}a_{77})+a_{52}(a_{61}a_{77}-a_{67}a_{71})+a_{57}(a_{62}a_{71}-a_{61}a_{72})]+$$
$$+a_{13}a_{24}a_{37}a_{42}[a_{51}(a_{66}a_{75}-a_{65}a_{76})+a_{55}(a_{61}a_{76}-a_{66}a_{71})+a_{56}(a_{65}a_{71}-a_{61}a_{75})]+$$
$$+a_{13}a_{24}a_{37}a_{45}[a_{51}(a_{62}a_{76}-a_{66}a_{72})+a_{52}(a_{66}a_{71}-a_{61}a_{76})+a_{56}(a_{61}a_{72}-a_{62}a_{71})]+$$
$$+a_{13}a_{25}a_{31}a_{42}[a_{54}(a_{67}a_{76}-a_{66}a_{77})+a_{56}(a_{64}a_{77}-a_{67}a_{74})+a_{57}(a_{66}a_{74}-a_{64}a_{76})]+$$
$$+a_{13}a_{25}a_{32}a_{41}[a_{54}(a_{66}a_{77}-a_{67}a_{76})+a_{56}(a_{67}a_{74}-a_{64}a_{77})+a_{57}(a_{64}a_{76}-a_{66}a_{74})]+$$
$$+a_{13}a_{25}a_{32}a_{44}[a_{51}(a_{67}a_{76}-a_{66}a_{77})+a_{56}(a_{61}a_{77}-a_{67}a_{71})+a_{57}(a_{66}a_{71}-a_{61}a_{76})]+$$
$$+a_{13}a_{25}a_{34}a_{42}[a_{51}(a_{66}a_{77}-a_{67}a_{76})+a_{56}(a_{67}a_{71}-a_{61}a_{77})+a_{57}(a_{61}a_{76}-a_{66}a_{71})]+$$
$$+a_{13}a_{25}a_{34}a_{46}[a_{51}(a_{62}a_{77}-a_{67}a_{72})+a_{52}(a_{61}a_{77}-a_{67}a_{71})+a_{57}(a_{62}a_{71}-a_{61}a_{72})]+$$
$$+a_{13}a_{25}a_{34}a_{47}[a_{51}(a_{62}a_{76}-a_{66}a_{72})+a_{52}(a_{66}a_{71}-a_{61}a_{76})+a_{56}(a_{61}a_{72}-a_{62}a_{71})]+$$
$$+a_{13}a_{25}a_{36}a_{44}[a_{51}(a_{62}a_{77}-a_{67}a_{72})+a_{52}(a_{67}a_{71}-a_{61}a_{77})+a_{57}(a_{61}a_{72}-a_{62}a_{71})]+$$
$$+a_{13}a_{25}a_{37}a_{44}[a_{51}(a_{66}a_{72}-a_{62}a_{76})+a_{52}(a_{61}a_{76}-a_{66}a_{71})+a_{56}(a_{62}a_{71}-a_{61}a_{72})]+$$
$$+a_{13}a_{26}a_{31}a_{42}[a_{54}(a_{65}a_{77}-a_{67}a_{75})+a_{55}(a_{67}a_{74}-a_{64}a_{77})+a_{57}(a_{64}a_{75}-a_{65}a_{74})]+$$
$$+a_{13}a_{26}a_{32}a_{41}[a_{54}(a_{67}a_{75}-a_{65}a_{77})+a_{55}(a_{64}a_{77}-a_{67}a_{74})+a_{57}(a_{65}a_{74}-a_{64}a_{75})]+$$

$$+a_{13}a_{26}a_{32}a_{44}[a_{51}(a_{65}a_{77}-a_{67}a_{75})+a_{55}(a_{67}a_{71}-a_{61}a_{77})+a_{57}(a_{61}a_{75}-a_{65}a_{71})]+$$
$$+a_{13}a_{26}a_{34}a_{42}[a_{51}(a_{67}a_{75}-a_{65}a_{77})+a_{55}(a_{61}a_{77}-a_{67}a_{71})+a_{57}(a_{65}a_{71}-a_{61}a_{75})]+$$
$$+a_{13}a_{26}a_{34}a_{45}[a_{51}(a_{62}a_{77}-a_{67}a_{72})+a_{52}(a_{67}a_{71}-a_{61}a_{77})+a_{57}(a_{61}a_{72}-a_{62}a_{71})]+$$
$$+a_{13}a_{26}a_{35}a_{44}[a_{51}(a_{67}a_{72}-a_{62}a_{77})+a_{52}(a_{61}a_{77}-a_{67}a_{71})+a_{57}(a_{62}a_{71}-a_{61}a_{72})]+$$
$$+a_{13}a_{27}a_{31}a_{42}[a_{54}(a_{66}a_{75}-a_{65}a_{76})+a_{55}(a_{64}a_{76}-a_{66}a_{74})+a_{56}(a_{65}a_{74}-a_{64}a_{75})]+$$
$$+a_{13}a_{27}a_{32}a_{41}[a_{54}(a_{65}a_{76}-a_{66}a_{75})+a_{55}(a_{66}a_{74}-a_{64}a_{76})+a_{56}(a_{64}a_{75}-a_{65}a_{74})]+$$
$$+a_{13}a_{27}a_{32}a_{44}[a_{51}(a_{66}a_{75}-a_{65}a_{76})+a_{55}(a_{61}a_{76}-a_{66}a_{71})+a_{56}(a_{65}a_{71}-a_{61}a_{75})]+$$
$$+a_{13}a_{27}a_{34}a_{42}[a_{51}(a_{65}a_{76}-a_{66}a_{75})+a_{55}(a_{66}a_{71}-a_{61}a_{76})+a_{56}(a_{61}a_{75}-a_{65}a_{71})]+$$
$$+a_{13}a_{27}a_{34}a_{45}[a_{51}(a_{66}a_{72}-a_{62}a_{76})+a_{52}(a_{61}a_{76}-a_{66}a_{71})+a_{56}(a_{62}a_{71}-a_{61}a_{72})]+$$
$$+a_{13}a_{27}a_{35}a_{44}[a_{51}(a_{62}a_{76}-a_{66}a_{72})+a_{52}(a_{66}a_{71}-a_{61}a_{76})+a_{56}(a_{61}a_{72}-a_{62}a_{71})]+$$

$$+a_{14}a_{21}a_{32}a_{43}[a_{55}(a_{67}a_{76}-a_{66}a_{77})+a_{56}(a_{65}a_{77}-a_{67}a_{75})+a_{57}(a_{66}a_{75}-a_{65}a_{76})]+$$
$$+a_{14}a_{21}a_{33}a_{42}[a_{55}(a_{66}a_{77}-a_{67}a_{76})+a_{56}(a_{67}a_{75}-a_{65}a_{77})+a_{57}(a_{65}a_{76}-a_{66}a_{75})]+$$
$$+a_{14}a_{22}a_{31}a_{43}[a_{55}(a_{66}a_{77}-a_{67}a_{76})+a_{56}(a_{67}a_{75}-a_{65}a_{77})+a_{57}(a_{65}a_{76}-a_{66}a_{75})]+$$
$$+a_{14}a_{22}a_{33}a_{41}[a_{55}(a_{67}a_{76}-a_{66}a_{77})+a_{56}(a_{65}a_{77}-a_{67}a_{75})+a_{57}(a_{66}a_{75}-a_{65}a_{76})]+$$
$$+a_{14}a_{22}a_{33}a_{45}[a_{51}(a_{66}a_{77}-a_{67}a_{76})+a_{56}(a_{67}a_{71}-a_{61}a_{77})+a_{57}(a_{61}a_{76}-a_{66}a_{71})]+$$
$$+a_{14}a_{22}a_{33}a_{46}[a_{51}(a_{67}a_{75}-a_{65}a_{77})+a_{55}(a_{61}a_{77}-a_{67}a_{71})+a_{57}(a_{65}a_{71}-a_{61}a_{75})]+$$
$$+a_{14}a_{22}a_{33}a_{47}[a_{51}(a_{65}a_{76}-a_{66}a_{75})+a_{55}(a_{66}a_{71}-a_{61}a_{76})+a_{56}(a_{61}a_{75}-a_{65}a_{71})]+$$
$$+a_{14}a_{22}a_{35}a_{43}[a_{51}(a_{67}a_{76}-a_{66}a_{77})+a_{56}(a_{61}a_{77}-a_{67}a_{71})+a_{57}(a_{66}a_{71}-a_{61}a_{76})]+$$
$$+a_{14}a_{22}a_{36}a_{43}[a_{51}(a_{65}a_{77}-a_{67}a_{75})+a_{55}(a_{67}a_{71}-a_{61}a_{77})+a_{57}(a_{61}a_{75}-a_{65}a_{71})]+$$
$$+a_{14}a_{22}a_{37}a_{43}[a_{51}(a_{66}a_{75}-a_{65}a_{76})+a_{55}(a_{61}a_{76}-a_{66}a_{71})+a_{56}(a_{65}a_{71}-a_{61}a_{75})]+$$
$$+a_{14}a_{23}a_{31}a_{42}[a_{55}(a_{67}a_{76}-a_{66}a_{77})+a_{56}(a_{65}a_{77}-a_{67}a_{75})+a_{57}(a_{66}a_{75}-a_{65}a_{76})]+$$
$$+a_{14}a_{23}a_{32}a_{41}[a_{55}(a_{66}a_{77}-a_{67}a_{76})+a_{56}(a_{67}a_{75}-a_{65}a_{77})+a_{57}(a_{65}a_{76}-a_{66}a_{75})]+$$
$$+a_{14}a_{23}a_{32}a_{45}[a_{51}(a_{67}a_{76}-a_{66}a_{77})+a_{56}(a_{61}a_{77}-a_{67}a_{71})+a_{57}(a_{66}a_{71}-a_{61}a_{76})]+$$
$$+a_{14}a_{23}a_{32}a_{46}[a_{51}(a_{65}a_{77}-a_{67}a_{75})+a_{55}(a_{67}a_{71}-a_{61}a_{77})+a_{57}(a_{61}a_{75}-a_{65}a_{71})]+$$
$$+a_{14}a_{23}a_{32}a_{47}[a_{51}(a_{66}a_{75}-a_{65}a_{76})+a_{55}(a_{61}a_{76}-a_{66}a_{71})+a_{56}(a_{65}a_{71}-a_{61}a_{75})]+$$
$$+a_{14}a_{23}a_{35}a_{42}[a_{51}(a_{66}a_{77}-a_{67}a_{76})+a_{56}(a_{67}a_{71}-a_{61}a_{77})+a_{57}(a_{61}a_{76}-a_{66}a_{71})]+$$
$$+a_{14}a_{23}a_{35}a_{46}[a_{51}(a_{67}a_{72}-a_{62}a_{77})+a_{52}(a_{61}a_{77}-a_{67}a_{71})+a_{57}(a_{62}a_{71}-a_{61}a_{72})]+$$
$$+a_{14}a_{23}a_{35}a_{47}[a_{51}(a_{62}a_{76}-a_{66}a_{72})+a_{52}(a_{66}a_{71}-a_{61}a_{76})+a_{56}(a_{61}a_{72}-a_{62}a_{71})]+$$
$$+a_{14}a_{23}a_{36}a_{42}[a_{51}(a_{67}a_{75}-a_{65}a_{77})+a_{55}(a_{61}a_{77}-a_{67}a_{71})+a_{57}(a_{65}a_{71}-a_{61}a_{75})]+$$
$$+a_{14}a_{23}a_{36}a_{45}[a_{51}(a_{62}a_{77}-a_{67}a_{72})+a_{52}(a_{67}a_{71}-a_{61}a_{77})+a_{57}(a_{61}a_{72}-a_{62}a_{71})]+$$
$$+a_{14}a_{23}a_{37}a_{42}[a_{51}(a_{65}a_{76}-a_{66}a_{75})+a_{55}(a_{66}a_{71}-a_{61}a_{76})+a_{56}(a_{61}a_{75}-a_{65}a_{71})]+$$
$$+a_{14}a_{23}a_{37}a_{45}[a_{51}(a_{66}a_{72}-a_{62}a_{76})+a_{52}(a_{61}a_{76}-a_{66}a_{71})+a_{56}(a_{62}a_{71}-a_{61}a_{72})]+$$
$$+a_{14}a_{25}a_{32}a_{43}[a_{51}(a_{66}a_{77}-a_{67}a_{76})+a_{56}(a_{67}a_{71}-a_{61}a_{77})+a_{57}(a_{61}a_{76}-a_{66}a_{71})]+$$
$$+a_{14}a_{25}a_{33}a_{42}[a_{51}(a_{67}a_{76}-a_{66}a_{77})+a_{56}(a_{61}a_{77}-a_{67}a_{71})+a_{57}(a_{66}a_{71}-a_{61}a_{76})]+$$
$$+a_{14}a_{25}a_{33}a_{46}[a_{51}(a_{62}a_{77}-a_{67}a_{72})+a_{52}(a_{67}a_{71}-a_{61}a_{77})+a_{57}(a_{61}a_{72}-a_{62}a_{71})]+$$
$$+a_{14}a_{25}a_{33}a_{47}[a_{51}(a_{66}a_{72}-a_{62}a_{76})+a_{52}(a_{61}a_{76}-a_{66}a_{71})+a_{56}(a_{62}a_{71}-a_{61}a_{72})]+$$
$$+a_{14}a_{25}a_{36}a_{43}[a_{51}(a_{67}a_{72}-a_{62}a_{77})+a_{52}(a_{61}a_{77}-a_{67}a_{71})+a_{57}(a_{62}a_{71}-a_{61}a_{72})]+$$
$$+a_{14}a_{25}a_{36}a_{47}[a_{51}(a_{62}a_{73}-a_{63}a_{72})+a_{52}(a_{63}a_{71}-a_{61}a_{73})+a_{53}(a_{61}a_{72}-a_{62}a_{71})]+$$
$$+a_{14}a_{25}a_{37}a_{43}[a_{51}(a_{62}a_{76}-a_{66}a_{72})+a_{52}(a_{66}a_{71}-a_{61}a_{76})+a_{56}(a_{61}a_{72}-a_{62}a_{71})]+$$
$$+a_{14}a_{25}a_{37}a_{46}[a_{51}(a_{63}a_{72}-a_{62}a_{73})+a_{52}(a_{61}a_{73}-a_{63}a_{71})+a_{53}(a_{62}a_{71}-a_{61}a_{72})]+$$
$$+a_{14}a_{26}a_{32}a_{43}[a_{51}(a_{67}a_{75}-a_{65}a_{77})+a_{55}(a_{61}a_{77}-a_{67}a_{71})+a_{57}(a_{65}a_{71}-a_{61}a_{75})]+$$
$$+a_{14}a_{26}a_{33}a_{42}[a_{51}(a_{65}a_{77}-a_{67}a_{75})+a_{55}(a_{67}a_{71}-a_{61}a_{77})+a_{57}(a_{61}a_{75}-a_{65}a_{71})]+$$
$$+a_{14}a_{26}a_{33}a_{42}[a_{51}(a_{65}a_{77}-a_{67}a_{75})+a_{55}(a_{67}a_{71}-a_{61}a_{77})+a_{57}(a_{61}a_{75}-a_{65}a_{71})]+$$
$$+a_{14}a_{26}a_{33}a_{45}[a_{51}(a_{67}a_{72}-a_{62}a_{77})+a_{52}(a_{61}a_{77}-a_{67}a_{71})+a_{57}(a_{62}a_{71}-a_{61}a_{72})]+$$

324

$$+a_{14}a_{26}a_{35}a_{43}[a_{51}(a_{62}a_{77}-a_{67}a_{72})+a_{52}(a_{67}a_{71}-a_{61}a_{77})+a_{57}(a_{61}a_{72}-a_{62}a_{71})]+$$
$$+a_{14}a_{26}a_{35}a_{47}[a_{51}(a_{63}a_{72}-a_{62}a_{73})+a_{52}(a_{61}a_{73}-a_{63}a_{71})+a_{53}(a_{62}a_{71}-a_{61}a_{72})]+$$
$$+a_{14}a_{26}a_{37}a_{45}[a_{51}(a_{62}a_{73}-a_{63}a_{72})+a_{52}(a_{63}a_{71}-a_{61}a_{73})+a_{53}(a_{61}a_{72}-a_{62}a_{71})]+$$
$$+a_{14}a_{27}a_{32}a_{43}[a_{51}(a_{65}a_{76}-a_{66}a_{75})+a_{55}(a_{66}a_{71}-a_{61}a_{76})+a_{56}(a_{61}a_{75}-a_{65}a_{71})]+$$
$$+a_{14}a_{27}a_{33}a_{42}[a_{51}(a_{66}a_{75}-a_{65}a_{76})+a_{55}(a_{61}a_{76}-a_{66}a_{71})+a_{56}(a_{65}a_{71}-a_{61}a_{75})]+$$
$$+a_{14}a_{27}a_{33}a_{45}[a_{51}(a_{62}a_{76}-a_{66}a_{72})+a_{52}(a_{66}a_{71}-a_{61}a_{76})+a_{56}(a_{61}a_{72}-a_{62}a_{71})]+$$
$$+a_{14}a_{27}a_{35}a_{43}[a_{51}(a_{66}a_{72}-a_{62}a_{76})+a_{52}(a_{61}a_{76}-a_{66}a_{71})+a_{56}(a_{62}a_{71}-a_{61}a_{72})]+$$
$$+a_{14}a_{27}a_{35}a_{46}[a_{51}(a_{62}a_{73}-a_{63}a_{72})+a_{52}(a_{63}a_{71}-a_{61}a_{73})+a_{53}(a_{61}a_{72}-a_{62}a_{71})]+$$
$$+a_{14}a_{27}a_{36}a_{45}[a_{51}(a_{63}a_{72}-a_{62}a_{73})+a_{52}(a_{61}a_{73}-a_{63}a_{71})+a_{53}(a_{61}a_{72}-a_{62}a_{71})]+$$

$$+a_{15}a_{21}a_{32}a_{43}[a_{54}(a_{66}a_{77}-a_{67}a_{76})+a_{56}(a_{67}a_{74}-a_{64}a_{77})+a_{57}(a_{64}a_{76}-a_{66}a_{74})]+$$
$$+a_{15}a_{21}a_{33}a_{42}[a_{54}(a_{67}a_{76}-a_{66}a_{77})+a_{56}(a_{64}a_{77}-a_{67}a_{74})+a_{57}(a_{66}a_{74}-a_{64}a_{76})]+$$
$$+a_{15}a_{22}a_{31}a_{45}[a_{54}(a_{67}a_{76}-a_{66}a_{77})+a_{56}(a_{64}a_{77}-a_{67}a_{74})+a_{57}(a_{66}a_{74}-a_{64}a_{76})]+$$
$$+a_{15}a_{22}a_{33}a_{41}[a_{54}(a_{66}a_{77}-a_{67}a_{76})+a_{56}(a_{67}a_{74}-a_{64}a_{77})+a_{57}(a_{64}a_{76}-a_{66}a_{74})]+$$
$$+a_{15}a_{22}a_{33}a_{44}[a_{51}(a_{67}a_{76}-a_{66}a_{77})+a_{56}(a_{61}a_{77}-a_{67}a_{71})+a_{57}(a_{66}a_{71}-a_{61}a_{76})]+$$
$$+a_{15}a_{22}a_{34}a_{43}[a_{51}(a_{66}a_{77}-a_{67}a_{76})+a_{56}(a_{67}a_{71}-a_{61}a_{77})+a_{57}(a_{61}a_{76}-a_{66}a_{71})]+$$
$$+a_{15}a_{23}a_{31}a_{42}[a_{54}(a_{66}a_{77}-a_{67}a_{76})+a_{56}(a_{67}a_{74}-a_{64}a_{77})+a_{57}(a_{64}a_{76}-a_{66}a_{74})]+$$
$$+a_{15}a_{23}a_{32}a_{41}[a_{54}(a_{67}a_{76}-a_{66}a_{77})+a_{56}(a_{64}a_{77}-a_{67}a_{74})+a_{57}(a_{66}a_{74}-a_{64}a_{76})]+$$
$$+a_{15}a_{23}a_{32}a_{44}[a_{51}(a_{66}a_{77}-a_{67}a_{76})+a_{56}(a_{67}a_{71}-a_{61}a_{77})+a_{57}(a_{61}a_{76}-a_{66}a_{71})]+$$
$$+a_{15}a_{23}a_{34}a_{42}[a_{51}(a_{67}a_{76}-a_{66}a_{77})+a_{56}(a_{61}a_{77}-a_{67}a_{71})+a_{57}(a_{66}a_{71}-a_{61}a_{76})]+$$
$$+a_{15}a_{23}a_{34}a_{46}[a_{51}(a_{62}a_{77}-a_{67}a_{72})+a_{52}(a_{67}a_{71}-a_{61}a_{77})+a_{57}(a_{61}a_{72}-a_{62}a_{71})]+$$
$$+a_{15}a_{23}a_{34}a_{47}[a_{51}(a_{66}a_{72}-a_{62}a_{76})+a_{52}(a_{61}a_{76}-a_{66}a_{71})+a_{56}(a_{62}a_{71}-a_{61}a_{72})]+$$
$$+a_{15}a_{23}a_{36}a_{44}[a_{51}(a_{67}a_{72}-a_{62}a_{77})+a_{52}(a_{61}a_{77}-a_{67}a_{71})+a_{57}(a_{62}a_{71}-a_{61}a_{72})]+$$
$$+a_{15}a_{23}a_{37}a_{44}[a_{51}(a_{62}a_{76}-a_{66}a_{72})+a_{52}(a_{66}a_{71}-a_{61}a_{76})+a_{56}(a_{61}a_{72}-a_{62}a_{71})]+$$
$$+a_{15}a_{24}a_{32}a_{43}[a_{51}(a_{67}a_{76}-a_{66}a_{77})+a_{56}(a_{61}a_{77}-a_{67}a_{71})+a_{57}(a_{66}a_{71}-a_{61}a_{76})]+$$
$$+a_{15}a_{24}a_{33}a_{42}[a_{51}(a_{66}a_{77}-a_{67}a_{76})+a_{56}(a_{67}a_{71}-a_{61}a_{77})+a_{57}(a_{61}a_{76}-a_{66}a_{71})]+$$
$$+a_{15}a_{24}a_{33}a_{46}[a_{51}(a_{67}a_{72}-a_{62}a_{77})+a_{52}(a_{61}a_{77}-a_{67}a_{71})+a_{57}(a_{62}a_{71}-a_{61}a_{72})]+$$
$$+a_{15}a_{24}a_{33}a_{47}[a_{51}(a_{62}a_{76}-a_{66}a_{72})+a_{52}(a_{66}a_{71}-a_{61}a_{76})+a_{56}(a_{61}a_{72}-a_{62}a_{71})]+$$
$$+a_{15}a_{24}a_{36}a_{43}[a_{51}(a_{62}a_{77}-a_{67}a_{72})+a_{52}(a_{67}a_{71}-a_{61}a_{77})+a_{57}(a_{61}a_{72}-a_{62}a_{71})]+$$
$$+a_{15}a_{24}a_{36}a_{47}[a_{51}(a_{63}a_{72}-a_{62}a_{73})+a_{52}(a_{61}a_{73}-a_{63}a_{71})+a_{53}(a_{62}a_{71}-a_{61}a_{72})]+$$
$$+a_{15}a_{24}a_{37}a_{43}[a_{51}(a_{66}a_{72}-a_{62}a_{76})+a_{52}(a_{61}a_{76}-a_{66}a_{71})+a_{56}(a_{62}a_{71}-a_{61}a_{72})]+$$
$$+a_{15}a_{24}a_{37}a_{46}[a_{51}(a_{62}a_{73}-a_{63}a_{72})+a_{52}(a_{63}a_{71}-a_{61}a_{73})+a_{53}(a_{61}a_{72}-a_{62}a_{71})]+$$
$$+a_{15}a_{26}a_{33}a_{44}[a_{51}(a_{62}a_{77}-a_{67}a_{72})+a_{52}(a_{67}a_{71}-a_{61}a_{77})+a_{57}(a_{61}a_{72}-a_{62}a_{71})]+$$
$$+a_{15}a_{26}a_{34}a_{43}[a_{51}(a_{67}a_{72}-a_{62}a_{77})+a_{52}(a_{61}a_{77}-a_{67}a_{71})+a_{57}(a_{62}a_{71}-a_{61}a_{72})]+$$
$$+a_{15}a_{26}a_{34}a_{47}[a_{51}(a_{62}a_{73}-a_{63}a_{72})+a_{52}(a_{63}a_{71}-a_{61}a_{73})+a_{53}(a_{61}a_{72}-a_{62}a_{71})]+$$
$$+a_{15}a_{26}a_{37}a_{44}[a_{51}(a_{63}a_{72}-a_{62}a_{73})+a_{52}(a_{61}a_{73}-a_{63}a_{71})+a_{53}(a_{62}a_{71}-a_{61}a_{72})]+$$
$$+a_{15}a_{27}a_{33}a_{44}[a_{51}(a_{66}a_{72}-a_{62}a_{76})+a_{52}(a_{61}a_{76}-a_{66}a_{71})+a_{56}(a_{62}a_{71}-a_{61}a_{72})]+$$
$$+a_{15}a_{27}a_{34}a_{43}[a_{51}(a_{62}a_{76}-a_{66}a_{72})+a_{52}(a_{66}a_{71}-a_{61}a_{76})+a_{56}(a_{61}a_{72}-a_{62}a_{71})]+$$
$$+a_{15}a_{27}a_{34}a_{46}[a_{51}(a_{63}a_{72}-a_{62}a_{73})+a_{52}(a_{61}a_{73}-a_{63}a_{71})+a_{53}(a_{62}a_{71}-a_{61}a_{72})]+$$
$$+a_{15}a_{27}a_{36}a_{44}[a_{51}(a_{62}a_{73}-a_{63}a_{72})+a_{52}(a_{63}a_{71}-a_{61}a_{73})+a_{53}(a_{61}a_{72}-a_{62}a_{71})]+$$

$$+a_{16}a_{21}a_{32}a_{43}[a_{54}(a_{67}a_{75}-a_{65}a_{77})+a_{55}(a_{64}a_{77}-a_{67}a_{74})+a_{57}(a_{65}a_{74}-a_{64}a_{75})]+$$
$$+a_{16}a_{21}a_{33}a_{42}[a_{54}(a_{65}a_{77}-a_{67}a_{75})+a_{55}(a_{67}a_{74}-a_{64}a_{77})+a_{57}(a_{64}a_{75}-a_{65}a_{74})]+$$
$$+a_{16}a_{22}a_{31}a_{43}[a_{54}(a_{65}a_{77}-a_{67}a_{75})+a_{55}(a_{67}a_{74}-a_{64}a_{77})+a_{57}(a_{64}a_{75}-a_{65}a_{74})]+$$
$$+a_{16}a_{22}a_{33}a_{41}[a_{54}(a_{67}a_{75}-a_{65}a_{77})+a_{55}(a_{64}a_{77}-a_{67}a_{74})+a_{57}(a_{65}a_{74}-a_{64}a_{75})]+$$

$$+a_{16}a_{22}a_{33}a_{44}[a_{51}(a_{65}a_{77}-a_{67}a_{75})+a_{55}(a_{67}a_{71}-a_{61}a_{77})+a_{57}(a_{61}a_{75}-a_{65}a_{71})]+$$
$$+a_{16}a_{22}a_{34}a_{43}[a_{51}(a_{67}a_{75}-a_{65}a_{77})+a_{55}(a_{61}a_{77}-a_{67}a_{71})+a_{57}(a_{65}a_{71}-a_{61}a_{75})]+$$
$$+a_{16}a_{23}a_{31}a_{42}[a_{54}(a_{67}a_{75}-a_{65}a_{77})+a_{55}(a_{64}a_{77}-a_{67}a_{74})+a_{57}(a_{65}a_{74}-a_{64}a_{75})]+$$
$$+a_{16}a_{23}a_{32}a_{41}[a_{54}(a_{65}a_{77}-a_{67}a_{75})+a_{55}(a_{67}a_{74}-a_{64}a_{77})+a_{57}(a_{64}a_{75}-a_{65}a_{74})]+$$
$$+a_{16}a_{23}a_{32}a_{44}[a_{51}(a_{67}a_{75}-a_{65}a_{77})+a_{55}(a_{61}a_{77}-a_{67}a_{71})+a_{57}(a_{65}a_{71}-a_{61}a_{75})]+$$
$$+a_{16}a_{23}a_{34}a_{42}[a_{51}(a_{65}a_{77}-a_{67}a_{75})+a_{55}(a_{67}a_{71}-a_{61}a_{77})+a_{57}(a_{61}a_{75}-a_{65}a_{71})]+$$
$$+a_{16}a_{23}a_{34}a_{45}[a_{51}(a_{67}a_{72}-a_{62}a_{77})+a_{52}(a_{61}a_{77}-a_{67}a_{71})+a_{57}(a_{62}a_{71}-a_{61}a_{72})]+$$
$$+a_{16}a_{23}a_{35}a_{44}[a_{51}(a_{62}a_{77}-a_{67}a_{72})+a_{52}(a_{67}a_{71}-a_{61}a_{77})+a_{57}(a_{61}a_{72}-a_{62}a_{71})]+$$
$$+a_{16}a_{24}a_{32}a_{43}[a_{51}(a_{65}a_{77}-a_{67}a_{75})+a_{55}(a_{67}a_{71}-a_{61}a_{77})+a_{57}(a_{61}a_{75}-a_{65}a_{71})]+$$
$$+a_{16}a_{24}a_{33}a_{42}[a_{51}(a_{67}a_{75}-a_{65}a_{77})+a_{55}(a_{61}a_{77}-a_{67}a_{71})+a_{57}(a_{65}a_{71}-a_{61}a_{75})]+$$
$$+a_{16}a_{24}a_{33}a_{45}[a_{51}(a_{62}a_{77}-a_{67}a_{72})+a_{52}(a_{67}a_{71}-a_{61}a_{77})+a_{57}(a_{61}a_{72}-a_{62}a_{71})]+$$
$$+a_{16}a_{24}a_{35}a_{43}[a_{51}(a_{67}a_{72}-a_{62}a_{77})+a_{52}(a_{61}a_{77}-a_{67}a_{71})+a_{57}(a_{62}a_{71}-a_{61}a_{72})]+$$
$$+a_{16}a_{24}a_{35}a_{47}[a_{51}(a_{62}a_{73}-a_{63}a_{72})+a_{52}(a_{63}a_{71}-a_{61}a_{73})+a_{53}(a_{61}a_{72}-a_{62}a_{71})]+$$
$$+a_{16}a_{24}a_{37}a_{45}[a_{51}(a_{63}a_{72}-a_{62}a_{73})+a_{52}(a_{61}a_{73}-a_{63}a_{71})+a_{53}(a_{62}a_{71}-a_{61}a_{72})]+$$
$$+a_{16}a_{25}a_{33}a_{44}[a_{51}(a_{67}a_{72}-a_{62}a_{77})+a_{52}(a_{61}a_{77}-a_{67}a_{71})+a_{57}(a_{62}a_{71}-a_{61}a_{72})]+$$
$$+a_{16}a_{25}a_{34}a_{43}[a_{51}(a_{62}a_{77}-a_{67}a_{72})+a_{52}(a_{67}a_{71}-a_{61}a_{77})+a_{57}(a_{61}a_{72}-a_{62}a_{71})]+$$
$$+a_{16}a_{25}a_{34}a_{47}[a_{51}(a_{63}a_{72}-a_{62}a_{73})+a_{52}(a_{61}a_{73}-a_{63}a_{71})+a_{53}(a_{62}a_{71}-a_{61}a_{72})]+$$
$$+a_{16}a_{25}a_{37}a_{44}[a_{51}(a_{62}a_{73}-a_{63}a_{72})+a_{52}(a_{63}a_{71}-a_{61}a_{73})+a_{53}(a_{61}a_{72}-a_{62}a_{71})]+$$
$$+a_{16}a_{27}a_{34}a_{45}[a_{51}(a_{62}a_{73}-a_{63}a_{72})+a_{52}(a_{63}a_{71}-a_{61}a_{73})+a_{53}(a_{61}a_{72}-a_{62}a_{71})]+$$
$$+a_{16}a_{27}a_{35}a_{44}[a_{51}(a_{63}a_{72}-a_{62}a_{73})+a_{52}(a_{61}a_{73}-a_{63}a_{71})+a_{53}(a_{62}a_{71}-a_{61}a_{72})]+$$

$$+a_{17}a_{21}a_{32}a_{43}[a_{54}(a_{65}a_{76}-a_{66}a_{75})+a_{55}(a_{66}a_{74}-a_{64}a_{76})+a_{56}(a_{64}a_{75}-a_{65}a_{74})]+$$
$$+a_{17}a_{21}a_{33}a_{42}[a_{54}(a_{66}a_{75}-a_{65}a_{76})+a_{55}(a_{64}a_{76}-a_{66}a_{74})+a_{56}(a_{65}a_{74}-a_{64}a_{75})]+$$
$$+a_{17}a_{22}a_{31}a_{43}[a_{54}(a_{66}a_{75}-a_{65}a_{76})+a_{55}(a_{64}a_{76}-a_{66}a_{74})+a_{56}(a_{65}a_{74}-a_{64}a_{75})]+$$
$$+a_{17}a_{22}a_{33}a_{41}[a_{54}(a_{65}a_{76}-a_{66}a_{75})+a_{55}(a_{66}a_{74}-a_{64}a_{76})+a_{56}(a_{64}a_{75}-a_{65}a_{74})]+$$
$$+a_{17}a_{22}a_{33}a_{44}[a_{51}(a_{66}a_{75}-a_{65}a_{76})+a_{55}(a_{61}a_{76}-a_{66}a_{71})+a_{56}(a_{65}a_{71}-a_{61}a_{75})]+$$
$$+a_{17}a_{22}a_{34}a_{43}[a_{51}(a_{65}a_{76}-a_{66}a_{75})+a_{55}(a_{66}a_{71}-a_{61}a_{76})+a_{56}(a_{61}a_{75}-a_{65}a_{71})]+$$
$$+a_{17}a_{23}a_{31}a_{42}[a_{54}(a_{65}a_{76}-a_{66}a_{75})+a_{55}(a_{66}a_{74}-a_{64}a_{76})+a_{56}(a_{64}a_{75}-a_{65}a_{74})]+$$
$$+a_{17}a_{23}a_{32}a_{41}[a_{54}(a_{66}a_{75}-a_{65}a_{76})+a_{55}(a_{64}a_{76}-a_{66}a_{74})+a_{56}(a_{65}a_{74}-a_{64}a_{75})]+$$
$$+a_{17}a_{23}a_{32}a_{44}[a_{51}(a_{65}a_{76}-a_{66}a_{75})+a_{55}(a_{66}a_{71}-a_{61}a_{76})+a_{56}(a_{61}a_{75}-a_{65}a_{71})]+$$
$$+a_{17}a_{23}a_{34}a_{42}[a_{51}(a_{66}a_{75}-a_{65}a_{76})+a_{55}(a_{61}a_{76}-a_{66}a_{71})+a_{56}(a_{65}a_{71}-a_{61}a_{75})]+$$
$$+a_{17}a_{23}a_{34}a_{45}[a_{51}(a_{62}a_{76}-a_{66}a_{72})+a_{52}(a_{66}a_{71}-a_{61}a_{76})+a_{56}(a_{61}a_{72}-a_{62}a_{71})]+$$
$$+a_{17}a_{23}a_{35}a_{44}[a_{51}(a_{66}a_{72}-a_{62}a_{76})+a_{52}(a_{61}a_{76}-a_{66}a_{71})+a_{56}(a_{62}a_{71}-a_{61}a_{72})]+$$
$$+a_{17}a_{24}a_{32}a_{43}[a_{51}(a_{66}a_{75}-a_{65}a_{76})+a_{55}(a_{61}a_{76}-a_{66}a_{71})+a_{56}(a_{65}a_{71}-a_{61}a_{75})]+$$
$$+a_{17}a_{24}a_{33}a_{42}[a_{51}(a_{65}a_{76}-a_{66}a_{75})+a_{55}(a_{66}a_{71}-a_{61}a_{76})+a_{56}(a_{61}a_{75}-a_{65}a_{71})]+$$
$$+a_{17}a_{24}a_{33}a_{45}[a_{51}(a_{66}a_{72}-a_{62}a_{76})+a_{52}(a_{61}a_{76}-a_{66}a_{71})+a_{56}(a_{62}a_{71}-a_{61}a_{72})]+$$
$$+a_{17}a_{24}a_{35}a_{43}[a_{51}(a_{62}a_{76}-a_{66}a_{72})+a_{52}(a_{66}a_{71}-a_{61}a_{76})+a_{56}(a_{61}a_{72}-a_{62}a_{71})]+$$
$$+a_{17}a_{24}a_{35}a_{46}[a_{51}(a_{63}a_{72}-a_{62}a_{73})+a_{52}(a_{61}a_{73}-a_{63}a_{71})+a_{53}(a_{62}a_{71}-a_{61}a_{72})]+$$
$$+a_{17}a_{24}a_{36}a_{45}[a_{51}(a_{62}a_{73}-a_{63}a_{72})+a_{52}(a_{63}a_{71}-a_{61}a_{73})+a_{53}(a_{61}a_{72}-a_{62}a_{71})]+$$
$$+a_{17}a_{25}a_{33}a_{44}[a_{51}(a_{62}a_{76}-a_{66}a_{72})+a_{52}(a_{66}a_{71}-a_{61}a_{76})+a_{56}(a_{61}a_{72}-a_{62}a_{71})]+$$
$$+a_{17}a_{25}a_{34}a_{43}[a_{51}(a_{66}a_{72}-a_{62}a_{76})+a_{52}(a_{61}a_{76}-a_{66}a_{71})+a_{56}(a_{62}a_{71}-a_{61}a_{72})]+$$
$$+a_{17}a_{26}a_{34}a_{45}[a_{51}(a_{63}a_{72}-a_{62}a_{73})+a_{52}(a_{61}a_{73}-a_{63}a_{71})+a_{53}(a_{62}a_{71}-a_{61}a_{72})]+$$
$$+a_{17}a_{25}a_{34}a_{46}[a_{51}(a_{62}a_{73}-a_{63}a_{72})+a_{52}(a_{63}a_{71}-a_{61}a_{73})+a_{53}(a_{61}a_{72}-a_{62}a_{71})]+$$
$$+a_{17}a_{26}a_{35}a_{44}[a_{51}(a_{62}a_{73}-a_{63}a_{72})+a_{52}(a_{63}a_{71}-a_{61}a_{73})+a_{53}(a_{61}a_{72}-a_{62}a_{71})]+$$
$$+a_{17}a_{25}a_{36}a_{44}[a_{51}(a_{63}a_{72}-a_{62}a_{73})+a_{52}(a_{61}a_{73}-a_{63}a_{71})+a_{53}(a_{62}a_{71}-a_{61}a_{72})] \quad (\&\&)$$

In Figure A is shown the algorithm for evaluation of matrix determinant 7x7 by Lapalce's theorem for the $1^{st}, 2^{nd}, 3^{rd}$ and 4th rows in total. And in Table G are considered the main steps for evaluation matrix determinant 7x7 in accordance with Laplace's theorem the the first, second ,third and fourth rows in total.

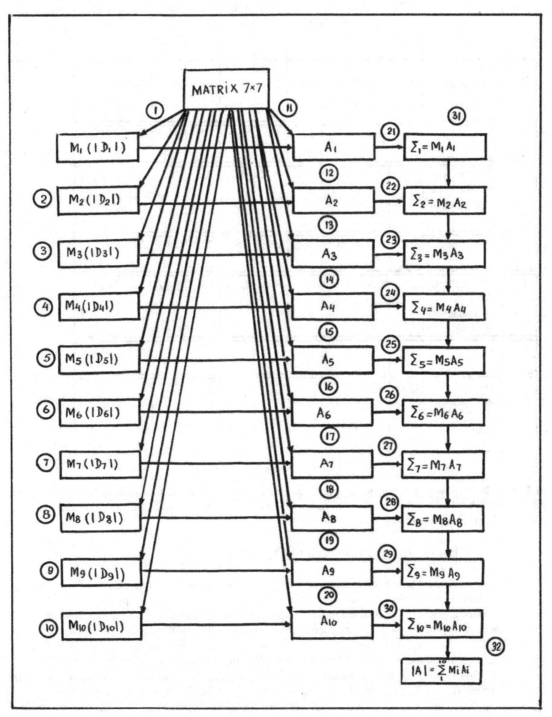

Figure A   Algorithm for evaluation of matrix determinant   7x7 by Laplace's theorem for the $1^{st}, 2^{nd}, 3^{rd}$ and 4th rows in total

Table G The main steps for evaluation of matrix determinant 7x7 in accordance with Lapalce's theorem for the first,second,third and fourth rows in total

n/n	Description	The main steps for solving of matrix determinant 7x7										Matrix determinant $\lvert A \rvert$
		1	2	3	4	5	6	7	8	9	10	
1	To mark out the minors (M_i)	M_1	M_2	M_3	M_4	M_5	M_6	M_7	M_8	M_9	M_{10}	
		11	12	13	14	15	16	17	18	19	20	
2	To mark out the algebraic supplemental (A_i)	A_1	A_2	A_3	A_4	A_5	A_6	A_7	A_8	A_9	A_{10}	
		21	22	23	24	25	26	27	28	29	30	
3	To multiple two parameters of M_i*A_i	M_1A_1	M_2A_2	M_3A_3	M_4A_4	M_5A_5	M_6A_6	M_7A_7	M_8A_8	M_9A_9	$M_{10}A_{10}$	
		31									32	
4	To summarize all parameters $\sum M_iA_i$	$\sum_1$	$\sum_2$	$\sum_3$	$\sum_4$	$\sum_5$	$\sum_6$	$\sum_7$	$\sum_8$	$\sum_9$	$\sum_{10}$	$\lvert A \rvert = $ $\sum_{1}^{10} M_iA_i$

In Figure B is shown the functional graphs between elements of the first minors M_1 and algebraic supplemental A_1 of matrix determinant 7x7 at Laplace's method. Analysis of Figure B shows the following facts for matrix determinant 7x7:

For the first element (a_{11}) of the first row

- The first element (a_{11}) has functional ties with the second element (a_{22}) of the second row and third element (a_{33})of the third row and fourth element (a_{44}) of the fourth row for positive area;
- The first element (a_{11}) has functional ties with the third element (a_{23}) of the second row and fourth element (a_{34}) of the third row and second element (a_{42}) of the fourth row for positive area;
- The first element(a_{11}) has functional ties with the fourth element(a_{24}) of the second row and second element (a_{32}) of the third row and third element (a_{43}) of the fourth row for positive area;
- The first element (a_{11})has functional ties with the fourth element (a_{24}) of the second row and third element (a_{33}) of the third row and second element (a_{42}) of the fourth row for negative area;
- The first element (a_{11}) has functional ties with the second element (a_{22}) of the second row and fourth element (a_{34}) of the third row and third element (a_{43}) of the fourth row for negative area;
- The first element(a_{11}) has functional ties with the third element (a_{23}) of the second row and second element (a_{32}) of the third row and fourth element (a_{44}) of the fourth row for negative area.

328

In Figure B is shown the functional graphs between elements of the first minor M_1 and algebraic supplemental A_1 of matrix determinant 7x7 at Laplace's method.

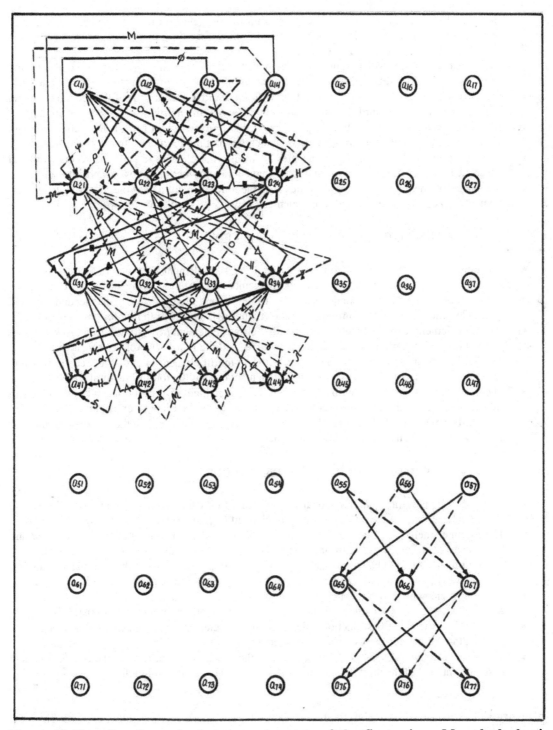

Figure B Functional graphs between elements of the first minor M_1 and algebraic supplemental A_1 of matrix determinant 7x7 at Laplace's method

For the second element (a₁₂) of the first row

- The second element (a_{12}) has functional ties with the first element (a_{21}) of the second row and third element (a_{33}) of the third row and negative fourth element (a_{44}) of the fourth row for negative area;
- The second element (a_{12}) has functional ties with the third element (a_{23}) of the second row and fourth element (a_{34}) of the third row and first element (a_{41}) of the fourth row for negative area;
- The second element (a_{12}) has functional ties with the fourth element (a_{24}) of the second row and first element (a_{31}) of the third row and third element (a_{43}) of the fourth row for negative area;
- The second element (a_{12}) has functional ties with the fourth element (a_{24}) of the second row and third element (a_{33}) of the third row and first element (a_{41}) of the fourth row for positive area;
- The second element (a_{12}) has functional ties with the first element (a_{21}) of the second row and fourth element (a_{34}) of the third row and third element (a_{43}) of the fourth row for positive area;
- The second element (a_{12}) has functional ties with the third element (a_{23}) of the second row and first element (a_{31}) of the third row and fourth element (a_{44}) of the fourth row for positive area;

For the third element (a₁₃) of the first row

- The third element (a_{13}) has functional ties with the first element (a_{21}) of the second row and second element (a_{32}) of the third row and fourth element (a_{44}) of the fourth row for positive area;
- The third element (a_{13}) has functional ties with the second element (a_{22}) of the second row and fourth element (a_{34}) of the third row and first element (a_{41}) of the fourth row for positive area;
- The third element (a_{13}) has functional ties with the fourth element (a_{24}) of the second row and first element (a_{31}) of the third row and second element (a_{42}) of the fourth row for positive area;
- The third element (a_{13}) has functional ties with the fourth element (a_{24}) of the second row and second element (a_{32}) of the third row and first element (a_{41}) of the fourth row for negative area;
- The third element (a_{13}) has functional ties with the first element (a_{21}) of the second row and fourth element (a_{34}) of the third row and second element (a_{42}) of the fourth row for negative area:
- The third element (a_{13}) has functional ties with the second element (a_{22}) of the second row and first element (a_{31}) of the third row and fourth element (a_{44}) of the fourth row for negative area.

For the fourth element (a₁₄) of the first row

- The fourth element (a_{14}) has functional ties with the first element (a_{21}) of the second row and second element (a_{32}) of the third row and third element (a_{43}) of the fourth row for negative area;
- The fourth element (a_{14}) has functional ties with the second element (a_{22}) of the second row and third element (a_{33}) of the third row and first element (a_{41}) of the fourth row for negative area;
- The fourth element (a_{14}) has functional ties with the third element (a_{23}) of the second row and first element (a_{31}) of the third row and second element (a_{42}) of the fourth row for negative area;
- The fourth element (a_{14}) has functional ties with the third element (a_{23}) of the second row and second element (a_{32}) of the third row and first element (a_{41}) of the fourth row for positive area;
- The fourth element (a_{14}) has functional ties with the first element (a_{21}) of the second row and third element (a_{33}) of the third row and second element (a_{42}) of the fourth row for positive area;
- The fourth element (a_{14}) has functional ties with the second element (a_{22}) of the second row and first element (a_{31}) of the third row and third element (a_{43}) of the fourth row for positive area.

In **Figure C** is shown the functional graphs between elements of the second minor M_2 and algebraic supplemental A_2 of matrix determinant 7x7 at Laplace's method.

Analysis of Figure C shows the following facts relatively of matrix determinant 7x7:

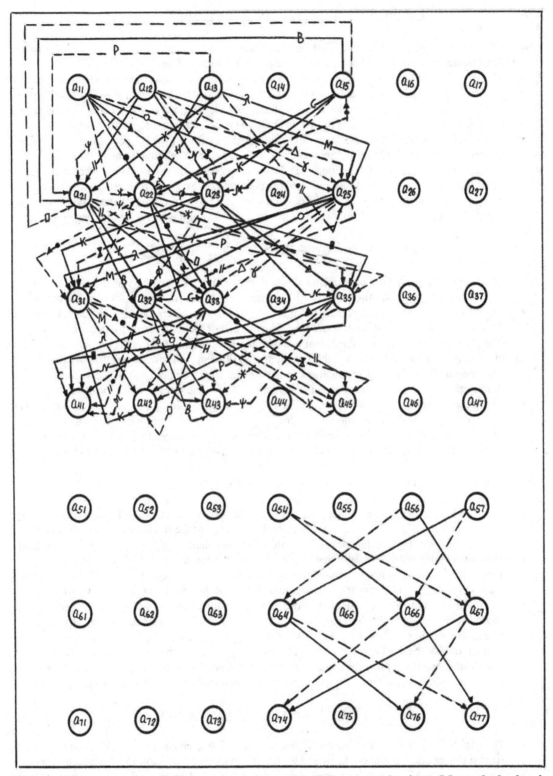

Figure C Functional graphs between elements of the second minor M₂ and algebraic supplemental A₂ of matrix determinant 7x7 at Laplace's method

For the first element (a_{11}) of the first row

- The first element (a_{11}) has functional ties with the second element (a_{22}) of the second row and third element (a_{33}) of the third row and fifth element (a_{45}) of the fourth row for positive area;
- The first element (a_{11}) has functional ties with the third element (a_{23}) of the second row and fifth element (a_{35}) of the third row and second element (a_{42}) of the fourth row for positive area;
- The first element (a_{11}) has functional ties with the fifth element(a_{25}) of the second row and second element (a_{32}) of the third row and third element (a_{43}) of the fourth row for positive area;
- The first element (a_{11}) has functional ties with the fifth element (a_{25}) of the second row and third element (a_{33}) of the third row and second element (a_{42}) of the fourth row for negative area;
- The first element (a_{11}) has functional ties with the second element (a_{22}) of the second row and fifth element (a_{35}) of the third row and third element (a_{43}) of the fourth row for negative area;
- The first element (a_{11}) has functional ties with the third element (a_{23}) of the second row and second element (a_{32}) of the third row and fifth element (a_{45}) of the fourth row for negative area.

For the second element (a_{12}) of the first row

- The second element (a_{12}) has functional ties with the first element (a_{21}) of the second row and third element (a_{33}) of the third row and fifth element (a_{45}) of the fourth row for negative area;
- The second element (a_{12}) has functional ties with the third element (a_{23}) of the second row and fifth element (a_{35}) of the third row and first element (a_{41}) of the fourth row for negative area;
- The second element (a_{12}) has functional ties with the fifth element (a_{25}) of the second row and first element (a_{31}) of the third row and third element (a_{43}) of the fourth row for negative area;
- The second element (a_{12}) has functional ties with the fifth element (a_{25}) of the second row and third element (a_{33}) of the third row and first element (a_{41}) of the fourth row for positive area;
- The second element (a_{12}) has functional ties with the first element (a_{21}) of the second row and fifth element (a_{35}) of the third row and third element (a_{43}) of the fourth row for ositive area;
- The second element (a_{12}) has functional ties with the third element (a_{23}) of the second row and first element (a_{31}) of the third row and fifth element (a_{45}) of the fourth row for positive area;

For the third element (a_{13}) of the first row

- The third element (a_{13}) has functional ties with the first element (a_{21}) of the second row and second element (a_{32}) of the third row and fifth element (a_{45}) of the fourth row for positive area;
- The third element (a_{13})has functional ties with the second element (a_{22}) of the second row and fifth element (a_{35}) of the third row and first element (a_{41}) of the fourth row for positive area;
- The third element (a_{13}) has functional ties with the fifth element (a_{25}) of the second row and first element (a_{31}) of the third row and second element (a_{42}) of the fourth row for positive area;
- The third element (a_{13}) has functional ties with the fifth element (a_{25}) of the second row and second element (a_{32}) of the third row and first element (a_{41}) of the fourth row for negative area;
- The third element (a_{13}) has functional ties with the first element (a_{21}) of the second row and fifth element (a_{35}) of the third row and second element (a_{42}) of the fourth row for negative area;
- The third element (a_{13}) has functional ties with the second element (a_{22}) of the second row and first element (a_{31}) of the third row and fifth element (a_{45}) of the fourth row for negative area.

For the fifth element (a_{15}) of the first row

- The fifth element (a_{15}) has functional ties with the first element (a_{21}) of the second row and second element (a_{32}) of the third row and third element (a_{43}) of the fourth row for negative area;
- The fifth element (a_{15}) has functional ties with the second element (a_{22}) of the second row and third element (a_{33}) of the third row and first element (a_{41}) of the fourth row for negative area;

- The fifth element (a_{15}) has functional ties with the third element (a_{23}) of the second row and first element (a_{31}) of the third row and second element (a_{42}) of the fourth row for negative area;
- The fifth element (a_{15}) has functional ties with the third element (a_{23}) of the second row and second element (a_{32}) of the third row and first element (a_{41}) of the fourth row for positive area;
- The fifth element (a_{15}) has functional ties with the first element (a_{21}) of the second row and third element (a_{33}) of the third row and second element (a_{42}) of the fourth row for positive area;
- The fifth element (a_{15}) has functional ties with the second element (a_{22}) of the second row and first element (a_{31}) of the third row and third element (a_{43}) of the fourth row for positive area.

In Figure D is shown the functional graphs between elements of the third minor M_3 and algebraic supplemental A_3 of matrix determinant 7x7 at Laplace's method. Analysis of Figure D shows the following facts relatively of matrix determinant 7x7:

For the first element (a_{11}) of the first row

- The first element (a_{11}) has functional ties with the second element (a_{22}) of the second row and third element (a_{33}) of the third row and sixth element (a_{46}) of the fourth row for positive area;
- The first element (a_{11}) has functional ties with the third element (a_{23}) of the second row and sixth element (a_{36}) of the third row and second element (a_{42}) of the fourth row for positive area;
- The first element (a_{11}) has functional ties with the sixth element (a_{26}) of the second row and second element (a_{32}) of the third row and third element (a_{43}) of the fourth row for positive area;
- The first element (a_{11}) has functional ties with the sixth element (a_{26}) of the second row and third element (a_{33}) of the third row and second element (a_{42}) of the fourth row for negative area;
- The first element (a_{11}) has functional ties with the second row and sixth element (a_{36}) of the third row and third element (a_{43}) of the fourth row for negative area;
- The first element (a_{11}) has functional ties with the third element (a_{23}) of the second row and second element (a_{32}) of the third row and sixth element (a_{46}) of the fourth row for negative area.

For the second element (a_{12}) of the first row

- The second element (a_{12}) has functional ties with the first element (a_{21}) of the second row and third element (a_{33}) of the third row and sixth element (a_{46}) of the fourth row for negative area;
- The second element (a_{12}) has functional ties with the third element (a_{23}) of the second row and six element (a_{36}) of the third row and first element (a_{41}) of the fourth row for negative area;
- The second element (a_{12}) has functional ties with the sixth element (a_{26}) of the second row and first element (a_{31}) of the third row and third element (a_{43}) of the fourth row for negative area;
- The second element (a_{12}) has functional ties with the sixth element (a_{26}) of the second row and third element (a_{33}) of the third row and first element (a_{41}) of the fourth row for positive area;
- The second element (a_{12}) has functional ties with the first element (a_{21}) of the second row and sixth element (a_{36}) of the third row and third element (a_{43}) of the fourth row for positive area;
- The second element (a_{12}) has functional ties with the third element (a_{23}) of the second row and first element (a_{31}) of the third row and sixth element (a_{46}) of the fourth row for positive area.

For the third element (a_{13}) of the first row

- The third element (a_{13}) has functional ties with the first element (a_{21}) of the second row and second element (a_{32}) of the third row and sixth element (a_{46}) of the fourth row for positive area;
- The third element (a_{13}) has functional ties with the second element (a_{22}) of the second row and sixth element (a_{36}) of the third row and first element (a_{41}) of the fourth row for positive area;
- The third element (a_{13}) has functional ties with the sixth element (a_{26}) of the second row and first element (a_{31}) of the third row and second element (a_{42}) of the fourth row for positive area;

333

- The third element (a_{13}) has functional ties with the sixth element (a_{26}) of the second row and second element (a_{32}) of the third row and first element (a_{41}) of the fourth row for negative area;
- The third element (a_{13}) has functional ties with the first element (a_{21}) of the second row and sixth element (a_{36}) of the third row and second element (a_{42}) of the fourth row for negative area;
- The third element (a_{13}) has functional ties with the second element (a_{22}) of the second row and first element (a_{31}) of the third row and sixth element (a_{46}) of the fourth row for negative area.

For the sixth element (a_{16}) of the first row

- The sixth element (a_{16}) has functional ties with the first element (a_{21}) of the second row and second element (a_{32}) of the third row and third element (a_{43}) of the fourth row for negative area;
- The sixth element (a_{16}) has functional ties with the second element (a_{22}) of the second row and third element (a_{33}) of the third row and first element (a_{41}) of the fourth row for negative area;
- The sixth element (a_{16}) has functional ties with the third element (a_{23}) of the second row and first element (a_{31}) of the third row and second element (a_{42}) of the fourth row for negative area;
- The sixth element (a_{16}) has functional ties with the third element (a_{23}) of the second row and second element (a_{32}) of the third row and first element (a_{41}) of the fourth row for positive area;
- The sixth element (a_{16}) has functional ties with the first element (a_{21}) of the second row and third element (a_{33}) of the third row and second element (a_{42}) of the fourth row for positive area;
- The sixth element (a_{16}) has functional ties with the second element (a_{22}) of the second row and first element (a_{31}) of the third row and third element (a_{43}) of the fourth row for positive area.

In Figure E is shown the functional graphs between elements of the fourth minor M_4 and algebraic supplemental A_4 of matrix determinant 7x7 at Laplace's method. Analysis of Figure E shows the following facts relatively to matrix determinant 7x7:

For the first element (a_{11}) of the first row

- The first element (a_{11}) has functional ties with the second element (a_{22}) of the second row and third element (a_{33}) of the third row and seventh element (a_{47}) of the fourth row for positive area;
- The first element (a_{11}) has functional ties with the third element (a_{23}) of the second row and seventh element(a_{37}) of the third row and second element(a_{42}) of the fourth row for positive area;
- The first element (a_{11}) has functional ties with the seventh element (a_{27}) of the second row and second element (a_{32}) of the third row and third element (a_{43}) of the fourth row for positive area;
- The first element (a_{11}) has functional ties with the seventh element (a_{27})of the second row and third element (a_{33}) of the third row and second element (a_{42}) of the fourth row for negative area;
- The first element (a_{11}) has functional ties with the second element (a_{22}) of the second row and seventh element (a_{37}) of the third row and third element (a_{43}) of the fourth row for negative area;
- The first element (a_{11}) has functional ties with the third element (a_{23}) of the second row and second element (a_{32}) of the third row and seventh element (a_{47}) of the fourth row for negative area.

For the second element (a_{12}) of the first row

- The second element (a_{12}) has functional ties with the first element (a_{21}) of the second row and third element (a_{33}) of the third row and seventh element (a_{47}) of the fourth row for negative area;
- The second element (a_{12}) has functional ties with the third element (a_{23}) of the second row and seventh element(a_{37}) of the third row and first element (a_{41}) of the fourth row for negative area;
- The second element (a_{12}) has functional ties with the seventh element (a_{27}) of the second row and first element (a_{31}) of the third row and third element (a_{43}) of the fourth row for negative area;
- The second element (a_{12}) has functional ties with the seventh element (a_{27}) of the second row and third element (a_{33}) of the third row and first element (a_{41})of the fourth row for positive area;

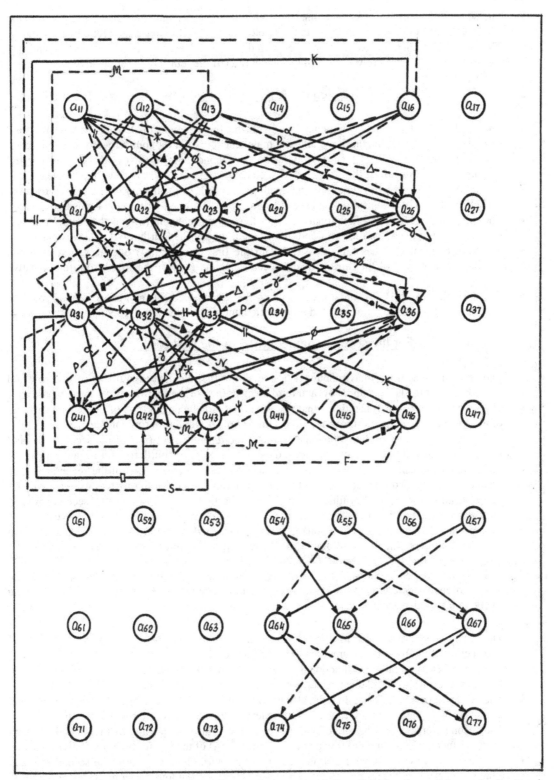

Figure D Functional graphs between elements of the third minor M_3 and algebraic supplemental A_3 of matrix determinant 7x7 at Laplace's method

335

- The second element (a_{12}) has functional ties with the first element (a_{21}) of the second row and seventh element (a_{37}) of the third row and third element (a_{43}) of the fourth row for positive area;
- The second element (a_{12}) has functional ties with the third element (a_{23}) of the second row and first element (a_{31}) of the third row and seventh element (a_{47}) of the fourth row for positive area.

For the third element (a_{13}) of the first row

- The third element (a_{13}) has functional ties with the first element (a_{21}) of the second row and second element (a_{32}) of the third row and seventh element (a_{47}) of the fourth row for positive area;
- The third element (a_{13}) has functional ties with the second element (a_{22}) of the second row and seventh element (a_{37}) of the third row and first element (a_{41}) of the fourth row for positive area;
- The third element (a_{13}) has functional ties with the seventh element (a_{27}) of the second row and first element (a_{31}) of the third row and second element (a_{42}) of the fourth row for positive area;
- The third element (a_{13}) has functional ties with the seventh element (a_{27}) of the second row and second element (a_{32}) of the third row and first element (a_{41}) of the fourth row for negative area;
- The third element (a_{13}) has functional ties with the first element (a_{21}) of the second row and seventh element (a_{37}) of the third row and second element (a_{42}) of the fourth row for negative area;
- The third element (a_{13}) has functional ties with the second element (a_{22}) of the second row and first element (a_{31}) of the third row and seventh element (a_{47}) of the fourth row for negative area.

For the seventh element (a_{17}) of the first row

- The seventh element (a_{17}) has functional ties with the first element (a_{21}) of the second row and second element (a_{32}) of the third row and third element (a_{43}) of the fourth row for negative area;
- The seventh element (a_{17}) has functional ties with the second element (a_{22}) of the second row and third element (a_{33}) of the third row and first element (a_{41}) of the fourth row for negative area;
- The seventh element (a_{17}) has functional ties with the third element (a_{23}) of the second row and first element (a_{31}) of the third row and second element (a_{42}) of the fourth row for negative area;
- The seventh element (a_{17}) has functional ties with the third element (a_{23}) of the second row and second element (a_{32}) of the third row and first element (a_{41}) of the fourth row for positive area;
- The seventh element (a_{17}) has functional ties with the first element (a_{21}) of the second row and third element (a_{33}) of the third row and second element (a_{42}) of the fourth row for positive area;
- The seventh element (a_{17}) has functional ties with the second element (a_{22}) of the second row and first element (a_{31}) of the third row and third element (a_{43}) of the fourth row for positive area.

In Figure F is shown the functional graphs between elements of the fifth minor M_5 and algebraic supplemental A_5 of matrix determinant 7x7 at Laplace's method. Analysis of Figure F shows the following facts relatively to matrix determinant 7x7:

- The second element (a_{12}) has functional ties with the third element (a_{23}) of the second row and fourth element (a_{34}) and fifth element (a_{45}) of the fourth row for positive area;
- The second element (a_{12}) has functional ties with the fourth element (a_{24}) of the second row and fifth element (a_{35}) and third element (a_{43}) of the fourth row for positive area;
- The second element (a_{12}) has functional ties with the fifth element (a_{25}) of the second row and third element (a_{33}) of the third row and fourth element (a_{44}) of the fourth row for positive area;
- The second element (a_{12}) has functional ties with the fifth element (a_{25}) of the second row and fourth element (a_{34}) of the third row and third element (a_{43}) of the fourth row for negative area;
- The second element (a_{12}) has functional ties with the third element (a_{23}) of the second row and fifth element (a_{35}) of the third row and fourth element (a_{44}) of the fourth row for negative area;
- The second element (a_{12}) has functional ties with the fourth element (a_{24}) of the second row and third element (a_{33}) of the third row and fifth element (a_{45}) of the fourth row for negative area.

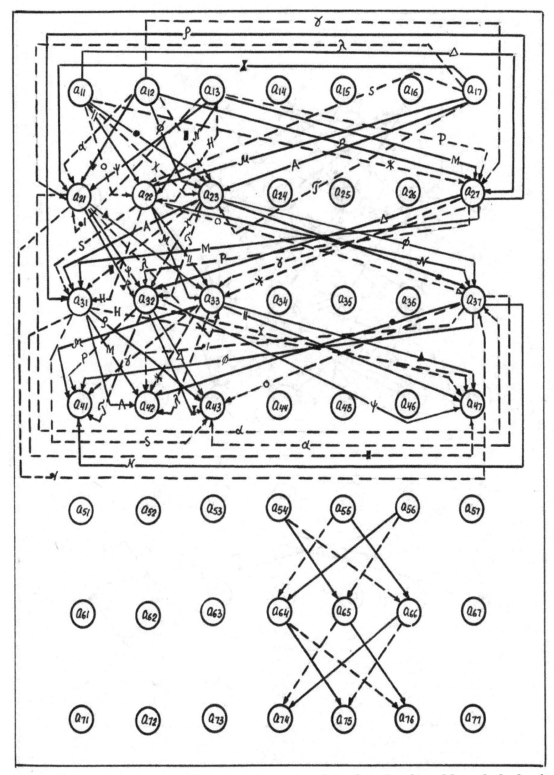

Figure E Functional graphs between elements of the fourth minor M₄ and algebraic supplemental A₄ of matrix determinant 7x7 at Laplace's method

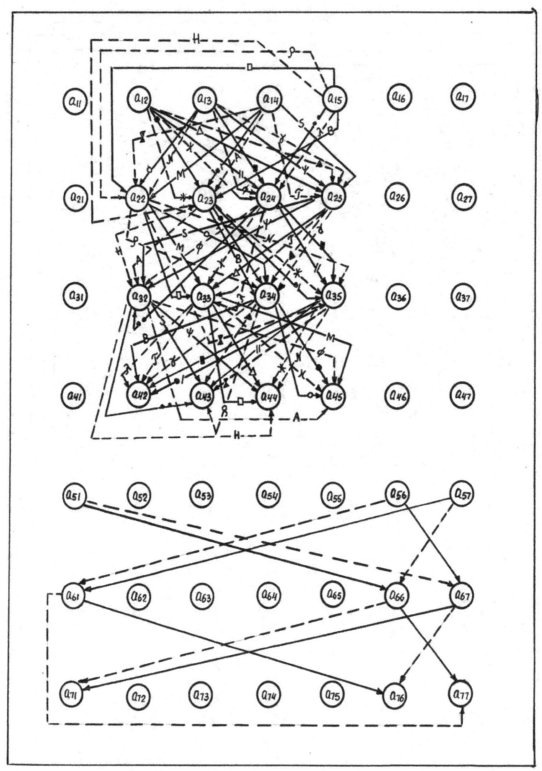

Figure F Functional graphs between elements of the fifth minor M₅ and algebraic supplemental A₅ of matrix determinant 7x7 at Laplace's method

For the third element (a_{13}) of the first row

- The third element (a_{13}) has functional ties with the second element (a_{22}) of the second row and fourth element (a_{34}) of the third row and fifth element (a_{45}) of the fourth row for negative area;
- The third element (a_{13}) has functional ties with the fourth element (a_{24}) of the second row and fifth element (a_{35})of the third row and second element (a_{42}) of the fourth row for negative area;
- The third element (a_{13})has functional ties with the fifth element (a_{25}) of the second row and second element (a_{32}) of the third row and fourth element (a_{44}) of the fourth row for negative area;
- The third element (a_{13}) has functional ties with the fifth element (a_{25}) of the second row and fourth element (a_{34}) of the third row and second element (a_{42}) of the fourth row for positive area;
- The third element (a_{13}) has functional ties with the second element (a_{22}) of the second row and fifth element (a_{35}) of the third row and fourth element (a_{44}) of the fourth row for positive area;
- The third element (a_{13}) has functional ties with the fourth element (a_{24}) of the second row and second element (a_{32}) of the third row and fifth element (a_{45}) of the fourth row for positive area.

For the fourth element (a_{14}) of the first row

- The fourth element (a_{14}) has functional ties with the second element (a_{22}) of the second row and third element (a_{33}) of the third row and fifth element (a_{45}) of the fourth row for positive area;
- The fourth element (a_{14}) has functional ties with the third element (a_{23}) of the second row and fifth element (a_{35}) of the third row and second element (a_{42}) of the fourth row for positive area;
- The fourth element (a_{14}) has functional ties with the fifth element (a_{25}) of the second row and secons element (a_{32}) of the third row and third element (a_{43})of the fourth row for positive area;
- The fourth element (a_{14}) has functional ties with the fifth element (a_{25}) of the second row and third element (a_{33}) of the third row and second element (a_{42}) of the fourth row for negative area;
- The fourth element (a_{14}) has functional ties with the second element (a_{22})of the second row and fifth element (a_{35}) of the third row and third element (a_{43}) of the fourth row for negative area;
- The fourth element (a_{14}) has functional ties with the third element (a_{23}) of the second row and second element (a_{32}) of the third row and fifth element (a_{45}) of the fourth row for negative area.

For the fifth element (a_{15}) of the first row

- The fifth element (a_{15}) has functional ties with the second element (a_{22}) of the second row and third element (a_{33}) of the third row and fourth element (a_{44}) of the fourth row for negative area;
- The fifth element (a_{15})has functional ties with the third element (a_{23}) of the second row and fourth element (a_{34}) of the third row and second element (a_{42}) of the fourth row for negative area;
- The fifth element (a_{15})has functional ties with the fourth element (a_{24}) of the second row and second element (a_{32}) of the third row and third element (a_{43}) of the fourth row for negative area;
- The fifth element (a_{15}) has functional ties with the fourth element (a_{24}) of the second row and third element (a_{33}) of the third row and second element (a_{42}) of the fourth row for positive area;
- The fifth element (a_{15}) has functional ties with the second element (a_{22}) of the second row and fourth element(a_{34}) of the third row and third element (a_{43}) of the fourth row for positive area;
- The fifth element (a_{15}) has functional ties with the third element (a_{23}) of the second row and second element (a_{32}) of the third row and fourth element (a_{44}) of the fourth row for positive area.

In **Figure G** is shown the functional graphs between elements of the sixth minor M_6 and algebraic supplemental A_6 of matrix determinant 7x7 at Laplace's method. Analysis of Figure G shows the following facts relatively to matrix determinant 7x7:

For the second element (a_{12}) of the first row

- The second element (a_{12}) has functional ties with the third element (a_{23}) of the second row and fourth element (a_{34}) of the third row and sixth element (a_{46}) of the fourth row for positive area;

- The second element (a_{12}) has functional ties with the fourth element (a_{24}) of the second row and sixth element (a_{36}) of the third row and third element (a_{43}) of the fourth row for positive area;
- The second element (a_{12}) has functional ties with the sixth element (a_{26}) of the second row and third element (a_{33}) of the third row and fourth element (a_{44}) of the fourth row for positive area;
- The second element (a_{12}) has functional ties with the sixth element (a_{26}) of the second row and fourth element (a_{34}) of the third row and third element (a_{43}) of the fourth row for negative area;
- The second element (a_{12}) has functional ties with the third element (a_{23}) of the second row and sixth element (a_{36}) of the third row and fourth element (a_{44}) of the fourth row for negative area;
- The second element (a_{12}) has functional ties with the fourth element (a_{24}) of the second row and third element (a_{33}) of the third row and sixth element (a_{46}) of the fourth row for negative area.

For the third element (a_{13}) of the first row

- The third element (a_{13}) has functional ties with the second element (a_{22}) of the second row and fourth element (a_{34}) of the third row and sixth element (a_{46}) of the fourth row for negative area;
- The third element (a_{13}) has functional ties with the fourth element (a_{24}) of the second row and sixth element (a_{36}) of the third row and second element (a_{42}) of the fourth row for negative area;
- The third element (a_{13}) has functional ties with the sixth element (a_{26}) of the second row and second element (a_{32}) of the third row and fourth element (a_{44}) of the fourth row for negative area;
- The third element (a_{13}) has functional ties with the sixth element (a_{26}) of the second row and fourth element (a_{34}) of the third row and second element (a_{42}) of the fourth row for positive area;
- The third element (a_{13}) has functional ties with the second element (a_{22}) of the second row and sixth element (a_{36}) of the third row and fourth element (a_{44}) of the fourth row for positive area;
- The third element (a_{13}) has functional ties with the fourth element (a_{24}) of the second row and second element (a_{32}) of the third row and sixth element (a_{46}) of the fourth row for positive area.

For the fourth element (a_{14}) of the first row

- The fourth element (a_{14}) has functional ties with the second element (a_{22}) of the second row and third element (a_{33}) of the third row and sixth element (a_{46}) of the fourth row for positive area;
- The fourth element (a_{14}) has functional ties with the third element (a_{23}) of the second row and sixth element (a_{36}) of the third row and second element (a_{42}) of the fourth row for positive area;
- The fourth element (a_{14}) has functional ties with the sixth element (a_{26}) of the second row and second element (a_{32}) of the third row and third element (a_{43}) of the fourth row for positive area;
- The fourth element (a_{14}) has functional ties with the sixth element (a_{26}) of the second row and third element (a_{33}) of the third row and second element (a_{42}) of the fourth row for negative area;
- The fourth element (a_{14}) has functional ties with the second element (a_{22}) of the second row and sixth element (a_{36}) of the third row and third element (a_{43}) of the fourth row for negative area;
- The fourth element (a_{14}) has functional ties with the third element (a_{23}) of the second row and second element (a_{32}) of the third row and sixth element (a_{46}) of the fourth row for negative area.

For the sixth element (a_{16}) of the first row

- The sixth element (a_{16}) has functional ties with the second element (a_{22}) of the second row and third element (a_{33}) of the third row and fourth element (a_{44}) of the fourth row for negative area;
- The sixth element (a_{16}) has functional ties with the third element (a_{23}) of the second row and fourth element (a_{34}) of the third row and second element (a_{42}) of the fourth row for negative area;
- The sixth element (a_{16}) has functional ties with the fourth element (a_{24}) of the second row and second element (a_{32}) of the third row and third element (a_{43}) of the fourth row for negative area;
- The sixth element (a_{16}) has functional ties with the fourth element (a_{24}) of the second row and third element (a_{33}) of the third row and second element (a_{42}) of the fourth row for positive area;
- The sixth element (a_{16}) has functional ties with the second element (a_{22}) of the second row and fourth element (a_{34}) of the third row and third element (a_{43}) of the fourth row for positive area;

- The sixth element (a_{16}) has functional ties with the third element (a_{23}) of the second row and second element (a_{32}) of the third row and fourth element (a_{44}) of the fourth row for positive area.

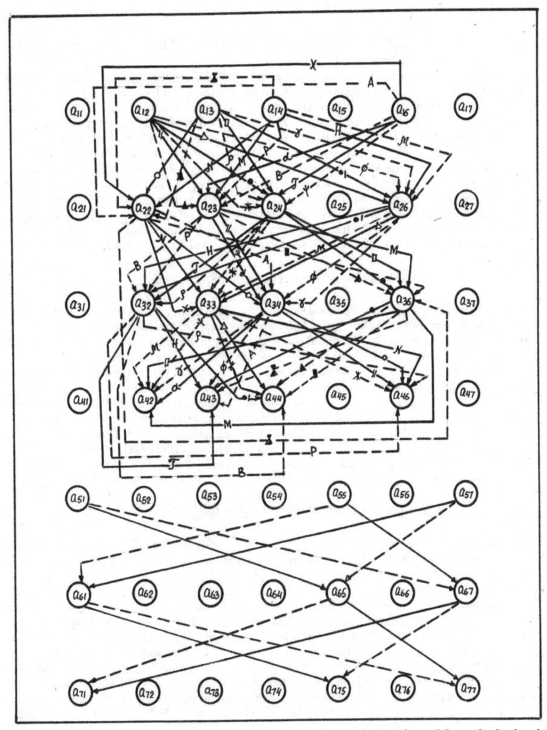

Figure G Functional graphs between elements of the sixth minor M_6 and algebraic supplemental A_6 of matrix determinant 7x7 at Laplace's method

In **Figure H** is shown the functional graphs between elements of the seventh minor M_7 and algebraic supplemental A_7 of matrix determinant 7x7 at Laplace's method. Analysis of Figure H shows the following facts relatively to matrix determinant 7x7:

For the second element (a_{12}) of the first row

- The second element (a_{12}) has functional ties with the third element (a_{23})of the second row and fourth element (a_{34}) of the third row and seventh element (a_{47})of the fourth row for positive area;
- The second element (a_{12}) has functional ties with the fourth element (a_{24})of the second row and seventh element (a_{37}) of the third row and third element (a_{43}) of the fourth row for positive area;
- The second element (a_{12}) has functional ties with the seventh element(a_{27}) of the second row and third element (a_{33}) of the third row and fourth element (a_{44}) of the fourth row for positive area;
- The second element (a_{12}) has functional ties with the seventh element (a_{27}) of the second row and fourth element (a_{34}) of the third row and third element (a_{43}) of the fourth row for negative area;
- The second element (a_{12}) has functional ties with the third element (a_{23}) of the second row and seventh element (a_{37}) of the third row and fourth element (a_{44}) of the fourth row for negative area;
- The second element (a_{12}) has functional ties with the fourth element (a_{24}) of the second row and third element (a_{33}) of the third row and seventh element(a_{47}) of the fourth row for negative area.

For the third element (a_{13}) of the first row

- The third element (a_{13}) has functional ties with the second element (a_{22}) of the second row and fourth element (a_{34}) of the third row and seventh element (a_{47}) of the fourth row for negative area;
- The third element (a_{13}) has functional ties with the fourth element (a_{24}) of the second row and seventh element (a_{37}) of the third row and second element (a_{42}) of the fourth row for negative area;
- The third element (a_{13}) has functional ties with the seventh element (a_{27}) of the second row and second element (a_{32}) of the third row and fourth element (a_{44}) of the fourth row for negative area;
- The third element (a_{13}) has functional ties with the seventh element (a_{27}) of the second row and fourth element (a_{34}) of the third row and second element (a_{42}) of the fourth row for positive area;
- The third element (a_{13}) has functional ties with the second element (a_{22}) of the second row and seventh element (a_{37}) of the third row and fourth element (a_{44}) of the fourth row for positive area;
- The third element (a_{13}) has functional ties with the fourth element (a_{24}) of the second row and second element (a_{32}) of the third row and seventh element (a_{47}) of the fourth row for positive area.

For the fourth element (a_{14}) of the first row

- The fourth element (a_{14}) has functional ties with the second element (a_{22}) of the second row and third element (a_{33}) of the third row and seventh element (a_{47}) of the fourth row for positive area;
- The fourth element (a_{14}) has functional ties with the third element (a_{23}) of the second row and seventh element(a_{37}) of the third row and second element (a_{42}) of the fourth row for positive area;
- The fourth element (a_{14}) has functional ties with the seventh element (a_{27}) of the second row and second element (a_{32}) of the third row and third element (a_{43}) of the fourth row for positive area;
- The fourth element (a_{14}) has functional ties with the seventh element (a_{27}) of the second row and third element (a_{33}) of the third row and second element (a_{42}) of the fourth row for negative area;
- The fourth element (a_{14}) has functional ties with the second element (a_{22}) of the second row and seventh element (a_{37}) of the third row and third element (a_{43}) of the fourth row for negative area;
- The fourth element (a_{14}) has functional ties with the third element (a_{23}) of the second row and second element (a_{32}) of the third row and seventh element (a_{47}) of the fourth row for negative area;

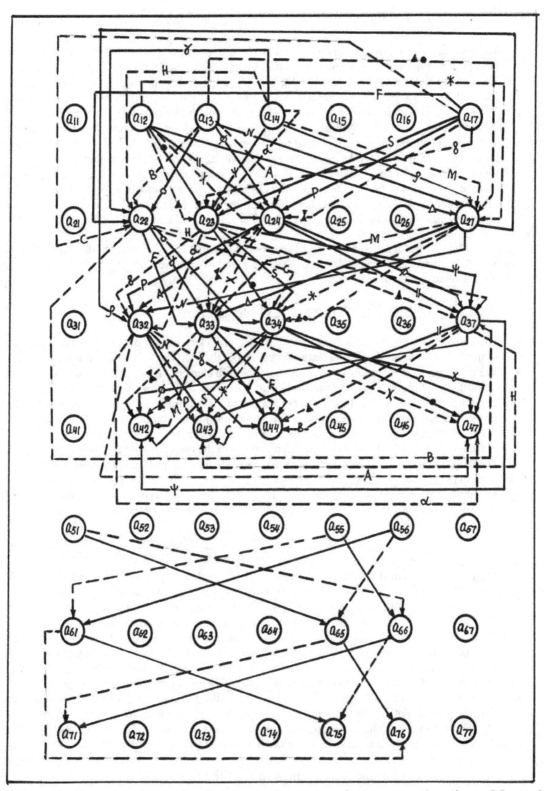

Figure H Functional graphs between elements of the seventh minor M₇ and algebraic supplemental of matrix determinant 7x7 at Laplace's method

For the seventh element (a_{17}) of the first row

- The seventh element (a_{17}) has functional ties with the second element (a_{22}) of the second row and third element (a_{33}) of the third row and fourth element (a_{44}) of the fourth row for negative area;
- The seventh element (a_{17}) has functional ties with the third element (a_{23}) of the second row and fourth element (a_{34}) of the third row and second element (a_{42}) of the second row for negative area;
- The seventh element (a_{17}) has functional ties with the fourth element (a_{24}) of the fourth row and second element (a_{32}) of the third row and third element (a_{43}) of the fourth row for negative area;
- The seventh element (a_{17}) has functional ties with the fourth element (a_{24}) of the second row and third element (a_{33}) of the third row and second element (a_{42}) of the fourth row for positive area;
- The seventh element (a_{17}) has functional ties with the second element (a_{22}) of the second row and fourth element (a_{34}) of the third row and third element (a_{43}) of the fourth row for positive area;
- The seventh element (a_{17}) has functional ties with the third element (a_{23}) of the second row and second element (a_{32}) of the third row and fourth element (a_{44}) of the fourth row for positive area.

In Figure J is shown the functional graphs between elements of the eighth minor M_8 and algebraic supplemental A_8 of matrix determinant 7x7 at Laplace's method. Analysis of Figure J shows the following facts relatively to matrix determinant 7x7:

For the third element (a_{13}) of the first row

- The third element (a_{13})has functional ties with the fourth element (a_{24}) of the second row and fifth element (a_{35}) of the third row and sixth element (a_{46})of the fourth row for positive area;
- The third element (a_{13}) has functional ties with the fifth element (a_{25}) of the second row and sixth element (a_{36}) of the third row and fourth element (a_{44}) of the fourth row for positive area;
- The third element (a_{13}) has functional ties with the sixth element (a_{26}) of the second row and fourth element (a_{34}) of the third row and fifth element (a_{45}) of the fourth row for positive area;
- The third element (a_{13}) has functional ties with the sixth element (a_{26}) of the second row and fifth element (a_{35}) of the third row and fourth element (a_{44}) of the fourth row for negative area;
- The third element (a_{13}) has functional ties with the fourth element (a_{24}) of the second row and sixth element (a_{36}) of the third row and fifth element (a_{45}) of the fourth row for negative area;
- The third element (a_{13}) has functional ties with the fifth element (a_{25}) of the second row and fourth element (a_{34}) of the third row and sixth element (a_{46}) of the fourth row for negative area.

For the fourth element (a_{14}) of the first row

- The fourth element (a_{14}) has functional ties with the third element (a_{23}) of the second row and fifth element (a_{35}) of the third row and sixth element (a_{46}) of the fourth row for negative area;
- The fourth element (a_{14}) has functional ties with the fifth element (a_{25}) of the second row and sixth element (a_{36}) of the third row and third element (a_{43}) of the fourth row for negative area;
- The fourth element (a_{14}) has functional ties with the sixth element (a_{26}) of the second row and third element (a_{33}) of the third row and fifth element (a_{45}) of the fourth row for negative area;
- The fourth element (a_{14}) has functional ties with the sixth element (a_{26}) of the second row and fifth element (a_{35}) of the third row and third element (a_{43}) of the fourth row for positive area;
- The fourth element (a_{14}) has functional ties with the third element (a_{23}) of the second row and sixth element (a_{36}) of the third row and fifth element (a_{45}) of the fourth row for positive area;
- The fourth element (a_{14}) has functional ties with the fifth element (a_{25}) of the second row and third element (a_{33}) of the third row and sixth element (a_{46}) of the fourth row for positive area.

For the fifth element (a_{15}) of the first row

- The fifth element (a_{15}) has functional ties with the third element (a_{23}) of the second row and fourth element (a_{34}) of the third row and sixth element (a_{46}) of the fourth row for positive area;

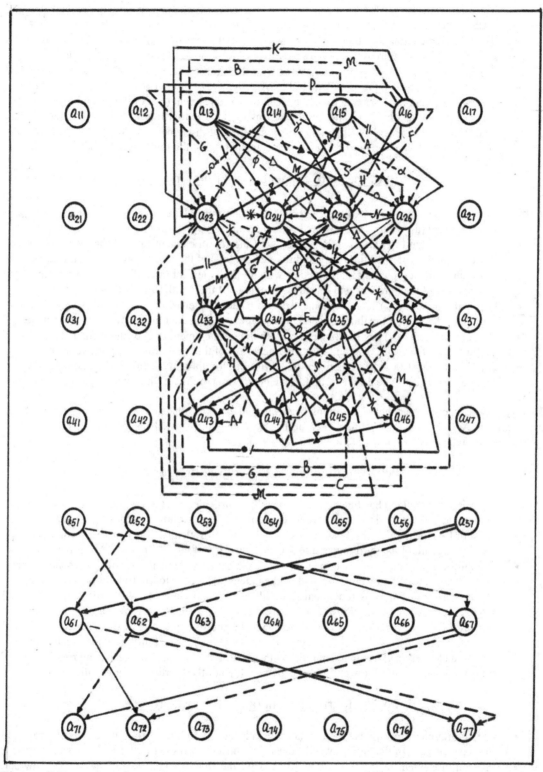

Figure J Functional graphs between elements of the eighth minor M$_8$ and algebraic supplemental A$_8$ of matrix determinant 7x7 at Laplace's method

- The fifth element (a_{15}) has functional ties with the fourth element (a_{24}) of the second row and sixth element (a_{36}) of the third row and third element (a_{43}) of the fourth row for positive area;
- The fifth element (a_{15}) has functional ties with the sixth element (a_{26}) of the second row and third element (a_{33}) of the third row and fourth element (a_{44}) of the fourth row for positive area;
- The fifth element (a_{15}) has functional ties with the sixth element (a_{26}) of the second row and fourth element (a_{34}) of the third row and third element (a_{43}) of the fourth row for negative area;
- The fifth element (a_{15}) has functional ties with the third element (a_{23}) of the second row and sixth element (a_{36}) of the third row and fourth element (a_{44}) of the fourth row for positive area;
- The fifth element (a_{15}) has functional ties with the fourth element (a_{24}) of the second row and third element (a_{33}) of the third row and sixth element (a_{46}) of the fourth row for positive area.

For the sixth element (a_{16}) of the first row

- The sixth element (a_{16}) has functional ties with the third element (a_{23}) of the second row and fourth element (a_{34}) of the third row and fifth element (a_{45}) of the fourth row for negative area;
- The sixth element (a_{16}) has functional ties with the fourth element (a_{24}) of the second row and fifth element (a_{35}) of the third row and third element (a_{43}) of the fourth row for negative area;
- The sixth element (a_{16}) has functional ties with the fifth element (a_{25}) of the second row and third element (a_{33}) of the third row and fourth element (a_{44}) of the fourth row for negative area;
- The sixth element (a_{16}) has functional ties with the fifth element (a_{25}) of the second row and fourth element (a_{34}) of the third row and third element (a_{43}) of the fourth row for positive area;
- The sixth element (a_{16}) has functional ties with the third element (a_{23}) of the second row and fifth element (a_{35}) of the third row and fourth element (a_{44}) of the fourth row for positive area;
- The sixth element (a_{16}) has functional ties with the fourth element (a_{24}) of the second row and third element (a_{33}) of the third row and fifth element (a_{45}) of the fourth row for positive area.

In Figure K is shown the functional graphs between elements of the nineth minor M_9 and algebraic supplemental A_9 of matrix determinant 7x7 at Laplace's method. Analysis of Figure K shows the following facts relatively to matrix determinant 7x7:

For the third element (a_{13}) of the first row

- The third element (a_{13}) has functional ties with the fourth element (a_{24}) of the second row and fifth element (a_{35}) of the third row and seventh element (a_{47}) of the fourth row for positive area;
- The third element (a_{13}) has functional ties with the fifth element (a_{25}) of the second row and seventh element of the third row and fourth element (a_{44}) of the fourth row for positive area;
- The third element (a_{13}) has functional ties with the seventh element (a_{27}) of the second row and fourth element (a_{34}) of the third row and fifth element (a_{45}) of the fourth row for positive area;
- The third element (a_{13}) has functional ties with the seventh element (a_{27}) of the second row and fifth element (a_{35}) of the third row and fourth element (a_{44}) of the fourth row for negative area;
- The third element (a_{13}) has functional ties with the fourth element (a_{24}) of the second row and seventh element (a_{37}) of the third row and fifth element (a_{45}) of the fourth row for negative area;
- The third element (a_{13}) has functional ties with the fifth element (a_{25}) of the second row and fourth element (a_{34}) of the third row and seventh element (a_{47}) of the fourth row for negative area.

For the fourth element (a_{14}) of the first row

- The fourth element (a_{14}) has functional ties with the third element (a_{23}) of the second row and fifth element (a_{35}) of the third row and seventh element (a_{47}) of the fourth row for negative area;
- The fourth element (a_{14}) has functional ties with the fifth element (a_{25}) of the second row and seventh element (a_{37}) of the third row and third element (a_{43}) of the fourth row for negative area;
- The fourth element (a_{14}) has functional ties with the seventh element (a_{27}) of the second row and third element (a_{33}) of the third row and fifth element (a_{45}) of the fourth row for positive area;

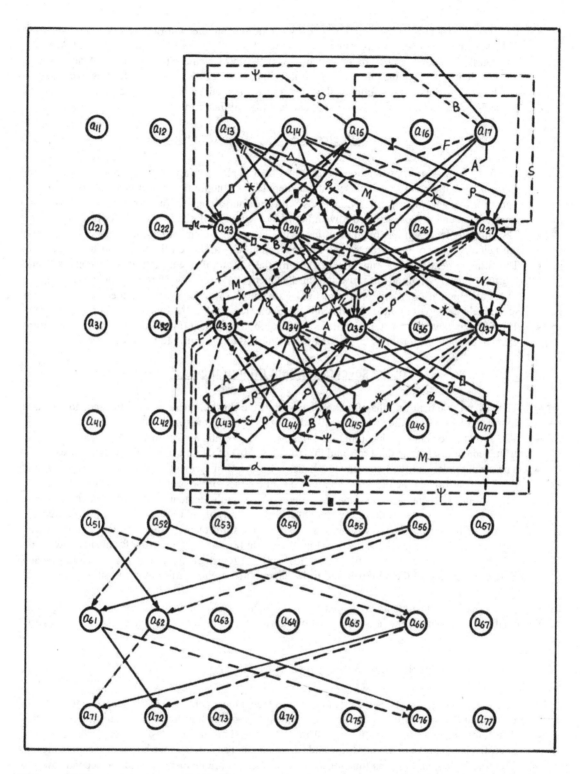

Figure K Functional graphs between element of the nineth minor M₉ and algebraic supplemental A₉ of matrix determinant 7x7 at Laplace's method

- The fourth element (a_{14}) has functional ties with the seventh element (a_{27}) of the second row and fifth element (a_{35}) of the third row and third element (a_{43}) of the fourth row for positive area;
- The fourth element (a_{14}) has functional ties with the third element (a_{23}) of the second row and seventh element (a_{37}) of the third row and fifth element (a_{45}) of the fourth row for positive area;
- The fourth element (a_{14}) has functional ties with the fifth element (a_{25}) of the second row and third element (a_{33}) of the third row and seventh element (a_{47}) of the fourth row for positive area.

For the fifth element (a_{15}) of the first row

- The fifth element (a_{15}) has functional ties with the third element (a_{23}) of the second row and fourth element (a_{34}) of the third row and seventh element (a_{47}) of the fourth row for positive area;
- The fifth element (a_{15}) has functional ties with the fourth element (a_{24}) of the second row and seventh element (a_{37}) of the third row and third element (a_{43}) of the fourth row for positive area;
- The fifth element (a_{15}) has functional ties with the seventh element (a_{27}) of the second row and third element (a_{33}) of the third row and fourth element (a_{44}) of the fourth row for positive area;
- The fifth element (a_{15}) has functional ties with the seventh element (a_{27}) of the second row and fourth element (a_{34}) of the third row and third element (a_{43}) of the fourth row for negative area;
- The fifth element (a_{15}) has functional ties with the third element (a_{23}) of the second row and seventh element (a_{37}) of the third row and fourth element (a_{44}) of the fourth row for negative area;
- The fifth element (a_{15}) has functional ties with the fourth element (a_{24}) of the second row and third element (a_{33}) of the third row and seventh element (a_{47}) of the fourth row for negative area.

For the seventh element (a_{17}) of the first row

- The seventh element (a_{17}) has functional ties with the third element (a_{23}) of the second row and fourth element (a_{34}) of the third row and fifth element (a_{45}) of the fourth row for negative area;
- The seventh element (a_{17}) has functional ties with the fourth element (a_{24}) of the second row and fifth element (a_{35}) of the third row and third element (a_{43}) of the fourth row for negative area;
- The seventh element (a_{17}) has functional ties with the fifth element (a_{25}) of the second row and third element (a_{33}) of the third row and fourth element (a_{44}) of the fourth row for negative area;
- The seventh element (a_{17}) has functional ties with the fifth element (a_{25}) of the second row and fourth element (a_{34}) of the third row and third element (a_{43}) of the fourth row for positive area;
- The seventh element (a_{17}) has functional ties with the third element (a_{23}) of the second row and fifth element (a_{35}) of the third row and fourth element (a_{44}) of the fourth row for positive area;
- The seventh element (a_{17}) has functional ties with the fourth element (a_{24}) of the second row and third element (a_{33}) of the third row and fifth element (a_{45}) of the fourth row for positive area.

In **Figure L** is shown the functional graphs between elements of the tenth minor M_{10} and algebraic supplemental A_{10} of Laplace's method. Analysis of Figure L shos the following facts relatively to matrix determinant 7x7:

For the fourth element (a_{14}) of the first row

- The fourth element (a_{14}) has functional ties with the fifth element (a_{25}) of the second row and sixth element (a_{36}) of the third row and seventh element (a_{47}) of the fourth row for positive area;
- The fourth element (a_{14}) has functional ties with the sixth element (a_{26}) of the second row and seventh element (a_{37}) of the third row and fifth element (a_{45}) of the fourth row for positive area;
- The fourth element (a_{14}) has functional ties with the seventh element (a_{27}) of the second row and fifth element (a_{35}) of the third row and sixth element (a_{46}) of the fourth row for positive area;
- The fourth element (a_{14}) has functional ties with the seventh element (a_{27}) of the second row and sixth element (a_{36}) of the third row and fifth element (a_{45}) of the fourth row for negative area;

348

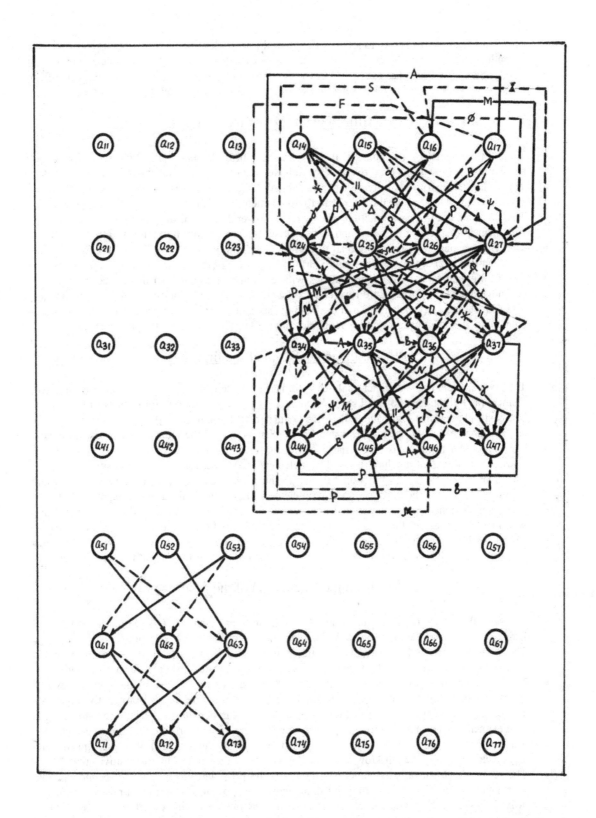

Figure L Funcional graphs between elements of the tenth minor M₁₀and algebraic supplemental A₁₀ of matrix determinant 7x7 at Laplace's method

- The fourth element (a_{14}) has functional ties with the fifth element (a_{25}) of the second row and seventh element (a_{37}) of the third row and sixth element (a_{46}) of the fourth row for negative area;
- The fourth element (a_{14}) has functional ties with the sixth element (a_{26}) of the second row and fifth element (a_{35}) of the third row and seventh element (a_{47}) of the fourth row for negative area.

For the fifth element (a_{15}) of the first row

- The fifth element (a_{15}) has functional ties with the fourth element (a_{24}) of the second row and sisxth element (a_{36}) of the third row and seventh element (a_{47}) of the fourth row for negative area;
- The fifth element (a_{15}) has functional ties with the sixth element (a_{26}) of the second row and seventh element (a_{37}) of the third row and fourth element (a_{44}) of the fourth row for negative area;
- The fifth element (a_{15}) has functional ties with the seventh element (a_{27}) of the second row and fifth element (a_{34}) of the third row and sixth element (a_{46}) of the fourth row for negative area;
- The fifth element (a_{15}) has functional ties with the seventh element (a_{27}) of the second row and sixth element (a_{36}) of the third row and fourth element (a_{44}) of the fourth row for positive area;
- The fifth element (a_{15}) has functional ties with the fourth element (a_{24}) of the second row and seventh element (a_{37}) of the third row and sixth element (a_{46}) of the fourth row for positive area;
- The fifth element (a_{15}) has functional ties with the sixth element (a_{26}) of the second row and fourth element (a_{34}) of the third row and seventh element (a_{47}) of the fourth row for positive area.

For the sixth element (a_{16}) of the first row

- The sixth element (a_{16}) has functional ties with the fourth element (a_{24}) of the second row and fifth element (a_{35}) of the third row and seventh element (a_{47}) of the fourth row for positive area;
- The sixth element (a_{16}) has functional ties with the fifth element (a_{25}) of the second row and seventh element (a_{37}) of the third row and fourth element (a_{44}) of the fourth row for positive area;
- The sixth element (a_{16}) has functional ties with the seventh element (a_{27}) of the second row and fourth element (a_{34}) of the third row and fifth element (a_{45}) of the fourth row for positive area;
- The sixth element (a_{16}) has functional ties with the seventh element (a_{27}) of the second row and fifth element (a_{35}) of the third row and fourth element (a_{44}) of the fourth row for negative area;
- The sixth element (a_{16}) has functional ties with the fourth element (a_{24}) of the second row and seventh element (a_{37}) of the third row and fifth element (a_{45}) of the fourth row for negative area;
- The sixth element (a_{16}) has functional ties with the fifth element (a_{25}) of the second row and fourth element (a_{34}) of the third row and seventh element (a_{47}) of the fourth element for negative area.

For the seventh element (a_{17}) of the first row

- The seventh element (a_{17}) has functional ties with the fourth element (a_{24}) of the second row and fifth element (a_{35}) of the third row and sixth element (a_{46}) of the fourth row for negative area;
- The seventh element (a_{17}) has functional ties with the fifth element (a_{25}) of the second row and sixth element (a_{36}) of the third row and fourth element (a_{44}) of the fourth row for negative area;
- The seventh element (a_{17}) has functional ties with the sixth element (a_{26}) of the second row and fourth element (a_{34}) of the third row and fifth element (a_{45}) of the fourth row for negative area;
- The seventh element (a_{17}) has functional ties with the sixth element (a_{26}) of the second row and fifth element (a_{35}) of the third row and fourth element (a_{44}) of the fourth row for positive area;
- The seventh element (a_{17}) has functional ties with the fourth element (a_{24}) of the second row and sixth element (a_{36}) of the third row and fifth element ($_{45}$) of the fourth row for positive area;
- The seventh element (a_{17}) has functional ties with the fifth element (a_{25}) of the second row and fourth element (a_{34}) of the third row and sixth element (a_{46}) of the fourth row for positive area.

In Appendix R is shown the evaluation of system normal equations with using of Cramer's rule and also the Laplace's theorem at conditions when in 7x7 matrix marked out joinly the first,second ,third and fourth rows.

APPENDICES

Appendix 1

Problem:

Evaluate the determinant $|A|$ of the 4x4 matrix with using of *Expansion method* for the rows and columns at such numerical data as:

$$A = \begin{vmatrix} 0 & 1 & -2 & -3 \\ 1 & 2 & -1 & 0 \\ -3 & 0 & 2 & 1 \\ 4 & 0 & -2 & 1 \end{vmatrix}$$

Solving:

1) Expansion method for the rows

- *To the first row*

The value of determinant of the 4x4 matrix elements of which have expansion to the first row is equal in general view:

$$|A| = a_{11}A_{11} - a_{12}A_{12} + a_{13}A_{13} - a_{14}A_{14}$$

Applicably to formula (17) we have the value of determinant for the above-indicated 4x4 matrix in numerical data which is equal $|A| = -13$.

- *To the second row*

The value of determinant of the 4x4 matrix elements of which have expansion to the second row is equal in general view:

$$|A| = -a_{21}A_{21} + a_{22}A_{22} - a_{23}A_{23} + a_{24}A_{24}$$

Applicably to formula (19) we have the value of determinant for the above-indicated 4x4 matrix in numerical data which is equal $|A| = -13$.

- *To the third row*

The value of determinant of the 4x4 matrix elements of which have expansion to the third row is equal in general view:

$$|A| = a_{31}A_{31} - a_{32}A_{32} + a_{33}A_{33} - a_{34}A_{34}$$

Applicably to formula (21) we have the value of determinant for the above-indicated 4x4 matrix in numerical data which is equal $|A| = -13$.

- *To the fourth row*

The value of determinant of the 4x4 matrix elements of which have expansion to the fourth row is equal in general view:

$$|A| = -a_{41}A_{41} + a_{42}A_{42} - a_{43}A_{43} + a_{44}A_{44}$$

Applicably to formula (23) we have the value of determinant for the above-indicated 4x4 matrix in numerical data which is equal $|A| = -13$.

2) Expansion method for the columns

351

- *To the first column*

The value of determinant of the 4x4 matrix elements of which have expansion to the first column is equal in general view:

$$|A|=a_{11}A_{11}-a_{21}A_{21}+a_{31}A_{31}-a_{41}A_{41}$$

Applicably to formula (25) we have the value of determinant for the above-indicated 4x4 matrix in numerical data which is equal $|A|=-13$.

- *To the second column*

The value of determinant of the 4x4 matrix elements of which have expansion to the second column is equal in general view:

$$|A|=-a_{12}A_{12}+a_{22}A_{22}-a_{32}A_{32}+a_{42}A_{42}$$

Applicably to formula (27) we have the value of determinant for the above-indicated 4x4 matrix in numerical data which is equal $|A|=-13$.

- *To the third column*

The value of determinant of the 4x4 matrix elements of which have expansion to the third column is equal in general view:

$$|A|=a_{13}A_{13}-a_{23}A_{23}+a_{33}A_{33}-a_{43}A_{43}$$

Applicably to formula (29) we have the value of determinant for the above-indicated 4x4 matrix in numerical data which is equal $|A|=-13$.

- *To the fourth column*

The value of determinant of the 4x4 matrix elements of which have expansion to the fourth column is equal in general view:

$$|A|=-a_{14}A_{14}+a_{24}A_{24}-a_{34}A_{34}+a_{44}A_{44}$$

Applicably to formula (31) we have the value of determinant for the above-indicated 4x4 matrix in numerical data which is equal $|A|=-13$. As we see that value of determinant of the 4x4 matrix at expansion method is the same for any rows and columns. And for above-indicated reasons we can now to define the algorithm for evaluation of the determinant of the 4x4 matrix *applicably to Expansion method for any rows or columns* .

Algorithm for evaluation of the determinant of the 4x4 matrix applicably to the *Expansion method* for any rows or columns

1) Primary should to analyze the given matrix in numerical data on subject of presenting elements *which have the null values* and then mark out of them and take into consideration in process of further calculations;
2) And then should to make the right decision regarding on which *row or column should to expanse elements* of given matrix;
3) After of choosing row or column ,necessary *to consider the above-considered formula* with goal of reducing the calculation procedures and finally to define the formula for given conditions;
4) And then *to draw the functional ties between elements of matrix* on base of simplified formula;
5) With goal of visual solving of present problem *should to built the diagram of elements matrix* in accordance with approved formula and numerical data *and calculate sum $\sum S_1$ for positive area and sum $\sum S_2$ for negative area;*
6) And finally to define the value of the determinant of the matrix by formula $|A|=\sum S_1-\sum S_2$.

Realization of suggested algorithm is shown **in Figure 1** of Appendix 1 for the above-considered conditions, for a instance, when elements of 4x4 matrix have expansion to the first row. As we see the evaluation of the determinant of the 4x4 matrix can be defined from the functional drawing or using the simplified formula for the given 4x4 matrix in numerical data .

352

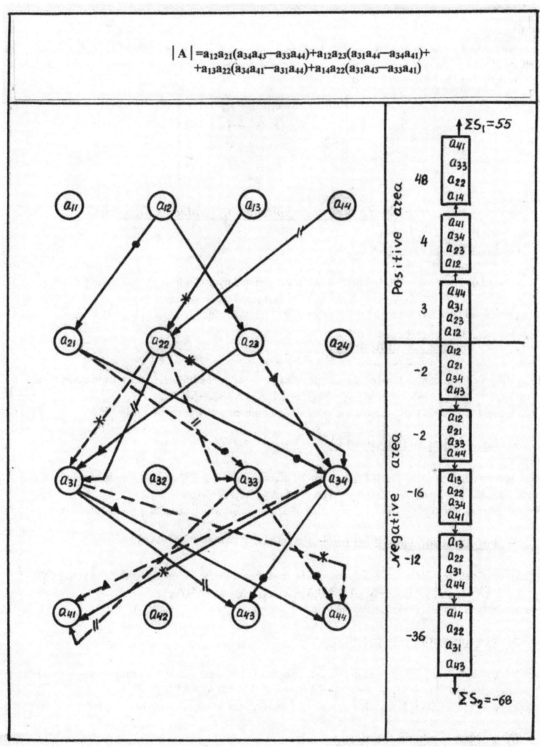

Figure 1 Evaluation of the determinant of the given 4x4 matrix of Appendix 1 applicably to the Expansion method for the first row
(— — —negative ties; ● elements of matrix equal zero)

Appendix 2

Problem:

Evaluate the value of determinant of the 4x4 matrix in numerical data with using of Laplace's theorem for each jointly rows and columns. And then compare these results with results for determinants evaluated by the Expansion method for any row and column.

$$A = \begin{vmatrix} 2 & -8 & 10 & 5 \\ 0 & 3 & -1 & 2 \\ -2 & -6 & 0 & 1 \\ 5 & -1 & -7 & 0 \end{vmatrix}$$

Solving:

❖ *To the rows applicably to Laplace's theorem:*

1) For the first and second rows

The value of determinant of the 4x4 matrix can be determined by the following formula in general view:
$$|A| = M_1 A_1 - M_2 A_2 + M_3 A_3 + M_4 A_4 - M_5 A_5 + M_6 A_6$$
or applicably to formula (33) we have the value of determinant which is equal $|A| = 1556$

2) For the first and third rows

The value of determinant of the 4x4 matrix can be determined by the following formula in general view:
$$|A| = -M_1 A_1 + M_2 A_2 - M_3 A_3 - M_4 A_4 + M_5 A_5 - M_6 A_6$$
or applicably to formula (35) we have the value of determinant which is equal $|A| = 1556$

3) For the first and fourth rows

The value of determinant of the 4x4 matrix can be determined by the following formula in general view:
$$|A| = M_1 A_1 - M_2 A_2 + M_3 A_3 + M_4 A_4 - M_5 A_5 + M_6 A_6$$
or applicably to formula(37) we have the value of determinant which is equal $|A| = 1556$

4) For the second and third rows

The value of determinant of the 4x4 matrix can be determined by the following formula in general view:
$$|A| = M_1 A_1 - M_2 A_2 + M_3 A_3 + M_4 A_4 - M_5 A_5 + M_6 A_6$$
or applicably to formula (39) we have the value of determinant which is equal $|A| = 1556$

5) For the second and fourth rows

The value of determinant of the 4x4 matrix can be determined by the following formula in general view:
$$|A| = -M_1 A_1 + M_2 A_2 - M_3 A_3 - M_4 A_4 + M_5 A_5 - M_6 A_6$$
or applicably to formula (41) we have the value of determinant which is equal $|A| = 1556$

6) For the third and fourth rows

The value of determinant of the 4x4 matrix can be determined by the following formula in general view:
$$|A| = M_1 A_1 - M_2 A_2 + M_3 A_3 + M_4 A_4 - M_5 A_5 + M_6 A_6$$

or applicably to formula (43) we have the value of determinant which is equal $|A|$ =1556

❖ *To the columns applicably to Laplace's theorem:*

1) For the first and second columns

The value of determinant of the 4x4 matrix can be determined by the following formula in general view:
$$|A| = M_1A_1 - M_2A_2 + M_3A_3 + M_4A_4 - M_5A_5 + M_6A_6$$
or applicably to formula (45) we have the value of determinant which is equal $|A|$ =1556

2) For the first and third columns

The value of determinant of the 4x4 matrix can be determined by the following formula in general view:
$$|A| = -M_1A_1 + M_2A_2 - M_3A_3 - M_4A_4 + M_5A_5 - M_6A_6$$
or applicably to formula (47) we have the value of determinant which is equal $|A|$ =1556

3) For the first and fourth columns

The value of determinant of the 4x4 matrix can be determined by the following formula in general view:
$$|A| = M_1A_1 - M_2A_2 + M_3A_3 + M_4A_4 - M_5A_5 + M_6A_6$$
or applicably to formula (49) we have the value of determinant which is equal $|A|$ =1556

4) For the second and third columns

The value of determinant of the 4x4 matrix can be determined by the following formula in general view:
$$|A| = M_1A_1 - M_2A_2 + M_3A_3 + M_4A_4 - M_5A_5 + M_6A_6$$
or applicably to formula (51) we have the value of determinant which is equal $|A|$ =1556

5) For the second and fourth columns

The value of determinant of the 4x4 matrix can be determined by the following formula in general view:
$$|A| = -M_1A_1 + M_2A_2 - M_3A_3 - M_4A_4 + M_5A_5 - M_6A_6$$
or applicably to formula (53) we have the value of determinant which is equal $|A|$ =1556

6) For the third and fourth columns

The value of determinant of the 4x4 matrix can be determined by the following formula in general view:
$$|A| = M_1A_1 - M_2A_2 + M_3A_3 + M_4A_4 - M_5A_5 + M_6A_6$$
or applicably to formula (55) we have the value of determinant which is equal $|A|$ =1556

❖ *To the rows applicably to Expansion(cofactor) method:*

The value of determinant of the 4x4 matrix elements of which have expansion, for a instance , to the first row can be determined by the following formula in general view:
$$|A| = a_{11}A_{11} + a_{12}A_{12} + a_{13}A_{13} + a_{14}A_{14}$$
or applicably to formula (17) we have the value of determinant which is equal $|A|$ =1556

❖ *To the columns applicably to Expansion(cofactor) method:*
The value of determinant of the 4x4 matrix elements of which have expansion, for a instance, to the third column can be determined by the following formula in general view:
$$|A| = a_{13}A_{13} + a_{23}A_{23} + a_{33}A_{33} + a_{43}A_{43}$$

or applicably to formula (29) we have the value of determinant which is equal $\left|A\right|=1556$.

In Figure 2 for the above-considered condition is shown the functional graph between elements of the 5x5 given matrix in numerical data of Appendix 2 applicably to Laplace's theorem for the first and second rows in general and calculations processes.

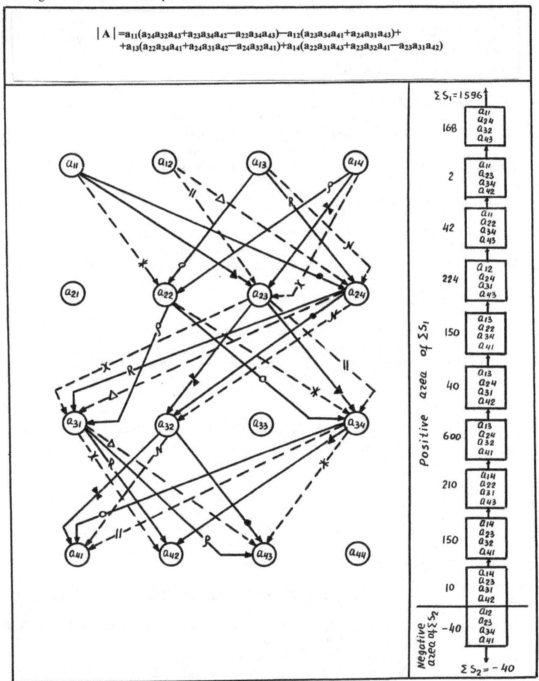

$$\left|A\right|=a_{11}(a_{24}a_{32}a_{43}+a_{23}a_{34}a_{42}-a_{22}a_{34}a_{43})-a_{12}(a_{23}a_{34}a_{41}+a_{24}a_{31}a_{43})+$$
$$+a_{13}(a_{22}a_{34}a_{41}+a_{24}a_{31}a_{42}-a_{24}a_{32}a_{41})+a_{14}(a_{22}a_{31}a_{43}+a_{23}a_{32}a_{41}-a_{23}a_{31}a_{42})$$

Figure 2 Functional graph between elements of the 5x5 matrix of Appendix 2 applicably to Laplace's theorem for the first and second rows
(------ negative ties; • elements of matrix equal zero)

356

Conclusion:

The determinant $|A|$ of the 4x4 matrix is obtained by the Expansion and Laplace's methods have the same value in both calculations.

Appendix 3

Problem:
Evaluate the determinant of the 5x5 matrix with using of the Laplace's theorem for marked out jointly the first and second rows. And compare that result with determinant evaluated by the Expansion (cofactor) method for the first row.

$$A = \begin{vmatrix} 0 & 0 & 1 & 4 & 0 \\ 0 & 6 & -7 & 8 & 0 \\ -2 & -5 & 1 & -1 & 3 \\ 4 & -4 & 6 & -2 & 2 \\ -7 & 2 & 3 & 1 & 6 \end{vmatrix}$$

Solving:

- **Lapalce's theorem applicably to the first and second rows**

The value of determinant of the 5x5 matrix at conditions when marked out jointly the first and second rows can be evaluated by the following formula in general view:

$$|A| = M_{12}A_{12} - M_{13}A_{13} + M_{14}A_{14} - M_{15}A_{15} + M_{23}A_{23} - M_{24}A_{24} + M_{25}A_{25} + M_{34}A_{34} - M_{35}A_{35} + M_{45}A_{45}$$

or applicably to formula (86) the value of determinant of the 5x5 matrix is equal $|A| = 8016$.

- **Expansion(cofactor) method applicably to the first row**

The value of determinant of the 5x5 matrix at conditions when its elements have expansion to the first row can be evaluated by the following formula in general view :

$$|A| = a_{11}A_{11} - a_{12}A_{12} + a_{13}A_{13} - a_{14}A_{14} + a_{15}A_{15}$$

or applicably to formula (58) the value of determinant of the 5x5 matrix is equal $|A| = 8016$.
As we see, the determinant $|A|$ of the 5x5 matrix is obtained by the Expansion and Laplace's methods have the same value in both calculations.

Appendix 4

Problem:
Evaluate the determinant of the 5x5 matrix with using of the Laplace's theorem for marked out jointly the first and third rows. And compare that result with determinant evaluated by the Expansion (cofactor) method for the third row.

$$A = \begin{vmatrix} 0 & 0 & 5 & -1 & 2 \\ 2 & -2 & 1 & 6 & -4 \\ 7 & 0 & -4 & 0 & 0 \\ 6 & -6 & 1 & 1 & -3 \\ 4 & 5 & -2 & 2 & 7 \end{vmatrix}$$

357

<u>Solving:</u>

- <u>Laplace's theorem applicably to the first and third rows</u>

The value of determinant of the 5x5 matrix at conditions when marked out jointly the first and third rows can be evaluated by the following formula in general view:

$$|A| = -M_1A_1 + M_2A_2 - M_3A_3 + M_4A_4 - M_5A_5 + M_6A_6 - M_7A_7 - M_8A_8 + M_9A_9 - M_{10}A_{10}$$

or applicably to formula (87) the value of determinant of the 5x5 matrix is equal $|A| = -6033$.

- <u>Expansion(cofactor) method applicably to the third row</u>

The value of determinant of the 5x5 matrix at conditions when its elements have expansion to the third row can be evaluated by the following formula in general view :

$$|A| = a_{31}A_{31} - a_{32}A_{32} + a_{33}A_{33} - a_{34}A_{34} + a_{35}A_{35}$$

or applicably to formula (64) the value of determinant of the 5x5 matrix is equal $|A| = -6033$.
As we see , the determinant $|A|$ of the 5x5 matrix is obtained by the Expansion and Laplace's methods have the same value in both calculations.

Appendix 5

<u>Problem:</u>
Evaluate the determinant of the 5x5 matrix with using of the Laplace's theorem for marked out jointly the first and fourth rows. And compare that result with determinant evaluated by the Expansion (cofactor) method for the fourth row.

$$A = \begin{vmatrix} 0 & 0 & -1 & 2 & 4 \\ 5 & -1 & 2 & -3 & 3 \\ 4 & 6 & -6 & 1 & -2 \\ -4 & 0 & 0 & -5 & 0 \\ 7 & 8 & -9 & 1 & 4 \end{vmatrix}$$

<u>Solving:</u>

- <u>Laplace's theorem applicably to the first and fourth rows</u>

The value of determinant of the 5x5 matrix at conditions when marked out jointly the first and fourth rows can be evaluated by the following formula in general view:

$$|A| = M_1A_1 - M_2A_2 + M_3A_3 - M_4A_4 + M_5A_5 - M_6A_6 + M_7A_7 + M_8A_8 - M_9A_9 + M_{10}A_{10}$$

or applicably to formula (88) the value of determinant of the 5x5 matrix is equal $|A| = -200$.

- <u>Expansion(cofactor) method applicably to the fourth row</u>

The value of determinant of the 5x5 matrix at conditions when its elements have expansion to the fourth row can be evaluated by the following formula in general view :

$$|A| = -a_{41}A_{41} + a_{42}A_{42} - a_{43}A_{43} + a_{44}A_{44} - a_{45}A_{45}$$

or applicably to formula (68) the value of determinant of the 5x5 matrix is equal $|A| = -200$.

Appendix 6

<u>Problem:</u>
Evaluate the determinant of the 5x5 matrix with using of the Laplace's theorem for marked out jointly

358

the first and fifth rows. And compare that result with determinant evaluated by the Expansion (cofactor) method for the fifth row.

$$A=\begin{vmatrix} 0 & -3 & 2 & 1 & 0 \\ 2 & -2 & 4 & 4 & -7 \\ 3 & -3 & 5 & -8 & 1 \\ -6 & 6 & 7 & -7 & 3 \\ 5 & 0 & 0 & 0 & 1 \end{vmatrix}$$

Solving:

- Laplace's theorem applicably to the first and fifth rows

The value of determinant of the 5x5 matrix at conditions when marked out jointly the first and fifth rows can be evaluated by the following formula in general view:
$$|A| = -M_1A_1+M_2A_2-M_3A_3+M_4A_4-M_5A_5+M_6A_6-M_7A_7-M_8A_8+M_9A_9-M_{10}A_{10}$$
or applicably to formula (89) the value of determinant of the 5x5 matrix is equal $|A| = 11166$

- Expansion method applicably to the fifth row

The value of determinant of the 5x5 matrix at conditions when its elements have expansion to the fifth row can be determined by the following formula in general view :
$$|A| = a_{51}A_{51}-a_{52}A_{52}+a_{53}A_{53}-a_{54}A_{54}+a_{55}A_{55}$$
or applicably to formula (71)the value of determinant of the 5x5 matrix is equal $|A| = 11166$.

Appendix 7

Problem:
Evaluate the determinant of the 5x5 matrix with using of the Laplace's theorem for marked out jointly the second and third rows. And compare that result with determinant evaluated by the Expansion method for the second row.

$$A=\begin{vmatrix} 10 & -11 & 13 & 7 & -6 \\ 0 & -1 & 0 & 2 & 0 \\ 5 & 0 & 8 & 0 & 10 \\ 12 & -3 & 4 & 6 & -8 \\ -7 & 8 & 9 & 10 & -13 \end{vmatrix}$$

Solving:

- Laplace's theorem applicably to the second and third rows

The value of determinant of the 5x5 matrix at conditions when marked out jointly the second and third rows can be evaluated by the following formula in general view:
$$|A| = M_1A_1-M_2A_2+M_3A_3-M_4A_4+M_5A_5-M_6A_6+M_7A_7+M_8A_8-M_9A_9+M_{10}A_{10}$$
or applicably to formula (90) the value of determinant of the 5x5 matrix is equal $|A| = 86236$.

- Expansion method applicably to the second row

The value of determinant of the 5x5 matrix at conditions when its elements have expansion to the second row can be determined by the following formula in general view :

$$|A| = -a_{21}A_{21} + a_{22}A_{22} - a_{23}A_{23} + a_{24}A_{24} - a_{25}A_{25}$$
or applicably to formula (61) the value of determinant of the 5x5 matrix is equal $|A| = 86236$.

Appendix 8

Problem:
Evaluate the determinant of the 5x5 matrix with using of the Laplace's theorem for marked out jointly the second and fourth rows. And compare that result with determinant evaluated by the Expansion method for the fourth row.

$$A = \begin{vmatrix} -1 & 2 & 1 & 3 & 4 \\ 0 & 4 & 0 & -1 & -2 \\ 6 & -4 & 2 & -3 & -6 \\ 7 & 0 & -1 & 0 & 0 \\ 2 & -2 & 4 & -4 & 1 \end{vmatrix}$$

Solving:

- Laplace's theorem applicably to the second and fourth rows

The value of determinant of the 5x5 matrix at conditions when marked out jointly the second and fourth rows can be evaluated by the following formula in general view:
$$|A| = -M_1A_1 + M_2A_2 - M_3A_3 + M_4A_4 - M_5A_5 + M_6A_6 - M_7A_7 - M_8A_8 + M_9A_9 - M_{10}A_{10}$$
or applicably to formula (91) the value of determinant of the 5x5 matrix is equal $|A| = -3624$.

- Expansion method applicably to the fourth row

The value of determinant of the 5x5 matrix at conditions when its elements have expansion to the fourth row can be determined by the following formula in general view:
$$|A| = -a_{41}A_{41} + a_{42}A_{42} - a_{43}A_{43} + a_{44}A_{44} - a_{45}A_{45}$$
or applicably to formula (47) the value of determinant of the 5x5 matrix is equal $|A| = -3624$.

Appendix 9

Problem:
Evaluate the determinant of the 5x5 matrix with using of the Laplace's theorem for marked out jointly the second and fifth rows. And compare that result with determinant evaluated by the Expansion method for the fifth row.

$$A = \begin{vmatrix} 3 & 4 & -1 & 1 & 4 \\ 0 & 0 & 6 & -7 & 8 \\ 10 & 6 & -5 & 5 & -3 \\ -4 & 2 & -2 & -1 & 4 \\ 0 & 4 & 0 & 10 & 0 \end{vmatrix}$$

Solving:

- Laplace's theorem applicably to the second and fifth rows

The value of determinant of the 5x5 matrix at conditions when marked out jointly the second and fifth rows can be evaluated by the following formula in general view:

$$|A|=M_1A_1-M_2A_2+M_3A_3-M_4A_4+M_5A_5-M_6A_6+M_7A_7+M_8A_8-M_9A_9+M_{10}A_{10}$$

or applicably to formula (92) the value of determinant of the 5x5 matrix is equal $|A|=4328$.

- ## Expansion method applicably to the fifth row

The value of determinant of the 5x5 matrix at conditions when its elements have expansion to the fifth row can be determined by the following formula in general view :

$$|A|=a_{51}A_{51}-a_{52}A_{52}+a_{53}A_{53}-a_{54}A_{54}+a_{55}A_{55}$$

or applicably to formula (71) the value of determinant of the 5x5 matrix is equal $|A|=4328$.

Appendix 10

Problem:
Evaluate the determinant of the 5x5 matrix with using of the Laplace's theorem for marked out jointly the third and fourth rows. And compare that result with determinant evaluated by the Expansion method for the fourth row.

$$A=\begin{vmatrix} 6 & -5 & 4 & 2 & -1 \\ -2 & 1 & 1 & -4 & 5 \\ 0 & 0 & 0 & -3 & 4 \\ -7 & 5 & 0 & 0 & 0 \\ 3 & -3 & 2 & -2 & 1 \end{vmatrix}$$

Solving:

- ## Laplace's theorem applicably to the third and fourth rows

The value of determinant of the 5x5 matrix at conditions when marked out jointly the third and fourth rows can be evaluated by the following formula in general view:

$$|A|=M_1A_1-M_2A_2+M_3A_3-M_4A_4+M_5A_5-M_6A_6+M_7A_7+M_8A_8-M_9A_9+M_{10}A_{10}$$

or applicably to formula (93) the value of determinant of the 5x5 matrix is equal $|A|=21$.

- ## Expansion method applicably to the fourth row

The value of determinant of the 5x5 matrix at conditions when its elements have expansion to the fourth row can be evaluated by the following formula in general view :

$$|A|=-a_{41}A_{41}+a_{42}A_{42}-a_{43}A_{43}+a_{44}A_{44}-a_{45}A_{45}$$

or applicably to formula (68) the value of determinant of the 5x5 matrix is equal $|A|=21$.

Appendix 11

Problem:
Evaluate the determinant of the 5x5 matrix with using of the Laplace;s theorem for marked out jointly the third and fifth rows. And compare that result with determinant evaluated by the Expansion method for the fifth row.

$$A=\begin{vmatrix} 6 & 4 & -4 & 5 & -5 \\ 3 & -3 & 1 & 2 & -3 \\ 2 & 0 & 0 & 0 & 1 \\ 6 & -6 & 1 & -3 & 4 \\ 0 & 0 & -7 & 8 & 0 \end{vmatrix}$$

Solving:

- Laplace's theorem applicably to the third and fifth rows

The value of determinant of the 5x5 matrix at conditions when marked out jointly the third and fifth rows can be evaluated by the following formula in general view:
$$|A| = -M_1A_1 + M_2A_2 - M_3A_3 + M_4A_4 - M_5A_5 + M_6A_6 - M_7A_7 - M_8A_8 + M_9A_9 - M_{10}A_{10}$$
or applicably to formula (94) the value of determinant of the 5x5 matrix is equal $|A| = 2848$

- Expansion method applicably to the fifth row

The value of determinant of the 5x5 matrix at conditions when its elements have expansion to the fifth row can be evaluated by the following formula in general view :
$$|A| = a_{51}A_{51} - a_{52}A_{52} + a_{53}A_{53} - a_{54}A_{54} + a_{55}A_{55}$$
or applicably to formula (71) the value of determinant of the 5x5 matrix is equal $|A| = 2848$.

Appendix 12

Problem:
Evaluate the determinant of the 5x5 matrix with using of the Laplace's theorem for marked out jointly the fourth and fifth rows. And compare that result with determinant evaluated by the Expansion method for the fourth row.

$$A = \begin{vmatrix} 3 & 4 & 1 & 2 & -4 \\ 5 & -5 & 6 & 7 & -1 \\ 2 & 2 & -3 & 4 & 6 \\ 4 & -4 & 0 & 0 & 0 \\ 0 & 0 & 3 & 4 & 5 \end{vmatrix}$$

Solving:

- Laplace's theorem applicably to the fourth and fifth rows

The value of determinant of the 5x5 matrix at conditions when marked out jointly the fourth and fifth rows can be evaluated by the following formula in general view:
$$|A| = M_1A_1 - M_2A_2 + M_3A_3 - M_4A_4 + M_5A_5 - M_6A_6 + M_7A_7 + M_8A_8 - M_9A_9 + M_{10}A_{10}$$
or applicably to formula (95) the value of the 5x5 matrix is equal $|A| = -5844$.

- Expansion method applicably to the fourth row

The value of determinant of the 5x5 matrix at conditions when its elements have expansion to the fourth row can be determined by the following formula in general view :
$$|A| = -a_{41}A_{41} + a_{42}A_{42} - a_{43}A_{43} + a_{44}A_{44} - a_{45}A_{45}$$
or applicably to formula (68) the value of determinant of the 5x5 matrix is equal $|A| = -5844$.

Appendix 13

Problem:
Evaluate the determinant of the 5x5 matrix with using of the Laplace's theorem for marked out jointly the first and second columns. And compare that result with determinant evaluated by the Expansion method for the column.

362

$$A = \begin{vmatrix} 0 & 0 & 4 & 2 & -5 \\ -5 & 0 & 5 & 6 & 7 \\ 0 & -9 & -6 & 8 & 9 \\ 0 & 4 & 7 & 11 & 6 \\ 2 & 9 & 3 & -4 & -5 \end{vmatrix}$$

Solving:

- ### Laplace's theorem applicably to the first and second columns

The value of determinant of the 5x5 matrix at conditions when marked out jointly the first and second columns can be evaluated by the following formula in general view:
$$|A| = M_1A_1 - M_2A_2 + M_3A_3 - M_4A_4 + M_5A_5 - M_6A_6 + M_7A_7 + M_8A_8 - M_9A_9 + M_{10}A_{10}$$
or applicably to formula (96) the value of determinant of the 5x5 matrix is equal $|A| = -6771$.

- ### Expansion method applicably to the first column

The value of determinant of the 5x5 matrix at conditions when its elements have expansion to the first column can be determined by the following formula in general view:
$$|A| = a_{11}A_{11} - a_{21}A_{21} + a_{31}A_{31} - a_{41}A_{41} + a_{51}A_{51}$$
or applicably to formula (74) the value of determinant of the 5x5 matrix is equal $|A| = -6771$.

Appendix 14

Problem:
Evaluate the determinant of the 5x5 matrix with using of the Laplace's theorem for marked out jointly the first and third columns. And compare that result with determinant evaluated by the Expansion method for the third column.

$$A = \begin{vmatrix} 0 & 4 & 0 & -4 & 5 \\ -6 & -3 & 0 & 1 & 3 \\ 8 & 2 & 0 & 4 & 1 \\ 0 & 1 & -2 & 5 & -6 \\ 0 & -1 & 3 & 7 & -7 \end{vmatrix}$$

Solving:

- ### Laplace's theorem applicably to the first and third columns

The value of determinant of the 5x5 matrix at conditions when marked out jointly the first and third columns can be evaluated by the following formula in general view:
$$|A| = -M_1A_1 + M_2A_2 - M_3A_3 + M_4A_4 - M_5A_5 + M_6A_6 - M_7A_7 - M_8A_8 + M_9A_9 - M_{10}A_{10}$$
or applicably to formula (97) the value of determinant of the 5x5 matrix is equal $|A| = -8060$.

- ### Expansion method applicably to the third column

The value of determinant of the 5x5 matrix at conditions when its elements have expansion to the third column can be evaluated by the following formula in general view:
$$|A| = a_{13}A_{13} - a_{23}A_{23} + a_{33}A_{33} - a_{43}A_{43} + a_{53}A_{53}$$
or applicably to formula (79) the value of determinant of the 5x5 matrix is equal $|A| = -8060$.

Appendix 15

Problem:

Evaluate the determinant of the 5x5 matrix with using of the Laplace's theorem for marked out jointly the first and fourth columns. And compare that result with the determinant evaluated by the Expansion method for the fourth column.

$$A = \begin{vmatrix} 0 & 6 & 2 & 0 & -1 \\ -3 & 1 & -2 & 0 & 1 \\ 0 & -1 & 3 & -5 & 2 \\ 4 & 5 & -4 & 0 & 4 \\ 0 & -7 & 6 & 8 & -3 \end{vmatrix}$$

Solving:

- #### Laplace's theorem applicably to the first and fourth columns

The value of determinant of the 5x5 matrix at conditions when marked out jointly the first and fourth columns can be evaluated by the following formula in general view:

$$|A| = M_1A_1 - M_2A_2 + M_3A_3 - M_4A_4 + M_5A_5 - M_6A_6 + M_7A_7 + M_8A_8 - M_9A_9 + M_{10}A_{10}$$

or applicably to formula (98) the value of determinant of the 5x5 matrix is equal $|A| = -6884$.

- #### Expansion method applicably to the fourth column:

The value of determinant of the 5x5 matrix at conditions when its elements have expansion to the fourth column can be evaluated by the following formula in general view:

$$|A| = -a_{14}A_{14} + a_{24}A_{24} - a_{34}A_{34} + a_{44}A_{44} - a_{54}A_{54}$$

or applicably to formula (82) the value of determinant of the 5x5 matrix is equal $|A| = -6884$.

Appendix 16

Problem:

Evaluate the determinant of the 5x5 matrix with using of the Laplace's theorem for marked out jointly the first and fifth columns. And compare that result with determinant evaluated by the Expansion method for the fifth column.

$$A = \begin{vmatrix} 0 & -4 & 2 & 4 & 7 \\ 0 & 5 & -2 & 6 & 0 \\ 3 & -6 & 1 & 7 & 8 \\ -2 & 7 & -1 & -8 & 9 \\ 0 & -1 & 3 & -1 & 0 \end{vmatrix}$$

Solving:

- #### Laplace's theorem applicably to the first and fifth columns

The value of determinant of the 5x5 matrix at conditions when marked out jointly the first and fifth column can be evaluated by the following formula in general view:

$$|A| = -M_1A_1 + M_2A_2 - M_3A_3 + M_4A_4 - M_5A_5 + M_6A_6 - M_7A_7 - M_8A_8 + M_9A_9 - M_{10}A_{10}$$

or applicably to formula (99) the value of determinant of the 5x5 matrix is equal $|A| = -6813$.

The value of determinant of the 5x5 matrix at conditions when its elements have expansion to the fifth column can be evaluated by the following formula in general view :

$$|A| = a_{15}A_{15} - a_{25}A_{25} + a_{35}A_{35} - a_{45}A_{45} + a_{55}A_{55}$$

or applicably to formula (85) the value of determinant of the 5x5 matrix is equal $|A| = -6813$.

Appendix 17

Problem:

Evaluate the determinant of the 5x5 matrix with using of the Laplace's theorem for marked out jointly the second and third columns. And compare that result with determinant evaluated by Expansion method for the second column.

$$A = \begin{vmatrix} 4 & 0 & 0 & 1 & 2 \\ 3 & -7 & 0 & 1 & -2 \\ -2 & 7 & 0 & -4 & 7 \\ 2 & 0 & 3 & 5 & -7 \\ 5 & 0 & -4 & 6 & 8 \end{vmatrix}$$

Solving:

- Laplace's theorem applicably to the second and third columns

The value of determinant of the 5x5 matrix at conditions when marked out jointly the second and third columns can be evaluated by the following formula in general view:

$$|A| = M_1A_1 - M_2A_2 + M_3A_3 - M_4A_4 + M_5A_5 - M_6A_6 + M_7A_7 + M_8A_8 - M_9A_9 + M_{10}A_{10}$$

or applicably to formula (100) the value of determinant of the 5x5 matrix is equal $|A| = -2653$

- Expansion method applicably to the second column

The value of determinant of the 5x5 matrix at conditions when its elements have expansion to the second column can be evaluated by the following formula in general view:

$$|A| = -a_{12}A_{12} + a_{22}A_{22} - a_{32}A_{32} + a_{42}A_{42} - a_{52}A_{52}$$

or applicably to formula (76) the value of determinant of the 5x5 matrix is equal $|A| = -2653$.

Appendix 18

Problem:

Evaluate the determinant of the 5x5 matrix with using of the Laplace's theorem for marked out jointly the second and fourth columns. And compare that result with determinant evaluated by the Expansion method for the third column.

$$A = \begin{vmatrix} -4 & 0 & 7 & 0 & 2 \\ 2 & 5 & -2 & 0 & -2 \\ 1 & 0 & 4 & 0 & 3 \\ -1 & 2 & -4 & 3 & -6 \\ 6 & -3 & 7 & -4 & 7 \end{vmatrix}$$

Solving:

- Laplace's theorem applicably to the second and fourth columns

The value of determinant of the 5x5 matrix at conditions when marked out jointly the second and fourth columns can be evaluated by the following formula in general view:

$$|A| = -M_1A_1 + M_2A_2 - M_3A_3 + M_4A_4 - M_5A_5 + M_6A_6 - M_7A_7 - M_8A_8 + M_9A_9 - M_{10}A_{10}$$

or applicably to formula (101) the value of determinant of the 5x5 matrix is equal $|A| = 1649$.

- ## Expansion method applicably to the third column

The value of determinant of the 5x5 matrix at conditions when its elements have expansion to the third column can be evaluated by the following formula in general view:

$$|A| = a_{13}A_{13} - a_{23}A_{23} + a_{33}A_{33} - a_{43}A_{43} + a_{53}A_{53}$$

or applicably to formula (79) the value of determinant of the 5x5 matrix is equal $|A| = 1649$.

Appendix 19

Problem:
Evaluate the determinant of the 5x5 matrix with using of the Laplace's theorem for marked out jointly the second and fifth columns. And compare that result with determinant evaluated by the Expansion method for the second column.

$$A = \begin{vmatrix} 4 & 0 & -6 & 7 & 0 \\ 2 & -3 & 1 & 3 & 5 \\ -2 & 0 & -1 & 4 & 0 \\ -6 & 0 & 5 & -2 & 0 \\ 7 & 0 & -6 & 1 & 6 \end{vmatrix}$$

Solving:
- ## Laplace's theorem applicably to the second and fifth columns

The value of determinant of the 5x5 matrix at conditions when marked out jointly the second and fifth columns can be evaluated by the following formula in general view:

$$|A| = M_1A_1 - M_2A_2 + M_3A_3 - M_4A_4 + M_5A_5 - M_6A_6 + M_7A_7 + M_8A_8 - M_9A_9 + M_{10}A_{10}$$

or applicably to formula (102) the value of determinant of the 5x5 matrix is equal $|A| = 288$.

- ## Expansion method applicably to the second column
The value of determinant of the 5x5 matrix at conditions when its elements have expansion to the second column can be evaluated by the following formula in general view:

$$|A| = -a_{12}A_{12} + a_{22}A_{22} - a_{32}A_{32} + a_{42}A_{42} - a_{52}A_{52}$$

or applicably to formula (76) the value of determinant of the 5x5 matrix is equal $|A| = 288$.

Appendix 20

Problem:
Evaluate the determinant of the 5x5 matrix with using of the Laplace's theorem for marked out jointly the third and fourth columns. And compare that result with determinant evaluated by the Expansion method for the third column.

$$A = \begin{vmatrix} -2 & 1 & 0 & 4 & 5 \\ 4 & -1 & 0 & -3 & -5 \\ 5 & 6 & 2 & 0 & 6 \\ 4 & 2 & -3 & 0 & -6 \\ -3 & 2 & 0 & 5 & 3 \end{vmatrix}$$

Solving:

- Laplace's theorem applicably to the third and fourth columns

The value of determinant of the 5x5 matrix at conditions when marked out jointly the third and fourth columns can be evaluated by the following formula in general view:
$$|A| = M_1A_1 - M_2A_2 + M_3A_3 - M_4A_4 + M_5A_5 - M_6A_6 + M_7A_7 + M_8A_8 - M_9A_9 + M_{10}A_{10}$$
or applicably to formula (103) the value of determinant of the 5x5 matrix is equal $|A| = -973$.

- Expansion method applicably to the third column

The value of determinant of the 5x5 matrix at conditions when its elements have expansion to the third column can be evaluated by the following formula in general view:
$$|A| = a_{13}A_{13} - a_{23}A_{23} + a_{33}A_{33} - a_{43}A_{43} + a_{53}A_{53}$$
or applicably to formula (79) the value of determinant of the 5x5 matrix is equal $|A| = -973$.

Appendix 21

Problem:
Evaluate the determinant of the 5x5 matrix with using of the Laplace's theorem for marked out jointly the third and fifth columns. And compare that result with determinant evaluated by the Expansion method for the second column.

$$A = \begin{vmatrix} 6 & -5 & 0 & 3 & 0 \\ -2 & 1 & 0 & -1 & 2 \\ 3 & -4 & 5 & 6 & 0 \\ 7 & -7 & -6 & 8 & 0 \\ 2 & 3 & 0 & -8 & -4 \end{vmatrix}$$

Solving:

- Laplace's theorem applicably to the third and fifth columns

The value of determinant of the 5x5 matrix at conditions when marked out jointly the third and fifth columns can be evaluated by the following formula in general view:
$$|A| = -M_1A_1 + M_2A_2 - M_3A_3 + M_4A_4 - M_5A_5 + M_6A_6 - M_7A_7 - M_8A_8 + M_9A_9 - M_{10}A_{10}$$
or applicably to formula (104) the value of determinant of the 5x5 matrix is equal $|A| = -378$.

- Expansion method applicably to the second column

The value of determinant of the 5x5 matrix at conditions when its elements have expansion to the second column can be evaluated by the following formula in general view:
$$|A| = -a_{12}A_{12} + a_{22}A_{22} - a_{32}A_{32} + a_{42}A_{42} - a_{52}A_{52}$$
or applicably to formula (76) the value of determinant of the 5x5 matrix is equal $|A| = -378$.

Appendix 22

Problem:
Evaluate the determinant of the 5x5 matrix with using of the Laplace's theorem for marked out jointly the fourth and fifth columns. And compare that result with determinant evaluated by the Expansion method for the first column.

$$A = \begin{vmatrix} 6 & 2 & -3 & 0 & 0 \\ -3 & -2 & 3 & 0 & -4 \\ 3 & 4 & 6 & 2 & 5 \\ 1 & -4 & -6 & 0 & 0 \\ -1 & 5 & 7 & 3 & 0 \end{vmatrix}$$

Solving:

- ## Laplace's theorem applicably to the fourth and fifth columns

The value of determinant of the 5x5 matrix at conditions when marked out jointly the fourth and fifth columns can be evaluated by the following formula in general view:
$$|A| = M_1A_1 - M_2A_2 + M_3A_3 - M_4A_4 + M_5A_5 - M_6A_6 + M_7A_7 + M_8A_8 - M_9A_9 + M_{10}A_{10}$$
or applicably to formula (105) the value of determinant of the 5x5 matrix is equal $|A| = 2288$.

- ## Expansion method applicably to the first column

The value of determinant of the 5x5 matrix at conditions when its elements have expansion to the first column can be evaluated by the following formula in general view:
$$|A| = a_{11}A_{11} - a_{21}A_{21} + a_{31}A_{31} - a_{41}A_{41} + a_{51}A_{51}$$
or applicably to formula (74) the value of determinant of the 5x5 matrix is equal $|A| = 2288$.

Conclusion:

So, analyzing the numerical data of 5x5 matrices shown in Appendices 3 to 22 for evaluation of the determinants by the Lapalce's and Expansion methods, we can definitely to conclude the following arguments:
1) The determinant $|A|$ for each given 5x5 matrix in numerical data is obtained by the above-indicated methods ,in view of Laplace's and Expansion methods, *has the same value* in both calculations;
2) And of which method is the best in choosing for calculation of the determinant depends advantageously from the given conditions of 5x5 matrix in numerical data;
3) Comparative analysis of evaluation the determinant of the 5x5 matrix shown in Appendices 3 to 22 discovers fact *that Laplace's method has advantages* against of the Expansion method. And these conclusions are confirmed by the numerical data shown in Table 1 also by the functional graphs shown **in Figure 3** in process of evaluation the determinant of the 5x5 matrix by the Laplace's theorem ,and **in Figure 4** for evaluation of the determinant of the 5x5 matrix by the Expansion method applicably to numerical data of 5x5 matrix in Appendix 19.

Table 1 Comparative analysis of evaluation the determinant of the 5x5 matrix by the Laplace's and Expansion methods

Number of appendix	Quantity of used elements in formulas for determinant $|A|$ of the 5x5 matrix	
	Laplace's theorem	Expansion method
3	68	68
4	85	91
5	85	100
6	102	102
7	102	102
8	102	124
9	85	107
10	68	112

11	68	90
12	102	102
13	85	85
14	68	100
15	68	101
16	68	68
17	68	68
18	68	100
19	17	68
20	102	102
21	68	104
22	51	195
Total:	1530	1963

Analysis of data shown in Table 1 indicates on fact that evaluation of the determinant of the 5x5 matrix applicably to Laplace's theorem in compare with Expansion method *has of some advantages* in question of reducing of quantity used elements in calculation procedures.

And besides the analysis of numerical data shown **in Figure 3** of functional graphs for evaluation of the determinant of the 5x5 matrix *by the Laplace's theorem* applicably to data shown in Appendix 19 indicates on the following results:

❖ **For positive area:**

- The second element(a_{22}) of the second column is tied with the fifth element(a_{55}) of the fifth column and then;
- The fifth element (a_{55}) of the fifth column is tied with the first element (a_{11})of the first column and then;
- The first element(a_{11}) of the first column is tied with the third element(a_{33}) of the third column and then;
- The third element (a_{33}) of the third column is tied with the fourth element (a_{44}) of the fourth column and then;
- The fifth element(a_{55}) of the fifth column is tied with the first element(a_{13}) of the third column and then;
- The first element (a_{13}) of the third column is tied with the third element(a_{34}) of the fourth column and then;
- The third element(a_{34})of the fourth column is tied with the fourth element(a_{41}) of the first column and then;
- The fifth element(a_{55}) of the fifth column is tied with the first element(a_{14}) of the fourth column and then;
- The first element(a_{14}) of the fourth column is tied with the third element (a_{31}) of the first column and then;
- The third element (a_{31}) of the first column is tied with the fourth element (a_{43}) of the third column.

❖ **For negative area:**

- The second element(a_{22}) of the second column is tied with the fifth element (a_{55}) of the fifth column and then;
- The fifth element(a_{55}) of the fifth column is tied with the first element(a_{11}) of the first column and then;
- The first element(a_{11}) of the first column is tied with the third element(a_{34}) of the fourth column and then;

369

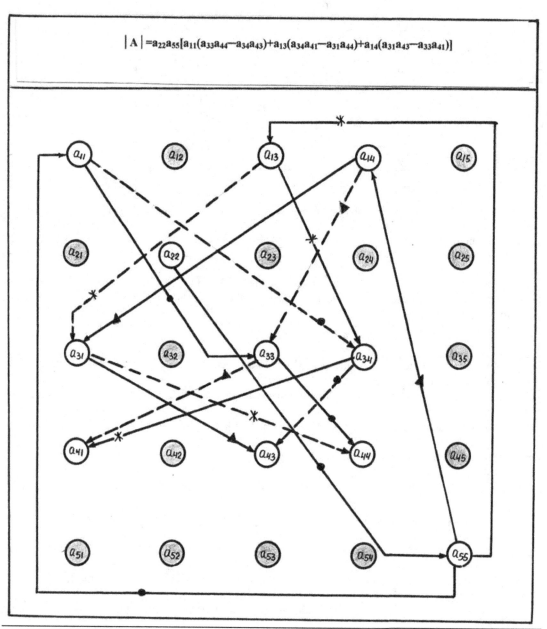

$$|A| = a_{22}a_{55}[a_{11}(a_{33}a_{44} - a_{34}a_{43}) + a_{13}(a_{34}a_{41} - a_{31}a_{44}) + a_{14}(a_{31}a_{43} - a_{33}a_{41})]$$

Figure 3 Functional graphs for evaluation of the determinant of the 5x5 matrix by Laplace's theorem applicably to numerical data of Appendix 19

(— — —negative ties;● elements of matrix equal zero)

- The third element (a_{34}) of the fifth column is tied with the fourth element (a_{43}) of the third column and then;
- The fifth element(a_{55}) of the fifth column is tied with the first element(a_{13}) of the third column and then;
- The first element(a_{13}) of the third column is tied with the third element (a_{31}) of the first column and then;

- The third element(a_{31}) of the first column is tied with the fourth element(a_{44}) of the fourth column and then;
- The fifth element(a_{55}) of the fifth column is tied with the first element(a_{14}) of the fourth column and then;
- The first element(a_{14}) of the fourth column is tied with the third element(a_{33}) of the third column and then;
- The third element (a_{33}) of the third column is tied with the fourth element(a_{41}) of the first column.

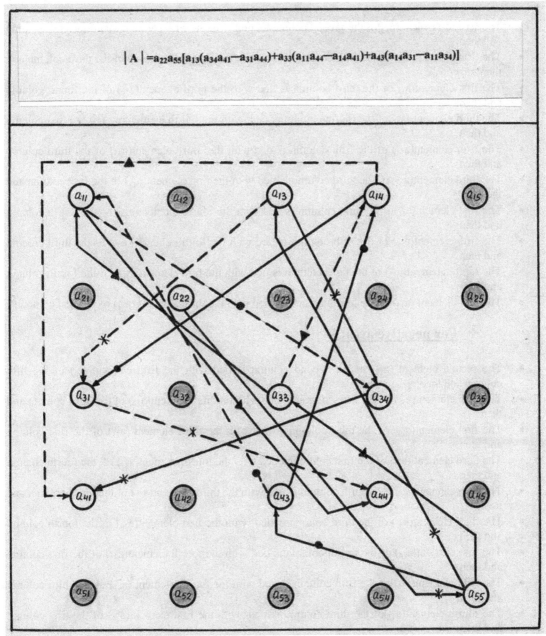

$$|A| = a_{22}a_{55}[a_{13}(a_{34}a_{41} - a_{31}a_{44}) + a_{33}(a_{11}a_{44} - a_{14}a_{41}) + a_{43}(a_{14}a_{31} - a_{11}a_{34})]$$

Figure 4 Functional graphs for evaluation of the determinant of the 5x5 matrix by Expansion method applicably to numerical data of Appendix 19
(— — —negative ties; ● elements of matrix equal zero)

And besides from Figure 3 we see that such 14 elements as $a_{12}, a_{15}, a_{21}, a_{23}, a_{24}, a_{25}, a_{32}, a_{35}, a_{42}, a_{45}, a_{51}, a_{52}, a_{53}, a_{54}$ of the 5x5 matrix practically do not participate in evaluation of the determinant by Laplace's theorem. And for this reason these above-indicated elements can ignore for the given conditions shown in Appendix 19.

Analysis of data shown in Figure 4 of functional graphs for evaluation of the determinant of the 5x5 matrix by *Expansion method* applicably to numerical data of Appendix 19 indicates on the following results:

❖ For positive area:

- The second element (a_{22})of the second column is tied with the fifth element(a_{55})of the fifth column and then;
- The fifth element(a_{55}) of the fifth column is tied with the first element(a_{13}) of the third column and then;
- The first element(a_{13})of the third column is tied with the third element(a_{34}) of the fourth column and then;
- The third element (a_{34})of the fourth column is tied with the fourth element(a_{41}) of the first column and then;
- The fifth element(a_{55}) of the fifth column is tied with the third element(a_{33}) of the third column and then;
- The third element(a_{33}) of the third column is tied with the first element(a_{11}) of the first column and then;
- The first element(a_{11})of the first column is tied with the fourth element(a_{44}) of the fourth column and then;
- The fifth element(a_{55}) of the fifth column is tied with the fourth element(a_{43}) of the third column and then;
- The fourth element(a_{43}) of the third column is tied with the first element (a_{14}) of the fourth column and then;
- The first element (a_{14}) of the fourth column is tied with the third element (a_{31}) of the first column.

❖ For negative area:

- The second element (a_{22}) of the second column is tied with the fifth element(a_{55}) of the fifth column and then;
- The fifth element(a_{55}) of the fifth column is tied with the first element(a_{13}) of the third column and then;
- The first element(a_{13}) of the third column is tied with the third element (a_{31}) of the first column and then;
- The third element (a_{31}) of the first column is tied with the fourth element(a_{44}) of the fourth column and then;
- The fifth element(a_{55})of the fifth column is tied with the third element(a_{33}) of the third column and then;
- The third element(a_{33}) of the third column is tied with the first element(a_{14}) of the fourth column and then;
- The first element(a_{14}) of the fourth column is tied with the fourth element(a_{41}) of the first column and then;
- The fifth element(a_{55}) of the fifth column is tied with the fourth element (a_{43}) of the third column and then;
- The fourth element(a_{43})of the third column is tied with the first element(a_{11}) of the first column and then;
- The first element (a_{11}) of the first column is tied with the third element (a_{34}) of the fourth column.

And besides from Figure 4 we see that such 14 elements as $a_{12}, a_{15}, a_{21}, a_{23}, a_{24}, a_{25}, a_{32}, a_{35}, a_{42}, a_{45}, a_{51}, a_{52}, a_{53}, a_{54}$ of the 5x5 matrix also practically do not participate in evaluation of the determinant $|A|$ by the Expansion method. And for this reason these above-indicated elements also can ignore for the given conditions shown in Appendix 19 for 5x5matrix in numerical data.

Appendix 23

Problem:
Compute the determinant of the 6x6 matrix applicably to the Expansion (cofactor) method for the first row elements. And express the functional ties between elements of the 6x6 matrix in graphs.

$$A = \begin{vmatrix} 1 & 2 & 1 & 0 & 2 & 0 \\ 3 & 0 & 2 & 0 & 1 & 0 \\ -1 & 0 & 2 & 0 & 6 & 0 \\ 0 & 0 & 0 & 2 & 4 & 1 \\ 0 & 0 & 0 & 1 & 0 & 2 \\ 0 & 0 & 0 & 1 & 1 & 1 \end{vmatrix}$$

Solution:
Applicably to the above-indicated 6x6 matrix in numerical data, the combined determinant $\sum |A| = |A_{12}|$ can be determined by the following formula (111)in general view as:

$$\sum |A| = |A_{12}| = a_{12}a_{21}a_{33}[a_{44}(a_{56}a_{65} - a_{55}a_{66}) + a_{45}(a_{54}a_{66} - a_{56}a_{64}) + a_{46}(a_{55}a_{64} - a_{54}a_{65})] +$$
$$+ a_{12}a_{23}a_{31}[a_{44}(a_{55}a_{66} - a_{56}a_{65}) + a_{45}(a_{56}a_{64} - a_{54}a_{66}) + a_{46}(a_{54}a_{65} - a_{55}a_{64})]$$

For the above-given numerical data of the 6x6 matrix, we have the solution of determinant value of which is equal $\sum |A| = |A_{12}| = -16$. **In Figure 5** is shown the functional graphs between elements of the 6x6 matrix for above- considered conditions.

Appendix 24

Problem:
Solve the systems of normal linear six equations with using of the Cramer's rule and Expansion (cofactor) method for evaluation of the determinants of the 6x6 matrices. And express the functional ties between elements of the 6x 6 matrices in graphs.

$$X_1 + 2X_2 + X_3 + 2X_5 \leq 2$$
$$3X_1 + 2X_3 + X_5 = -4$$
$$-X_1 + 2X_3 + 6X_5 = 0$$
$$2X_4 + 4X_5 + X_6 \geq 5$$
$$X_4 + 2X_6 = 6$$
$$X_4 + X_5 + X_6 \leq -7$$

Solution:
1) The above-indicated system, having six equations, can be written in view of the 6x6 matrix in numerical data:

$$A = \begin{vmatrix} 1 & 2 & 1 & 0 & 2 & 0 \\ 3 & 0 & 2 & 0 & 1 & 0 \\ -1 & 0 & 2 & 0 & 6 & 0 \\ 0 & 0 & 0 & 2 & 4 & 1 \\ 0 & 0 & 0 & 1 & 0 & 2 \\ 0 & 0 & 0 & 1 & 1 & 1 \end{vmatrix} \qquad B = \begin{vmatrix} 2 \\ -4 \\ 0 \\ 5 \\ 6 \\ -7 \end{vmatrix}$$

2) Applicably to Cramer's rule ,we have the following 6x6 matrices for evaluation of the determinants:

$$A_1=\begin{vmatrix} 2 & 2 & 1 & 0 & 2 & 0 \\ -4 & 0 & 2 & 0 & 1 & 0 \\ 0 & 0 & 2 & 0 & 6 & 0 \\ 5 & 0 & 0 & 2 & 4 & 1 \\ 6 & 0 & 0 & 1 & 0 & 2 \\ -7 & 0 & 0 & 1 & 1 & 1 \end{vmatrix} \qquad A_2=\begin{vmatrix} 1 & 2 & 1 & 0 & 2 & 0 \\ 3 & -4 & 2 & 0 & 1 & 0 \\ -1 & 0 & 2 & 0 & 6 & 0 \\ 0 & 5 & 0 & 2 & 4 & 1 \\ 0 & 6 & 0 & 1 & 0 & 2 \\ 0 & -7 & 0 & 1 & 1 & 1 \end{vmatrix}$$

$$A_3=\begin{vmatrix} 1 & 2 & 2 & 0 & 2 & 0 \\ 3 & 0 & -4 & 0 & 1 & 0 \\ -1 & 0 & 0 & 0 & 6 & 0 \\ 0 & 0 & 5 & 2 & 4 & 1 \\ 0 & 0 & 6 & 1 & 0 & 2 \\ 0 & 0 & -7 & 1 & 1 & 1 \end{vmatrix} \qquad A_4=\begin{vmatrix} 0 & 2 & 1 & 2 & 2 & 0 \\ 3 & 0 & 2 & -4 & 1 & 0 \\ -1 & 0 & 2 & 0 & 6 & 0 \\ 0 & 0 & 0 & 5 & 4 & 1 \\ 0 & 0 & 0 & 6 & 0 & 2 \\ 0 & 0 & 0 & -7 & 1 & 1 \end{vmatrix}$$

$$A_5=\begin{vmatrix} 1 & 2 & 1 & 0 & 2 & 0 \\ 3 & 0 & 2 & 0 & -4 & 0 \\ -1 & 0 & 2 & 0 & 0 & 0 \\ 0 & 0 & 0 & 2 & 5 & 1 \\ 0 & 0 & 0 & 1 & 6 & 2 \\ 0 & 0 & 0 & 1 & -7 & 1 \end{vmatrix} \qquad A_6=\begin{vmatrix} 1 & 2 & 1 & 0 & 2 & 2 \\ 3 & 0 & 2 & 0 & 1 & -4 \\ -1 & 0 & 2 & 0 & 6 & 0 \\ 0 & 0 & 0 & 2 & 4 & 5 \\ 0 & 0 & 0 & 1 & 0 & 6 \\ 0 & 0 & 0 & 1 & 1 & -7 \end{vmatrix}$$

3) The unknown parameters $X_1;X_2;X_3;X_4;X_4;X_5$ and X_5 can be evaluated by the Cramer's rule from formulas:

$$X_1=|A_1|/|A|\ ;X_2=|A_2|/|A|\ ;X_3=|A_3|/|A|\ ;X_4=|A_4|/|A|\ ;X_5=|A_5|/|A|\ ;$$
$$X_6=|A_6|/|A|\ .$$

4) The value of above-indicated determinants of the 6x6 matrices can be determined by the Expansion(cofactor) method at conditions when elements of each 6x6 matrix have expansion to the first row for the following considered formulas ,such as:

- Determinant $|A|$ for the 6x6 matrix (A) from formula shown in Appendix 23 equal $|A|=-16$;
- Determinant $|A_1|$ for the 6x6 matrix(A_1) is defined from the suggested formula and equal $|A_1|=-624$:

$$|A_1|=|A^*{}_{12}|=a_{12}a_{21}a_{33}[a_{44}(a_{56}a_{65}-a_{55}a_{66})+a_{45}(a_{54}a_{66}-a_{56}a_{64})+a_{46}(a_{55}a_{64}-a_{54}a_{65})]+$$
$$+a_{12}a_{23}a_{35}[a_{41}(a_{54}a_{66}-a_{56}a_{64})+a_{44}(a_{56}a_{61}-a_{51}a_{66})+a_{46}(a_{51}a_{64}-a_{54}a_{61})]+$$
$$+a_{12}a_{25}a_{33}[a_{41}(a_{56}a_{64}-a_{54}a_{66})+a_{44}(a_{51}a_{66}-a_{56}a_{61})+a_{46}(a_{54}a_{61}-a_{51}a_{64})]$$

The functional graphs between elements of the 6x6 matrix (A_1), participating in definition of the determinant $|A_1|$ is shown **in Figure 6.**

- Combined determinant $|A_2|$ for the 6x6 matrix (A_2) is sum of the different determinant which are equal $|A_2|=|A'_{11}|+|A''_{12}|+|A'_{13}|+|A'_{15}|$

where,

$|A'_{11}|$ -determinant which is defined from elements of the 6x6 matrix (A_2) which participate in evaluation of the determinant by formula and equal $|A'_{11}|=312$:

$$|A'_{11}|=a_{11}a_{22}a_{33}[a_{44}(a_{55}a_{66}-a_{56}a_{65})+a_{45}(a_{56}a_{64}-a_{54}a_{66})+a_{46}(a_{54}a_{65}-a_{55}a_{64})]+$$
$$+a_{11}a_{23}a_{35}[a_{42}(a_{56}a_{64}-a_{54}a_{66})+a_{44}(a_{52}a_{66}-a_{56}a_{62})+a_{46}(a_{54}a_{62}-a_{52}a_{64})]+$$
$$+a_{11}a_{25}a_{33}[a_{42}(a_{54}a_{66}-a_{56}a_{64})+a_{44}(a_{56}a_{62}-a_{52}a_{66})+a_{46}(a_{52}a_{64}-a_{54}a_{62})]$$

374

In Figure 7 is shown the functional graphs between elements of the above-considered formula $|A'_{11}|$.

$|A''_{12}|$ -determinant which is defined from elements of the 6x6 matrix (A_2) ,participating in evaluation of the determinant by formula and equal $|A''_{12}| = -16$:

$$|A''_{12}| = a_{12}a_{21}a_{33}[a_{44}(a_{56}a_{65}-a_{55}a_{66})+a_{45}(a_{54}a_{66}-a_{56}a_{64})+a_{46}(a_{55}a_{64}-a_{54}a_{65})]+$$
$$+a_{12}a_{23}a_{31}[a_{44}(a_{55}a_{66}-a_{56}a_{65})+a_{45}(a_{56}a_{64}-a_{54}a_{66})+a_{46}(a_{54}a_{65}-a_{55}a_{64})]$$

In Figure 8 is shown the functional graphs between elements of the above-considered formula for $|A''_{12}|$.

$|A'_{13}|$ -determinant which is defined from elements of the 6x6 matrix (A_2) ,participating in evaluation of the determinant by formula and equal $|A'_{13}| = -612$:

$$|A'_{13}| = a_{13}a_{21}a_{35}[a_{42}(a_{54}a_{66}-a_{56}a_{64})+a_{44}(a_{56}a_{62}-a_{52}a_{66})+a_{46}(a_{52}a_{64}-a_{54}a_{62})]+$$
$$+a_{13}a_{22}a_{31}[a_{44}(a_{56}a_{65}-a_{55}a_{66})+a_{45}(a_{54}a_{66}-a_{56}a_{64})+a_{46}(a_{55}a_{64}-a_{54}a_{65})]+$$
$$+a_{13}a_{25}a_{31}[a_{42}(a_{56}a_{64}-a_{54}a_{66})+a_{44}(a_{52}a_{66}-a_{56}a_{62})+a_{46}(a_{54}a_{62}-a_{52}a_{64})]$$

In Figure 9 is shown the functional graphs between elements of the above-considered formula for $|A'_{13}|$.

$|A'_{15}|$ -determinant which is defined from elements of the 6x6 matrix (A_2), participating in evaluation of the determinant by formula and equal $|A'_{15}| = 512$:

$$|A'_{15}| = a_{15}a_{21}a_{33}[a_{42}(a_{56}a_{64}-a_{54}a_{66})+a_{44}(a_{52}a_{66}-a_{56}a_{62})+a_{46}(a_{54}a_{62}-a_{52}a_{64})]+$$
$$+a_{15}a_{23}a_{31}[a_{42}(a_{54}a_{66}-a_{56}a_{64})+a_{44}(a_{56}a_{62}-a_{52}a_{66})+a_{46}(a_{52}a_{64}-a_{54}a_{62})]$$

In Figure 10 is shown the functional graphs between elements of the above-considered formula for $|A'_{15}|$.

So, we can define the combined determinant which is equal as $|A_2| = 196$.

- Determinant $|A_3|$ for the 6x6 matrix (A_3) is defined from elements ,participating in evaluation of its value from the suggested formula and equal $|A_3| = 1224$:

$$|A_3| = a_{12}a_{21}a_{35}[a_{43}(a_{56}a_{64}-a_{54}a_{66})+a_{44}(a_{53}a_{66}-a_{56}a_{63})+a_{46}(a_{54}a_{63}-a_{53}a_{64})]+$$
$$+a_{12}a_{23}a_{31}[a_{44}(a_{55}a_{66}-a_{56}a_{65})+a_{45}(a_{56}a_{64}-a_{54}a_{66})+a_{46}(a_{54}a_{65}-a_{55}a_{64})]+$$
$$+a_{12}a_{25}a_{31}[a_{43}(a_{54}a_{66}-a_{56}a_{64})+a_{44}(a_{56}a_{63}-a_{53}a_{66})+a_{46}(a_{53}a_{64}-a_{54}a_{63})]$$

In Figure 11 is shown the functional graphs between elements of the above-considered formula for $|A_3|$.

- Determinant $|A_4|$ for the 6x6 matrix (A_4) is defined from elements ,participating in evaluation of its value from the suggested formula and equal $|A_4| = 1344$:

$$|A_4| = a_{12}a_{21}a_{33}[a_{44}(a_{56}a_{65}-a_{55}a_{66})+a_{45}(a_{54}a_{66}-a_{56}a_{64})+a_{46}(a_{55}a_{64}-a_{54}a_{65})]+$$
$$+a_{12}a_{23}a_{31}[a_{44}(a_{55}a_{66}-a_{56}a_{65})+a_{45}(a_{56}a_{64}-a_{54}a_{66})+a_{46}(a_{54}a_{65}-a_{55}a_{64})]$$

The functional graphs between elements of the above-considered formula is shown in Figure 5 of Appendix 23 in general view ,as for the determinant of $|A|$. The same formula we have for definition of the determinant $|A_5|$ and $|A_6|$, where determinant $|A_5| = -512$:

$$|A_5| = a_{12}a_{21}a_{33}[a_{44}(a_{56}a_{65}-a_{55}a_{66})+a_{45}(a_{54}a_{66}-a_{56}a_{64})+a_{46}(a_{55}a_{64}-a_{54}a_{65})]+$$
$$+a_{12}a_{23}a_{31}[a_{44}(a_{55}a_{66}-a_{56}a_{65})+a_{45}(a_{56}a_{64}-a_{54}a_{66})+a_{46}(a_{54}a_{65}-a_{55}a_{64})]$$

and determinant $|A_6|$ is equal $|A_6| = -720$:

$$|A_6| = a_{12}a_{21}a_{33}[a_{44}(a_{56}a_{65}-a_{55}a_{66})+a_{45}(a_{54}a_{66}-a_{56}a_{64})+a_{46}(a_{55}a_{64}-a_{54}a_{65})]+$$
$$+a_{12}a_{23}a_{31}[a_{44}(a_{55}a_{66}-a_{56}a_{65})+a_{45}(a_{56}a_{64}-a_{54}a_{66})+a_{46}(a_{54}a_{65}-a_{55}a_{64})]$$

The functional graphs between elements of the above-considered formulas for $|A_5|$ and $|A_6|$ are shown in Figure 5 of Appendix 23 in general view ,as for the determinant of $|A|$.

5) So, at given numerical data for the 6x6 matrices ,we have the value for the determinants which are equal:

$$|A| = -16; \quad |A_1| = -624; \quad |A_2| = 196; \quad |A_3| = 1224; \quad |A_4| = 1344; \quad |A_5| = -512;$$
$$|A_6| = -720.$$

And then the coefficients of system for six linear equations are equal:

$$X_1 = 39 \; ; X_2 = -12.25; \; X_3 = -76.5 \; ; X_4 = -84; \; X_5 = 32; \; X_6 = 45$$

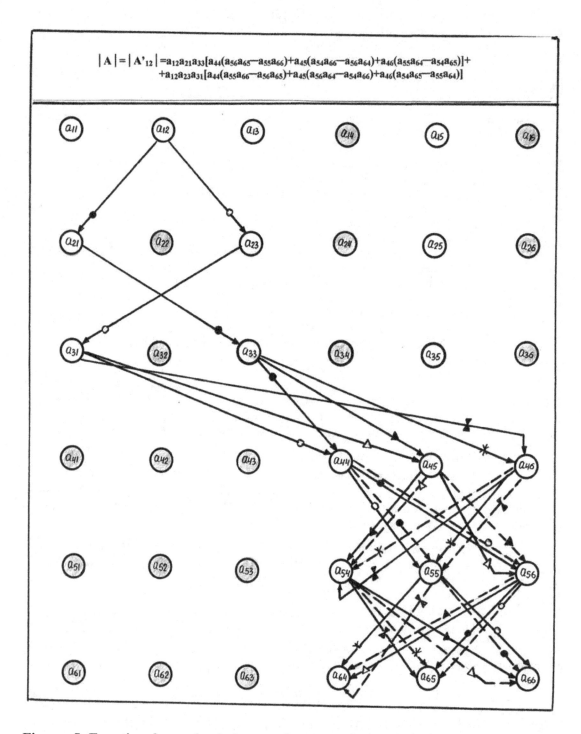

$$|A| = |A'_{12}| = a_{12}a_{21}a_{33}[a_{44}(a_{56}a_{65} - a_{55}a_{66}) + a_{45}(a_{54}a_{66} - a_{56}a_{64}) + a_{46}(a_{55}a_{64} - a_{54}a_{65})] +$$
$$+ a_{12}a_{23}a_{31}[a_{44}(a_{55}a_{66} - a_{56}a_{65}) + a_{45}(a_{56}a_{64} - a_{54}a_{66}) + a_{46}(a_{54}a_{65} - a_{55}a_{64})]$$

Figure 5 Functional graphs between elements of the 6x6 matrix applicably to numerical data of Appendix 23 in general view for determinant $|A|$
(— — —negative ties; ● elements of matrix equal zero)

376

$$|A_1| = |A^*_{12}| = a_{12}a_{21}a_{33}[a_{44}(a_{56}a_{65}-a_{55}a_{66})+a_{45}(a_{54}a_{66}-a_{56}a_{64})+a_{46}(a_{55}a_{64}-a_{54}a_{65})]+$$
$$+a_{12}a_{23}a_{35}[a_{41}(a_{54}a_{66}-a_{56}a_{64})+a_{44}(a_{56}a_{61}-a_{51}a_{66})+a_{46}(a_{51}a_{64}-a_{54}a_{61})]+$$
$$+a_{12}a_{25}a_{33}[a_{41}(a_{56}a_{64}-a_{54}a_{66})+a_{44}(a_{51}a_{66}-a_{56}a_{61})+a_{46}(a_{54}a_{61}-a_{51}a_{64})]$$

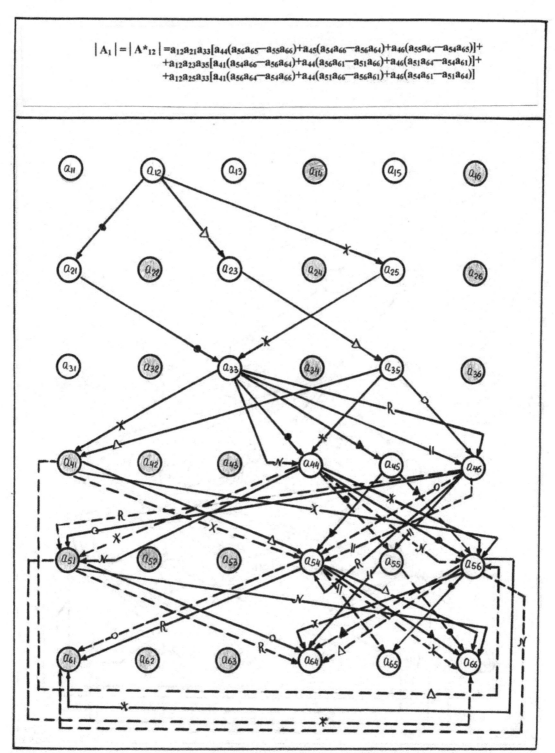

Figure 6 Functional graphs between elements of the determinant $|A_1|$ of the 6x6 matrix applicably to numerical data in general view of Appendix 24 (— — — negative ties; • elements of matrix equal zero)

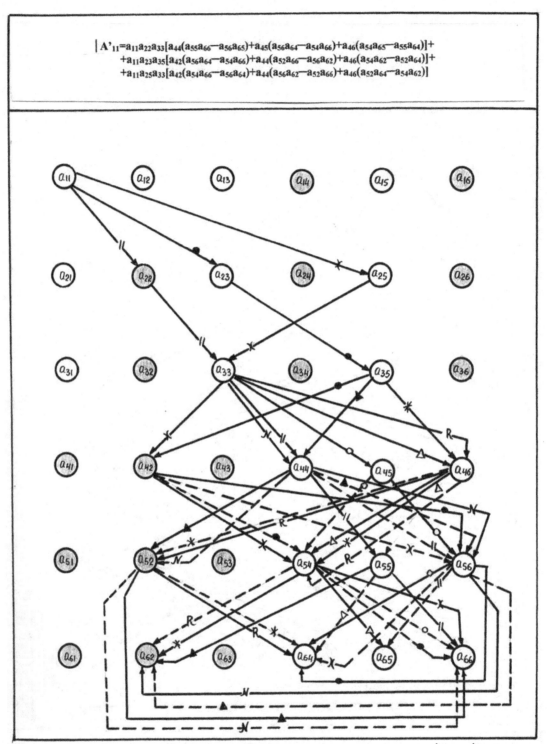

$$|A'_{11}=a_{11}a_{22}a_{33}[a_{44}(a_{55}a_{66}-a_{56}a_{65})+a_{45}(a_{56}a_{64}-a_{54}a_{66})+a_{46}(a_{54}a_{65}-a_{55}a_{64})]+$$
$$+a_{11}a_{23}a_{35}[a_{42}(a_{56}a_{64}-a_{54}a_{66})+a_{44}(a_{52}a_{66}-a_{56}a_{62})+a_{46}(a_{54}a_{62}-a_{52}a_{64})]+$$
$$+a_{11}a_{25}a_{33}[a_{42}(a_{54}a_{66}-a_{56}a_{64})+a_{44}(a_{56}a_{62}-a_{52}a_{66})+a_{46}(a_{52}a_{64}-a_{54}a_{62})]$$

Figure 7 Functional graphs between elements of the determinant $|A'_{11}|$ of the 6x6 matrix applicably to numerical data in general view of Appendix 24 (— — — negative ties; ● elements of matrix equal zero)

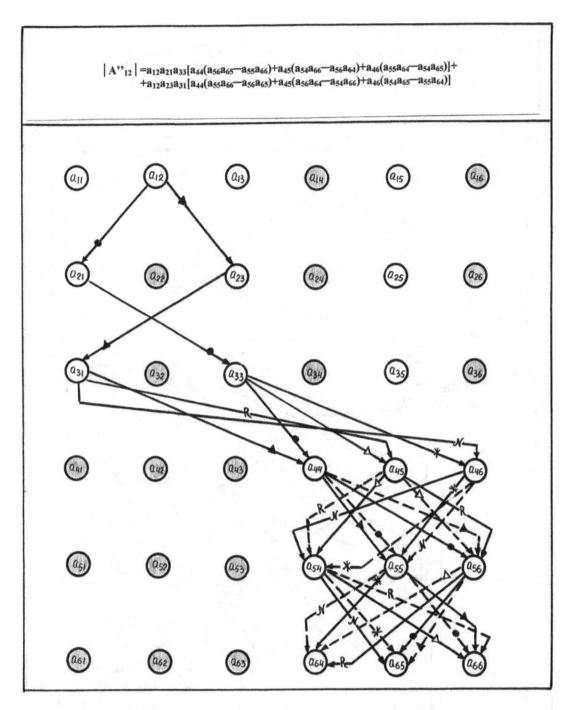

$$\left| A''_{12} \right| = a_{12}a_{21}a_{33}[a_{44}(a_{56}a_{65}-a_{55}a_{66})+a_{45}(a_{54}a_{66}-a_{56}a_{64})+a_{46}(a_{55}a_{64}-a_{54}a_{65})]+$$
$$+a_{12}a_{23}a_{31}[a_{44}(a_{55}a_{66}-a_{56}a_{65})+a_{45}(a_{56}a_{64}-a_{54}a_{66})+a_{46}(a_{54}a_{65}-a_{55}a_{64})]$$

Figure 8 Functional graphs between elements of the determinant $\left| A''_{12} \right|$ of the 6x6 matrix applicably to numerical data in general view of Appendix 24 ($-----$ negative ties; • elements of matrix equal zero)

$$|A'_{13}| = a_{13}a_{21}a_{35}[a_{42}(a_{54}a_{66}-a_{56}a_{64})+a_{44}(a_{56}a_{62}-a_{52}a_{66})+a_{46}(a_{52}a_{64}-a_{54}a_{62})]+$$
$$+a_{13}a_{22}a_{31}[a_{44}(a_{56}a_{65}-a_{55}a_{66})+a_{45}(a_{54}a_{66}-a_{56}a_{64})+a_{46}(a_{55}a_{64}-a_{54}a_{65})]+$$
$$+a_{13}a_{25}a_{31}[a_{42}(a_{56}a_{64}-a_{54}a_{66})+a_{44}(a_{52}a_{66}-a_{56}a_{62})+a_{46}(a_{54}a_{62}-a_{52}a_{64})]$$

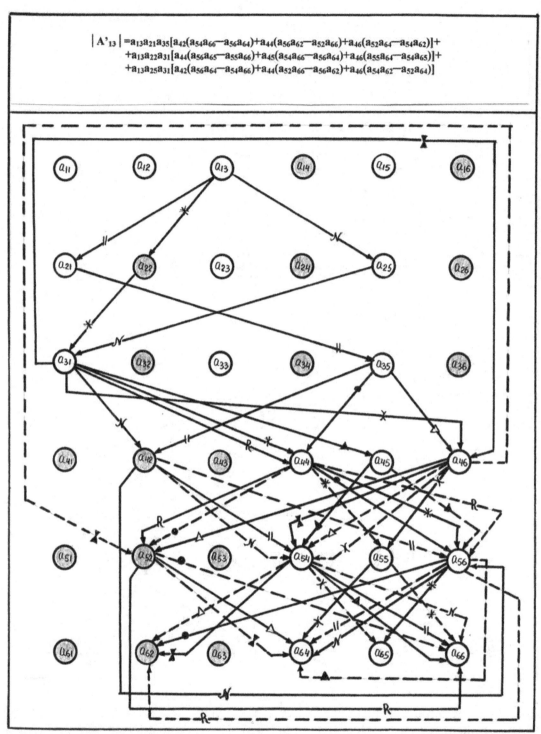

Figure 9 Functional graphs between elements of the determinant $|A'_{13}|$ of the 6x6 matrix applicably to numerical data in general view of Appendix 24 (— — — negative ties; • elements of matrix equal zero)

$$|A'_{15}| = a_{15}a_{21}a_{33}[a_{42}(a_{56}a_{64}-a_{54}a_{66})+a_{44}(a_{52}a_{66}-a_{56}a_{62})+a_{46}(a_{54}a_{62}-a_{52}a_{64})]+$$
$$+a_{15}a_{23}a_{31}[a_{42}(a_{54}a_{66}-a_{56}a_{64})+a_{44}(a_{56}a_{62}-a_{52}a_{66})+a_{46}(a_{52}a_{64}-a_{54}a_{62})]$$

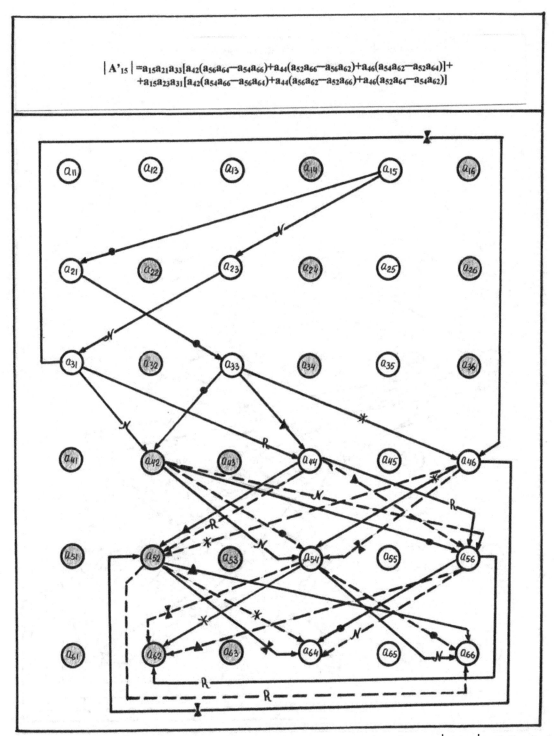

Figure 10 Functional graphs between elements of the determinant $|A'_{15}|$ of the 6x6 matrix applicably to numerical data in general view of Appendix 24
(— — — negative ties; ●elements of matrix equal zero)

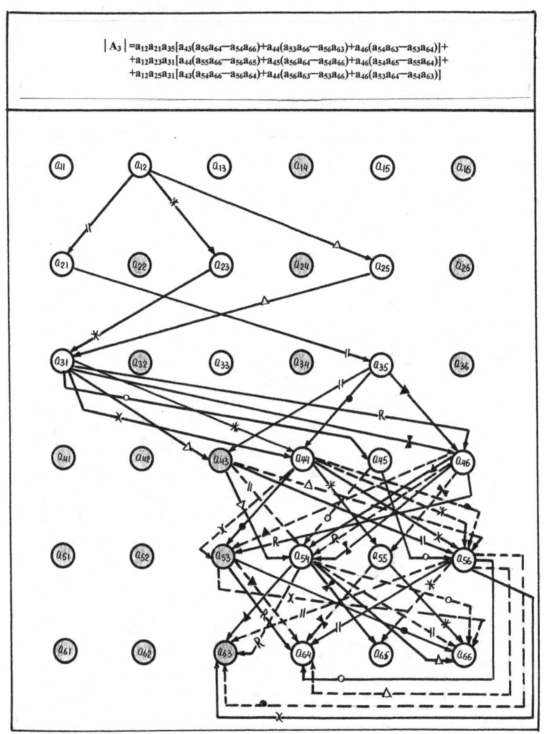

$$|A_3| = a_{12}a_{21}a_{35}[a_{43}(a_{56}a_{64}-a_{54}a_{66})+a_{44}(a_{53}a_{66}-a_{56}a_{63})+a_{46}(a_{54}a_{63}-a_{53}a_{64})]+$$
$$+a_{12}a_{23}a_{31}[a_{44}(a_{55}a_{66}-a_{56}a_{65})+a_{45}(a_{56}a_{64}-a_{54}a_{66})+a_{46}(a_{54}a_{65}-a_{55}a_{64})]+$$
$$+a_{12}a_{25}a_{31}[a_{43}(a_{54}a_{66}-a_{56}a_{64})+a_{44}(a_{56}a_{63}-a_{53}a_{66})+a_{46}(a_{53}a_{64}-a_{54}a_{63})]$$

Figure 11 Functional graphs between elements of the determinant $|A_3|$ of the 6x6 matrix applicably to numerical data in general view of Appendix 24 (— — —negative ties; • elements of matrix equal zero)

Appendix 25

Problem:

Solve the system of normal linear six equations with using of the Expansion (cofactor) and Lapalace's methods for evaluation of the determinants of the 6x6 matrices applicably *for marked out jointly the first and second rows* and also the Cramer's rule for definition of the unknown coefficients:

$$2X_2-X_4=4$$
$$5X_1-4X_2-3X_3+6X_5+4X_6 \leq -6$$
$$2X_2-2X_3-3X_6=0$$
$$6X_1+5X_2-5X_3+2X_4-2X_5+4X_6 \geq 3$$
$$4X_1-X_2+X_3+2X_4-2X_5+3X_6=1$$
$$6X_1-6X_2+7X_3-7X_4-X_5+2X_6 \leq 5$$

Solution:

1) The above-indicated system, having six equations, can be written in view of the 6x6 matrix in numerical data:

$$A=\begin{vmatrix} 0 & 2 & 0 & -1 & 0 & 0 \\ 5 & -4 & -3 & 0 & 6 & 4 \\ 0 & 2 & -2 & 0 & 0 & -3 \\ 6 & 5 & -5 & 2 & -2 & 4 \\ 4 & -1 & 1 & 2 & -2 & 3 \\ 6 & -6 & 7 & -7 & -1 & 2 \end{vmatrix} \quad B=\begin{vmatrix} 4 \\ -6 \\ 0 \\ 3 \\ 1 \\ 5 \end{vmatrix}$$

2) Applicably to Cramer's rule, we have the following formula for evaluation of unknown coefficients:

$$X_1=|A_1|/|A| \; ; X_2=|A_2|/|A| \; ; X_3=|A_3|/|A| \; ; X_4=|A_4|/|A| \; ;$$
$$X_5=|A_5|/|A| \; ; X_6=|A_6|/|A|$$

where,

$|A_1| \ldots \ldots |A_6|$ -the determinants of the 6x6 matrices which are equal:

$$A_1=\begin{vmatrix} 4 & 2 & 0 & -1 & 0 & 0 \\ -6 & -4 & -3 & 0 & 6 & 4 \\ 0 & 2 & -2 & 0 & 0 & -3 \\ 3 & 5 & -5 & 2 & -2 & 4 \\ 1 & -1 & 1 & 2 & -2 & 3 \\ 5 & -6 & 7 & -7 & -1 & 2 \end{vmatrix} \quad A_2=\begin{vmatrix} 0 & 4 & 0 & -1 & 0 & 0 \\ 5 & -6 & -3 & 0 & 6 & 4 \\ 0 & 0 & -2 & 0 & 0 & -3 \\ 6 & 3 & -5 & 2 & -2 & 4 \\ 4 & 1 & 1 & 2 & -2 & 3 \\ 6 & 5 & 7 & -7 & -1 & 2 \end{vmatrix}$$

$$A_3=\begin{vmatrix} 0 & 2 & 4 & 1 & 0 & 0 \\ 5 & -4 & -6 & 0 & 6 & 4 \\ 0 & 2 & 0 & 0 & 0 & -3 \\ 6 & 5 & 3 & 2 & -2 & 4 \\ 4 & -1 & 1 & 2 & -2 & 3 \\ 6 & -6 & 5 & -7 & -1 & 2 \end{vmatrix} \quad A_4=\begin{vmatrix} 0 & 2 & 0 & 4 & 0 & 0 \\ 5 & -4 & -3 & -6 & 6 & 4 \\ 0 & 2 & -2 & 0 & 0 & -3 \\ 6 & 5 & -5 & 3 & -2 & 4 \\ 4 & -1 & 1 & 1 & -2 & 3 \\ 6 & -6 & 7 & 5 & -1 & 2 \end{vmatrix}$$

$$A_5=\begin{vmatrix} 0 & 2 & 0 & -1 & 4 & 0 \\ 5 & -4 & -3 & 0 & -6 & 4 \\ 0 & 2 & -2 & 0 & 0 & -3 \\ 6 & 5 & -5 & 2 & 3 & 4 \\ 4 & -1 & 1 & 2 & 1 & 3 \\ 6 & -6 & 7 & -7 & 5 & 2 \end{vmatrix} \quad A_6=\begin{vmatrix} 0 & 2 & 0 & -1 & 0 & 4 \\ 5 & -4 & -3 & 0 & 6 & -6 \\ 0 & 2 & -2 & 0 & 0 & 0 \\ 6 & 5 & -5 & 2 & -2 & 3 \\ 4 & -1 & 1 & 2 & -2 & 1 \\ 6 & -6 & 7 & -7 & -1 & 5 \end{vmatrix}$$

3) The determinants $|A|, |A_1|, |A_2|, |A_3|, |A_4|, |A_5|$ and $|A_6|$ of the 6x6 matrices can be defined by the Expansion(cofactor) and also by the Laplace's methods with using of the above-indicated formula(108).

So , in numerical data, we have the value for the determinants: $|A| = -9810$; $|A_1| = -6180$; $|A_2| = -19080$; $|A_3| = -17756$; $|A_4| = 2200$; $|A_5| = -6420$; $|A_6| = -780$ and then the coefficients are equal:

$$X_1 = 0.63; \ X_2 = 1.945; \ X_3 = 1.81; \ X_4 = -0.224; \ X_5 = 0.654; \ X_6 = 0.08$$

Appendix 26

Problem:

Evaluate of the determinant of the 6x6 matrix by Expansion and Laplace's methods. And express the functional ties between elements of the 6x6 matrix in graphs for both methods:

$$A = \begin{vmatrix} 3 & 2 & 0 & 0 & 4 & 0 \\ -2 & 4 & 0 & 0 & 0 & 5 \\ 0 & 0 & 3 & 2 & -1 & 0 \\ 0 & 0 & -2 & 4 & 0 & -7 \\ 6 & -6 & 0 & 0 & 0 & 0 \\ 0 & 0 & 6 & -6 & 0 & 0 \end{vmatrix}$$

Solution:

1) Evaluation of the determinant of the 6x6 matrix can be defined in comparative form by two suggested methods ,such as:
    a) With using of Laplace's method, for a instance, when in 6x6 matrix *marked out jointly three columns, such as the third, fourth and fifth columns;*
    b) With using of Expansion method, for a instance , when elements of the 6x6 matrix have expansion to the first row.

A. Definition of the determinant $|A|$ by Laplace's method

Combinations of used rows and minors of the third order (M) and algebraic(A) supplemental(cofactors) for the above-considered conditions have the following view in numerical data and shown in Table 2:

Table 2 Definition of the determinant $|A|$ by Laplace's method

n/n	Combinations of rows	Minors of the third order(M)	Algebraic(A) supplemental
1	$i_1+i_2+i_3$	$M_1=0$	$A_1=0$
2	$i_1+i_2+i_4$	$M_2=0$	$A_2=0$
3	$i_1+i_2+i_5$	$M_3=0$	$A_3=0$
4	$i_1+i_2+i_6$	$M_4=0$	$A_4=0$
5	$i_2+i_3+i_4$	$M_5=0$	$A_5=0$
6	$i_2+i_3+i_5$	$M_6=0$	$A_6=0$
7	$i_2+i_3+i_6$	$M_7=0$	$A_7=210$
8	$i_3+i_4+i_5$	$M_8=12$	$A_8=0$
9	$i_3+i_4+i_6$	$M_9=12$	$A_9=-150$
10	$i_4+i_5+i_6$	$M_{10}=0$	$A_{10}=0$
11	$i_1+i_3+i_4$	$M_{11}=64$	$A_{11}=0$
12	$i_1+i_3+i_5$	$M_{12}=0$	$A_{12}=0$
13	$i_1+i_3+i_6$	$M_{13}=-120$	$A_{13}=-84$

384

So, analysis of numerical data shown in Table 8 indicates on fact that value of the determinant $|A|$ of the 6x6 matrix can be defined by formula in general view as:

$$|A| = -M_9A_9 + M_{13}A_{13} \quad \text{or in detail we have}$$

$$|A| = -(a_{35}a_{43}a_{64} - a_{35}a_{44}a_{63})(a_{12}a_{26}a_{51} - a_{11}a_{26}a_{52}) +$$
$$+ (a_{15}a_{33}a_{64} - a_{15}a_{34}a_{63})(a_{22}a_{46}a_{51} - a_{21}a_{46}a_{52}) =$$
$$= a_{35}a_{26}(a_{43}a_{64} - a_{44}a_{63})(a_{11}a_{52} - a_{12}a_{51}) +$$
$$+ a_{15}a_{46}(a_{33}a_{64} - a_{34}a_{63})(a_{22}a_{51} - a_{21}a_{52})$$

In numerical data the determinant of the above-indicated 6x6 matrix is equal $|A|$=8280. In Figure 12 is shown the functional ties between elements of the determinant $|A|$ of the 6x6 matrix applicably to Laplace's method.

B. Definition of the determinant $|A|$ by Expansion method

In accordance with above-considered conditions, we can define the value of determinant $|A|$ of the 6x6 matrix by formula (108) applicably to Expansion method :

$$|A| = |A_1| + |A_2| + |A_3|$$

where,

$$|A_1| = a_{11}a_{26}a_{35}[a_{42}(a_{54}a_{63} - a_{53}a_{64}) + a_{43}(a_{52}a_{64} - a_{54}a_{62}) + a_{44}(a_{53}a_{62} - a_{52}a_{63})] +$$
$$+ a_{12}a_{26}a_{35}[a_{41}(a_{53}a_{64} - a_{54}a_{63}) + a_{43}(a_{54}a_{61} - a_{51}a_{64}) + a_{44}(a_{51}a_{63} - a_{53}a_{61})]$$

and

$$|A_2| = a_{15}a_{21}a_{33}[a_{42}(a_{56}a_{64} - a_{54}a_{66}) + a_{44}(a_{52}a_{66} - a_{56}a_{62}) + a_{46}(a_{54}a_{62} - a_{52}a_{64})] +$$
$$+ a_{15}a_{21}a_{34}[a_{42}(a_{53}a_{66} - a_{56}a_{63}) + a_{43}(a_{56}a_{62} - a_{52}a_{66}) + a_{46}(a_{52}a_{63} - a_{53}a_{62})]$$

and

$$|A_3| = a_{15}a_{22}a_{33}[a_{41}(a_{54}a_{66} - a_{56}a_{64}) + a_{44}(a_{56}a_{61} - a_{51}a_{66}) + a_{46}(a_{51}a_{64} - a_{54}a_{61})] +$$
$$+ a_{15}a_{22}a_{34}[a_{41}(a_{56}a_{63} - a_{53}a_{66}) + a_{43}(a_{51}a_{66} - a_{56}a_{61}) + a_{46}(a_{53}a_{61} - a_{51}a_{63})]$$

So, in numerical data , the value of determinant of the 6x6 matrix evaluated by Expansion(cofactor) method has the same result and equal $|A|$=8280.
In Figure 13 is shown the functional ties between elements of the determinant $|A_1|$ of the 6x6 matrix applicably to Expansion method. In Figure 14 is shown the functional ties between elements of the determinant $|A_2|$ of the 6x6 matrix applicably to Expansion method. In Figure 15 is shown the functional ties between elements of the determinant $|A_3|$ of the 6x6 matrix applicably to Expansion method.

Conclusion:
1) Analysis of both methods, Expansion and Laplace's ,in question of evaluation of the determinant $|A|$ of the 6x6 matrix shows of their differences at the same value equal $|A|$=8280;
2) These differences at evaluation of the determinant $|A|$ of the 6x6 matrix are expressed in the following statements shown in Table 3:

Table 3 Comparative analysis of both methods (Expansion and Laplace's) in evaluation of the determinant $|A|$ of the 6x6 matrix

n/n	Criteria of analysis	Laplace's method	Expansion method
1	The value of determinant	8280	8280
2	Quantity of used elements of 6x6 matrix	48	216
3	Relative time for calculation of $\lvert A \rvert$	1	4.5

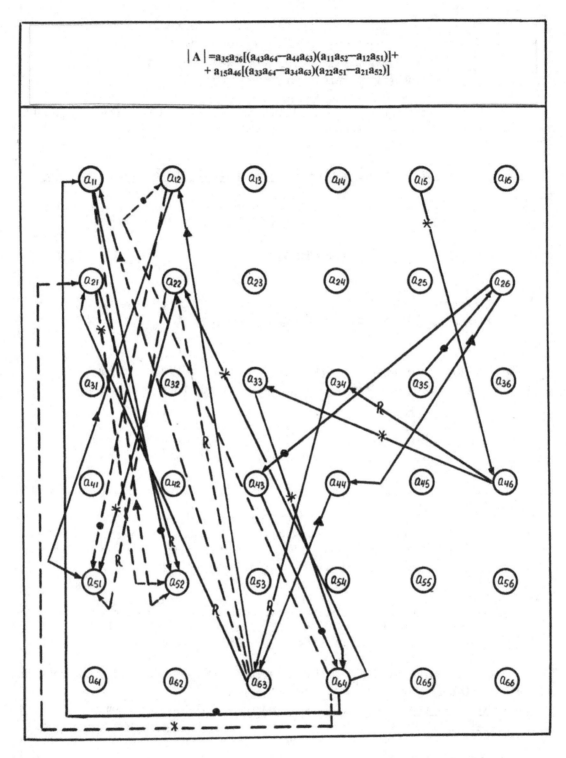

$$|A| = a_{35}a_{26}[(a_{43}a_{64}-a_{44}a_{63})(a_{11}a_{52}-a_{12}a_{51})]+$$
$$+ a_{15}a_{46}[(a_{33}a_{64}-a_{34}a_{63})(a_{22}a_{51}-a_{21}a_{52})]$$

Figure 12 Functional ties between elements of the determinant $|A|$ of the 6x6 matrix applicably to Laplace's method

386

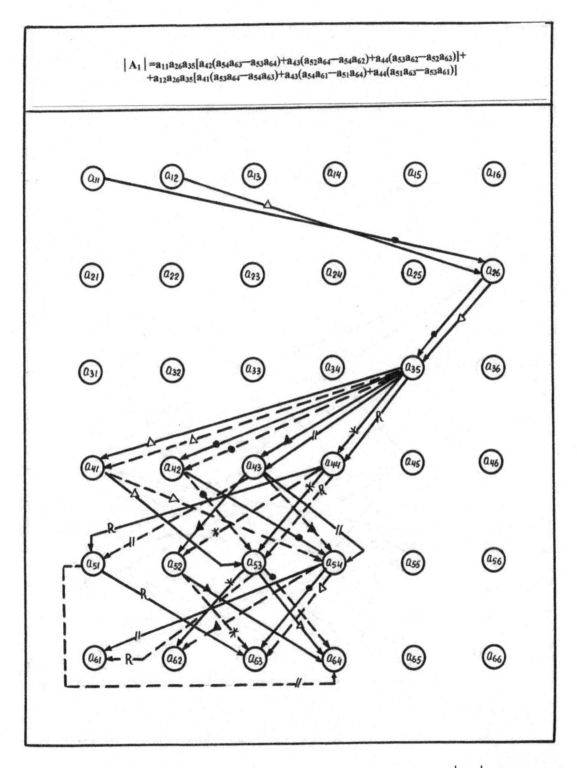

$$|A_1| = a_{11}a_{26}a_{35}[a_{42}(a_{54}a_{63} - a_{53}a_{64}) + a_{43}(a_{52}a_{64} - a_{54}a_{62}) + a_{44}(a_{53}a_{62} - a_{52}a_{63})] +$$
$$+ a_{12}a_{26}a_{35}[a_{41}(a_{53}a_{64} - a_{54}a_{63}) + a_{43}(a_{54}a_{61} - a_{51}a_{64}) + a_{44}(a_{51}a_{63} - a_{53}a_{61})]$$

Figure 13 Functional ties between elements of the determinant $|A_1|$ of the 6x6 matrix applicably to Expansion(cofactor) method

$$|\mathbf{A}_2| = a_{15}a_{21}a_{33}[a_{42}(a_{56}a_{64} - a_{54}a_{66}) + a_{44}(a_{52}a_{66} - a_{56}a_{62}) + a_{46}(a_{54}a_{62} - a_{52}a_{64})] +$$
$$+ a_{15}a_{21}a_{34}[a_{42}(a_{53}a_{66} - a_{56}a_{63}) + a_{43}(a_{56}a_{62} - a_{52}a_{66}) + a_{46}(a_{52}a_{63} - a_{53}a_{62})]$$

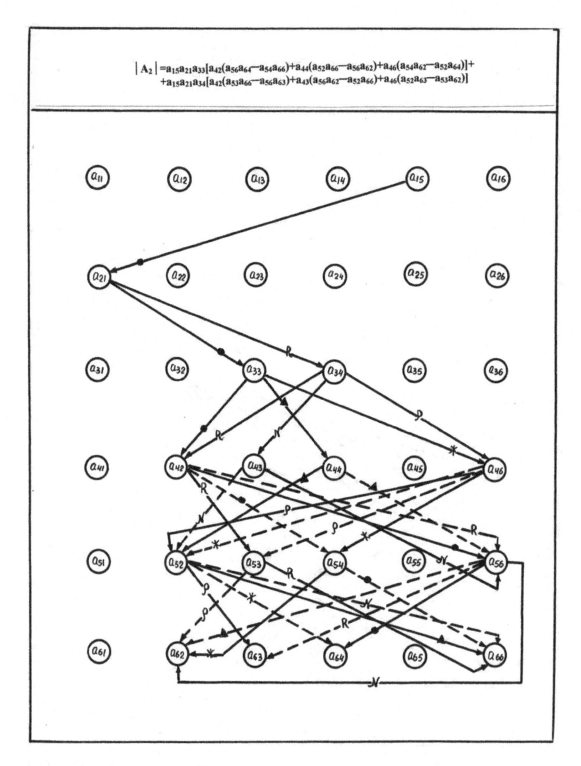

Figure 14 Functional ties between elements of the determinant $|\mathbf{A}_2|$ **of the 6x6 matrix applicably to Expansion(cofactor) method**

388

$$|A_3| = a_{15}a_{22}a_{33}[a_{41}(a_{54}a_{66}-a_{56}a_{64})+a_{44}(a_{56}a_{61}-a_{51}a_{66})+a_{46}(a_{51}a_{64}-a_{54}a_{61})]+$$
$$+a_{15}a_{22}a_{34}[a_{41}(a_{56}a_{63}-a_{53}a_{66})+a_{43}(a_{51}a_{66}-a_{56}a_{61})+a_{46}(a_{53}a_{61}-a_{51}a_{63})]$$

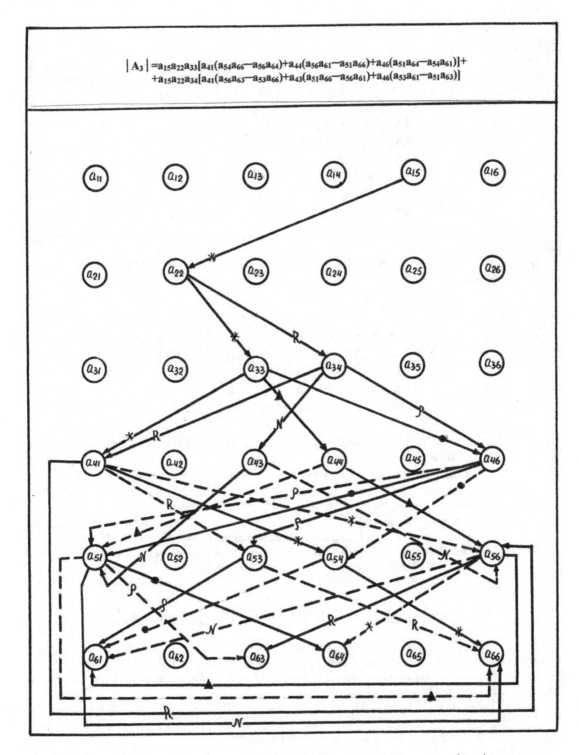

Figure 15 Functional ties between elements of the determinant $|A_3|$ of the 6x6 matrix applicably to Expansion (cofactor) method

389

Appendix 27

Problem:
Solve the system of normal linear six equations with using of Laplace's theorem for evaluation of the determinants of the 6x6 matrices and Cramer's rule for finding of the unknown coefficients. And express the functional ties between elements of the determinant of the 6x6 matrix in graphs.

$$3X_1+2X_2+4X_5 \geq -6$$
$$-2X_1+4X_2+5X_6 \leq 3$$
$$3X_3+2X_4-X_5=0$$
$$-2X_3+4X_4-7X_6=2$$
$$6X_1-6X_2 \leq -4$$
$$6X_3-6X_4 \geq 8$$

Solution:
1) The above-indicated system, having six equations, can be written in view of the 6x6 matrix in numerical data:

$$A = \begin{vmatrix} 3 & 2 & 0 & 0 & 4 & 0 \\ -2 & 4 & 0 & 0 & 0 & 5 \\ 0 & 0 & 3 & 2 & -1 & 0 \\ 0 & 0 & -2 & 4 & 0 & -7 \\ 6 & -6 & 0 & 0 & 0 & 0 \\ 0 & 0 & 6 & -6 & 0 & 0 \end{vmatrix} \qquad B = \begin{vmatrix} -6 \\ 3 \\ 0 \\ 2 \\ -4 \\ 8 \end{vmatrix} \qquad (*)$$

2) Evaluation of the determinants of the 6x6 matrices we can define for the comparative analysis at such conditions as:

a) In the different 6x6 matrices for evaluation of the determinants $|A_1| \ldots \ldots |A_6|$ in accordance with Cramer's rule uses the Laplace's theorem for *marked out jointly three columns ,such as the third, fourth and fifth columns (Version #1)*;

b) And in primary numerical 6x6 matrix (A) (*) uses the Expansion method *(Version #2)* applicably to formula (111), elements of which have expansion to the first row, **as this was shown in Appendix 26 ,where the value of primary determinant equal** : $|A| = 8280$. This Expansion method , as Version #2 , applicably to formula (111) can be used for another determinants $|A_1| \ldots |A_6|$ in accordance with Cramer's rule.

A. **Version #1: Definition of the determinant $|A_1|$:**

Combinations of used rows and minors of the third order (M_{ij}) and algebraic (A_{ij}) supplemental (cofactors) for above-considered conditions have the following view in numerical data and shown in Table 4.
So, analysis of numerical data shown in Table 2 indicates on fact that value of the determinant $|A_1|$ of the 6x6 matrix can be defined by formula in general view as:
$$|A_1| = -M_9A_9 + M_{11}A_{11} + M_{13}A_{13} \text{ or at numerical data we have the value } |A_1| = 26880.$$

In general view the value of combined determinant $|A_1|$ is equal:
$$|A_1| = -(a_{35}a_{43}a_{64} - a_{35}a_{44}a_{63})(a_{12}a_{26}a_{51} - a_{11}a_{26}a_{52}) +$$
$$+ (a_{15}a_{33}a_{44} - a_{15}a_{34}a_{43})(a_{26}a_{51}a_{62} - a_{26}a_{52}a_{61}) +$$
$$+ (a_{15}a_{33}a_{64} - a_{15}a_{34}a_{63})(a_{26}a_{41}a_{52} + a_{22}a_{46}a_{51} - a_{21}a_{46}a_{52});$$

390

After of some calculations and transformations we have finally formula for evaluation of combined determinant $|A_1|$:

$$|A_1| = |A'_1| + |A''_1| + |A'''_1|$$

where,

$|A'_1|$ -partial combined determinant of the 6x6 matrix in general view which is expressed by the following formula:

$$|A'_1| = a_{35}a_{26}a_{44}a_{63}a_{11}a_{52} - a_{35}a_{26}a_{43}a_{64}a_{11}a_{52} - a_{35}a_{26}a_{44}a_{63}a_{12}a_{51} + a_{35}a_{26}a_{43}a_{64}a_{12}a_{51}$$

$|A''_1|$ - partial combined determinant of the 6x6 matrix in general view which is expressed by the following formula:

$$|A''_1| = a_{15}a_{26}a_{33}a_{44}a_{51}a_{62} - a_{15}a_{26}a_{33}a_{44}a_{52}a_{61} - a_{15}a_{26}a_{34}a_{43}a_{51}a_{62} + a_{15}a_{26}a_{34}a_{43}a_{52}a_{61}$$

$|A'''_1|$ -partial combined determinant of the 6x6 matrix in general view which is expressed by the following formula:

$$|A'''_1| = a_{15}a_{33}a_{64}a_{26}a_{41}a_{52} - a_{15}a_{34}a_{63}a_{26}a_{41}a_{52} + a_{15}a_{33}a_{64}a_{22}a_{46}a_{51} - a_{15}a_{34}a_{63}a_{22}a_{46}a_{51} - $$
$$- a_{15}a_{33}a_{64}a_{21}a_{46}a_{52} + a_{15}a_{34}a_{63}a_{21}a_{46}a_{52}$$

In Figure 16 is shown the functional ties between elements of the combined partial determinant $|A'_1|$ of the 6x6 matrix in general view. In Figure 17 is shown the functional ties between elements of the combined partial determinant $|A''_1|$ of the 6x6 matrix in general view. In Figure 18 is shown the functional ties between elements of the combined partial determinant $|A'''_1|$ of the 6x6 matrix in general view.

Table 4 Combinations of used rows and other parameters of the determinant of the 6x6 matrix

n/n	Combination of used rows	Parameters of the determinant of the 6x6 matrix	
		(M_{ij})	(A_{ij})
1	$i_1+i_2+i_3$	0	-336
2	$i_1+i_2+i_4$	0	0
3	$i_1+i_2+i_5$	0	0
4	$i_1+i_2+i_6$	0	0
5	$i_2+i_3+i_4$	0	0
6	$i_2+i_3+i_5$	0	-112
7	$i_2+i_3+i_6$	0	308
8	$i_3+i_4+i_5$	0	80
9	$i_3+i_4+i_6$	12	-220
10	$i_4+i_5+i_6$	0	0
11	$i_1+i_3+i_4$	64	240
12	$i_1+i_3+i_5$	0	-224
13	$i_1+i_3+i_6$	-120	-74
14	$i_1+i_4+i_5$	0	0
15	$i_1+i_4+i_6$	-48	0
16	$i_3+i_5+i_6$	0	230

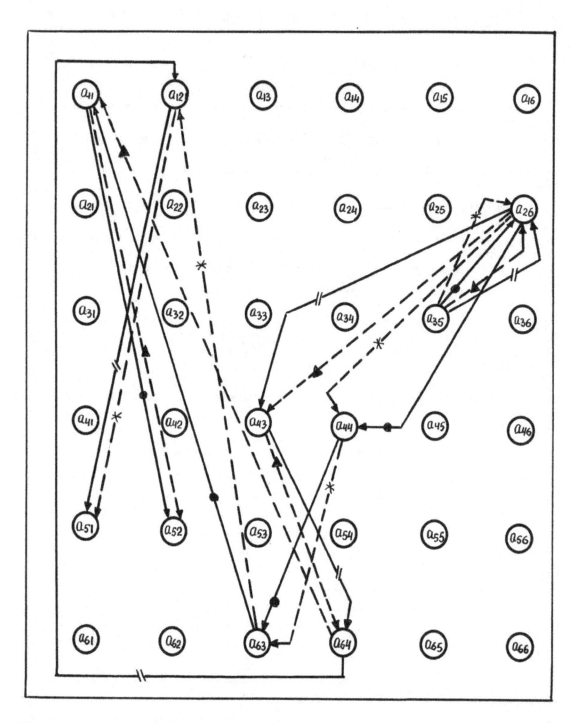

Figure 16 Functional ties between elements of the 6x6 matrix of the combined partial determinant $|A'_1|$ for positive(————) and negative (-------) areas applicably to Laplace's theorem at marked out jointly the third ,fourth and fifth columns

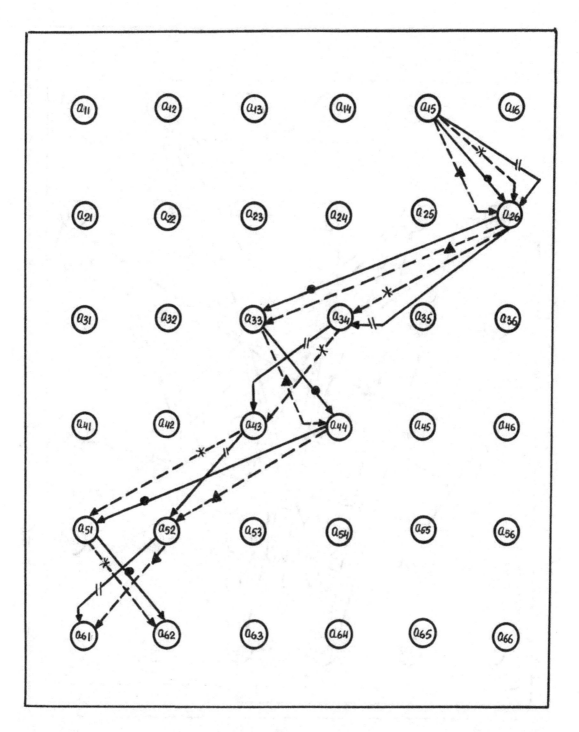

Figure 17 Functional ties between elements of the 6x6 matrix of the combined partial determinant $\left| A''_1 \right|$ for positive (——) and negative (------) areas applicably to Laplace's theorem at marked out jointly the third, fourth and fifth columns

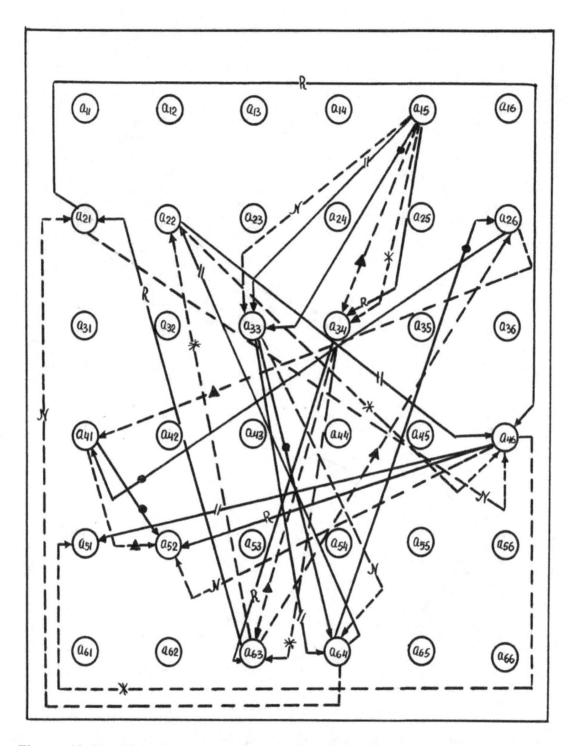

Figure 18 Functional ties between of the 6x6 matrix of the combined partial determinant $\left|A'''_1\right|$ for positive (———) and negative (------) areas applicably to Laplace's theorem at marked out jointly the third, fourth and fifth columns

B. Version #1: The Definition of combined determinant $|A_2|$:

Combinations of used rows and minors of the third order(M_{ij}) and algebraic A_{ij} supplemental(cofactors) fo the above-considered conditions have the following view in numerical data and shown in Table 5.

Table 5 Combination of used rows and other parameters of the determinant of the 6x6 matrix

n/n	Combination of used rows	Parameters of the determinant of the 6x6 matrix	
		(M_{ij})	(A_{ij})
1	$i_1+i_2+i_3$	0	-336
2	$i_1+i_2+i_4$	0	0
3	$i_1+i_2+i_5$	0	0
4	$i_1+i_2+i_6$	0	0
5	$i_1+i_3+i_4$	64	240
6	$i_1+i_3+i_5$	0	-112
7	$i_1+i_3+i_6$	-120	-130
8	$i_2+i_3+i_4$	0	0
9	$i_2+i_3+i_5$	0	168
10	$i_2+i_3+i_6$	0	252
11	$i_3+i_4+i_5$	0	-120
12	$i_3+i_4+i_6$	12	-120
13	$i_4+i_5+i_6$	0	0

So, analysis of numerical data shown in Table 5 indicates on the essential fact that value of the determinant $|A_2|$ of the 6x6 matrix can be defined by the following formula in general view as:

$$|A_2| = M_5A_5 + M_7A_7 - M_{12}A_{12}$$ or in numerical data the value of determinant $|A_2| = 32400$

In general form the value of the combined determinant $|A_2|$ of the 6x6 matrix for the above-considered conditions can be defined by the following formula:

$$|A_2| = (a_{15}a_{33}a_{44} - a_{15}a_{34}a_{43})(a_{26}a_{51}a_{62}) +$$
$$+ (a_{15}a_{33}a_{64} - a_{15}a_{34}a_{63})(a_{22}a_{46}a_{51} - a_{26}a_{42}a_{51} - a_{21}a_{46}a_{52}) -$$
$$- (a_{35}a_{43}a_{64} - a_{35}a_{44}a_{63})(a_{12}a_{26}a_{51} - a_{11}a_{26}a_{52})$$

After of some calculations and transformations the value of the combined determinant $|A_2|$ of the 6x6 matrix is equal $|A_2| = |A'_2| + |A''_2| + |A'''_2|$:

where,

$|A'_2|$ -combined partial determinant of the 6x6 matrix equal:

$$|A'_2| = a_{26}a_{35}(a_{11}a_{52} - a_{12}a_{51})(a_{43}a_{64} - a_{44}a_{63})$$

$|A''_2|$ -combined partial determinant of the 6x6 matrix equal:

$$|A''_2| = a_{15}a_{46}(a_{22}a_{51} - a_{21}a_{52})(a_{33}a_{64} - a_{34}a_{63})$$

$|A'''_2|$ -combined partial determinant of the 6x6 matrix equal:

$$|A'''_2| = a_{15}a_{26}a_{51}[a_{33}(a_{44}a_{62} - a_{42}a_{64}) + a_{34}(a_{42}a_{63} - a_{43}a_{62})]$$

In Figure 19 is shown the functional ties between elements of the 6x6 matrix of the combined partial determinant $|A'_2|$ for positive and negative areas applicably to Laplace's theorem. In Figure 20 is shown the functional ties between elements of the 6x6 matrix of the combined partial determinant $|A''_2|$ for positive and negative areas applicably to Laplace's theorem. And in Figure 21 is shown the functional ties between elements of the 6x 6 matrix of the combined partial determinant $|A'''_2|$ for positive and negative areas applicably to Laplace's theorem.

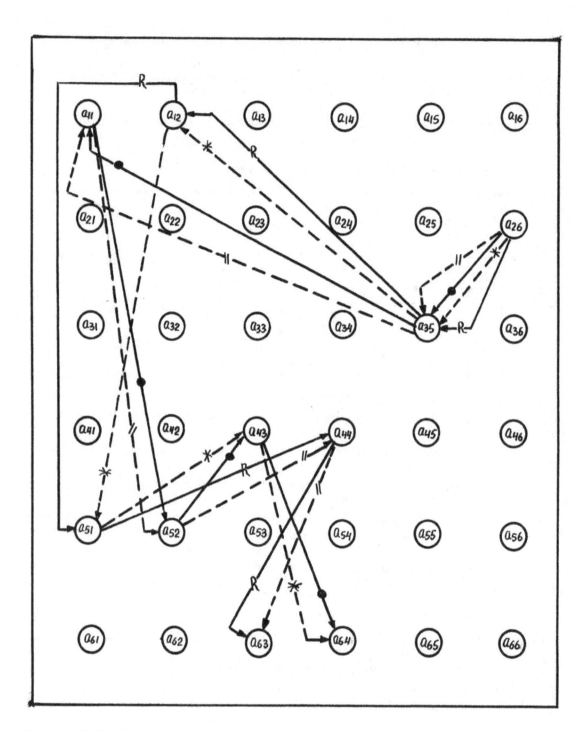

Figure 19 Functional ties between elements of the 6x6 matrix of the combined partial determinant $|A'_2|$ for positive (———) and negative (------) areas applicably to Laplace's theorem at marked out jointly the third, fourth and fifth columns

396

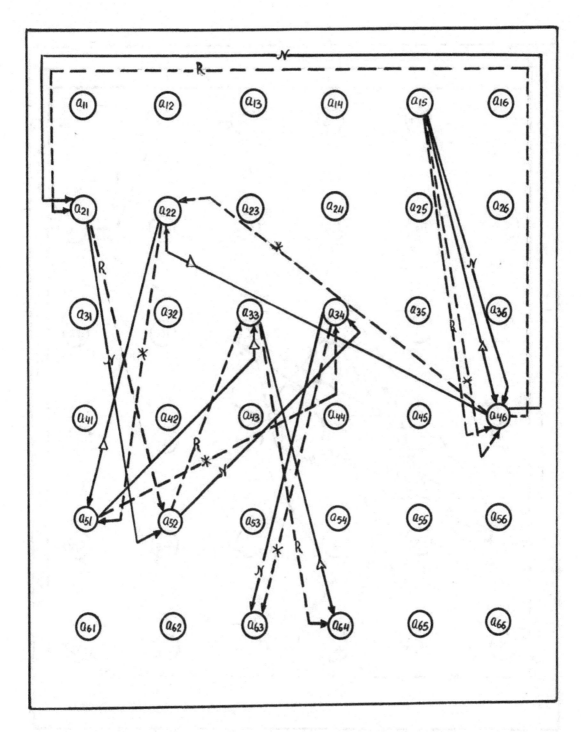

Figure 20 Functional ties between elements of the 6x6 matrix of the combined partial determinant $|A''_2|$ for positive(———) and negative(------) areas applicably to Laplace's theorem at marked out jointly the third, fourth and fifth columns

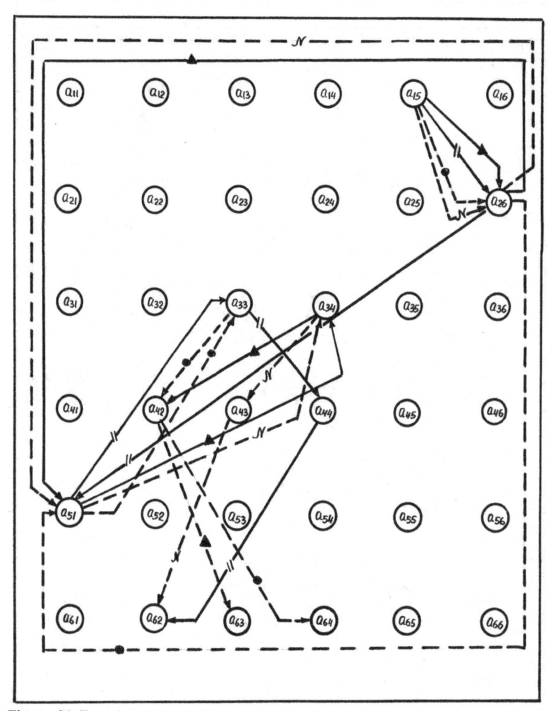

Figure 21 Functional ties between elements of the 6x6 matrix of the combined partial determinant $\left|A'''_2\right|$ for positive (———) and negative (------) areas applicably to Laplace's theorem at marked out jointly the third, fourth and fifth columns

C. Version#1: Definition of the determinant $|A_3|$:

Combination of used rows and minors of the third order (M_{ij}) and algebraic A_{ij} supplemental (cofactors) for above-considered conditions have the following view in numerical data and shown in Table 6.

Table 6 Combination of used rows and other parameters of the determinant $|A_3|$ of the 6x 6 matrix

n/n	Combination of used rows	Parameters of the determinant $|A_3|$ of the 6x6 matrix	
		(M_{ij})	(A_{ij})
1	$i_1+i_2+i_3$	24	0
2	$i_1+i_2+i_4$	48	0
3	$i_1+i_2+i_5$	0	0
4	$i_1+i_2+i_6$	-72	0
5	$i_1+i_3+i_4$	-40	0
6	$i_1+i_3+i_5$	32	0
7	$i_1+i_3+i_6$	−28	−84
8	$i_1+i_4+i_5$	64	0
9	$i_1+i_4+i_6$	−80	0
10	$i_2+i_3+i_4$	12	0
11	$i_2+i_3+i_5$	0	0
12	$i_2+i_3+i_6$	−18	−210
13	$i_3+i_4+i_5$	−16	0
14	$i_3+i_4+i_6$	44	150
15	$i_3+i_5+i_6$	−24	−112
16	$i_4+i_5+i_6$	0	0

So, analysis of numerical data shown in Table 6 indicates on the essential fact that value of the determinant $|A_3|$ of the 6x6 matrix can be defined by formula in general view as:

$$|A_3| = M_7A_7 - M_{12}A_{12} - M_{14}A_{14} + M_{15}A_{15}$$

In numerical data the determinant $|A_3|$ of the 6x6 matrix is equal: $|A_3| = -5340$. In general view the above-indicated formula has the following view:

$$|A_3| = (-a_{15}a_{34}a_{63} - a_{13}a_{35}a_{64})(a_{22}a_{46}a_{51} - a_{21}a_{46}a_{52}) - (-a_{23}a_{35}a_{64})(a_{12}a_{46}a_{51} - a_{11}a_{46}a_{52}) - (a_{35}a_{43}a_{64} - a_{35}a_{44}a_{63})(a_{12}a_{26}a_{51} - a_{11}a_{26}a_{52}) + (a_{35}a_{53}a_{64})(a_{11}a_{22}a_{46} - a_{12}a_{21}a_{46})$$

After of some calculations and transformations, the above-indicated formula for determinant $|A_3|$ of the 6x6 matrix has the following view in general form:

$$|A_3| = a_{46}(a_{21}a_{52} - a_{22}a_{51})(a_{13}a_{35}a_{64} + a_{15}a_{34}a_{63}) + a_{35}(a_{11}a_{52} - a_{12}a_{51})(a_{43}a_{64}a_{26} - a_{23}a_{64}a_{46}) + a_{35}a_{44}a_{63}[a_{26}(a_{12}a_{51} - a_{11}a_{52})] + a_{35}a_{53}a_{64}[a_{46}(a_{11}a_{22} - a_{12}a_{21})]$$

where,

combined determinant $|A_3|$ includes three partial determinants $|A_3| = |A'_3| + |A''_3| + |A'''_3|$:

- $|A'_3| = a_{46}a_{21}a_{52}a_{13}a_{35}a_{64} - a_{46}a_{22}a_{51}a_{13}a_{35}a_{64} + a_{46}a_{21}a_{52}a_{15}a_{34}a_{63} - a_{46}a_{22}a_{51}a_{15}a_{34}a_{63}$

- $|A''_3| = a_{35}a_{11}a_{52}a_{43}a_{64}a_{26} - a_{35}a_{12}a_{51}a_{43}a_{64}a_{26} - a_{35}a_{11}a_{52}a_{23}a_{64}a_{46} + a_{35}a_{12}a_{51}a_{23}a_{64}a_{46}$

- $|A'''_3| = a_{35}a_{44}a_{63}a_{26}a_{12}a_{51} - a_{35}a_{44}a_{63}a_{26}a_{11}a_{52} + a_{35}a_{53}a_{64}a_{46}a_{11}a_{22} - a_{35}a_{53}a_{64}a_{46}a_{12}a_{21}$

In Figure 22is shown the functional ties between elements of the combined partial determinant $|A'_3|$ of the 6x6 matrix . In Figure 23 is shown the functional ties between elements of the combined partial determinant $|A''_3|$ of the 6x6 matrix. In Figure 24 is shown the functional ties between elements of the combined partial determinant $|A'''_3|$ of the 6x6 matrix.

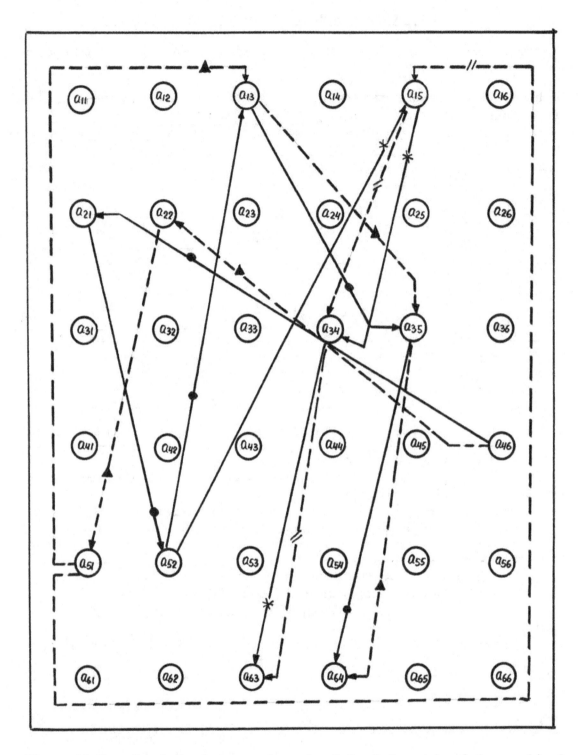

Figure 22 Functional ties between elements of the 6x6 matrix of the combined partial determinant $\left| A'_3 \right|$ for positive (——) and negative (------) areas applicably to Laplace's theorem at marked out jointly the third, fourth and fifth columns

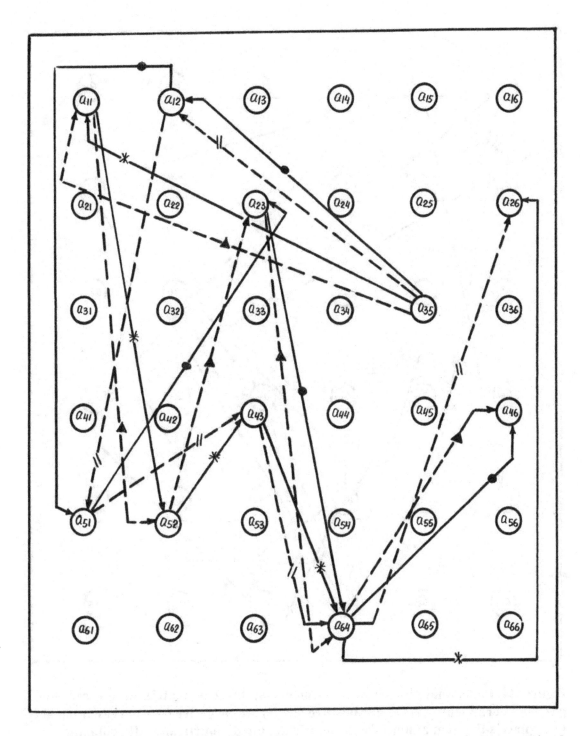

Figure 23 Functional ties between elements of the 6x6 matrix of the combined partial determinant $\left|A''_3\right|$ for positive (——) and negative(------) areas applicably to Laplace's theorem at marked out jointly the third, fourth and fifth columns

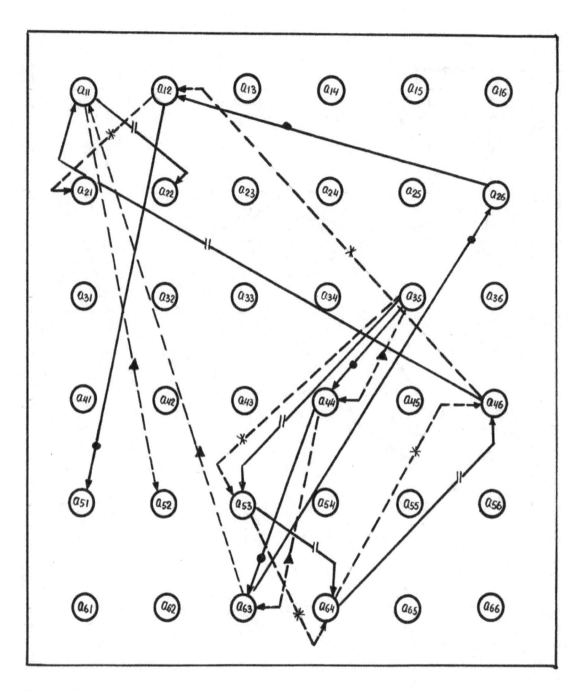

Figure 24 Functional ties between elements of the 6x6 matrix of the combined partial determinant $\left| A'''_3 \right|$ for positive (———) and negative (-----) areas applicably to Laplace's theorem at marked out jointly the third , fourth and fifth columns

D. Version #1:Definition of the determinant $|A_4|$:

Combination of used rows and minors of the third order (M_{ij}) and algebraic A_{ij} supplemental(cofactors) for above-considered conditions have the following view in numerical data and shown in Table 7.

Table 7 Combination of used rows and other parameters of the determinant $|A_4|$

n/n	Combination of used rows	Parameters of the determinant of the 6x6 matrix	
		(M_{ij})	(A_{ij})
1	$i_1+i_2+i_3$	-36	0
2	$i_1+i_2+i_4$	24	0
3	$i_1+i_2+i_5$	0	0
4	$i_1+i_2+i_6$	-72	0
5	$i_2+i_3+i_4$	6	0
6	$i_2+i_3+i_5$	0	0
7	$i_2+i_3+i_6$	−18	−210
8	$i_3+i_4+i_5$	-8	0
9	$i_3+i_4+i_6$	28	150
10	$i_4+i_5+i_6$	0	0
11	$i_1+i_3+i_4$	-12	0
12	$i_1+i_3+i_5$	-48	0
13	$i_1+i_3+i_6$	132	−84
14	$i_1+i_4+i_5$	32	0
15	$i_1+i_4+i_6$	-112	0
16	$i_3+i_5+i_6$	−24	−112

So, analysis of numerical data shown in Table 7 indicates on the essential fact that value of the determinant $|A_4|$ of the 6x6 matrix can be defined by formula in general view as:

$$|A_4|=-M_7A_7-M_9A_9+M_{13}A_{13}+M_{16}A_{16}$$

In numerical data the value of determinant $|A_4|$ is equal: $|A_4|=-16380$ and in general view we have the following formula for evaluation of the determinant $|A_4|$ for above –considered conditions applicably to Laplace's theorem:

$$|A_4| = -(a_{24}a_{35}a_{63})(a_{12}a_{46}a_{51}-a_{11}a_{46}a_{52})-$$
$$-(a_{35}a_{43}a_{64}-a_{35}a_{44}a_{63})(a_{12}a_{26}a_{51}-a_{11}a_{26}a_{52})+$$
$$+(a_{14}a_{35}a_{63}+a_{15}a_{33}a_{64})(a_{22}a_{46}a_{51}-a_{21}a_{46}a_{52})+$$
$$+ (-a_{35}a_{54}a_{63})(a_{11}a_{22}a_{46}-a_{12}a_{21}a_{46})$$

After of some calculations and transformations we have the value for combined determinant $|A_4|$ of the 6x6 matrix in the following view which consists from three combined partial determinants:

$$|A_4|=|A'_4|+|A''_4|+|A'''_4|$$

where,

$|A'_4|$ -combined partial determinant equal:

$$|A'_4|=(a_{11}a_{52}-a_{12}a_{51})(a_{24}a_{35}a_{63}a_{46}+a_{26}a_{35}a_{64}a_{43})$$

$|A''_4|$ -combined partial determinant equal:

$$|A''_4|=a_{26}a_{35}a_{63}[a_{44}(a_{12}a_{51}-a_{11}a_{52})]+a_{35}a_{54}a_{63}[a_{46}(a_{12}a_{21}-a_{11}a_{22})]$$

$|A'''_4|$ -combined partial determinant equal:

$$|A'''_4|=(a_{22}a_{51}-a_{21}a_{52})(a_{14}a_{35}a_{63}a_{46}+a_{15}a_{33}a_{64}a_{46})$$

In Figure 25 is shown the functional ties between elements of the combined partial determinant $|A'_4|$ of the 6x6 matrix. In Figure 26 is shown the functional ties between elements of the combined partial determinant $|A''_4|$ of the 6x6 matrix. In Figure 27 is shown the functional ties between elements of the combined partial determinant $|A'''_4|$ of the 6x6 matrix.

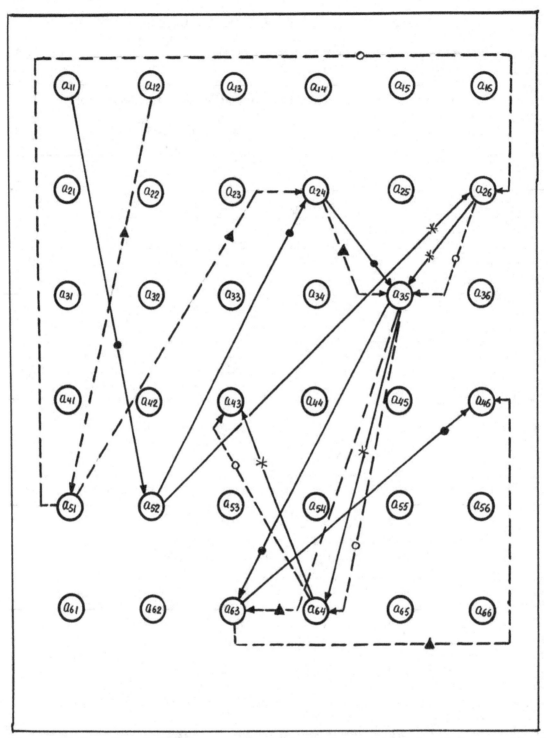

Figure 25 Functional ties between elements of the 6x6 matrix of the combined partial determinant $\left| A'_4 \right|$ for positive (———) and negative(------) areas applicably to Laplace's theorem at marked out jointly the third, fourth and fifth columns

404

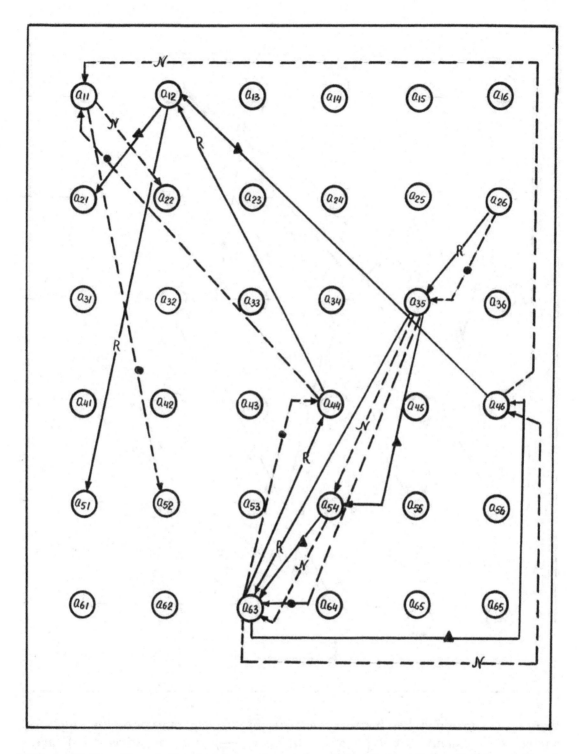

Figure 26 Functional ties between elements of the 6x6 matrix of the combined partial determinant $\left| A''_4 \right|$ for positive (——) and negative areas applicably to Laplace's theorem at marked out jointly the third, fourth and fifth columns

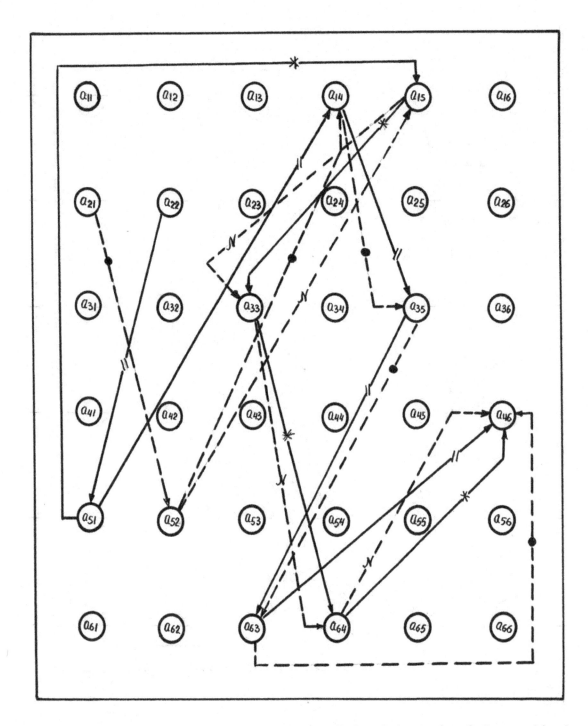

Figure 27 Functional ties between elements of the 6x6 matrix of the combined partial determinant $\left|A'''_4\right|$ for positive (——) and negative(-------) areas applicably to Laplace's theorem at marked out jointly the third, fourth and fifth columns

Combination of used rows and minors of the third order(M_{ij}) and algebraic A_{ij} supplemental(cofactors) for above-considered conditions have the following view in numerical data and shown in Table 8.

Table 8 Combination of used rows and other parameters of the determinant $|A_5|$

n/n	Combination of used rows	Parameters of the determinant of the 6x6 matrix	
1	$i_1+i_2+i_3$	0	0
2	$i_1+i_2+i_4$	0	0
3	$i_1+i_2+i_5$	0	0
4	$i_1+i_2+i_6$	0	0
5	$i_1+i_3+i_4$	-96	0
6	$i_1+i_3+i_5$	0	0
7	$i_1+i_3+i_6$	180	−84
8	$i_1+i_4+i_5$	0	0
9	$i_1+i_4+i_6$	72	0
10	$i_2+i_3+i_4$	48	0
11	$i_2+i_3+i_5$	0	0
12	$i_2+i_3+i_6$	−90	−210
13	$i_3+i_4+i_5$	−64	0
14	$i_3+i_4+i_6$	188	150
15	$i_3+i_5+i_6$	−120	−112
16	$i_4+i_5+i_6$	−48	0

So, analysis of numerical data shown in Table 8 indicates on the essential fact that value of the determinant $|A_5|$ of the 6x6 matrix can be defined by the following formula in general view as :

$$|A_5| = M_7A_7 - M_{12}A_{12} - M_{14}A_{14} + M_{15}A_{15} \text{ or}$$

$$|A_5| = (a_{15}a_{33}a_{64} - a_{15}a_{34}a_{63})(a_{22}a_{46}a_{51} - a_{21}a_{46}a_{52}) - $$
$$- (a_{25}a_{33}a_{64} - a_{25}a_{34}a_{63})(a_{12}a_{46}a_{51} - a_{11}a_{46}a_{52}) - $$
$$- (a_{33}a_{44}a_{65} + a_{34}a_{45}a_{63} - a_{33}a_{45}a_{64} - a_{34}a_{43}a_{65})(a_{12}a_{26}a_{51} - a_{11}a_{26}a_{52}) + $$
$$+ (a_{34}a_{55}a_{63} - a_{33}a_{55}a_{64})(a_{11}a_{22}a_{46} - a_{12}a_{21}a_{46})$$

In numerical data the value of determinant $|A_5|$ of the 6x 6 matrix is equal $|A_5| = -48780$.
After of some calculations and transformations, we can conclude that combined determinant $|A_5|$ consists from some combined partial determinants of the 6x6 matrices in general view:

$$|A_5| = |A'_5| + |A^{iv}_5| = (|A''_5| + |A'''_5|) + |A^{iv}_5|$$

where,
$|A'_5|$ -combined partial determinant, including two partial determinants $|A''_5|$ and $|A'''_5|$ which is equal as:

$$|A'_5| = a_{46}(a_{33}a_{64} - a_{34}a_{63})[a_{15}(a_{22}a_{51} - a_{21}a_{52}) + a_{25}(a_{11}a_{52} - a_{12}a_{51}) - a_{55}(a_{11}a_{22} - a_{12}a_{21})]$$

and $|A^{iv}_5| = (a_{26}a_{11}a_{52} - a_{26}a_{12}a_{51})[(a_{33}a_{44}a_{65} - a_{33}a_{45}a_{64}) + (a_{34}a_{45}a_{63} - a_{34}a_{43}a_{65})]$

In Figure 28 are shown the functional ties between elements of the partial determinant $|A''_5|$ of the combined determinant $|A'_5|$ and in Figure 29 are shown the functional ties between elements of the partial determinant $|A'''_5|$ of the combined determinant $|A'_5|$ of the 6x6 matrix. In Figure 30 are shown the functional ties between elements of the partial determinant $|A^{iv}_5|$ of the 6x6 matrix.

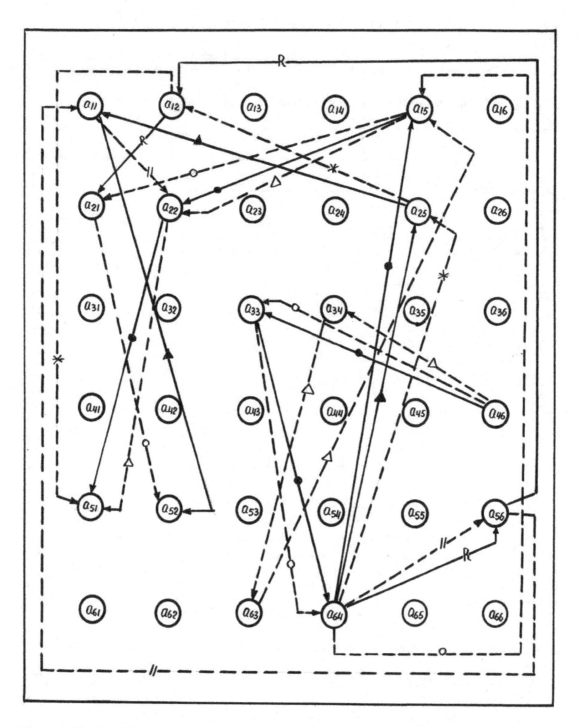

Figure 28 Functional ties between elements of the 6x6 matrix of the partial determinant $|A''_5|$ of the combined partial determinant $|A'_5|$ for positive (——) and negative(------) areas applicably to Laplace's theorem at marked out jointly the third ,fourth and fifth columns

408

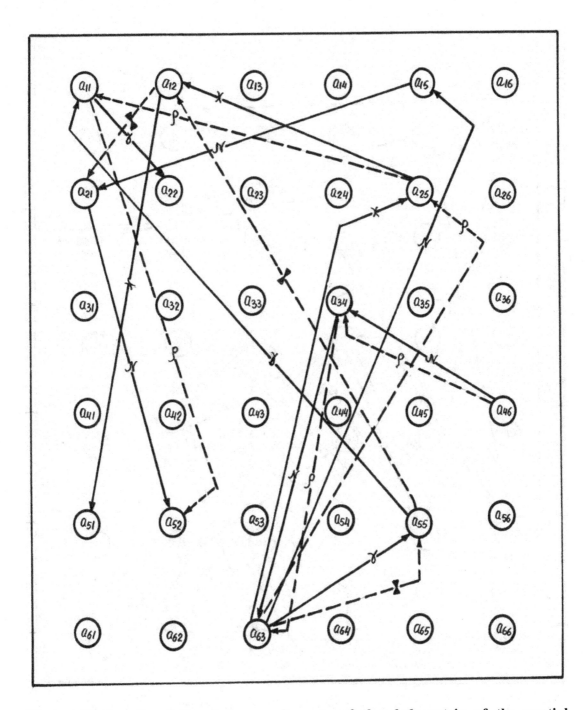

Figure 29 Functional ties between elements of the 6x6 matrix of the partial determinant $|A'''_5|$ of the combined partial determinant $|A'_5|$ for positive(——) and negative(------) areas applicably to Lapalce's theorem at marked out jointly the third ,fourth and fifth columns

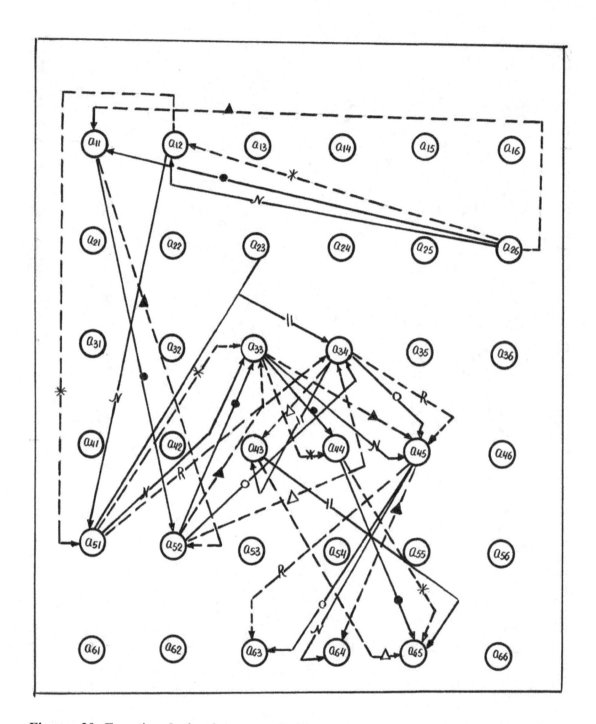

Figure 30 Functional ties between elements of the 6x6 matrix of the partial determinant $|A^{iv}_5|$ of the combined determinant $|A_5|$ for positive (———) and negative(-----) areas applicably to Laplace's theorem at marked out jointly the third, fourth and fifth columns

F. Version #1:Definition of the determinant $|A_6|$:

Combination of used rows and minors of the third order (M_{ij}) and algebraic A_{ij} supplemental (cofactors) for above-considered conditions have the following view in numerical data and shown in Table 9.

Table 9 Combination of used rows and other parameters of the 6x 6 matrix

n/n	Combination of used rows	Parameters of the determinant of the 6x 6 matrix	
		(M_{ij})	(A_{ij})
1	$i_1+i_2+i_3$	0	0
2	$i_1+i_2+i_4$	0	0
3	$i_1+i_2+i_5$	0	0
4	$i_1+i_2+i_6$	0	0
5	$i_1+i_3+i_4$	64	−96
6	$i_1+i_3+i_5$	0	0
7	$i_1+i_3+i_6$	−120	24
8	$i_1+i_4+i_5$	0	0
9	$i_1+i_4+i_6$	−48	0
10	$i_2+i_3+i_4$	0	−240
11	$i_2+i_3+i_5$	0	0
12	$i_2+i_3+i_6$	0	60
13	$i_3+i_4+i_5$	0	128
14	$i_3+i_4+i_6$	12	98
15	$i_3+i_5+i_6$	0	32
16	$i_4+i_5+i_6$	0	0

So, analysis of numerical data shown in Table 9 indicates on the essential fact that value of the determinant $|A_6|$ of the 6x6 matrix can be defined by the following formula in general view as :

$$|A_6|=M_5A_5+M_7A_7-M_{14}A_{14} \text{ or}$$

$$|A_6|=(a_{15}a_{33}a_{44}-a_{15}a_{34}a_{43})(a_{21}a_{52}a_{66}-a_{22}a_{51}a_{66})+$$
$$+(a_{15}a_{33}a_{64}-a_{15}a_{34}a_{63})(a_{22}a_{46}a_{51}-a_{21}a_{46}a_{52})-$$
$$-(a_{35}a_{43}a_{64}-a_{35}a_{44}a_{63})[(a_{11}a_{22}a_{56}+a_{12}a_{26}a_{51}+a_{16}a_{21}a_{52}-a_{16}a_{22}a_{51}-a_{11}a_{26}a_{52}-a_{12}a_{21}a_{56})]$$

In numerical data the value of the determinant $|A_6|$ of the 6x6 matrix is equal: $|A_6|=-10200$.
After of some calculations and transformations, we have the value of the combined determinant $|A_6|$ of the 6x6 matrix consists from some partial determinants and is equal:

$$|A_6|=|A'_6|+|A^{iv}_6|=(|A''_6|+|A'''_6|)+|A^{iv}_6|$$

where,
$|A'_6|$ -combined partial determinant, having two partial determinants which is equal :
$$|A'_6|=|A''_6|+|A'''_6|$$
$|A'_6|=(a_{21}a_{52}-a_{22}a_{51})[(a_{15}a_{33}(a_{44}a_{66}-a_{64}a_{46})+a_{15}a_{34}(a_{63}a_{46}-a_{43}a_{66})+a_{35}a_{16}(a_{44}a_{63}-a_{43}a_{64})]$,where:

- $|A''_6|=a_{21}a_{52}a_{15}a_{33}a_{44}a_{66}-a_{21}a_{52}a_{15}a_{33}a_{64}a_{46}+a_{21}a_{52}a_{15}a_{34}a_{63}a_{46}-$
$-a_{21}a_{52}a_{15}a_{34}a_{43}a_{66}+a_{21}a_{52}a_{35}a_{16}a_{44}a_{63}-a_{21}a_{52}a_{35}a_{16}a_{43}a_{64}$

- $|A'''_6|=-a_{22}a_{51}a_{33}a_{44}a_{66}+a_{22}a_{51}a_{15}a_{33}a_{64}a_{46}-a_{22}a_{51}a_{15}a_{34}a_{63}a_{46}+$
$+a_{22}a_{51}a_{15}a_{34}a_{43}a_{66}-a_{22}a_{51}a_{35}a_{16}a_{44}a_{63}+a_{22}a_{51}a_{35}a_{16}a_{43}a_{64}$

$|A^{iv}_6|$ -partial determinant of the combined determinant $|A_6|$ equal :
$|A^{iv}_6|=a_{35}(a_{43}a_{64}-a_{44}a_{63})[a_{11}(a_{26}a_{52}-a_{22}a_{56})+a_{12}(a_{21}a_{56}-a_{26}a_{51})$

In Figure 31 is shown the functional ties between elements of the 6x 6 matrix of the partial determinant $|A''_6|$ of the combined partial determinant $|A'_6|$ for positive and negative areas applicably to Laplace's theorem. In Figure 32 is shown the functional ties between elements of the 6x6 matrix of the partial determinant $|A'''_6|$ of the combined partial determinant $|A'_6|$ and in Figure 33 is shown the functional ties between elements of the 6x6 matrix of partial determinant $|A^{iv}_6|$ of the combined determinant $|A_6|$.

411

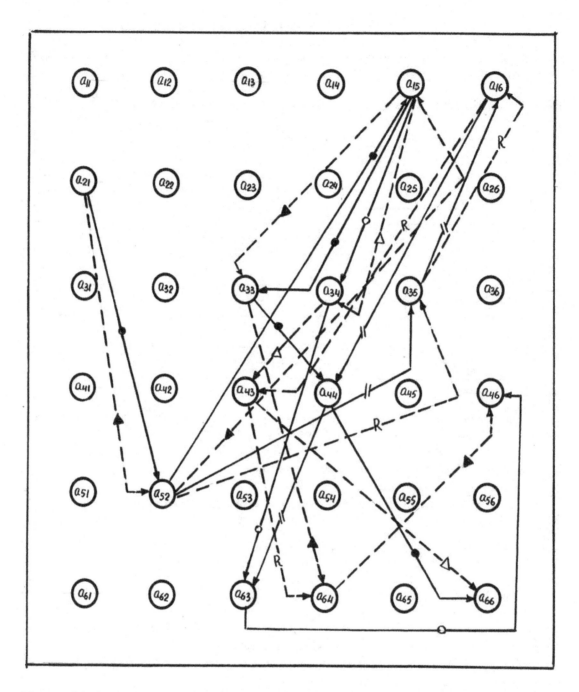

Figure 31 Functional ties between elements of the 6x6 matrix of the partial determinant $\left|A''_6\right|$ of the combined partial determinant $\left|A'_6\right|$ for positive (———) and negative (-------) areas applicably to Laplace's theorem at marked out jointly the third, fourth and fifth columns

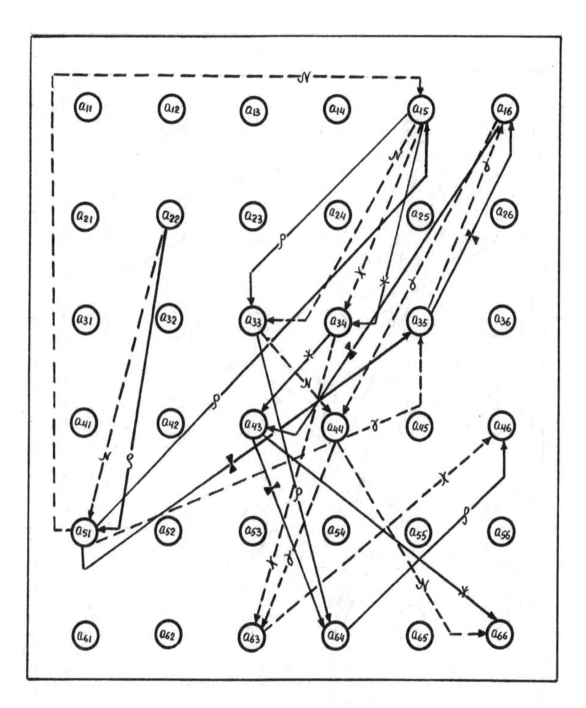

Figure 32 Functional ties between elements of the 6x6 matrix of the partial determinant $\left|A'''_6\right|$ of the combined partial determinant $\left|A'_6\right|$ for positive (———) and negative(--------) areas applicably to Laplace's theorem at marked out jointly the third ,fourth and fifth columns

413

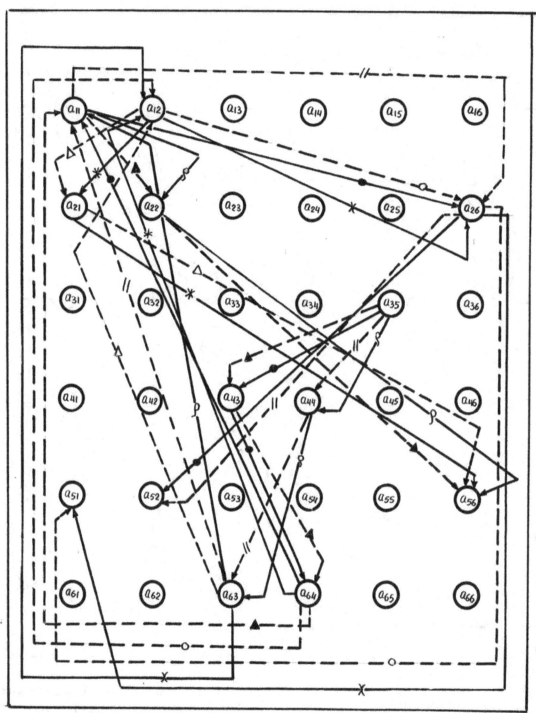

Figure 33 Functional ties between elements of the 6x6 matrix of the combined partial determinant $\left| A^{iv}_6 \right|$ **of the combined determinant** $\left| A_6 \right|$ **for positive (——) and negative(------) areas applicably to Laplace's theorem at marked out jointly the third, fourth and fifth columns**

414

In accordance with Cramer's rule the unknown coefficients of system normal six equations can be defined by the following formulas:

$$X_1=|A_1|/|A|; X_2=|A_2|/|A|; X_3=|A_3|/|A|; X_4=|A_4|/|A|; X_5=|A_5|/|A|$$

and $X_6=|A_6|/|A|$.

The values of above-considered determinants of the 6x6 matrices have the following data:

$$|A|=8280; |A_1|=26880; |A_2|=32400; |A_3|=-5340; |A_4|=-16380;$$
$$|A_5|=-48780; |A_6|=-10200$$

and unknown coefficients of the system six normal equations are equal:

$$X_1=3.246; X_2=3.913; X_3=-0.645; X_4=-1.978; X_5=-5.891; X_6=-1.232$$

Appendix 28

Problem:

Evaluate the system of six normal equations with using of Cramer's rule and also the Laplace's theorem at conditions when in 6x6 matrix marked out jointly the third, fourth and fifth columns.

$$2X_1-3X_2+4X_6=0$$
$$4X_1-4X_2+X_3-X_4+3X_5+2X_6=0$$
$$X_1-3X_2-2X_3+5X_5-5X_6=-1$$
$$2X_2-3X_4=3$$
$$X_1+X_2+3X_3-3X_4+4X_5+5X_6=-2$$
$$4X_1-4X_2+X_3-X_4+2X_5+3X_6=1$$

Solution:

1) The above-indicated system of six normal equations can be written in view of the 6x6 matrix in numerical data:

$$A=\begin{vmatrix} 2 & -3 & 0 & 0 & 0 & 4 \\ 4 & -4 & 1 & -1 & 3 & 2 \\ 1 & -3 & -2 & 0 & 5 & -5 \\ 0 & 2 & 0 & -3 & 0 & 0 \\ 1 & 1 & 3 & -3 & 4 & 5 \\ 4 & -4 & 1 & -1 & 2 & 3 \end{vmatrix} \quad B=\begin{vmatrix} 0 \\ 0 \\ -1 \\ 3 \\ -2 \\ 1 \end{vmatrix}$$

2) The unknown parameters X_1, X_2, X_3, X_4, X_5 and X_6 can be evaluated by the Cramer's rule from formulas:

$$X_1=|A_1|/|A|; X_2=|A_2|/|A|; X_3=|A_3|/|A|; X_4=|A_4|/|A|; X_5=|A_5|/|A|;$$
$$X_6=|A_6|/|A|.$$

3) The determinants $|A|; |A_1|; |A_2|; |A_3|; |A_4|; |A_5|$ and $|A_6|$ of the 6x6 matrices for the above-considered conditions can be evaluated by the formulas (117).

4) So, at given numerical data for the 6x6 matrices ,we have the value for determinants which are equal: $|A|=655; |A_1|=688; |A_2|=932; |A_3|=-570; |A_4|=33; |A_5|=283 ; |A_6|=251$ and then the coefficients of system for six linear equations are equal:

$$X_1=1.051; X_2=1.423; X_3=-0.869; X_4=0.052; X_5=0.432; X_6=0.383$$

415

Appendix R

<u>Problem:</u>

Evaluate the system of seven normal equations with using of Cramer's rule and also the Laplace's theorem at conditions when in 7x7 matrix marked out jointly the first ,second,third and fourth rows.

$$-2X_6+3X_7=0$$
$$-X_1+3X_2+2X_3+4X_4-3X_5+5X_6+X_7=-4$$
$$2X_1-2X_2+4X_3-5X_4+2X_5+5X_6-X_7=3$$
$$-3X_1+X_2-X_3+5X_4+2X_5-3X_6+2X_7=5$$
$$4X_1+X_2-2X_3+6X_4-3X_5+4X_6-2X_7=6$$
$$5X_1-3X_2+4X_3-X_4-4X_5+6X_6-3X_7=-3$$
$$2X_1+2X_2-5X_3+2X_4+X_5-X_6-3X_7=2$$

<u>Solution:</u>

1) The above-indicated system of seven normal equations can be written in view of the 7x7 matrix in numerical data:

$$A=\begin{vmatrix} 0 & 0 & 0 & 0 & 0 & -2 & 3 \\ -1 & 3 & 2 & 4 & -3 & 5 & 1 \\ 2 & -2 & 4 & -5 & 2 & 5 & -1 \\ -3 & 1 & -1 & 5 & 2 & -3 & 2 \\ 4 & 1 & -2 & 6 & -3 & 4 & -2 \\ 5 & -3 & 4 & -1 & -4 & 6 & -3 \\ 1 & 2 & -5 & 2 & 1 & -1 & -3 \end{vmatrix} \quad B=\begin{vmatrix} 0 \\ -4 \\ 3 \\ 5 \\ 6 \\ -3 \\ 2 \end{vmatrix}$$

2) The unknown parameters $X_1 \ldots \ldots X_7$ can be evaluated by the Cramer's rule from formulas:

$$X_1=|A_1|/|A|; X_2=|A_2|/|A|; X_3=|A_3|/|A|; X_4=|A_4|/|A|; X_5=|A_5|/|A|$$
$$X_6=|A_6|/|A|; X_7=|A_7|/|A|.$$

3) The determinants $|A|, |A_1| \ldots \ldots \ldots |A_7|$ of the 7x7 matrices for the above-considered conditions can be evaluated by the formula (&&) or other methods;

4) So, at given numerical data for the 7x7 matrices ,we have the value for determinants which are equal :

$$|A|=-8241; |A_1|=21682.07; |A_2|=34043.57; |A_3|=30590.59;$$
$$|A_4|=-6971.89; |A_5|=-26709.08; |A_6|=-32263.52; |A_7|=-21682.07$$

5) The coefficients $X_1 \ldots \ldots X_7$ of system for seven normal equations are equal:

$$X_1=-2.983; X_2=-4.131; X_3=-3.712; X_4=0.846; X_5=3.241; X_6=3.915; X_7=2.631$$

416

CHAPTER FIVE   FUNCTIONAL GRAPHS AND ALGORITHMS FOR SYMMETRIC AND OBLIQUE(SKEW) DETERMINANTS

1. Evaluation of some matrices for symmetric determinants

1.1 For 3x3 matrix

In general view we have the symmetric determinant of 3x3 matrix at conditions when all of its elements submit to $a_{ij}=a_{ji}$. And of its evaluation can be defined by the following methods:

A. *Sarrus rule*

In total view we have the following results for 3x3 matrix for the symmetric determinant:

$$A= \begin{vmatrix} a_{11} & a_{12} & a_{13} \\ a_{12} & a_{22} & a_{23} \\ a_{13} & a_{23} & a_{33} \end{vmatrix} \qquad (*)$$

The symmetric determinant for 3x3 matrix is equal in accordance with Sarrus rule:

$$\begin{aligned} |A| &= a_{13}(a_{12}a_{23}-a_{13}a_{22})+a_{23}(a_{13}a_{12}-a_{23}a_{11})+a_{33}(a_{11}a_{22}-a_{12}a_{12})= \\ &= a_{11}(a_{22}a_{33}-a_{23}a_{23})+a_{12}(a_{23}a_{13}-a_{12}a_{33})+a_{13}(a_{12}a_{23}-a_{22}a_{13}) \end{aligned} \qquad (118)$$

Example 2:

The value of the symmetric determinant for 3x3 matrix in numerical data in accordance of formula (118) is equal:

$$A= \begin{vmatrix} 2 & 6 & 5 \\ 6 & -3 & 4 \\ 5 & 4 & 0 \end{vmatrix} \qquad |A| = (0+120+120)-(-75+32+0)=283$$

In Figure 192 is shown the functional ties between elements of 3x3 matrix for evaluation of the symmetric determinant. From Figure 192 we see the following facts:

❖ For positive area

- The first element (a_{11})of the first row has functional ties with the second element (a_{22}) of the second row and third element (a_{33}) of the third row;
- The second element (a_{12})of the first row has functional ties with the third element(a_{23}) of the second row and first element (a_{13}) of the third row;
- The third element(a_{13}) of the first row has functional ties with the first element (a_{12}) of the second row and second element (a_{23}) of the third row.

❖ For negative area

- The first element(a_{11}) of the first row has functional ties with the third element (a_{23}) of the second row and second element(a_{23}) of the third row;
- The second element(a_{12}) of the first row has functional ties with the first element (a_{12}) of the second row and third element (a_{33}) of the third row;
- The third element (a_{13}) of the first row has functional ties with the second element (a_{22}) of the second row and first element (a_{13}) of the third row, forming for all elements of 3x3 matrix such

functional ties as:

$$a_{11} \rightarrow a_{22}, a_{23}, a_{23}, a_{23}$$
$$a_{12} \rightarrow a_{23}, a_{13}, a_{12}, a_{33}$$
$$a_{13} \rightarrow a_{12}, a_{23}, a_{22}, a_{13}$$

having the frequency of ties equal f=2 for all elements of 3x3 matrix ,besides of elements a_{11}, a_{12}, a_{13} for the first row.

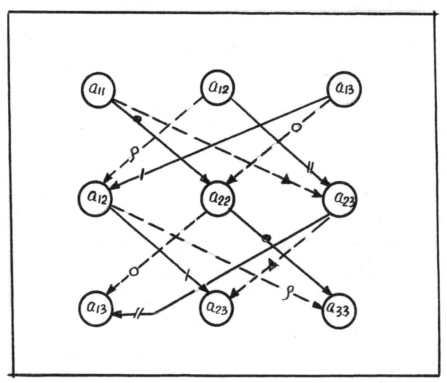

Figure 192 Symmetric determinant of 3x3 matrix in functional graphs
(------negative area)

B. *Expansion (cofactor method)*

For above- indicated 3x3 matrix (*) ,we can take any row or column that to evaluate the matrix determinant**[2].** So ,expanding determinant of 3x3 matrix to elements ,for a instance of the first row ,we have the following results:

$$|\,A\,| = a_{11}A_{11} + a_{12}A_{12} + a_{13}A_{13}$$

where,

a_{11}-the first element of the first row;

a_{12}-the second element of the first row;

a_{13}-the third element of the first row;

A_{11}- algebraic supplemental of the first row and first column equal:

$$A_{11} = (-1)^{1+1}(\,a_{22}a_{33} - a_{23}a_{23}) = (a_{22}a_{33} - a_{23}a_{23})$$

A_{12}-algebraic supplemental of the first row and second column equal:

$$A_{12} = (-1)^{1+2}(a_{12}a_{33} - a_{23}a_{13}) = -(a_{12}a_{33} - a_{23}a_{13}) = (a_{23}a_{13} - a_{12}a_{33})$$

A_{13}-algebraic supplemental of the first row and third column equal:

$$A_{13} = (-1)^{1+3}(a_{12}a_{23} - a_{22}a_{13})$$

And then the symmetric determinant of 3x3 matrix in general view is equal:

$$|\,A\,| = a_{11}(a_{22}a_{33} - a_{23}a_{23}) + a_{12}(a_{23}a_{13} - a_{12}a_{33}) + a_{13}(a_{12}a_{23} - a_{22}a_{13})$$

418

C. *Laplace's theorem*

In above-considered 3x3 matrix (*) mark out randomly two rows, for a instance the first and second rows for symmetric determinant, writing the minors(M_{ij}) of the elements in two considered rows. And then define the corresponding algebraic supplemental of these minors(A_{ij}) with writing the value of symmetric determinant in view of formula :

$$|A| = M_1A_1 + M_2A_2 + M_3A_3$$

where,

$M_1 = (a_{11}a_{22} - a_{12}a_{12}); \quad A_1 = (-1)^{(1+2)+(1+2)} \; |a_{33}| = a_{33}$

$M_2 = (a_{11}a_{23} - a_{13}a_{12}); \quad A_2 = (-1)^{(1+2)+(1+3)} \; |a_{23}| = -a_{23}$

$M_3 = (a_{12}a_{23} - a_{13}a_{22}); \quad A_3 = (-1)^{(1+2)+(2+3)} \; |a_{13}| = a_{13}$

So, in accordance with Laplace's theorem ,we have the following results for symmetric determinant of 3x3 matrix:

$$|A| = a_{33}(a_{11}a_{22} - a_{12}a_{12}) + a_{23}(a_{13}a_{12} - a_{23}a_{11}) + a_{13}(a_{12}a_{23} - a_{13}a_{22})$$

In Appendix 29 is shown evaluation of symmetric determinant of 3x3 matrix with using of Sarrus rule and solving of system of linear equation with Cramer's rule.

1.2 For 4x4 matrix

In general view the symmetric determinant of 4x4 matrix can be introduced by the following form :

$$A = \begin{vmatrix} a_{11} & a_{12} & a_{13} & a_{14} \\ a_{12} & a_{22} & a_{23} & a_{24} \\ a_{13} & a_{23} & a_{33} & a_{34} \\ a_{14} & a_{24} & a_{34} & a_{44} \end{vmatrix} \quad (**)$$

and of its evaluation can be solved by the following methods ,such as:

A. *Expansion method*

Referring to above- indicated 4x4 matrix, we have the following results ,for a instance, when expansion method use at conditions for elements of the first row of this symmetric determinant $|A|$:

$$|A| = a_{11}A_{11} + a_{12}A_{12} + a_{13}A_{13} + a_{14}A_{14}$$

where,

$a_{11}, a_{12}; a_{13}; a_{14}$-elements of the first row for 4x4 matrix;

A_{11}-algebraic supplemental of the first row and first column equal:

$A_{11} = (a_{22}a_{33}a_{44} + a_{23}a_{34}a_{24} + a_{24}a_{23}a_{34} - a_{24}a_{33}a_{24} - a_{22}a_{34}a_{34} - a_{23}a_{23}a_{35})$

A_{12}-algebraic supplemental of the first and second column equal:

$A_{12} = -(a_{12}a_{33}a_{44} + a_{23}a_{34}a_{14} + a_{24}a_{13}a_{34} - a_{24}a_{33}a_{14} - a_{12}a_{34}a_{34} - a_{23}a_{13}a_{44})$

A_{13}-algebraic supplemental of the first row and third column equal:

$A_{13} = (a_{12}a_{23}a_{44} + a_{22}a_{34}a_{14} + a_{24}a_{13}a_{24} - a_{24}a_{23}a_{14} - a_{12}a_{34}a_{24} - a_{22}a_{13}a_{44}$

A_{14}-algebraic supplemental of the first row and fourth column equal:

$A_{14} = -(a_{12}a_{23}a_{34} + a_{22}a_{33}a_{14} + a_{23}a_{13}a_{24} - a_{23}a_{23}a_{14} - a_{12}a_{33}a_{24} - a_{22}a_{13}a_{34}$

And then the symmetric determinant of 4x4 matrix in general view for *Expansion method* is equal:

$$|A| = a_{11}[a_{22}(a_{33}a_{44}-a_{34}a_{34})+a_{23}(a_{34}a_{24}-a_{23}a_{44})+a_{24}(a_{23}a_{34}-a_{33}a_{24})]+$$
$$+a_{12}[a_{33}(a_{24}a_{14}-a_{12}a_{44})+a_{34}(a_{12}a_{34}-a_{23}a_{14})+a_{13}(a_{23}a_{44}-a_{24}a_{34})]+$$
$$+a_{13}[a_{23}(a_{12}a_{44}-a_{24}a_{14})+a_{34}(a_{22}a_{14}-a_{12}a_{24})+a_{13}(a_{24}a_{24}-a_{22}a_{44})]+$$
$$+a_{14}[a_{23}(a_{23}a_{14}-a_{12}a_{34})+a_{33}(a_{12}a_{24}-a_{22}a_{14})+a_{13}(a_{22}a_{34}-a_{23}a_{24})] \quad (119)$$

In Appendix 30 is shown evaluation of symmetric determinant of 4x4 matrix with using of *Expansion method* and also the solving of system linear equations with using of Cramer's rule. In Figure 193 is shown the functional ties for symmetric determinant at Expansion method of 4x4 matrix for positive and negative (----------) areas.

Analysis of functional ties between of elements 4x4 matrix for the symmetric determinant in accordance with *Expansion method* indicates on the following results:

For the first element(a_{11}) of the first row

This element has such functional ties for positive and negative areas as:
- Element(a_{11}) has connection with the second element (a_{22}) of the second row and third element(a_{33}) of the third row and fourth element(a_{44})of the fourth row for positive area;
- Element(a_{11}) has connection with the second element (a_{22}) of the second row and fourth element(a_{34}) of the third row and third element (a_{34}) of the fourth row for negative area;
- Element (a_{11}) has connection with the third element (a_{23}) of the second row and fourth element(a_{34}) of the third row and second element(a_{24}) of the fourth row for positive area;
- Element(a_{11}) has ties with the third element (a_{23}) of the second row and second element (a_{23}) of the third row and fourth element (a_{44}) of the fourt row for negative area;
- Element(a_{11}) has ties with the fourth element(a_{24}) of the second row and second element(a_{23}) of the third row and third element (a_{34}) of the fourth row for positive area;
- Element(a_{11}) has ties with the fourth element (a_{24})of the second row and third element (a_{33}) of the third row and second element(a_{24}) of the fourth row for negative area.

For the second element (a_{12}) of the first row

This element has such functional ties for positive and negative areas as:
- Element(a_{12}) has ties with the third element (a_{33}) of the third row and second element(a_{24}) of the fourth row and fourth element(a_{14}) of the first row for positive area;
- Element(a_{12}) has ties with the third element (a_{33}) of the third row and first element (a_{12}) of the second row and fourth element (a_{44}) of the fourth row for negative area;
- Element (a_{12}) has ties with the fourth element (a_{34}) of the third row and first element(a_{12}) of the second row and third element (a_{34}) of the fourth row for positive area;
- Element(a_{12}) has ties with the fourth element(a_{34}) of the third row and third element(a_{23}) of the second row and fourth element (a_{14}) of the first row for negative area;
- Element (a_{12}) has ties with the first element(a_{13})of the third row and third element(a_{23}) of the second row and fourth element (a_{44}) of the fourth row for positive area;
- Element (a_{12}) has ties with the third element (a_{13}) of the first row and fourth element (a_{24}) of the second row and third element (a_{34}) of the fourth row for negative area;

For the third element (a_{13}) of the first row

This element has such functional ties for positive and negative areas as:
- Element(a_{13}) has ties with the third element (a_{23}) of the second row and second element (a_{12}) of the first row and fourth element (a_{44}) of the fourth row for positive area;

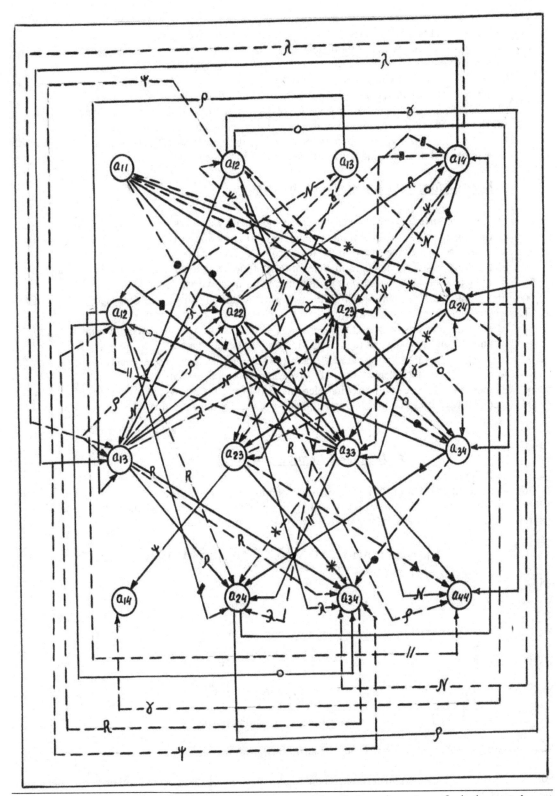

Figure 193 Functional graphs for symmetric determinant of 4x4 matrix at Expansion method for positive and negative areas

421

- Element(a_{13}) has ties with the second element (a_{23}) of the third row and fourth element (a_{24}) of the second row and first element (a_{14}) of the fourth row for negative area;
- Element(a_{13}) has ties with the third element(a_{34}) of the fourth row and second element(a_{22}) of the second row and fourth element (a_{14}) of the first row for positive area;
- Element(a_{13}) has ties with the third element(a_{34}) of the fourth row and first element(a_{12}) of the second row and second element(a_{24}) of the fourth row for negative area;
- Element(a_{13}) has ties with the first element (a_{13}) of the third row and second element(a_{24}) of the fourth row and fourth element(a_{24}) of the second row for positive area;
- Element(a_{13}) has ties with the first element (a_{13}) of the third row and second element (a_{22}) of the second row and fourth element (a_{44}) of the fourth row for negative area.

For the fourth element(a_{14}) of the first row

This element (a_{14}) has such functional ties for positive and negative areas as:
- Element(a_{14}) has ties with the third element(a_{23}) of the second row and second element(a_{23}) of the third row and first element (a_{14}) of the fourth row for positive area;
- Element(a_{14}) has ties with the third element (a_{23}) of the second row and second element (a_{12}) of the first row and third element (a_{34}) of the fourth row for negative area;
- Element(a_{14}) has ties with the third element (a_{33}) of the third row and first element (a_{12}) of the second row and second element (a_{24}) of the fourth row for positive area;
- Element(a_{14}) has ties with the third element (a_{33}) of the third row and second element (a_{22}) of the second row and first element (a_{14}) of the first row for negative area;
- Element(a_{14}) has ties with the first element (a_{13}) of the third row and second element(a_{22}) of the second row and third element (a_{34}) of the fourth row for positive area;
- Element (a_{14}) has ties with the first element (a_{13}) of the third row and third element (a_{23}) of the second row and second element (a_{24}) of the fourth row for negative area.

B. Laplace's theorem(method)

In above-considered 4x4 matrix (**) ,we mark out randomly two rows, for a instance the first and second columns, for symmetric determinant $|A|$, writing the minors (M_{ij}) of these elements in two considered columns. And then define the corresponding algebraic supplemental of these minors (A_{ij}):

$$|A|=M_1A_1+M_2A_2+M_3A_3+M_4A_4+M_5A_5+M_6A_6$$

where,
$M_1=(a_{11}a_{22}-a_{12}a_{12})$; $A_1=(-1)^{(1+2)+(1+2)} (a_{33}a_{44}-a_{34}a_{34})= (a_{33}a_{44}-a_{34}a_{34};$

$M_2=(a_{11}a_{23}-a_{12}a_{13})$; $A_2=(-1)^{(1+2)+(1+3)} (a_{23}a_{44}-a_{24}a_{34}) = -(a_{23}a_{44}-a_{24}a_{34});$

$M_3=(a_{11}a_{24}-a_{12}a_{14})$; $A_3=(-1)^{(1+2)+(1+4)} (a_{23}a_{34}-a_{24}a_{33})= (a_{23}a_{34}-a_{24}a_{33});$

$M_4=(a_{12}a_{23}-a_{22}a_{13})$; $A_4=(-1)^{(1+2)+(2+3)} (a_{13}a_{44}-a_{14}a_{34})= (a_{13}a_{44}-a_{14}a_{34});$

$M_5=(a_{12}a_{24}-a_{22}a_{14})$; $A_5=(-1)^{(1+2)+(2+4)} (a_{13}a_{34}-a_{14}a_{33})= -(a_{13}a_{34}-a_{14}a_{33});$

$M_6=(a_{13}a_{24}-a_{23}a_{14})$; $A_6=(-1)^{(1+2)+(3+4)} (a_{13}a_{24}-a_{14}a_{23})= (a_{13}a_{24}-a_{14}a_{23}).$

So ,we can conclude that symmetric determinant $|A|$ of the 4x4 matrix in accordance with *Laplace's theorem* is equal:

$$|A| = a_{11}[a_{22}(a_{33}a_{44}-a_{34}a_{34})+a_{23}(a_{24}a_{34}-a_{23}a_{44})+a_{24}(a_{23}a_{34}-a_{24}a_{33})] +$$
$$+a_{12}[a_{12}(a_{34}a_{34}-a_{33}a_{44})+a_{13}(a_{23}a_{44}-a_{24}a_{34})+a_{14}(a_{24}a_{33}-a_{23}a_{34})]+$$
$$+a_{13}[a_{22}(a_{14}a_{34}-a_{13}a_{44})+a_{23}(a_{12}a_{44}-a_{14}a_{24})+a_{24}(a_{13}a_{24}-a_{12}a_{34})]+$$
$$+a_{14}[a_{12}(a_{24}a_{33}-a_{23}a_{34})+a_{22}(a_{13}a_{34}-a_{14}a_{33})+a_{23}(a_{23}a_{14}-a_{13}a_{24})] \qquad (120)$$

In Appendix 31 is shown evaluation of the symmetric determinant of 4x4 matrix with using of Laplace's theorem and also the solving of system linear equations with using of Cramer's rule.

In Figure 194 is shown the functional ties for symmetric determinant $|A|$ of 4x4 matrix for positive and negative areas.

Analysis of functional ties between of elements of 4x4 matrix for the symmetric determinant shown in Figure 194 in accordance with Laplace's theorem indicates on the following results:

For the first element(a_{11}) of the first row

This element has such functional ties for positive and negative areas as:
- Element(a_{11})has ties with the second element (a_{22}) of the second row and third element (a_{33}) of the third row and fourth element (a_{44}) of the fourth row for positive area;
- Element(a_{11}) has ties with the second element(a_{22}) of the second row and fourth element(a_{34}) of the third row and third element (a_{34}) of the fourth row for negative area;
- Element(a_{11}) has ties with the third element(a_{23}) of the second row and second element(a_{24}) of the fourth row and fourth element(a_{34}) of the third row for positive area;
- Element(a_{11}) has ties with the third element (a_{23}) of the second row and second element(a_{23})of the third row and fourth element(a_{44})of the fourth row for negative area;
- Element(a_{11}) has ties with the fourth element(a_{24}) of the second row and second element(a_{23})of the third row and third element (a_{34}) of the fourth row for positive area;
- Element(a_{11}) has ties with the fourth element(a_{24}) of the second row and second element (a_{24}) of the fourth row and third element(a_{33}) of the third row for negative area.

For the second element(a_{12}) of the first row

This element has such functional ties for positive and negative areas as:
- Element(a_{12}) has ties with the first element(a_{12}) of the second row and fourth element (a_{34}) of the third row and third element(a_{34}) of the fourth row for positive area;
- Element(a_{12}) has ties with the first element (a_{12}) of the second row and third element (a_{33}) of the third row and fourth element (a_{44}) of the fourth row for negative area;
- Element(a_{12}) has ties with the first element(a_{13}) of the third row and third element (a_{23}) of the third row and fourth element (a_{44}) of the fourth row for positive area;
- Element (a_{12}) has ties with the first element(a_{13}) of the third row and second element(a_{24}) of the fourth row and fourth element(a_{34}) of the third row for negative area;
- Element(a_{12}) has ties with the first element (a_{14}) of the fourth row and fourth element(a_{24}) of the second row and third element (a_{33}) of the third row for positive area;
- Element(a_{12}) has ties with the first element (a_{14}) of the fourth row and second element(a_{23}) of the third row and third element (a_{34}) of the fourth row for negative areas.

For the third element(a_{13}) of the first row

This element has such functional ties for positive and negative areas as:
- Element(a_{13}) has ties with the second element(a_{22}) of the second row and fourth element (a_{14}) of the first row and third element(a_{34}) of the fourth row for positive area;
- Element(a_{13}) has ties with the second element(a_{22}) of the second row and first element(a_{13}) of the third row and fourth element (a_{44}) of the fourth row for negative area;

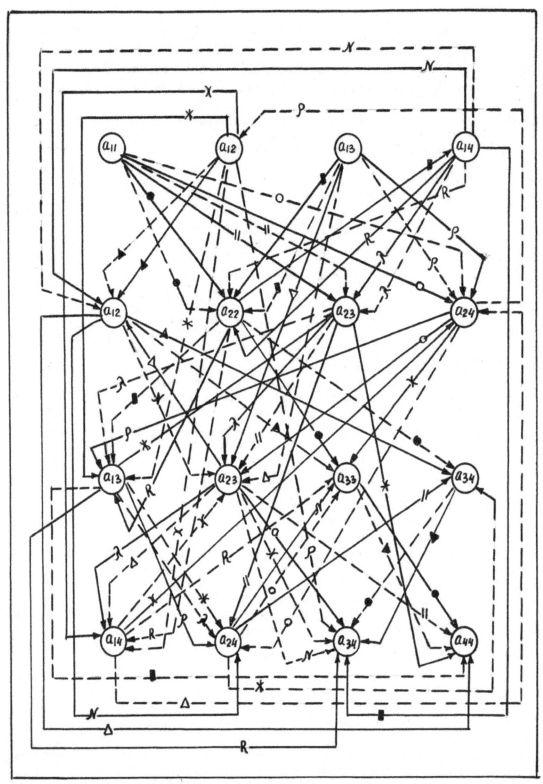

Figure 194 Functional ties for symmetric determinant of 4x4 matrix at Laplace's theorem for positive and negative areas

- Element (a_{13}) has ties with the second element(a_{23}) of the third row and first element (a_{12}) of the second row and fourth element (a_{44}) of the fourth row for positive area;
- Element(a_{13}) has ties with the second element (a_{23}) of the third row and first element (a_{14}) of the fourth row and fourth element(a_{24}) of the second row for negative area;
- Element(a_{13}) has ties with the fourth element (a_{24}) of the second row and first element (a_{13}) of the third row and second element(a_{24}) of the fourth row for positive area;
- Element(a_{13}) has ties with the fourth element(a_{24}) of the second row and second element(a_{12}) of the first row and third element (a_{34}) of the fourth row for negative area.

For the fourth element (a_{14}) of the first row

This element has such functional ties for positive and negative areas as:
- Element(a_{14})has ties with the first element (a_{12}) of the second row and second element (a_{24}) of the fourth row and third element (a_{33}) of the third row for positive area;
- Element(a_{14}) has ties with the first element(a_{12}) of the second row and second element (a_{23}) of the third row and third element(a_{34}) of the fourth row for negative area;
- Element(a_{14}) has ties with the second element (a_{22}) of the second row and first element (a_{13})of the third row and third element (a_{34}) of the fourth row for positive area;
- Element(a_{14}) has ties with the second element (a_{22}) of the second row and first element (a_{14}) of the fourth row and third element(a_{33}) of the third row for negative area;
- Element(a_{14}) has ties with the third element (a_{23}) of the second row and second element (a_{23}) of the third row and first element(a_{14}) of the fourth row for positive area;
- Element(a_{14}) has ties with the third element (a_{23}) of the second row and first element(a_{13}) of the third row and second element (a_{24}) of the fourth row for negative area.

1.3 For 5x5 matrix

In general view ,we have the symmetric determinant of 5x5 matrix:

$$A = \begin{vmatrix} a_{11} & a_{12} & a_{13} & a_{14} & a_{15} \\ a_{12} & a_{22} & a_{23} & a_{24} & a_{25} \\ a_{13} & a_{23} & a_{33} & a_{34} & a_{35} \\ a_{14} & a_{24} & a_{34} & a_{44} & a_{45} \\ a_{15} & a_{25} & a_{35} & a_{45} & a_{55} \end{vmatrix} \qquad (***)$$

and of its evaluation can be defined by the following methods:

A. *Expansion method*

Referring to 5x5 matrix(***) ,we have the following results ,for a instance ,when expansion method use at conditions for elements of the second row of this symmetric determinant $|A|$:

$$|A| = a_{12}A_{12} + a_{22}A_{22} + a_{23}A_{23} + a_{24}A_{24} + a_{25}A_{25}$$

where,
$a_{12}, a_{22}, a_{23}, a_{24}, a_{25}$-elements of the second row for 5x5 matrix;
A_{12}-algebraic supplemental of the second row and first column ;
A_{22}-algebraic supplemental of the second row and second column;

425

A_{23}- algebraic supplemental of the second row and third column;

A_{24}-algebraic supplemental of the second row and fourth column;

A_{25}- algebraic supplemental of the second row and fifth column which are expressed in such form as:

$$A_{12}=(-1)^{2+1}\begin{vmatrix} a_{12} & a_{13} & a_{14} & a_{15} \\ a_{23} & a_{33} & a_{34} & a_{35} \\ a_{24} & a_{34} & a_{44} & a_{45} \\ a_{25} & a_{35} & a_{45} & a_{55} \end{vmatrix}$$

$$A_{22}=(-1)^{2+2}\begin{vmatrix} a_{11} & a_{13} & a_{14} & a_{15} \\ a_{13} & a_{33} & a_{34} & a_{35} \\ a_{14} & a_{34} & a_{44} & a_{45} \\ a_{15} & a_{35} & a_{45} & a_{55} \end{vmatrix}$$

$$A_{23}=(-1)^{2+3}\begin{vmatrix} a_{11} & a_{12} & a_{14} & a_{15} \\ a_{13} & a_{23} & a_{34} & a_{35} \\ a_{14} & a_{24} & a_{44} & a_{45} \\ a_{15} & a_{25} & a_{45} & a_{55} \end{vmatrix}$$

$$A_{24}=(-1)^{2+4}\begin{vmatrix} a_{11} & a_{12} & a_{13} & a_{15} \\ a_{13} & a_{23} & a_{33} & a_{35} \\ a_{14} & a_{24} & a_{34} & a_{45} \\ a_{15} & a_{25} & a_{35} & a_{55} \end{vmatrix}$$

$$A_{25}=(-1)^{2+5}\begin{vmatrix} a_{11} & a_{12} & a_{13} & a_{14} \\ a_{13} & a_{23} & a_{33} & a_{34} \\ a_{14} & a_{24} & a_{34} & a_{44} \\ a_{15} & a_{25} & a_{35} & a_{45} \end{vmatrix}$$

The evaluation of above-indicated algebraic supplemental $A_{12} = \lvert A_{12} \rvert$; $A_{22}= \lvert A_{22} \rvert$; $A_{23}= \lvert A_{23} \rvert$; $A_{24}= \lvert A_{24} \rvert$ and $A_{25}= \lvert A_{25} \rvert$ as the determinants of 5x5 matrices can be solved, for a instance ,also by Expansion method and further with using of Sarrus rule for 3x3 matrices.

So ,after of some calculations and transformations ,we have the formula for evaluation of symmetric determinant of 5x5 matrix with using of *Expansion method.*

In Appendix 32 is shown evaluation of symmetric determinant of 5x5 matrix with using *of Expansion method* and also the solving of system linear equations with using of Cramer's rule.

$$|A| = a_{12}a_{12}[a_{33}(a_{45}a_{45}-a_{44}a_{55})+a_{34}(a_{34}a_{55}-a_{45}a_{35})+a_{35}(a_{44}a_{35}-a_{34}a_{45})]+$$
$$+a_{12}a_{13}[a_{23}(a_{44}a_{55}-a_{45}a_{45})+a_{34}(a_{45}a_{25}-a_{24}a_{55})+a_{35}(a_{24}a_{45}-a_{44}a_{25})]+$$
$$+a_{12}a_{14}[a_{23}(a_{45}a_{35}-a_{34}a_{55})+a_{33}(a_{24}a_{55}-a_{45}a_{25})+a_{35}(a_{34}a_{25}-a_{24}a_{35})]+$$
$$+a_{12}a_{15}[a_{23}(a_{34}a_{45}-a_{44}a_{35})+a_{33}(a_{44}a_{25}-a_{24}a_{45})+a_{34}(a_{24}a_{35}-a_{34}a_{25})]+$$

$$+a_{22}a_{11}[a_{33}(a_{44}a_{55}-a_{45}a_{45})+a_{34}(a_{45}a_{35}-a_{34}a_{55})+a_{35}(a_{34}a_{45}-a_{44}a_{35})]+$$
$$+a_{22}a_{13}[a_{13}(a_{45}a_{45}-a_{44}a_{55})+a_{34}(a_{14}a_{55}-a_{45}a_{15})+a_{35}(a_{44}a_{15}-a_{14}a_{45})]+$$
$$+a_{22}a_{14}[a_{13}(a_{34}a_{55}-a_{45}a_{35})+a_{33}(a_{45}a_{15}-a_{14}a_{55})+a_{35}(a_{14}a_{35}-a_{34}a_{15})]+$$
$$+a_{22}a_{15}[a_{13}(a_{44}a_{35}-a_{34}a_{45})+a_{33}(a_{14}a_{45}-a_{44}a_{15})+a_{34}(a_{34}a_{15}-a_{14}a_{35})]+$$

$$+a_{23}a_{11}[a_{23}(a_{45}a_{45}-a_{44}a_{55})+a_{34}(a_{24}a_{55}-a_{45}a_{25})+a_{35}(a_{44}a_{25}-a_{24}a_{45})]+$$
$$+a_{23}a_{12}[a_{13}(a_{44}a_{55}-a_{45}a_{45})+a_{34}(a_{45}a_{15}-a_{14}a_{55})+a_{35}(a_{14}a_{45}-a_{44}a_{15})]+$$
$$+a_{23}a_{14}[a_{13}(a_{45}a_{25}-a_{24}a_{55})+a_{23}(a_{14}a_{55}-a_{45}a_{15})+a_{35}(a_{24}a_{15}-a_{14}a_{25})]+$$
$$+a_{23}a_{15}[a_{13}(a_{24}a_{45}-a_{44}a_{25})+a_{23}(a_{44}a_{15}-a_{14}a_{45})+a_{34}(a_{14}a_{25}-a_{24}a_{15})]+$$

$$+a_{24}a_{11}[a_{23}(a_{34}a_{55}-a_{45}a_{35})+a_{33}(a_{45}a_{25}-a_{24}a_{55})+a_{35}(a_{24}a_{35}-a_{34}a_{25})]+$$
$$+a_{24}a_{12}[a_{13}(a_{45}a_{35}-a_{34}a_{55})+a_{33}(a_{14}a_{55}-a_{45}a_{15})+a_{35}(a_{34}a_{15}-a_{14}a_{35})]+$$
$$+a_{24}a_{13}[a_{13}(a_{24}a_{55}-a_{45}a_{25})+a_{23}(a_{45}a_{15}-a_{14}a_{55})+a_{35}(a_{14}a_{25}-a_{24}a_{15})]+$$
$$+a_{24}a_{15}[a_{13}(a_{34}a_{25}-a_{24}a_{35})+a_{23}(a_{14}a_{35}-a_{34}a_{15})+a_{33}(a_{24}a_{15}-a_{14}a_{25})]+$$

$$+a_{25}a_{11}[a_{23}(a_{44}a_{35}-a_{34}a_{45})+a_{33}(a_{24}a_{45}-a_{44}a_{25})+a_{34}(a_{34}a_{25}-a_{24}a_{35})]+$$
$$+a_{25}a_{12}[a_{13}(a_{34}a_{45}-a_{44}a_{35})+a_{33}(a_{44}a_{15}-a_{14}a_{45})+a_{34}(a_{14}a_{35}-a_{34}a_{15})]+$$
$$+a_{25}a_{13}[a_{13}(a_{44}a_{25}-a_{24}a_{45})+a_{23}(a_{14}a_{45}-a_{44}a_{15})+a_{34}(a_{24}a_{15}-a_{14}a_{25})]+$$
$$+a_{25}a_{14}[a_{13}(a_{24}a_{35}-a_{34}a_{25})+a_{23}(a_{34}a_{15}-a_{14}a_{35})+a_{33}(a_{14}a_{25}-a_{24}a_{15})] \qquad (121)$$

In Figure 195 is shown the functional ties at *Expansion method for partial part of* $|A_{12}|$ complex symmetric determinant $|A|$ of 5x5 matrix for positive and negative areas.

Analysis of functional ties between elements of 5x5 matrix for partial part $|A_{12}|$ of complex symmetric determinant $|A|$ for positive and negative areas in accordance with Expansion method shown in Figure 195 indicates on the following results:

For the second element (a_{12}) of the first row

This element has such functional ties for positive and negative areas as:
- Element(a_{12}) has ties with the first element (a_{12}) of the second row and third element(a_{33}) of the third row and fifth element(a_{45}) of the fourth row and fourth element (a_{45}) of the fifth row for positive area;
- Element(a_{12}) has ties with the first element(a_{12}) of the second row and third element(a_{33}) of the third row and fourth element(a_{44}) of the fourth row and fifth element (a_{55}) of the fifth row for negative area;
- Element(a_{12}) has ties with the first element(a_{12}) of the second row and fourth element(a_{34}) of the third row and third element (a_{34}) of the fourth row and fifth element(a_{55}) of the fifth row for positive area;
- Element(a_{12}) has ties with the first element (a_{12}) of the second row and fourth element (a_{34}) of the third row and fifth element(a_{45}) of the fourth row and third element (a_{35}) of the fifth row for negative area;

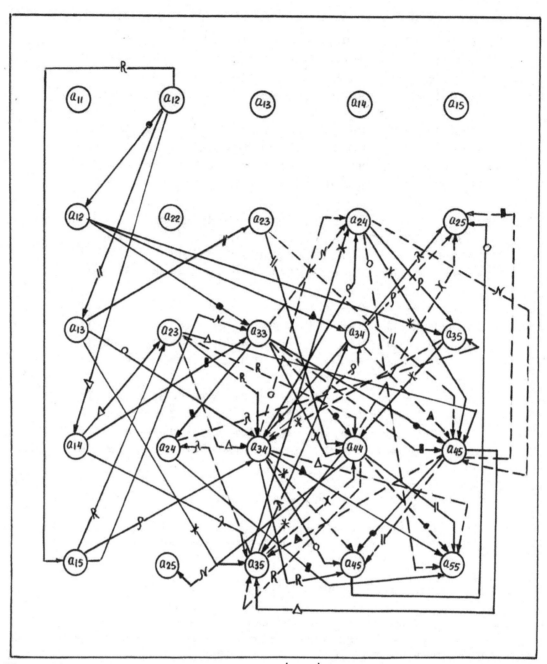

Figure 195 Functional ties for partial part $|A_{12}|$ complex symmetric determinant at Expansion method for positive and negative (---------) areas

- Element(a_{12}) has ties with the first element (a_{12}) of the second row and fifth element(a_{35}) of the third row and fourth element(a_{44}) of the fourth row and third element (a_{35}) of the fifth row for positive area;
- Element (a_{12}) has ties with the first element (a_{12}) of the second row and fifth element (a_{35}) of the third row and third element(a_{34}) of the fourth row and fourth element(a_{45}) of the fifth row for negative area;

- Element(a_{12}) has ties with the first element(a_{13})of the third row and third element(a_{23}) of the second row and fourth element (a_{44}) of the fourth row and fifth element (a_{55}) of the fifth row for positive area;
- Element(a_{12}) has ties with the first element (a_{13}) of the third row and third element(a_{23}) of the second row and fifth element(a_{45}) of the fourth row and fourth element(a_{45}) of the fifth row for negative area;
- Element(a_{12}) has ties with the first element(a_{13}) of the third row and third element (a_{34}) of the fourth row and fourth element (a_{45}) of the fifth row and fifth element(a_{25}) of the second row for positive area;
- Element(a_{12}) has ties with the first element (a_{13})of the third row and third element (a_{34}) of the fourth row and fourth element (a_{24}) of the second row and fifth element (a_{55}) of the fifth row for negative area;
- Element(a_{12}) has ties with the first element (a_{13}) of the third row and third element (a_{35}) of the fifth row and fourth element (a_{24}) of the second row and fifth element (a_{45}) of the fourth row for positive area;
- Element(a_{12}) has ties with the first element (a_{13}) of the third row and third element(a_{35}) of the fifth row and fourth element (a_{44}) of the fourth row and fifth element (a_{25}) of the second row for negative area;
- Element(a_{12}) has ties with the first element (a_{14}) of the fourth row and second element(a_{23}) of the third row and fifth element (a_{45}) of the fourth row and third element(a_{35}0 of the fifth row for positive area;
- Element(a_{12}) has ties with the first element (a_{14}) of the fourth row and second element(a_{23}) of the third row and third element(a_{34}) of the fourth row and fifth element(a_{55}) of the fifth row for negative area;
- Element (a_{12})has ties with the first element (a_{14}) of the fourth row and third element(a_{33}) of the third row and second element(a_{24}) of the fourth row and fifth element(a_{55}) of the fifth row for positive area;
- Element(a_{12}) has ties with the first element (a_{14}) of the fourth row and third element (a_{33}) of the third row and fifth element (a_{45}) of the fourth row and fifth element (a_{25}) of the second row for negative area;
- Element(a_{12}) has ties with the first element (a_{14}) of the fourth row and third element (a_{35}) of the fifth row and fourth element (a_{34}) of the third row and fifth element (a_{25}) of the second row for positive area;
- Element (a_{12}) has ties with the first element(a_{14}) of the fourth row and third element (a_{35}) of the fifth row and second element (a_{24})of the fourth row and fifth element (a_{35}) of the third row for negative area;
- Element(a_{12}) has ties with the first element (a_{15}) of the fifth row and second element(a_{23}) of the third row and third element (a_{34}) of the fourth row and fourth element (a_{45}) of the fifth row for positive area;
- Element (a_{12}) has ties with the first element (a_{15}) of the fifth row and second element (a_{23}) of the third row and fourth element (a_{44}) of the fourth row and third element (a_{35}) of the fifth row for negative area;
- Element(a_{12}) has ties with the first element (a_{15}) of the fifth row and third element (a_{33}) of the third row and fourth element (a_{44}) of the fourth row and second element(a_{25}) of the fifth row for positive area;
- Element(a_{12}) has ties with the first element(a_{15}) of the fifth row and third element (a_{33})of the third row and third element (a_{24}) of the second row and fifth element(a_{45}) of the fourth row for negative area;
- Element(a_{12}) has ties with the first element(a_{15}) of the fifth row and third element(a_{34}) of the fourth row and fourth element(a_{24}) of the second row and fourth element (a_{35}) of the third row for positive area;
- Element(a_{12})has ties with the first element(a_{15} of the fifth row and third element(a_{34}) of the fourth row and fourth element(a_{34}) of the third row and fifth element (a_{25}) of the second row for negative area.

In Figure 196　is shown the functional ties at *Expansion method for partial part of* $|A_{22}|$ complex symmetric determinant $|A|$ of 5x5 matrix for positive and negative areas.

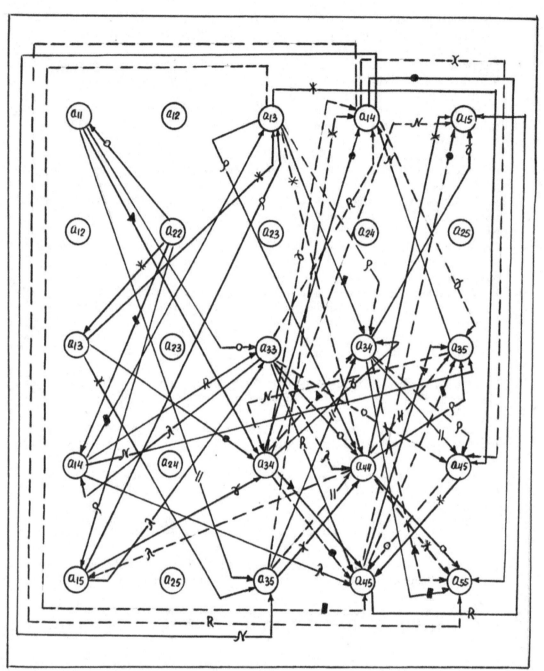

Figure 196 Functional ties for partial part $|A_{22}|$ complex symmetric determinant at Expansion method for positive and negative (---------) areas

Analysis of functional ties between of elements of 5x5 matrix for partial part $|A_{22}|$ of complex symmetric determinant $|A|$ for positive and negative areas in accordance with Figure 196 indicates on the following results:

For the second element (a₂₂) of the second row

This element has such functional ties for positive and negative areas as:

- Element(a₂₂) has ties with the first element(a₁₁) of the first row and third element (a₃₃) of the third row and fourth element(a₄₄) of the fourth row and fifth element(a₅₅) of the fifth row for positive area;
- Element(a₂₂) has ties with the first element (a₁₁)of the first row and third element(a₃₃) of the third row and fifth element(a₄₅) of the fourth row and fifth element(a₄₅) of the fifth row for negative area;
- Element(a₂₂) has ties with the first element (a₁₁) of the first row and third element(a₃₄) of the fourth row and fourth element (a₄₅) of the fifth row and fifth element(a₃₅) of the third row for positive area;
- Element(a₂₂)has ties with the first element(a₁₁) of the first row and third element(a₃₄) of the fourth row and fourth element(a₃₄) of the third row and fifth element(a₅₅) of the fifth row for negative area;
- Element (a₂₂) has ties with the first element (a₁₁)of the first row and fifth element(a₃₅) of the fifth row and fourth element(a₃₄) of the third row and fifth element(a₄₅) of the fourth row for positive area;
- Element(a₂₂) has ties with the first element(a₁₁) of the first row and fifth element(a₃₅) of the fifth row and fourth element(a₄₄) of the fourth row and fifth element (a₃₅) of the third row for negative area;
- Element(a₂₂) has ties with the first element(a₁₃) of the third row and third element(a₁₃) of the first row and fifth element (a₄₅) of the fourth row and fourth element (a₄₅) of the fifth row for positive area;
- Element(a₂₂) has ties with the first element (a₁₃) of the third row and third element(a₁₃) of the first row and fourth element(a₄₄) of the fourth row and fifth element (a₅₅) of the fifth row for negative area;
- Element(a₂₂) has ties with the first element(a₁₃) of the third row and third element(a₃₄) of the fourth row and fourth element(a₁₄) of the first row and fifth element (a₅₅) the fifth row for positive area;
- Element(a₂₂) has ties with the first element (a₁₃) of the third row and third element (a₃₄) of the fourth row and fourth element(a₄₅) of the fifth row and fifth element(a₁₅) of the first row for negative area;
- Element(a₂₂) has ties with the first element(a₁₃)of the third row and third element(a₃₅) of the fifth row and fourth element (a₄₄) of the fourth row and fifth element(a₁₅) of the first row for positive area;
- Element(a₂₂) has ties with the first element(a₁₃) of the third row and third element(a₃₅) and fourth element (a₁₄) of the first row and fifth element(a₄₅) of the fourth row for negative area;
- Element(a₂₂) has ties with the first element (a₁₄)of the fourth row and third element(a₁₃) of the first row and fourth element(a₃₄) of the third row and fifth element(a₅₅) of the fifth row for positive area;
- Element(a₂₂) has ties with the first element(a₁₄) of the fourth row and third element(a₁₃) of the first row and fourth element(a₄₅) of the fifth row and fifth element(a₃₅) of the third row for negative area;
- Element(a₂₂) has ties with the first element (a₁₄) of the fourth row and third element(a₃₃)of the third row and fourth element (a₄₅) of the fifth row and fifth element(a₁₅) of the first row for positive area;
- Element(a₂₂) has ties with the first element(a₁₄) of the fourth row and third element (a₃₃) of the third row and fourth element(a₁₄) of the first row and fifth element(a₅₅) of the fifth row for negative area;
- Element(a₂₂) has ties with the first element(a₁₄) of the fourth row and fifth element(a₃₅) of the third row and fourth element(a₁₄) of the first row and third element (a₃₅) of the fifth row for positive area;

- Element(a_{22}) has ties with the first element (a_{14}) of the fourth row and fifth element (a_{35}) of the third row and third element(a_{34}) of the fourth row and fifth element (a_{15}) of the first row for negative area;
- Element(a_{22})has ties with the first element(a_{15}) of the fifth row and third element (a_{13}) of the first row and fourth element (a_{44}) of the fourth row and fifth element (a_{35}) of the third row for positive area;
- Element (a_{22}) has ties with the first element (a_{15}) of the fifth row and third element (a_{13}) of the first row and fourth element (a_{34}) of the third row and fifth element(a_{45}) of the fourth row for negative area;
- Element(a_{22}) has ties with the first element(a_{15}) of the fifth row and third element(a_{33}) of the third row and first element(a_{14}) of the fourth row and fourth element (a_{45}) of the fifth row for the fifth row for positive area;
- Element(a_{22}) has ties with the first element (a_{15}) of the fifth row and third element(a_{33}) of the thirds row and fourth element(a_{44}) of the fourth row and first element (a_{15}) of the fifth row for negative area;
- Element(a_{22}) has ties with the first element(a_{15}) of the fifth row and third element(a_{34}) of the fourth row and fourth element (a_{34}) of the third row and fifth element (a_{15}) of the first row for positive area;
- Element (a_{22}) has ties with the first element(a_{15}) of the fifth row and third element(a_{34}) of the fourth row and fourth element(a_{14}) of the first row and fifth element (a_{35}) of the third row for negative area.

In Figure 197 is shown the functional ties at Expansion method for partial part of $|A_{23}|$ complex symmetric determinant $|A|$ of 5x5 matrix for positive and negative areas.

Analysis of functional ties between of elements of 5x5 matrix for partial part $|A_{23}|$ of complex symmetric determinant $|A|$ for positive and negative areas in accordance with Expansion method shown in Figure 197 indicates on the following results:

For the second element (a_{23}) of the third row

This element has such functional ties for positive and negative areas as:
- Element (a_{23}) has ties with the first element (a_{11}) of the first row and third element (a_{23}) of the second row and fifth element (a_{45}) of the fourth row and fourth element (a_{45}) of the fifth row for positive area;
- Element(a_{23}) has ties with the first element(a_{11}) of the first row and third element(a_{23}) of the second row and fourth element (a_{44}) of the fourth row and fifth element(a_{55}) of the fifth row for negative area;
- Element (a_{23}) has ties with the first element(a_{11}) of the first row and third element (a_{34}) of the fourth row and fourth element (a_{24}) of the second row and fifth element(a_{55}) of the fifth row for positive area;
- Element(a_{23}) has ties with the first element(a_{11}) of the first row and third element(a_{34}) of the fourth row and fourth element(a_{45}) of the fifth row and fifth element(a_{25}) of the second row for negative area;
- Element(a_{23}) has ties with the first element(a_{11}) of the first row and third element(a_{35}) of the fifth row and fourth element(a_{44}) of the fourth row and fifth element (a_{25}) of the second row for positive area;
- Element(a_{23}) has ties with the first element(a_{11}) of the first row and third element(a_{35}) of the fifth row and second element (a_{24}) of the fourth row and fourth element(a_{45}) of the fifth row for negative area;
- Element(a_{23}) has ties with the first element (a_{12}) of the second row and third element(a_{13}) of the first row and fourth element(a_{44}) of the fourth row and fifth element(a_{55}) of the fifth row for positive area;
- Element (a_{23}) has ties with the first element (a_{12}) of the second row and third element(a_{13}) of the first row and fifth element (a_{45}) of the fourth row and fourth element(a_{45}) of the fifth row for negative area;

432

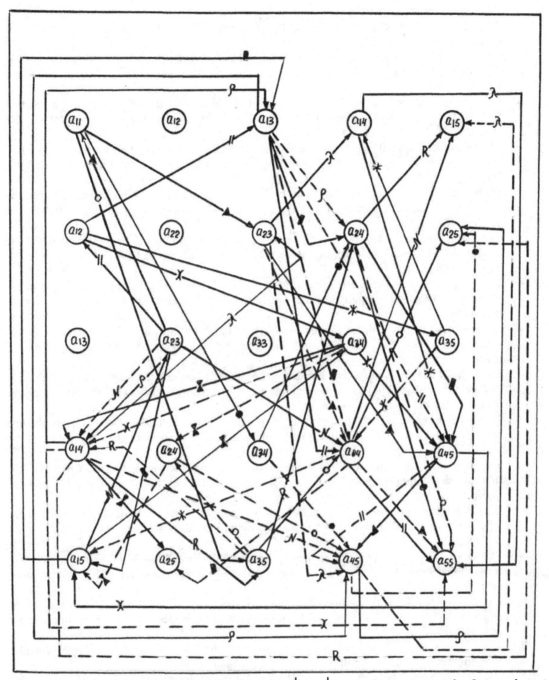

Figure 197 Functional ties for partial part $|A_{23}|$ complex symmetric determinant at Expansion method for positive and negative(---------) areas

- Element(a_{23})has ties with the first element(a_{12})of the second row and fourth element(a_{34}) of the third row and fifth element(a_{45}) of the fourth row and first element(a_{15}) of the fifth row for positive area;
- Element(a_{23}) has ties with the first element (a_{12}) of the second row and fourth element(a_{34}) of the third row and first element(a_{14}) of the fourth row and fifth element(a_{55}) of the fifth row for negative area;

- Element(a_{23}) has ties with the first element(a_{12}) of the second row and fifth element(a_{35}) of the third row and fourth element(a_{14}) of the first row and fifth element(a_{45}) of the fourth row for positive area;
- Element(a_{23}) has ties with the first element (a_{12}) of the second row and fifth element(a_{35}) of the third row and fourth element(a_{44}) of the fourth row and first element(a_{15}) of the fifth row for negative area;
- Element(a_{23}) has ties with the first element(a_{14}) of the third row and third element(a_{13}) of the first row and fourth element(a_{45}) of the fifth row and fifth element(a_{25}) of the second row for positive area;
- Element(a_{23}) has ties with the first element (a_{14}) of the third row and third element(a_{13}) of the first row and fourth element (a_{24}) of the second row and fifth element (a_{55}) of the fifth row for negative area;
- Element(a_{23}) has ties with the first element(a_{14}) of the fourth row and third element(a_{23}) of the second row and fourth element(a_{14}) of the first row and fifth element(a_{55}) of the fifth row for positive area;
- Element(a_{23}) has ties with the first element (a_{14}) of the fourth row and third element(a_{23}) of the second row and fourth element (a_{45}) of the fifth row and fifth element (a_{15}) of the first row for negative area;
- Element(a_{23}) has ties with the first element (a_{14}) of the fourth row and third element (a_{35}) of the fifth row and fourth element(a_{24}) of the second row and fifth element (a_{15}) of the first row for positive area;
- Element(a_{23}) has ties with the first element(a_{14}) of the fourth row and third element(a_{35}) of the fifth row and first element(a_{14}) of the fourth row and fifth element(a_{25}) of the second row of negative area;
- Element(a_{23}) has ties with the first element(a_{15}) of the fifth row and third element(a_{13}) of the first row and fourth element(a_{24}) of the second row and fifth element(a_{45}) of the fourth row for positive area;
- Element(a_{23}) has ties with the first element (a_{15}) of the fifth row and third element (a_{13}) of the first row and fourth element(a_{44}) of the fourth row and second element (a_{25}) of the fifth row for negative area;
- Element(a_{23}) has ties with the first element (a_{15}) of the fifth row and second element(a_{23}) of the third row and fourth element(a_{44}) of the fourth row and fifth element (a_{15}) of the first row for positive area;
- Element(a_{23})has ties with the first element (a_{15}) of the fifth row and second element(a_{23}) of the third row and first element(a_{14}) of the fourth row and fourth element(a_{45}) of the fifth row for negative area;
- Element(a_{23}) has ties with the first element(a_{15}) of the fifth row and fourth element(a_{34}) of the third row and first element(a_{14}) of the fourth row and second element(a_{25}) of the fifth row for positive area;
- Element(a_{23}) has ties with the first element(a_{15}) of the fifth row and fourth element (a_{34}) of the third row and second element (a_{24}) of the fourth row and first element (a_{15}) of the fifth row for negative area.

In Figure 198 is shown the functional ties at *Expansion method* for partial part of $|A_{24}|$ complex symmetric determinant $|A|$ of 5x5 matrix for positive and negative areas.

Analysis of functional ties between of elements of 5x5 matrix for partial part $|A_{24}|$ of complex symmetric determinant $|A|$ for positive and negative areas in accordance with Expansion method shown in Figure 198 indicates on the following results:

For the second element(a_{24}) of the fourth row

This element has such functional ties for positive and negative areas as:
- Element (a_{24}) has ties with the first element (a_{11}) of the first row and second element (a_{23}) of the third row and third element (a_{34}) of the fourth row and fifth element (a_{55})of the fifth row for positive area;

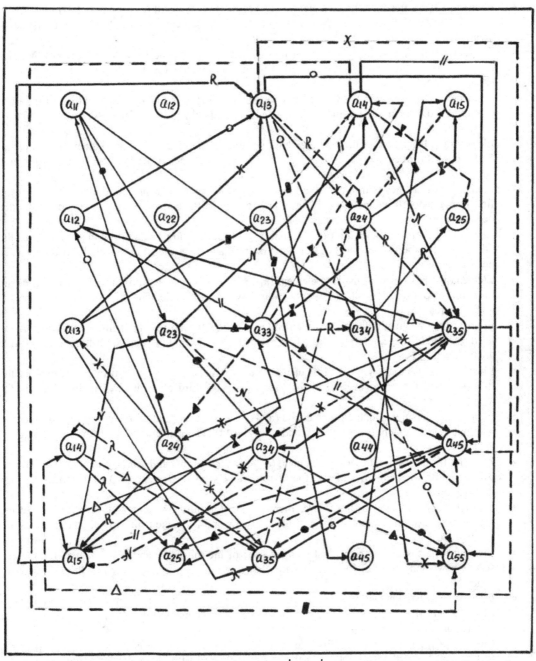

Figure 198 Functional ties for partial part $\left|A_{24}\right|$ complex symmetric determinant at Expansion method for positive and negative(--------) areas

- Element (a_{24}) has ties with the first element(a_{11}) of the first row and second element (a_{23})of the third row and fifth element (a_{45}) of the fourth row and third element(a_{35}) of the fifth row for negative area;

- Element(a_{24}) has ties with the first element (a_{11}) of the first row and third element (a_{33}) of the third row and fifth element (a_{45}) of the fourth row and second element (a_{25}) of the fifth row for positive area;

435

- Element(a_{24}) has ties with the first element(a_{11}) of the first row and third element (a_{33}) of the third row and second element(a_{24}) of the fourth row and fifth element(a_{55}) of the fifth row for negative area;

- Element(a_{24}) has ties with the first element(a_{11}) of the first row and fifth element (a_{35}) of the third row and second element(a_{24}) of the fourth row and third element(a_{35}) of the fifth row for positive area;

- Element(a_{24}) has ties with the first element(a_{11}) of the first row and fifth element(a_{35}) of the third row and third element (a_{34}) of the fourth row and second element(a_{25}) of the fifth row for negative area;

- Element(a_{24}) has ties with the first element (a_{12}) of the second row and third element (a_{13}) of the first row and fifth element (a_{45}) of the four row and third element (a_{35}) of the fifth row for positive area;

- Element (a_{24}) has ties with the first element(a_{12}) of the second row and third element (a_{13}) of the first row and fourth element(a_{34}) of the third row and fifth element (a_{55}) of the fifth row for negative area;

- Element (a_{24}) has ties with the first element (a_{12}) of the second row and third element (a_{33}) of the third row and fourth element (a_{14}) of the first row and fifth element(a_{55}) of the fifth row for positive area;

- Element(a_{24}) has ties with the first element (a_{12}) of the second row and third element (a_{33}) of the third row and fifth element (a_{45}) of the four row and first element (a_{15}) of the fifth row for negative area;

- Element (a_{24}) has ties with the first element(a_{12}) of the second row and fifth element (a_{35})of the third row and third element(a_{34}) of the four row and first element (a_{15}) of the fifth row for positive area;

- Element (a_{24}) has ties with the first element (a_{12}) of the second row and fifth element (a_{35})of the third row and first element (a_{14}) of the fourth row and third element (a_{35}) of the fifth row for negative area ;

- Element(a_{24}) has ties with the first element (a_{13}) of the third row and third element (a_{13}) of the first row and fourth element (a_{24}) of the second row and fifth element (a_{55}) of the fifth row for positive area;

- Element(a_{24}) has ties with the first element (a_{13}) of the third row and third element (a_{13}) of the first row and fifth element (a_{45}) of the fourth row and second element (a_{25}) of the fifth row for negative area;

- Element (a_{24}) has ties with the first element (a_{13}) of the third row and third element (a_{23}) of the second row and fourth element (a_{45}) of the fifth row and fifth element (a_{15}) of the first row for positive area;

- Element (a_{24}) has ties with the first element (a_{13}) of the third row and third element (a_{23}) of the second row and fourth element (a_{14})of the first row and fifth element (a_{55}) of the fifth row for negative area;

- Element (a_{24}) has ties with the first element (a_{13}) of the third row and third element (a_{35}) of the fifth row and first element (a_{14}) of the fourth row and second element (a_{25}) of the fifth row for positive area;

- Element (a_{24}) has ties with the first element (a_{13}) of the third row and third element (a_{35}) of the fifth row and fourth element (a_{24}) of the second row and fifth element (a_{15}) of the first row for negative area;

- Element(a_{24}) has ties with the first element (a_{15})of the fifth row and third element (a_{13}) of the first row and fourth element (a_{34}) of the third row and fifth element (a_{25}) of the second row for positive area;

- Element (a_{24}) has ties with the first element (a_{15}) of the fifth row and third element (a_{13}) of the first row and fourth element (a_{24}) of the second row and fifth element (a_{35}) of the third row for negative area;

- Element (a_{24}) has ties with the first element (a_{15}) of the fifth row and second element (a_{23}) of the third row and fourth element (a_{14}) of the first row and fifth element (a_{35}) of the third row for positive area;

- Element (a_{24}) has ties with the first element (a_{15}) of the fifth row and second element (a_{23}) of the third row and third element (a_{34}) of the fourth row and first element (a_{15}) of the fifth row for negative area;
- Element (a_{24}) has ties with the first element (a_{15}) of the fifth row and third element (a_{33}) of the third row and fourth element (a_{24}) of the second row and fifth element (a_{15}) of the first row for positive area;
- Element (a_{24}) has ties with the first element (a_{15}) of the fifth row and third element (a_{33}) of the third row and fourth element (a_{14}) of the first row and fifth element (a_{25}) of the second row for negative area.

In Figure 199 is shown the functional ties at *Expansion method* for partial part of $|A_{25}|$ complex symmetric determinant $|A|$ of 5x5 matrix for positive and negative areas. Analysis of functional ties between elements of 5x5 matrix for partial part $|A_{25}|$ of complex symmetric determinant $|A|$ for positive and negative areas in accordance with Expansion method shown in Figure 199 indicates on the following results:

For the second element (a_{25}) of the fifth row

This element has such functional ties for positive and negative areas as:
- Element (a_{25}) has ties with the first element (a_{11}) of the first row and second element (a_{23}) of the third row and fourth element (a_{44}) of the fourth row and third element (a_{35}) of the fifth row for positive area;
- Element (a_{25}) has ties with the first element (a_{11}) of the first row and second element (a_{23}) of the third row and third element (a_{34}) of the fourth row and fourth element(a_{45}) of the fifth row for negative area;
- Element(a_{25}) has ties with the first element (a_{11}) of the first row and third element (a_{33}) of the third row and second element (a_{24}) of the fourth row and fourth element (a_{45}) of the fifth row for positive area;
- Element (a_{25}) has ties with the first element (a_{11}) of the first row and third element (a_{33}) of the third row and fourth element (a_{44}) of the fourth row and second element (a_{25}) of the fifth row for negative area;
- Element(a_{25}) has ties with the first element (a_{11}) of the first row and fourth element (a_{34}) of the third row and third element(a_{34}) of the fourth row and second element (a_{25}) of the fifth row for positive area;
- Element(a_{25}) has ties with the first element (a_{11}) of the first row and fourth element (a_{34}) of the third row and second element (a_{24}) of the fourth row and third element (a_{35}) of the fifth row for negative area;
- Element(a_{25}) has ties with the first element (a_{12}) of the second row and third element (a_{13}) of the first row and fourth element (a_{34}) of the third row and fifth element (a_{45}) of the fourth row for positive area;
- Element(a_{25}) has ties with the first element (a_{12}) of the second row and third element (a_{13}) of the first row and fourth element (a_{44}) of the fourth row and third element (a_{35})of the fifth row for negative area;
- Element(a_{25}) has ties with the first element (a_{12}) of the second row and third element (a_{33}) of the third row and fourth element (a_{44}) of the fourth row and first element (a_{15}) of the fifth row for positive area;
- Element (a_{25}) has ties with the first element (a_{12}) of the second row and third element (a_{33}) of the third row and first element (a_{14}) of the fourth row and fourth element (a_{45}) of the fifth row for negative area;
- Element (a_{25}) has ties with the first element(a_{12}) of the second row and fourth element (a_{34}) of the third row and first element (a_{14}) of the fourth row and third element (a_{35}) of the fifth row for positive area;

437

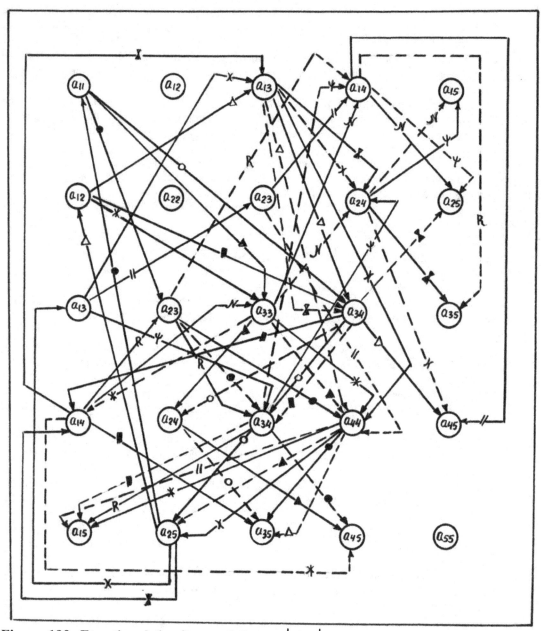

Figure 199 Functional ties for partial part │A₂₅│ complex symmetric determinant at Expansion method for positive and negative(----------) areas

- Element (a₂₅) has ties with the first element (a₁₂) of the second row and fourth element(a₃₄) of the third row and third element (a₃₄) of the fourth row and first element (a₁₅) of the fifth row for negative area;

- Element (a₂₅) has ties with the first element (a₁₃) of the third row and third element (a₁₃) of the first row   and fourth element (a₄₄) of the fourth row and second element (a₂₅) of the fifth row for positive area;

- Element (a₂₅) has ties with the first element (a₁₃) of the third row and third element (a₁₃) of the first row and fourth element (a₂₄) of the second row and fifth element (a₄₅) of the fourth row for negative area;

438

- Element (a_{25}) has ties with the first element (a_{13}) of the third row and third element (a_{23})of the second row and fourth element (a_{14}) of the first row and fifth element (a_{45}) of the fourth row for positive area;
- Element(a_{25}) has ties with the first element (a_{13}) of the third row and third element (a_{23}) of the second row and fourth element(a_{44}) of the fourth row and first element (a_{15}) of the fifth row for negative area;
- Element (a_{25}) has ties with the first element (a_{13}) of the third row and third element (a_{34}) of the fourth row and fourth element (a_{24})of the second row and fifth element(a_{15}) of the first row for positive area;
- Element (a_{25}) has ties with the first element (a_{13}) of the third row and third element(a_{34}) of the fourth row and fourth element (a_{14}) of the first row and fifth element(a_{25}) of the second row for negative area;
- Element (a_{25})has ties with the first element (a_{14}) of the fourth row and third element (a_{13}) of the first row and fourth element (a_{24}) of the second row and fifth element (a_{35}) of the third row for positive area;
- Element (a_{25}) has ties with the first element (a_{14}) of the fourth row and third element (a_{13}) of the first row and fourth element (a_{34}) of the third row and fifth element (a_{25}) of the second row for negative area;
- Element(a_{25}) has ties with the first element (a_{14}) of the fourth row and second element (a_{23}) of the third row and third element (a_{34}) of the fourth row and first element (a_{15}) of the fifth row for positive area;
- Element(a_{25}) has ties with the first element (a_{14}) of the fourth row and second element(a_{23}) of the third row and fourth element (a_{14}) of the first row and fifth element (a_{35}) of the third row for negative area;
- Element(a_{25}) has ties with the first element (a_{14}) of the fourth row and third element (a_{33}) of the third row and fourth element (a_{14}) of the first row and fifth element (a_{25}) of the second row for positive area;
- Element(a_{25}) has ties with the first element (a_{14}) of the fourth row and third element (a_{33}) of the third row and fourth element (a_{24})of the second row and fifth element (a_{15}) of the first row for negative area.

B. Laplace theorem (method)

Referring to 5x5 matrix(***) ,we marked out randomly two rows, for a instance, the first and second rows for symmetric determinant ,writing the minors (M_{ij}) of these elements in two considered rows. And then define the corresponding algebraic supplemental of these minors (A_{ij}):

$M_1 = (a_{11}a_{22} - a_{12}a_{12})$; $A_1 = (a_{33}a_{44}a_{55} + a_{34}a_{45}a_{35} + a_{35}a_{34}a_{45} - a_{35}a_{44}a_{35} - a_{33}a_{45}a_{45} - a_{34}a_{34}a_{55})$

$M_2 = (a_{11}a_{23} - a_{13}a_{12})$; $A_2 = -(a_{23}a_{44}a_{55} + a_{34}a_{45}a_{25} + a_{35}a_{24}a_{45} - a_{35}a_{44}a_{25} - a_{23}a_{45}a_{45} - a_{34}a_{24}a_{55})$

$M_3 = (a_{11}a_{24} - a_{14}a_{12})$; $A_3 = (a_{23}a_{34}a_{55} + a_{33}a_{45}a_{25} + a_{35}a_{24}a_{35} - a_{35}a_{34}a_{25} - a_{23}a_{45}a_{35} - a_{33}a_{24}a_{55})$

$M_4 = (a_{11}a_{25} - a_{15}a_{12})$; $A_4 = -(a_{23}a_{34}a_{45} + a_{33}a_{44}a_{25} + a_{34}a_{24}a_{35} - a_{34}a_{34}a_{25} - a_{23}a_{44}a_{35} - a_{33}a_{24}a_{45})$

$M_5 = (a_{12}a_{23} - a_{13}a_{22})$; $A_5 = (a_{13}a_{44}a_{55} + a_{34}a_{45}a_{15} + a_{35}a_{14}a_{45} - a_{35}a_{44}a_{15} - a_{13}a_{45}a_{45} - a_{34}a_{14}a_{55})$

$M_6 = (a_{12}a_{24} - a_{14}a_{22})$; $A_6 = -(a_{13}a_{34}a_{55} + a_{33}a_{45}a_{15} + a_{35}a_{14}a_{35} - a_{35}a_{34}a_{15} - a_{13}a_{45}a_{35} - a_{33}a_{14}a_{55})$

$M_7 = (a_{12}a_{25} - a_{15}a_{22})$; $A_7 = (a_{13}a_{34}a_{45} + a_{33}a_{44}a_{15} + a_{34}a_{14}a_{35} - a_{34}a_{34}a_{15} - a_{13}a_{44}a_{35} - a_{33}a_{14}a_{45})$

$M_8 = (a_{13}a_{24} - a_{14}a_{23})$; $A_8 = (a_{13}a_{24}a_{55} + a_{23}a_{45}a_{15} + a_{35}a_{14}a_{25} - a_{35}a_{24}a_{15} - a_{13}a_{45}a_{25} - a_{23}a_{14}a_{55})$

$M_9=(a_{13}a_{25}{-}a_{15}a_{23})$; $A_9={-}(a_{13}a_{24}a_{45}{+}a_{23}a_{44}a_{15}{+}a_{34}a_{14}a_{25}{-}a_{34}a_{24}a_{15}{-}a_{13}a_{44}a_{25}{-}a_{23}a_{14}a_{45})$

$M_{10}=(a_{14}a_{25}{-}a_{15}a_{24})$; $A_{10}=(a_{13}a_{24}a_{35}{+}a_{23}a_{34}a_{15}{+}a_{33}a_{14}a_{25}{-}a_{33}a_{24}a_{15}{-}a_{13}a_{34}a_{25}{-}a_{23}a_{14}a_{35})$.

So, we have the value for symmetric determinant of matrix 5x5 equal :

$$|A|=M_1A_1+M_2A_2+M_3A_3+M_4A_4+M_5A_5+M_6A_6+M_7A_7+M_8A_8+M_9A_9+M_{10}A_{10} \quad (122)$$

After of some calculations and transformations ,we have the following formula for evaluation of symmetric matrix determinant $|A|$ of 5x5 matrix at Laplace's method evaluated ,for a instance , for the first and second rows:

$$|A|=a_{11}a_{22}[a_{33}(a_{44}a_{55}{-}a_{45}a_{45}){+}a_{34}(a_{45}a_{35}{-}a_{34}a_{55}){+}a_{35}(a_{34}a_{45}{-}a_{44}a_{35})]+$$
$$+a_{11}a_{23}[a_{23}(a_{45}a_{45}{-}a_{44}a_{55}){+}a_{34}(a_{24}a_{55}{-}a_{45}a_{25}){+}a_{35}(a_{44}a_{25}{-}a_{24}a_{45})]+$$
$$+a_{11}a_{24}[a_{23}(a_{34}a_{55}{-}a_{45}a_{35}){+}a_{33}(a_{45}a_{25}{-}a_{24}a_{55}){+}a_{35}(a_{24}a_{35}{-}a_{34}a_{25})]+$$
$$+a_{11}a_{25}[a_{23}(a_{44}a_{35}{-}a_{34}a_{45}){+}a_{33}(a_{24}a_{45}{-}a_{44}a_{25}){+}a_{34}(a_{34}a_{25}{-}a_{24}a_{35})]+$$

$$+a_{12}a_{12}[a_{33}(a_{45}a_{45}{-}a_{44}a_{55}){+}a_{34}(a_{34}a_{55}{-}a_{45}a_{35}){+}a_{35}(a_{44}a_{35}{-}a_{34}a_{45})]+$$
$$+a_{12}a_{23}[a_{13}(a_{44}a_{55}{-}a_{45}a_{45}){+}a_{34}(a_{45}a_{15}{-}a_{14}a_{55}){+}a_{35}(a_{14}a_{45}{-}a_{44}a_{15})]+$$
$$+a_{12}a_{24}[a_{13}(a_{45}a_{35}{-}a_{34}a_{55}){+}a_{33}(a_{14}a_{55}{-}a_{45}a_{15}){+}a_{35}(a_{34}a_{15}{-}a_{14}a_{35})]+$$
$$+a_{12}a_{25}[a_{13}(a_{34}a_{45}{-}a_{44}a_{35}){+}a_{33}(a_{44}a_{15}{-}a_{14}a_{45}){+}a_{34}(a_{14}a_{35}{-}a_{34}a_{15})]+$$

$$+a_{13}a_{12}[a_{23}(a_{44}a_{55}{-}a_{45}a_{45}){+}a_{34}(a_{45}a_{25}{-}a_{24}a_{55}){+}a_{35}(a_{24}a_{45}{-}a_{44}a_{25})]+$$
$$+a_{13}a_{22}[a_{13}(a_{45}a_{45}{-}a_{14}a_{55}){+}a_{34}(a_{14}a_{55}{-}a_{45}a_{15}){+}a_{35}(a_{44}a_{15}{-}a_{14}a_{45})]+$$
$$+a_{13}a_{24}[a_{13}(a_{24}a_{55}{-}a_{45}a_{25}){+}a_{23}(a_{45}a_{15}{-}a_{14}a_{55}){+}a_{35}(a_{14}a_{25}{-}a_{24}a_{15})]+$$
$$+a_{13}a_{25}[a_{13}(a_{44}a_{25}{-}a_{24}a_{45}){+}a_{23}(a_{14}a_{45}{-}a_{44}a_{15}){+}a_{34}(a_{24}a_{15}{-}a_{14}a_{25})]+$$

$$+a_{14}a_{12}[a_{23}(a_{45}a_{35}{-}a_{34}a_{55}){+}a_{33}(a_{24}a_{55}{-}a_{45}a_{25}){+}a_{35}(a_{34}a_{25}{-}a_{24}a_{35})]+$$
$$+a_{14}a_{22}[a_{13}(a_{34}a_{55}{-}a_{45}a_{35}){+}a_{33}(a_{45}a_{15}{-}a_{14}a_{35}){+}a_{35}(a_{14}a_{35}{-}a_{34}a_{15})]+$$
$$+a_{14}a_{23}[a_{13}(a_{45}a_{25}{-}a_{24}a_{15}){+}a_{23}(a_{14}a_{55}{-}a_{45}a_{15}){+}a_{35}(a_{24}a_{15}{-}a_{14}a_{25})]+$$
$$+a_{14}a_{25}[a_{13}(a_{24}a_{35}{-}a_{34}a_{25}){+}a_{23}(a_{34}a_{15}{-}a_{14}a_{35}){+}a_{33}(a_{14}a_{25}{-}a_{24}a_{15})]+$$

$$+a_{15}a_{12}[a_{23}(a_{34}a_{45}{-}a_{44}a_{35}){+}a_{33}(a_{44}a_{25}{-}a_{24}a_{45}){+}a_{34}(a_{24}a_{35}{-}a_{34}a_{25})]+$$
$$+a_{15}a_{22}[a_{13}(a_{44}a_{35}{-}a_{34}a_{45}){+}a_{33}(a_{14}a_{45}{-}a_{44}a_{15}){+}a_{34}(a_{34}a_{15}{-}a_{14}a_{35})]+$$
$$+a_{15}a_{23}[a_{13}(a_{24}a_{45}{-}a_{44}a_{25}){+}a_{23}(a_{44}a_{15}{-}a_{14}a_{45}){+}a_{34}(a_{14}a_{25}{-}a_{24}a_{15})]+$$
$$+a_{15}a_{24}[a_{13}(a_{34}a_{25}{-}a_{24}a_{35}){+}a_{23}(a_{14}a_{35}{-}a_{34}a_{15}){+}a_{33}(a_{24}a_{15}{-}a_{14}a_{25})] \quad (123)$$

In Figure 200 is shown the functional ties at Laplace's method (theorem) for partial part of $|A_{11}|$ complex symmetric determinant $|A|$ of 5x5 matrix for positive and negative areas. Analysis of functional ties between elements of 5x5 matrix for partial part $|A_{11}|$ of complex symmetric determinant $|A|$ for positive and negative areas in accordance with Laplace's method shown in Figure 200 indicates on the following results:

For the first element (a₁₁)of the first row

This element has such functional ties for positive and negative areas as:
- Element (a_{11}) has ties with the second element(a_{22}) of the second row and third element (a_{33}) of the third row and fourth element (a_{44}) of the fourth row and fifth element(a_{55}) of the fifth row for positive area;

- Element(a_{11}) has ties with the second element (a_{22}) of the second row and third element(a_{33}) of the third row and fifth element (a_{45}) of the fourth row and fourth element (a_{45}) of the fifth row for negative area;

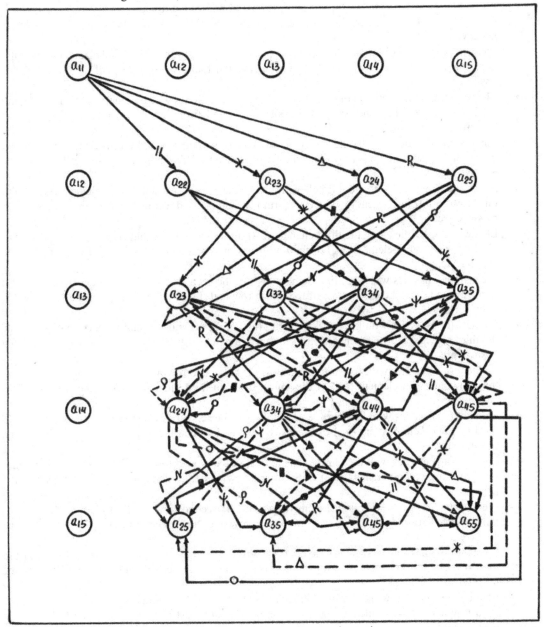

Figure 200 Functional ties for partial part $\left|A_{11}\right|$ of complex symmetric determinant at Laplace's method for positive and negative(--------) areas

- Element(a_{11}) has ties with the second element (a_{22}) of the second row and fourth element (a_{34}) of the third row and fifth element (a_{45}) of the fourth row and third element (a_{35}) of the fifth row for positive area;
- Element(a_{11}) has ties with the second element (a_{22}) of the second row and fourth element (a_{34}) of the third row and third element (a_{34}) of the fourth row and fifth element (a_{55}) of the fifth row for negative area;

441

- Element(a_{11})has ties with the second element (a_{22}) of the second row and fifth element(a_{35}) of the third row and third element (a_{34}) of the fourth row and fourth element (a_{45}) of the fifth row for positive area;
- Element(a_{11}) has ties with the second element (a_{22}) of the second row and fifth element (a_{35}) of the third row and fourth element (a_{44}) of the fourth row and third element (a_{35}) of the fifth row for negative area;
- Element (a_{11}) has ties with the third element (a_{23})of the second row and second element (a_{23}) of the third row and fifth element (a_{45}) of the fourth row and fourth element (a_{45}) of the fifth row for positive area;
- Element(a_{11})has ties with the third element (a_{23}) of the second row and second element (a_{23}) of the third row and fourth element(a_{44}) of the fourth row and fifth element (a_{55}) of the fifth row for negative area;
- Element (a_{11})has ties with the third element (a_{23}) of the second row and fourth element (a_{34}) of the third row and second element (a_{24}) of the fourth row and fifth element (a_{55}) of the fifth row for positive area;
- Element(a_{11}) has ties with the third element (a_{23})of the second row and fourth element (a_{34}) of the third row and fifth element (a_{45}) of the fourth row and second element (a_{25}) of the fifth row for negative area;
- Element (a_{11}) has ties with the third element (a_{23}) of the second row and fifth element (a_{35}) of the third row and fourth element (a_{44}) of the fourth row and second element (a_{25}) of the fifth row for positive area;
- Element(a_{11}) has ties with the third element (a_{23}) of the second row and fifth element (a_{35}) of the third row and second element (a_{24}) of the fourth row and fourth element (a_{45}) of the fifth row for negative area;
- Element (a_{11}) has ties with the fourth element (a_{24}) of the second row and second element (a_{23}) of the third row and third element (a_{34}) of the fourth row and fifth element (a_{55}) of the fifth row for positive area;
- Element (a_{11})has ties with fourth element (a_{24}) of the second row and second element (a_{23}) of the third row and fifth element (a_{45}) of the fourth row and third element (a_{35}) of the fifth row for negative area;
- Element (a_{11})has ties with the fourth element (a_{24}) of the second row and third element (a_{33}) of the third row and fifth element (a_{45}) of the fourth row and second element (a_{25}) of the fifth row for positive area;
- Element (a_{11}) has ties with the fourth element (a_{24}) of the second row and third element (a_{33}) of the third row and second element (a_{24}) of the fourth row and fifth element (a_{55}) of the fifth row for negative area;
- Element (a_{11}) has ties with fourth element (a_{24}) of the second row and fifth element (a_{35}) of the third row and second element (a_{24}) of the fourth row and third element (a_{35}) of the fifth row for positive area;
- Element (a_{11}) has ties with fourth element (a_{24}) of the second row and fifth element (a_{35}) of the third row and third element (a_{34}) of the fourth row and second element (a_{25}) of the fifth row for negative area;
- Element (a_{11}) has ties with the fifth element (a_{25}) of the second row and second element (a_{23}) of the third row and fourth element (a_{44}) of the fourth row and third element (a_{35})of the fifth row for positive area;
- Element (a_{11}) has ties with the fifth element (a_{25})of the second row and second element (a_{23}) of the third row and third element (a_{34}) of the fourth row and fourth element(a_{45}) of the fifth row for negative area;
- Element (a_{11}) has ties with the fifth element (a_{25}) of the second row and third element (a_{33}) of the third row and second element (a_{24}) of the fourth row and fourth element (a_{45}) of the fifth row for positive area;
- Element(a_{11}) has ties with the fifth element (a_{25}) of the second row and third element (a_{33}) of the third row and fourth element (a_{44}) of the fourth row and second element (a_{25}) of the fifth row for negative area;

- Element (a_{11}) has ties with the fifth element (a_{25}) of the second row and fourth element (a_{34}) of the third row and third element (a_{34}) of the fourth row and second element (a_{25}) of the fifth row for positive area;
- Element (a_{11}) has ties with the fifth element(a_{25}) of the second row and fourth element (a_{34})of the third row and second element (a_{24}) of the fourth row and third element (a_{35})of the fifth row for negative area.

In Figure 201 is shown the functional ties at Laplace's theorem (method) for partial part $|A_{12}|$ of complex symmetric determinant $|A|$ of 5x5 matrix for positive and negative areas. Analysis of functional ties between elements of 5x5 matrix for partial part $|A_{12}|$ of complex symmetric determinant $|A|$ for positive and negative areas in accordance with Laplace's method shown in Figure 201 indicates on the following results:

For the second element (a_{12}) of the first row

This element has such functional ties for positive and negative areas as :
- Element (a_{12})has ties with the first element(a_{12})of the second row and third element (a_{33})of the third row and fifth element (a_{45}) of the fourth row and fourth element (a_{45}) of the fifth row for positive area;
- Element (a_{12}) has ties with the first element (a_{12}) of the second row and third element (a_{33}) of the third row and fourth element (a_{44}) of the fourth row and fifth element (a_{55}) of the fifth row for negative area;
- Element (a_{12}) has ties with the first element (a_{12}) of the second row and fourth element (a_{34}) of the third row and third element (a_{34}) of the fourth row and fifth element (a_{55}) of the fifth row for positive area;
- Element (a_{12}) has ties with the first element (a_{12}) of the second row and fourth element (a_{34}) of the third row and fifth element (a_{45}) of the fourth row and third element(a_{35}) of the fifth row for negative area;
- Element (a_{12}) has ties with the first element (a_{12}) of the second row and fifth element (a_{35}) of the third row and fourth element (a_{44})of the fourth row and third element (a_{35}) of the fifth row for positive area;
- Element (a_{12})has ties with the first element (a_{12}) of the second row and fifth element (a_{35}) of the third row and third element (a_{34}) of the fourth row and fourth element (a_{45}) of the fifth row for negative area;
- Element (a_{12}) has ties with the third element (a_{23}) of the second row and first element (a_{13}) of the third row and fourth element (a_{44}) of the fourth row and fifth element (a_{55}) of the fifth row for positive area;
- Element(a_{12}) has ties with the third element (a_{23}) of the second row and first element (a_{13})of the third row and fifth element (a_{45}) of the fourth row and fourth element (a_{45}) of the fifth row for negative area;
- Element (a_{12}) has ties with the third element (a_{23}) of the second row and fourth element (a_{34}) of the third row and fifth element (a_{45}) of the fourth row and first element (a_{15}) of the fifth row for positive area;
- Element (a_{12}) has ties with the third element (a_{23}) of the second row and fourth element (a_{34}) of the third row and first element (a_{14}) of the fourth row and fifth element (a_{55}) of the fifth row for negative area;
- Element (a_{12}) has ties with the third element (a_{23})of the second row and fifth element (a_{35}) of the third row and first element (a_{14}) of the fourth row and fourth element (a_{45}) of the fifth row for positive area;
- Element (a_{12}) has ties with third element (a_{23}) of the second row and fifth element (a_{35}) of the third row and fourth element (a_{44}) of the fourth row and first element (a_{15}) of the fifth row for negative area;

- Element (a_{12}) has ties with the fourth element (a_{24}) of the second row and first element (a_{13}) of the third row and fifth element (a_{45}) of the fourth row and third element (a_{35}) of the fifth row for positive area;
- Element(a_{12})has ties with the fourth element (a_{24}) of the second row and first element (a_{13}) of the third row and third element (a_{34}) of the fourth row and fifth element (a_{55}) of the fifth row for negative area;

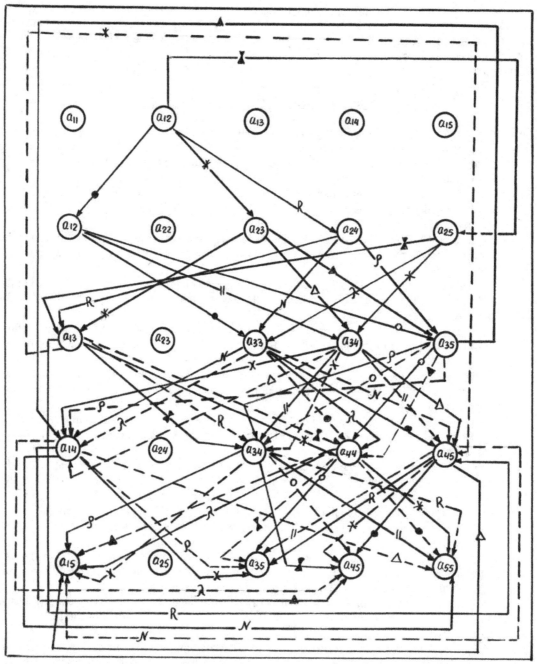

Figure 201   Functional ties for partial part $|A_{12}|$ of complex symmetric determinant at Laplace's method for positive and negative(-------) areas

- Element (a_{12}) has ties with the fourth element (a_{24}) of the second row and third element (a_{33}) of the third row and first element (a_{14}) of the fourth row and fifth element (a_{55}) of the fifth row for positive area;
- Element (a_{12}) has ties with fourth element (a_{24}) of the second row and third element (a_{33}) of the third row and fifth element (a_{45}) of the fourth row and first element (a_{15}) of the fifth row for negative area;
- Element (a_{12}) has ties with fourth element (a_{24}) of the second row and fifth element (a_{35}) of the third row and third element (a_{34}) of the fourth row and first element (a_{15}) of the fifth row for positive area;
- Element (a_{12}) has ties with the fourth element (a_{24}) of the second row and fifth element (a_{35}) of the third row and first element (a_{14}) of the fourth row and third element (a_{35}) of the fifth row for negative area;
- Element (a_{12}) has ties with the fifth element (a_{25}) of the second row and first element (a_{13}) of the third row and third element (a_{34}) of the fourth row and fourth element (a_{45}) of the fifth row for positive area;
- Element (a_{12}) has ties with the fifth element (a_{25}) of the second row and first element (a_{13}) of the third row and fourth element (a_{44}) of the fourth row and third element (a_{35}) of the fifth row for negative area;
- Element (a_{12}) has ties with the fifth element (a_{25}) of the second row and third element (a_{33}) of the third row and fourth element (a_{44}) of the fourth row and first element (a_{15}) of the first row for positive area;
- Element (a_{12}) has ties with the fifth element (a_{25}) of the second row and third element (a_{33}) of the third row and first element (a_{14}) of the fourth row and fourth element (a_{45}) of the fifth row for negative area;
- Element (a_{12}) has ties with the fifth element (a_{25}) of the second row and fourth element (a_{34}) of the third row and first element (a_{14}) of the fourth row and third element (a_{35}) of the fifth row for positive area;
- Element (a_{12}) has ties with the fifth element (a_{25}) of the second row and fourth element (a_{34}) of the third row and third element (a_{34}) of the fourth row and first element (a_{15}) of the fifth row for negative area.

In Figure 202 is shown the functional ties at Laplace's theorem(method) for partial part of $\left| A_{13} \right|$ complex symmetric determinant $\left| A \right|$ of 5x5 matrix for positive and negative areas. Analysis of functional ties between elements of 5x5 matrix for partial part $\left| A_{13} \right|$ of complex symmetric determinant $\left| A \right|$ for positive and negative areas in accordance with Laplace's theorem shown in Figure 202 indicates on the following results:

For the third element (a_{13}) of the first row

This element has such functional ties for positive and negative areas as:
- Element (a_{13}) has ties with the first element (a_{12}) of the second row and second element (a_{23}) of the third row and fourth element (a_{44}) of the fourth row and fifth element (a_{55}) of the fifth row for positive area;
- Element (a_{13}) has ties with the first element (a_{12}) of the second row and second element (a_{23}) of the third row and fifth element(a_{45}) of the fourth row and fourth element (a_{45}) of the fifth row for negative area;
- Element (a_{13}) has ties with the first element (a_{12}) of the second row and fourth element (a_{34}) of the third row and fifth element (a_{45}) of the fourth row and second element (a_{25}) of the fifth row for positive area;
- Element (a_{13}) has ties with the first element (a_{12}) of the second row and fourth element (a_{34}) of the third row and second element (a_{24}) of the fourth row and fifth element (a_{55}) of the fifth row for negative area;

445

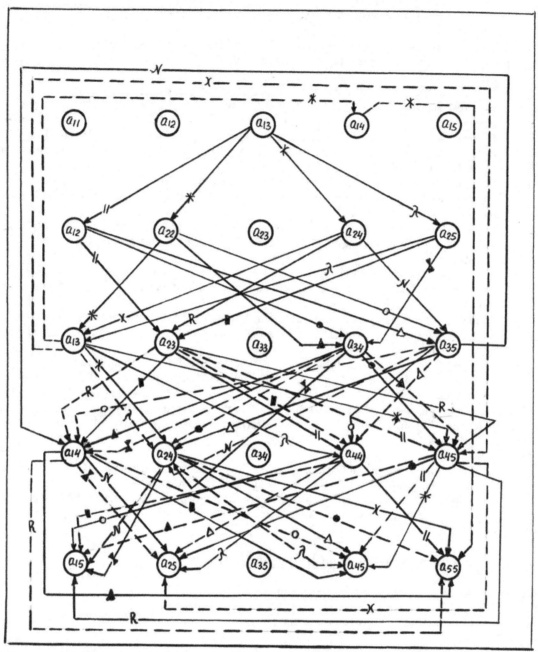

Figure 202 Functional ties for partial part $|A_{13}|$ of complex symmetric determinant at Laplace's method for positive and negative(--------) areas

- Element (a_{13})has ties with the first element (a_{12}) of the second row and fifth element (a_{35}) of the third row and second element (a_{24}) of the fourth row and fourth element (a_{45}) of the fifth row for positive area;
- Element (a_{13}) has ties with the first element (a_{12}) of the second row and fifth element (a_{35}) of the third row and fourth element (a_{44}) of the fourth row and second element (a_{25}) of the fifth row for negative area;

446

- Element (a_{13}) has ties with the second element (a_{22}) of the second row and first element (a_{13}) of the third row and fifth element (a_{45}) of the fourth row and fourth element (a_{45}) of the fifth row for positive area;
- Element (a_{13}) has ties with the second element (a_{22}) of the second row and first element (a_{13}) of the third row and fourth element (a_{14}) of the first row and fifth element (a_{55}) of the fifth row for negative area;
- Element (a_{13}) has ties with the second element (a_{22}) of the the second row and fourth element (a_{34}) of the third row and first element (a_{14}) of the fourth row and fifth element (a_{55}) of the fifth row for positive area;
- Element (a_{13}) has ties with the second element (a_{22}) of the second row and fourth element(a_{34}) of the third row and fifth element (a_{45}) of the fourth row and fifth element (a_{15}) of the fifth row for negative area;
- Element (a_{13}) has ties with the second element (a_{22}) of the second row and fifth element (a_{35}) of the third row and fourth element (a_{44}) of the fourth row and first element (a_{15}) of the fifth row for positive area;
- Element (a_{13}) has ties with the second element (a_{22}) of the second row and fifth element (a_{35}) of the third row and first element (a_{14}) of the fourth row and fourth element (a_{45}) of the fifth row for negative area;
- Element (a_{13}) has ties with the fourth element (a_{24}) of the second row and first element (a_{13}) of the third row and second element (a_{24}) of the fourth row and fifth element (a_{55}) of the fifth row for the positive area;
- Element (a_{13}) has ties with the fourth element (a_{24}) of the second row and first element (a_{13}) of the third row and fifth element (a_{45}) of the fourth row and second element (a_{25}) of the fifth row for negative area;
- Element (a_{13}) has ties with the fourth element (a_{24}) of the second row and second element (a_{23}) of the third row and fifth element (a_{45}) of the fourth row and first element (a_{15}) of the fifth row for positive area;
- Element (a_{13}) has ties with the fourth element (a_{24}) of the second row and second element (a_{23}) of the third row and first element (a_{14}) of the fourth row and fifth element (a_{55}) of the fifth row for negative area;
- Element (a_{13}) has ties with the fourth element (a_{24}) of the second row and fifth element (a_{35}) of the third row and first element (a_{14}) of the fourth row and second element (a_{25}) of the fifth row for positive area;
- Element (a_{13}) has ties with the fourth element (a_{24}) of the second row and fifth element (a_{35}) of the third row and second element (a_{24}) of the fourth row and first element (a_{15}) of the fifth row for negative area;
- Element (a_{13}) has ties with the fifth element (a_{25}) of the second row and first element (a_{13}) of the third row and fourth element (a_{44}) of the fourth row and second element (a_{25}) of the fifth row for positive area;
- Element (a_{13}) has ties with the fifth element (a_{25}) of the second row and first element (a_{13}) of the third row and second element (a_{24}) of the fourth row and fourth element (a_{45}) of the fifth row for negative area;
- Element (a_{13}) has ties with the fifth element (a_{25}) of the second row and second element (a_{23}) of the third row and first element (a_{14}) of the fourth row and fourth element (a_{45}) of the fifth row for positive area;
- Element(a_{13}) has ties with the fifth element (a_{25}) of the second row and second element (a_{23}) of the third row and fourth element (a_{44}) of the fourth row and first element (a_{15}) of the fifth row for negative area;
- Element(a_{13}) has ties with the fifth element (a_{25}) of the second row and fourth element (a_{34}) of the third row and second element (a_{24}) of the fourth row and first element (a_{15}) of the fifth row for positive area;
- Element(a_{13}) has ties with the fifth element (a_{25}) of the second row and fourth element (a_{34}) of the third row and first element (a_{14}) of the fourth row and second element (a_{25}) of the fifth row for negative area.

In Figure 203 is shown the functional ties at Laplace's theorem(method) for partial part of $\left|A_{14}\right|$ for symmetric determinant $\left|A\right|$ of 5x5 matrix for positive and negative areas. Analysis of functional ties between elements of 5x5 matrix for partial part $\left|A_{14}\right|$ of complex symmetric determinant $\left|A\right|$ for positive and negative areas in accordance with Laplace's method shown in Figure 203 indicates on the following results:

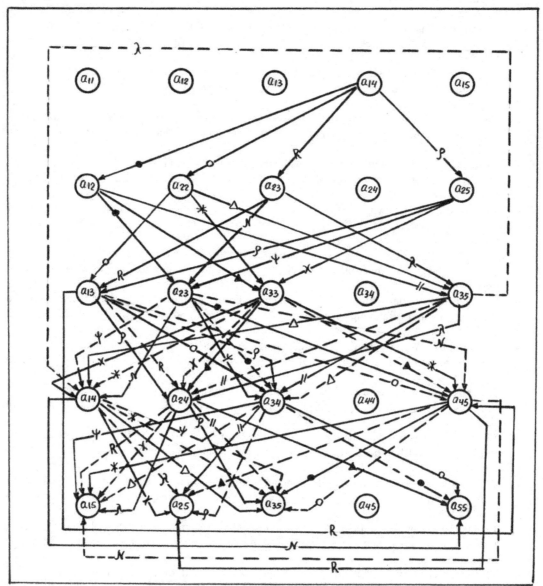

Figure 203   Functional ties   for partial part $\left|A_{14}\right|$ of complex symmetric determinant at Laplace's method for positive and negative (--------)areas

For the fourth element (a₁₄) of the first row

This element has such functional ties for positive and negative areas as:
- Element (a₁₄) has ties with the first element (a₁₂) of the second row and second element (a₂₃) of the third row and fifth element (a₄₅) of the fourth row and third element (a₃₅) of the fifth row for positive area;

448

- Element (a_{14}) has ties with the first element (a_{12}) of the second row and second element (a_{23}) of the third row and third element (a_{34}) of the fourth row and fifth element (a_{55}) of the fifth row for negative area;
- Element (a_{14})has ties with the first element (a_{12}) of the second row and third element (a_{33}) of the third row and second element(a_{24}) of the fourth row and fifth element (a_{55}) of the ffith row for positive area;
- Element (a_{14}) has ties with the first element (a_{12}) of the second row and third element (a_{33}) of the third row and fifth element (a_{45}) of the fourth row and second element (a_{25}) of the ffith row for negative area;
- Element (a_{14}) has ties with the first element (a_{12}) of the second row and fifth element (a_{35}) of the third row and third element (a_{34}) of the fourth row and second element (a_{25}) of the ffith row for positive area;
- Element(a_{14}) has ties with the first element (a_{12}) of the second row and fifth element (a_{35}) of the third row and second element (a_{24}) of the fourth row and third element (a_{35}) of the fifth row for negative area;
- Element(a_{14}) has ties with the second element (a_{22}) of the second row and first element (a_{13}) of the third row and third element (a_{34}) of the fourth row and fifth element (a_{55}) of the fifth row for positive area;
- Element(a_{14}) has ties with the second element (a_{22}) of the second row and first element (a_{13}) of the third row and fifth element (a_{45}) of the fourth row and third element (a_{35}) of the fifth row for negative area;
- Element (a_{14}) has ties with the second element (a_{22}) of the second row and third element (a_{33}) of the third row and fifth element (a_{45}) of the fourth row and first element (a_{15}) of the fifth row for positive area;
- Element(a_{14}) has ties with the second element (a_{22}) of the second row and third element (a_{33}) of the third row and first element (a_{14}) of the fourth row and third element (a_{35}) of the fifth row for negative area;
- Element (a_{14}) has ties with the second element (a_{22}) of the second row and fifth element (a_{35}) of the third row and first element (a_{14}) of the fourth row and third element (a_{35}) of the fifth row for positive area;
- Element(a_{14}) has ties with the second element (a_{22})of the second row and fifth element(a_{35}) of the third row and third element (a_{34})of the fourth row and first element (a_{15}) of the fifth row for negative area;
- Element (a_{14}) has ties with the third element (a_{23}) of the second row and first element (a_{13}) of the third row and fifth element (a_{45}) of the fourth row and second element (a_{25}) of the fifth row for positive area;
- Element (a_{14}) has ties with the third element (a_{23}) of the second row and first element (a_{13}) of the third row and second element (a_{24}) of the fourth row and first element (a_{15}) of the ffith row for negative area;
- Element (a_{14})has ties with the third element (a_{23}) of the second row and second element (a_{23}0 of the third row and first element (a_{14}) of the fourth row and fifth element (a_{15}) of the fifth row for positive area;
- Element(a_{14}) has ties with the third element (a_{23}) of the second row and second element (a_{23}) of the third row and fifth element (a_{45}) of the fourth row and first element (a_{15}) of the fifth row for negative area;
- Element (a_{14}) has ties with the third element (a_{23}) of the second row and fifth element(a_{35}) of the third row and second element (a_{24}) of the fourth row and first element (a_{15}) of the fifth row for positive area;
- Element (a_{14}) has ties with the third element (a_{23}) of the second row and fifth element (a_{35}) of the third row and first element (a_{14}) of the fourth row and second element (a_{25}) of the fifth row for negative area;
- Element (a_{14}) has ties with the fifth element (a_{25}) of the second row and first element (a_{13}) of the third row and second element (a_{24}) of the fourth row and third element (a_{35}) of the fifth row for positive area;

- Element(a_{14}) has ties with the fifth element (a_{25}) of the second row and first element (a_{13}) of the third row and third element (a_{34}) of the fourth row and second element (a_{25}) of the fifth row for negative area;
- Element (a_{14}) has ties with the fifth element (a_{25}) of the second row and second element (a_{23}) of the third row and third element (a_{34}) of the fourth row and first element (a_{15}) of the fifth row for positive area;
- Element(a_{14}) has ties with the fifth element (a_{25}) of the second row and second element (a_{23}) of the third row and first element (a_{14}) of the fourth row and third element (a_{35}) of the fifth row for negative area;
- Element (a_{14}) has ties with the fifth element (a_{25}) of the second row and third element (a_{33}) of the third row and first element (a_{14}) of the fourth row and second element (a_{25}) of the fifth row for positive area;
- Element (a_{14}) has ties with the fifth element (a_{25}) of the second row and third element (a_{33}) of the third row and second element (a_{24}) of the fourth row and first element (a_{15}) of the fifth row for negative area.

In Figure 204 is shown the functional ties at Laplace's theorem(method) for partial part of $\left|A_{15}\right|$ for symmetric determinant $\left|A\right|$ of 5x5 matrix for positive and negative areas. Analysis of functional ties between of elements of 5x 5 matrix for partial part $\left|A_{15}\right|$ of complex symmetric determinant $\left|A\right|$ for positive and negative areas in accordance with Laplace's method shown in Figure 204 indicates on the following results:

For the fifth element (a_{15}) of the first row

This element has such functional ties for positive and negative areas as:
- Element(a_{15}) has ties with the first element (a_{12}) of the second row and second element (a_{23}) of the third row and third element (a_{34}) of the fourth row and fourth element (a_{45}) of the fifth row for positive area;
- Element (a_{15}) has ties with the first element (a_{12}) of the second row and second element (a_{23}) of the third row and fourth element (a_{44}) of the fourth row and third element (a_{35}) of the fith row for negative area;
- Element(a_{15}) has ties with the first element (a_{12}) of the second row and third element (a_{33}) of the third row and fourth element (a_{44}) of the fourth row and second element (a_{25}) of the fifth row for positive area;
- Element (a_{15}) has ties with the first element (a_{12}) of the second row and third element (a_{33}) of the third row and second element (a_{24}) of the fourth row and fourth element (a_{45}) of the fifth row for negative area;
- Element (a_{15}) has ties with the first element (a_{12}) of the second row and fourth element (a_{34}) of the third row and second element (a_{24})of the third row and third element (a_{35}) of the fifth row for positive area;
- Element (a_{15}) has ties with the first element (a_{12}) of the second row and fourth element (a_{34}) of the third row and third element (a_{34}) of the fourth row and second element (a_{25}) of the fifth row for negative area;
- Element (a_{15}) has ties with the second element(a_{22}) of the second row and first element (a_{13}) of the third row and fourth element (a_{44}) of the fourth row and third element (a_{35}) of the fifth row for positive area;
- Element (a_{15}) has ties with the second element (a_{22}) of the second row and first element (a_{13}) of the third row and third element (a_{34}) of the fourth row and fourth element (a_{45}) of the fifth row for negative area;
- Element (a_{15}) has ties with the second element (a_{22}) of the second row and third element (a_{33}) of the third row and first element (a_{14}) of the fourth row and fourth element (a_{45}) of the fifth row for positive area;
- Element (a_{15}) has ties with the second element (a_{22}) of the second row and third element (a_{33}) of the third row and fourth element (a_{44}) of the fourth row and first element (a_{15}) of the fifth row for negative area;

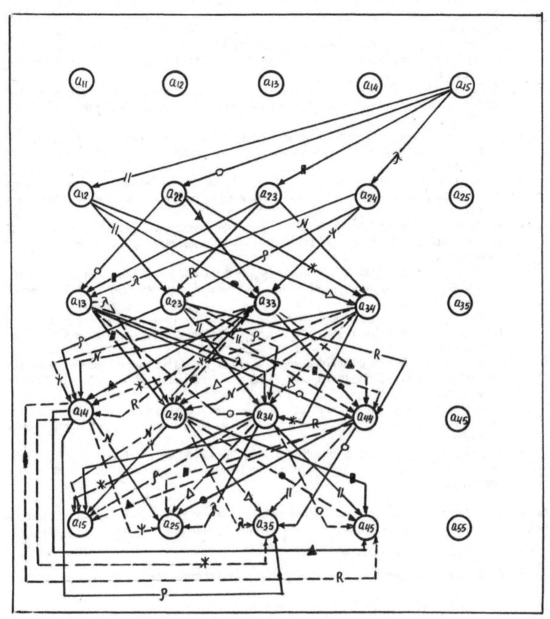

Figure 204 Functional ties for partial part $\left|A_{15}\right|$ of complex symmetric determinant at Laplace's method for positive and negative (-------) areas

- Element (a_{15})has ties with the second element (a_{22}) of the second row and fourth element (a_{34}) of the third row and third element (a_{34}) of the fourth row and first element (a_{15}) of the fifth row for positive area;
- Element (a_{15}) has ties with the second element (a_{22}) of the second row and fourth element (a_{34}) of the third row and first element (a_{14}) of the fourth row and third element (a_{35}) of the fifth row for negative area;
- Element (a_{15}) has ties with the third element (a_{23}) of the second row and fifth element (a_{13}) of the third row and second element (a_{24}) of the fourth row and fourth element (a_{45}) of the fifth row for positive area;

451

- Element (a_{15}) has ties with the third element (a_{23}) of the second row and first element (a_{13}) of the third row and fourth element(a_{44}) of the fourth row and second element (a_{25}) of the fifth row for negative area;
- Element (a_{15}) has ties with the third element (a_{23}) of the second row and second element (a_{23}) of the third row and fourth element (a_{44}) of the fourth row and first element (a_{15}) of the fifth row for positive area;
- Element (a_{15}) has ties with the third element (a_{23}) of the second row and second element (a_{23}) of the third row and first element (a_{14}) of the fourth row and fourth element (a_{45})of the fifth row for negative area;
- Element (a_{15}) has ties with the third element (a_{23}) of the second row and fourth element (a_{34}) of the third row and first element (a_{14}) of the fourth row and second element (a_{25}) of the fifth row for negative area;
- Element (a_{15}) has ties with the third element (a_{23}) of the second row and fourth element (a_{34}) of the third row and second element (a_{24}) of the fourth row and first element (a_{15}) of the fifth row for negative area;
- Element (a_{15}) has ties with the fourth element (a_{24}) of the second row and first element (a_{13}) of the third row and third element (a_{34}) of the fourth row and second element (a_{25}) of the fifth row for positive area;
- Element(a_{15}) has ties with the fourth element (a_{24}) of the second row and first element (a_{13}) of the third row and second element (a_{24}) of the fourth row and third element (a_{35}) of the fifth row for negative area;
- Element (a_{15}) has ties with the fourth element (a_{24}) of the second row and second element (a_{23}) of the third row and first element (a_{14}) of the fourth row and third element (a_{35}) of the fifth row for positive area;
- Element (a_{15}) has ties with the fourth element (a_{24}) of the second row and second element (a_{23}) of the third row and third element (a_{34}) of the fourth row and first element (a_{15}) of the fifth row for negative area;
- Element (a_{15}) has ties with the fourth element (a_{24}) of the second row and third element (a_{33}) of the third row and second element (a_{24}) of the fourth row and first element (a_{15}) of the fifth row for positive area;
- Element (a_{15}) has ties with the fourth element (a_{24}) of the second row and third element (a_{33}) of the third row and first element(a_{14}) of the fourth row and second element (a_{25}) of the ffith row for negative area.

In Appendix 33 is shown the evaluation of symmetric determinant of 5x 5 matrix with using of Laplace's method and also the solving of system linear equations with application of Cramer's rule.

2 Evaluation of some matrices for oblique(skew)determinants

2.1 For 3x3 matrix

In general view we have the oblique(skew) determinant of 3x3 matrix ,i.e when all of its elements must submit to the conditions of $a_{ij=-aji}$ [2] :

$$A= \begin{vmatrix} a_{11} & a_{12} & a_{13} \\ -a_{12} & a_{22} & a_{23} \\ -a_{13} & -a_{23} & a_{33} \end{vmatrix} \quad (\blacktriangle)$$

452

And evaluation of this determinant can be defined by the following methods:

❖ Sarrus rule

In total view we have the following results for oblique(skew)determinant $|A|$ of above-indicated 3x3 matrix (▲) :

$$|A| = (a_{11}a_{22}a_{33} - a_{12}a_{23}a_{13} + a_{13}a_{12}a_{23} + a_{13}a_{13}a_{22} + a_{11}a_{23}a_{23} + a_{12}a_{12}a_{33});$$

where,
$$|A| = a_{11}(a_{22}a_{33} + a^2_{23}) + a^2_{12}a_{33} + a^2_{13}a_{22} \qquad (124)$$

In Appendix 34 is shown evaluation of the oblique (skew)determinant of 3x3 matrix, for a instance, by Sarrus rule with using of formula (124) and also solving of system linear equations with using od Cramer's rule.

❖ Expansion(cofactor) method for oblique(skew) determinant of 3x3 matrix

For above-indicated 3x3 matrix(▲) ,we can take any row or column that to evaluate the matrix oblique(skew) determinant.
So, expanding the oblique(skew) determinant of 3x3 matrix to elements ,for a instance, to the second row ,we have the following results:

$$|A| = -a_{12}A_{12} + a_{22}A_{22} + a_{23}A_{23}A_{23}$$

where,
$(-a_{12})$ –the first element of the second row;
(a_{22}) –the second element of the second row;
(a_{23}) -the third element of the second row;
A_{12}-algebraic supplemental of the second row and first column equal:
$$A_{12} = -(a_{12}a_{33} + a_{13}a_{23});$$
A_{22}-algebraic supplemental of the second row and second column equal:
$$A_{22} = (a_{11}a_{33} + a^2_{13})$$
A_{23}-algebraic supplemental of the second row and third column equal:
$$A_{23} = (a_{11}a_{23} - a_{12}a_{13})$$
And then the oblique(skew)determinant in accordance *with Expansion method* is equal the same of formula (124),i.e we have :

$$|A| = a_{11}(a_{22}a_{33} + a^2_{23}) + a^2_{12}a_{33} + a^2_{13}a_{22}$$

❖ Laplace's theorem(method)

In above-considered 3x3 matrix(▲) ,we mark out randomly two columns, for a instance ,the first and second columns for oblique(skew determinant ,writing the minors (M_{ij}) of these elements in two considered columns. And then define the corresponding algebraic supplemental of these minors(A_{ij}) with writing the value of oblique(skew) determinant:

$$|A| = M_1A_1 + M_2A_2 + M_3A_3$$

where,
$M_1 = (a_{11}a_{22} + a_{12}a_{12})$; $A_1 = a_{33}$;

$M_2 = (a_{12}a_{13} - a_{11}a_{23})$; $A_2 = -a_{23}$;

$M_3 = (a_{12}a_{23} + a_{22}a_{13})$; $A_3 = a_{13}$.
So, in accordance with Laplace's theorem ,we have the following results for oblique(skew) determinant of 3x3 matrix:

$$|A| = a_{11}(a_{22}a_{33} + a^2_{23}) + a^2_{12}a_{33} + a^2_{13}a_{22}$$

453

Analysis of formula (124) shows that it has the same value for calculation of oblique(skew) determinant of 3x3 matrix and can be fitted for any considered version ,such as Sarrus rule, Expansion and Laplace's methods.

2.2 For 4x4 matrix

In general view ,we have the oblique (skew) determinant of 4x4 matrix ,i.e when all of its elements must submit to the conditions of $a_{ij}= -a_{ji}$:

$$A=\begin{vmatrix} a_{11} & a_{12} & a_{13} & a_{14} \\ a_{12}' & a_{22} & a_{23} & a_{24} \\ a'_{13} & a_{23}' & a_{33} & a_{34} \\ a'_{14} & a'_{24} & a_{34}' & a_{44} \end{vmatrix} \qquad A=\begin{vmatrix} 6 & -2 & -1 & -7 \\ 2 & 3 & -8 & 10 \\ 1 & 8 & -3 & 1 \\ 7 & -10 & -1 & 5 \end{vmatrix}$$

a) In general view (•)   b) In numerical data

and of its evaluation can be defined by the following methods:

❖ Expansion (cofactor)method

Referring to above-indicated 4x4 matrix in general view ,we have the following results ,for a instance ,when expansion method use at conditions for elements of the first row of this oblique(skew) determinant $|A|$, i.e we have the following formula:

$$|A|=a_{11}A_{11}+a_{12}A_{12}+a_{13}A_{13}+a_{14}A_{14}$$

where,
$a_{11};a_{12};a_{13};a_{14}$-elements of the first row for 4x4 matrix

A_{11}-algebraic supplemental of the first row and first column equal:
$$A_{11}=a_{22}(a_{33}a_{44}-a_{34}a'_{34})+a_{23}(a_{34}a'_{24}-a'_{23}a_{44})+a_{24}(a'_{23}a'_{34}-a_{33}a'_{24})$$

A_{12}-algebraic supplemental of the first row and second column equal:
$$A_{12}=a_{23}(a'_{13}a_{44}-a_{34}a'_{14})+a_{33}(a_{24}a'_{14}-a'_{12}a_{44})+a'_{34}(a'_{12}a_{34}-a_{24}a'_{13})$$

A_{13}-algebraic supplemental of the first row and third column equal:
$$A_{13}=a'_{12}(a'_{23}a_{44}-a_{34}a'_{24})+a'_{13}(a_{24}a'_{24}-a_{22}a_{44})+a'_{14}(a_{22}a_{34}-a_{24}a'_{23})$$

A_{14}-algebraic supplemental of the first row and fourth column equal:
$$A_{14}=a'_{12}(a_{33}a'_{24}-a'_{23}a'_{34})+a'_{13}(a_{22}a'_{34}-a_{23}a'_{24})+a'_{14}(a_{23}a'_{23}-a_{22}a_{33})$$

And then the oblique(skew) determinant of 4x4 matrix in general view *for Expansion method* is equal:

$$\begin{aligned}|A|=&a_{11}[a_{22}(a_{33}a_{44}-a_{34}a'_{34})+a_{23}(a_{34}a'_{24}-a'_{23}a_{44})+a_{24}(a'_{23}a'_{34}-a_{33}a'_{24})]+\\ &+a_{12}[a_{23}(a'_{13}a_{44}-a_{34}a'_{14})+a_{33}(a_{24}a'_{14}-a'_{12}a_{44})+a'_{34}(a'_{12}a_{34}-a_{24}a'_{13})]+\\ &+a_{13}[a'_{12}(a'_{23}a_{44}-a_{34}a'_{24})+a'_{13}(a_{24}a'_{24}-a_{22}a_{44})+a'_{14}(a_{22}a_{34}-a_{24}a'_{23})]+\\ &+a_{14}[a'_{12}(a_{33}a'_{24}-a'_{23}a'_{34})+a'_{13}(a_{22}a'_{34}-a_{23}a'_{24})+a'_{14}(a_{23}a'_{23}-a_{22}a_{33})]\quad(125)\end{aligned}$$

In Appendix 35 is shown the evaluation of oblique(skew) determinant with using of Expansion method and also the solving of system linear equations with using of Cramer's rule.

454

In Figure 205 is shown the functional ties between elements of oblique(skew) determinant for 4x4 matrix at Expansion method for positive and negative areas.

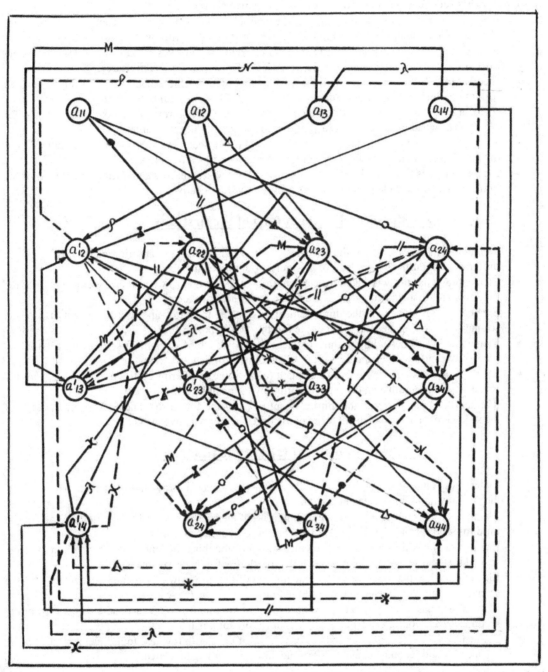

Figure 205 Functional ties between elements of oblique(skew) determinant for 4x4 matrix at Expansion method for positive and negative(--------) areas

Analysis of functional ties between of elements 4x4 matrix for oblique(skew) determinant shown in Figure 205 in accordance with Expansion method indicates on the following results:

455

For the first element (a_{11}) of the first row

This element has such functional ties for positive and negative areas as:

- Element (a_{11}) has connection with the second element (a_{22}) of the second row and third element(a_{33}) of the third row and fourth element (a_{44}) of the fourth row for positive area;
- Element (a_{11} has connections with the second element (a_{22}) of the second row and fourth element (a_{34}) of the third row and third element (a'_{34}) of the fourth row for negative area;
- Element (a_{11}) has connections with the third element (a_{23}) of the second row and fourth element (a_{34}) of the third row and second element (a'_{24}) of the fourth row for positive area;
- Element (a_{11}) has ties with the third element (a_{23}) of the second row and second element (a'_{23}) of the third row and fourth element (a_{44}) of the fourth row for negative area;
- Element (a_{11}) has ties with the fourth element (a_{24}) of the second row and second element (a'_{23}) of the third row and third element (a'_{34}) of the fourth row for positive area;
- Element(a_{11}) has ties with the fourth element (a_{24}) of the second row and third element (a_{33}) of the third row and second element (a'_{24}) of the fourth row for negative area.

For the second element (a_{12}) of the first row

- Element(a_{12})has ties with the third element (a_{23}) of the second row and first element (a'_{13}) of the third row and fourth element (a_{44}) of the fourth row for positive area;
- Element (a_{12}) has ties with the third element (a_{23}) of the second row and fourth element (a_{34}) of the third row and first element (a'_{14}) of the fourth ro for negative area;
- Element (a_{12}) has ties with the third element (a_{33}) of the third row and fourth element (a_{24}) of the second row and first element (a'_{14}) of the fourth row for positive area;
- Element (a_{12}) has ties with the third element (a_{33}) of the third row and first element (a'_{12}) of the second row and fourth element (a_{44}) of the fourth row for negative area;
- Element (a_{12}) has ties with the third element (a'_{34}) of the fourth row and first element (a'_{12})of the second row and fourth element (a_{34}) of third row for positive area;
- Element (a_{12}) has ties with the third element (a'_{34}) of the fourth row and fourth element (a_{24}) of the second row and first element (a'_{13}) of the third row for negative areas;

For the third element (a_{13}) of the first row

- Element (a_{13}) has ties with the first element (a'_{12}) of the second row and second element (a'_{23}) of the third row and fourth element (a_{44}) of the positive area;
- Element (a_{13}) has ties with the first element (a'_{12}) of the second row and fourth element (a_{34}) of the third row and second element (a'_{24}) of the fourth row for negative area;
- Element (a_{13}) has ties with the first element (a'_{13}) of the third row and fourth element (a_{24}) of the second row and second element (a'_{24}) of the fourth row for positive area;
- Element(a_{13}) has ties with the first element (a'_{13}) of the third row and second element (a_{22}) of the second row and fourth element (a_{44}) of the fourth row for negative area;
- Element(a_{13}) has ties with the first element (a'_{14}) of the fourth row and second element (a_{22}) of the second row and fourth element (a_{34}) of the third row for positive area;
- Element (a_{13}) has ties with the first element (a'_{14}) of the fourth row and fourth element (a_{24}) of the second row and second element (a'_{23}) of the third row for negative area.

For the fourth element (a_{14}) of the first row

- Element (a_{14}) has ties with the first element (a'_{12}) of the second row and third element (a_{33}) of the third row and second element (a'_{24}) of the fourth row for positive area;
- Element (a_{14}) has ties with the first element (a'_{12}) of the second row and second element (a'_{23}) of the third row and third element (a'_{34}) of the fourth row for negative area;

456

- Element (a_{14}) has ties with the first element (a'_{13}) of the third row and second element (a_{22}) of the second row and third element (a'_{34}) of the fourth row for positive area;
- Element (a_{14}) has ties with the first element (a'_{13}) of the third row and third element (a_{23}) of the second row and second element (a'_{24}) of the fourth row for negative area;
- Element (a_{14}) has ties with the first element (a'_{14}) of the fourth row and third element (a_{23}) of the second row and second element (a'_{23}) of the third row for positive area;
- Element (a_{14}) has ties with the first element (a'_{14}) of the fourth row and second element (a_{22}) of the second row and third element (a_{33}) of the third row for negative area.

❖ Laplace's theorem(method)

In above-considered 4x4 matrix (●) for general view ,we mark out randomly two rows, for a instance ,the first and second columns for oblique(skew) determinant $|A|$,writing the minors(M_{ij}) of these elements in two considered columns. And then define the corresponding algebraic supplemental of these minors(A_{ij}), writing the formula for evaluation of oblique(skew) determinant in view of :

$$|A|=M_1A_1+M_2A_2+M_3A_3+M_4A_4+M_5A_5+M_6A_6$$

where,

$M_1=(a_{11}a_{22}-a_{12}a'_{12})$; $A_1=(a_{33}a_{44}-a_{34}a'_{34})$;

$M_2=(a_{11}a_{23}-a_{13}a'_{12})$; $A_2=-(a'_{23}a_{44}-a_{34}a'_{24})$;

$M_3=(a_{11}a_{24}-a_{14}a'_{12})$; $A_3=(a'_{23}a'_{34}-a_{33}a'_{24})$;

$M_4=(a_{12}a_{23}-a_{13}a_{22})$; $A_4=(a'_{13}a_{44}-a_{34}a'_{14})$;

$M_5=(a_{12}a_{24}-a_{14}a_{22})$; $A_5=-(a'_{13}a'_{34}-a_{33}a'_{14})$;

$M_6=(a_{13}a_{24}-a_{14}a_{23})$; $A_6=(a'_{13}a'_{24}-a'_{23}a'_{14})$.

So, we can conclude that oblique(skew) determinant $|A|$ of the 4x4 matrix in accordance with Laplace's theorem is equal:

$$|A|=a_{11}[a_{22}(a_{33}a_{44}-a_{34}a'_{34})+a_{23}(a_{34}a'_{24}-a'_{23}a_{44})+a_{24}(a'_{23}a'_{34}-a_{33}a'_{24})]+$$
$$+a_{12}[a_{23}(a'_{13}a_{44}-a_{34}a'_{14})+a_{24}(a_{33}a'_{14}-a'_{13}a'_{34})+a'_{12}(a_{34}a'_{34}-a_{33}a_{34})]+ \quad (*)$$
$$+a_{13}[a'_{12}(a'_{23}a_{44}-a_{34}a'_{24})+a'_{13}(a_{24}a'_{24}-a_{22}a_{44})+a'_{14}(a_{22}a_{34}-a_{24}a'_{23})]+$$
$$+a_{14}[a'_{12}(a_{33}a'_{24}-a'_{23}a'_{34})+a'_{13}(a_{22}a'_{34}-a_{23}a'_{24})+a'_{14}(a_{23}a'_{23}-a_{22}a_{33})] \quad (126)$$

where(*) modification is equal,

$[a_{24}(a_{33}a'_{14}-a'_{13}a'_{34})+a'_{12}(a_{34}a'_{34}-a_{33}a_{34})]=a_{33}(a_{24}a'_{14}-a'_{12}a_{44})+a'_{34}(a'_{12}a_{34}-a_{24}a'_{13})$

and this meant that formulas (125) and (126) have the similarity.

In Appendix 36 is shown the evaluation of the oblique(skew) determinant of 4x4 matrix with using of Laplace's theorem(method) and also the solving of system linear equations with using of Cramer's rule.

In Figure 206 is shown the functional ties at Laplace's method for oblique (skew) determinant of 4x4 matrix for positive and negative areas.

Analysis of functional ties between of elements of 4x4 matrix for the oblique(skew) determinant shown in Figure 206 in accordance with Laplace's method indicates on the following results:

457

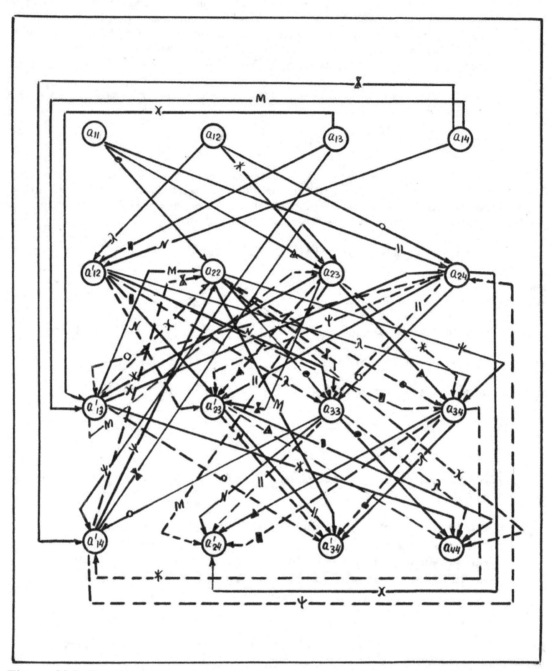

Figure 206 Functional ties between elements of the oblique(skew) determinant at Laplace's method for positive and negative (--------) areas

For the first element (a₁₁) of the first row

This element has such functional ties for positive and negative areas as:
- Element (a₁₁) has ties with the second element (a₂₂) of the second row and third element (a₃₃) of the third row and fourth element (a₄₄) of the fourth row for positive area;

458

- Element(a_{11}) has ties with the second element (a_{22}) of the second row and fourth element (a_{34}) of the third row and third element (a'_{34}) of the fourth row for negative area;
- Element (a_{11}) has ties with the third element (a_{23}) of the second row and fourth element (a_{34}) of the third row and second element (a'_{24}) of the fourth row for positive area;
- Element(a_{11})has ties with the third element (a_{23}) of the second row and second element (a'_{23}) of the third row and fourth element (a_{44}) of the fourth row for negative area;
- Element(a_{11}) has ties with the fourth element (a_{24}) of the second row and second element (a'_{23}) of the third row and third element (a'_{34}) of the fourth row for positive area;
- Element (a_{11}) has ties with the fourth element (a_{24}) of the second row and third element (a_{33}) of the third row and second element (a'_{24}) of the fourth row for negative area.

For the second element (a_{12}) of the first row

- Element(a_{12}) has ties with the third element(a_{23})of the second row and first element (a'_{13}) of the third row and fourth element(a_{44}) of the fourth row for positive area;
- Element (a_{12}) has ties with the third element (a_{23}) of the second row and fourth element(a_{34}) of the third row and first element (a'_{14}) of the fourth row for negative area;
- Element (a_{12}) has ties with the fourth element (a_{24}) of the second row and third element (a_{33}) of the third row and first element (a'_{14}) of the fourth row for positive area;
- Element(a_{12}) has ties with the fourth element (a_{24}) of the second row and first element (a'_{13}) of the third row and third element (a'_{34}) of the fourth row for negative area;
- Element (a_{12}) has ties with the first element (a'_{12}) of the second row and fourth element (a_{34}) of the third row and third element (a'_{34}) of the fourth row for positive area;
- Element(a_{12}) has ties with the first element (a'_{12}) of the second row and third element (a_{33}) of the third row and fourth element (a_{44}) of the fourth row for negative area.

For the third element (a_{13}) of the first row

- Element(a_{13}) has ties with the first element (a'_{12}) of the second row and second element (a'_{23}) of the third row and fourth element (a_{44}) of the fourth row for positive area;
- Element (a_{13}) has ties with the first element (a'_{12}) of the second row and fourth element (a_{34}) of the third row and second element (a'_{24}) of the fourth row for negative area;
- Element(a_{13}) has ties with the first element (a'_{13}) of the third row and fourth element (a_{24}) of the second row and second element (a'_{24}) of the fourth row for positive area;
- Element (a_{13}) has ties with the first element (a'_{13}) of the third row and second element (a_{22}) of the second row and fourth element (a_{44}) of the fourth row for negative area;
- Element (a_{13}) has ties with the first element (a'_{14}) of the fourth row and second element (a_{22}) of the second row and fourth element (a_{34}) of the third row for positive area;
- Element (a_{13}) has ties with the first element (a'_{14}) of the fourth row and fourth element (a_{24}) of the second row and second element (a'_{23}) of the third row for negative area.

For the fourth element (a_{14}) of the first row

- Element (a_{14}) has ties with the first element (a'_{12})of the second row and third element (a_{33}) of the third row and second element(a'_{24}) of the fourth row for positive area;
- Element (a_{14}) has ties with the first element(a'_{12}) of the second row and second element (a'_{23}) of the third row and third element (a'_{34}) of the fourth row for negative area;
- Element (a_{14}) has ties with the first element (a'_{13}) of the third row and second element (a_{22}) of the second row and third element (a'_{34}) of the fourth row for positive area;
- Element (a_{14}) has ties with the first element (a'_{13}) of the third row and third element (a_{23}) of the second row and second element (a'_{24}) of the fourth row for negative area;

- Element (a_{14}) has ties with the first element (a'_{14}) of the fourth row and third element (a_{23}) of the second row and second element (a'_{23}) of the third row for positive area;
- Element(a_{14}) has ties with the first element (a'_{14}) of the fourth row and second element (a_{22}) of the second row and third element (a_{33}) of the third row for negative area.

2.3 For 5x5 matrix

In general view ,we have the oblique (skew) determinant of 5x5 matrix(*):

$$A=\begin{vmatrix} a_{11} & a_{12} & a_{13} & a_{14} & a_{15} \\ a_{12}' & a_{22} & a_{23} & a_{24} & a_{25} \\ a_{13}' & a_{23}' & a_{33} & a_{34} & a_{35} \\ a_{14}' & a_{24}' & a_{34}' & a_{44} & a_{45} \\ a_{15}' & a_{25}' & a_{35}' & a_{45}' & a_{55} \end{vmatrix} \qquad A=\begin{vmatrix} 5 & -2 & 1 & 4 & 6 \\ 2 & 3 & -5 & 6 & -7 \\ -1 & 5 & -2 & 1 & 2 \\ -4 & -6 & -1 & -1 & -2 \\ -6 & 7 & -2 & 2 & 1 \end{vmatrix}$$

a) In general view (*)   b) In numerical data

The oblique(skew) determinant can be defined ,for a instance ,by the Expansion and Laplace's methods.

❖ <u>Expansion method for oblique (skew) determinant of 5x5 matrix</u>

Referring to 5x5 matrix(*),we have the following results ,for a instance ,when Expansion method use at conditions for elements of the first row of this oblique(skew) determinant $|A|$:

$$|A| = a_{11}A_{11} - a_{12}A_{12} + a_{13}A_{13} - a_{14}A_{14} + a_{15}A_{15}$$

where,

$a_{11}, a_{12}, a_{13}, a_{14}, a_{15}$ —elements of the first row for 5x5 matrix

A_{11}-algebraic supplemental of the first row and first column
A_{12}-algebraic supplemental of the first row and second column
A_{13}-algebraic supplemental of the first row and third column
A_{14}-algebraic supplemental of the first row and fourth column
A_{15}-algebraic supplemental of the first row and fifth column.

So, we have the following formula for evaluation of the oblique(skew) determinant of 5x5 matrix:

$$|A| = a_{11}a_{22}[a_{33}(a_{44}a_{55} - a_{45}a_{45}') + a_{34}(a_{45}a_{35}' - a_{34}'a_{55}) + a_{35}(a_{34}'a_{45}' - a_{44}a_{35}')] +$$
$$+ a_{11}a_{23}[a_{23}'(a_{45}a_{45}' - a_{44}a_{55}) + a_{34}(a_{24}'a_{55} - a_{45}a_{25}') + a_{35}(a_{44}a_{25}' - a_{24}'a_{45}')] +$$
$$+ a_{11}a_{24}[a_{23}'(a_{34}'a_{55} - a_{45}a_{35}') + a_{33}(a_{45}a_{25}' - a_{24}'a_{55}) + a_{35}(a_{24}'a_{35}' - a_{34}'a_{25}')] +$$
$$+ a_{11}a_{25}[a_{23}'(a_{44}a_{35}' - a_{34}'a_{45}') + a_{33}(a_{24}'a_{45}' - a_{44}a_{25}') + a_{34}(a_{34}'a_{25}' - a_{24}'a_{35}')] +$$

$$+ a_{12}a_{12}'[a_{33}(a_{45}a_{45}' - a_{44}a_{55}) + a_{34}(a_{34}'a_{55} - a_{45}a_{35}') + a_{35}(a_{44}a_{35}' - a_{34}'a_{45}')] +$$
$$+ a_{12}a_{23}[a_{13}'(a_{44}a_{55} - a_{45}a_{45}') + a_{34}(a_{45}a_{15}' - a_{14}'a_{55}) + a_{35}(a_{14}'a_{45}' - a_{44}a_{15}')] +$$
$$+ a_{12}a_{24}[a_{13}'(a_{45}a_{35}' - a_{34}'a_{55}) + a_{33}(a_{14}'a_{55} - a_{45}a_{15}') + a_{35}(a_{34}'a_{15}' - a_{14}'a_{35}')] +$$
$$+ a_{12}a_{25}[a_{13}'(a_{34}'a_{45}' - a_{44}a_{35}') + a_{33}(a_{44}a_{15}' - a_{14}'a_{45}') + a_{34}(a_{14}'a_{35}' - a_{34}'a_{15}')] +$$

$+a_{13}a_{12}'[a_{23}'(a_{44}a_{55}-a_{45}a_{45}')+a_{34}(a_{45}a_{25}'-a_{24}'a_{55})+a_{35}(a_{24}'a_{45}'-a_{44}a_{25}')]+$
$+a_{13}a_{22}[a_{13}'(a_{45}a_{45}'-a_{44}a_{55})+a_{34}(a_{14}'a_{55}-a_{45}a_{15}')+a_{35}(a_{44}a_{15}'-a_{14}'a_{45})]+$
$+a_{13}a_{24}[a_{13}'(a_{24}'a_{55}-a_{45}a_{25}')+a_{23}'(a_{45}a_{15}'-a_{14}'a_{55})+a_{35}(a_{14}'a_{25}'-a_{24}'a_{15}')]+$
$+a_{13}a_{25}[a_{13}'(a_{44}a_{25}'-a_{24}'a_{45})+a_{23}'(a_{14}'a_{45}-a_{44}a_{15}')+a_{34}(a_{24}'a_{15}'-a_{14}'a_{25}')]+$

$+a_{14}a_{12}'[a_{23}'(a_{45}a_{35}'-a_{34}'a_{55})+a_{33}(a_{24}'a_{55}-a_{45}a_{25}')+a_{35}(a_{34}'a_{25}'-a_{24}'a_{35}')]+$
$+a_{14}a_{22}[a_{13}'(a_{34}'a_{55}-a_{45}a_{35}')+a_{33}(a_{45}a_{15}'-a_{14}'a_{55})+a_{35}(a_{14}'a_{35}'-a_{34}'a_{15}')]+$
$+a_{14}a_{23}[a_{13}'(a_{45}a_{25}'-a_{24}'a_{55})+a_{23}'(a_{14}'a_{55}-a_{45}a_{15}')+a_{35}(a_{24}'a_{15}'-a_{14}'a_{25}')]+$
$+a_{14}a_{25}[a_{13}'(a_{24}'a_{35}'-a_{34}'a_{25}')+a_{23}'(a_{34}'a_{15}'-a_{14}'a_{35}')+a_{33}(a_{14}'a_{25}'-a_{24}'a_{15}')]+$

$+a_{15}a_{12}'[a_{23}'(a_{34}'a_{45}'-a_{44}a_{35}')+a_{33}(a_{44}a_{25}'-a_{24}'a_{45}')+a_{34}(a_{24}'a_{35}'-a_{34}'a_{25}')]+$
$+a_{15}a_{22}[a_{13}'(a_{44}a_{35}'-a_{34}'a_{45})+a_{33}(a_{14}'a_{45}-a_{44}a_{15}')+a_{34}(a_{34}'a_{15}'-a_{14}'a_{35}')]+$
$+a_{15}a_{23}[a_{13}'(a_{24}'a_{45}'-a_{44}a_{25}')+a_{23}'(a_{44}a_{15}'-a_{14}'a_{45})+a_{34}(a_{14}'a_{25}'-a_{24}'a_{15}')]+$
$+a_{15}a_{24}[a_{13}'(a_{34}'a_{25}'-a_{24}'a_{35}')+a_{23}'(a_{14}'a_{35}'-a_{34}'a_{15}')+a_{33}(a_{24}'a_{15}'-a_{14}'a_{25}')]$

$$(127)$$

$$A_{11}=(-1)^{1+1}\begin{vmatrix} a_{22} & a_{23} & a_{24} & a_{25} \\ a_{23}' & a_{33} & a_{34} & a_{35} \\ a_{24} & a_{34} & a_{44} & a_{45} \\ a_{25} & a_{35} & a_{45} & a_{55} \end{vmatrix} \qquad A_{12}=(-1)^{1+2}\begin{vmatrix} a_{12}' & a_{23} & a_{24} & a_{25} \\ a_{13}' & a_{33} & a_{34} & a_{35} \\ a_{14}' & a_{34} & a_{44} & a_{45} \\ a_{15} & a_{35} & a_{45} & a_{55} \end{vmatrix}$$

$$A_{13}=(-1)^{1+3}\begin{vmatrix} a_{12}' & a_{22} & a_{24} & a_{25} \\ a_{13}' & a_{23} & a_{34} & a_{35} \\ a_{14}' & a_{24} & a_{44} & a_{45} \\ a_{15} & a_{25} & a_{45} & a_{55} \end{vmatrix} \qquad A_{14}=(-1)^{1+4}\begin{vmatrix} a_{12}' & a_{22} & a_{23} & a_{25} \\ a_{13}' & a_{23}' & a_{33} & a_{35} \\ a_{14}' & a_{24} & a_{34}' & a_{45} \\ a_{15} & a_{25} & a_{35} & a_{55} \end{vmatrix}$$

$$A_{15}=(-1)^{1+5}\begin{vmatrix} a_{12}' & a_{22} & a_{23} & a_{24} \\ a_{13}' & a_{23}' & a_{33} & a_{34} \\ a_{14}' & a_{24} & a_{34}' & a_{44} \\ a_{15} & a_{25} & a_{35} & a_{45} \end{vmatrix}$$

In Appendices 37 and 38 are shown the evaluation of the oblique(skew)determinant of 5x5 matrix with using of Expansion method and also solving of system linear equations with using of Cramer's rule.

In Figures 207 to 211 are shown the functional ties at Expansion method for oblique(skew) determinant of 5x5 matrix for positive and negative areas.

Analysis of functional ties between of elements of 5x5 matrix for the complex oblique(skew) determinant shown in partial part $|A_{11}|$ of Figure 207 at Expansion method indicates on the following results:

For the first partial part $|A_{11}|$ of element (a_{11})

This element has such functional ties for positive and negative areas as:
- Element (a_{11}) has ties with the second element (a_{22}) of the second row and third element (a_{33})of the third row and fourth element (a_{44}) of the fourth row and fifth row (a_{55}) of the fifth row for positive area;

461

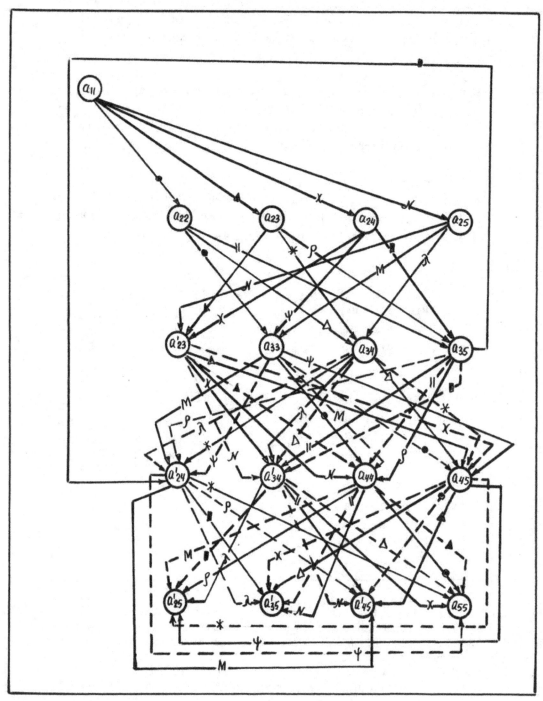

Figure 207 Functional ties for partial part $|A_{11}|$ of complex oblique(skew) determinant at Expansion method for positive and negative (------) areas

- Element (a_{11}) has ties with the second element (a_{22}) of the second row and third element (a_{33}) of the third row and fifth element(a_{45}) of the fourth row and fourth element (a_{45}') of the fifth row for negative area;

- Element(a_{11}) has ties with the second element (a_{22}) of the second row and fourth element (a_{34}) of the third row and fifth element (a_{45}) of the fourth row and third element (a_{35}') of the fifth row for positive area;
- Element (a_{11}) has ties with the second element (a_{22}) of the second row and fourth element (a_{34}) of the third row and third element (a_{34}') of the fourth row and fifth element (a_{55}) of the fifth row for negative area;
- Element (a_{11}) has ties with the second element (a_{22}) of the second row and fifth element (a_{35}) of the third row and third row and third row (a_{34}') of the fourth row and fourth element (a_{45}')of the fifth row for positive area;
- Element (a_{11}) has ties with the second element (a_{22}) of the second row and fifth element (a_{35}) of the third row and fourth element (a_{44}) of the fourth row and third element (a_{35}') of the fifth row for negative area;
- Element (a_{11}) has ties with the third element (a_{23}) of the second roe and second element(a_{23}') of the third row and fifth element (a_{45}) of the fourth row and fourth element (a_{45}') of the fifth row for positive area;
- Element (a_{11}) has ties with the third element (a_{23}) of the second row and second element (a_{23}') of the third row and fourth element (a_{44}) of the fourth row and fifth element (a_{55}) of the fifth row for negative area;
- Element(a_{11}) has ties with the third element (a_{23}) of the second row and fourth element (a_{34}) of the third row and second element (a_{24}') of the fourth row and fifth element (a_{55}) of the fifth row for positive area;
- Element (a_{11}) has ties with the third element (a_{23}) of the second row and fourth element (a_{34}) of the third row and fifth element (a_{45}) of the fourth row and second element (a_{25}') of the fifth row for negative area;
- Element(a_{11})has ties with the third element (a_{23}) of the second row and fifth element (a_{35}) of the third row and fourth element (a_{44}) of the fourth row and second element (a_{25}') of the fifth row for positive area;
- Element (a_{11}) has ties with the third element (a_{23}) of the second row and fifth element (a_{35}) of the third row and second element (a_{24}') of the fourth row and fourth element (a_{45}') of the fifth row for negative area;
- Element (a_{11}) has ties with the fourth element (a_{24}) of the second row and second element (a_{23}') of the third row and third element (a_{34}')of the fourth row and fifth element(a_{55}) of the fifth row for positive area;
- Element(a_{11}) has ties with the fourth element (a_{24}) of the second row and second element(a_{23}') of the third row and fifth element (a_{45}) of the fourth row and third element (a_{35}') of the fifth row for negative area;
- Element (a_{11}) has ties with the fourth element (a_{24}) of the second row and third element (a_{33}) of the third row and fifth element (a_{45})of the fourth row and second element (a_{25}') of the fifth row for positive area;
- Element (a_{11}) has ties with the fourth element (a_{24}) of the second row and third element (a_{33}) of the third row and second element (a_{24}') of the fourth row and fifth element (a_{55}) of the fifth row for negative area;
- Element (a_{11}) has ties with the fourth element(a_{24}) of the second row and fifth element (a_{35}) of the third row and second element (a_{24}') of the fourth row and third element (a_{35}') of the fifth row for positive area;
- Element (a_{11})has ties with the fourth element (a_{24}) of the second row and fifth element (a_{35}) of the third row and third element (a_{34}') of the fourth row and second element (a_{25}') of the fifth row for negative area;
- Element (a_{11}) has ties with the fifth element (a_{25}) of the second row and second element (a_{23}') of the third row and fourth element (a_{44}) of the fourth row and third element (a_{35}') of the fifth row for positive area;
- Element (a_{11}) has ties with the fifth element (a_{25}) of the second row and second element (a_{23}') of the third row and third element (a_{34}') of the fourth row and fourth element (a_{45}') of the fifth row for negative area;

- Element (a_{11}) has ties with the fifth element(a_{25}) of the second row and third element (a_{33}) of the third row and second element (a_{24}') of the fourth row and fourth element (a_{45}') of the fifth row for positive area;
- Element (a_{11}) has ties with the fifth element (a_{25}) of the second row and third element (a_{33}) of the third row and fourth element (a_{44}) of the fourth row and second element (a_{25}') of the fifth row for the negative area;
- Element(a_{11}) has ties with the fifth element (a_{25}) of the second row and fourth element (a_{34}) of the third row and third element (a_{34}') of the fourth row and second element (a_{25}') of the fifth row for positive area;
- Element(a_{11}) has ties with the fifth element (a_{25}) of the second row and fourth element (a_{34}) of the third row and second element (a_{24}') of the fourth row and third element (a_{35}') of the fifth row for negative area.

For the second partial part $\left|A_{12}\right|$ of element (a_{12})

Analysis of functional ties between of elements 5x 5 matrix for the complex oblique(skew) determinant shown in partial part of Figure 208 at Expansion method indicates on the following results:

This element has such functional ties for positive and negative areas as:
- Element (a_{12}) has ties with the first element (a_{12}')of the second row and third element (a_{33}) of the third row and fifth element (a_{45}) of the fourth row and fourth element (a_{45}) of the fifth row for positive area;
- Element (a_{12}) has ties with the first element (a_{12}')of the second row and third element (a_{33}) of the third row and fourth element (a_{44}) of the fourth row and fifth element (a_{55}) of the fifth row for negative area;
- Element (a_{12}) has ties with the first element (a_{12}') of the second row and fourth element (a_{34}) of the third row and third element (a_{34}') of the fourth row and fifth element (a_{55}) of the fifth row for positive area;
- Element (a_{12}) has ties with the first element (a_{12}') of the second row and fourth element (a_{34}) of the third row and fifth element (a_{45}) of the fourth row and third element (a_{35}')of the fifth row for negative area;
- Element (a_{12})has ties with the first element (a_{12}') of the second row and fifth element (a_{35}) of the third row and fourth element (a_{44}) of the fourth row and third element (a_{35}')of the fifth row for positive area;
- Element(a_{12}) has ties with the first element (a_{12}') of the second row and fifth element (a_{35}) of the third row and third element (a_{34}') of the fourth row and fourth element (a_{45}') of the fifth row for negative area;
- Element (a_{12}) has ties with the third element (a_{23}) of the second row and first element (a_{13}') of the third row and fourth element(a_{44}) of the fourth row and fifth element (a_{55}) of the fifth row for positive area;
- Element (a_{12}) has ties with the third element (a_{23}) of the second row and the first element (a_{13}') of the third row and fifth element (a_{45}) of the fourth row and fourth element (a_{45}') of the fifth row for negative area;
- Element(a_{12}) has ties with the third element (a_{23}) of the second row and fourth element (a_{34}) of the third row and fifth element (a_{45}) of the fourth row and first element (a_{15}') of the fifth row for positive area;
- Element (a_{12}) has ties with the third element (a_{23}) of the second row and fourth element (a_{34}) of the third row and first element (a_{14}') of the fourth row and fifth element (a_{55}) of the fifth row for negative area;
- Element (a_{12}) has ties with the third element (a_{23}) of the second row and fifth element (a_{35}) of the third row and first element (a_{14}') of the fourth row and fourth element (a_{45}') of the fifth row for positive area;

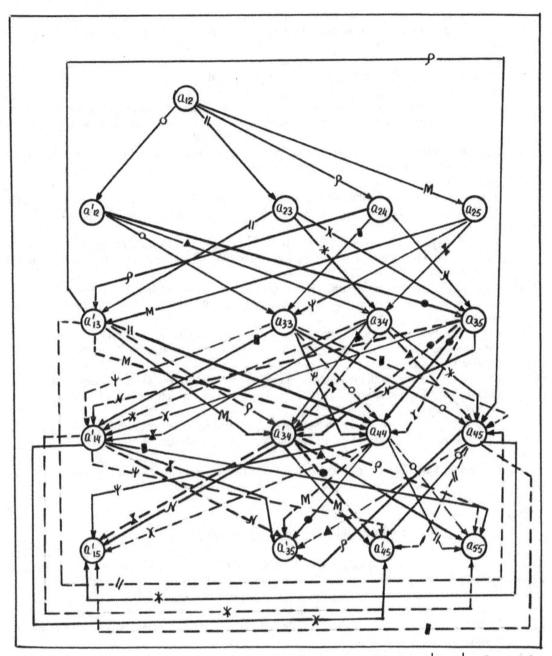

Figure 208 Functional ties between elements for partial part $|A_{12}|$ of complex oblique(skew) determinant at Expansion method for positive and negative(----) areas

- Element (a_{12}) has ties with the third element (a_{23}) of the second row and fifth element (a_{35}) of the third row and fourth element (a_{44}) of the fourth row and first element (a_{15}') of the fifth row for negative area;
- Element (a_{12}) has ties with the fourth element (a_{24}) of the second row and first element (a_{13}') of the third row and fifth element (a_{45}) of the fourth row and third element (a_{35}') of the fifth row for positive area;

465

- Element (a_{12}) has ties with the fourth element (a_{24}) of the second row and first element (a_{13}') of the third row and third element (a_{34}') of the third row and fifth element (a_{55}) of the fifth row for negative area;
- Element (a_{12}) has ties with the fourth element(a_{24}) of the second row and third element (a_{33}) of the third row and first element (a_{14}') of the fourth row and fifth element (a_{55}) of the fifth row for positive area;
- Element (a_{12}) has ties with the fourth element (a_{24}) of the second row and third element (a_{33}) of the third row and fifth element (a_{45}) of the fourth row and first element (a_{15}') of the fifth row for negative area;
- Element (a_{12}) has ties with the fourth element (a_{24}) of the second row and fifth element (a_{35}) of the third row and third element (a_{34}') of the fourth row and first element (a_{15}') of the ffith row for positive area;
- Element (a_{12}) has ties with the fourth element (a_{24}) of the second row and fifth element (a_{35}) of the third row and first element (a_{14}') of the fourth row and third element (a_{35}') of the fifth row for negative area;
- Element (a_{12}) has ties with the fifth element (a_{25}) of the second row and first element (a_{13}') of the third row and third element (a_{34}') of the fourth row and fourth element (a_{45}') of the fifth row for positive area;
- Element (a_{12}) has ties with the fifth element (a_{25}) of the second row and first element (a_{13}') of the third row and fourth element (a_{44}) of the fourth row and third element (a_{35}') of the fifth row for negative area;
- Element (a_{12}) has ties with the fifth element (a_{25}) of the second row and third element (a_{33}) of the third row and fourth element (a_{44}) of the fourth row and first element (a_{15}') of the fifth row for positive area;
- Element (a_{12}) has ties with the fifth element (a_{25}) of the second row and third element (a_{33}) of the third row and first element (a_{14}') of the fourth row and fourth element (a_{45}') of the fifth row for negative area;
- Element (a_{12}) has ties with the fifth element (a_{25}) of the second row and fourth element (a_{34}) of the third row and first element (a_{14}') of the fourth row and third element (a_{35}') of the ffith row for positive area;
- Element (a_{12}) has ties with the fifth element (a_{25}) of the second row and fourth element (a_{34}) of the third row and fourth element (a_{34}') of the fourth row and first element (a_{15}') of the ffith row for negative area.

For the third partial part $|A_{13}|$ of element (a_{13})

Analysis of functional ties between of elements 5x5 matrix for the complex oblique(skew) determinant shown in partial part $|A_{13}|$ of Figure 209 at Expansion method indicates on the following results:
This element has such functional ties for positive and negative areas as:
- Element(a_{13}) has ties with the first element (a_{12}') of the second row and second element (a_{23}') 0f the third row and fourth element (a_{44}) of the fourth row and fifth element (a_{55}) of the ffith row for positive area;
- Element (a_{13})has ties with the first element (a_{12}') of the second row and second element (a_{23}') of the third row and fifth element (a_{45}) of the fourth row and fourth element (a_{45}') of the fifth row for negative area;
- Element(a_{13}) has ties with the first element (a_{12}') of the second row and fourth element (a_{34})of the third row and fifth element (a_{45}) of the fourth row and second element (a_{25}') of the fifth row for positive area;
- Element (a_{13}) has ties with the first element (a_{12}') of the second row and fourth element (a_{34}) of the third row and second element (a_{24}') of the fourth row and fifth element (a_{55})of the fifth row for negative area;
- Element (a_{13})has ties with the first element (a_{12}') of the second row and fifth element (a_{35}) of the third row and second element (a_{24}') of the fourth row and fourth element (a_{45}') of the fifth row for positive area;

466

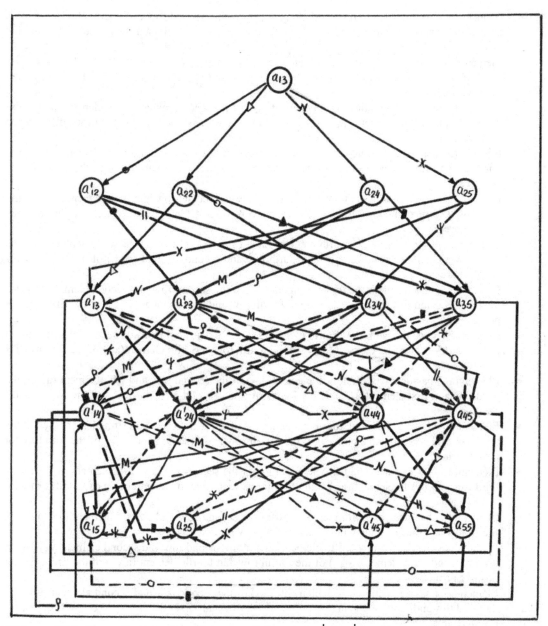

Figure 209 Functional ties for partial part $|A_{13}|$ of complex oblique(skew) determinant at Expansion method for positive and negative (------) areas

- Element (a_{13}) has ties with the first element (a_{12}') of the second row and fifth element (a_{35})of the third row and fourth element (a_{44}) of the fourth row and second element (a_{25}') of the fifth row for negative area;
- Element (a_{13}) has ties with the second element (a_{22}) of the second row and first element (a_{13}') of the third row and fifth element (a_{45}) of the fourth row and fourth element (a_{45}') of the fifth row for positive area;
- Element (a_{13}) has ties with the second element (a_{22}) of the second row and first element (a_{13}') of the third row and fourth element (a_{44}) of the fourth row and fifth element (a_{55}) of the fifth row for negative area;

- Element (a_{13}) has ties with the second element (a_{22}) of the second row and fourth element (a_{34}) of the third row and first element (a_{14}') of the fourth row and fifth element(a_{55}) of the fifth row for positive area;
- Element (a_{13}) has ties with the second element (a_{22}) of the second row and fourth element (a_{34})of the third row and fifth element (a_{45}) of the fourth row and first element (a_{15}') of the fifth row for negative area;
- Element (a_{13}) has ties with the second element (a_{22}) of the second row and fifth element (a_{35}) of the third row and fourth element (a_{44}) of the fourth row and first element (a_{15}')of the fifth row for positive area;
- Element (a_{13}) has ties with the second element (a_{22}) of the second row and fifth element (a_{35}) of the third row and first element (a_{14}')of the fourth row and fourth element (a_{45}') of the fifth row for negative area;
- Element (a_{13}) has ties with the fourth element (a_{24}) of the second row and first element (a_{13}')of the third row and second element (a_{24}') of the fourth row and fifth element (a_{55}) of the fifth row for positive area;
- Element(a_{13})has ties with the fourth element (a_{24})of the second row and first element(a_{13}') of the third row and fifth element (a_{45}) of the fourth row and second element (a_{25}') of the fifth row for negative area;
- Element (a_{13}) has ties with the fourth element (a_{24}) of the second row and second element (a_{23}')of the third row and fifth element (a_{45}) of the fourth row and first element (a_{15}') of the fifth row for positive area;
- Element (a_{13}) has ties with the fourth element (a_{24}) of the second row and second element (a_{23}') of the third row and first element (a_{14}') of the fourth row and fifth element(a_{55}) of the fifth row for negative area;
- Element(a_{13}) has ties with the fourth element (a_{24}) of the second row and fifth element (a_{35})of the third row and first element(a_{14}') of the fourth row and second element (a_{25}') of the fifth row for positive area;
- Element (a_{13}) has ties with the fourth element (a_{24}) of the second row and fifth element (a_{35}) of the third row and second element (a_{24}') of the fourth row and first element (a_{15}') of the fifth row for negative area;
- Element(a_{13})has ties with the fifth element (a_{25}) of the second row and first element (a_{13}') of the third row and fourth element (a_{44})of the fourth row and second element (a_{25}') of the fifth row for positive area;
- Element(a_{13}) has ties with fifth element (a_{25})of the second row and second element (a_{24}') of the fourth row and fourth element (a_{45}') of the fifth row for negative area;
- Element (a_{13}) has ties with the fifth element (a_{25}) of the second row and second element (a_{23}' of the third row and first element (a_{14}') of the fourth row and fourth element (a_{45}') of the fifth row for positive area;
- Element(a_{13}) has ties with the fifth element (a_{25}) of the second row and second element (a_{23}') of the third row and fourth element (a_{44}) of the fourth row and first element (a_{15}') of the fifth row for negative area;
- Element (a_{13}) has ties with the fifth element (a_{25}) of the second row and fourth element (a_{34}) of the third row and second element (a_{24}') of the fourth row and first element (a_{15}') of the fifth row for positive area;
- Element(a_{14}) has ties with the fifth element (a_{25}) of the second row and fourth element (a_{34}) of the third row and first element (a_{14}') of the fourth row and second element (a_{25}') of the fifth row for negative area.

For the fourth partial part $\left| A_{14} \right|$ of element (a_{14})

Analysis of functional ties between of elements 5x5 matrix for the complex oblique(skew) determinant shown in partial part $\left| A_{14} \right|$ of Figure 210 at Expansion method indicates on the following results:
This element has such functional ties for positive and negative areas as:

468

- Element (a_{14})has ties with the first element (a_{12}') of the second row and second element (a_{23}') of the third row and fifth element (a_{45}) of the fourth row and third element (a_{35}') of the fifth row for positive area;
- Element(a_{14}) has ties with the first element (a_{12}') of the second row and second element (a_{23}') of the third row and third element (a_{34}') of the fourth row and fifth element (a_{55}) of the fifth row for negative area;
- Element (a_{14}) has ties with the first element (a_{12}') of the second row and third element (a_{33}) of the third row and second element (a_{24}') of the fourth row and fifth element (a_{55}) of the fifth row for positive area;
- Element (a_{14}) has ties with the first element (a_{12}') of the second row and third element (a_{33}) of the third row and fifth element (a_{45}) of the fourth row and second element (a_{25}') of the fifth row for negative area;
- Element (a_{14}) has ties with the first element (a_{12}')of the second row and fifth element (a_{35}) of the third row and third element (a_{34}') of the fourth row and second element (a_{25}') of the fifth row for positive area;
- Element (a_{14}) has ties with the first element (a_{12}') of the second row and fifth element (a_{35}) of the third row and second element (a_{24}') of the fourth row and third element (a_{35}') of the fifth row for negative area;
- Element (a_{14}) has ties with the second element (a_{22}) of the second row and first element (a_{13}') of the third row and third element (a_{34}') of the fourth row and fifth element (a_{55}) of the fifth row for positive area;
- Element (a_{14}) has ties with the second element (a_{22}) of the second row and first element (a_{13}') of the third row and fifth element (a_{45}) of the fourth row and third element (a_{35}') of the fifth row for negative area;
- Element (a_{14}) has ties with the second element (a_{22}) of the second row and third element (a_{33}) of the third row and fifth element (a_{45}) of the fourth row and first element (a_{15}') of the fifth row for positive area;
- Element(a_{14}) has ties with the second element (a_{22}) of the second row and third element (a_{33}) of the third row and first element (a_{14}') of the fourth row and fifth element (a_{55}) of the fifth row for negative area;
- Element (a_{14}) has ties with the second element (a_{22}) of the second row and fifth element (a_{35}) of the third row and first element (a_{14}') of the fourth row and thirds element (a_{35}') of the fifth row for positive area;
- Element (a_{14}) has ties with the second element (a_{22}) of the second row and fifth element (a_{35}) of the third row and third element (a_{34}') of the fourth row and first element (a_{15}') of the fifth row for negative area;
- Element (a_{14})has ties with the third element (a_{23})of the second row and first element (a_{13}') of the third row and fifth element (a_{45}) of the fourth row and second element (a_{25}') of the fifth row for positive area;
- Element(a_{14}) has ties with the third element (a_{23}) of the second row and first element (a_{13}') of the third row and second element (a_{24}') of the fourth row and fifth element (a_{55}) of the fifth row for negative area;
- Element(a_{14}) has ties with the third element (a_{23}) of the second row and second element (a_{23}') of the third row and first element (a_{14}') of the fourth row and fifth element (a_{55}) of the fifth row for positive area;
- Element (a_{14}) has ties with the third element (a_{23}) of the second row and second element (a_{23}') of the third row and fifth element (a_{45}) of the fourth row and first element (a_{15}') of the fifth row for negative area;
- Element(a_{14}) has ties with the third element (a_{23}) of the second row and fifth element (a_{35}) of the third row and second element (a_{24}') of the fourth row and first element (a_{15}') of the fifth row for positive area;
- Element(a_{14}) has ties with the third element (a_{23}) of the second row and fifth element (a_{35}) of the third row and first element (a_{14}') of the fourth row and second element (a_{25}') of the fifth row for negative area;

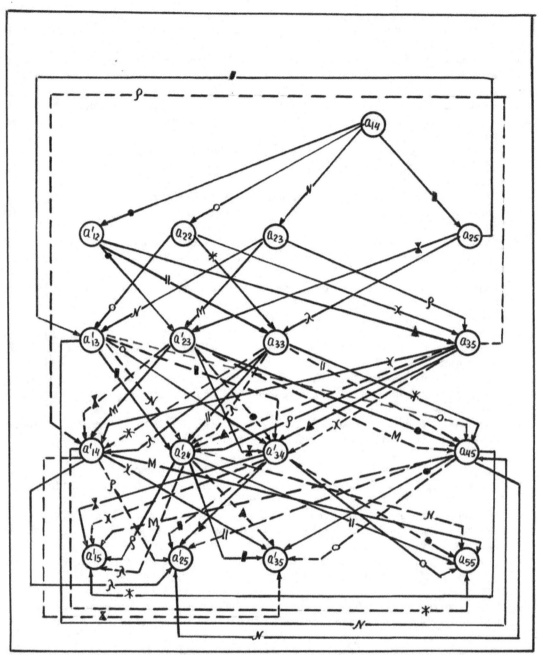

Figure 210 Functional ties between elements for partial part $\left|A_{14}\right|$ of complex oblique(skew) determinant at Expansion method for positive and negative (------) areas

- Element (a_{14}) has ties with the fifth element (a_{25}) of the second row and first element $(a_{13}{}')$ of the third row and second element $(a_{24}{}')$ of the fourth row and third element $(a_{35}{}')$ of the fifth row for positive area;
- Element (a_{14}) has ties with the fifth element (a_{25}) of the second row and first element $(a_{13}{}')$ of the third row and third element $(a_{34}{}')$ of the fourth row and second element $(a_{25}{}')$ of the fifth row for negative area;

- Element (a_{14})has ties with the fifth element (a_{25})of the second row and second element (a_{23}') of the third row and third element (a_{34}') of the fourth row and first element (a_{15}') of the fifth row for positive area;
- Element(a_{14}) has ties with the fifth element (a_{25}) of the second row and second element (a_{23}') of the third row and first element (a_{14}') of the fourth row and third element (a_{35}') of the fifth row for negative area;
- Element(a_{14}) has ties with the fifth element (a_{25}) of the second row and third element (a_{33}) of the third row and first element (a_{14}') of the fourth row and second element (a_{25}') of the fifth row for positive area;
- Element (a_{14}) has ties with the fifth element (a_{25}) of the second row and third element (a_{33}) of the third row and second element (a_{24}') of the fourth row and first element (a_{15}') of the fifth row for negative area.

For the fifth partial part $\left|A_{15}\right|$ of element (a_{15})

Analysis of functional ties between of elements 5x5 matrix for the complex oblique(skew) determinant shown in partial part $\left|A_{15}\right|$ of Figure 211 at Expansion method indicates on the following results:

This element has such functional ties for positive and negative areas as:
- Element (a_{15}) has ties with the first element (a_{12}') of the second row and second element (a_{23}') of the third row and third element (a_{34}') of the fourth row and fourth element (a_{45}')of the fifth row for positive area;
- Element (a_{15}) has ties with the first element (a_{12}') of the second row and second element (a_{23}') of the third row and fourth element (a_{44}) of the fourth row and third element (a_{35}') of the fifth row for negative area;
- Element (a_{15}) has ties with the first element (a_{12}') of the second row and third element (a_{33}) of the third row and fourth element (a_{44}) of the fourth row and second element (a_{25}')of the fifth row for positive area;
- Element (a_{15}) has ties with the first element (a_{12}') of the second row and third element (a_{33}) of the third row and second element (a_{24}') of the fourth row and fourth element (a_{45}') of the fifth row for negative area;
- Element (a_{15})has ties with the first element (a_{12}') of the second row and fourth element (a_{34}) of the third row and second element (a_{24}') of the fourth row and third element (a_{35}') of the fifth row for positive area;
- Element (a_{15}) has ties with the first element (a_{12}') of the second row and fourth element (a_{34}) of the third row and third element (a_{34}') of the fourth row and second element (a_{25}') of the fifth row for negative area;
- Element (a_{15}) has ties with the second element (a_{22}) of the second row and first element (a_{13}') of the third row and fourth element (a_{44}) of the fourth row and third element (a_{35}') of the fifth row for positive area;
- Element (a_{15}) has ties with the second element (a_{22}) of the second row and first element (a_{13}') of the third row and third element (a_{34}') of the fourth row and fourth element (a_{45}') of the fifth row for negative area;
- Element(a_{15}) has ties with the second element (a_{22}) of the second row and third element (a_{33}) of the third row and first element (a_{14}') of the fourth row and fourth element (a_{45}') of the fifth row for positive area;
- Element (a_{15}) has ties with the second element (a_{22}) of the second row and third element (a_{33}) of the third row and fourth element (a_{44}) of the fourth row and first element (a_{15}') of the fifth row for negative area;
- Element(a_{15}) has ties with the second element (a_{22}) of the second row and fourth element (a_{34}) of the third row and third element (a_{34}') of the fourth row and first element (a_{15}') of the fifth row for positive area;

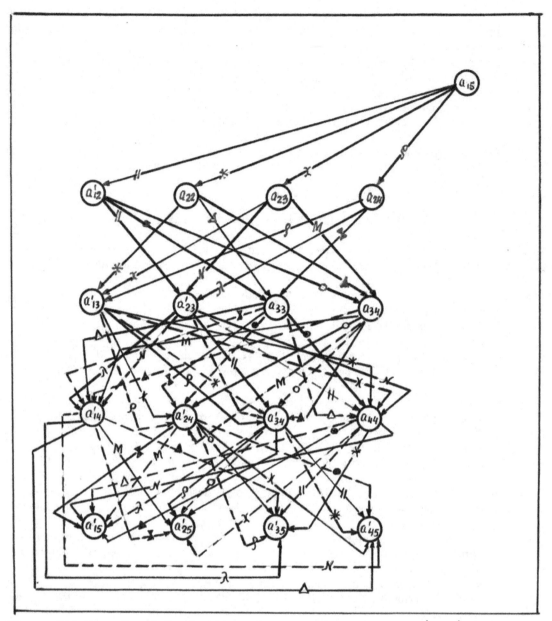

Figure 211 Functional ties between elements of partial part │A₁₅│ of complex oblique(skew) determinant at Expansion method for positive and negative (---) area

- Element (a₁₅) has ties with the second element (a₂₂) of the second row and fourth element (a₃₄) of the third row and first element (a₁₄') of the fourth row and third element (a₃₅') of the fifth row for negative area;
- Element (a₁₅) has ties with the third element (a₂₃) of the second row and first element (a₁₃') of the third row and second element (a₂₄') of the fourth row and fourth element (a₄₅') of the fifth row for positive area;
- Element (a₁₅) has ties with the third element (a₂₃) of the second row and first element (a₁₃') of the third row and fourth element (a₄₄) of the fourth row and second element (a₂₅') of the fifth row for negative area;

472

- Element (a_{15}) has ties with the third element (a_{23}) of the second row and second element (a_{23}') of the third row and fourth element (a_{44}) of the fourth row and first element (a_{15}') of the fifth row for positive area;
- Element (a_{15}) has ties with the third element (a_{23}) of the second row and second element (a_{23}') of the third row and first element (a_{14}') of the fourth row and fourth element (a_{45}') of the fifth row for negative area;
- Element (a_{15}) has ties with the third element (a_{23}) of the second row and fourth element (a_{34}) of the third row and first element (a_{14}') of the fourth row and second element (a_{25}') of the fifth row for positive area;
- Element(a_{15}) has ties with the third element (a_{23}) of the second row and fourth element (a_{34}) of the third row and second element (a_{24}') of the fourth row and first element (a_{15}') of the fifth row for negative area;
- Element (a_{15}) has ties with the fourth element (a_{24}) of the second row and first element (a_{13}') of the third row and third element (a_{34}') of the fourth row and second element (a_{25}') of the fifth row for positive area;
- Element (a_{15}) has ties with the fourth element (a_{24}) of the second row and first element (a_{13}') of the third row and second element (a_{24}') of the fourth row and third element (a_{35}') of the fifth row for negative area;
- Element(a_{15}) has ties with the fourth element (a_{24}) of the second row and second element (a_{23}') of the third row and first element (a_{14}') of the fourth row and third element (a_{35}') of the fifth row for positive area;
- Element (a_{15}) has ties with the fourth element (a_{24}) of the second row and second element (a_{23}') of the third row and third element (a_{34}') of the fourth row and first element (a_{15}') of the fifth row for negative area;
- Element(a_{15}) has ties with the fourth element (a_{24}) of the second row and thirds element (a_{33}) of the third row and second element (a_{24}') of the fourth row and first element (a_{15}') of the fifth row for positive area;
- Element(a_{15}) has ties with the fourth element (a_{24}) of the second row and third element (a_{33}) of the third row and first element (a_{14}') of the fourth row and second element (a_{25}') of the fifth row for negative area.

❖ Laplace's method for oblique(skew) determinant of 5x5 matrix

In general and numerical views ,we have the oblique (skew) determinant of 5x5 matrix and of its evaluation can be defined by the following Laplace's method:

$$
A=\begin{vmatrix} a_{11} & a_{12} & a_{13} & a_{14} & a_{15} \\ a_{12}' & a_{22} & a_{23} & a_{24} & a_{25} \\ a'_{13} & a_{23}' & a_{33} & a_{34} & a_{35} \\ a'_{14} & a'_{24} & a_{34}' & a_{44} & a_{45} \\ a'_{15} & a'_{25} & a'_{35} & a'_{45} & a_{55} \end{vmatrix}
\qquad
A=\begin{vmatrix} 6 & -4 & -1 & -5 & 7 \\ 4 & 3 & -5 & 8 & 10 \\ 1 & 5 & -3 & 1 & -5 \\ 5 & -8 & -1 & 5 & 9 \\ -7 & -10 & 5 & -9 & 11 \end{vmatrix}
$$

a) In general view   b) In numerical data

473

In above-considered 5x5 matrix , we mark out randomly two columns ,for a instance the first and second columns for oblique (skew) determinant $|A|$, writing the minors (M_{ij}) of these elements in two considered columns. And then define the corresponding algebraic supplemental of these minors(A_{ij}):

M_1 —minor and A_1 — algebraic supplemental (cofactor) for marked out jointly the first and second columns and also the first and second rows:

$M_1=(a_{11}a_{22}-a_{12}a'_{12})$; $A_1=(a_{33}a_{44}a_{55}+a_{34}a_{45}a'_{35}+a_{35}a'_{34}a'_{45}-a_{35}a_{44}a'_{35}-a_{33}a_{45}a'_{45}-a_{34}a'_{34}a_{55})$;

M_2-minor and A_2- algebraic supplemental(cofactor)for marked out jointly the first and second columns and also the first and third rows:

$M_2=(a_{11}a'_{23}-a_{12}a'_{13})$; $A_2=-(a_{23}a_{44}a_{55}+a_{24}a_{45}a'_{35}+a_{25}a'_{34}a'_{45}-a_{25}a_{44}a'_{35}-a_{23}a_{45}a'_{45}-a_{24}a'_{34}a_{55})$;

M_3-minor and A_3-algebraic supplemental (cofactor)for marked out jointly the first and second columns and also the first and fourth rows:

$M_3=(a_{11}a'_{25}-a_{12}a'_{15})$; $A_3=(a_{23}a_{34}a_{55}+a_{24}a_{35}a'_{35}+a_{25}a_{33}a'_{45}-a_{25}a_{34}a'_{35}-a_{23}a_{35}a'_{45}-a_{24}a_{33}a_{55}$;

M_4-minor and A_4-algebraic supplemental (cofactor) for marked out jointly the first and second columns and also the first and fifth rows:

$M_4=(a_{11}a'_{24}-a_{12}a'_{15})$; $A_4=-(a_{23}a_{34}a_{45}+a_{24}a_{35}a'_{34}+a_{25}a_{33}a_{44}-a_{25}a_{34}a'_{34}-a_{23}a_{35}a_{44}-a_{24}a_{33}a_{45})$;

M_5-minor and A_5-algebraic supplemental (cofactor) for marked out jointly the first and second columns and also the second and third rows:

$M_5=(a'_{12}a'_{23}-a_{22}a'_{13})$; $A_5=(a_{13}a_{44}a_{55}+a_{14}a_{45}a'_{35}+a_{15}a'_{34}a'_{45}-a_{15}a_{44}a'_{35}-a_{13}a_{45}a'_{45}-a_{14}a'_{34}a_{55})$;

M_6-minor and A_6-algebraic supplemental (cofactor) for marked out jointly the first and second columns and also the second and fourth rows:

$M_6=(a'_{12}a'_{24}-a_{22}a'_{14})$; $A_6=-(a_{13}a_{34}a_{55}+a_{14}a_{35}a'_{35}+a_{15}a_{33}a'_{45}-a_{15}a_{34}a'_{35}-a_{13}a_{35}a'_{45}-a_{14}a_{33}a_{55})$;

M_7-minor and A_7-algebraic supplemental (cofactor) for marked out jointly the first and second columns and also the second and fifth rows:

$M_7=(a'_{12}a'_{24}-a_{22}a'_{15})$; $A_7=(a_{13}a_{34}a_{45}+a_{14}a_{35}a'_{34}+a_{15}a_{33}a_{44}-a_{15}a_{34}a'_{34}-a_{13}a_{35}a_{44}-a_{14}a_{33}a_{45})$;

M_8-minor and A_8-algebraic supplemental(cofactor) for marked out jointly the first and second columns and also the third and fourth rows:

$M_8=(a'_{13}a'_{24}-a'_{23}a'_{14})$; $A_8=(a_{13}a_{24}a_{55}+a_{14}a_{25}a'_{35}+a_{15}a_{23}a'_{45}-a_{15}a_{24}a'_{35}-a_{13}a_{25}a'_{45}-a_{14}a_{23}a_{55})$;

M_9-minor and A_9-algebraic supplemental (cofactor) for marked out jointly the first and second columns and also the third and fifth rows:

$M_9=(a'_{13}a'_{25}-a'_{23}a'_{15})$; $A_9=-(a_{13}a_{24}a_{45}+a_{14}a_{25}a'_{34}+a_{15}a_{23}a_{44}-a_{15}a_{24}a'_{34}-a_{13}a_{25}a_{44}-a_{14}a_{23}a_{45})$;

M_{10}-minor and A_{10}-algebraic supplemental(cofactor) for marked out jointly the first and second columns and also the fourth and fifth rows:

$M_{10}=(a'_{14}a'_{25}-a'_{24}a'_{15})$; $A_{10}=(a_{13}a_{24}a_{35}+a_{14}a_{25}a_{33}+a_{15}a_{23}a_{34}-a_{15}a_{24}a_{33}-a_{13}a_{25}a_{34}-a_{14}a_{23}a_{35})$.

So ,after of some analysis we can conclude that the oblique(skew) determinant of 5x5 matrix regarding of Laplace's method is equal:

$$|A|=M_1A_1-M_2A_2+M_3A_3-M_4A_4+M_5A_5-M_6A_6+M_7A_7+M_8A_8-M_9A_9+M_{10}A_{10}$$

After of some calculations and transformations ,we have finally formula(128) for evaluation of complex oblique (skew) determinant of 5x5 matrix for using of Laplace's theorem in such general view as:

$$|A| = |A_{11}| + |A_{12}| + |A'_{12}| + |A'_{13}| + |A_{22}| + |A'_{23}|$$

$$|A| = a_{11}a_{22}[a_{33}(a_{44}a_{55}-a_{45}a'_{45})+a_{34}(a_{45}a'_{35}-a'_{34}a_{55})+a_{35}(a'_{34}a'_{45}-a_{44}a'_{35})] + \quad |A_{11}|$$
$$+a_{11}a'_{23}[a_{23}(a_{45}a'_{45}-a_{44}a_{55})+a_{24}(a'_{34}a_{55}-a_{45}a'_{35})+a_{25}(a_{44}a'_{35}-a'_{34}a'_{45})]+$$
$$+a_{11}a'_{24}[a_{23}(a_{35}a_{44}-a_{34}a_{45})+a_{24}(a_{33}a_{45}-a_{35}a'_{34})+a_{25}(a_{34}a'_{34}-a_{33}a_{44})]+$$
$$+a_{11}a'_{25}[a_{23}(a_{34}a_{55}-a_{35}a'_{45})+a_{24}(a_{35}a'_{35}-a_{33}a_{55})+a_{25}(a_{33}a'_{45}-a_{34}a'_{35})]+$$

$$+a_{12}a'_{12}[a_{33}(a_{45}a'_{45}-a_{44}a_{55})+a_{34}(a'_{34}a_{55}-a_{45}a'_{35})+a_{35}(a_{44}a'_{35}-a'_{34}a'_{45})]+ \quad |A_{12}|$$
$$+a_{12}a'_{13}[a_{23}(a_{44}a_{55}-a_{45}a'_{45})+a_{24}(a_{45}a'_{35}-a'_{34}a_{55})+a_{25}(a'_{34}a'_{45}-a_{44}a'_{35})]+$$
$$+a_{12}a'_{14}[a_{23}(a_{35}a'_{45}-a_{34}a_{55})+a_{24}(a_{33}a_{55}-a_{35}a'_{35})+a_{25}(a_{34}a'_{35}-a_{33}a'_{45})]+$$
$$+a_{12}a'_{15}[a_{23}(a_{34}a_{45}-a_{35}a_{44})+a_{24}(a_{35}a'_{34}-a_{33}a_{45})+a_{25}(a_{33}a_{44}-a_{34}a'_{34})]+$$

$$+a'_{12}a'_{23}[a_{13}(a_{44}a_{55}-a_{45}a'_{45})+a_{14}(a_{45}a_{35}-a'_{34}a_{55})+a_{15}(a'_{34}a'_{45}-a_{44}a'_{35})]+ \quad |A'_{12}|$$
$$+a'_{12}a'_{24}[a_{13}(a_{35}a'_{45}-a_{34}a_{55})+a_{14}(a_{33}a_{55}-a_{35}a'_{35})+a_{15}(a_{34}a'_{35}-a_{33}a'_{45})]+$$
$$+a'_{12}a'_{25}[a_{13}(a_{34}a_{45}-a_{35}a_{44})+a_{14}(a_{35}a'_{34}-a_{33}a_{45})+a_{15}(a_{33}a_{44}-a_{34}a'_{34})]+$$

$$+a'_{13}a'_{24}[a_{13}(a_{24}a_{55}-a_{25}a'_{45})+a_{14}(a_{25}a'_{35}-a_{23}a_{55})+a_{15}(a_{23}a'_{45}-a_{24}a'_{35})]+ \quad |A'_{13}|$$
$$+a'_{13}a'_{25}[a_{13}(a_{25}a_{44}-a_{24}a_{45})+a_{14}(a_{23}a_{45}-a_{25}a'_{34})+a_{15}(a_{24}a'_{34}-a_{23}a_{44})]+$$
$$+a'_{14}a'_{25}[a_{13}(a_{24}a_{35}-a_{25}a_{34})+a_{14}(a_{25}a_{33}-a_{23}a_{35})+a_{15}(a_{23}a_{34}-a_{24}a_{33})]+$$

$$+a_{22}a'_{13}[a_{13}(a_{45}a'_{45}-a_{44}a_{55})+a_{14}(a'_{34}a_{55}-a_{45}a'_{35})+a_{15}(a_{44}a'_{35}-a'_{34}a'_{45})]+ \quad |A_{22}|$$
$$+a_{22}a'_{14}[a_{13}(a_{34}a_{55}-a_{35}a'_{45})+a_{14}(a_{35}a'_{35}-a_{33}a_{55})+a_{15}(a_{33}a'_{45}-a_{34}a'_{35})]+$$
$$+a_{22}a'_{15}[a_{13}(a_{35}a_{44}-a_{34}a_{45})+a_{14}(a_{33}a_{45}-a_{35}a'_{34})+a_{15}(a_{34}a'_{34}-a_{33}a_{44})]+$$

$$+a'_{23}a'_{14}[a_{13}(a_{25}a'_{45}-a_{24}a_{55})+a_{14}(a_{23}a_{55}-a_{25}a'_{35})+a_{15}(a_{24}a'_{35}-a_{23}a_{45})]+ \quad |A'_{23}|$$
$$+a'_{23}a'_{15}[a_{13}(a_{24}a_{45}-a_{25}a_{44})+a_{14}(a_{25}a'_{34}-a_{23}a'_{45})+a_{15}(a_{23}a_{44}-a_{24}a'_{34})]+$$
$$+a'_{24}a'_{15}[a_{13}(a_{25}a_{34}-a_{24}a_{35})+a_{14}(a_{23}a_{35}-a_{25}a_{33})+a_{15}(a_{24}a_{33}-a_{23}a_{34})] \qquad (128)$$

In Figures 212 to 217 are shown the functional ties for combined oblique (skew)determinant $|A|$ of 5x5 matrix for positive and negative areas applicably to formula (128)of Laplace's theorem(method):

- ❖ *For Figure 212 of the partial part $|A_{11}|$ of combined oblique (skew)determinant $|A| = |A_{11}| + |A_{12}| + |A'_{12}| + |A'_{13}| + |A_{22}| + |A'_{23}|$ of 5x5 matrix for Laplace's method*

Analysis of data shown in Figure 212 applicably to formula (128) of the partial part $|A_{11}|$ of combined oblique(skew) determinant $|A|$ of 5x5 matrix for positive and negative areas with Laplace's method indicates on the following results in general view:
- Element (a_{11}) has functional ties with the second element (a_{22}) of the second row and third element (a_{33}) of the third row and fourth element (a_{44}) of the fourth row and fifth element (a_{55}) of the fifth row for positive and negative areas;
- Element (a_{11}) has functional ties with the second element (a_{22}) of the second row and third element (a_{33}) of the third row and fifth element (a_{45}) of the fourth row and fourth element (a'_{45}) of the fifth row for negative area;
- Element (a_{11}) has functional ties with the second element (a_{22}) of the second row and fourth element (a_{34}) of the third row and fifth element (a_{45}) of the fourth row and third element (a_{35}) of the fifth row for positive area;
- Element (a_{11}) has functional ties with the second element (a_{22}) of the second row and fourth element (a_{34}) of the third row and third element (a'_{34}) of the fourth row and fifth element (a_{55}) of the fifth row for negative area;

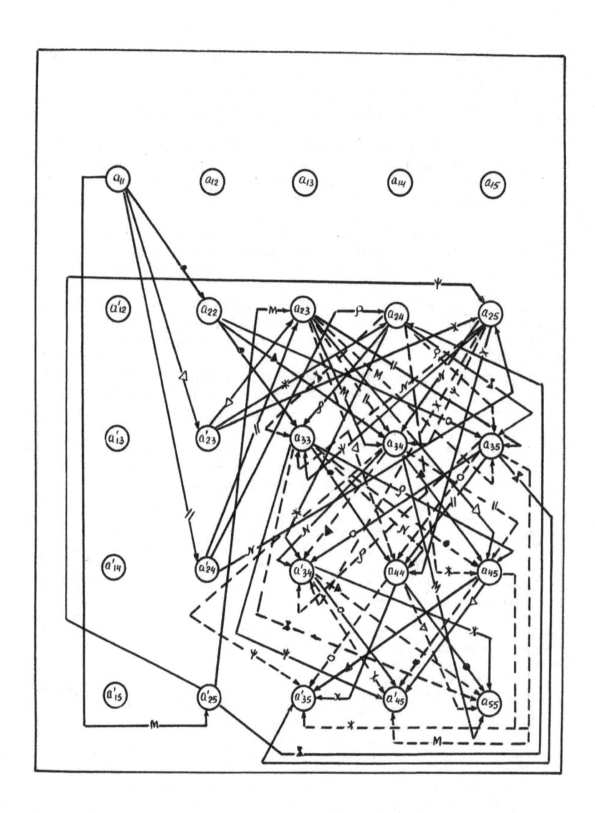

Figure 212 Functional ties for partial part $\left|A_{11}\right|$ of combined oblique(skew) determinant $\left|A\right|$ of 5x5 matrix at Laplace's method

476

- Element (a_{11})has functional ties with the second element (a_{22}) of the second row and fifth element (a_{35}) of the third row and third element (a'_{34}) of the fourth row and fourth element (a'_{45}) of the fifth row for positive area;
- Element (a_{11}) has functional ties with the second element (a_{22}) of the second row and fifth element (a_{35}) of the third row and fourth element (a_{44})of the fourth row and third element (a'_{35}) of the fifth row for negative area;
- Element (a_{11}) has functional ties with the second element (a'_{23}) of the third row and third element (a_{23})of the second row and fifth element (a_{45}) of the fourth row and fourth element (a'_{45}) of the fifth row for positive area;
- Element (a_{11}) has functional ties with the second element (a'_{23}) of the third row and third element (a_{23}) of the second row and fourth element (a_{44}) of the fourth row and fifth element (a_{55}) of the fifth row for negative area;
- Element (a_{11}) has functional ties with the second element (a'_{23}) of the third row and fourth element (a_{24}) of the second row and third element (a'_{34}) of the fourth row and fifth element (a_{55}) of the fifth row for positive area;
- Element (a_{11}) has functional ties with the second element (a'_{23}) of the third row and fourth element (a_{24}) of the second row and fifth element (a_{45}) of the fourth row and third element (a'_{35}) of the fifth row for negative area;
- Element (a_{11}) has functional ties with the second element (a'_{23}) of the third row and fifth element (a_{25}) of the second row and fourth element (a_{44}) of the fourth row and third element (a'_{35}) of the fifth row for positive area;
- Element (a_{11}) has functional ties with the second element (a'_{23}) of the third row and fifth element (a_{25}) of the second row and third element (a'_{34}) of the fourth row and fourth element (a'_{45}) of the fifth row for negative area;
- Element (a_{11}) has functional ties with the second element (a'_{24}) of the fourth row and third element (a_{23}) of the second row and fifth element (a_{35})of the third row and fourth element (a_{44}) of the fourth row for positive area;
- Element(a_{11}) has functional ties with the second element (a'_{24}) of the fourth row and third element (a_{23}) of the second row and fourth element (a_{34}) of the third row and fifth element (a_{45}) of the fourth row for negative area;
- Element (a_{11}) has functional ties with the second element (a'_{24}) of the fourth row and fourth element (a_{24}) of the second row and third element (a_{33}) of the third row and fifth element (a_{45}) of the fourth row for positive area;
- Element (a_{11}) has functional ties with the second element (a'_{24}) of the fourth row and fourth element (a_{24}) of the second row and fifth element (a_{35}) of the third row and third element (a'_{34}) of the fourth row for negative area;
- Element (a_{11}) has functional ties with the second element (a'_{24})of the fourth row and fifth element (a_{25}) of the second row and fourth element (a_{34}) of the third row and third element (a'_{34}) of the fourth row for positive area;
- Element (a_{11}) has functional ties with the second element (a'_{24}) of the fourth row and fifth element (a_{25}) of the second row and third element (a_{33}) of the third row and fourth element (a_{44}) of the fourth row for negative area;
- Element (a_{11}) has functional ties with the second element (a'_{25}) of the fifth row and third element (a_{23}) of the second row and fourth element (a_{34}) of the third row and fifth element (a_{55}) of the fifth row for positive area;
- Element (a_{11}) has functional ties with the second element (a'_{25}) of the fifth row and third element (a_{23}) of the second row and fifth element (a_{35}) of the third row and fourth element (a'_{45}) of the fifth row for negative area;
- Element (a_{11})has functional ties with the second element (a'_{25}) of the fifth row and fourth element (a_{24}) of the second row and fifth element (a_{35}) of the third row and third element (a'_{35}) of the fifth row for positive area;
- Element (a_{11}) has functional ties with the second element (a'_{25}) of the fifth row and fourth element (a_{24}) of the second row and third element (a_{33}) of the third row and fifth element (a_{55}) of the fifth row for negative area;

- Element (a_{11}) has functional ties with the second element (a'_{25}) of the fifth row and fifth element (a_{25}) of the second row and third element (a_{33}) of the third row and fourth element (a'_{45}) of the fifth row for positive area;
- Element (a_{11}) has functional ties with the second element (a'_{25}) of the fifth row and fifth element (a_{25}) of the second row and fourth element (a_{34}) of the third row and third element (a'_{35}) of the fifth row for negative area.

❖ For Figure 213 of the partial part $|A_{12}|$ of combined oblique(skew) determinant $|A|$ of 5x5 matrix for Laplace's method

Analysis of data shown in Figure 213 applicably to formula (128) of the partial part $|A_{12}|$ of combined oblique(skew)determinant $|A|$ of 5x5 matrix for positive and negative area with Laplace's method indicates of the following results in general view:
- Element (a_{12}) has functional ties with the first element (a'_{12})of the second row and third element (a_{33}) of the third row and fifth element (a_{45}) of the fourth row and fourth element (a'_{45}) of the fifth row for positive area;
- Element (a_{12}) has functional ties with the first element (a'_{12})of the second row and third element (a_{33}) of the third row and fourth element (a_{44}) of the fourth row and fifth element (a_{55}) of the ffith row for negative area;
- Element (a_{12}) has functional ties with the first element (a'_{12}) of the second row and fourth element (a_{34}) of the third row and third element (a'_{34}) of the fourth row and fifth element (a_{55}) of the fifth row for positive area;
- Element (a_{12}) has functional ties with the first element (a'_{12})of the second row and fourth element (a_{34}) of the third row and fifth element (a_{45}) of the fourth row and third element (a'_{35}) of the fifth row for negative area;
- Element (a_{12}) has functional ties with the first element (a'_{12}) of the second row and fifth element (a_{35}) of the third row and fourth element (a_{44}) of the fourth row and third element (a'_{35}) of the fifth row for positive area;
- Element (a_{12}) has functional ties with the first element (a'_{12}) of the second row and fifth element (a_{35}) of the third row and third element (a'_{34}) of the fourth row and fourth element (a'_{45}) of the fifth row for negative area;
- Element (a_{12})has functional ties with the first element (a'_{13}) of the third row and third element (a_{23}) of the second row and fourth element (a_{44}) of the fourth row and fifth element (a_{55}) of the fifth row for positive area;
- Element (a_{12}) has functional ties with the first element (a'_{13}) of the third row and third element (a_{23}) of the second row and fifth element (a_{45}) of the fourth row and fourth element (a'_{45}) of the fifth row for negative area;
- Element (a_{12}) has functional ties with the first element (a'_{13}) of the third row and fourth element (a_{24})of the second row and fifth element (a_{45}) of the fourth row and third element (a'_{35}) of the fifth row for positive area;
- Element (a_{12}) has functional ties with the first element (a'_{13}) of the third row and fourth element (a_{24})of the second row and third element (a'_{34}) of the fourth row and fifth element (a_{55}) of the fifth row for negative area;
- Element (a_{12}) has functional ties with the first element (a'_{13}) of the third row and fifth element (a_{25}) of the second row and third element (a'_{34}) of the fourth row and fourth element (a'_{45}) of the fifth row for positive area;
- Element (a_{12}) has functional ties with the first element (a'_{13}) of the third row and fifth element (a_{25}) of the second row and fourth element (a_{44}) of the fourth row and third element (a'_{35}) of the fifth row for negative area;
- Element (a_{12}) has functional ties with the first element (a'_{14}) of the fourth row and third element (a_{23}) of the second row and fifth element (a_{35}) of the third row and fourth element (a'_{45}) of the fifth row for positive area;

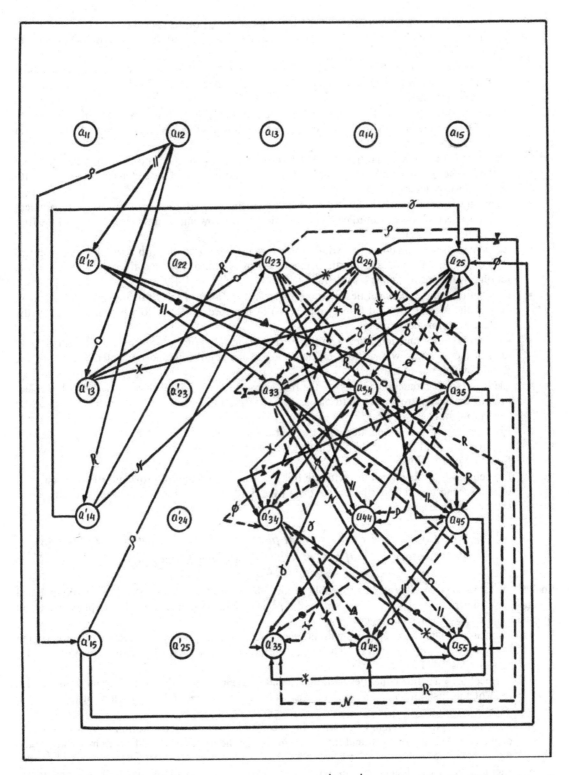

Figure 213 Functional ties for partial part $|A_{12}|$ of combined oblique (skew) determinant $|A|$ of 5x5 matrix at Laplace's method

479

- Element (a_{12}) has functional ties with the first element (a'_{14}) of the fourth row and third element (a_{23}) of the second row and fourth element (a_{34}) of the third row and fifth element (a_{55}) of the fifth row for negative area;
- Element (a_{12}) has functional ties with the first element (a'_{14}) of the fourth row and fourth element (a_{24}) of the second row and third element (a_{33}) of the third row and fifth element (a_{55}) of the fifth row for positive area;
- Element(a_{12}) has functional ties with the first element (a'_{14}) of the fourth row and fourth element (a_{24}) of the second row and fifth element (a_{35})of the third row and third element (a'_{35}) of the fifth row for negative area;
- Element (a_{12}) has functional ties with the first element (a'_{14}) of the fourth row and fifth element (a_{25}) of the second row and fourth element (a_{34}) of the third row and third element (a'_{35}) of the fifth row for positive area;
- Element(a_{12}) has functional ties with the first element (a'_{14}) of the fourth row and fifth element (a_{25})of the second row and third element (a_{33}) of the third row and fourth element (a'_{45}) of the fifth row for negative area;
- Element (a_{12}) has functional ties with the first element (a'_{15}) of the fifth row and third element (a_{23}) of the second row and fourth element (a_{34}) of the third row and fifth element (a_{45})of the fourth row for positive area;
- Element (a_{12}) has functional ties with the first element (a'_{15}) of the fifth row and third element (a_{23}) of the second row and fifth element (a_{35}) of the third row and fourth element (a_{44}) of the fourth row for negative area;
- Element (a_{12}) has functional ties with the first element (a'_{15}) of the fifth row and fourth element (a_{24}) of the second row and fifth element (a_{35}) of the third row and third element(a'_{34}) of the fourth row for positive area;
- Element (a_{12}) has functional ties with the first element (a'_{15}) of the fifth row and fourth element (a_{24}) of the second row and third element (a_{33}) of the third row and fifth element (a_{45}) of the fourth row for negative area;
- Element (a_{12}) has functional ties with the first element (a'_{15}) of the fifth row and fifth element (a_{25})of the second row and third element (a_{33}) of the third row and fourth element (a_{44}) of the fourth row for positive area;
- Element (a_{12}) has functional ties with the first element (a'_{15})of the fifth row and fifth element (a_{25}) of the second row and fourth element (a_{34} of the third row and third element (a'_{34}) of the fourth row for negative area.

❖ *For Figure 214 of the partial part $\left| A'_{12} \right|$ of combined oblique(skew) of determinant $\left| A \right|$ of 5x5 matrix for Laplace's method*

Analysis of data shown in Figure 214 applicably to formula (128) of the partial part $\left| A'_{12} \right|$ of combined oblique(skew) determinant $\left| A \right|$ of 5x5 matrix for positive and negative areas for Laplace's method indicates on the following results in general view:
- Element (a'_{12}) has functional ties with the second element (a'_{23}) of the third row and third element (a_{13}) of the first row and fourth element (a_{44}) of the fourth row and fifth element (a_{55}) of the fifth row for positive area;
- Element(a'_{12}) has functional ties with the second element (a'_{23}) of the third row and third element (a_{13}) of the first row and fifth element (a_{45}) of the fourth row and fourth element (a'_{45}) of the negative area;
- Element (a'_{12})has functional ties with the second element (a'_{23}) of the third row and fourth element (a_{14}) of the first row and fifth element (a_{45}) of the fourth row and fifth element (a_{35}) of the third row for positive area;
- Element (a'_{12}) has functional ties with the second element (a'_{23}) of the third row and fourth element (a_{14}) of the first row and third element (a'_{34}) of the fourth row and fifth element (a_{55}) of the fifth row for negative area;

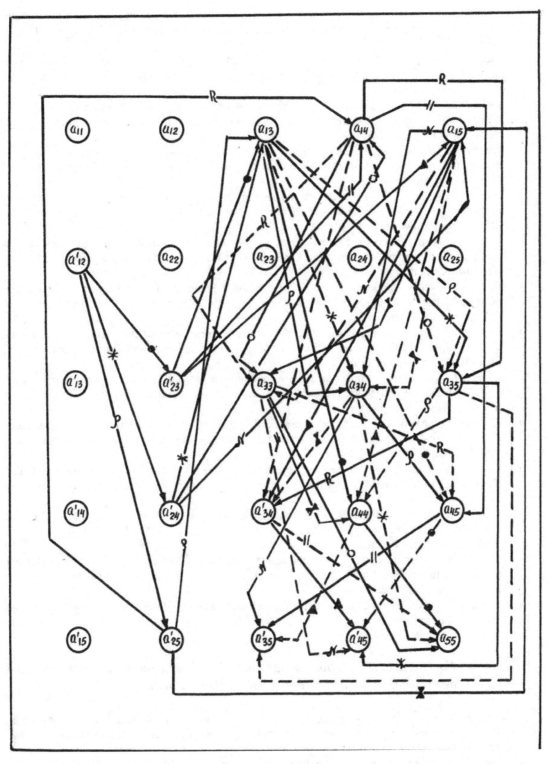

Figure 214 Functional ties for partial part $|A'_{12}|$ of combined oblique(skew) determinant $|A|$ of 5x5 matrix at Laplace's method

- Element (a'$_{12}$) has functional ties with the second element (a'$_{23}$) of the third row and fifth element (a$_{15}$) of the first row and third element of the fourth row and fourth element (a'$_{45}$) of the fifth row for positive area;
- Element (a'$_{12}$) has functional ties with the second element (a'$_{23}$) of the third row and fifth element (a$_{15}$) of the first row and fourth element (a$_{44}$) of the fourth row and third element (a'$_{35}$) of the fifth row for negative area;
- Element (a'$_{12}$) has functional ties with the second element (a'$_{24}$) of the fourth row and third element (a$_{13}$) of the first row and fifth element (a$_{35}$) of the third row and fourth element (a'$_{45}$) of the fifth row for positive area;
- Element (a'$_{12}$) has functional ties with the second element (a'$_{24}$) of the fourth row and third element (a$_{13}$) of the first row and fourth element (a$_{34}$) of the third row and fifth element (a$_{55}$) of the fifth row for negative area;
- Element (a'$_{12}$) has functional ties with the second element (a'$_{24}$) of the fourth row and fourth element (a$_{14}$) of the first row and third element (a$_{33}$) of the third row and fifth element (a$_{55}$) of the fifth row for positive area;
- Element (a'$_{12}$) has functional ties with the second element (a'$_{24}$) of the fourth row and fourth element (a$_{14}$) of the first row and fifth element (a$_{35}$) of the third row and third element (a'$_{35}$) of the fifth row for negative area;
- Element (a'$_{12}$) has functional ties with the second element (a'$_{24}$) of the fourth row and fifth element (a$_{15}$) of the first row and fourth element (a$_{34}$) of the third row and third element (a'$_{35}$) of the fifth row for positive area;
- Element (a'$_{12}$) has functional ties with the second element (a'$_{24}$) of the fourth row and fifth element (a$_{15}$) of the first row and third element (a$_{33}$) of the third row and fourth element (a'$_{45}$) of the fifth row for negative area;
- Element (a'$_{12}$) has functional ties with the second element (a'$_{25}$) of the fifth row and third element (a$_{13}$) of the first row and fourth element (a$_{34}$) of the third row and fifth element (a$_{45}$) of the fourth row for positive area;
- Element (a'$_{12}$) has functional ties with the second element (a'$_{25}$) of the fifth row and third element (a$_{13}$) of the first row and fifth element (a$_{35}$) of the third row and fourth element (a$_{44}$) of the fourth row for negative area;
- Element (a'$_{12}$) has functional ties with the second element (a'$_{25}$) of the fifth row and fourth element (a$_{14}$) of the first row and fifth element (a$_{35}$) of the third row and third element (a'$_{34}$) of the fourth row for positive area;
- Element (a'$_{12}$) has functional ties with the second element (a'$_{25}$) of the fifth row and fourth element (a$_{14}$) of the first row and third element (a$_{33}$) of the third row and fifth element (a$_{45}$) of the fourth row for negative area;
- Element (a'$_{12}$) has functional ties with the second element (a'$_{25}$) of the fifth row and fifth element (a$_{15}$) of the first row and third element (a$_{33}$) of the third row and fourth element (a$_{44}$) of the fourth row for positive area;
- Element (a'$_{12}$) has functional ties with the second element (a'$_{25}$) of the fifth row and fifth element (a$_{15}$) of the first row and fourth element (a$_{34}$) of the third row and third element (a'$_{34}$) of the fourth row for negative area.

❖ **For Figure 215 of the partial part $|A'_{13}|$ of combined oblique(skew) determinant $|A|$ of 5x5 matrix at Laplace's method**

Analysis of data shown in Figure 215 applicably to formula (128) of the partial part $|A'_{13}|$ of combined oblique(skew) determinant $|A|$ of 5x5 matrix for positive and negative areas with Laplace's method indicates on the following results in general view:
- Element(a'$_{13}$) has functional ties with the second element (a'$_{24}$) of the fourth row and third element (a$_{13}$) of the first row and fourth element (a$_{24}$) of the second row and fifth element (a$_{55}$) of the fifth row for positive area;

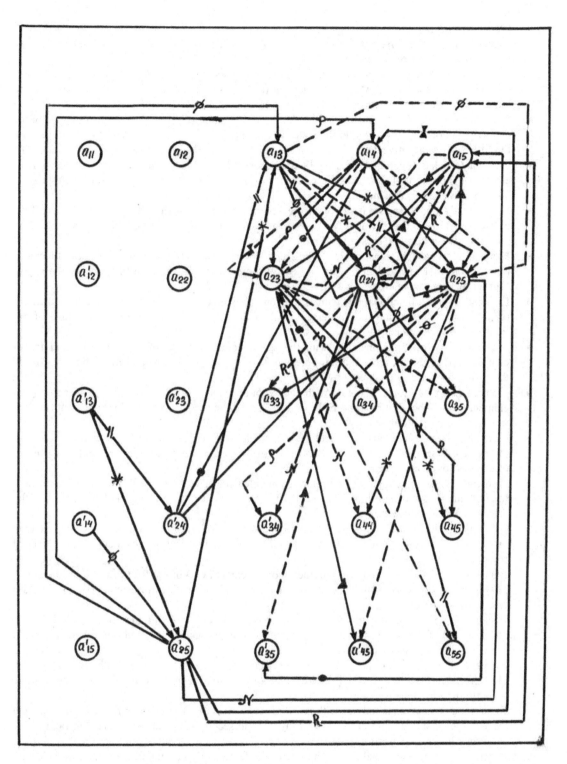

Figure 215 Functional ties for partial part $\left|A'_{13}\right|$ of combined oblique(skew)determinant $\left|A\right|$ of 5x5 matrix at Laplace's method

- Element (a'_{13}) has functional ties with the second element (a'_{24}) of the fourth row and third element (a_{13}) of the first row and fifth element (a_{25}) of the second row and fourth element (a'_{45}) of the fifth row for negative area;
- Element (a'_{13}) has functional ties with the second element (a'_{24}) of the fourth row and fourth element (a_{14}) of the first row and fifth element (a_{25}) of the second row and third element (a'_{35}) of the fifth row for positive area;
- Element (a'_{13}) has functional ties with the second element (a'_{24}) of the fourth row and fourth element (a_{14}) of the first row and third element (a_{23}) of the second row and fifth element (a_{55}) of the fifth row for negative area;
- Element (a'_{13})has functional ties with the second element (a'_{24}) of the fourth row and fifth element(a_{15}) of the first row and third element (a_{23}) of the second row and fourth element (a'_{45})of the fifth row for positive area;
- Element (a'_{13}) has functional ties with the second element (a'_{24}) of the fourth row and fifth element (a_{15}) of the first row and fourth element (a_{24}) of the second row and third element (a'_{35}) of the fifth row for negative area;
- Element (a'_{13}) has functional ties with the second element (a'_{25}) of the fifth row and third element (a_{13}) of the first row and fifth element (a_{25}) of the second row and fourth element (a_{44}) of the fourth row for positive area;
- Element (a'_{13}) has functional ties with the second element (a'_{25})of the fifth row and third element (a_{13}) of the first row and fourth element (a_{24}) of the second row and fifth element (a_{45}) of the fourth row for negative area;
- Element (a'_{13}) has functional ties with the second element (a'_{25}) of the fifth row and fourth element (a_{14}) of the first row and third element (a_{23}) of the second row and fifth element (a_{45}) of the fourth row for positive area;
- Element (a'_{13}) has functional ties with the second element (a'_{25}) of the fifth row and fourth element (a_{14}) of the first row and fifth element (a_{25}) of the second row and third element (a'_{34}) of the fourth row for negative area;
- Element (a'_{13}) has functional ties with the second element (a'_{25}) of the fifth row and fifth element (a_{15}) of the first row and fourth element (a_{24}) of the second row and third element (a'_{34}) of the fourth row for positive area;
- Element(a'_{13}) has functional ties with the second element (a'_{25}) of the fifth row and fifth element (a_{15}) of the first row and third element (a_{23}) of the second row and fourth element (a_{44}) of the fourth row for negative area;
- Element(a'_{14}) has functional ties with the second element (a'_{25}) of the fifth row and third element (a_{13}) of the first row and fourth element (a_{24})of the second row and fifth element (a_{35}) of the third row for positive area;
- Element (a'_{14}) has functional ties with the second element (a'_{25}) of the fifth row and third element (a_{13}) of the first row and fifth element (a_{25}) of the second row and fourth element (a_{34}) of the third row for negative area;
- Element (a'_{14}) has functional ties with the second element (a'_{25}) of the fifth row and fourth element (a_{14}) of the first row and fifth element (a_{25}) of the second row and third element (a_{33}) of the third row for positive area;
- Element (a'_{14}) has functional ties with the second element (a'_{25}) of the fifth row and fourth element (a_{14}) of the first row and third element (a_{23}) of the second row and fifth element (a_{35}) of the third row for negative area;
- Element (a'_{14}) has functional ties with the second element (a'_{25}) of the fifth row and fifth element (a_{15}) of the first row and third element (a_{23}) of the second row and fourth element (a_{34}) of the third row for positive area;
- Element (a'_{14}) has functional ties with the second element (a'_{25}) of the fifth row and fifth element (a_{15}) of the first row and fourth element (a_{24}) of the second row and third element (a_{33}) of the third row for negative area.

❖ *For Figure 216 of the partial part* $\left|A_{22}\right|$ *of combined oblique(skew) determinant* $\left|A\right|$ *of 5x5 matrix at Laplace's method*

Analysis of data shown in Figure 216 applicably to formula (128) of the partial part $\left|A_{22}\right|$ of combined oblique(skew) determinant $\left|A\right|$ of 5x5 matrix for positive and negative areas with Laplace's method indicates on the following results in general view:

- Element(a_{22}) has functional ties with the first element (a'_{13}) of the third row and third element (a_{13}) of the first row and fifth element (a_{45}) of the fourth row and fourth element (a'_{45}) of the fifth row for positive area;
- Element (a_{22}) has functional ties with the first element (a'_{13}) of the third row and third element (a_{13}) of the first row and fourth element (a_{44}) of the fourth row and fifth element (a_{55}) of the fifth row for negative area;
- Element (a_{22}) has functional ties with the first element (a'_{13})of the third row and fourth element (a_{14}) of the first row and third element (a'_{34}) of the fourth row and fifth element (a_{55}) of the fifth row for positive area;
- Element (a_{22}) has functional ties with the first element (a'_{13}) of the third row and fourth element (a_{14}) of the first row and fifth element (a_{45}) of the fourth row and third element (a'_{35}) of the fifth row for negative area;
- Element (a_{22}) has functional ties with the first element (a'_{13}) of the third row and fifth element (a_{15}) of the first row and fourth element (a_{44}) of the fourth row and third element (a'_{35}) of the fifth row for positive area;
- Element (a_{22}) has functional ties with the first element (a'_{13}) of the third row and fifth element (a_{15}) of the first row and third element (a'_{34}) of the fourth row and fourth element (a'_{45}) of the fifth row for negative area;
- Element(a_{22}) has functional ties with the first element (a'_{14}) of the fourth row and third element (a_{13}) of the first row and fourth element (a_{34}) of the third row and fifth element (a_{55}) of the ffith row for positive area;
- Element (a_{22}) has functional ties with the first element (a'_{14}) of the fourth row and third element (a_{13}) of the first row and fifth element (a_{35}) of the third row and fourth element (a'_{45}) of the ffith row for negative area;
- Element (a_{22}) has functional ties with the first element (a'_{14}) of the fourth row and fourth element (a_{14}) of the first row and fifth element (a_{35}) of the third row and third element (a'_{35}) of the fifth row for positive area;
- Element (a_{22}) has functional ties with the first element (a'_{14}) of the fourth row and fourth element (a_{14}) of the first row and third element (a_{33}) of the third row and fifth element (a_{55}) of the fifth row for negative area;
- Element (a_{22}) has functional ties with the first element (a'_{14}) of the fourth row and fifth element (a_{15}) of the first row and third element (a_{33}) of the third row and fourth element (a'_{45}) of the fifth row for positive area;
- Element(a_{22}) has functional ties with the first element (a'_{14}) of the fourth row and fifth element (a_{15}) of the first row and fourth element (a_{34}) of the third row and third element (a'_{35}) of the fifth row for negative area;
- Element(a_{22}) has functional ties with the first element (a'_{15}) of the fifth row and third element (a_{13}) of the first row and fifth element (a_{35}) of the third row and fourth element (a_{44}) of the fourth row for positive area;
- Element (a_{22}) has functional ties with the first element (a'_{15}) of the fifth row and third element (a_{13}) of the first row and fourth element (a_{34}) of the third row and fifth element (a_{45}) of the fourth row for negative area;
- Element (a_{22}) has functional ties with the first element (a'_{15}) of the fifth row and fourth element (a_{14})of the first row and third element (a_{33}) of the third row and fifth element (a_{45}) of the fourth row for positive area;

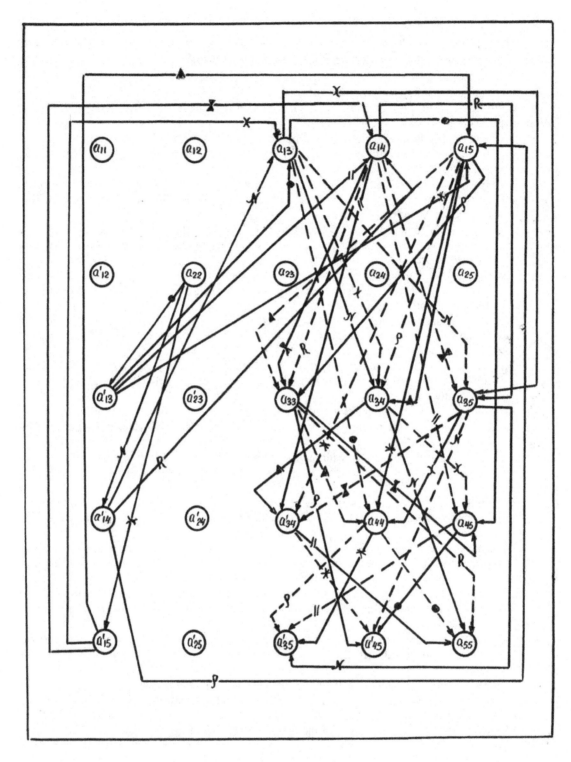

Figure 216 Functional ties for partial part $\left|A_{22}\right|$ of combined oblique(skew) determinant $\left|A\right|$ of 5x5 matrix at Laplace's method

- Element (a_{22}) has functional ties with the first element (a'_{15}) of the fifth row and fourth element (a_{14}) of the fifth row and fifth element (a_{35}) of the third row and third element (a'_{34}) of the third row for negative area;
- Element (a_{22}) has functional ties with the first element (a'_{15}) of the fifth row and fifth element (a_{15}) of the first row and fourth element (a_{34}) of the third row and third element (a'_{34}) of the fourth row for positive area;
- Element(a_{22}) has functional ties with the first element (a'_{15}) of the fifth row and fifth element (a_{15}) of the first row and third element (a_{33}) of the third row and fourth element (a_{44}) of the fourth row for negative area.

❖ *For Figure 217 of the partial part $|A'_{23}|$ of combined oblique(skew) determinant $|A|$ of 5x5 matrix at Laplace's method*

Analysis of data shown in Figure 217 applicably to formula(128) of the partial part $|A'_{23}|$ of combined oblique (skew) determinant $|A|$ of 5x5 matrix for positive and negative areas with Laplace's method indicates on the following results in general view:
- Element (a'_{23}) has functional ties with the first element (a'_{14}) of the fourth row and third element (a_{13}) of the first row and fifth element (a_{25}) of the second row and fourth element (a'_{45}) of the fifth row for positive area;
- Element(a'_{23}) has functional ties with the first element (a'_{14}) of the fourth row and third element (a_{13}) of the first row and fourth element (a_{24}) of the second row and fifth element (a_{55}) of the fifth row for negative area;
- Element (a'_{23}) has functional ties with the first element (a'_{14}) of the fourth row and fourth element (a_{14}) of the first row and third element (a_{23}) of the second row and fifth element (a_{55}) of the fifth row for positive area;
- Element (a'_{23}) has functional ties with the first element (a'_{14}) of the fourth row and fourth element (a_{14}) of the first row and fifth element (a_{25}) of the second row and third element (a'_{35}) of the fifth row for negative area;
- Element (a'_{23}) has functional ties with the first element (a'_{14}) of the fourth row and fifth element (a_{15}) of the first row and fourth element (a_{24}) of the second row and third element (a'_{35}) of the fifth row for positive area;
- Element (a'_{23}) has functional ties with the first element (a'_{14}) of the fourth row and fifth element (a_{15}) of the first row and third element (a_{23}) of the second row and fifth element (a_{45}) of the fourth row for negative area;
- Element (a'_{23}) has functional ties with the first element (a'_{15}) of the fifth row and third element (a_{13}) of the first row and fourth element (a_{24}) of the second row and fifth element (a_{45}) of the fourth row for positive area;
- Element (a'_{23}) has functional ties with the first element (a'_{15}) of the fifth row and third element (a_{13}) of the first row and fifth element (a_{25}) of the second row and fourth element (a_{44}) of the fourth row for negative area;
- Element (a'_{23}) has functional ties with the first element (a'_{15}) of the fifth row and fourth element (a_{14}) of the first row and fifth element (a_{25}) of the second row and third element (a'_{34}) of the fourth row for positive area;
- Element (a'_{23}) has functional ties with the first element (a'_{15}) of the fifth row and fourth element (a_{14}) of the first row and third element (a_{23}) of the second row and fourth element (a'_{45}) of the fifth row for negative area;
- Element (a'_{23}) has functional ties with the first element (a'_{15})of the fifth row and fifth element(a_{15}) of the first row and third element (a_{23}) of the second row and fourth element (a_{44}) of the fourth row for positive area;
- Element (a'_{23}) has functional ties with the first element (a'_{15}) of the fifth row and fifth element (a_{15}) of the first row and fourth element (a_{24}) of the second row and third element (a'_{34}) of the fourth row for negative area;

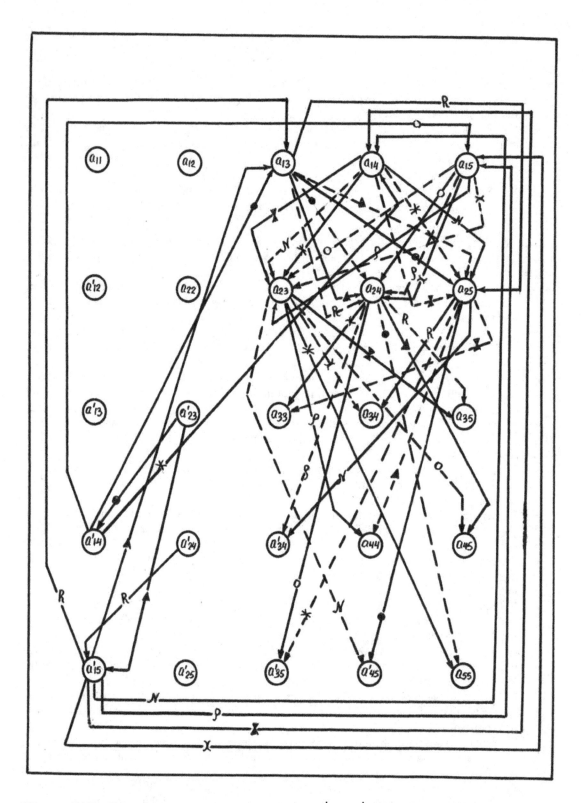

Figure 217 Functional ties for partial part $\left| A'_{23} \right|$ of combined oblique(skew) determinant $\left| A \right|$ of 5x5 matrix at Laplace's method

488

- Element (a'$_{24}$) has functional ties with the first element (a'$_{15}$) of the fifth row and third element (a$_{13}$) f the first row and fifth element (a$_{25}$) of the second row and fourth element (a$_{34}$) of the third row for positive area;
- Element (a'$_{24}$) has functional ties with the first element (a'$_{15}$)of the fifth row and third element (a$_{13}$) of the first row and fourth element (a$_{24}$) of the second row and fifth element (a$_{35}$) of the third row for negative area;
- Element (a'$_{24}$) has functional ties with the first element (a'$_{15}$)of the fifth row and fourth element (a$_{14}$) of the first row and third element (a$_{23}$) of the second row and fifth element (a$_{35}$) of the third row for positive area;
- Element(a'$_{24}$) has functional ties with the first element (a'$_{15}$) of the fifth row and fourth element (a$_{14}$) of the first row and fifth element (a$_{25}$) of the second row and third element (a$_{33}$) of the third row for negative area;
- Element (a'$_{24}$) has functional ties with the first element (a'$_{15}$) of the fifth row and fifth element (a$_{15}$) of the first row and fourth element (a$_{24}$) of the second row and third element (a$_{33}$) of the third row for positive area;
- Element (a'$_{24}$) has functional ties with the first element (a'$_{15}$) of the fifth row and fifth element (a$_{15}$) of the first row and third element (a$_{23}$) of the second row and fourth element (a$_{34}$) of the third row for negative area.

In Appendix 39 is shown the evaluation of oblique(skew) determinant of 5x5 matrix in accordance with Laplace's theorem and also the solving of system linear equations with using of Cramer's rule.

In Appendices 40 to 64 are shown the algorithms for 3x3 ,4x4 and 5x5 matrices by solving of Sarrus rule and Expansion methods .

APPENDICES

Appendix 29

Problem
Solve the symmetric determinant of 3x3 matrix and system of linear equations with using of Sarrus and Cramer's rules:

$$3X_1+2X_2+5X_3=4$$
$$2X_1+4X_2+X_3=2$$
$$5X_1+X_2+6X_3=1$$

Solving
1) The above-indicated system of linear equations can be introduced by matrix form:

$$A=\begin{vmatrix} 3 & 2 & 5 \\ 2 & 4 & 1 \\ 5 & 1 & 6 \end{vmatrix} \qquad B=\begin{vmatrix} 4 \\ 2 \\ 1 \end{vmatrix}$$

2) The value of symmetric determinant of 3x3 matrix and other determinants can be evaluated by formula (118) or Sarrus rule ,where :

$|A|=-35; |A_1|=60; |A_2|=-35; |A_3|=-50$

In Figure 1 is shown the solving of value symmetric determinant $|A|$ with using of functional graphs in accordance with formula (118) :

$|A|=a_{11}(a_{22}a_{33}-a_{23}a_{23})+a_{12}(a_{23}a_{13}-a_{12}a_{33})+a_{13}(a_{12}a_{23}-a_{22}a_{13})$

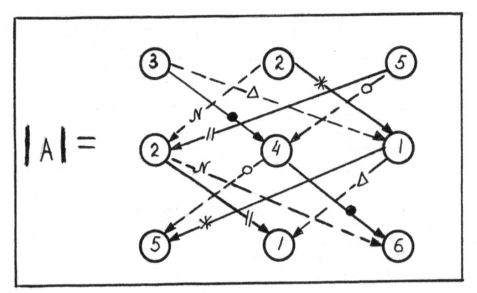

Figure 1 Solving of symmetric determinant with using of functional graphs

3) Evaluation of unknown coefficients X_1; X_2 and X_3 of system linear equations can be determined by the Cramer's rule:

$$X_1= |A_1|\,/\,|A|\,;\, X_2= |A_2|\,/\,|A|\,;\, X_3= |A_3|\,/\,|A|$$

So, the coefficients X_1; X_2; and X_3 of above-indicated system of linear equations are equal:

$$X_1= -1.714 \,;\, X_2=1.012;\, X_3= 1.429$$

Appendix 30

Problem

Solve the symmetric determinant of the 4x4 matrix *by Expansion method* and also the system of linear equations with using of Cramer's rule:

$$3X_1-2X_2+5X_3 -4X_4= -5$$
$$-2X_1+6X_2-4X_3+3X_4=2$$
$$5X_1-4X_2+6X_3-X_4= -3$$
$$-4X_1+3X_2-X_3+5X_4=0$$

Solving

1) The above-indicated system of linear equations can be introduced by matrix form:

$$A = \begin{vmatrix} 3 & -2 & 5 & -4 \\ -2 & 6 & -4 & 3 \\ 5 & -4 & 6 & -1 \\ -4 & 3 & -1 & 5 \end{vmatrix} \quad B= \begin{vmatrix} -5 \\ 2 \\ -3 \\ 0 \end{vmatrix}$$

2) The value of symmetric determinant $|A|$ can be solved by formula (119) or any other methods, for a instance ,by Expansion method, where :

$|A| = -445;\, |A_1| = -336;\, |A_2| =158;\, |A_3| =500;\, |A_4| = -188$.

490

Evaluation of symmetric determinant $|A|$ is shown in Figure 2 with using of functional graphs.

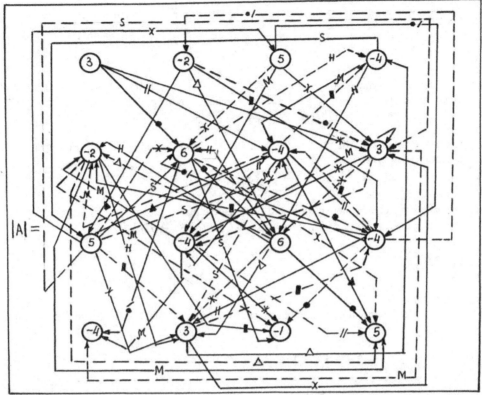

$$|A| =$$

Figure 2 Evaluation of symmetric determinant $|A|$ of 4x4 matrix in graphs

3) Determination of unknown coefficients of system linear equation can be evaluated by Cramer's rule with the following formulas:

$$X_1 = |A_1| / |A| \; ; X_2 = |A_2| / |A| \; ; X_3 = |A_3| / |A| \; ; X_4 = |A_4| / |A|$$

So, the value of coefficients for above-indicated system are equal:
$$X_1 = 0.755; \; X_2 = -0.356; \; X_3 = -1.124; \; X_4 = 0.422$$

Appendix 31

Problem

Solve the symmetric determinant of the 4x4 matrix by Laplace's method and also to evaluate the system of linear equations with using of Cramer's rule:

$$-4X_1 + 2X_2 - 3X_3 + 4X_4 = 5$$
$$2X_1 - 7X_2 + 4X_3 + X_4 = -3$$
$$-3X_1 + 4X_2 - 3X_3 + 5X_4 = 2$$
$$4X_1 + X_2 + 5X_3 - 7X_4 = -8$$

Solving

1) The above-indicated system of linear equations can be introduced by matrix form:

491

$$A = \begin{vmatrix} -4 & 2 & -3 & 4 \\ 2 & -7 & 4 & 1 \\ -3 & 4 & -3 & 5 \\ 4 & 1 & 5 & -7 \end{vmatrix} \qquad B = \begin{vmatrix} 5 \\ -3 \\ 2 \\ -8 \end{vmatrix}$$

2) The value of symmetric determinant $|A|$ we can evaluate by formula (120) and other determinants $|A_1|, |A_2|, |A_3|, |A_4|$ by Expansion method, where:

$|A| = -274; |A_1| = 306; |A_2| = 178; |A_3| = 324; |A_4| = 114$

3) Evaluation of unknown coefficients X_1; X_2; X_3 and X_4 of system linear equations can be determined by the Cramer's rule:

$$X_1 = |A_1| / |A|; X_2 = |A_2| / |A|; X_3 = |A_3| / |A|; X_4 = |A_4| / |A|$$

So, the value of coefficients for above-indicated system are equal:

$$X_1 = -1.117; X_2 = -0.649; X_3 = -1.182; X_4 = -0.416$$

Appendix 32

Problem

Solve the symmetric determinant of the 5x5 matrix by *Expansion method* and also the system of linear equations with using of Cramer's rule:

$$-3X_2 + 4X_5 = 2$$
$$-3X_1 + X_2 - 4X_3 = -4$$
$$-4X_2 + 5X_3 + 3X_4 - 6X_5 = 0$$
$$3X_3 - 2X_4 + 5X_5 = -3$$
$$4X_1 - 6X_3 + 5X_4 + X_5 = 7$$

Solving

1) The above- indicated system of linear equations can be introduced by matrix form:

$$A = \begin{vmatrix} 0 & -3 & 0 & 0 & 4 \\ -3 & 1 & -4 & 0 & 0 \\ 0 & -4 & 5 & 3 & -6 \\ 0 & 0 & 3 & -2 & 5 \\ 4 & 0 & -6 & 5 & 1 \end{vmatrix} \qquad B = \begin{vmatrix} 2 \\ -4 \\ 0 \\ -3 \\ 7 \end{vmatrix}$$

2) The value of symmetric determinant $|A|$ we can evaluate by formula (121) and determinants $|A_1|; |A_2|; |A_3|; |A_4|; |A_5|$ with using ,for a instance ,of Expansion method, where:

$|A| = 2348; |A_1| = 4936; |A_2| = -3352; |A_3| = -2165; |A_4| = -3296; |A_5| = -1340$

3) Determination of unknown coefficients of system linear equations can be evaluated by Cramer's rule with the following formulas:

$$X_1 = |A_1| / |A|; X_2 = |A_2| / |A|; X_3 = |A_3| / |A|; X_4 = |A_4| / |A|; X_5 = |A_5| / |A|.$$

So, the value of coefficients for above-indicated system of linear equations are equal:

$$X_1 = 2.102; X_2 = -1.428; X_3 = -0.922; X_4 = -1.404; X_5 = -0.571$$

Appendix 33

Problem

Solve the symmetric determinant of 5x5 matrix by Laplace's method and also the system of linear equation with using of Cramer's rule:

$$4X_1 - 2X_4 + 3X_5 = 5$$
$$5X_2 + 4X_3 = -3$$
$$4X_2 - 3X_3 + 2X_4 - X_5 = 4$$
$$-2X_1 + 2X_3 + 6X_4 + 3X_5 = 1$$
$$3X_1 - X_3 + 3X_4 - 4X_5 = -2$$

Solving

1) The above-indicated system of linear equations can be introduced by matrix form:

$$A = \begin{vmatrix} 4 & 0 & 0 & -2 & 3 \\ 0 & 5 & 4 & 0 & 0 \\ 0 & 4 & -3 & 2 & -1 \\ -2 & 0 & 2 & 6 & 3 \\ 3 & 0 & -1 & 3 & -4 \end{vmatrix} \qquad B = \begin{vmatrix} 5 \\ -3 \\ 4 \\ 1 \\ -2 \end{vmatrix}$$

2) The value of symmetric determinant $|A|$ can be solved by formula (123) or any other methods, for a instance by Expansion method ,where :

$|A| = 6486;\ |A_1| = 2654;\ |A_2| = 2582;\ |A_3| = -7582;\ |A_4| = 688;\ |A_5| = 7790;$

3) Determination of unknown coefficients of system linear equations can be evaluated by Cramer's rule with the following formulas:

$$X_1 = |A_1| / |A|;\ X_2 = |A_2| / |A|;\ X_3 = |A_3| / |A|;\ X_4 = |A_4| / |A|;\ X_5 = |A_5| / |A|$$

So, the value of coefficients for above-indicated system of linear equation are equal:

$$X_1 = 0.409;\ X_2 = 0.398;\ X_3 = -1.169;\ X_4 = 0.106;\ X_5 = 1.201$$

Appendix 34

Problem

Solve the oblique (skew) determinant of 3x3 matrix by Sarrus rule and also the system of linear equations with using of Cramer's rule:

$$7X_1 - 3X_2 - 6X_3 = 2$$
$$3X_1 - 4X_2 - X_3 = -5$$
$$6X_1 + X_2 - 2X_3 = 3$$

Solving

1) The above-indicated system of linear equations can be introduced by matrix for:

493

$$A = \begin{vmatrix} 7 & -3 & -6 \\ 3 & -4 & -1 \\ 6 & 1 & -2 \end{vmatrix} \qquad B = \begin{vmatrix} 2 \\ -5 \\ 3 \end{vmatrix}$$

2) The value of oblique(skew) determinant $|A|$ can be solved by formula (124) or other determinants by any above-indicated methods, where:

$|A| = -99$; $|A_1| = 15$; $|A_2| = -143$; $|A_3| = 122$;

3) Evaluation of unknown coefficients $X_1; X_2; X_3$ of system linear equations can be solved by the Cramer's rule:

$$X_1 = |A_1| / |A|; X_2 = |A_2| / |A|; X_3 = |A_3| / |A|.$$

So, the coefficients $X_1; X_2$ and X_3 of system linear equations are equal:

$$X_1 = -0.152; X_2 = 1.444; X_3 = -1.232$$

Appendix 35

Problem

Solve the oblique(skew) determinant of the 4x4 matrix by Expansion method and also the system of linear equations with using of Cramer's rule:

$$4X_1 - 3X_2 + X_3 + 2X_4 = 0$$
$$3X_1 + 5X_2 - 2X_3 - X_4 = -4$$
$$-X_1 + 2X_2 - 4X_3 + 6X_4 = 2$$
$$-2X_1 + X_2 - 6X_3 + 3X_4 = 7$$

Solving

1) The above-indicated system of linear equations can be introduced by matrix form:

$$A = \begin{vmatrix} 4 & -3 & 1 & 2 \\ 3 & 5 & -2 & -1 \\ -1 & 2 & -4 & 6 \\ -2 & 1 & -6 & 3 \end{vmatrix} \qquad B = \begin{vmatrix} 0 \\ -4 \\ 2 \\ 7 \end{vmatrix}$$

2) The value of oblique(skew)determinant $|A|$ we can evaluate by formula (125) and other determinants ,for a instance ,by Expansion method, where:

$|A| = 780$; $|A_1| = -312$; $|A_2| = -897$; $|A_3| = -945$; $|A_4| = -243$;

3) Determination of unknown coefficients $X_1; X_2; X_3$ and X_4 of system linear equations can be evaluated by Cramer's rule with the following formula:

$$X_1 = |A_1| / |A|; X_2 = |A_2| / |A|; X_3 = |A_3| / |A|; X_4 = |A_4| / |A|.$$

So, the value of coefficients for above-indicated system of linear equations are equal:

$$X_1 = -0.4; X_2 = -1.15; X_3 = -1.35; X_4 = -0.312$$

Appendix 36

Problem

Solve the oblique(skew) determinant of the 4x4 matrix by Laplace's method and also the system of linear equations with using of Cramer's rule:

$$2X_1 + 4X_2 - 3X_3 + 6X_4 = -5$$
$$-4X_1 + 2X_2 + X_3 - 5X_4 = 0$$
$$3X_1 - X_2 + 4X_3 + 7X_4 = 6$$
$$-6X_1 + 5X_2 - 7X_3 + 10X_4 = -7$$

Solving

1) The above-indicated system of linear equations can be introduced my matrix form:

$$A = \begin{vmatrix} 2 & 4 & -3 & 6 \\ -4 & 2 & 1 & -5 \\ 3 & -1 & 4 & 7 \\ -6 & 5 & -7 & 10 \end{vmatrix} \qquad B = \begin{vmatrix} -5 \\ 0 \\ 6 \\ -7 \end{vmatrix}$$

2) The value of oblique(skew) determinant $|A|$ we can evaluate by formula (126)and determinants $|A_1|; |A_2|; |A_3|$ and $|A_4|$ by using ,for a instance the Laplace's method:

$$|A| = 2045; \quad |A_1| = -591; \quad |A_2| = -214; \quad |A_3| = 2383; \quad |A_4| = 475;$$

3) Determination of unknown coefficients of system linear equations can be evaluated by Cramer's rule by the following formulas:

$$X_1 = |A_1| / |A|; X_2 = |A_2| / |A|; X_3 = |A_3| / |A|; X_4 = |A_4| / |A|.$$

So, the value of coefficients for above-indicated system of linear equation are equal:

$$X_1 = -0.289; X_2 = -0.105; X_3 = 1.165; X_4 = 0.232$$

Appendix 37

Problem

Solve the oblique(skew)determinant of 5x5 matrix by Expansion method and also the system of linear equations with using of Cramer's rule:

$$2X_1 + 3X_2 + X_3 + 4X_4 + 6X_5 = 0$$
$$-3X_1 + 5X_2 + 4X_3 + 6X_4 + 7X_5 = -3$$
$$-X_1 - 4X_2 + 8X_3 + 2X_4 + X_5 = 2$$
$$-4X_1 - 6X_2 - 2X_3 + 5X_4 + 3X_5 = 4$$
$$-6X_1 - 7X_2 - X_3 - 3X_4 + 6X_5 = 0$$

Solving

- The above-indicated system of linear equations can be introduced by matrix form:

495

$$A = \begin{vmatrix} 2 & 3 & 1 & 4 & 6 \\ -3 & 5 & 4 & 6 & 7 \\ -1 & -4 & 8 & 2 & 1 \\ -4 & -6 & -2 & 5 & 3 \\ -6 & -7 & -1 & -3 & 6 \end{vmatrix} \quad B = \begin{vmatrix} 0 \\ -3 \\ 2 \\ 4 \\ 0 \end{vmatrix}$$

- The value of oblique(skew) determinant $|A|$ we can evaluate by formula (127) for Expansion method and other determinants $|A_1|; |A_2|; |A_3|; |A_4|; |A_5|$ by any above-shown methods, where:

  $|A| = 34653; |A_1| = 9002; |A_2| = -20201; |A_3| = -1189; |A_4| = 13572; |A_5| = -12400;$

- Determination of unknown coefficients of system linear equations can be evaluated by Cramer's rule with the following formulas:

$$X_1 = |A_1| / |A| ; X_2 = |A_2| / |A| ; X_3 = |A_3| / |A| ; X_4 = |A_4| / |A| ; X_5 = |A_5| / |A|$$

So, the value of coefficients for above- considered system are equal:

$$X_1 = 0.259; X_2 = -0.583; X_3 = -0.034; X_4 = 0.392; X_5 = -0.358$$

Appendix 38

Problem

Solve the oblique (skew) determinant of 5x5 matrix by Expansion method and also the system of linear equations with using of Cramer's rule:

$$4X_1 - 2X_2 + X_3 + 3X_4 - 5X_5 = 2$$
$$2X_1 - 3X_2 + 4X_3 - X_4 + 3X_5 = -4$$
$$-X_1 - 4X_2 + 5X_3 - 6X_4 + 7X_5 = 0$$
$$-3X_1 + X_2 + 6X_3 - X_4 - 3X_5 = 1$$
$$5X_1 - 3X_2 - 7X_3 + 3X_4 + 2X_5 = 3$$

Solving

1) The above-indicated system of linear equations can be introduced by matrix form:

$$A = \begin{vmatrix} 4 & -2 & 1 & 3 & -5 \\ 2 & -3 & 4 & -1 & 3 \\ -1 & -4 & 5 & -6 & 7 \\ -3 & 1 & 6 & -1 & -3 \\ 5 & -3 & -7 & 3 & 2 \end{vmatrix} \quad B = \begin{vmatrix} 2 \\ -4 \\ 0 \\ 1 \\ 3 \end{vmatrix}$$

2) The value of oblique(skew) determinant $|A|$ we can evaluate by formula (58) for Expansion method and other determinants $|A_1|; |A_2|; |A_3|; |A_4|$ and $|A_5|$, where:
$|A| = 1015; |A_1| = -2516; |A_2| = -3768; |A_3| = -1195; |A_4| = -982; |A_5| = -1838.$

3) Determination of unknown coefficients of system linear equations can be evaluated by Cramer's rule with the following formulas:

496

$$X_1 = |A_1| / |A|; \; X_2 = |A_2| / |A|; \; X_3 = |A_3| / |A|; \; X_4 = |A_4| / |A|; X_5 = |A_5| / |A|$$

So, the value of coefficients for above-considered system are equal:

$$X_1 = -2.479; \; X_2 = -3.712; \; X_3 = -1.177; \; X_4 = -0.967; \; X_5 = -1.811.$$

Appendix 39

Problem

Solve the oblique(skew) determinant of 5x5 matrix with using of Laplace's method and also the system of linear equations with Cramer's rule:

$$-2X_1 + 5X_2 - 4X_3 + X_4 + 2X_5 = 0$$
$$-5X_1 + 3X_2 + 6X_3 + 4X_4 - X_5 = 4$$
$$4X_1 - 6X_2 - 4X_3 + 5X_4 + 3X_5 = -5$$
$$-X_1 - 4X_2 - 5X_3 - 7X_4 - 2X_5 = 6$$
$$-2X_1 + X_2 - 3X_3 + 2X_4 + 6X_5 = -3$$

Solving

1) The above-indicated system of linear equations can be introduced by matrix form:

$$A = \begin{vmatrix} -2 & 5 & -4 & 1 & 2 \\ -5 & 3 & 6 & 4 & -1 \\ 4 & -6 & -4 & 5 & 3 \\ -1 & -4 & -5 & -7 & -2 \\ -2 & 1 & -3 & 2 & 6 \end{vmatrix} \qquad B = \begin{vmatrix} 0 \\ 4 \\ -5 \\ 6 \\ -3 \end{vmatrix}$$

2) The value of oblique(skew) determinant $|A|$ of 5x5 matrix can be evaluated by formula (128) or ,for a instance, by Expansion method. And also another determinants $|A_1|$; $|A_2|$; $|A_3|$; $|A_4|$ and $|A_5|$ -by above-named methods, where:

$|A| = 14530$; $|A_1| = -15418$; $|A_2| = -918$; $|A_3| = -2880$; $|A_4| = -4384$; $|A_5| = -12000$.

3) Determination of unknown coefficients of above-shown system linear equations can be evaluated by Cramer's rule with the following formulas:

$$X_1 = |A_1| / |A|; \; X_2 = |A_2| / |A|; \; X_3 = |A_3| / |A|; \; X_4 = |A_4| / |A|; \; X_5 = |A_5| / |A|$$

So, the value of coefficients for above- considered system are equal:
$$X_1 = -1.061; \; X_2 = -0.063; \; X_3 = -0.198; \; X_4 = -0.302; \; X_5 = -0.826$$

Appendix 40

Algorithm for evaluation of determinant $\left| A \right|$ of 3x3 matrix by Sarrus rule

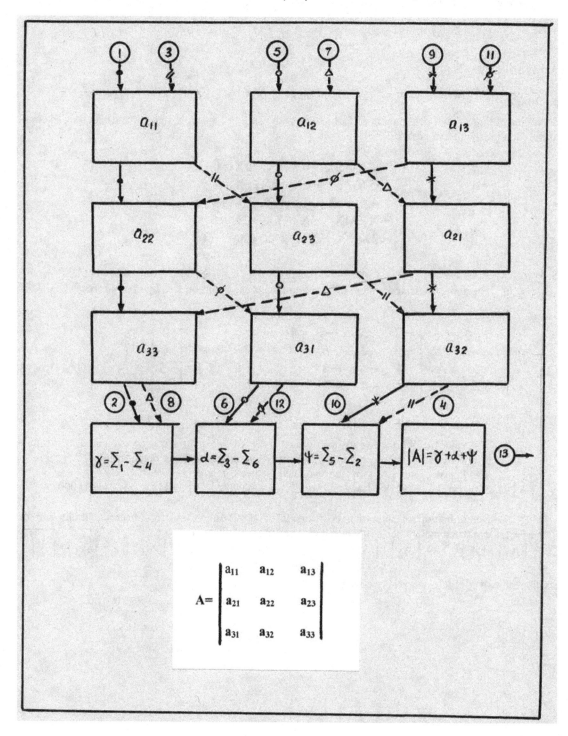

In Table 1 is shown the general steps for evaluation of determinant $|A|$ of 3x3 matrix by Sarrus rule.

Table 1 The general steps for evaluation of determinant $|A|$ of 3x3 matrix by Sarrus rule [2]

Number of steps		Elements involved in evaluation of determinant $|A|$									Sum
in	out	a_{11}	a_{12}	a_{13}	a_{21}	a_{22}	a_{23}	a_{31}	a_{32}	a_{33}	
1	2	x				x				x	$\sum_1$
3	4	x					x	x			$\sum_2$
5	6		x				x	x			$\sum_3$
7	8		x		x					x	$\sum_4$
9	10			x	x				x		$\sum_5$
11	12			x		x		x			$\sum_6$
2	8										γ
6	12										α
10	4										ψ
$|A|=$ $\gamma+\alpha+\psi$											$|A|$

Example 1:

Problem

Solve the determinant $|A|$ of 3x 3 matrix by Sarrus rule using of above-indicated algorithm.

$$A= \begin{vmatrix} 2 & -1 & 4 \\ 5 & 3 & -2 \\ 4 & 1 & -3 \end{vmatrix}$$

Solving

In accordance with above-indicated algorithm and Table 1 we have the following data:

$\sum_1=a_{11}a_{22}a_{33}=-18$; $\sum_2=a_{11}a_{23}a_{32}=-4$; $\sum_3=a_{12}a_{23}a_{31}=8$; $\sum_4=a_{12}a_{21}a_{33}=15$; $\sum_5=a_{13}a_{21}a_{32}=20$; $\sum_6=a_{13}a_{22}a_{31}=48$; $\gamma=-33$; $\alpha=-40$; $\psi=24$; $|A|=-49$.

Appendix 41
Algorithm for evaluation of partial part $|A_{11}|$ of complex determinant $|A|=|A_{11}|+|A_{12}|+|A_{13}|+|A_{14}|$ of 4x4 matrix by Expansion method

Example 2:

Problem

Solve the partial part $|A_{11}|$ of the first row (i_1) formula (17) of 4x4 matrix for complex determinant $|A|$ by Expansion method:

$$A= \begin{vmatrix} 2 & -2 & 1 & 0 \\ 3 & 1 & -1 & 4 \\ -4 & 0 & 2 & -3 \\ 0 & 1 & -3 & 4 \end{vmatrix} \qquad (*)$$

499

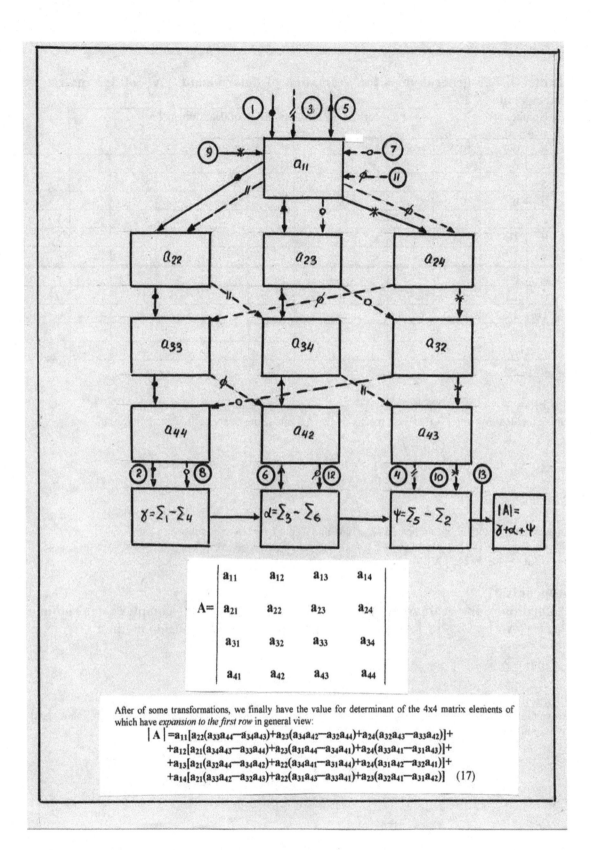

$$A = \begin{vmatrix} a_{11} & a_{12} & a_{13} & a_{14} \\ a_{21} & a_{22} & a_{23} & a_{24} \\ a_{31} & a_{32} & a_{33} & a_{34} \\ a_{41} & a_{42} & a_{43} & a_{44} \end{vmatrix}$$

After of some transformations, we finally have the value for determinant of the 4x4 matrix elements of which have *expansion to the first row* in general view:

$$\begin{aligned} |A| &= a_{11}[a_{22}(a_{33}a_{44}-a_{34}a_{43})+a_{23}(a_{34}a_{42}-a_{32}a_{44})+a_{24}(a_{32}a_{43}-a_{33}a_{42})]+ \\ &+ a_{12}[a_{21}(a_{34}a_{43}-a_{33}a_{44})+a_{23}(a_{31}a_{44}-a_{34}a_{41})+a_{24}(a_{33}a_{41}-a_{31}a_{43})]+ \\ &+ a_{13}[a_{21}(a_{32}a_{44}-a_{34}a_{42})+a_{22}(a_{34}a_{41}-a_{31}a_{44})+a_{24}(a_{31}a_{42}-a_{32}a_{41})]+ \\ &+ a_{14}[a_{21}(a_{33}a_{42}-a_{32}a_{43})+a_{22}(a_{31}a_{43}-a_{33}a_{41})+a_{23}(a_{32}a_{41}-a_{31}a_{42})] \quad (17) \end{aligned}$$

Solving:

In accordance with above-indicated algorithm of formula (17) for the first row (i_1) and Table 2, we have the following data for partial part $|A_{11}|$ of complex determinant $|A|$ of matrix 4x4:

Table 2 The general steps for evaluation of partial part $|A_{11}|$ of complex determinant $|A|$ by Expansion method

Number of steps in	out	Elements involved in evaluation process for partial part $|A_{11}|$ of determinant $|A|$																Sum
		a_{11}	a_{12}	a_{13}	a_{14}	a_{21}	a_{22}	a_{23}	a_{24}	a_{31}	a_{32}	a_{33}	a_{34}	a_{41}	a_{42}	a_{43}	a_{44}	
1	2	x					x					x					x	$\sum_1$
3	4	x					x						x			x		$\sum_2$
5	6	x						x					x		x			$\sum_3$
7	8	x						x			x						x	$\sum_4$
9	10	x							x		x					x		$\sum_5$
11	12	x							x			x			x			$\sum_6$
2	8																	γ
6	12																	α
10	4																	ψ
Partial part is equal:																		$|A|=$ $\gamma+\alpha+\psi$

where,

$\sum_1=a_{11}a_{22}a_{33}a_{44}=16$; $\sum_2=a_{11}a_{22}a_{34}a_{43}=18$; $\sum a_{11}a_{23}a_{34}a_{42}=6$; $\sum_4=a_{11}a_{23}a_{32}a_{44}=0$; $\sum_5=a_{11}a_{24}a_{32}a_{43}=0$; $\sum_6=a_{11}a_{24}a_{33}a_{42}=16$; $\gamma=\sum_1-\sum_4=16$; $\alpha=\sum_3-\sum_6=-10$; $\psi=\sum_5-\sum_2=-18$; $|A|=-12$.

Appendix 42

Algorithm for evaluation of partial part $|A_{12}|$ of complex determinant $|A|$ of 4x4 matrix by Expansion method

Example 3:

Problem

Solve the partial part $|A_{12}|$ of the second row (i_2) formula (17) of 4x4 matrix (*) for complex determinant $|A|$ by Expansion method.

Solving

In accordance with above-indicated algorithm of formula(17) for the second row(i_2) and Table 3 "The general steps for evaluation of partial part $|A_{12}|$ by Expansion method", we have the following data for partial part $|A_{12}|$ of complex determinant $|A|$ of matrix 4x4 (*) :

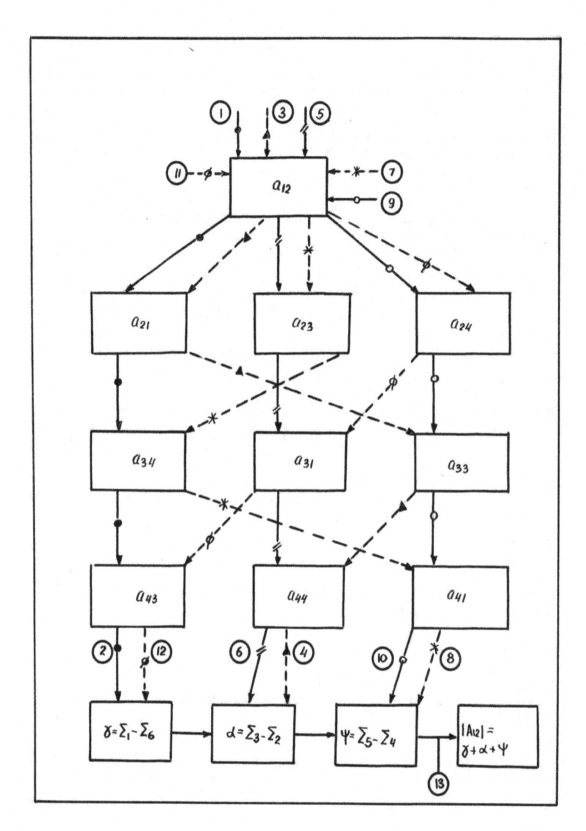

502

Table 3 The general steps for evaluation of partial part $|A_{12}|$ of complex determinant $|A|$ of 4x4 matrix by Expansion method

Number of steps		Elements involved in calculation process of partial part $|A_{12}|$																Sum
in	out	a_{11}	a_{12}	a_{13}	a_{14}	a_{21}	a_{22}	a_{23}	a_{24}	a_{31}	a_{32}	a_{33}	a_{34}	a_{41}	a_{42}	a_{43}	a_{44}	
1	2		x			x							x			x		Σ_1
3	4		x			x						x					x	Σ_2
5	6		x					x		x							x	Σ_3
7	8		x					x				x	x					Σ_4
9	10		x						x			x	x					Σ_5
11	12		x						x	x						x		Σ_6
2	12																	γ
6	4												`					α
10	8																	ψ
$|A_{12}| =$ γ+α+ψ																		$|A_{12}|$

where,

$\Sigma_1=a_{12}a_{21}a_{34}a_{43}=-54$; $\Sigma_2=a_{12}a_{21}a_{33}a_{44}=-48$; $\Sigma_3=a_{12}a_{23}a_{31}a_{44}=-32$; $\Sigma_4=a_{12}a_{23}a_{34}a_{44}=0$; $\Sigma_5=a_{12}a_{24}a_{33}a_{41}=0$;$\Sigma_6=a_{12}a_{24}a_{31}a_{43}=-96$ and $|A_{12}|=58$.

Appendix 43

Algorithm for evaluation of partial part $|A_{13}|$ of complex determinant $|A|$ of 4x4 matrix by Expansion method

Example 4:

Problem

Solve the partial part $|A_{13}|$ of the third row (i_3) formula (17) of 4x4 matrix for complex determinant $|A|$ by Expansion method.

Solving

In accordance with above-indicated algorithm of formula (17) for the third row (i_3) and Table 4 *"The general steps for evaluation of partial part $|A_{13}|$ by Expansion method"* ,we have the following data for partial part $|A_{13}|$ of complex determinant $|A|$ of matrix 4x4 at numerical data (*):

$$A=\begin{vmatrix} a_{11} & a_{12} & a_{13} & a_{14} \\ a_{21} & a_{22} & a_{23} & a_{24} \\ a_{31} & a_{32} & a_{33} & a_{34} \\ a_{41} & a_{42} & a_{43} & a_{44} \end{vmatrix}$$

503

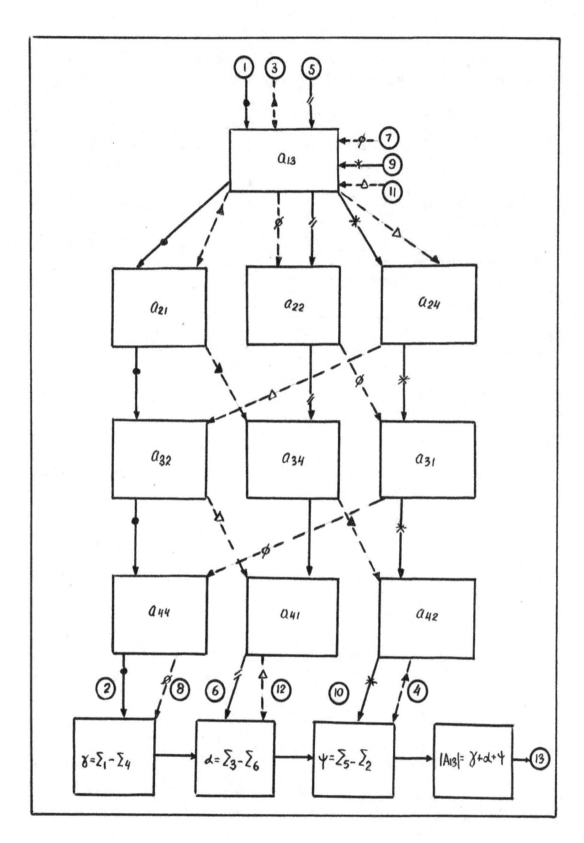

Table 4   The general steps for evaluation of partial part $|A_{13}|$ of complex determinant $|A|$ of 4x4 matrix by Expansion method

Number of steps in	out	a_{11}	a_{12}	a_{13}	a_{14}	a_{21}	a_{22}	a_{23}	a_{24}	a_{31}	a_{32}	a_{33}	a_{34}	a_{41}	a_{42}	a_{43}	a_{44}	Sum
1	2			X		X					X						X	$\sum_1$
3	4			X		X							X		X			$\sum_2$
5	6			X			X						X	X				$\sum_3$
7	8			X			X			X							X	$\sum_4$
9	10			X					X	X					X			$\sum_5$
11	12			X					X		X			X				$\sum_6$
2	8																	γ
6	12																	α
10	4																	ψ
$|A_{13}|=$ $\gamma+\alpha+\psi$																		A_{13}

where,

$\sum_1=a_{13}a_{21}a_{32}a_{44}=0$; $\sum_2=a_{13}a_{21}a_{34}a_{42}=-9$; $\sum_3=a_{13}a_{22}a_{34}a_{41}=0$; $\sum_4=a_{13}a_{22}a_{31}a_{44}=-16$; $\sum_5=a_{13}a_{24}a_{31}a_{42}=-16$; $\sum_6=a_{13}a_{24}a_{32}a_{41}=0$; $\gamma=16$; $\alpha=0$; $\psi=-7$; $|A_{13}|=9$.

Appendix 44

Algorithm for evaluation of partial part $|A_{14}|$ of complex determinant $|A|$ of 4x4 matrix by Expansion method

Example 5:

Problem

Solve the partial part $|A_{14}|$ of the fourth row (i_4) formula (17) of 4x4 matrix for complex determinant $|A|$ by Expansion method.

Solving

In accordance with above-indicated algorithm of formula(17) for the fourth row (i_4) and Table 5 *"The general steps for evaluation of partial part $|A_{14}|$ by Expansion method"*, we have the following data for partial part $|A_{14}|$ of complex determinant $|A|$ of matrix 4x4 with numerical data (*):

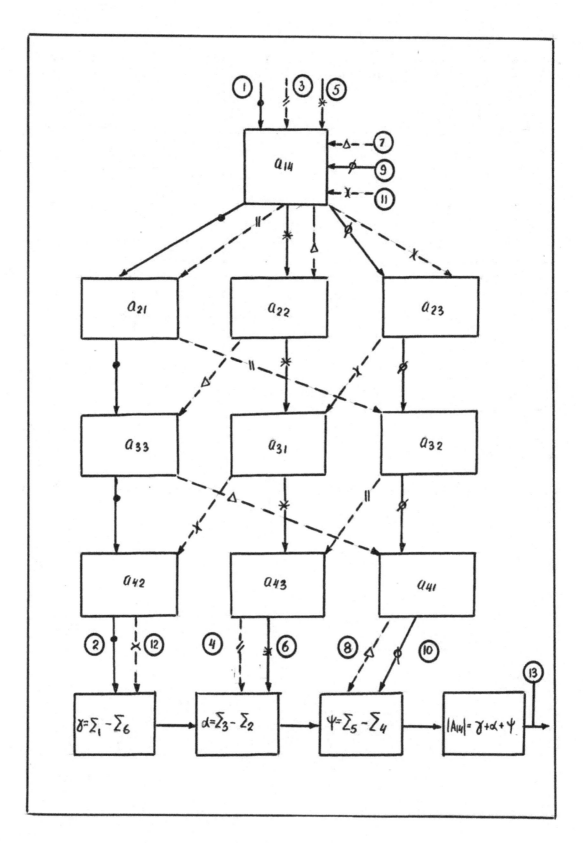

506

Table 5   **The general steps for evaluation of partial part** $|A_{14}|$ **of complex determinant** $|A|$ **of matrix 4x4 by Expansion method**

Number of steps		Elements involved in calculation process of partial part $|A_{14}|$																Sum
in	out	a_{11}	a_{12}	a_{13}	a_{14}	a_{21}	a_{22}	a_{23}	a_{24}	a_{31}	a_{32}	a_{33}	a_{34}	a_{41}	a_{42}	a_{43}	a_{44}	
1	2				X	X						X			X			$\sum_1$
3	4				X	X					X					X		$\sum_2$
5	6				X		X			X						X		$\sum_3$
7	8				X		X					X		X				$\sum_4$
9	10				X			X			X			X				$\sum_5$
11	12				X			X		X					X			$\sum_6$
2	12																	γ
6	4																	α
10	8																	ψ
$|A_{14}|=$ $\gamma+\alpha+\psi$																		$|A_{14}|$

where,

$\sum_1 = a_{14}a_{21}a_{33}a_{42} = 0$; $\sum_2 = a_{14}a_{21}a_{32}a_{43} = 0$; $\sum_3 = a_{14}a_{22}a_{31}a_{43} = 0$; $\sum_4 = a_{14}a_{22}a_{33}a_{41} = 0$; $\sum_5 = a_{14}a_{23}a_{32}a_{41} = 0$;

$\sum_6 = a_{14}a_{23}a_{31}a_{42} = 0$; $\gamma = \sum_1 - \sum_6 = 0$; $\alpha = \sum_3 - \sum_2 = 0$; $\psi = \sum_5 - \sum_4 =$; $|A_{14}| = 0$

So, in total we have the value of determinant $|A|$ of 4x4 matrix $|A| = |A_{11}| + |A_{12}| + |A_{13}| + |A_{14}|$, $|A| = 55$.

Appendix 45

Algorithm for evaluation of partial part $|A^1_{11}|$ **of complex determinant** $|A| = \sum|A_{11}| + \sum|A_{12}| + \sum|A_{13}| + \sum|A_{14}| + \sum|A_{15}|$ **of 5x5 matrix by Expansion method for row(i_1) of formula (58).**

<u>Example 6:</u>

<u>Problem</u>
Solve the partial part $|A^1_{11}|$ of the first row (i_1) formula (58) of 5x5 matrix for complex determinant $|A|$ by Expansion method.

<u>Solving</u>
In accordance with above-indicated algorithm of formula (58) for the first row(i_1) and Table 6 *"The general steps for evaluation of partial part* $|A^1_{11}|$ *of complex determinant* $|A|$ *of 5x5 matrix by Expansion method"* ,we have the following data for partial part $|A^1_{11}|$ of complex determinant $|A|$ of matrix 5x5 at numerical data (**):

$$A = \begin{vmatrix} 1 & 2 & -1 & 4 & 5 \\ 4 & 1 & -3 & 5 & 6 \\ 3 & 3 & -4 & 2 & 1 \\ 6 & -1 & 2 & 4 & -1 \\ -3 & 2 & -1 & -4 & 1 \end{vmatrix} \quad (**)$$

507

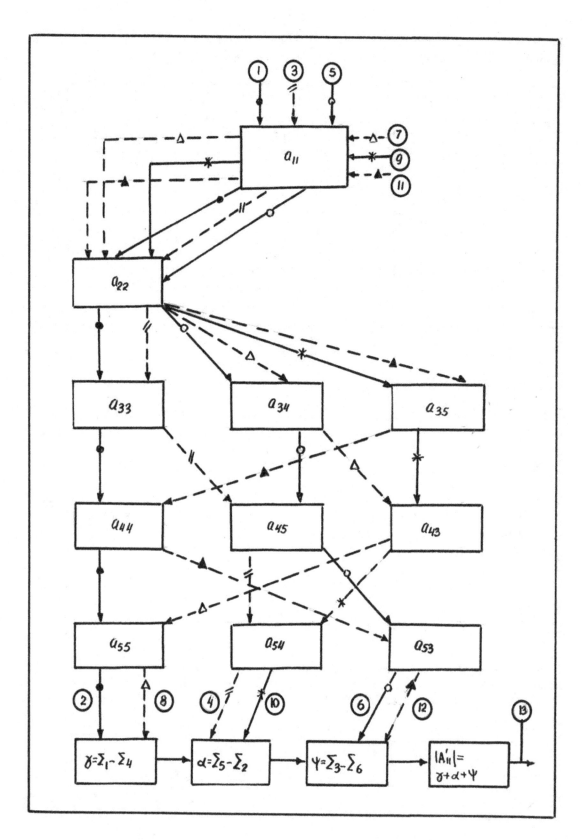

508

$$A=\begin{vmatrix} a_{11} & a_{12} & a_{13} & a_{14} & a_{15} \\ a_{21} & a_{22} & a_{23} & a_{24} & a_{25} \\ a_{31} & a_{32} & a_{33} & a_{34} & a_{35} \\ a_{41} & a_{42} & a_{43} & a_{44} & a_{45} \\ a_{51} & a_{52} & a_{53} & a_{54} & a_{55} \end{vmatrix}$$

So, the value of determinant of the 5x5 matrix *at expansion of its elements to the first row*, after of some transformations, can be defined by the following formula in general view:

$$
\begin{aligned}
|A| =\, & a_{11}a_{22}[a_{33}(a_{44}a_{55}-a_{45}a_{54})+a_{34}(a_{45}a_{53}-a_{43}a_{55})+a_{35}(a_{43}a_{54}-a_{44}a_{53})]+ \\
& +a_{11}a_{23}[a_{32}(a_{45}a_{54}-a_{44}a_{55})+a_{34}(a_{42}a_{55}-a_{45}a_{52})+a_{35}(a_{44}a_{52}-a_{42}a_{54})]+ \\
& +a_{11}a_{24}[a_{32}(a_{43}a_{55}-a_{45}a_{53})+a_{33}(a_{45}a_{52}-a_{42}a_{55})+a_{35}(a_{42}a_{53}-a_{43}a_{52})]+ \\
& +a_{11}a_{25}[a_{32}(a_{44}a_{53}-a_{43}a_{54})+a_{33}(a_{42}a_{54}-a_{44}a_{52})+a_{34}(a_{43}a_{52}-a_{42}a_{53})]+ \\[4pt]
& +a_{12}a_{21}[a_{33}(a_{45}a_{54}-a_{44}a_{55})+a_{34}(a_{43}a_{55}-a_{45}a_{53})+a_{35}(a_{44}a_{53}-a_{43}a_{54})]+ \\
& +a_{12}a_{23}[a_{31}(a_{44}a_{55}-a_{45}a_{54})+a_{34}(a_{45}a_{51}-a_{41}a_{55})+a_{35}(a_{41}a_{54}-a_{44}a_{51})]+ \\
& +a_{12}a_{24}[a_{31}(a_{45}a_{53}-a_{43}a_{55})+a_{33}(a_{41}a_{55}-a_{45}a_{51})+a_{35}(a_{43}a_{51}-a_{41}a_{53})]+ \\
& +a_{12}a_{25}[a_{31}(a_{43}a_{54}-a_{44}a_{53})+a_{33}(a_{44}a_{51}-a_{41}a_{54})+a_{34}(a_{41}a_{53}-a_{43}a_{51})]+ \\[4pt]
& +a_{13}a_{21}[a_{32}(a_{44}a_{55}-a_{45}a_{54})+a_{34}(a_{45}a_{52}-a_{42}a_{55})+a_{35}(a_{42}a_{54}-a_{44}a_{52})]+ \\
& +a_{13}a_{22}[a_{31}(a_{45}a_{54}-a_{44}a_{55})+a_{34}(a_{41}a_{55}-a_{45}a_{51})+a_{35}(a_{44}a_{51}-a_{41}a_{54})]+ \\
& +a_{13}a_{24}[a_{31}(a_{42}a_{55}-a_{45}a_{52})+a_{32}(a_{45}a_{51}-a_{41}a_{55})+a_{35}(a_{41}a_{52}-a_{42}a_{51})]+ \\
& +a_{13}a_{25}[a_{31}(a_{44}a_{52}-a_{42}a_{54})+a_{32}(a_{41}a_{54}-a_{44}a_{51})+a_{34}(a_{42}a_{51}-a_{41}a_{52})]+ \\[4pt]
& +a_{14}a_{21}[a_{32}(a_{45}a_{53}-a_{43}a_{55})+a_{33}(a_{42}a_{55}-a_{45}a_{52})+a_{35}(a_{43}a_{52}-a_{42}a_{53})]+ \\
& +a_{14}a_{22}[a_{31}(a_{43}a_{55}-a_{45}a_{53})+a_{33}(a_{45}a_{51}-a_{41}a_{55})+a_{35}(a_{41}a_{53}-a_{43}a_{51})]+ \\
& +a_{14}a_{23}[a_{31}(a_{45}a_{52}-a_{42}a_{55})+a_{32}(a_{41}a_{55}-a_{45}a_{51})+a_{35}(a_{42}a_{51}-a_{41}a_{52})]+ \\
& +a_{14}a_{25}[a_{31}(a_{42}a_{53}-a_{43}a_{52})+a_{32}(a_{43}a_{51}-a_{41}a_{53})+a_{33}(a_{41}a_{52}-a_{42}a_{51})]+ \\[4pt]
& +a_{15}a_{21}[a_{32}(a_{43}a_{54}-a_{44}a_{53})+a_{33}(a_{44}a_{52}-a_{42}a_{54})+a_{34}(a_{42}a_{53}-a_{43}a_{52})]+ \\
& +a_{15}a_{22}[a_{31}(a_{44}a_{53}-a_{43}a_{54})+a_{33}(a_{41}a_{54}-a_{44}a_{51})+a_{34}(a_{43}a_{51}-a_{41}a_{53})]+ \\
& +a_{15}a_{23}[a_{31}(a_{42}a_{54}-a_{44}a_{52})+a_{32}(a_{44}a_{51}-a_{41}a_{54})+a_{34}(a_{41}a_{52}-a_{42}a_{51})]+ \\
& +a_{15}a_{24}[a_{31}(a_{43}a_{52}-a_{42}a_{53})+a_{32}(a_{41}a_{53}-a_{43}a_{51})+a_{33}(a_{42}a_{51}-a_{41}a_{52})]
\end{aligned} \qquad (58)
$$

509

Table 6 The general steps for evaluation of partial part $|A^1_{11}|$ of complex determinant $|A|$ of 5x5 matrix by Expansion method for row(i_1) of formula (58)

Number of steps		Elements involved in calculation process																										Sum
in	out	a_{11}	a_{12}	a_{13}	a_{14}	a_{15}	a_{21}	a_{22}	a_{23}	a_{24}	a_{25}	a_{31}	a_{32}	a_{33}	a_{34}	a_{35}	a_{41}	a_{42}	a_{43}	a_{44}	a_{45}	a_{51}	a_{52}	a_{53}	a_{54}	a_{55}	
1	2	X						X						X						X						X	Σ_1
3	4	X						X						X							X				X		Σ_2
5	6	X						X							X						X			X			Σ_3
7	8	X						X							X				X							X	Σ_4
9	10	X						X								X			X						X		Σ_5
11	12	X						X								X				X				X			Σ_6
2	8																										γ
4	10																										α
6	12																										ψ
$|A^1_{11}|=$ $\gamma+\alpha+\psi$																											$|A^1_{11}|$

where,

$\Sigma_1=a_{11}a_{22}a_{33}a_{44}a_{55}=-16$; $\Sigma_2=a_{11}a_{22}a_{33}a_{45}a_{54}=-16$; $\Sigma_3=a_{11}a_{22}a_{34}a_{45}a_{53}=2$; $\Sigma_4=a_{11}a_{22}a_{34}a_{43}a_{55}=4$;
$\Sigma_5=a_{11}a_{22}a_{35}a_{43}a_{54}=-8$; $\Sigma_6=a_{11}a_{22}a_{35}a_{44}a_{53}=-8$; $\gamma=\Sigma_1-\Sigma_4=-20$; $\alpha=\Sigma_5-\Sigma_2=8$; $\psi=\Sigma_3-\Sigma_6=10$;
$|A^1_{11}|=-2$.

Appendix 46

Algorithm for evaluation of partial part $|A^2_{11}|$ of complex determinant $|A|$ of 5x5 matrix by Expansion method for the second row(i_2) of formula (58)

Example 7:

Problem
Solve the partial part $|A^2_{11}|$ of the second row (i_2) of formula (58) of matrix 5x5 for complex determinant $|A|$ by Expansion method.

Solving
In accordance with above-indicated algorithm of formula (58) for the second row (i_2) and Table 7 "*The general steps for evaluation of partial part $|A^2_{11}|$ by Expansion method*", we have the following data for partial part $|A^2_{11}|$ of complex determinant $|A|$ of matrix 5x5 at numerical data (**):

510

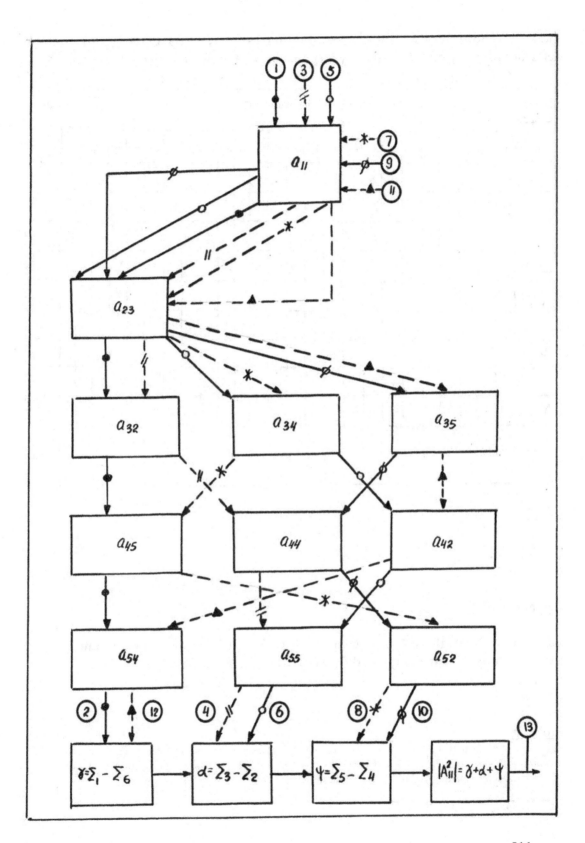

Table 7 The general steps for evaluation for partial part $|A^2_{11}|$ of complex determinant $|A|$ of 5x5 matrix by Expansion method for row(i_2) of formula (58)

Number of steps in	out	a11	a12	a13	a14	a15	a21	a22	a23	a24	a25	a31	a32	a33	a34	a35	a41	a42	a43	a44	a45	a51	a52	a53	a54	a55	Sum
1	2	X							X				X								X				X		Σ_1
3	4	X							X				X							X					X		Σ_2
5	6	X							X						X			X							X		Σ_3
7	8	X							X						X						X		X				Σ_4
9	10	X							X							X				X			X				Σ_5
11	12	X							X							X		X							X		Σ_6
2	12																										γ
6	4																										α
10	8																										ψ
$|A^2_{11}|=$ $\gamma+\alpha+\psi$																											$|A^2_{11}|$

where ,

$$\Sigma_1=a_{11}a_{23}a_{32}a_{45}a_{54}=-36; \ \Sigma_2=a_{11}a_{23}a_{32}a_{44}a_{55}=-36; \Sigma_3=a_{11}a_{23}a_{34}a_{42}a_{55}=6; \ \Sigma_4=a_{11}a_{23}a_{34}a_{45}a_{52}=12;$$
$$\Sigma_5=a_{11}a_{23}a_{35}a_{44}a_{52}=-24; \ \Sigma_6=a_{11}a_{23}a_{35}a_{42}a_{54}=-12; \ \gamma=\Sigma_1-\Sigma_6=-24; \ \alpha=\Sigma_3-\Sigma_2=42; \ \psi=\Sigma_5-\Sigma_4=-36;$$
$$|A^2_{11}|=-18.$$

Appendix 47

Algorithm for evaluation of partial part $|A^3_{11}|$ of complex determinant $|A|$ of 5x5 matrix by Expansion method for row (i_3) of formula (58)

Example 8:

Problem

Solve the partial part $|A^3_{11}|$ of the third row(i_3) of formula (58) and Table 8 "*The general steps for evaluation of partial part $|A^3_{11}|$ by Expansion method*, we have the following data for partial part $|A^3_{11}|$ of complex determinant $|A|$ of matrix 5x5 at numerical data (**):

512

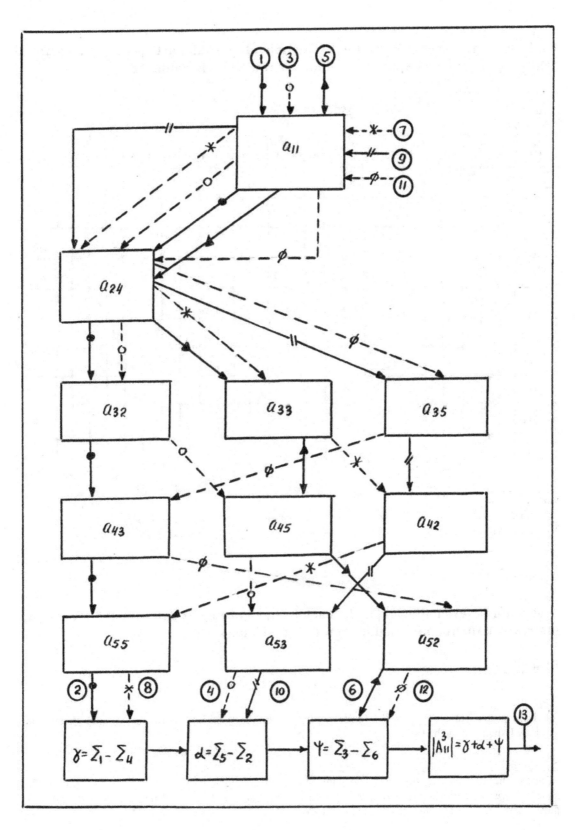

513

Table 8 The general steps for evaluation of partial part $\left|A^3_{11}\right|$ of complex determinant $\left|A\right|$ by Expansion method for row (i_3) of formula (58)

Number of steps		Elements involved in calculation process																									Sum
in	out	a_{11}	a_{12}	a_{13}	a_{14}	a_{15}	a_{21}	a_{22}	a_{23}	a_{24}	a_{25}	a_{31}	a_{32}	a_{33}	a_{34}	a_{35}	a_{41}	a_{42}	a_{43}	a_{44}	a_{45}	a_{51}	a_{52}	a_{53}	a_{54}	a_{55}	
1	2	×								×			×						×							×	Σ_1
3	4	×								×			×								×			×			Σ_2
5	6	×								×				×							×		×				Σ_3
7	8	×								×				×				×								×	Σ_4
9	10	×								×						×		×						×			Σ_5
11	12	×								×						×			×				×				Σ_6
2	8																										γ
4	10																										α
6	12																										ψ
$\left|A^3_{11}\right|=$ $\gamma+\alpha+\psi$																											$\left|A^3_{11}\right|$

where,

$\Sigma_1 = a_{11}a_{24}a_{32}a_{43}a_{55} = 30$; $\Sigma_2 = a_{11}a_{24}a_{32}a_{45}a_{53} = 15$; $\Sigma_3 = a_{11}a_{24}a_{33}a_{45}a_{52} = 40$; $\Sigma_4 = a_{11}a_{24}a_{33}a_{42}a_{55} = 20$; $\Sigma_5 = a_{11}a_{24}a_{35}a_{42}a_{53} = 5$; $\Sigma_6 = a_{11}a_{24}a_{35}a_{43}a_{52} = 20$; $\gamma = \Sigma_1 - \Sigma_4 = 10$; $\alpha = \Sigma_5 - \Sigma_2 = -10$; $\psi = \Sigma_3 - \Sigma_6 = 20$; $\left|A^3_{11}\right| = 20$.

Appendix 48

Algorithm for evaluation of partial part $\left|A^4_{11}\right|$ of complex determinant $\left|A\right|$ of 5x5 matrix by Expansion method for row (i_4) of formula (58)

Example 9:

Problem

Solve the partial part $\left|A^4_{11}\right|$ of the fourth row(i_4) formula (58) of 5x5 matrix for complex determinant $\left|A\right|$ by Expansion method.

Solving

In accordance with above-indicated algorithm of formula (58 for the fourth row (i_4) and Table 9 "*The general steps for evaluation of partial part $\left|A^4_{11}\right|$ by Expansion method*", we have the following data for partial part $\left|A^4_{11}\right|$ of complex determinant $\left|A\right|$ of matrix 5x5 at numerical data(**):

514

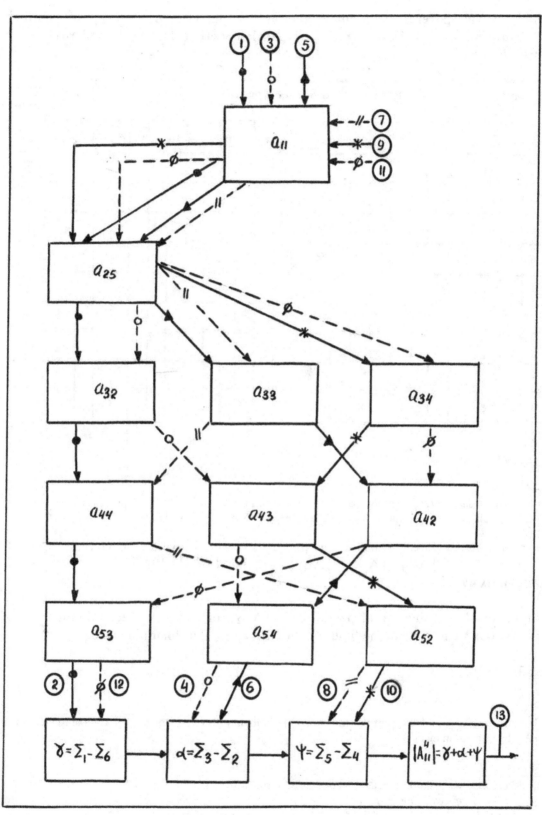

515

Table 9 The general steps for evaluation of partial part $|A^4_{11}|$ of complex determinant $|A|$ of 5x5 matrix by Expansion method for the fourth row (i_4) of formula (58)

Number of steps in	out	a_{11}	a_{12}	a_{13}	a_{14}	a_{15}	a_{21}	a_{22}	a_{23}	a_{24}	a_{25}	a_{31}	a_{32}	a_{33}	a_{34}	a_{35}	a_{41}	a_{42}	a_{43}	a_{44}	a_{45}	a_{51}	a_{52}	a_{53}	a_{54}	a_{55}	Sum
1	2	X									X		X							X				X			Σ_1
3	4	X									X		X						X						X		Σ_2
5	6	X									X			X				X							X		Σ_3
7	8	X									X			X						X			X				Σ_4
9	10	X									X				X				X				X				Σ_5
11	12	X									X				X			X						X			Σ_6
2	12																										γ
4	6																										α
8	10																										ψ
$|A^4_{11}|=$ $\gamma+\alpha+\psi$																											$|A^4_{11}|$

where,

$\Sigma_1=a_{11}a_{25}a_{32}a_{44}a_{53}=-72$; $\Sigma_2=a_{11}a_{25}a_{32}a_{43}a_{54}=-144$; $\Sigma_3=a_{11}a_{25}a_{33}a_{42}a_{54}=-96$; $\Sigma_4=a_{11}a_{25}a_{33}a_{44}a_{52}=-192$; $\Sigma_5=a_{11}a_{25}a_{34}a_{43}a_{52}=48$; $\Sigma_6=a_{11}a_{25}a_{34}a_{42}a_{53}=12$; $\gamma=\Sigma_1-\Sigma_6=-84$; $\alpha=\Sigma_3-\Sigma_2=38$; $\psi=\Sigma_5-\Sigma_4=240$; $|A^4_{11}|=194$.

So, in total we have for summary partial part $|A_{11}|$ = the following value which is equal:

$$|A_{11}|=|A^1_{11}|+|A^2_{11}|+|A^3_{11}|+|A^4_{11}|=194$$

Appendix 49

Algorithm for evaluation of partial part $|A^5_{12}|$ of complex determinant $|A|$ of 5x5 matrix by Expansion method for the fifth row (i_5) of formula (58):

Example 10:

Problem
Solve the partial part $|A^5_{12}|$ of the fifth row (i_5) formula (58) of 5x5 matrix for complex determinant $|A|$ by Expansion method.

Solving
In accordance with above-indicated algorithm of formula (58) for the fifth row(i_5) and Table 10 *"The general steps for evaluation of partial part $|A^5_{12}|$ by Expansion method*, we have the following data for partial part $|A^5_{12}|$ of complex determinant $|A|$ of 5x5 matrix at numerical data (**):

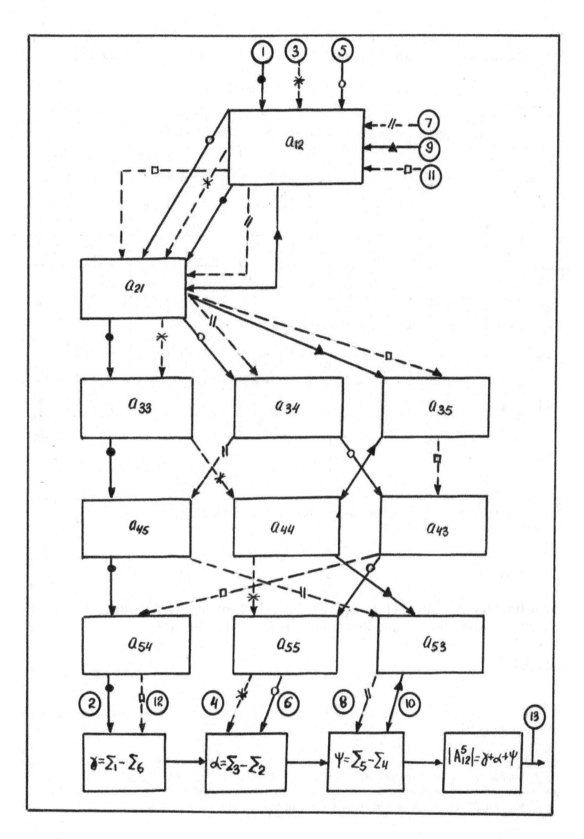

517

Number of steps		Elements involved in calculation process																									Sum
in	out	a_{11}	a_{12}	a_{13}	a_{14}	a_{15}	a_{21}	a_{22}	a_{23}	a_{24}	a_{25}	a_{31}	a_{32}	a_{33}	a_{34}	a_{35}	a_{41}	a_{42}	a_{43}	a_{44}	a_{45}	a_{51}	a_{52}	a_{53}	a_{54}	a_{55}	
1	2		X				X							X							X				X		Σ_1
3	4		X				X							X						X						X	Σ_2
5	6		X				X								X				X							X	Σ_3
7	8		X				X								X						X			X			Σ_4
9	10		X				X									X				X				X			Σ_5
11	12		X				X									X			X						X		Σ_6
2	12																										γ
6	4																										α
10	8																										ψ
$\left|A^5_{11}\right|=$ $\gamma+\alpha+\psi$																											$\left|A^5_{11}\right|$

where,

$\Sigma_1 = a_{12}a_{21}a_{33}a_{45}a_{54} = -128$; $\Sigma_2 = a_{12}a_{21}a_{33}a_{44}a_{55} = -128$; $\Sigma_3 = a_{12}a_{21}a_{34}a_{43}a_{55} = 32$; $\Sigma_4 = a_{12}a_{21}a_{34}a_{45}a_{53} = 16$; $\Sigma_5 = a_{12}a_{21}a_{35}a_{44}a_{53} = -32$; $\Sigma_6 = a_{12}a_{21}a_{35}a_{43}a_{54} = -64$; $\gamma = \Sigma_1 - \Sigma_6 = -64$; $\alpha = \Sigma_3 - \Sigma_2 = 160$; $\psi = \Sigma_5 - \Sigma_4 = -48$; $\left|A^5_{12}\right| = 48$.

Appendix 50

Algorithm for evaluation of partial part $\left|A^6_{12}\right|$ of complex determinant $|A|$ of 5x5 matrix by Expansion method for the sixth row (i_6) of formula (58)

Example 11:

Problem
Solve the partial part $\left|A^6_{12}\right|$ of the sixth row(i_6) formula (58) of 5x5 matrix for complex determinant $|A|$ by Expansion method

Solving
In accordance with above-indicated algorithm of formula (58) for the sixth row(i_6) and Table 11 "*The general steps for evaluation of partial part $\left|A^6_{12}\right|$ by Expansion method* ", we have the following data for partial part $\left|A^6_{12}\right|$ of complex determinant $|A|$ of matrix 5x5 at numerical data(**):

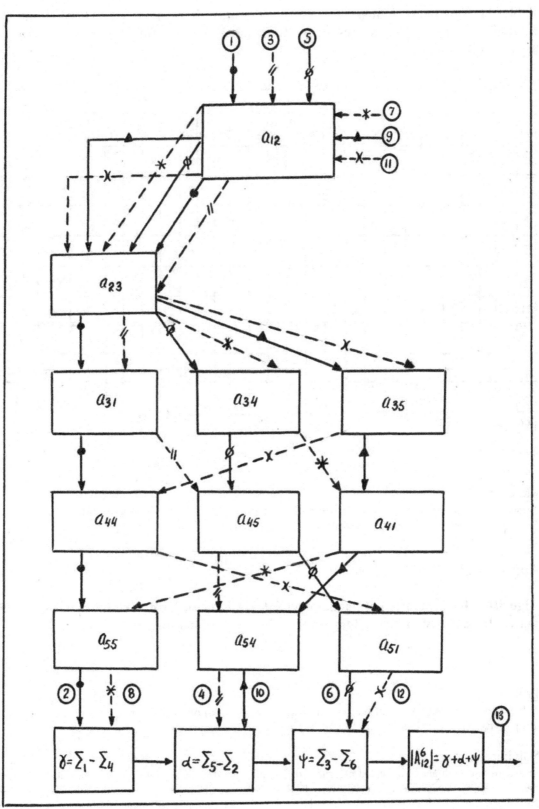

519

Table 11 The general steps for evaluation of partial part $|A^6_{12}|$ of complex determinant $|A|$ of 5x5 matrix by Expansion method for the sixth row(i_6) of formula (58)

Number of steps		Elements involved in calculation process																									Sum
in	out	a_{11}	a_{12}	a_{13}	a_{14}	a_{15}	a_{21}	a_{22}	a_{23}	a_{24}	a_{25}	a_{31}	a_{32}	a_{33}	a_{34}	a_{35}	a_{41}	a_{42}	a_{43}	a_{44}	a_{45}	a_{51}	a_{52}	a_{53}	a_{54}	a_{55}	
1	2		X						X			X								X						X	Σ_1
3	4		X						X			X									X				X		Σ_2
5	6		X						X						X						X	X					Σ_3
7	8		X						X						X		X									X	Σ_4
9	10		X						X							X	X								X		Σ_5
11	12		X						X							X				X		X					Σ_6
2	8																										γ
10	4																										α
6	12																										ψ
$\|A^6_{12}\|=$ $\gamma+\alpha+\psi$																											$\|A^6_{12}\|$

where,

$\Sigma_1=a_{12}a_{23}a_{31}a_{44}a_{55}=-72$; $\Sigma_2=a_{12}a_{23}a_{31}a_{45}a_{54}=-72$; $\Sigma_3=a_{12}a_{23}a_{34}a_{45}a_{51}=-36$; $\Sigma_4=a_{12}a_{23}a_{34}a_{41}a_{55}=-72$;
$\Sigma_5=a_{12}a_{23}a_{35}a_{41}a_{54}=144$; $\Sigma_6=a_{12}a_{23}a_{35}a_{44}a_{51}=72$; $\gamma=\Sigma_1-\Sigma_4=0$; $\alpha=\Sigma_5-\Sigma_2=216$; $\psi=\Sigma_3-\Sigma_6=-108$;
$|A^6_{12}|=108$.

Appendix 51

Algorithm for evaluation of partial part $|A^7_{12}|$ of complex determinant $|A|$ of 5x5 matrix by Expansion method for seventh row (i_7) of formula (58)

Example 12:

Problem
Solve the partial part $|A^7_{12}|$ of the seventh row(i_7) formula (58) of 5x5 matrix for complex determinant $|A|$ by Expansion method.

Solving
In accordance with above-indicated algorithm of formula (58) for the seventh row(i_7) and Table 12 "The general steps for evaluation of partial part $|A^7_{12}|$ by Expansion method", we have the following data for partial part $|A^7_{12}|$ of complex determinant $|A|$ of matrix 5x5 at numerical data(**):

520

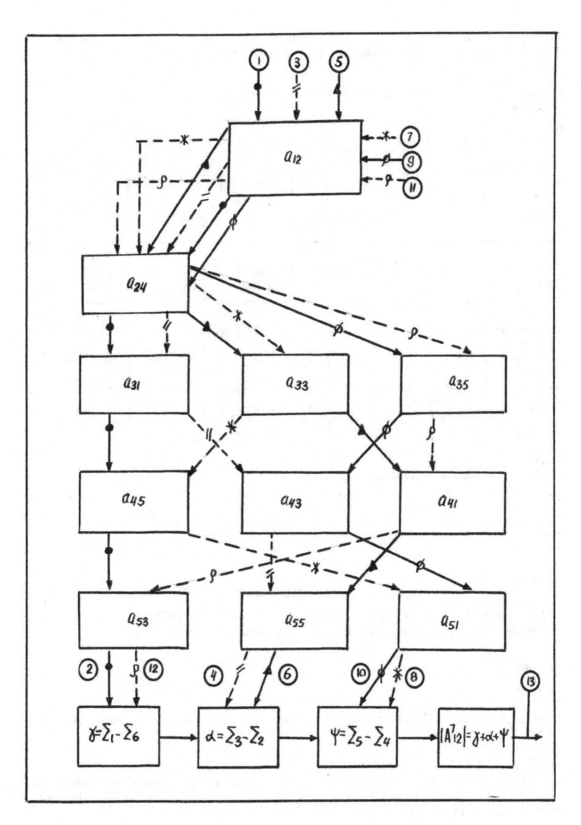

Table 12 The general steps for evaluation of partial part $|A^7{}_{12}|$ of complex data $|A|$ of 5x5 matrix by Expansion method for seventh row(i_7) of formula (58)

Number of steps		Elements involved in calculation process																									Sum
in	out	a_{11}	a_{12}	a_{13}	a_{14}	a_{15}	a_{21}	a_{22}	a_{23}	a_{24}	a_{25}	a_{31}	a_{32}	a_{33}	a_{34}	a_{35}	a_{41}	a_{42}	a_{43}	a_{44}	a_{45}	a_{51}	a_{52}	a_{53}	a_{54}	a_{55}	
1	2	X								X		X									X			X			Σ_1
3	4	X								X		X							X							X	Σ_2
5	6	X								X				X			X									X	Σ_3
7	8	X								X				X							X	X					Σ_4
9	10	X								X						X			X			X					Σ_5
11	12	X								X						X	X							X			Σ_6
2	12																										γ
6	4																										α
10	8																										ψ
$\|A^7{}_{12}\|=$ $\gamma+\alpha+\psi$																											$\|A^7{}_{12}\|$

where,

$\Sigma_1=a_{12}a_{24}a_{31}a_{45}a_{53}=30$; $\Sigma_2=a_{12}a_{24}a_{31}a_{43}a_{55}=60$; $\Sigma_3=a_{12}a_{24}a_{33}a_{41}a_{55}=-240$; $\Sigma_4=a_{12}a_{24}a_{33}a_{45}a_{51}=-120$; $\Sigma_5=a_{12}a_{24}a_{35}a_{43}a_{51}=-60$; $\Sigma_6=a_{12}a_{24}a_{35}a_{41}a_{53}=-60$; $\gamma=\Sigma_1-\Sigma_6=90$; $\alpha=\Sigma_3-\Sigma_2=-300$; $\psi=\Sigma_5-\Sigma_4=60$; $|A^7{}_{12}|=-150$.

Appendix 52

Algorithm for evaluation of partial part $|A^8{}_{12}|$ of complex determinant $|A|$ of 5x5 matrix by Expansion method for the eighth row(i_8) of formula (58)

Example 13:

Problem

Solve the partial part $|A^8{}_{12}|$ of the eighth row(i_8) formula (58) of 5x5 matrix for complex determinant $|A|$ by Expansion method.

Solving

In accordance with above-indicated algorithm of formula (58) for the eighth row (i_8) and Table 13 " *The general steps for evaluation of partial part $|A^8{}_{12}|$ by Expansion method*", we have the following data for partial part $|A^8{}_{12}|$ of complex determinant $|A|$ of matrix 5x5 at numerical data(**):

522

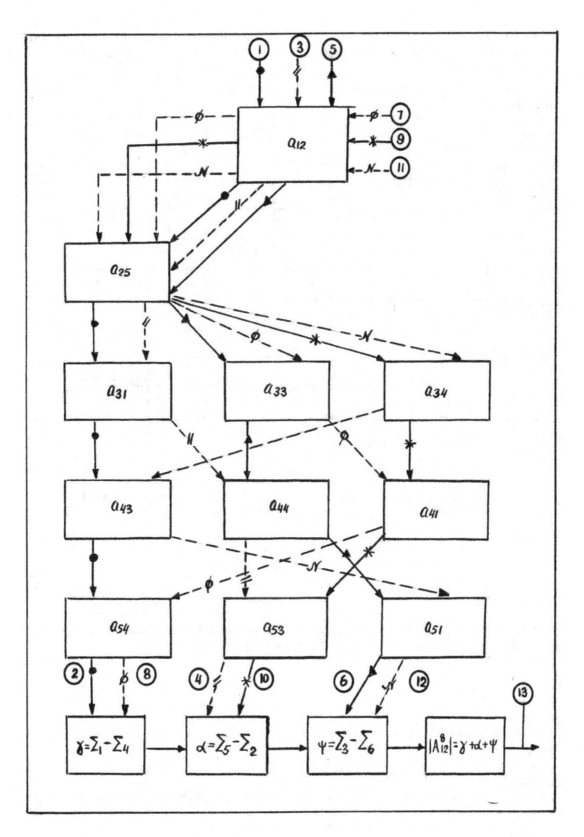

Table 13 The general steps for evaluation of partial part $|A^8_{12}|$ of complex data $|A|$ of 5x5 matrix by Expansion method for eighth row (i_8) of formula (58)

Number of steps		Elements involved in calculation process																								Sum
in	out	a_{11}	a_{12}	a_{13}	a_{14}	a_{15}	a_{21}	a_{22}	a_{23}	a_{24}	a_{25}	a_{31}	a_{32}	a_{33}	a_{34}	a_{35}	a_{41}	a_{42}	a_{43}	a_{44}	a_{45}	a_{51}	a_{52}	a_{53}	a_{54}	a_{55}	
1	2		✕								✕	✕							✕						✕		Σ_1
3	4		✕								✕	✕								✕				✕			Σ_2
5	6		✕								✕			✕						✕		✕					Σ_3
7	8		✕								✕			✕			✕								✕		Σ_4
9	10		✕								✕				✕		✕							✕			Σ_5
11	12		✕								✕				✕				✕			✕					Σ_6
2	8																										γ
10	4																										α
6	12																										ψ
$|A^8_{12}|=$ $\gamma+\alpha+\psi$																										$|A^8_{12}|$

where,

$\Sigma_1 = a_{12}a_{25}a_{31}a_{43}a_{54} = -288; \Sigma_2 = a_{12}a_{25}a_{31}a_{44}a_{53} = -144; \Sigma_3 = a_{12}a_{25}a_{33}a_{44}a_{51} = 576; \Sigma_4 = a_{12}a_{25}a_{33}a_{41}a_{54} = 1152;$
$\Sigma_5 = a_{12}a_{25}a_{34}a_{41}a_{53} = -144; \Sigma_6 = a_{12}a_{25}a_{34}a_{43}a_{51} = -144; \gamma = \Sigma_1 - \Sigma_4 = -1440; \alpha = \Sigma_5 - \Sigma_2 = 0; \psi = \Sigma_3 - \Sigma_6 = 720;$
$|A^8_{12}| = -720.$

So ,the summary value of complex determinant $|A_{12}|$ is equal:
$$|A_{12}| = |A^5_{12}| + |A^6_{12}| + |A^7_{12}| + |A^8_{12}| = -714$$

Appendix 53

Algorithm for evaluation of partial part $|A^9_{13}|$ of complex determinant $|A|$ of 5x5 matrix by Expansion method for the ninth row(i_9) of formula (58)

<u>Example 14:</u>

<u>Problem</u>
Solve the partial part $|A^9_{13}|$ of the ninth row (i_9) of 5x5 matrix for complex determinant $|A|$ by Expansion method.

<u>Solving</u>
In accordance with above-indicated algorithm of formula(58) and Table 14 "The general steps for evaluation of partial part $|A^9_{13}|$ by Expansion method" ,we have the following data for partial part $|A^9_{13}|$ of complex determinant $|A|$ of matrix 5x5 at numerical data(**):

524

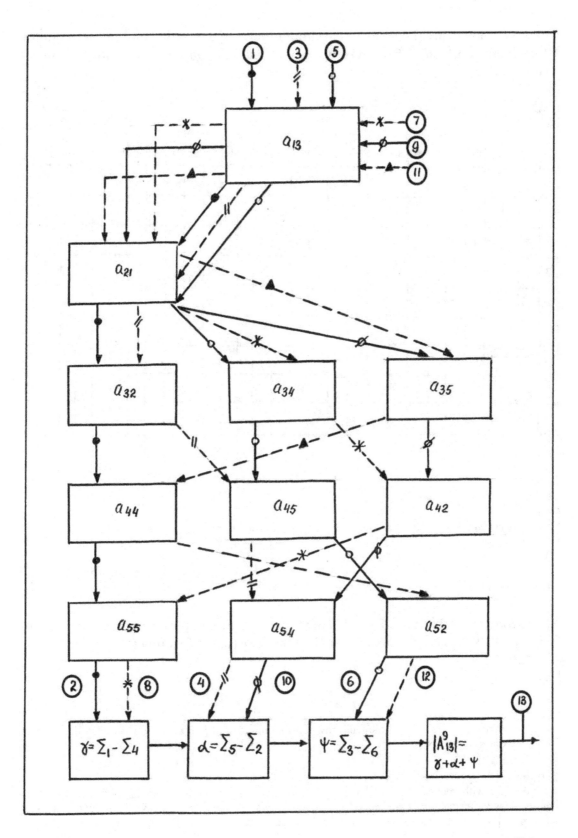

Table 14 The general steps for evaluation of partial part $\left|A^9_{13}\right|$ of complex determinant $\left|A\right|$ of 5x5 matrix by Expansion method for the ninth(i_9) row of formula (58)

Number of steps		Elements involved in calculation process																									Sum
in	out	a_{11}	a_{12}	a_{13}	a_{14}	a_{15}	a_{21}	a_{22}	a_{23}	a_{24}	a_{25}	a_{31}	a_{32}	a_{33}	a_{34}	a_{35}	a_{41}	a_{42}	a_{43}	a_{44}	a_{45}	a_{51}	a_{52}	a_{53}	a_{54}	a_{55}	
1	2			X			X						X							X					X		Σ_1
3	4			X			X						X								X				X		Σ_2
5	6			X			X								X						X		X				Σ_3
7	8			X			X								X			X								X	Σ_4
9	10			X			X									X		X							X		Σ_5
11	12			X			X									X				X			X				Σ_6
2	8																										γ
10	4																										α
6	12																										ψ
$\left|A^9_{13}\right|=$ $\gamma+\alpha+\psi$																											$\left|A^9_{13}\right|$

where,

$\Sigma_1=a_{13}a_{21}a_{32}a_{44}a_{55}=-48$; $\Sigma_2=a_{13}a_{21}a_{32}a_{45}a_{54}=-48$; $\Sigma_3=a_{13}a_{21}a_{34}a_{45}a_{52}=16$; $\Sigma_4=a_{13}a_{21}a_{34}a_{42}a_{55}=8$;
$\Sigma_5=a_{13}a_{21}a_{35}a_{42}a_{54}=-16$; $\Sigma_6=a_{13}a_{21}a_{35}a_{44}a_{52}=-32$; $\gamma=\Sigma_1-\Sigma_4=-56$; $\alpha=\Sigma_5-\Sigma_2=32$; $\psi=\Sigma_3-\Sigma_6=48$;
$\left|A^9_{13}\right|=24$.

Appendix 54

Algorithm for evaluation of partial part $\left|A^{10}_{13}\right|$ of complex determinant $\left|A\right|$ of 5x5 matrix by Expansion method for the tenth row(i_{10}) of formula (58)

Example 15:

Problem

Solve the partial part $\left|A^{10}_{13}\right|$ of the tenth row(i_{10}) formula (58) of 5x5 matrix for complex determinant $\left|A\right|$ by Expansion method.

Solving

In accordance with above-indicated algorithm of formula (58) for the tenth row(i_{10}) and Table 15 "*The general steps for evaluation of partial part $\left|A^{10}_{13}\right|$ by Expansion method*", we have the following data for partial part $\left|A^{10}_{13}\right|$ of complex determinant $\left|A\right|$ of matrix 5x5 at numerical data (**):

526

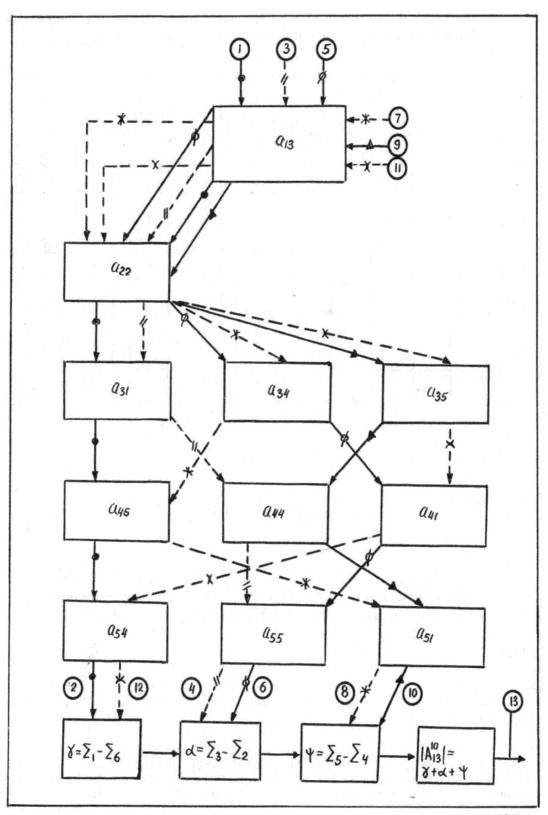

527

Table 15 The general steps for evaluation of partial part $\left|A^{10}_{13}\right|$ of complex data $\left|A\right|$ of 5x5 matrix by Expansion method for the tenth row(i_{10}) of formula (58)

Number of steps		a_{11}	a_{12}	a_{13}	a_{14}	a_{15}	a_{21}	a_{22}	a_{23}	a_{24}	a_{25}	a_{31}	a_{32}	a_{33}	a_{34}	a_{35}	a_{41}	a_{42}	a_{43}	a_{44}	a_{45}	a_{51}	a_{52}	a_{53}	a_{54}	a_{55}	Sum
in	out																										
1	2			X				X				X									X				X		Σ_1
3	4			X				X				X								X						X	Σ_2
5	6			X				X							X		X									X	Σ_3
7	8			X				X							X						X	X					Σ_4
9	10			X				X								X				X		X					Σ_5
11	12			X				X								X	X								X		Σ_6
2	12																										γ
6	4																										α
10	8																										ψ
$\left|A^{10}_{13}\right|=$ $\gamma+\alpha+\psi$																											$\left|A^{10}_{13}\right|$

where,

$\Sigma_1=a_{13}a_{22}a_{31}a_{45}a_{54}=-12$; $\Sigma_2=a_{13}a_{22}a_{31}a_{44}a_{55}=-12$; $\Sigma_3=a_{13}a_{22}a_{34}a_{41}a_{55}=-12$; $\Sigma_4=a_{13}a_{22}a_{34}a_{45}a_{51}=-6$;
$\Sigma_5=a_{13}a_{22}a_{35}a_{44}a_{51}=12$; $\Sigma_6=a_{13}a_{22}a_{35}a_{41}a_{54}=24$; $\gamma=\Sigma_1-\Sigma_6=-36$; $\alpha=\Sigma_3-\Sigma_2=0$; $\psi=\Sigma_5-\Sigma_4=18$; $\left|A^{10}_{13}\right|=-18$.

Appendix 55

Algorithm for evaluation of partial part $\left|A^{11}_{13}\right|$ of complex determinant $\left|A\right|$ of 5x5 matrix by Expansion method for the eleventh row(i_{11}) of formula (58)

Example 16:

Problem
Solve the partial part $\left|A^{11}_{13}\right|$ of the eleventh row(i_{11}) formula (58) of 5x5 matrix for complex determinant $\left|A\right|$ by Expansion method.

Solving
In accordance with above-indicated algorithm of formula (58) for the eleventh row(i_{11}) and Table 16 *"The general steps for evaluation of partial part $\left|A^{11}_{13}\right|$ by Expansion method "*, we have the following data for partial part $\left|A^{11}_{13}\right|$ of complex determinant $\left|A\right|$ of matrix 5x5 at numerical data(**):

528

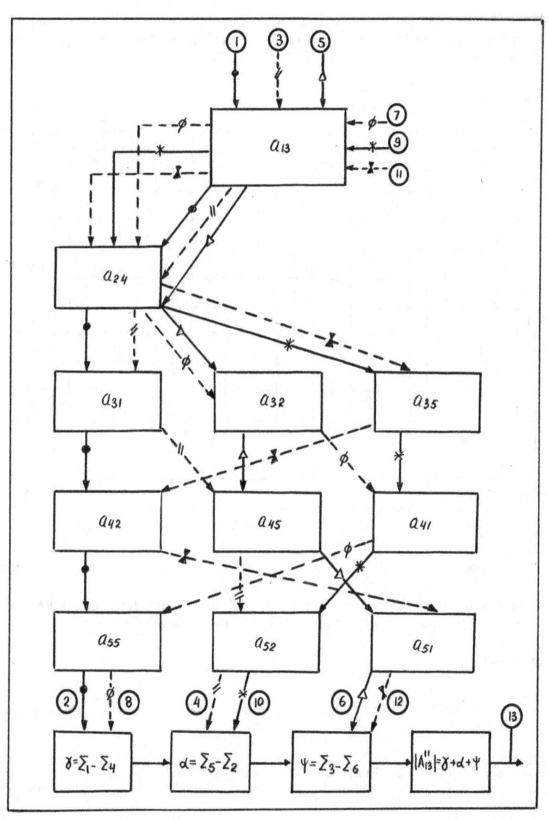

Table 15 The general steps for evaluation of partial part $\left|A^{11}_{13}\right|$ of complex determinant $\left|A\right|$ of 5x5 matrix by Expansion method for the eleventh row(i_{11}) of formula (58)

Number of steps		Elements involved in calculation process																										Sum
in	out	a_{11}	a_{12}	a_{13}	a_{14}	a_{15}	a_{21}	a_{22}	a_{23}	a_{24}	a_{25}	a_{31}	a_{32}	a_{33}	a_{34}	a_{35}	a_{41}	a_{42}	a_{43}	a_{44}	a_{45}	a_{51}	a_{52}	a_{53}	a_{54}	a_{55}	
1	2			X						X		X						X								X	Σ_1
3	4			X						X		X										X		X				Σ_2
5	6			X						X			X									X	X					Σ_3
7	8			X						X			X				X									X		Σ_4
9	10			X						X						X	X							X				Σ_5
11	12			X						X						X		X					X					Σ_6
2	8																											γ
10	4																											α
6	12																											ψ
$\left|A^{11}_{13}\right|=$ $\gamma+\alpha+\psi$																												$\left|A^{11}_{13}\right|$

where,

$\Sigma_1=a_{13}a_{24}a_{31}a_{42}a_{55}=15$; $\Sigma_2=a_{13}a_{24}a_{31}a_{45}a_{52}=30$; $\Sigma_3=a_{13}a_{24}a_{32}a_{45}a_{51}=-45$; $\Sigma_4=a_{13}a_{24}a_{32}a_{41}a_{55}=-90$;
$\Sigma_5=a_{13}a_{24}a_{35}a_{41}a_{52}=-60$; $\Sigma_6=a_{13}a_{24}a_{35}a_{42}a_{51}=-15$; $\gamma=\Sigma_1-\Sigma_4=105$; $\alpha=\Sigma_5-\Sigma_2=-90$; $\psi=\Sigma_3-\Sigma_6=-30$;
$\left|A^{11}_{13}\right|=-15$.

Appendix 56

Algorithm for evaluation of partial part $\left|A^{12}_{13}\right|$ of complex determinant $\left|A\right|$ of 5x5 matrix by Expansion method for the twelfth row(i_{12}) of formula (58)

Example 16:

Problem
Solve the partial part $\left|A^{12}_{13}\right|$ of the twelfth row(i_{12}) formula (58) of matrix 5x5 for complex determinant $\left|A\right|$ by Expansion method.

Solving
In accordance with above-indicated algorithm of formula (58) for the twelfth row (i_{12}) and Table 16 *"The general steps for evaluation of partial part $\left|A^{12}_{13}\right|$ by Expansion method "*, we have the following data for partial part $\left|A^{12}_{13}\right|$ of complex determinant $\left|A\right|$ of matrix 5x5 at numerical data(**):

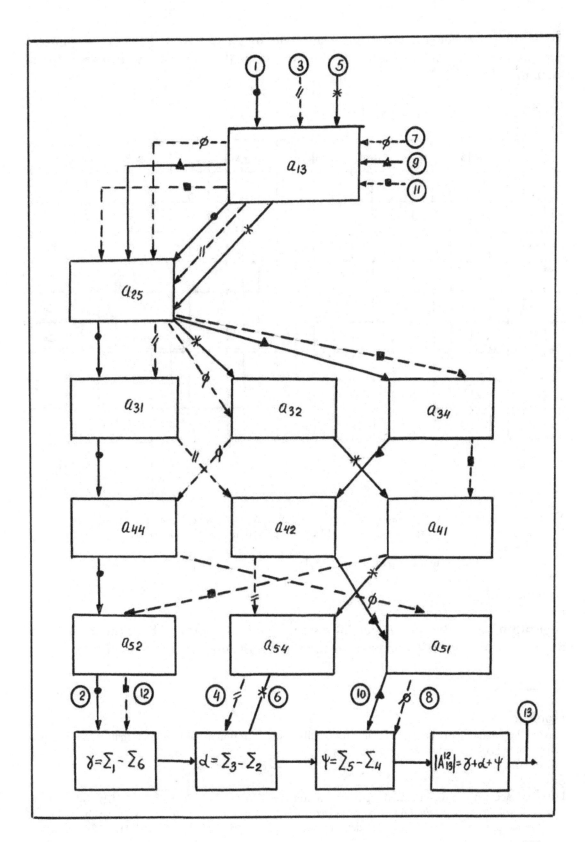

Table 16 **The general steps for evaluation of partial part** $\left|A^{12}_{13}\right|$ **of complex determinant** $\left|A\right|$ **of 5x5 matrix by Expansion method for the twelfth row (i$_{12}$) of formula (58)**

Number of steps		Elements involved in calculation process																									Sum
in	out	a_{11}	a_{12}	a_{13}	a_{14}	a_{15}	a_{21}	a_{22}	a_{23}	a_{24}	a_{25}	a_{31}	a_{32}	a_{33}	a_{34}	a_{35}	a_{41}	a_{42}	a_{43}	a_{44}	a_{45}	a_{51}	a_{52}	a_{53}	a_{54}	a_{55}	
1	2			X							X	X								X			X				Σ_1
3	4			X							X	X						X							X		Σ_2
5	6			X							X		X				X								X		Σ_3
7	8			X							X		X							X		X					Σ_4
9	10			X							X				X			X				X					Σ_5
11	12			X							X				X		X						X				Σ_6
2	12																										γ
6	4																										α
10	8																										ψ
$\left|A^{12}_{13}\right|=$ $\gamma+\alpha+\psi$																											$\left|A^{12}_{13}\right|$

where,

$\Sigma_1= a_{13}a_{25}a_{31}a_{44}a_{52}= -144;\ \Sigma_2=a_{13}a_{25}a_{31}a_{42}a_{54}= -72;\ \Sigma_3=a_{13}a_{25}a_{32}a_{41}a_{54}=432;\ \Sigma_4=a_{13}a_{25}a_{32}a_{44}a_{51}=216;$
$\Sigma_5=a_{13}a_{25}a_{34}a_{42}a_{51}= -36;\ \Sigma_6=a_{13}a_{25}a_{34}a_{41}a_{52}= -144;\ \gamma=\Sigma_1-\Sigma_6=0;\ \alpha=\Sigma_3-\Sigma_2=504;\ \psi=\Sigma_5-\Sigma_4=-252;$
$\left|A^{12}_{13}\right|=252.$

So, the value of summary partial part $\left|A_{13}\right|=\left|A^{9}_{13}\right|+\left|A^{10}_{13}\right|+\left|A^{11}_{13}\right|+\left|A^{12}_{13}\right|=243.$

Appendix 57

Algorithm for evaluation of partial part $\left|A^{13}_{14}\right|$ **of complex determinant** $\left|A\right|$ **of 5x5 matrix by Expansion method for the thirteenth row(i$_{13}$) of formula (58)**

Example 17:

Problem
Solve the partial part $\left|A^{13}_{14}\right|$ of the thirteenth row(i$_{13}$) formula (58) of 5x5 matrix for complex determinant $\left|A\right|$ by Expansion method.

Solving
In accordance with above-indicated algorithm of formula (58) for the thirteenth row (i$_{13}$) and Table 17 *"The general steps for evaluation of partial part* $\left|A^{13}_{14}\right|$ *by Expansion method"*, we have the following data for partial part $\left|A^{13}_{14}\right|$ of complex determinant $\left|A\right|$ of matrix 5x5 at numerical data (**):

532

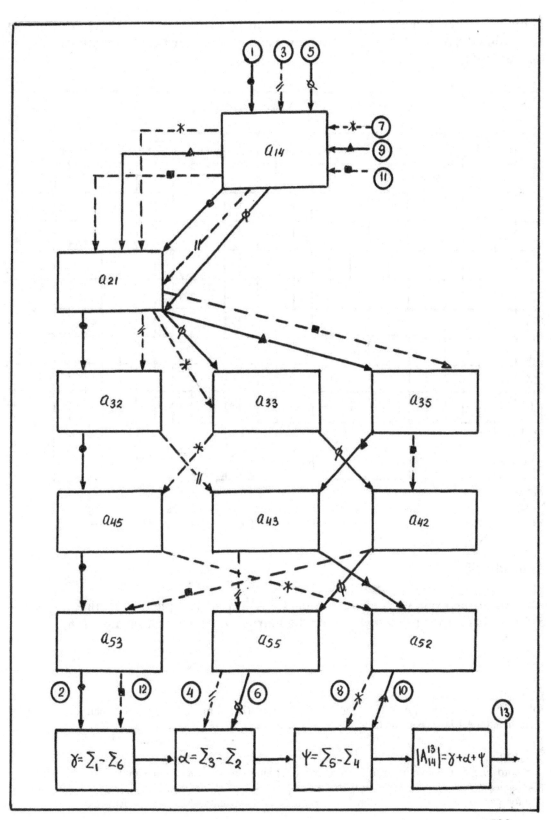

Table 17 The general steps for evaluation of partial part $\left|A^{13}_{14}\right|$ of complex determinant $\left|A\right|$ of 5x5 matrix by Expansion method for the thirteenth row (i_{13}) of formula (58)

Number of steps		Elements involved in calculation process																									Sum
in	out	a11	a12	a13	a14	a15	a21	a22	a23	a24	a25	a31	a32	a33	a34	a35	a41	a42	a43	a44	a45	a51	a52	a53	a54	a55	
1	2				X		X						X								X			X			Σ_1
3	4				X		X						X						X							X	Σ_2
5	6				X		X							X				X								X	Σ_3
7	8				X		X							X							X		X				Σ_4
9	10				X		X									X			X				X				Σ_5
11	12				X		X									X		X						X			Σ_6
2	12																										γ
6	4																										α
10	8																										ψ
$\left|A^{13}_{14}\right|=$ $\gamma+\alpha+\psi$																											$\left|A^{13}_{14}\right|$

where,

$\Sigma_1 = a_{14}a_{21}a_{32}a_{45}a_{53} = 48$; $\Sigma_2 = a_{14}a_{21}a_{32}a_{43}a_{55} = 96$; $\Sigma_3 = a_{14}a_{21}a_{33}a_{42}a_{55} = 64$; $\Sigma_4 = a_{14}a_{21}a_{33}a_{45}a_{52} = 128$;
$\Sigma_5 = a_{14}a_{21}a_{35}a_{43}a_{52} = 64$; $\Sigma_6 = a_{14}a_{21}a_{35}a_{42}a_{53} = 16$; $\gamma = \Sigma_1 - \Sigma_6 = 32$; $\alpha = \Sigma_3 - \Sigma_2 = -32$; $\psi = \Sigma_5 - \Sigma_4 = -64$;
$\left|A^{13}_{14}\right| = -64$.

Appendix 58

Algorithm for evaluation of partial part $\left|A^{14}_{14}\right|$ of complex determinant $\left|A\right|$ of 5x5 matrix by Expansion method for the fourteenth row(i_{14}) of formula (58)

Example 18:

Problem

Solve the partial part $\left|A^{14}_{14}\right|$ of the fourteenth row (i_{14}) formula (58) of 5x5 matrix for complex determinant $\left|A\right|$ by Expansion method.

Solving

In accordance with above-indicated algorithm of formula (58) for the fourteenth row (i_{14}) and Table 18 *"The general steps for evaluation of partial part $\left|A^{14}_{14}\right|$ by Expansion method "*, we have the following data for partial part $\left|A^{14}_{14}\right|$ of complex determinant $\left|A\right|$ of matrix 5x5 at numerical data (**):

534

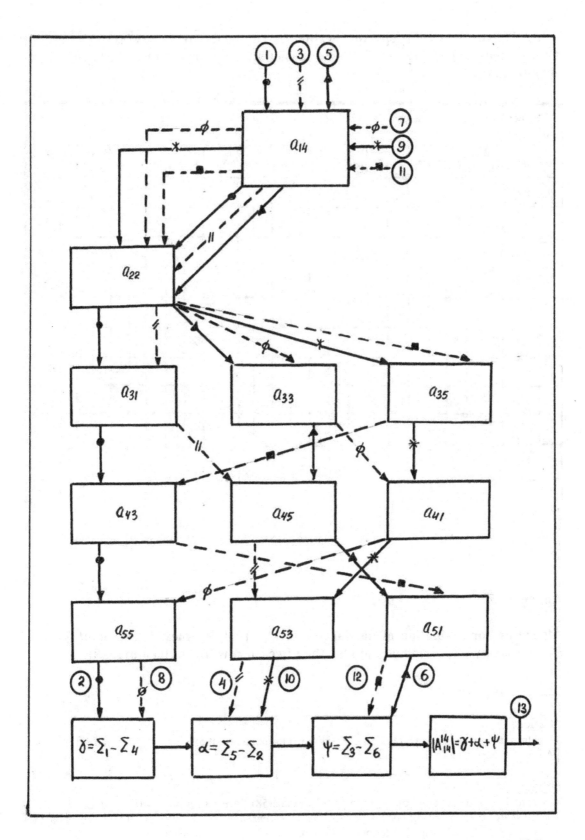

Table 17 The general steps for evaluation of partial part $\left|A^{14}_{14}\right|$ of complex determinant $\left|A\right|$ of 5x5 matrix by Expansion method for the fourteenth row(i_{14}) of formula (58)

in	out	a_{11}	a_{12}	a_{13}	a_{14}	a_{15}	a_{21}	a_{22}	a_{23}	a_{24}	a_{25}	a_{31}	a_{32}	a_{33}	a_{34}	a_{35}	a_{41}	a_{42}	a_{43}	a_{44}	a_{45}	a_{51}	a_{52}	a_{53}	a_{54}	a_{55}	Sum
1	2				X			X				X							X							X	Σ_1
3	4				X			X				X									X			X			Σ_2
5	6				X			X						X							X	X					Σ_3
7	8				X			X						X			X									X	Σ_4
9	10				X			X								X	X							X			Σ_5
11	12				X			X								X			X		X						Σ_6
2	8																										γ
10	4																										α
6	12																										ψ
$\left|A^{14}_{14}\right|=$ $\gamma+\alpha+\psi$																											$\left|A^{14}_{14}\right|$

where,

$\Sigma_1=a_{14}a_{22}a_{31}a_{43}a_{55}=24$; $\Sigma_2=a_{14}a_{22}a_{31}a_{45}a_{53}=12$; $\Sigma_3=a_{14}a_{22}a_{33}a_{45}a_{51}=-48$; $\Sigma_4=a_{14}a_{22}a_{33}a_{41}a_{55}=-96$; $\Sigma_5=a_{14}a_{22}a_{35}a_{41}a_{53}=-24$; $\Sigma_6=a_{14}a_{22}a_{35}a_{43}a_{51}=-24$; $\gamma=\Sigma_1-\Sigma_4=120$; $\alpha=\Sigma_5-\Sigma_2=-36$; $\psi=\Sigma_3-\Sigma_6=-24$; $\left|A^{14}_{14}\right|=60$.

Appendix 59

Algorithm for evaluation of partial part $\left|A^{15}_{14}\right|$ of complex determinant $\left|A\right|$ of 5x5 matrix by Expansion method for the fifteenth row (i_{15}) of formula (58)

<u>Example 18:</u>

<u>Problem</u>
Solve the partial part $\left|A^{15}_{14}\right|$ of the fifteenth row(i_{15}) formula (58) of 5x5 matrix for complex determinant $\left|A\right|$ by Expansion method.

<u>Solving</u>
In accordance with above-indicated algorithm of formula(58) for the fifteenth row(i_{15}) and Table 18 "*The general steps for evaluation of partial part $\left|A^{15}_{14}\right|$ by Expansion method*" ,we have the following data for partial part $\left|A^{15}_{14}\right|$ of complex determinant $\left|A\right|$ of matrix 5x5 at numerical data(**):

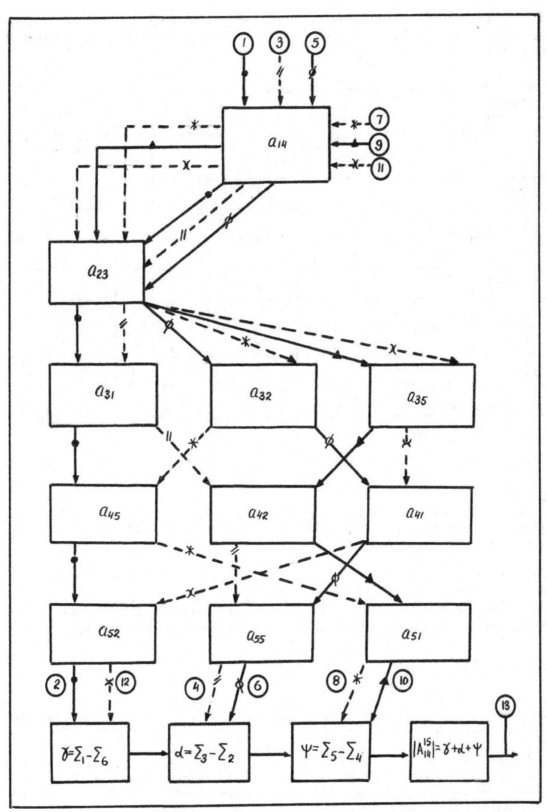

Table 18 The general steps for evaluation of partial part $|A^{15}_{14}|$ of complex determinant $|A|$ of 5x5 matrix by Expansion method for the fifteenth row(i_{15}) of formula (58)

Number of steps in	out	a_{11}	a_{12}	a_{13}	a_{14}	a_{15}	a_{21}	a_{22}	a_{23}	a_{24}	a_{25}	a_{31}	a_{32}	a_{33}	a_{34}	a_{35}	a_{41}	a_{42}	a_{43}	a_{44}	a_{45}	a_{51}	a_{52}	a_{53}	a_{54}	a_{55}	Sum
1	2				X				X			X									X		X				Σ_1
3	4				X				X			X						X								X	Σ_2
5	6				X				X				X				X									X	Σ_3
7	8				X				X				X								X	X					Σ_4
9	10				X				X							X		X				X					Σ_5
11	12				X				X							X	X						X				Σ_6
2	12																										γ
6	4																										α
10	8																										ψ
$|A^{15}_{14}|=$ $\gamma+\alpha+\psi$																											$|A^{15}_{14}|$

where,

$\Sigma_1=a_{14}a_{23}a_{31}a_{45}a_{52}=72$; $\Sigma_2=a_{14}a_{23}a_{31}a_{42}a_{55}=36$; $\Sigma_3=a_{14}a_{23}a_{32}a_{41}a_{55}=-216$; $\Sigma_4=a_{14}a_{23}a_{32}a_{45}a_{51}=-108$; $\Sigma_5=a_{14}a_{23}a_{35}a_{42}a_{51}=-36$; $\Sigma_6=a_{14}a_{23}a_{35}a_{41}a_{52}=-144$; $\gamma=\Sigma_1-\Sigma_6=216$; $\alpha=\Sigma_3-\Sigma_2=-252$; $\psi=\Sigma_5-\Sigma_4=72$; $|A^{15}_{14}|=36$.

Appendix 60

Algorithm for evaluation of partial part $|A^{16}_{14}|$ of complex determinant $|A|$ of 5x5 matrix by Expansion method for the sixteenth row(i_{16}) of formula (58)

Example 19:

Problem
Solve the partial part $|A^{16}_{14}|$ of the sixteenth row(i_{16}) formula (58) of 5x5 matrix for complex determinant $|A|$ by Expansion method.

Solving
In accordance with above-indicated algorithm of formula (58) for the sixteenth row (i_{16}) and Table 19 "*The general steps for evaluation of partial part $|A^{16}_{14}|$ by Expansion method*" ,we have the following data for partial part $|A^{16}_{14}|$ of complex determinant $|A|$ of 5x5 matrix at numerical data(**):

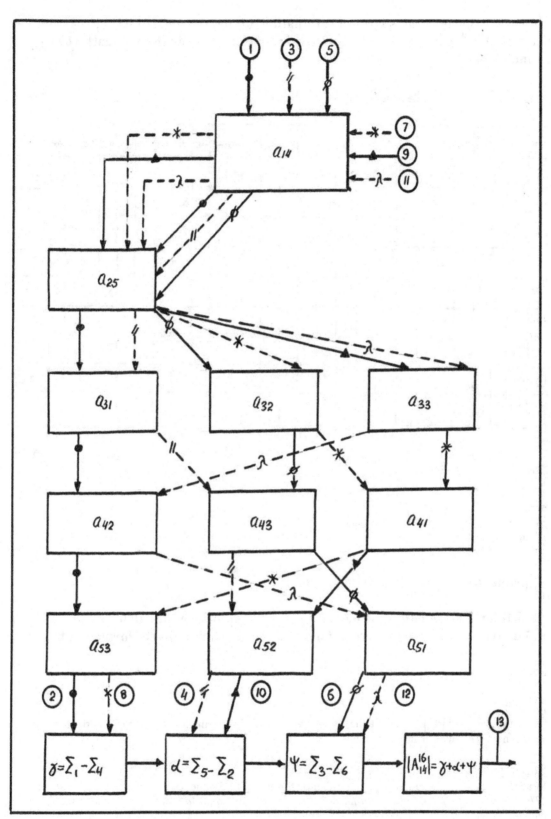

Table 19 The general steps for evaluation of partial part $|A^{16}_{14}|$ of complex determinant $|A|$ of 5x5 matrix by Expansion method for the sixteenth row(i_{16}) of formula (58)

Number of steps in	out	a_{11}	a_{12}	a_{13}	a_{14}	a_{15}	a_{21}	a_{22}	a_{23}	a_{24}	a_{25}	a_{31}	a_{32}	a_{33}	a_{34}	a_{35}	a_{41}	a_{42}	a_{43}	a_{44}	a_{45}	a_{51}	a_{52}	a_{53}	a_{54}	a_{55}	Sum
1	2				X						X	X						X						X			Σ_1
3	4				X						X	X							X				X				Σ_2
5	6				X						X		X						X			X					Σ_3
7	8				X						X		X				X							X			Σ_4
9	10				X						X			X			X						X				Σ_5
11	12				X						X			X				X				X					Σ_6
2	8																										γ
10	4																										α
6	12																										ψ
$|A^{16}_{14}| = \gamma+\alpha+\psi$																											$|A^{16}_{14}|$

where,

$\Sigma_1=a_{14}a_{25}a_{31}a_{42}a_{53}=72$; $\Sigma_2=a_{14}a_{25}a_{31}a_{43}a_{52}=288$; $\Sigma_3=a_{14}a_{25}a_{32}a_{43}a_{51}=-432$; $\Sigma_4=a_{14}a_{25}a_{32}a_{41}a_{53}=-432$; $\Sigma_5=a_{14}a_{25}a_{33}a_{41}a_{52}=-1152$; $\Sigma_6=a_{14}a_{25}a_{33}a_{42}a_{51}=-288$; $\gamma=\Sigma_1-\Sigma_4=504$; $\alpha=\Sigma_5-\Sigma_2=-1440$; $\psi=\Sigma_3-\Sigma_6=-144$; $|A^{16}_{14}|=-1080$.

Appendix 61

Algorithm for evaluation of partial part $|A^{17}_{15}|$ of complex determinant $|A|$ of 5x5 matrix by Expansion method for the seventeenth row (i_{17}) of formula (58)

<u>Example 20:</u>

<u>Problem</u>
Solve the partial part $|A^{17}_{15}|$ of the seventeenth row (i_{17}) formula (58) of 5x5 matrix for complex determinant $|A|$ by Expansion method.

<u>Solving</u>
In accordance with above-indicated algorithm of formula (58) for the seventeenth row (i_{17}) and Table 20 *"The general steps for evaluation of partial part $|A^{17}_{15}|$ by Expansion method"* ,we have the following data for partial part $|A^{17}_{15}|$ of complex determinant $|A|$ of matrix 5x5 at numerical data(**):

540

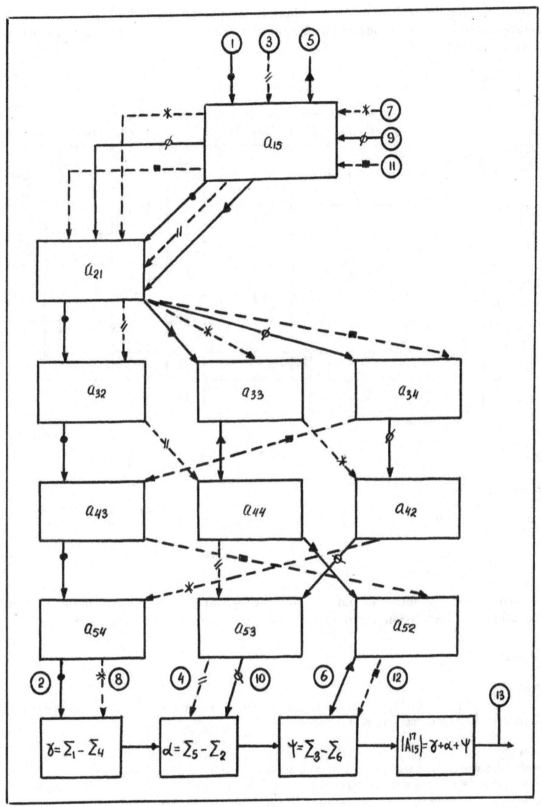

Table 20 The general steps for evaluation of partial part $\left|A^{17}_{15}\right|$ of complex determinant $\left|A\right|$ of 5x5 matrix by Expansion method for the seventeenth row(i_{17}) of formula (58)

Number of steps		Elements involved in calculation process																									Sum
in	out	a_{11}	a_{12}	a_{13}	a_{14}	a_{15}	a_{21}	a_{22}	a_{23}	a_{24}	a_{25}	a_{31}	a_{32}	a_{33}	a_{34}	a_{35}	a_{41}	a_{42}	a_{43}	a_{44}	a_{45}	a_{51}	a_{52}	a_{53}	a_{54}	a_{55}	
1	2					×	×						×						×						×		Σ_1
3	4					×	×						×							×				×			Σ_2
5	6					×	×							×						×			×				Σ_3
7	8					×	×							×				×							×		Σ_4
9	10					×	×								×			×						×			Σ_5
11	12					×	×								×				×				×				Σ_6
2	8																										γ
10	4																										α
6	12																										ψ
$\left|A^{17}_{15}\right| =$ $\gamma+\alpha+\psi$																											$\left|A^{17}_{15}\right|$

where,

$\Sigma_1=a_{15}a_{21}a_{32}a_{43}a_{54}=-480$; $\Sigma_2=a_{15}a_{21}a_{32}a_{44}a_{53}=-240$; $\Sigma_3=a_{15}a_{21}a_{33}a_{44}a_{52}=-640$; $\Sigma_4=a_{15}a_{21}a_{33}a_{42}a_{54}=-320$; $\Sigma_5=a_{15}a_{21}a_{34}a_{42}a_{53}=40$; $\Sigma_6=a_{15}a_{21}a_{34}a_{43}a_{52}=160$; $\gamma=\Sigma_1-\Sigma_4=-160$; $\alpha=\Sigma_5-\Sigma_2=280$; $\psi=\Sigma_3-\Sigma_6=-800$; $\left|A^{17}_{15}\right|=-680$.

Appendix 62

Algorithm for evaluation of partial part $\left|A^{18}_{15}\right|$ of complex determinant $\left|A\right|$ of 5x5 matrix by Expansion method for the eighteenth row(i_{18}) of formula (58)

<u>Example 21</u>:

<u>Problem</u>

Solve the partial part $\left|A^{18}_{15}\right|$ of the eighteenth row(i_{18}) formula (58) of 5x5 matrix for complex determinant $\left|A\right|$ by Expansion method.

<u>Solving</u>

In accordance with above-indicated algorithm of formula (58) for the eighteenth row(i_{18}) and Table 21 "*The general steps for evaluation of partial part $\left|A^{18}_{15}\right|$ by Expansion method*" ,we have the following data for partial part $\left|A^{18}_{15}\right|$ of complex determinant $\left|A\right|$ of matrix 5x5 at numerical data(**):

542

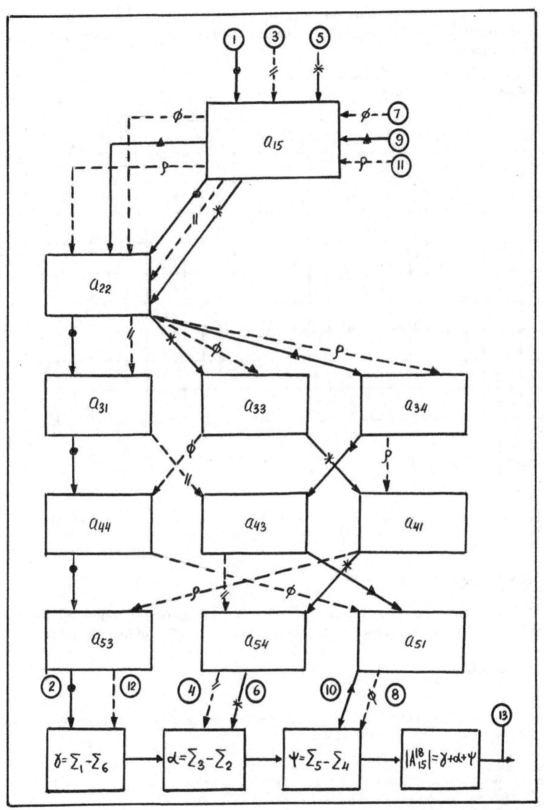

543

Table 21 The general steps for evaluation of partial part $\left|A^{18}_{15}\right|$ of complex determinant $\left|A\right|$ of 5x5 matrix by Expansion method for the eighteenth row (i_{18}) of formula (58)

Number of steps in	out	a_{11}	a_{12}	a_{13}	a_{14}	a_{15}	a_{21}	a_{22}	a_{23}	a_{24}	a_{25}	a_{31}	a_{32}	a_{33}	a_{34}	a_{35}	a_{41}	a_{42}	a_{43}	a_{44}	a_{45}	a_{51}	a_{52}	a_{53}	a_{54}	a_{55}	Sum
1	2					✕		✕				✕								✕				✕			Σ_1
3	4					✕		✕				✕							✕						✕		Σ_2
5	6					✕		✕						✕			✕								✕		Σ_3
7	8					✕		✕						✕						✕		✕					Σ_4
9	10					✕		✕							✕				✕			✕					Σ_5
11	12					✕		✕							✕		✕							✕			Σ_6
2	12																										γ
6	4																										α
10	8																										ψ
$\left|A^{18}_{15}\right| = \gamma + \alpha + \psi$																											$\left|A^{18}_{15}\right|$

where,

$\Sigma_1 = a_{15}a_{22}a_{31}a_{44}a_{53} = -60$; $\Sigma_2 = a_{15}a_{22}a_{31}a_{43}a_{54} = -120$; $\Sigma_3 = a_{15}a_{22}a_{33}a_{41}a_{54} = 480$; $\Sigma_4 = a_{15}a_{22}a_{33}a_{44}a_{51} = 240$;
$\Sigma_5 = a_{15}a_{22}a_{34}a_{43}a_{51} = -60$; $\Sigma_6 = a_{15}a_{22}a_{34}a_{41}a_{53} = -60$; $\gamma = \Sigma_1 - \Sigma_6 = 0$; $\alpha = \Sigma_3 - \Sigma_2 = 600$; $\psi = \Sigma_5 - \Sigma_4 = -300$;
$\left|A^{18}_{15}\right| = 300$.

Appendix 63

Algorithm for evaluation of partial part $\left|A^{19}_{15}\right|$ of complex determinant $\left|A\right|$ of 5x5 matrix by Expansion method for the nineteenth row (i_{19}) of formula (58)

Example 22:

Problem
Solve the partial part $\left|A^{19}_{15}\right|$ of the nineteenth row (i_{19}) formula (58) of 5x5 matrix for complex determinant $\left|A\right|$ by Expansion method.

Solving
In accordance with above-indicated algorithm of formula (58) for the nineteenth row (i_{19}) and Table 22 *"The general steps for evaluation of partial part $\left|A^{19}_{15}\right|$ by Expansion method"* ,we have the following data for partial part $\left|A^{19}_{15}\right|$ of complex determinant $\left|A\right|$ of matrix 5x5 at numerical data(**):

544

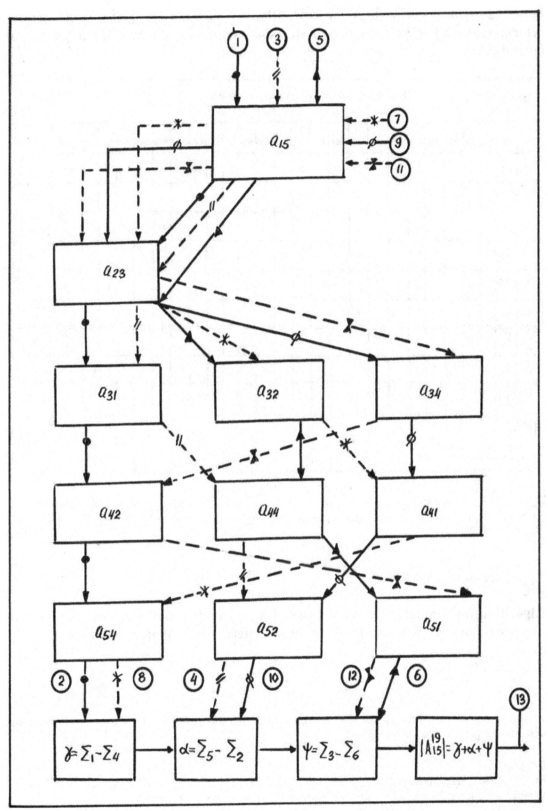

Table 22 The general steps for evaluation of partial part $|A^{19}_{15}|$ of complex determinant $|A|$ of 5x5 matrix by Expansion method for the nineteenth row (i_{19}) of formula (58)

Number of steps in	out	a_{11}	a_{12}	a_{13}	a_{14}	a_{15}	a_{21}	a_{22}	a_{23}	a_{24}	a_{25}	a_{31}	a_{32}	a_{33}	a_{34}	a_{35}	a_{41}	a_{42}	a_{43}	a_{44}	a_{45}	a_{51}	a_{52}	a_{53}	a_{54}	a_{55}	Sum
1	2					X			X			X						X							X		Σ_1
3	4					X			X			X								X			X				Σ_2
5	6					X			X				X							X		X					Σ_3
7	8					X			X				X				X								X		Σ_4
9	10					X			X						X		X						X				Σ_5
11	12					X			X						X			X				X					Σ_6
2	8																										γ
10	4																										α
6	12																										ψ
$|A^{19}_{15}|=$ $\gamma+\alpha+\psi$																											$|A^{19}_{15}|$

where,

$\Sigma_1 = a_{15}a_{23}a_{31}a_{42}a_{54} = -180$; $\Sigma_2 = a_{15}a_{23}a_{31}a_{44}a_{52} = -360$; $\Sigma_3 = a_{15}a_{23}a_{32}a_{44}a_{51} = 540$; $\Sigma_4 = a_{15}a_{23}a_{32}a_{41}a_{54} = 1080$; $\Sigma_5 = a_{15}a_{23}a_{34}a_{41}a_{52} = -360$; $\Sigma_6 = a_{15}a_{23}a_{34}a_{42}a_{51} = -90$; $\gamma = \Sigma_1 - \Sigma_4 = -1260$; $\alpha = \Sigma_5 - \Sigma_2 = 0$; $\psi = \Sigma_3 - \Sigma_6 = 630$; $|A^{19}_{15}| = -630$.

Appendix 64

Algorithm for evaluation of partial part $|A^{20}_{15}|$ of complex determinant $|A|$ of 5x5 matrix by Expansion method for the twentieth row (i_{20}) of formula (58)

Example 23:

Problem

Solve the partial part $|A^{20}_{15}|$ of the twentieth row(i_{20}) formula (58) of 5x5 matrix for complex determinant $|A|$ by Expansion method.

Solving

In accordance with above-indicated algorithm of formula (58) for the twentieth row (i_{20}) and Table 23 "The general steps for evaluation of partial part $|A^{20}_{15}|$ by Expansion method", we have the following data for partial part $|A^{20}_{15}|$ of complex determinant $|A|$ of matrix 5x5 at numerical data(**):

546

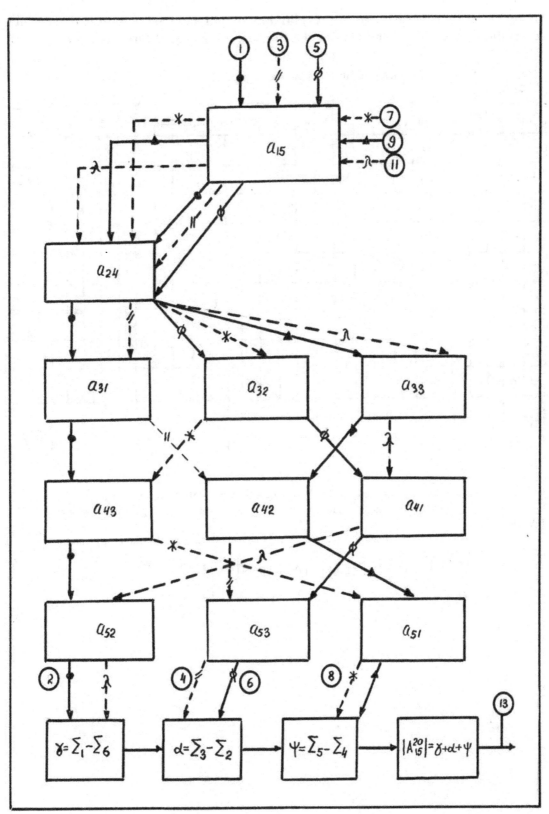

Table 23 The general steps for evaluation of partial part $|A^{20}_{15}|$ of complex determinant $|A|$ of 5x5 matrix for the twentieth row (i_{20}) of formula (58)

Number of steps		Elements involved in calculation process																									Sum
in	out	a_{11}	a_{12}	a_{13}	a_{14}	a_{15}	a_{21}	a_{22}	a_{23}	a_{24}	a_{25}	a_{31}	a_{32}	a_{33}	a_{34}	a_{35}	a_{41}	a_{42}	a_{43}	a_{44}	a_{45}	a_{51}	a_{52}	a_{53}	a_{54}	a_{55}	
1	2					X				X		X						X						X				Σ_1
3	4					X				X		X						X							X			Σ_2
5	6					X				X			X				X						X		X			Σ_3
7	8					X				X			X						X				X	X				Σ_4
9	10					X				X				X				X					X	X				Σ_5
11	12					X				X				X			X						X		X			Σ_6
2	12																											γ
6	4																											α
10	8																											ψ
$\|A^{20}_{15}\|$ = $\gamma+\alpha+\psi$																											$\|A^{20}_{15}\|$

where,

$\Sigma_1 = a_{15}a_{24}a_{31}a_{43}a_{52} = 300$; $\Sigma_2 = a_{15}a_{24}a_{31}a_{42}a_{53} = 75$; $\Sigma_3 = a_{15}a_{24}a_{32}a_{41}a_{53} = -450$; $\Sigma_4 = a_{15}a_{24}a_{32}a_{43}a_{51} = -450$; $\Sigma_5 = a_{15}a_{24}a_{33}a_{42}a_{51} = -300$; $\Sigma_6 = a_{15}a_{24}a_{33}a_{41}a_{52} = -1200$; $\gamma = \Sigma_1 - \Sigma_6 = 1500$; $\alpha = \Sigma_3 - \Sigma_2 = -525$; $\psi = \Sigma_5 - \Sigma_4 = 150$; $|A^{20}_{15}| = 1125$.

So, the complex determinant $|A|$ of 5x5 matrix in accordance with formula (58) can be evaluated by the following value equal:

$$|A| = |A_{11}| + |A_{12}| + |A_{13}| + |A_{14}| + |A_{15}| = -1210$$

where:

$|A_{11}| = 194$; $|A_{12}| = -714$; $|A_{13}| = 243$; $|A_{14}| = -1048$; $|A_{15}| = 115$.

References and bibliography:

[1] The theory determinants in the historical order of development, volumes I-IV ,1906-1923
- Dover Publications, Inc. New York ,New York

[2] Окунев Л.Я .Высшая алгебра.,1937
-Объединённое Научно-техническое издательство НКТП СССР,Москва

[3] Nicholson, W. Keith. Elementary linear algebra with applications .,1990
-PWS-Kent Publish Company

[4] Nakos, George .Linear algebra with applications.,1998
--Brooks /Cole Publishing Company

[5] College Algebra through modeling and visualization by Gary K. Rockswold, John Hornsby, Margaret L.Lial,1999
-Addison-Wesley

[6] Lay, David C. Linear Algebra and its applications.,2006
-Pearson Addison Wesley, 2006

[7] Handbook of Linear Algebra by Leslie Hogben , 2007
- Chapman &Hall/CRC (Taylor &Francis Group, LLC),New York

[8] Rozenblat, A. Matrix Determinant with Graphs for Laplace's and Expansion methods.,2013
-AuthorHouse Inc, Bloomington ,USA

[9] Бронштейн И.Н.,Семендяев К.А. Справочник пл математике для инженеров и учащихся   для втузов,1980
-Москва "Наука " ,Главная редакция физико-математической литературы.